RECENT PROGRESS IN GENERAL TOPOLOGY II

RECENT PROGRESS IN GENERAL TOPOLOGY II

Edited by

MIROSLAV HUŠEK
Charles University
Prague, Czech Republic

JAN van MILL
Vrije Universiteit
Amsterdam, The Netherlands

2002

ELSEVIER
Amsterdam – Boston – New York – Oxford – Paris
San Diego – San Francisco – Singapore – Sydney – Tokyo

ELSEVIER SCIENCE B.V.
Sara Burgerhartstraat 25
P.O. Box 211, 1000 AE Amsterdam, The Netherlands

First edition 2002

Library of Congress Cataloging in Publication Data
A catalog record from the Library of Congress has been applied for.

British Library Cataloguing in Publication Data
A catalogue record from the British Library has been applied for.

ISBN: 0-444-50980-1

♾ The paper used in this publication meets the requirements of ANSI/NISO Z39.48-1992 (Permanence of Paper).
Printed in The Netherlands.

Preface

Ten years ago Elsevier published the volume Recent Progress in General Topology. The idea behind that book was to present surveys describing recent developments in most of the primary subfields of General Topology and its applications to Algebra and Analysis. It was our belief that the book could be of help to researchers in General Topology as a background for the development of their own research. There were two similar predecessors, namely, the Handbook of Set-Theoretic Topology (North Holland, Amsterdam 1984, J.E. Vaughan and K. Kunen, eds.) and Open Problems in Topology (North Holland, Amsterdam 1990, J. van Mill and G.M. Reed, eds.).

It seems that these three books were well received by the community and suggestions to publish an update of Recent Progress in General Topology came from several directions. So, the editors agreed to prepare Recent Progress in General Topology II, again in connection with the Prague Topological Symposium, held in 2001. We asked a number of invited speakers to prepare a survey that would be suitable for the book. We are pleased that virtually everyone contributed a paper. Two contributions were written by authors who could not attend the Symposium. We would like to express our appreciation to all authors for their valuable work.

During the last 10 years the focus in General Topology changed and therefore our selection of topics differs slightly from those chosen in 1992. The following areas experienced significant developments: Topological Groups, Function Spaces, Dimension Theory, Hyperspaces, Selections, Geometric Topology (including Infinite-Dimensional Topology and the Geometry of Banach Spaces). Of course, not every important topic could be included in this book. For instance, we regret that a contribution on Continua Theory is missing (but the reader can find many important results from Continua Theory in the included contributions).

Apart from the survey articles on the progress of the past decade, the reader will find several historical essays at the end of the book. We asked a number of senior topologists to write a short essay expressing their personal view on the developments in General Topology in the last century. We expect that the reader will find it interesting to read the personal opinions of such eminent topologists as R.D. Anderson, W.W. Comfort, M. Henriksen, S. Mardešić, J. Nagata, M.E. Rudin, J.M. Smirnov and L. Vietoris. The essays were not refereed.

Another novelty in comparison to the 1992 edition is the author index and a combined list of problems and questions posed in the papers in this volume. The first named editor is responsible for the selection of problems and questions, and of items in both indexes.

June 2002 Mirek Hušek and Jan van Mill

Contents

CHAPTER 1

Topological Invariants in Algebraic Environment

Alexander Arhangel'skii
Ohio University, Athens, OH 45701, U.S.A
E-mail: arhangel@bing.math.ohiou.edu

Contents

RECENT PROGRESS IN GENERAL TOPOLOGY II
Edited by Miroslav Hušek and Jan van Mill

1. Introduction

This is a survey of that domain of topological algebra which studies the influence of algebraic structures on topologies properly tailored for them. This domain could be called "Topological invariants in topological algebra". The survey is by no means complete, since this vast area is energetically developing in various directions. On several occasions, when a good survey on a certain topic is already available, the reader is referred to such a survey. This is the case of minimal topological groups, we refer the reader to the works of DIKRANJAN, including his survey [1998a]. However, the author has attempted to provide a rather representative sample of relatively recent results in the domain, not restricting himself just to two or three topics. The areas covered to a lesser or greater extent are cardinal invariants in topological algebra, separate and joint continuity of group operation, extremally disconnected and related topologies on groups, free topological groups, completions of topological groups, Bohr topologies, duality theory. I must confess that treating these subjects I paid more attention to the research which is closer to my own. For results on compactness type conditions in topological groups, see TKAČENKO's survey in this book and the memoir on pseudocompact topological groups by DIKRANJAN and SHAKHMATOV [1993]. The last fundamental paper reflects a trend in topological algebra dual to the one we survey in this article: it mostly studies the influence of topological properties of topological groups on their algebraic structure. Few results of this kind we discuss below.

The survey covers the period from 1990 to 2001, though occasionally we cite some classical older results. For a systematic survey of the results in topological algebra obtained before 1990 we refer the reader to COMFORT [1990] and COMFORT, HOFMANN and REMUS [1992]. See also recent TKAČENKO's surveys [1999], [2000], ARHANGEL'SKII's old survey [1981] and SHAKHMATOV's survey [1999]. All topologies considered below are assumed to satisfy T_1 separation axiom. The standard reference book for general topology is ENGELKING [1977]. For a general background on topological groups, see PONTRYAGIN [1939] and ROELKE and DIEROLF [1981].

2. Topologies on groups

One of generic questions in topological algebra is how the relationships between topological properties depend on underlying algebraic structure. And, clearly, the answer to this question should strongly depend on the way algebraic structure is related to topology. The weaker the restrictions on the connection between topology and algebraic structure are, the larger is the class of objects entering the theory. Because of that, even when our main interest is, for example, in topological groups, it is natural to consider more general objects with not so rigid connection between topology and algebra. And examples we encounter in such a theory would help us to better understand and appreciate the fruits of the theory of topological groups. In this section we present a rich collection of results on generalizations of topological groups.

First, we should recall definitions of the objects we are going to discuss. Let S be a semigroup. The mapping of $S \times S$ into S associating with arbitrary $(x, y) \in S \times S$ the product $xy \in S$ is called the *product mapping*. If the semigroup S is endowed with a topology such that the product mapping of the space $S \times S$ into the space S is (jointly) continuous, we say that S is a *topological semigroup*. A *topological group* G is a group G with a (Hausdorff) topology such that the product mapping of $G \times G$ into G is (jointly) continuous and the inverse mapping of G onto itself associating x^{-1} with arbitrary $x \in G$ is continuous. A *paratopological group* G is a group G with a topology such that the product mapping of $G \times G$ into G is (jointly) continuous. A *semitopological group* G is a group G with a topology such that the product mapping of $G \times G$ into G is separately continuous. A *quasitopological group* G is a group G with a topology such that the product mapping of $G \times G$ into G is separately continuous and the inverse mapping of G onto itself is continuous. Finally, if G is a group with a topology such that the product mapping is continuous from the left (from the right), G is called a *left topological group* (a *right topological group*).

There are many natural examples showing that all these notions are distinct. A natural example of a paratopological group can be obtained by taking the group of autohomeomorphisms of a dense-in-itself locally compact zero-dimensional non-compact space, in the compact-open topology.

Obviously, it is important to know what restrictions on the topology of a paratopological group G imply that G is, in fact, a topological group. Similarly, we should ask under what restrictions on the topology of a semitopological group it becomes a paratopological group. Sorgenfrey line under the usual addition is a paratopological group which is hereditarily separable, hereditarily Lindelöf and has the Baire property. Thus, even this strong combination of restrictions on the topology of a paratopological group does not ensure the continuity of the inverse operation.

D. Montgomery in 1936 proved that every semitopological group metrizable by a complete metric is, in fact, a paratopological group. In [1957] ELLIS showed that every locally compact semitopological group is a topological group. In 1960 W. Zelazko established that each completely metrizable semitopological group is a topological group. Later, in 1982, N. Brand proved that every Čech-complete paratopological group is a topological group. Recently Ahmed Bouziad made a decisive contribution to this topic. He proved the next theorem which naturally covers and unifies both principal cases: of locally compact semitopological groups and of completely metrizable semitopological groups, BOUZIAD [1996a].

2.1. THEOREM. *Every Čech-complete semitopological group is a topological group.*

Since each Čech-complete topological group is paracompact, Bouziad's Theorem 2.1 implies the next interesting result:

2.2. COROLLARY. *Every Čech-complete semitopological group is paracompact.*

Recall that every Čech-complete space X is a Baire space, that is, for every countable family γ of dense open subsets of X, the intersection of γ is dense in X. Also, every Čech-complete space is a p-space. The class of p-spaces is much wider than the class of Čech-complete spaces, all metrizable spaces are p-spaces, which implies that p-spaces needn't be Baire spaces and needn't be complete. The next result was obtained by BOUZIAD [1996b]:

2.3. THEOREM. *If a semitopological group G is a Baire p-space, then G is a paratopological group.*

Theorem 2.1 follows from the last result and Brand's theorem that every Čech-complete paratopological group is a topological group. BOUZIAD deduced Theorem 2.3 from his slightly more general result on actions of groups on spaces, [1996b]:

2.4. THEOREM. *Suppose that G is a left topological group such that G is a Baire p-space. Then each separately continuous action of G on a p-space X is a continuous action.*

Let X and Y be topological spaces. A mapping $f : X \to Y$ is called *quasicontinuous at* $x \in X$ if for every open neighborhood V of $f(x)$ and each open neighborhood U of x there exists a non-empty open set $W \subset U$ such that $f(W) \subset V$. If f is quasicontinuous at every point of X, we say that f is *quasicontinuous*. A mapping $f : X \times Y \to Z$ is said to be *strongly quasicontinuous at* $(x, y) \in X \times Y$ if, for each open neighborhood V of $f(x, y)$ in Z and for each open neighborhood U of (x, y) in $X \times Y$, there exists a non-empty open subset W of X and an open neighborhood Oy of y in Y such that $W \times Oy \subset U$ and $f(W \times Oy) \subset V$. The next theorem was proved in a slightly more general setting by BOUZIAD [1996b]:

2.5. THEOREM. *Suppose that X is a Baire p-space, Y a space of point-countable type, and Z a Tychonoff space. Then every separately continuous mapping $f : X \times Y \to Z$ is strongly quasicontinuous (that is, strongly quasicontinuous at every point $(x, y) \in X \times Y$).*

Here another result of Bouziad (see KENDEROV, KORTEZOV and MOORS [2001]) should be mentioned:

2.6. LEMMA. *Suppose that G is a paratopological group. If the inversion is quasicontinuous at the neutral element e, then G is a topological group.*

The work of Bouziad was continued in KENDEROV, KORTEZOV and MOORS [2001], who introduced a somewhat technical notion of a strongly Baire space defined in terms of a topological game, and proved that if G is a semitopological group and a strongly Baire space, then the inversion is quasicontinuous. They also proved that if G is a semitopological group and a strongly Baire space, then G is a topological group. Hence, every regular countably compact semitopological group is a topological group.

In this direction some strong results were obtained by REZNICHENKO. In particular, in [1994b] he proved the following statement:

2.7. THEOREM. *Every pseudocompact paratopological group is a topological group.*

This result cannot be extended to pseudocompact semitopological groups. Indeed, A.V. Korovin has constructed an Abelian pseudocompact quasitopological group which is not a paratopological group (see KOROVIN [1992] and ARHANGEL'SKII and HUŠEK [2001]). However, Reznichenko has identified several special restrictions under which a pseudocompact semitopological group must be a topological group.

2.8. THEOREM (REZNICHENKO [1994b]). *Suppose that G is a Tychonoff pseudocompact semitopological group such that the space G satisfies at least one of the following conditions:*

a) *the tightness of G is countable;*
b) *G is separable;*
c) *G is a k-space.*
Then G is a topological group.

Recently ARHANGEL'SKII and HUŠEK in [2001], using a method of A.V. Korovin, constructed a Tychonoff pseudocompact quasitopological group G such that the Souslin number of G is countable and G is not a topological group.

3. Some relations between invariants of semitopological and paratopological groups

It is well known that many topological properties become much stronger in the presence of an algebraic structure nicely related to the topology. An important result of this kind is S. Kakutani's classical theorem on metrizability of every first countable topological group. Pontryagin's theorem on the equivalence of T_1 axiom to complete regularity for topological groups illustrates the same idea. There are many examples of this kind. However, much less is known on what happens to the relations among topological properties in the class of semitopological groups or in the class of paratopological groups. How strong is the influence of the algebraic structure on the topology in these cases?

In recent years certain progress was made in these directions, and new interesting open problems were formulated.

CHEN has proved the following statement in [1999]:

3.1. THEOREM. *For every Hausdorff first countable semitopological group G, the diagonal Δ_G is a G_δ (in $G \times G$).*

However, in contrast to the case of topological groups, even a Tychonoff first countable semitopological group needn't be metrizable, or Moore, or paracompact. As an example we could take the so called Kofner's plane. It is also possible to construct a Hausdorff paratopological group which is not regular (H.P.Künzi, a private communication). Thus, Pontryagin's theorem mentioned above does not generalize to the case of paratopological groups. However, the next question is still open:

3.2. PROBLEM. Is every regular paratopological group G Tychonoff? What if, in addition, G is first countable?

A proof of Chen's theorem can be based on some elementary facts worthy of being brought to the light.

Let G be a group and V a non-empty subset of it. Put $V_\Delta = \cup\{Vg \times Vg : g \in G\}$. Clearly, $\Delta_G \subset V_\Delta \subset G \times G$. We say that V_Δ is the *V-envelope* of the diagonal Δ in the product $G \times G$. The next easy to verify lemma is very useful.

3.3. LEMMA. *Suppose that G is a group and ξ is a family of non-empty subsets of G such that $\cap\{VV^{-1} : V \in \xi\} = \{e\}$. Then $\cap\{V_\Delta : V \in \xi\} = \Delta_G$.*

Let G be a group (just an algebraic group). A family $\mathcal{E}$ of subsets of G is called *discerning* or a *Hausdorff discernor on X* if all elements of $\mathcal{E}$ are non-empty and, for every $z \in G$ distinct from the neutral element e, there exists $P \in \mathcal{E}$ such that $zP \cap P = \varnothing$.

To provide an example of a discernor, we note that if G is a Hausdorff semitopological group and $\mathcal{B}$ is a base of G at some $a \in G$ then $\mathcal{B}$ is a Hausdorff discernor on G. A slightly less trivial and much more useful example of a Hausdorff discernor we obtain when we take an arbitrary π-network of G at e. A family $\mathcal{E}$ of subsets of a topological space X is called a *π-network* of X at a point $a \in X$ if all elements of $\mathcal{E}$ are non-empty and every open neighborhood of a in X contains an element of $\mathcal{E}$. If $\mathcal{E}$ is a π-network of X at $a \in X$ and all elements of $\mathcal{E}$ are open, we say that $\mathcal{E}$ is a *π-base* of X at a.

3.4. PROPOSITION (ARHANGEL'SKII and REZNICHENKO [2002]). *Suppose that G is a Hausdorff semitopological group. Then every π-network $\mathcal{E}$ of G at e is a Hausdorff discernor on G.*

3.5. PROPOSITION. *Suppose that G is a group and $\mathcal{E}$ a Hausdorff discernor on G. Then $\bigcap\{PP^{-1} : P \in \mathcal{E}\} = \{e\}$.*

The following definitions were introduced in ARHANGEL'SKII and REZNICHENKO [2002]. Let G be a left topological group. A *topological discernor $\mathcal{E}$ on G* is a Hausdorff discernor on G such that the interior of PP^{-1} contains e, for each $P \in \mathcal{E}$. A discernor is called *open* if all elements of it are open sets. Finally, a discernor $\mathcal{E}$ is said to be *coopen* if P^{-1} is open, for every $P \in \mathcal{E}$. It is clear that open discernors and coopen discernors are topological discernors.

3.6. PROPOSITION (ARHANGEL'SKII and REZNICHENKO [2002]). *Suppose that G is a semitopological group with a countable topological discernor. Then the diagonal is G_δ in $G \times G$ (and, hence, e is a G_δ-point in G).*

A Tychonoff space X is called "weakly pseudocompact" if there exists a Hausdorff compactification $b(X)$ of X such that X is G_δ-dense in $b(X)$. Clearly, every pseudocompact Tychonoff space is weakly pseudocompact. However, every uncountable discrete space is also weakly pseudocompact as well as the product of any countable family of such spaces. In ARHANGEL'SKII and REZNICHENKO [2002] the following improvements of Theorem 2.8 were obtained:

3.7. THEOREM. *Every weakly pseudocompact semitopological group G with a countable topological discernor is a topological group metrizable by a complete (invariant) metric.*

3.8. COROLLARY. *Every weakly pseudocompact semitopological group G of countable π-character is a topological group metrizable by a complete (invariant) metric.*

3.9. COROLLARY. *Every pseudocompact semitopological group G of countable π-character is a compact metrizable topological group.*

The following questions are open:

3.10. PROBLEM. Is every first countable semitopological (paratopological) group subparacompact?

3.11. PROBLEM. Can every first countable paratopological (semitopological) group be condensed onto a metrizable space?

3.12. EXAMPLE. In connection with the above results and problems, it is instructive to have a look at the Sorgenfrey line S. Indeed, S is a first countable paratopological group with the Baire property, and S is paracompact (even Lindelöf). On the other hand, S is not Čech complete, not metrizable, and not even a p-space. The square $S \times S$ is again a first countable paratopological group. However, $S \times S$ is no longer paracompact, but is subparacompact (see BURKE [1984]). See in this connection Problem 3.10.

The results below show that this combination of properties of S is typical for first countable paratopological groups. Here is a key statement:

3.13. LEMMA. *For every paratopological group G, there exists a topological group H homeomorphic to a closed subspace of $G \times G$ such that H can be mapped by a continuous isomorphism j onto G.*

The next result was established in ARHANGEL'SKII and REZNICHENKO [2002]:

3.14. THEOREM. *Suppose that G is a bisequential paratopological group. Then the following three conditions are equivalent:*
1) *$G \times G$ is Lindelöf;*
2) *$e(G \times G) \leq \omega$;*
3) *G has a countable network.*

RAVSKIJ [2001] noticed the following fact:

3.15. THEOREM. *Every first countable paratopological group with a countable network has a countable base.*

This allowed him to obtain the next result:

3.16. THEOREM. *Suppose that G is a first countable paratopological group. Then the following three conditions are equivalent:*
1) *$G \times G$ is Lindelöf;*
2) *$e(G \times G) \leq \omega$;*
3) *G has a countable base.*

3.17. PROBLEM. Suppose that G is a bisequential paratopological group such that $G \times G$ is Lindelöf. Must G have a countable base?

In view of Theorem 3.14, Problem 3.17 is equivalent to the following question, posed by Reznichenko:

3.18. PROBLEM. Is every regular bisequential paratopological group with a countable network first countable?

For any topological group G, the condition that G is of point-countable type (that is, G contains a non-empty compact subspace with a countable base of neighborhoods) is equivalent to the condition that G is a paracompact p-space (see ROELKE and DIEROLF

[1981]). This result does not generalize to paratopological groups, as the example of Sorgenfrey line shows. However, the next result about paratopological groups of point-countable type, which is parallel to Theorem 3.14, was established in ARHANGEL'SKII and REZNICHENKO [2002]:

3.19. THEOREM. *Suppose that G is a paratopological group of point-countable type. Then the following three conditions are equivalent:*
1) *$G \times G$ is Lindelöf;*
2) *$e(G \times G) \leq \omega$;*
3) *G is a Lindelöf Σ-space.*

CHOBAN [1977] has proved that if a compact subspace F of a topological group G is G_δ in G, then G is a dyadic compactum. I am not sure if something similar can be proved about compact subspaces of a paratopological group. However, we have the next statement (see ARHANGEL'SKII and REZNICHENKO [2002]):

3.20. THEOREM. *Suppose that G is a paratopological group, $e(G \times G) \leq \omega$, and F a non-empty compact subspace of G such that $t(F) \leq \omega$ and $\chi(F, G) \leq \omega$. Then G has a countable base.*

3.21. PROBLEM. Suppose that F is a compact G_δ subspace of a (regular, Tychonoff) paratopological group G. Is then F dyadic?

Reznichenko used Lemma 3.13 to establish the next generalization of Tkačenko's theorem on countability of Souslin number of arbitrary σ-compact topological group (TKAČENKO [1983a], extended by USPENSKIJ [1985] to all topological groups which are Lindelöf Σ-spaces).

3.22. THEOREM. *Every paratopological group which is a Lindelöf Σ space has the countable Souslin number.*

The following two results from ARHANGEL'SKII and REZNICHENKO [2002] depend on an additional assumption that $2^\omega < 2^{\omega_1}$.

3.23. THEOREM. ($2^\omega < 2^{\omega_1}$) *Suppose that G is a separable paratopological group such that $G \times G$ is normal. Suppose also that G is of point-countable type. Then G is a Lindelöf Σ-space.*

3.24. THEOREM. *($2^\omega < 2^{\omega_1}$) Suppose that G is a separable first countable paratopological group such that $G \times G$ is normal. Then G has a countable base.*

The above series of results clarify from the point of view of topological algebra why the square of Sorgenfrey line is neither Lindelöf, nor normal.

4. Some special algebraic structures and topologies

In this section we consider the influence on topology of algebraic structures different from the group structure. A *topological semigroup* is a topological space with a continuous associative binary operation. Any linearly ordered topological space with sup (or inf) operation can serve as an example of a topological semigroup. In particular, we can take the usual space of countable ordinals. A semigroup is called *cancellative* if the conditions $ab = ac$, $ca = ba$ and $b = c$ are equivalent for all a, b and c. The Sorgenfrey line is, of course, an example of a cancellative topological semigroup which is not a topological group. Indeed, every paratopological group is a cancellative topological semigroup. This example shows that, unlike topological groups, first countable cancellative topological semigroups needn't be metrizable.

HEATH [1992] has considered systematically how close to metrizable are first countable cancellative topological semigroups. We see from Heath's results that the class of cancellative topological semigroups is much larger than the class of paratopological groups, and that, in general, cancellative topological semigroups behave quite differently from paratopological groups.

4.1. EXAMPLE (HEATH [1992]). There exists a non-metrizable, connected, locally compact, cancellative Abelian topological semigroup which is a Moore space.

Note, that a paratopological group with the above properties cannot exist, since every locally compact paratopological group is a topological group, and a first countable topological group is metrizable. R.W. Heath has also constructed a cancellative topological semigroup that is a Moore space but is not quasi-metrizable (quasi-metrics are like metrics, but the symmetry axiom is dropped). Further, he has shown that a semimetrizable and stratifiable cancellative topological semigroup needn't be Moore. He also exhibited a paracompact, quasi-metrizable, quasi-developable cancellative topological semigroup that is not Moore (HEATH [1992]). See also a discussion of cancellative topological semigroups in Tkačenko's article in this book (in particular, about the Wallace Problem).

Another non-standard algebraic operation, which has an especially strong impact on compact topologies, was introduced by USPENSKIJ in [1989b]. He calls a topological space X *Mal'tsev* if there exists a continuous mapping $M : X^3 \to X$ such that $M(x, y, y) = M(y, y, x) = x$, for all x, y in X. Such a mapping M is called a *Mal'tsev operation* on X. It is easily seen that every topological group G is a Mal'tsev space: a Mal'tsev operation M on G can be defined as follows: $M(x, y, z) = xy^{-1}z$.

A Tychonoff space X is called *retral* if X is a retract of a topological group (ARHANGEL'SKII [1981]). Suppose that $X \subset G$, G is a topological group, and $r : G \to X$ is a retraction. Put $M(x, y, z) = r(xy^{-1}z)$. It is easily verified that M is a continuous Mal'tsev operation on X. Therefore, we have:

4.2. PROPOSITION. *Every retral space is Mal'tsev.*

SIPACHÖVA [1991] proved a partial converse:

4.3. THEOREM. *Every countably compact Tychonoff Mal'tsev space is retral.*

Later, REZNICHENKO and USPENSKIJ [1998] showed that even every pseudocompact Tychonoff Mal'tsev space is retral. This result was complemented by GARTSIDE, REZNICHENKO and SIPACHEVA who constructed an example of a non-retral Tychonoff Mal'tsev space in [1997]. It was proved there that if a Mal'tsev space is a k_ω-space, then it is retral. Therefore, every locally compact paracompact Mal'tsev space is retral. Every zero-dimensional separable metrizable space, as well as every Tychonoff space that condenses onto such a space, is Mal'tsev, GARTSIDE, REZNICHENKO and SIPACHEVA [1997]. On the basis of techniques developed for constructing Mal'tsev spaces, the following fact was established in GARTSIDE, REZNICHENKO and SIPACHEVA [1997]: there exists a Lindelöf topological group G such that the Souslin number of G is 2^ω. This answers a question of Ščepin. Several open questions are formulated in GARTSIDE, REZNICHENKO and SIPACHEVA [1997]:

4.4. PROBLEM. Is every compact Mal'tsev space a retract of a compact topological group?

4.5. PROBLEM. Is every metrizable Mal'tsev space retral? Is every countable Mal'tsev space retral?

4.6. PROBLEM. Is every Mal'tsev Lindelöf Σ-space retral?

However, many theorems on topological invariants of topological groups can be extended to Mal'tsev spaces. In particular, the next result was established by USPENSKIJ [1985]:

4.7. THEOREM. *For every Mal'tsev Lindelöf Σ-space X, the Souslin number of X is countable.*

This generalizes the famous theorem of Tkačenko about Souslin number of σ-compact topological groups.

4.8. THEOREM (USPENSKIJ [1989a]). *Every compact G_δ-subspace of a Mal'tsev space is a Dugundji compactum.*

This statement generalizes Choban's theorem that every compact G_δ-subset of a topological group is a dyadic compactum, CHOBAN [1977].

4.9. COROLLARY (USPENSKIJ [1989a]). *Every Mal'tsev compactum is a Dugundji compactum.*

In view of Sipacheva's Theorem 4.3, the last statement can be reformulated as follows:

4.10. COROLLARY (GARTSIDE, REZNICHENKO and SIPACHEVA [1997]). *Every retral compactum is a Dugundji compactum.*

One of the principal results of REZNICHENKO and USPENSKIJ [1998] is the next theorem:

4.11. THEOREM. *Suppose that X is a pseudocompact Mal'tsev space. Then:*

1) *βX is a Dugundji compactum;*
2) *every Mal'tsev operation on X extends to a Mal'tsev operation on βX;*
3) *X is a retract of a topological group, that is, X is retral.*

The proof of Theorem 4.11 is based on the next lemma which is interesting in itself:

4.12. LEMMA. *Suppose that X, Y, and Z are pseudocompact Tychonoff spaces. Then every continuous real-valued function on the product $X \times Y \times Z$ can be extended to a separately continuous function on $\beta X \times \beta Y \times \beta Z$.*

It is not clear if a similar statement holds for the product of four factors (Reznichenko). The proof of Sipacheva's Theorem 4.3 remains an important part of the argument of Reznichenko and Uspenskij. They also established another generalization of this theorem:

4.13. THEOREM. *If a countably compact Tychonoff space admits a separately continuous Mal'tsev operation $f : X^3 \to X$, then X is a Dugundji compactum.*

Using the above results, Uspenskij and Reznichenko have generalized the well known theorem of COMFORT and ROSS [1966] on the productivity of pseudocompactness for topological groups as follows:

4.14. THEOREM. *The product of any family of pseudocompact Mal'tsev spaces is pseudocompact.*

Some other interesting generalizations of Comfort-Ross theorem were obtained in HUŠEK [1992] and HERNÁNDEZ, SANCHIS and TKAČENKO [2000].

A.S. GUL'KO [1996] has studied another special algebraico-topological structure - that of a rectifiable space. The roots of the definition of the notion of a rectifiable space are in a paper SHAPIROVSKIJ [1978]. A space X is called *rectifiable*, or a *space with rectifiable diagonal*, if there exist a point $e \in X$ and a homeomorphism Ψ of X^2 onto X^2 such that:

(1) The first coordinate of $\Psi(x, y)$ is x, for each $(x, y) \in X^2$, and

(2) $\Psi(x, x) = (x, e)$, for every $x \in X$.

The first condition simply means that the "vertical" slices of X^2 are invariant under Ψ, and the second condition tells us that the diagonal in X^2 becomes "horizontal" under Ψ. M.M. Choban noticed that a topological space X is rectifiable if and only if there are continuous mappings $p : X^2 \to X$, $q : X^2 \to X$ and an element $e \in X$ such that, for all x and y in X, $p(x, q(x, y)) = q(x, p(x, y)) = y$ and $q(x, x) = e$. Using operations p and q, one can define a continuous Mal'tsev operation on an arbitrary rectifiable space as follows: $f(x, y, z) = p(x, q(y, z))$, for all x, y, z in X. Therefore, we have:

4.15. PROPOSITION. *Every rectifiable space is a Mal'tsev space.*

GUL'KO [1996] established the following result:

4.16. PROPOSITION. *Every topological group is rectifiable.*

To see this, we define a "rectifying" mapping Ψ on $G \times G$ as follows: $\Psi(x, y) = (x, x^{-1}y)$, for every $x \in G$.

Proposition 4.2 implies that the closed unit interval is a Mal'tsev space. Therefore, a compact Mal'tsev space needn't be homogeneous. On the other hand, we have (see GUL'KO [1996]):

4.17. PROPOSITION. *Every rectifiable space is homogeneous.*

It follows that not every Mal'tsev space is rectifiable, that is, the converse to Proposition 4.15 does not hold.

As in the case of topological groups, weak separation axioms in rectifiable spaces imply much stronger separation axioms. GUL'KO proved in [1996] that every rectifiable space satisfying T_0 axiom is a regular T_1 space. However, the next question remains open:

4.18. PROBLEM. Is every T_0 rectifiable space Tychonoff?

There is an interesting relationship between rectifiable spaces and the fixed point property, GUL'KO [1996]:

4.19. THEOREM. *Suppose that X is a rectifiable topological space such that every continuous mapping of X into itself has a fixed point. Then X consists of exactly one point, that is, $|X| = 1$.*

Therefore, the Tychonoff cube I^τ, for $\tau \geq 1$, is not rectifiable, though it is a homogeneous Dugundji compactum. It is not so easy to present an example of a rectifiable space which is not a topological group. The sphere S^7 is not homeomorphic to a topological group. However, using quaternions, one can show that S^7 is a rectifiable space. Multiplying S^7 by a zero-dimensional topological group, we again obtain a rectifiable space which is not homeomorphic to a topological group, GUL'KO [1996].

Kakutani's theorem on metrizability of first countable topological groups extends to rectifiable spaces, GUL'KO [1996]:

4.20. THEOREM. *Every first countable rectifiable T_0 space is metrizable.*

It follows that Sorgenfrey line S is not rectifiable. On the other hand, Sorgenfrey line is Mal'tsev (GARTSIDE, REZNICHENKO and SIPACHEVA [1997]). Thus, S is a homogeneous Mal'tsev space that is not rectifiable. The next question seems to be new and open:

4.21. PROBLEM. Is every rectifiable paratopological group a topological group?

A closely related question:

4.22. PROBLEM. Suppose that G is a paratopological group and a Mal'tsev space. Is then G a topological group? Is G homeomorphic to a topological group?

A.S. Gul'ko discovered many parallels in the behavior of cardinal invariants for topological groups and rectifiable spaces. In particular, the weak first countability for rectifiable spaces is equivalent to metrizability, the Fréchet-Urysohn property turns into the strong Fréchet-Urysohn property, bisequential rectifiable spaces are metrizable, the π-character coincides with the character, the pseudocharacter coincides with the diagonal number, GUL'KO [1996].

Since every rectifiable space is a Mal'tsev space, and every Mal'tsev compactum is Dugundji, we have (Uspenskij, Choban):

4.23. COROLLARY. *Every rectifiable compactum is Dugundji.*

The square of a pseudocompact rectifiable space X is pseudocompact, since the product of pseudocompact Mal'tsev spaces is pseudocompact. Therefore, by Glicksberg's theorem, $\beta X \times \beta X = \beta(X \times X)$, if X is rectifiable and pseudocompact. It follows that the operations on X providing for rectifiability of X can be extended continuously to βX, turning βX into a rectifiable compactum.

It seems, the next problem is formulated for the first time.

4.24. PROBLEM. Is every Tychonoff rectifiable space retral?

5. Extremal topologies and various algebraic structures

A topological space X is called *extremally disconnected*, if the closure of any open subset of X is open in X. It was noticed rather early that extremally disconnected homogeneous spaces with nice separation or compactness type properties are not easy to come by. In particular, an extremally disconnected compact Hausdorff space is homogeneous if and only if it is finite and discrete (Z. Frolík). And it is still an open question, formulated for the first time in ARHANGEL'SKII [1967], whether there exists in ZFC a non-discrete extremally disconnected topological group.

The first consistent example of a non-discrete extremally disconnected topological group was constructed, under CH, by SIROTA [1969]. Note that the next theorem holds in ZFC, ARHANGEL'SKII [1967].

5.1. THEOREM. *If G is an extremally disconnected topological group, then every compact subspace F of G is finite.*

Since a k-space, in which all compact subspaces are finite, must be discrete, we have:

5.2. COROLLARY. *Suppose that G is an extremally disconnected topological group and G is a k-space. Then G is discrete.*

Thus, a non-discrete extremally disconnected topological group is indeed a rare animal. To understand better where to look for a ZFC example of a non-discrete extremally disconnected topological group, a research was done on the influence of extremal disconnectedness on the algebraic structure of an extremally disconnected group. The next result of FROLÍK [1968] is instrumental in this connection.

5.3. THEOREM. *Let X be an extremally disconnected Hausdorff space, and h a homeomorphism of X onto itself. Then the set $M = \{x \in X : h(x) = x\}$ of all fixed points under h is an open and closed subset of X.*

This leads to the following theorem of MALYKHIN [1975]:

5.4. THEOREM. *Let G be an extremally disconnected topological group. Then there exists an open and closed Abelian subgroup H of G such that $a^2 = e$, for each $a \in H$.*

The proof of this theorem heavily depends on the assumption that G is a topological group, in particular, on the joint continuity of multiplication in G. If we replace this assumption with a weaker one that the multiplication is separately continuous, we cannot derive a conclusion as strong as in Theorem 5.4, but we still can obtain some interesting information on the topologico-algebraic structure of G. The next seven statements were proved in ARHANGEL'SKII [2000b].

5.5. THEOREM. *Let G be an extremally disconnected quasitopological group. Then the set W of all elements a of G such that $a^2 = e$ is an open (and closed) neighborhood of the neutral element e of G.*

If G is a group and $a \in G$, we denote by C_a the set of all $b \in G$ which commute with a (that is, satisfy the condition $ab = ba$).

5.6. THEOREM. *Let G be an extremally disconnected quasitopological group. Then, for any $a \in G$, the set C_a of all $b \in G$ that commute with a is an open and closed subgroup of G (containing a).*

Theorem 5.6 allows to strengthen Theorem 5.4 in the following way. Let G be an extremally disconnected topological group. Then, for any $a \in G$, there exists an open (and closed) Abelian subgroup A of G such that, for every element b of A, $ab = ba$ and $b^2 = e$.

5.7. THEOREM. *Let G be an extremally disconnected quasitopological group such that G is generated by every open neighborhood of the neutral element e. Then G is Abelian, and $a^2 = e$, for each $a \in G$.*

5.8. THEOREM. *Let G be a separable extremally disconnected quasitopological group. Then there exists an Abelian subgroup H of G such that H is a closed G_δ-subset of G. Moreover, H can be chosen so that every element of H commutes with every element of G.*

5.9. THEOREM. *Let G be an extremally disconnected quasitopological group, and b any element of G. Then the set $M_b = \{x \in G : x^2 = b\}$ is open and closed in G.*

We say that the *discrete Souslin number* of a space X is countable if every discrete in X family of non-empty open subsets of X is countable.

5.10. PROPOSITION. *Let G be an extremally disconnected quasitopological group such that the discrete Souslin number of the space G is countable. Then the set of all $b \in G$, for which there exists $a \in G$ such that $a^2 = b$, is countable.*

We will call a group G a *group with square roots*, if for each $b \in G$ there exists $a \in G$ such that $a^2 = b$.

5.11. THEOREM. *Let G be an extremally disconnected quasitopological group with square roots, and suppose that the discrete Souslin number of the space G is countable. Then G is countable.*

It follows that a pseudocompact extremally disconnected quasitopological group with square roots is finite, a Lindelöf extremally disconnected quasitopological group is countable, and an extremally disconnected quasitopological group with square roots and with the countable Souslin number is countable.

5.12. THEOREM. *Let G be an extremally disconnected quasitopological group with square roots such that the discrete Souslin number of the space G is countable and the space G has the Baire property. Then G is countable and discrete.*

This assertion follows from Theorem 5.11.

If G is an extremally disconnected group, then the set $L = \{x \in G : x^3 = e\}$ need not be open in G. Indeed, if L is open, then L is a neighborhood of e; therefore, $L \cap M_e$ is also an open neighborhood of the neutral element e in G (recall that $M_e = \{x \in G : x^2 = e\}$). On the other hand, it is clear that $M_e \cap L = \{e\}$; therefore, e is isolated in G, which implies that G is discrete.

The next old question remains open:

5.13. PROBLEM. Is there in ZFC an example of a non-discrete extremally disconnected topological group?

In connection with Theorem 5.4, it is natural to ask the following question:

5.14. PROBLEM. Let G be an extremally disconnected quasitopological group. Is then true that there exists an open and closed Abelian subgroup of G?

Theorem 5.3 immediately implies the following statement from ARHANGEL'SKII [2000b]:

5.15. THEOREM. *If a topological skew field F is extremally disconnected, then it is discrete.*

Theorem 5.15 remains true if we only assume that F is an extremally disconnected semitopological skew field.

A dense in itself non-empty Hausdorff space is called *maximal* if any strictly stronger topology on X has at least one isolated point. Every maximal space is extremally disconnected. E. van Douwen has shown that there exists a countable Tychonoff maximal infinite space (in ZFC). A natural question, whether a topological group can be a maximal space, was answered by V.I. Malykhin under Martin's Axiom: he established that under this assumption there exists a countable maximal topological group, MALYKHIN [1975].

Every maximal space has another interesting property: it is submaximal. A Hausdorff space X is called *submaximal* if every dense subset of X is open in X, or, equivalently, if every subset A of X is open in its closure. According to Malykhin's result, it is consistent that there exists a non-discrete submaximal topological group. However, the next question, formulated in ARHANGEL'SKII and COLLINS [1995] remains open:

5.16. PROBLEM. Is there in ZFC a non-discrete submaximal topological group?

The following simple facts were observed in ARHANGEL'SKII and COLLINS [1995]:

5.17. PROPOSITION. *Every subgroup of a submaximal topological group is closed.*

5.18. COROLLARY. *Every dense subset of a submaximal topological group G is a set of algebraic generators of G.*

5.19. COROLLARY. *Each separable submaximal topological group is countable.*

Later, it was shown by ALAS, PROTASOV, TKAČENKO, TKACHUK, WILSON and YASHCHENKO [1998] that separability in the last statement can be replaced by the countability of the Souslin number. If G is a submaximal topological group, then either the space G is submetrizable (that is, G condenses onto a metrizable space), or G is a P-space, that is, every G_δ-subset of G is open in G (ARHANGEL'SKII and COLLINS [1995]). Answering a question from the last mentioned paper, ALAS, PROTASOV, TKAČENKO, TKACHUK, WILSON and YASHCHENKO [1998] obtained the following strong result:

5.20. THEOREM. *Every submaximal topological group G of Ulam non-measurable cardinality is strongly σ-discrete, that is, the space G is the union of a countable family of closed discrete subspaces.*

According to an old result of MALYKHIN [1975], every maximal topological group satisfies the same conclusion in ZFC, since every such group must have an open countable subgroup. Here are another two very interesting results from ALAS, PROTASOV, TKAČENKO, TKACHUK, WILSON and YASHCHENKO [1998]:

5.21. THEOREM. *Every ω-bounded submaximal topological group is countable.*

5.22. THEOREM. *Every submaximal topological group of Ulam non-measurable cardinality is hereditarily paracompact.*

It follows that a submaximal topological group of Ulam non-measurable cardinality is zero-dimensional, see ALAS, PROTASOV, TKAČENKO, TKACHUK, WILSON and YASHCHENKO [1998]. In particular, under these assumptions G cannot be connected. This answers a question from ARHANGEL'SKII and COLLINS [1995].

Some time ago a method allowing to construct non-discrete maximal or extremally disconnected topologies on groups turning them into semitopological or quasitopological groups was developed (see PAPAZYAN [1991], PROTASOV [1993]). This method is based on the notion of idempotent. An element p of a semigroup O is called an *idempotent* if $p^2 = p$. ELLIS [1957] proved a very useful result: *every (non-empty) compact right topological semigroup has an idempotent.*

Let G be a discrete group with the discrete topology and βG the Čech-Stone compactification of the discrete space G. Then the product operation in G can be extended to a product operation in βG in such a way that βG becomes a right topological semigroup. This can be done so that the left action on βG by any element of G is continuous. Furthermore, under the last condition this extension is unique, and the neutral element e of G is also a left and right unit of the semigroup βG.

Indeed, for an element a of G, let λ_a be the left action by a on G, that is, $\lambda_a(b) = ab$, for each $b \in G$. Since λ_a is a continuous mapping of G into G, we can extend it to a continuous mapping of βG into βG. The latter mapping we also denote by λ_a, and put $aq = \lambda_a(q)$, for each $q \in \beta G$. Thus, the product aq in βG is defined for each $a \in G$ and each $q \in \beta G$.

Now fix $q \in \beta G$ and put $\rho_q(x) = xq$, for every $x \in G$. In this way a mapping ρ_q is defined on G, with values in G. Since G is discrete, ρ_q is continuous. Therefore, ρ_q can be extended to βG; we denote the extension also by ρ_q. Now, for any p in βG put $pq = \rho_q(p)$.

The definition of the product operation is complete. It can be shown that this operation is associative, HINDMAN and STRAUSS [1998]. Since the mapping ρ_q is continuous, for every $q \in \beta G$, βG with this product operation is a right topological semigroup. The statement about identities follows from the continuity of λ_a and ρ_a for each $a \in G$.

Whenever G is a discrete group, we consider βG as the right topological semigroup with the product operation defined in the argument above.

According to Ellis's theorem, there exists an idempotent in the right topological semigroup βG.

5.23. PROPOSITION. *Suppose that p is a free ultrafilter on a discrete group G. Then p is an idempotent in βG if and only if for each $A \in p$ there exists $B \in p$ such that $A \in bp$, for each $b \in B$.*

For any discrete group G, idempotents in βG can be used to produce natural topologies on the group G itself.

5.24. THEOREM. *Suppose that G is a discrete group with the identity e, and $p \in \beta G \setminus G$, p an idempotent of the compact semigroup βG. Put $\mathcal{F}_p = \{\{e\} \cup U : U \in p\}$. Then there exists a topology $\mathcal{T}_p$ on G such that:*

1) *G endowed with $\mathcal{T}_p$ is a left topological semigroup;*

2) *for each $a \in G$, the family $\eta_a = \{Int(aP) : P \in \mathcal{F}_p\} = \{\{a\} \cup Int(aU) : U \in p\}$ is a base of the space G at a;*

3) *the space G is Hausdorff;*

4) *G is extremally disconnected;*

5) *G is homogeneous;*

6) *there are no isolated points in G;*

7) *e belongs to the closure of a set $A \subset G \setminus \{e\}$ if and only if $A \in p$;*

8) *for every topology $\mathcal{T}$ on G that is strictly larger than the topology $\mathcal{T}_p$, there exists an isolated point in $(G, \mathcal{T})$ (this means that the topology $\mathcal{T}_p$ is maximal).*

An idempotent p in a semigroup S will be called a *Protasov idempotent* if the equation $xp = p$ has the unique solution $x = p$ in S. PROTASOV [1998b] established that the topology $\mathcal{T}_p$ in the preceding theorem is regular if and only if the idempotent p is Protasov. An equivalent condition is that the topology $\mathcal{T}_p$ is zero-dimensional. From Protasov's results it follows (see HINDMAN and STRAUSS [1998]) that if a countable infinite discrete group G can be algebraically embedded in a metrizable compact group, then there exists a Protasov idempotent in $\beta G \setminus G$. Therefore, we have the following result:

5.25. THEOREM (PROTASOV [1998b]). *On the group of integers there exists a non-discrete Tychonoff homogeneous maximal topology.*

In this way Protasov answered an old question of Eric van Douwen.

Protasov established several other interesting facts concerning maximal topological and paratopological groups.

5.26. THEOREM (PROTASOV [1998b]). *Every maximal topological group is complete in the left uniformity.*

5.27. THEOREM (PROTASOV [200?]). *Every maximal paratopological group is a topological group.*

I do not know if a similar statement holds for extremally disconnected paratopological groups. There is yet another natural way to use topology of βG to produce some natural topologies on G itself.

5.28. PROPOSITION. *Suppose that G is a group, $p \in \beta G$, and b, c are two distinct elements of G. Then $bp \neq cp$.*

5.29. COROLLARY. *Let G be a discrete group. Then, for any q in $\beta G \setminus G$, the restriction of the mapping ρ_q to G is one-to-one.*

If G is a discrete group, then we put $G_q = \rho_q(G)$ and $F_q = \rho_q(\beta G)$. Of course, G_q and F_q are taken with the topology generated from βG. The subspace G_q is called the *orbit* of q in βG under the action of G (or the G-*orbit* of q). Clearly, the orbit of q always contains q.

5.30. PROPOSITION. *Suppose that G is a discrete group and $q \in \beta G$. Then:*

1) *For each a in G, the mapping λ_a restricted to G_q is a homeomorphism of G_q onto itself;*

2) *For each a in G, $\rho_q \lambda_a | G = \lambda_a \rho_q | G$;*

3) $\lambda_a(\rho_q(b)) = \rho_q(ab) = abq$, *for each $b \in G$;*

4) *The space G_q is homogeneous.*

It is natural to ask if ρ_q restricted to G is actually a homomorphism of the group G into the semigroup βG. However, $\rho_q(a)\rho_q(b) = aqbq$ and $\rho_q(ab) = abq$. Since there is no reason to believe that $aqbq = abq$, we should not also expect that $\rho_q | G$ is a homomorphism.

The reasoning above also shows that the subspace $G_q = \{bq : b \in G\}$ is not, in general, a subgroup of the semigroup βG. Now we are going to show that we can introduce a new product operation on the subset G_q in such a way that, with the topology already defined on G_q, it will become a left topological group, and ρ_q will become an isomorphism of the group G onto the group G_q. In fact, if the last condition is to be satisfied, there is only one way to define the new operation $\times$ on G_q: we have to put $aq \times bq = abq$. Since ρ_q is one-to-one and $aq = \rho_q(a)$,with this operation G_q becomes a group, and ρ_q becomes an isomorphism of G onto G_q. The left action by the element aq on the group G_q so defined coincides with the restriction of λ_a to G_q and is therefore, by Proposition 5.30, a homeomorphism of the space G_q onto itself. Hence, G_q is a left topological group. We sum up the information obtained in the next statement.

5.31. THEOREM. *Suppose that G is a discrete group and $q \in \beta G$, and the product operation $\times$ on G_q is defined by the formula $aq \times bq = abq$. Then G_q, with this operation and with the subspace topology, is a left topological group, and the mapping $\rho_q | G$ is an isomorphism of the group G onto the group G_q. Furthermore, there exists a unique topology $\mathcal{T}'$ on G such that G with this topology is a left topological group and ρ_q is a topological isomorphism of it onto the left topological group G_q.*

Let us consider some particular cases of the construction described in Proposition 5.30 and Theorem 5.31.

5.32. THEOREM. *Suppose that G is a discrete group and $q \in \beta G$, q an idempotent. Then:*
1) *the subspace $G_q = \rho_q(G)$ of βG is extremally disconnected and dense-in-itself;*
2) *the closure F_q of G_q in βG is also extremally disconnected; and*
3) *$F_q = \overline{G_q}$ is the Čech-Stone compactification of the space G_q.*

Theorem 5.32 is based on the following general result.

5.33. THEOREM. *Suppose that S is a right topological semigroup and q an idempotent in S. Then the subspace Sq of S is a retract of S.*

The mapping ρ_q is continuous and is a retraction of S onto Sq. Indeed, take any $y \in Sq$. Then $y = xq$, for some $x \in S$, and we have $\rho_q(y) = yq = xqq = xq = y$.

5.34. THEOREM. *Suppose X is an extremally disconnected compact Hausdorff space, Y a retract of X, $r : X \to Y$ a retraction, and D a dense subspace of X such that X is the Čech-Stone compactification of D. Then:*
1) *Y is extremally disconnected; and*
2) *Y is the Čech-Stone compactification of the subspace $r(D)$.*

5.35. COROLLARY (PROTASOV [1998b]). *On every infinite discrete group G there exists an extremally disconnected Tychonoff topology $\mathcal{T}$ such that $(G, \mathcal{T})$ is a left topological group without isolated points.*

The topology in Corollary 5.35 is automatically regular (even Tychonoff) while the topology in Theorem 5.24 need not be regular.

Here is an important special case of Theorem 5.31:

5.36. THEOREM. *Suppose G is a discrete Abelian group, and $q \in \beta G \setminus G$, q an idempotent. Then $h_q = \rho_q|G$ is a monomorphism of the group G into the semigroup βG, and the image $G_q = h_q(G)$ is an extremally disconnected semitopological Abelian subgroup of the semigroup βG.*

The new element in this statement, compared to Theorem 5.31, is that to make the mapping $\rho_q|G$ into a homomorphism, we do not have to change the product operation on G_q: the multiplication which is already there fits well!

Indeed, $\rho_q(xy) = xyq = xyq^2 = xqyq = \rho_q(x)\rho_q(y)$ (since $yq = qy$, for each $y \in G$, whenever G is an Abelian group). Thus, h_q is a homomorphism of G into βG. Therefore, $G_q = h_q(G)$ is a subgroup of the semigroup βG. Clearly, $h_q = \rho_q|G$ is a monomorphism (see Theorem 5.31). By Theorem 5.32, G_q is extremally disconnected. Finally, G_q is a left topological group, since it is a subgroup of the left topological semigroup βG. However, G_q is commutative, since it is isomorphic to G. It follows that G_q is a semitopological Abelian subgroup of the semigroup βG.

Theorem 5.24 can be considerably strengthened in the following way.

5.37. THEOREM. *Let $(G, \mathcal{T})$ be a non-discrete paratopological group. Then there exists a maximal non-discrete Hausdorff topology $\mathcal{T}'$ on G such that $(G, \mathcal{T}')$ is a left topological group and $\mathcal{T} \subset \mathcal{T}'$.*

Using Theorem 5.36, we can easily construct a nondiscrete extremally disconnected quasitopological group. Indeed, let G be the σ-product of ω copies of the Boolean group

$D = \{0,1\}$. We endow G with the discrete topology, and apply Theorem 5.36 to it. Clearly, G and G_q are also Boolean groups, that is, the inverse operation is the identity mapping. Hence, the inverse operation in G_q is continuous, and G_q is the quasitopological group we are looking for.

The notion of a submaximal space is closely related to the notion of an irresolvable space. According to E. Hewitt, a space X is called *resolvable* if one can find two disjoint dense subsets in it. Otherwise, it is called *irresolvable*. Clearly, every submaximal space is irresolvable, since every dense subset in it is open. The theory of irresolvable spaces has a special flavor in the class of topological groups. COMFORT and VAN MILL [1994] proved that every non-discrete Abelian topological group with a finite number of elements of order 2 is resolvable. They introduced the notion of absolute resolvability in the context of topological groups which turned out to be very useful. A subset D of a group G is called *absolutely dense in* G if D is dense in G for every non-discrete group topology on G. ZELENYUK [1998] proved that every infinite countable Abelian group with a finite number of elements of order 2 contains an infinite disjoint family of absolutely dense subsets.

PROTASOV [1998e], answering a question of Comfort and van Mill, showed that every non-discrete Abelian irresolvable topological group contains a countable open subgroup consisting of elements of order 2. It is not clear if an irresolvable non-discrete group exists in ZFC. Protasov posed the following question which has good chances to get a positive answer:

5.38. PROBLEM. Is the product of two arbitrary non-discrete topological groups resolvable?

Curiously, the square of every non-discrete topological group is resolvable. This fact is cited in PROTASOV [1998e] and attributed to MASAVEU [1995]. Masaveu also established that the product of every two non-discrete Abelian topological groups is resolvable. Protasov proved that every totally bounded topological group is resolvable. A somewhat stronger result see in MALYKHIN and PROTASOV [1996].

Many interesting and deep results on the semigroup βG, with applications, contains the book HINDMAN and STRAUSS [1998]. In particular, ZELENYUK [1997], answering an old question, proved that if G is a countable discrete group without non-trivial finite subgroups, then the semigroup $\beta G \setminus G$ also does not have non-trivial finite subgroups.

5.39. PROBLEM. Is every extremally disconnected (regular) paratopological group a topological group?

5.40. PROBLEM. Is there an example in ZFC of a nondiscrete extremally disconnected regular paratopological group?

6. Topological groups and completions

A space X is called *Moscow*, ARHANGEL'SKII [1983], if for each open subset U of X, the closure of U in X is the union of a family of G_δ-subsets of X, that is, for each $x \in \overline{U}$ there exists a G_δ-subset P of X such that $x \in P \subset \overline{U}$.

The techniques based on the notion of Moscow space played a vital role in the recent solution in ARHANGEL'SKII [2000a] of the next problem posed by PESTOV and TKAČENKO [1985] (see also TKAČENKO [1991b]):

Let G be a topological group, and μG the Dieudonné completion of the space G. Can the operations on G be extended to μG in such a way that μG becomes a topological group containing G as a topological subgroup?

Recall, that the Dieudonné completion μG of G is the completion of G with respect to the maximal uniformity on G compatible with the topology of G. It is well known that the Dieudonné completion of a topological space X is always contained in the Hewitt-Nachbin completion υX of X. In fact, μX is the smallest Dieudonné complete subspace of υX containing X. Moreover, if there are no Ulam-measurable cardinals, then υX and μX coincide (see ENGELKING [1977]). Therefore, the next question, also belonging to Pestov and Tkačenko, is almost equivalent to the question above:

Let G be a topological group, and υG the Hewitt-Nachbin completion of the space G. Can the operations on G be extended to υG in such a way that υG becomes a topological group containing G as a topological subgroup?

Clearly, if there exists an Ulam-measurable cardinal τ, then for any discrete group G of cardinality τ the answer to the last question is in negative (since in this case the Hewitt-Nachbin completion υG is a non-discrete non-homogeneous space).

Below we call the first question the *PT-problem*. Until recently, even a consistent counterexample to the PT-problem was not known, though astonishingly large classes of topological groups were found in which the answer to it is positive (see USPENSKIJ [1989b], TKAČENKO [1991b]). The majority of these results are corollaries to the following general theorem obtained in ARHANGEL'SKII [2000a]:

6.1. THEOREM. *For every Moscow topological group G, the operations in G can be continuously extended to the Dieudonné completion μG of G so that μG becomes a topological group containing G as a (topological) subgroup.*

Because of this result it is important to find out which topological groups are Moscow. Such a study has been conducted in ARHANGEL'SKII [200?], [2000c]. Note, that quite a few important classes of topological spaces are subclasses of the class of Moscow spaces. Indeed, every space of countable pseudocharacter is, obviously, Moscow, every extremally disconnected space is Moscow, every perfectly κ-normal space ŠČEPIN [1976] is Moscow, and, hence, every κ-metrizable ŠČEPIN [1979] space is Moscow. Besides, the product of any family of first countable spaces is Moscow, the product of any family of spaces with a countable network is Moscow, and the product of any family of metrizable spaces is Moscow (see ARHANGEL'SKII [200?]). Also every dense subspace of a Moscow space is Moscow. Thus, the class of Moscow spaces is quite wide. However, it turned out that a topological group is Moscow even much more often than a topological space in general. In particular, the countability restriction on a cardinal invariant very often implies that a topological group satisfying this restriction is Moscow. For example, every topological group with the countable Souslin number is Moscow, USPENSKIJ [1989a]. It follows that all totally bounded topological groups (and, hence, all pseudocompact topological groups) are Moscow (see an elementary direct proof of this in ARHANGEL'SKII [1999a]). Since every pseudocompact group G is a Moscow space, it follows from Theorem 6.1 that the

Dieudonné completion μG of a pseudocompact topological group G is a topological group as well. Since G is pseudocompact, μG coincides with the Stone-Čech compactification βG of the space G. In this way we arrive at the classical result of COMFORT and ROSS [1966]: the Stone-Čech compactification βG of a pseudocompact topological group G is a compact topological group containing G as a subgroup.

In fact, not only topological groups, but paratopological groups and semitopological groups as well, are much more often Moscow than topological spaces in general. This can be seen from a very general result below.

Let G be a right topological group, and $U \subset G$. A subset A of G is called an *ω-deep subset of U* if there exists a G_δ-subset P of G such that $e \in P$ and $AP \subset U$. The *g-tightness* $t_g(G)$ of a right topological group G is countable ($t_g(G) \leq \omega$), if for each canonical (that is, regular) open subset U of G and each $x \in \overline{U}$, there exists an ω-deep subset A of U such that $x \in \overline{A}$, ARHANGEL'SKII [2000a]. If G is a paratopological group such that the Souslin number of G is countable, then the g-tightness of G is countable. Clearly, if G is an extremally disconnected topological group, then $t_g(G) \leq \omega$. The next result from ARHANGEL'SKII [2000c] demonstrates the remarkable phenomenon: a minor restriction on a topological group guarantees that this group is Moscow.

6.2. THEOREM. *Every right topological group G of countable g-tightness is a Moscow space.*

Theorem 6.2 covers very large classes of topological groups.

The *o-tightness* of a space X is countable (notation: $ot(X) \leq \omega$) if whenever a point a belongs to the closure of $\cup\gamma$, where γ is a family of open sets, there exists a countable subfamily η of γ such that a is in the closure of $\cup\eta$, TKAČENKO [1983b].

With the help of Theorem 6.2 it can be established (see ARHANGEL'SKII [2000c]) that a topological group G is Moscow in each of the following cases: 1) the pseudocharacter of G is countable; 2) the tightness of G is countable; 3) the Souslin number of G is countable; 4) G is extremally disconnected; 5) G is κ-metrizable; 6) G is a subgroup of a topological group F such that F is a k-space; 7) the o-tightness of G is countable.

If A is a subset of a space X, then the *G_δ-closure* of A in X is defined as the set of all points $x \in X$ such that every G_δ-subset of X containing x has a non-empty intersection with A. If X is the G_δ-closure of A, we say that A is *G_δ-dense in X*. If the G_δ-closure of A coincides with A, we say that A is *G_δ-closed.*

The abbreviation (MA+¬CH) stands for Martin's Axiom combined with the negation of the Continuum Hypothesis.

Let us describe the basic steps in the proof of Theorem 6.1. Recall that a subspace Y of a space X is said to be *C-embedded in X*, if every continuous real-valued function f on Y can be extended to a continuous real-valued function on X. It is well known that if a dense subspace Y of a space X is C-embedded in X, then Y is G_δ-dense in X. The converse to this statement is not true (to see this, take the Alexandroff one-point compactification of an uncountable discrete space). However, every G_δ-dense subspace Y of a Moscow space X is C-embedded in X, USPENSKIJ [1989a]. Another key fact was established in ARHANGEL'SKII [2000a]:

6.3. THEOREM. *If a Moscow space Y is a G_δ-dense subspace of a homogeneous space X, then X is also a Moscow space.*

A standard example of a non-Moscow space X such that every G_δ-dense subspace Y of X is C-embedded in X is the space $\omega_1 + 1$ of ordinals, since every G_δ-dense subspace Y of $\omega_1 + 1$ is either ω_1 or $\omega_1 + 1$. To see that $\omega_1 + 1$ is not Moscow, take two disjoint uncountable sets U and V consisting of isolated ordinals. Then the point ω_1 is in the intersection of their closures. Assume now that $\omega_1 + 1$ is Moscow. Then there are G_δ-subsets P_1 and P_2 in $\omega_1 + 1$ such that $\omega_1 \in P_1 \subset \overline{U}$ and $\omega_1 \in P_2 \subset \overline{V}$. Put $P = P_1 \cap P_2$. Then P is a G_δ-set and $\omega_1 \in P \subset \overline{U} \cap \overline{V}$. It follows that $P \cap (U \cup V) = \varnothing$. However, this is impossible, since every non-empty G_δ-set in $\omega_1 + 1$ containing the point ω_1 has a non-empty intersection with every uncountable subset of ω_1. Therefore, $\omega_1 + 1$ is not Moscow.

This example also shows that the Dieudonné completion of a Moscow space need not be a Moscow space.

The second step in the proof of Theorem 6.1 involves the Raĭkov completion of a topological group.

Recall that the *Raĭkov completion* ρG of a topological group G is the completion of G with respect to the natural two-sided uniformity of the topological group G. It is well known, that ρG can be interpreted as a Raĭkov complete topological group, containing G as a dense subgroup, ROELKE and DIEROLF [1981].

The G_δ-closure of G in ρG is denoted by $\rho_\omega G$. Clearly, $\rho_\omega G$ is a subgroup of ρG containing G. It is well known that, for every topological group G, the space $\rho_\omega G$ is Dieudonné complete.

6.4. THEOREM (ARHANGEL'SKII [2000c]). *Let G be a Moscow group. Then $\rho_\omega G$ is a Dieudonné complete Moscow group, in which the space G is C-embedded.*

To complete the outline of the proof of Theorem 6.1, it remains to establish that if Z is a Dieudonné complete topological group, and G a dense subgroup of Z C-embedded in Z, then there exists a subgroup M of Z such that $G \subset M$ and the space M is the Dieudonné completion μG of G.

For the sake of brevity, a topological group G is called below a *PT-group*, if the operations on G can be extended to the Dieudonné completion μG in such a way that G becomes a topological subgroup of the topological group μG.

Naturally, the following question arises: *when $\mu G = \rho_\omega G$?*

Notice, that every Raĭkov complete group trivially satisfies the above equality. However, the answer in ZFC to the next question is unknown:

6.5. PROBLEM. Is for every Moscow group G true that $\mu G = \rho_\omega G$?

Of course, Theorem 6.1 implies the next result for Hewitt-Nachbin completions:

6.6. THEOREM (ARHANGEL'SKII [2000a]). *Suppose that G is a Moscow group of Ulam non-measurable cardinality. Then the operations on G can be extended to the Hewitt-Nachbin completion υG of G in such a way that υG becomes a topological group containing G as a topological subgroup.*

Clearly, for every Moscow group G of Ulam non-measurable cardinality, $\mu G = \rho_\omega G$ $(= \upsilon G)$. In ARHANGEL'SKII [2000c] it was shown that if G is a topological group such that μG is a Lindelöf topological group (containing G as a topological subgroup), then

$\mu G = \rho_\omega G$. The argument runs as follows: μG can be interpreted as a subgroup of ρG such that $G \subset \mu G \subset \rho_\omega G$. It is well known that every Lindelöf space is G_δ-closed in each Tychonoff space in which it is dense. Therefore, μG is G_δ-closed in ρG, and $\mu G = \rho_\omega G$.

A construction of TKAČENKO [1991b] was applied to show that every Abelian topological group H can be represented as a closed subgroup of a Moscow group (see ARHANGEL'SKII [2000c]). Since not every Abelian topological group is a PT-group, it follows that not every closed subgroup of a Moscow group is a PT-group.

6.7. PROBLEM. Is every C-embedded subgroup of a Moscow group Moscow?

Clearly, if H is a C-embedded subgroup of a PT-group G, then H is also a PT-group.

6.8. PROBLEM. Can every topological group be embedded in a Moscow group?

Notice, that it follows from the above observations that if G is a Moscow group and the space G is normal, then every closed subgroup H of G is a PT-group.

It is an old result of V.G. Pestov that every topological group F can be represented as an image under an open continuous homomorphism of a topological group G of countable pseudocharacter. This group G can be also assumed to be Raĭkov complete, strongly σ-discrete, paracompact, and zero-dimensional in the sense of Lebesgue dimension dim. Since every topological group of countable pseudocharacter is Moscow, we have:

6.9. THEOREM. *Every topological group F can be represented as a quotient under an open continuous homomorphism of a paracompact Raĭkov complete Moscow group G. If F is Abelian, then G can also be chosen to be Abelian.*

However, if we take the quotient of a Moscow group with regard to a compact subgroup, then the result will be a Moscow space, ARHANGEL'SKII [200?]. Note that perfect mappings, in general, do not preserve the class of Moscow spaces.

Another approach to expanding our knowledge about the class of PT-groups is based on the notion of R-factorizability of topological group, introduced by TKAČENKO [1991b] (see also TKAČENKO [1991a] and Tkačenko's article in this book).

Let $\mathcal{P}$ be a class of topological groups. A topological group G is said to be *factorizable* over $\mathcal{P}$, or simply *$\mathcal{P}$-factorizable*, if for every continuous real-valued function f on G there exists a continuous homomorphism g of G into a topological group $H \in \mathcal{P}$ and a continuous real-valued function h on H such that $f = hg$. A topological group G is called *R-factorizable*, TKAČENKO [1991b], if it is factorizable over the class of separable metrizable groups.

Tkačenko established that every R-factorizable group is a PT-group. On the other hand, neither the class of Moscow groups contains the class of R-factorizable groups, nor the class of R-factorizable groups contains the class of Moscow groups. Indeed, every Lindelöf group is R-factorizable, TKAČENKO [1991b], while not every Lindelöf group is a Moscow space, as we will see below. Observe, that any discrete group is a Moscow space, while if a discrete group is R-factorizable, then it is countable. The following theorem, generalizing Tkačenko's result, was obtained in ARHANGEL'SKII [2000c].

6.10. THEOREM. *If a topological group G is factorizable over the class $\mathcal{PT}$ of all PT-groups, then G is a PT-group.*

6.11. THEOREM. *If a topological group G is factorizable over the class of Moscow groups, then G is C-embedded in $\rho_\omega G$, and therefore, G is a PT-group.*

A topological group G is *projectively Moscow*, ARHANGEL'SKII [2000c], if it is factorizable over the class of Moscow groups. The class of projectively Moscow groups is strictly smaller than the class of PT-groups.

6.12. PROBLEM. Is every Raĭkov complete group projectively Moscow?

Of course, it may be overoptimistic to expect the positive answer to this question. The next example was considered in ARHANGEL'SKII [2000c] and HERNÁNDEZ and TKAČENKO [1998].

Let $D(\omega_1)$ be a discrete space of cardinality ω_1 and $L = D(\omega_1) \cup \{a\}$, where a is not in $D(\omega_1)$, $D(\omega_1)$ is an open discrete subspace of L, and a subset U of L, containing a, is open if and only if the set $D(\omega_1) \setminus U$ is countable. It is well known that the space L, defined in this way, is a Lindelöf P-space; it is also Tychonoff. Let G be the free Abelian topological group $A(L)$ of the space L.

Since L is a Lindelöf P-space, L^n is a Lindelöf space, for each $n \in \omega$. It follows that the free Abelian topological group $G = A(L)$ of L is Lindelöf. The space G is also a P-space, since otherwise the topology generated on G by G_δ-sets would be a strictly stronger group topology on G inducing the original topology on L, which is impossible by the basic property of free topological groups.

Clearly, the cardinality of G is ω_1. Since G contains L as a subspace, G is not discrete. Being a topological group, G is a regular space. Therefore, since G is a P-space, G has a base $\mathcal{B}$ consisting of non-empty open and closed sets such that the intersection of any countable subfamily of $\mathcal{B}$ belongs to $\mathcal{B}$.

Since the pseudocharacter of G cannot not greater than the cardinality of G and G is a P-space of cardinality ω_1, the pseudocharacter of G is ω_1.

Put $Y = G \setminus \{e\}$. It can be shown that Y is not C-embedded in G, ARHANGEL'SKII [2000c]. On the other hand, Y is G_δ-dense in G. This, of course, implies that G is not Moscow. It follows that the g-tightness of G is uncountable.

Every proper dense subspace Z of G is not C-embedded in G, ARHANGEL'SKII [2000c]. Now, let H be the subgroup of G, algebraically generated by the set $D(\omega_1)$. Since $D(\omega_1)$ is dense in L, it follows that H is dense in G. Therefore, H is G_δ-dense in G, since G is a P-space. It is also clear that a is not in H; it follows that H is a proper subgroup of G. Therefore, H is not C-embedded in G, despite the fact that *H is a G_δ-dense subgroup of G.*

The character of the space G is ω_1. Since the cardinality of G is also ω_1, it follows that the weight of G is precisely ω_1. Now, it was observed in WILLIAMS [1984] that every P-space of the weight $\leq \omega_1$ is paracompact. It follows that every subspace of G is paracompact. In particular, H is paracompact and, therefore, H is Dieudonné complete (and Hewitt-Nachbin complete). Thus, *H is a PT-group.* Since H is not C-embedded in $\rho_\omega H$, we conclude that the main results on Moscow groups do not generalize to the class S of all topological groups G C-embedded in $\rho_\omega G$.

On the other hand, *G is Raĭkov complete.* This can be proved by an easy standard argument; it also follows from the deep theorem established by Tkačenko (see about it

TKAČENKO [1999], [2000]: the free Abelian topological group of a Dieudonné complete space is always Raĭkov complete.

The group H *is not factorizable over the class of Moscow groups, that is,* H *is not projectively Moscow.* This follows from the fact that H is not C-embedded in $\rho_\omega H$. On the other hand, *the group* G *is* R*-factorizable*, since it is Lindelöf, TKAČENKO [1991b].

The fact that the "bad" (not Moscow!) topological group we considered above turned out to be a P-space is rather suggestive. Indeed, if a space X is a P-space and a Moscow space, then, obviously, X is extremally disconnected. However, every extremally disconnected P-space of Ulam non-measurable cardinality is discrete, ISBELL [1955]. Therefore, in a search for a non-PT-group, we may start with any topological group G of uncountable pseudocharacter and of Ulam non-measurable cardinality, and introduce on G a new topology: the G_δ-modification of the original topology on G (G_δ-sets form a base of it). The topological group G^* so obtained is a P-space. Then G^* is not a Moscow space. Therefore, G^* may be a candidate for being a non-PT-group. However, the solution to the PT-problem in ARHANGEL'SKII [2000a] presented below is based on a different idea.

A topological group G is called $\aleph_0$*-bounded* provided that for each neighborhood V of the neutral element there exists a countable subset A of G such that $AV = G$. Tkačenko has shown that every R-factorizable group is $\aleph_0$-bounded and that not every $\aleph_0$-bounded group is R-factorizable (see TKAČENKO [1991b]). The groups G and H constructed above are $\aleph_0$-bounded, since every Lindelöf group is $\aleph_0$-bounded and every subgroup of an $\aleph_0$-bounded group is again $\aleph_0$-bounded. Since H is not projectively Moscow, H is not R-factorizable. Therefore, H can serve as another example of a non-R-factorizable $\aleph_0$-bounded group. Note that TKAČENKO [200?a] has shown that even a Raĭkov complete $\aleph_0$-bounded topological group need not be R-factorizable. However, it is not clear what is the answer to the following question:

6.13. PROBLEM (M.G. Tkačenko). Is every $\aleph_0$-bounded group a PT-group?

TKAČENKO [1991b] also asked another interesting question:

6.14. PROBLEM. Is every topological group with the countable Souslin number R-factorizable?

Rather unexpectedly, it turned out that Moscow groups are nicely related to the formula $\upsilon X \times \upsilon Y = \upsilon(X \times Y)$.

Here is one of classical questions in general topology: under what restrictions on spaces X and Y the formula $\upsilon X \times \upsilon Y = \upsilon(X \times Y)$ holds? Remarkable results in this direction were obtained in COMFORT and NEGREPONTIS [1966], HUŠEK [1970], COMFORT [1968], GLICKSBERG [1959], FROLÍK [1960]. In particular, the formula holds when X is any compact space of Ulam non-measurable cardinality, and if $X \times Y$ is pseudocompact – this was proved by GLICKSBERG [1959] and reproved, by another method, by FROLÍK [1960].

The case when the factors are topological groups was recently considered in ARHANGEL'SKII [2000c], where the following sufficient condition for the formula to hold was established. The key role in the formula again belongs to Moscow groups.

Let $\mathcal{S}$ be the class of topological groups G such that G is C-embedded in $\rho_\omega G$ (such groups could be called *strong* PT*-groups*).

6.15. THEOREM (ARHANGEL'SKII [2000c]). *Suppose that* $G = \Pi\{G_\alpha : \alpha \in A\}$ *is the topological product of topological groups* G_α *such that* $G \in \mathcal{S}$ *and the cardinality of* G *is Ulam non-measurable. Then* $\upsilon\Pi\{G_\alpha : \alpha \in A\} = \Pi\{\upsilon G_\alpha : \alpha \in A\}$.

With the help of this theorem the next result is easily established:

6.16. THEOREM (ARHANGEL'SKII [2000c]). *Let* $G = \Pi\{G_\alpha : \alpha \in A\}$ *be the product of topological groups* G_α *such that the space* G *is Moscow and* $|G|$ *is Ulam non-measurable. Then* $\upsilon\Pi\{G_\alpha : \alpha \in A\} = \Pi\{\upsilon G_\alpha : \alpha \in A\}$.

In particular, under the assumptions and in the notation of the above theorem, the formula

$$\upsilon\Pi\{G_\alpha : \alpha \in A\} = \Pi\{\upsilon G_\alpha : \alpha \in A\}.$$

holds if at least one of the following conditions is satisfied for $\mathcal{F} = \{G_\alpha : \alpha \in A\}$: 1) every group in $\mathcal{F}$ is totally bounded; 2) every group in $\mathcal{F}$ is k-separable; 3) ω_1 is a precaliber of every space in $\mathcal{F}$; 4) Souslin number of the product space $\Pi\{G_\alpha : \alpha \in A\}$ is countable; 5) Souslin number of every group in $\mathcal{F}$ is countable, and (MA+$\neg$CH) is satisfied; 6) every group in $\mathcal{F}$ is κ-metrizable; 7) the tightness of the product space $\Pi\{G_\alpha : \alpha \in A\}$ is countable; 8) the g-tightness of the product group $\Pi\{G_\alpha : \alpha \in A\}$ is countable; 9) the pseudocharacter of every group in $\mathcal{F}$ is countable, and each finite subproduct of the product has countable tightness; 10) every group in $\mathcal{F}$ has a countable network.

Indeed, if the Souslin number of the product group G is countable, then G is Moscow. This takes care of cases 1)–5). Similarly, in the cases 6), 7), and 8) the group G is also Moscow, since the g-tightness of it is countable. In the cases 9) and 10) G is also Moscow, ARHANGEL'SKII [200?].

6.17. THEOREM (ARHANGEL'SKII [2000c]). *Let* G_1 *and* G_2 *be two topological groups such that* $\mu G_i = \rho_\omega G_i$, *for* $i = 1, 2$. *Then the next conditions are equivalent:*

1) $G_1 \times G_2$ *is a* PT*-group;*
2) $\mu(G_1 \times G_2) = \mu G_1 \times \mu G_2$*;*
3) $\mu(G_1 \times G_2) = \rho_\omega(G_1 \times G_2)$*;*
4) $G_1 \times G_2$ *is* C*-embedded in* $\rho_\omega(G_1 \times G_2)$.

We will see below that the product of two Moscow groups may be a PT-group that is not a Moscow group.

6.18. PROBLEM. Let G_i be a topological group such that $\mu G_i = \rho_\omega G_i$, for each $i \in \omega$, and G the product of these groups. Assume also that G is a PT-group. Is then true that $\mu G = \Pi\{\mu G_i : i \in \omega\}$?

With the help of Theorem 6.17, the following facts were established in ARHANGEL'SKII [2000c]. If G is a Lindelöf group and H a pseudocompact group, then 1) $\mu(G \times H) = \rho_\omega(G \times H)$ and, therefore, $G \times H$ is a PT-group; and 2) $\mu G \times \mu H = \mu(G \times H)$. TKAČENKO [1991b] proved that under the restrictions on G and H in the above statement, the group $G \times H$ is R-factorizable. It follows that $G \times H$ is a PT-group. If G is a totally bounded group and H is a Lindelöf group which is a P-space, then $G \times H$ is C-embedded in $\rho_\omega(G \times H)$, $G \times H$ is a PT-group, and $\mu G \times \mu H = \mu(G \times H)$. In the proof we have to refer to the next fact, ARHANGEL'SKII [2000c]: every topological group G with the countable Souslin number satisfies the condition: $\mu G = \rho_\omega G$. Yet another result of this kind: For any totally bounded group H and any topological group G of countable o-tightness and of Ulam non-measurable cardinality, we have $\mu(G \times H) = \rho_\omega(G \times H)$, and $\mu G \times \mu H = \mu(G \times H)$.

The next statement slightly improves a theorem from COMFORT and NEGREPONTIS [1966]: If X is any space and Y a compact space, then $\mu(X \times Y) = \mu X \times Y$. Further, if G is a topological group which is a k-space and H a pseudocompact group, and the cardinal numbers $|G|$ and $|H|$ are Ulam non-measurable, then $\mu(G \times H) = \rho_\omega(G \times H)$, and the formula $\mu(G \times H) = \mu G \times \mu H$ holds, ARHANGEL'SKII [2000c]. In connection with Theorem 6.17, we note that the next result of ARHANGEL'SKII and HUŠEK [2001] holds:

6.19. THEOREM. *The product $G \times H$ of topological groups G and H is a PT-group if and only if G and H are PT-groups and the formula $\mu(G \times H) = \mu G \times \mu H$ holds.*

It was shown in ARHANGEL'SKII [2000c] that the product of a PT-group G and a compact group H is a PT-group. In particular, if G is a topological group satisfying at least one of the following conditions: 1) $c(G) \leq \omega$; 2) $t(G) \leq \omega$; 3) $t_g(G) \leq \omega$; 4) G is almost metrizable; 5) G is k-separable, then, for any compact group H, we have $\mu(G \times H) = \rho_\omega(G \times H)$ and therefore, $G \times H$ is a PT-group.

In ARHANGEL'SKII [2000a] an example of two PT-groups whose product is not a PT-group was presented. Here are some details. The construction below is a modification of HUŠEK's construction in [1972].

Let X be a non-Dieudonné complete topological group with the countable Souslin number. For example, let X be the Σ-product of ω_1 copies of the two element Boolean group $D = \{0, 1\}$. Then X is, in addition, countably compact and zero-dimensional. Note, that X is not compact and, therefore, X is not Dieudonné complete. Let us show that there exists a Moscow group G such that $X \times G$ is not a PT-group. In fact, G can be selected to be a topological group of countable pseudocharacter.

Fix an open covering η of X such that the union of any finite subfamily of η belongs to η, all elements of η are closed, and the closures of elements of η in μX do not cover μX.

Note, that μX is, in this case, the product of ω_1 copies of $D = \{0, 1\}$. Consider the space $G = C_\eta(X)$ of all continuous real-valued functions on X with the topology of uniform convergence on elements of η. Clearly, G is a topological group. It is also obvious that G is Raĭkov complete.

Since the Souslin number of X is countable, there exists a countable subfamily γ of η such that $\cup\gamma$ is dense in X. For each $P \in \gamma$ and each positive $n \in \omega$, the set $U_{P,n}$ of all $f \in G$ such that $|f(x)| < 1/n$ for every $x \in P$ is open in G and contains the zero-function θ on X which is the neutral element of G. It is obvious that θ is the only element in $\cap\{U_{P,n} : P \in \gamma, n \in \omega\}$. Therefore, θ is a G_δ-point in G. Since G is a topological group, it follows that the pseudocharacter of G is countable. Hence, G is a Moscow group. The group X is also Moscow. and, hence, μX is a topological group.

Consider the natural evaluation mapping ψ of the product space $X \times G$ into R. Clearly, ψ is continuous, since elements of η are open sets. Let us show that the group $H = X \times G$ is not C-embedded in $\mu X \times G$. Observe that $\mu X \times G$ is a topological group, and $X \times G$ is G_δ-dense in $\mu X \times G$, since X is G_δ-dense in μX.

Let us check that ψ cannot be extended to a continuous real-valued function on $\mu X \times G$. We can choose $a \in \mu X \setminus X$ such that a does not belong to the closure of any element of η. Take the point $(a, \theta) \in \mu X \times G$ and consider the subsets $B = \{(x, \theta) : x \in X\}$ and $C = \{(x, f) : f(x) = 1\}$ of $X \times G$. Clearly, $\psi(C) = \{1\}$ and $\psi(B) = \{0\}$. Obviously, the point (a, θ) is in the closure of B. Therefore, if ψ can be continuously extended to (a, θ), the value of this extension at (a, θ) has to be 0. On the other hand, a is not in the closure of any $P \in \eta$; therefore, (a, θ) is in the closure of C as well, and the extended function must take the value 1 at (a, θ), a contradiction.

Assume now that $X \times G$ is a PT-group. Then, according to Theorem 6.19, the formula $\mu(X \times G) = \mu X \times \mu G$ holds. Thus, since G is Raĭkov complete, it follows that $X \times G$ is C-embedded in $\mu X \times G$, a contradiction with the Claim. Hence, $H = X \times G$ is not a PT-group.

If the cardinality of G is Ulam non-measurable, then the space G is hereditarily Hewitt-Nachbin complete, since every topological group of countable pseudocharacter can be mapped by a one-to-one continuous mapping onto a metrizable space.

The group $\mu X \times G = D^{\omega_1} \times G$ is not Moscow, since otherwise $X \times G$, as a dense subspace of $\mu X \times G$, would have been a Moscow space and, therefore, a PT-group. On the other hand, $\mu X \times G$ is, obviously, Raĭkov complete.

Observe that the g-tightness of G, X and $\mu X = D^{\omega_1}$ is countable. However, the g-tightness of $\mu X \times G$ is not countable, since otherwise $\mu X \times G$ would be a Moscow group. Thus, we have the following result (see ARHANGEL'SKII [2000a]):

6.20. THEOREM. *There exist a countably compact group X and a Raĭkov complete group G of countable pseudocharacter with the following properties:*

1) *The product $X \times G$ is not a PT-group;*
2) *The product $\beta X \times G$ is not a Moscow group;*
3) $\mu X \times \mu G = \upsilon X \times \upsilon G \neq \upsilon(X \times G) = \mu(X \times G)$;
4) *The groups X, βX, and G are Moscow groups of countable g-tightness;*
5) *The g-tightness of $\beta X \times G$ is not countable;*
6) *$\beta X \times G$ is a Raĭkov complete group and, therefore, a PT-group.*

Here are a few interesting open questions, some of which are motivated by Theorem 6.20.

6.21. PROBLEM. Let G be a topological group of countable tightness. Is then $G \times G$ a Moscow group? A PT-group? Is then the g-tightness of $G \times G$ countable?

We still do not have a ZFC example of a topological group G of countable tightness such that the tightness of $G \times G$ is not countable (some non-ZFC examples of this kind can be found in HART and VAN MILL [1991b], MALYKHIN [1987]).

6.22. PROBLEM. Suppose G is an extremally disconnected topological group. Is then $G \times G$ Moscow? Is $G \times G$ a PT-group? Is the g-tightness of $G \times G$ countable?

6.23. PROBLEM. Suppose G is an extremally disconnected group and B a compact group. Is then $G \times B$ a Moscow group?

Note, that under the assumptions in Problem 6.23, $G \times B$ is a PT-group.

6.24. PROBLEM. (I.V. Yashchenko) Is the g-tightness of every Moscow group countable?

This question is related in a natural way to the next one:

6.25. PROBLEM. Suppose that G is a topological group of the countable g-tightness, and H is a dense subgroup of G. Is the g-tightness of H is countable?

The answer to the last question is "yes", if H is G_δ-dense in G.

7. Free topological groups

The free topological group $F(X)$ of a Tychonoff topological space X, from a purely algebraic point of view, is a rather standard object. Indeed, algebraically, $F(X)$ is determined just by the cardinality of X.

However, the passage from a space X to a new topological space $F(X)$ by means of the forces contained in the algebraic structure of $F(X)$ is a most non-trivial step, from the topological point of view. The existential nature of the definition of the topology of $F(X)$ is, probably, one of the reasons why the topological properties of $F(X)$ are so enigmatic. Very few things are clear here, almost everything is covered with clouds, almost every new step requires a lot of effort. Since the free topological groups constitute a very large and rich collection of topological groups, and every topological group can be represented as a quotient of a free topological group, it is important to study in depth these objects. And the first thing to learn is how the topological properties of $F(X)$ depend on properties of X. In this section we describe some progress made in the topological study of free topological groups in recent years. The spaces considered in this section are all assumed to be Tychonoff. By $A(X)$ we denote the free Abelian topological group of X.

It is well known that only when X is discrete, the space $F(X)$ can be first countable or locally compact. Therefore, only if X is discrete the space $F(X)$ is metrizable. Even more is true: the space $F(X)$ is Fréchet-Urysohn only if X is discrete, ARHANGEL'SKII [1981]. However, the space $F(X)$ can be sequential for a non-discrete X. Indeed, if X is a sequential compactum (in particular, a metrizable compactum), then $F(X)$ is sequential, that is, sequentially closed subsets of $F(X)$ are closed. M.I. Graev showed that $F(X)$ is a k-space, for any compactum X. Moving further in this direction, Kohzo Yamada considered systematically the question: when $F(X)$ or $A(X)$ is a k-space? (YAMADA [1994], [1996]). He has also studied from this point of view the standard subspaces $F_n(X)$ and $A_n(X)$ of $F(X)$ and $A(X)$, respectively, consisting of the "words" of length $\leq n$, for $n \in \omega$. These subspaces are closed in $F(X)$ (in $A(X)$). Below we call $F_n(X)$ ($A_n(X)$) an *n-kernel* of $F(X)$ (of $A(X)$, respectively).

The structure of $F(X)$ can be quite non-trivial even when X is a separable metrizable space. For example, the free topological group of the space Q of rational numbers is not a k-space. In fact, it was shown in FAY, ORDMAN and THOMAS [1979] that the subspace $F_3(Q)$ of $F(Q)$ is not a k-space.

In ARHANGEL'SKII, OKUNEV and PESTOV [1989] the following two theorems were obtained:

7.1. THEOREM. *For any metrizable space X, the following conditions are equivalent:*

1) *$F(X)$ is a k-space;*

2) *$F(X)$ is a k_ω-space or a discrete space;*
3) *X is a locally compact separable space or X is discrete.*

7.2. THEOREM. *Suppose that X is a metrizable space, and let X' be the set of all non-isolated points in X. Then the following conditions are equivalent:*
1) *$A(X)$ is a k-space;*
2) *$A(X)$ is homeomorphic to a product of a k_ω-space with a discrete space;*
3) *X is locally compact and X' is separable.*

Comparing these two theorems, we see that if $F(X)$ is a k-space then $A(X)$ is also a k-space, while there exists a metrizable space X such $A(X)$ is a k-space and $F(X)$ is not a k-space (take in the role of X the product of the space R of reals with an uncountable discrete space).

K. Yamada discovered that all n-kernels $A_n(X)$ of $A(X)$ can be k-spaces while $A(X)$ is not a k-space: it suffices to take as X the metrizable hedgehog with countably many spines, YAMADA [1996]. He proved that, for a metrizable space X, $A_4(X)$ is a k-space if and only if $A_n(X)$ is a k-space, for each $n \in \omega$.

This result is complemented by the next result of Yamada, which emphasizes the importance of the number 4 in the preceding theorem.

7.3. THEOREM.
a) *$A_2(X)$ is a k-space, for every metrizable space X;*
b) *there exists a metrizable space X such that $A_3(X)$ is not a k-space;*
c) *there exists a metrizable space X such that $A_3(X)$ is a k-space but $A_4(X)$ is not a k-space.*

Recall that $F_1(X) = X \cup \{e\} \cup X^{-1}$, where the subspace X^{-1} is naturally homeomorphic to the space X, and X, X^{-1} are open and closed in $F_1(X)$ and disjoint. For any positive $n \in \omega$, we denote by i_n the canonical mapping of the product $(F_1(X))^n$ onto the subspace $F_n(X)$ (thus, $i_n(h_1, h_2, ..., h_n) = h_1 h_2 ... h_n$). One of main theorems of Yamada is the next result:

7.4. THEOREM. *Suppose that X is a metrizable space. Then the following conditions are equivalent:*
1) *$A_n(X)$ is a k-space, for each positive $n \in \omega$;*
2) *$A_4(X)$ is a k-space;*
3) *i_n is a quotient mapping, for each positive $n \in \omega$;*
4) *i_4 is a quotient mapping;*
5) *either X is locally compact and the set X' of all non-isolated points of X is separable, or X' is compact.*

Yamada also noticed that the restriction of i_2 to $X \times X$ is a perfect mapping. Using this, he proved that $A_2(X)$ is a k-space if and only if $X \times X$ is a k-space (YAMADA [1996]), which has lead him to the following two statements:

7.5. THEOREM. *For each metrizable space X, $A_2(X)$ is a k-space and the mapping i_2 is quotient.*

7.6. EXAMPLE. $A_2(V(\omega_1))$ is not a k-space (where $V(\omega_1)$ is the Fréchet-Urysohn fan made of ω_1 usual sequences converging to the same point).

Yamada also gave a complete characterization of metrizable spaces for which the mapping i_3 is quotient:

7.7. THEOREM. *Suppose that X is a metrizable space. Then the following statements are equivalent:*
1) $A_3(X)$ *is a k-space;*
2) *the mapping i_3 is quotient;*
3) X *is locally compact or the set X' of all non-isolated points of X is compact.*

This result allowed Yamada to present an example of a metrizable space X such that $A_3(X)$ is a k-space but $A_4(X)$ is not a k-space.

Recall that a Tychonoff space is said to be a *k_R-space* if every continuous real-valued function f on X such that the restriction of f to arbitrary compact subset of X is continuous, must be continuous. Every Tychonoff k-space is obviously a k_R-space. After Yamada's results, the following questions are very natural:

7.8. PROBLEM. Is it true that, for every metrizable space X, $A(X)$ is a k_R-space? In particular, is the free Abelian topological group of the space Q of rational numbers a k_R-space?

7.9. PROBLEM. Is the o-tightness of the free (Abelian) topological group of a metrizable space countable?

One should expect a negative answer to this question. So we pose another problem:

7.10. PROBLEM. Characterize metrizable spaces X such that $A(X)$ is a k_R-space.

REZNICHENKO and SIPACHEVA [1994] obtained conditions for $F(X)$ to be a k-space for the case of non-metrizable X. In particular, they have proved the next theorem:

7.11. THEOREM. *Suppose that X is a non-discrete paracompact first countable space. Then the free topological group $F(X)$ is a k-space if and only if X is a locally compact σ-compact space.*

Pestov established in 1981 that the mapping i_2 is quotient if and only if the space X is strongly collectionwise normal, which explains the corresponding statement of Yamada about i_2 being always quotient for metrizable spaces.

YAMADA [1998] also proved the following result:

7.12. THEOREM. *For any metrizable space X, the following conditions are equivalent:*
1) A_2 *is first countable;*
2) A_2 *is metrizable;*
3) $A_n(X)$ *is first countable, for each $n \in \omega$;*
4) $A_n(X)$ *is metrizable, for each $n \in \omega$;*
5) *the set of all non-isolated points of X is compact.*

An important step in the proof of Theorem 7.12 is due to Gary Gruenhage who proved that if X is a metrizable space such that the set of all non-isolated points in X is compact, then $A_n(X)$ has a σ-disjoint base, for each $n \in \omega$. This nicely combines with a theorem in ARHANGEL'SKII [1981] which states that the free (Abelian) topological group of a metrizable space is a paracompact σ-space, and, hence, is perfectly normal. Indeed, it is well known that all perfectly normal spaces with a σ-disjoint base are metrizable.

Yamada also established that the situation for $F_n(X)$ changes:

7.13. THEOREM. *For any non-discrete metrizable space X, the following conditions are equivalent:*
1) $F_4(X)$ *is first countable;*
2) $F_4(X)$ *is metrizable;*
3) i_4 *is a closed mapping;*
4) $F_n(X)$ *is first countable, for each* $n \in \omega$*;*
5) $F_n(X)$ *is metrizable, for each* $n \in \omega$*;*
6) i_n *is a closed mapping, for each* $n \in \omega$*;*
7) X *is compact.*

However, the situation with $n = 2$ and $n = 3$ is quite different, as another theorem of Yamada shows (YAMADA [1998]):

7.14. THEOREM. *For any metrizable space X, the following conditions are equivalent:*
1) $F_2(X)$ *is first countable;*
2) $F_2(X)$ *is metrizable;*
3) $F_3(X)$ *is first countable;*
4) $F_3(X)$ *is metrizable;*
5) *the set of all non-isolated points of X is compact.*

The question, when $F_n(X)$ and $A_n(X)$ are Fréchet-Urysohn spaces, was considered in YAMADA [200?].

The next question naturally arises in the study of free topological groups: when $F(X)$ ($A(X)$) is the inductive topological limit of the sequence of spaces $\{F_n(X) : n \in \omega\}$ (respectively, $\{A_n(X) : n \in \omega\}$)? PESTOV and YAMADA [1999] answered this question for the case of metrizable X. They proved that in this case $A(X)$ has the inductive limit topology if and only if $A(X)$ is a k-space. Also, they proved that, for a metrizable non-discrete X, $F(X)$ has the inductive limit topology if and only if $F(X)$ is a k-space (that is, if and only if X is locally compact and separable).

In the same direction, TKAČENKO [1994] proved, that if X is a pseudocompact Tychonoff space, then $F(X)$ has the inductive limit topology if and only if X^n is normal and countably compact, for every $n \in \omega$.

Yamada, studying the seemingly unrelated question: when $F_n(X)$ is locally compact, for every $n \in \omega$, discovered a strong connection of this question with the above one. He proved the following statement:

7.15. THEOREM (YAMADA [1996]). *Suppose that X is a Tychonoff space such that $F_n(X)$ is locally compact for each $n \in \omega$. Then $F(X)$ has the inductive limit topology and is a k-space, and X^n is normal and countably compact for each $n \in \omega$.*

Alongside the k-space property in free topological groups, the tightness of subspaces of free topological groups was thoroughly investigated. In ARHANGEL'SKII, OKUNEV and PESTOV [1989] it was proved that, for a metrizable space X, the tightness of the free Abelian topological group $A(X)$ is countable if and only if the subspace X', consisting of all non-isolated points of X, is separable. On the other hand, YAMADA established the following three facts in [1997]:

7.16. THEOREM. *For any metrizable space X, the tightness of $A_3(X)$ is countable.*

7.17. THEOREM. *For any metrizable space X, the tightness of $A_4(X)$ is countable if and only if the tightness of $A(X)$ is countable.*

7.18. THEOREM. *Suppose that X is a metrizable space such that the weight of X' is ω_1. Then the tightness of $A_4(X)$ and the tightness of $A(X)$ are also equal ω_1.*

The last result provides us with a rich supply of metrizable spaces the free Abelian topological groups of which have uncountable tightness. The results above suggest the next general problem: find non-trivial tightness type properties such that the free topological group of every metrizable space has these properties. One very weak such property is the g-tightness, discussed in section 6. Indeed, we have:

7.19. THEOREM. *For every metrizable space X, the g-tightness of $F(X)$ and the g-tightness of $A(X)$ are countable.*

This is so, since the free topological group of a metrizable space is always submetrizable.

7.20. COROLLARY. *The free (Abelian) topological group of any metrizable space is Moscow.*

However, the general problem remains open:

7.21. PROBLEM. When the free topological group of a space X is Moscow?

7.22. PROBLEM. Is the free (Abelian) topological group of a first countable space Moscow?

SIPACHEVA [2000] has published a new proof of her following theorem:

7.23. THEOREM. *The free topological group $F(X)$ of any Dieudonné complete space X is Raĭkov complete (even Weil complete).*

On the other hand, it is an old result of PESTOV [1982] that the free topological group $F(X)$ of arbitrary Tychonoff space X is a topological subgroup of the free topological group $F(\mu X)$ of the Dieudonné completion μX of the space X. Combining this result with Theorem 7.23, we obtain the next statement:

7.24. THEOREM. *For any Tychonoff space X, the Raĭkov completion $\rho(F(X))$ of the free topological group $F(X)$ of the space X coincides with the free topological group of the Dieudonné completion μX of X.*

Since X is G_δ-dense in μX, it is clear that $F(X)$ is G_δ-dense in $F(\mu X)$. Therefore, the next statement follows from Theorem 7.24:

7.25. COROLLARY. *For every Tychonoff space X, the free topological group $F(X)$ is G_δ-dense in the Raĭkov completion $\rho(F(X))$ of the topological group $F(X)$.*

It is well known that X is C-embedded in μX, and that, for a dense subspace, to be G_δ-dense in the larger space is a necessary condition for being C-embedded in that larger space. So the next question naturally arises in connection with Corollary 7.25: is $F(X)$ always C-embedded in $F(\mu X)$? It turns out that the answer to this question is in negative. Indeed, we have:

7.26. THEOREM. *Suppose that G is a topological group such that the free topological group $F(G)$ of the space G is C-embedded in the free topological group $F(\mu G)$ of the Dieudonné completion μG of the space G. Then μG is also a topological group, that is, the operations on G can be continuously extended to μG.*

From this theorem it follows that to get a counterexample to the above question we only have to take in the role of X any non-PT topological group. For example, we can take the non-PT-group constructed in the preceding section.

7.27. PROBLEM. When $F(X)$ is C-embedded in $F(\mu X)$?

7.28. PROBLEM. Suppose that X is a first countable space. Is $F(X)$ C-embedded in $F(\mu X)$?

MORRIS and PESTOV [1993] proved that any open subgroup of the free Abelian topological group of any Tychonoff space X is again a free Abelian topological group, and the free topological bases of both groups have the same covering dimension. It is known for a long time that for non-open subgroups this is not necessarily true (see MORRIS and PESTOV [1993]).

LEIDERMAN, MORRIS and PESTOV [1997] characterized Tychonoff spaces X such that the free Abelian topological group $A(X)$ can be embedded as a topological subgroup into the free Abelian topological group $A(I)$ of the closed unit interval:

7.29. THEOREM. *For any Tychonoff space X, the following conditions are equivalent:*

1) *$A(X)$ embeds into $A(I)$ as a topological subgroup;*
2) *$F(X)$ embeds into $F(I)$ as a topological subgroup;*
3) *X is homeomorphic to a closed subspace of $A(I)$;*
4) *X is homeomorphic to a closed subspace of $F(I)$;*
5) *X is a k_ω-space such that every compact subspace of X is metrizable and finite-dimensional.*

The image under the embeddings in 1) and 2) above is automatically closed, since topological groups $A(X)$ and $F(X)$ are Raĭkov complete when X satisfies 3), 4), or 5). So we can replace "a topological subgroup" in conditions 1) and 2) with "a closed topological subgroup".

7.30. COROLLARY. *For any finite-dimensional metrizable compactum X, the free topological group $F(X)$ is topologically isomorphic to a closed subgroup of $F(I)$.*

Free topological G-groups over a (semi)group action of a topological (semi)group G on a topological space X were considered in MEGRELISHVILI [1996]. The situation here is quite different from the classical case. The difference can be already seen when X is S^n, the n-dimensional sphere, and G is the group of all homeomorphisms of S^n onto itself with the compact-open topology. See the details in MEGRELISHVILI [1996].

In that paper the following results were obtained. If X is a connected locally connected space, then the free topological groups $F(X)$ and $A(X)$ are locally connected. It follows that, under the same restrictions on X, every non-trivial continuous homomorphism of the free Abelian topological group $A(X)$ into the additive topological group of reals R is open.

We conclude this section with open questions, motivated by the following well known facts: the free topological group of a metrizable space is paracompact, ARHANGEL'SKII [1981], and the free topological group of a compact space is paracompact (since it is σ-compact and, therefore, Lindelöf).

7.31. PROBLEM. Is $F(X)$ ($A(X)$) paracompact, for every paracompact p-space? What if $X = M \times B$, where M is metrizable and B is compact?

8. The Bohr topology

Let G be an Abelian group. Then $G^\#$ stands for G with the *Bohr topology* which is the smallest topology on G that makes all homomorphisms of G into the circle group T continuous. It is easily seen that $G^\#$ is a totally bounded topological group, in fact, the topology of $G^\#$ is the maximal totally bounded group topology on G. Since $G^\#$ is totally bounded, the Raĭkov completion of $G^\#$ is a compact topological group bG containing $G^\#$ as a dense subgroup. The compact group bG is called the *Bohr compactification* of G. Since the definition of the topology of $G^\#$ is not an effective one, some natural questions about $G^\#$ are not easy to answer. One of such questions was asked by E. van Douwen: must $G^\#$ and $H^\#$ be homeomorphic as topological spaces whenever G and H are Abelian groups of the same cardinality?

Recently KUNEN [1998], using combinatorial methods, has shown that the answer is in negative, even for countable Abelian groups. Independently, the problem was solved by DIKRANJAN and WATSON [2001].

VAN DOUWEN [1990] has proved that if G is an Abelian group and A is any infinite subset of $G^\#$, then there exists a relatively discrete subset D of A such that $|D| = |A|$, D is C-embedded in $G^\#$, and D is C^*-embedded in bG. The proof of this theorem is technically quite involved. Recently GALINDO and HERNÁNDEZ [1998] gave a new proof of this theorem, more transparent.

Answering a question of E. van Douwen, HART and VAN MILL [1991a] obtained the following result:

8.1. THEOREM. *For any Abelian group G, every subset A of G contains a discrete and closed in $G^\#$ subset B of the same cardinality as A.*

It follows that every relatively countably compact subset of $G^\#$ is finite. In particular, $G^\#$ is a k-space if and only if the set G is finite. HART and VAN MILL [1991a] also presented an example showing that not every relatively discrete subset of $G^\#$ must be closed in $G^\#$. An essential role in their arguments belongs to the notion of an independent subset of a group. A subset A of a group G is said to be *independent* if, for every subset

B of A, the subgroups of G generated by B and by $A \setminus B$ have a trivial intersection (consisting of the neutral element only). Here is a useful lemma proved in HART and VAN MILL [1991a]:

8.2. LEMMA. *For any Abelian group G, every independent subset of G is closed and discrete in $G^{\#}$.*

GLADDINES [1995] showed that if G is any Abelian group containing an infinite Boolean subgroup, then $G^{\#}$ contains a countable infinite closed subset that is not a retract of $G^{\#}$. This answers a question of VAN DOUWEN. He proved in [1990] that, under the above restrictions on G and A, every mapping of A into the discrete space $\mathbf{N}$ of integers extends to a continuous mapping of $G^{\#}$ into the space $\mathbf{N}$. This motivated his question.

The notion of Bohr topology naturally extends from discrete Abelian groups to those topological groups that admit enough continuous homomorphisms into compact Hausdorff groups (such groups are called *maximally almost periodic* or *MAP* in the sense of J. von Neumann). In particular, it obviously extends to those topological Abelian groups which admit enough continuous characters to distinguish each element of the group from the neutral element. In this case, the Bohr topology on G coincides with the weak topology generated by the family of all continuous characters on G, and G with this topology is denoted by G^{+}.

In particular, all locally compact Abelian groups belong to the class of MAP groups. The Bohr topology of locally compact Abelian groups was studied in many articles, some of them quite recent. Here is a remarkable result of TRIGOS-ARRIETA [1991]:

8.3. THEOREM. *Suppose that G and H are any locally compact Abelian groups and $\phi : G \to H$ any homomorphism of G into H. Then ϕ is continuous if and only if the corresponding homomorphism ϕ of G^{+} into H^{+} is continuous.*

HERNÁNDEZ [1998] established that the covering dimension of G is the same as the covering dimension of G^{+}, for every locally compact Abelian group G. This generalizes the earlier result (see COMFORT and TRIGOS-ARRIETA [1991]) that $G^{\#}$ is zero-dimensional, for every discrete Abelian group G. Let us also mention an old result of I. Glicksberg that a subspace A of a locally compact Abelian group G is compact if and only if it is compact as a subspace of G^{+}, GLICKSBERG [1959]. In COMFORT and TRIGOS-ARRIETA [1991] this result was extended to countably compact and pseudocompact subsets of locally compact Abelian group, and a more elementary proof was given to Glicksberg's theorem.

COMFORT, HERNÁNDEZ and TRIGOS-ARRIETA [1996] have shown that a locally compact Abelian group G is Hewitt-Nachbin complete if and only if the space G^{+} is Hewitt-Nachbin complete. Recall that every locally compact Abelian group is paracompact and, therefore, Dieudonné complete. Under the same restrictions on G, every closed subgroup of G^{+} is C-embedded in G^{+}.

8.4. THEOREM (COMFORT, HERNÁNDEZ and TRIGOS-ARRIETA [1996]). *Suppose that G is a locally compact Abelian group. Then the following conditions are equivalent:*

1) *G^{+} is hereditarily realcompact;*

2) *G^{+} is submetrizable;*

3) *the space G is metrizable and $|G| \leq 2^{\omega}$.*

Trigos-Arrieta has shown that G^+ is rarely normal, TRIGOS-ARRIETA [1994]:

8.5. THEOREM. *Suppose that G is a locally compact Abelian group. Then G^+ is normal if and only if G is σ-compact.*

9. Duality theorems for topological groups

There was a series of papers in which Pontryagin's duality theory was extended to various classes of topological groups (see BANASZCZYK, CHASCO and MARTIN-PEINADOR [1994], CHASCO [1998], CHASCO and MARTIN-PEINADOR [2001], MARTIN-PEINADOR [1995], GALINDO and HERNÁNDEZ [1999], in particular). For a topological group G, G^* will stand for the group $Hom(G,T)$ of all continuous homomorphisms of G into the circumference T, endowed with the compact-open topology. We say that G satisfies Pontryagin's duality, or is *reflexive*, if the natural evaluation mapping ν of G into $(G^*)^*$ is a topological isomorphism. It is well known (see CHASCO and MARTIN-PEINADOR [2001] for precise references) that every locally compact Abelian group is reflexive and the additive group of arbitrary Banach space is reflexive. The products of reflexive groups are also reflexive. The class of reflexive groups doesn't behave nicely with regards to operations: a closed subgroup of a reflexive group need not be reflexive, and a quotient group of a reflexive group need not be reflexive. Because of this, it is natural to consider the following notion. A topological group is called *strongly reflexive* if every closed subgroup and every quotient group of G and G^* is reflexive (BANASZCZYK [1991]). BANASZCZYK [1990] proved that all countable products of locally compact Abelian topological groups are strongly reflexive. For further results on strongly reflexive groups see CHASCO and MARTIN-PEINADOR [2001].

In several papers the case of free Abelian topological groups was considered. Recall that a space X is said to be a *k_R-space* if a real-valued function f on X is continuous if and only if all restrictions of f to arbitrary compact subsets of X are continuous. A space X is called a *μ-space* if every bounded closed subset of X is compact. Suppose that $A(X)$ is the free Abelian topological group of a Tychonoff space X. PESTOV [1995] observed that the restriction mapping of the dual group $A(X)^*$ onto the space $C_c(X,T)$ of all continuous mappings of X into the space T, in the compact-open topology, is a topological isomorphism, for any μ-space X. Using this approach, he obtained the following three results:

9.1. THEOREM. *If X is both a μ-space and a k_R-space, then ν is a topological isomorphism of the topological group $A(X)$ onto a topological subgroup of $(A(X)^*)^*$.*

9.2. THEOREM. *If the mapping ν of $A(X)$ into $(A(X)^*)^*$ is onto, then X is totally linearly disconnected, that is, every continuous image of the closed unit interval in X is trivial.*

9.3. THEOREM. *If the covering dimension of X is zero, and X is both a μ-space and a k_R-space, then the free topological group $A(X)$ is reflexive.*

Note, that for every Tychonoff space X the continuous characters on $A(X)$ separate points of $A(X)$. This implies that the mapping $\nu : A(X) \to A(X)^{**}$ is always an injection. Pestov's approach was pursued further in GALINDO and HERNÁNDEZ [1999], where it was proved, in particular, that, for a compact Hausdorrf space X, $A(X)$ is reflexive if and only if X is zero-dimensional (or, equivalently, totally disconnected). Along these lines, a general criterion for reflexivity of $A(X)$ was also established in GALINDO and HERNÁNDEZ [1999].

In HERNÁNDEZ and USPENSKIJ [2000] it was investigated when the additive topological group $C_p(X)$ of all real-valued continuous functions on a Tychonoff space X, endowed with the topology of pointwise convergence, is reflexive. The authors established there that if this is the case, then X is a P-space, that is, every G_δ-subset of X is open. They also have shown that the last condition is not sufficient for $C_p(X)$ to be reflexive. On the other hand, the topological group $C_c(X)$ (with the compact-open topology) is reflexive for every metrizable space. Note, that the evaluation mapping ν of $C_p(X)$ into $C_p(X)^{**}$ is always a topological isomorphism of the topological group $C_p(X)$ onto its image (HERNÁNDEZ and USPENSKIJ [2000]).

Some other aspects of duality were considered in OHTA [1996] and EDA, KAMO and OHTA [1993], where spaces of continuous integer-valued functions on certain topological spaces X were used to construct Abelian groups with some special duality properties.

M. Bruguera, M.J. Chasco, E. Martin-Peinador and V. Tarieladze obtained new results on locally convex topological groups. This class of groups may be viewed as a bridge from the classical theory of linear topological spaces to the general theory of topological groups. In particular, their results concern certain duality questions and completeness problems. I refer the reader to the paper BRUGUERA, CHASCO, MARTIN-PEINADOR and TARIELADZE [2001] which also contains an introduction to this topic.

10. Some further results and problems on topological groups

As we have already seen on many examples, almost every question of general topology gets entirely new flavor when it is put in the setting of topological algebra, in particular, when it is formulated for topological groups. Because of that, practically every direction of research in general topology can have a non-trivial projection into topological algebra. Though until now not all these dormant domains have been activated, many classical lines of investigation in general topology have found their path in the domains of topological algebra in recent years. In this last section, I try to provide just a glimpse to these activities, not really aiming at achieving any kind of completeness at all.

A curious result was obtained in BRUGUERA, CHASCO, MARTIN-PEINADOR and TARIELADZE [2001]: the topological group R of reals, with the usual topology and operations, admits a strictly stronger connected group topology.

The topic of completeness and completions in topological algebra is far from being exhausted. Apart from topological groups and topological rings, we only have some partial results on when a natural completion of a topologico-algebraic structure exists. For example, this is not quite clear even for paratopological groups. One of the difficulties here is that paratopological groups and semitopological groups do not bear, in general, a natural uniform structure. Besides, in general, nothing similar to Raĭkov construction of a completion of a topological group exists. See about this KÜNZI, ROMAGUERA and SIPACHEVA [1998].

In this situation, it might be reasonable to declare a (regular, Tychonoff) paratopological group to be *Markov complete* if it is closed in every (regular, Tychonoff) paratopological group containing it as a subgroup and a subspace. After this definition, it is reasonable to ask, which results on completeness known in the theory of topological groups can be extended to paratopological groups. In particular, we have the following questions:

Is the free (Tychonoff) paratopological group of an arbitrary Dieudonné complete Tychonoff space Markov complete?

Is any regular (Tychonoff) paratopological group a (dense) subgroup of a Markov complete paratopological group?

Suppose a paratopological group is represented as a dense subgroup of two Markov complete paratopological groups M_1 and M_2. Are then M_1 and M_2 topologically isomorphic?

Similar questions can be formulated for Tychonoff (regular, Hausdorff) semitopological groups.

An important fact concerning the Raĭkov completion of a topological group G was established by GALINDO and HERNÁNDEZ [1999]. They constructed an Abelian topological group G such that every non-zero $x \in G$ can be separated from 0 by a continuous homomorphism to a compact Hausdorff group, while the Raĭkov completion ρG does not satisfy this property.

An interesting notion of sequential completeness of a topological group was introduced by DIKRANJAN and TKAČENKO [2000], [2001a]. A topological group G is called *sequentially complete* if it sequentially closed in its Raĭkov completion (that is, every Cauchy sequence in the uniform space G converges). It was shown in [2000] that every minimal sequentially complete connected Abelian topological group of Ulam non-measurable cardinality is compact. See also the paper [2001a] in which Dikranjan and Tkačenko study weakly complete free topological groups.

Some curious results on the transfinite dimension of topological groups were obtained by SHAKHMATOV [1993]. He proved that if the large transfinite inductive dimension trInd X of a homogeneous normal space X is defined, then either ind X is finite or X is countably compact. He also established that if G is a normal topological group such that the large transfinite dimension trInd G is defined, then ind G is finite.

There are two general problems on topological groups formulated as early as 1981 in ARHANGEL'SKII [1981] on which there was very little progress made in the 22 years since. They can be stated as follows.

10.1. PROBLEM. Given a class $\mathcal{P}$ of topological spaces, when every $X \in \mathcal{P}$ can be embedded as a closed subspace into a topological group $G \in \mathcal{P}$?

10.2. PROBLEM. Given a class $\mathcal{P}$ of topological spaces, when, for every topological group G in $\mathcal{P}$, the square $G \times G$ belongs to $\mathcal{P}$?

The last question is very close to the following one. Suppose that topological groups G and H have a certain topological property $\mathcal{P}$. Is then true that $G \times H$ also has the property $\mathcal{P}$? However, the two questions are not equivalent. While an example of two Lindelöf topological groups G and H such that $G \times H$ is not Lindelöf is known, we have only a consistent example of a Lindelöf topological group G such that the square of G is not Lindelöf (see MALYKHIN [1987]). Indeed, we do not know if any two Lindelöf topological groups can be represented as closed subgroups of one Lindelöf topological group. Note that any two Lindelöf spaces can be trivially represented as closed subspaces of one Lindelöf space: suffices to take their free topological sum.

Here are some interesting open concrete versions of the embedding Problem 10.1.

10.3. PROBLEM (ARHANGEL'SKII [1981]). Can Sorgenfrey line be embedded as a closed subspace into some Lindelöf topological group? Can every Lindelöf space be so embedded?

10.4. PROBLEM. Can arbitrary Tychonoff space of countable tightness be embedded as a closed subspace into a topological group of countable tightness? What if we drop the word "closed" in this question?

If the answer to question 10.3 is "yes" in ZFC, then we obviously obtain an absolute example of a Lindelöf topological group the square of which is not even normal. It is unknown at present, whether there exists in ZFC a normal topological group such that the square of it is not normal. This shows an important connection between general Problems 10.2 and 10.1. The next old question is also open in ZFC:

10.5. PROBLEM (ARHANGEL'SKII [1981]). Is there a countably compact topological group G such that the square $G \times G$ is not countably compact?

Comfort and Ross's theorem that the product of any family of pseudocompact topological groups is pseudocompact motivates this question. A consistent example of a countably compact topological group G such that $G \times G$ is not countably compact was first constructed by Malykhin (see MALYKHIN [1987], VAN DOUWEN [1980], and also HART and VAN MILL [1991b], TOMITA [1996], [1997], [1999]).

10.6. PROBLEM. Is the product of any two topological groups of countable tightness a Moscow group? A topological group of countable o-tightness?

10.7. PROBLEM. Is the square of a Moscow group a PT-group?

10.8. PROBLEM. Is the square of a PT-group a PT-group?

The next theorem was established in ARHANGEL'SKII and HUŠEK [2001]:

10.9. THEOREM. *Suppose that G is the product of a family $\{G_\alpha : \alpha \in A\}$ of topological groups. Then G is a PT-group if and only if G_α is a PT-group, for each $\alpha \in A$, and the formula*

$$\mu(\Pi\{G_\alpha : \alpha \in A\}) \subset \Pi\{\mu(G_\alpha) : \alpha \in A\}$$

holds.

It is not known whether "$\subset$" can be replaced with "$=$" in the last formula.

10.10. PROBLEM (ARHANGEL'SKII and HUŠEK [2001]). Is μG homogeneous for every topological group G?

It was shown in ARHANGEL'SKII and HUŠEK [2001] that there exists a pseudocompact Moscow quasitopological group the Dieudonné completion of which has points of different characters. Note, that not every pseudocompact quasitopological group is Moscow, ARHANGEL'SKII and HUŠEK [2001]. Besides, we do not have anything similar to Raĭkov completion for quasitopological groups. However, the next question remains open:

10.11. PROBLEM (ARHANGEL'SKII and HUŠEK [2001]). Is the product of arbitrary family of pseudocompact quasitopological groups pseudocompact?

Some interesting new questions on topological algebra were recently formulated by SHAKHMATOV [1999].

10.12. PROBLEM. Is there a Dowker topological field?

Note that a Dowker topological group does exist, HART, JUNILLA and VAN MILL [1985]. On the other hand, TKACHUK [1991] showed that every normal locally convex linear topological space is countably paracompact.

SHIBAKOV [1999] constructed, under CH, a countable Fréchet-Urysohn topological group G such that $G \times G$ is not Fréchet-Urysohn. Whether such a group can exist in ZFC is still unclear. The next problem, posed by V.I. Malykhin, is also open:

10.13. PROBLEM. Is there in ZFC a countable non-metrizable Fréchet-Urysohn topological group?

At the beginning of seventies, it became a part of the folklore that, under Martin's Axiom and not CH, there exists a countable non-metrizable Fréchet-Urysohn topological group (see ARHANGEL'SKII [1987], [1981]). The last question is, obviously, equivalent to the question: is every separable Fréchet-Urysohn topological group metrizable? Note, that for paratopological groups the answer to this question is in negative as is demonstrated by Sorgenfrey line. However, I do not know the answer to the next question:

10.14. PROBLEM. Is every regular Fréchet-Urysohn paratopological group first countable?

Answering an old question, SHIBAKOV showed that, under CH, there exists a countable sequential topological group of an intermediate sequential order ([1996], [1998]). On the other hand, DOLECKI and PEIRONE [1992] proved in ZFC that, for every countable ordinal α, there exists a sequential countable Hausdorff paratopological group of sequential order α. For further interesting results and problems on Fréchet-Urysohn topological groups, see NYIKOS [1992], REZNICHENKO and SIPACHEVA [1999], SHAKHMATOV [1990], and Shakhmatov's article in this book.

S. Todorčević has constructed in ZFC two countably tight σ-compact topological groups G and H such that $G \times H$ has uncountable tightness. OKUNEV [1996] improved this result by making one factor countable.

10.15. PROBLEM (SHAKHMATOV [1999]). Is every topological field with the countable Souslin number separable?

A. Shibakov has constructed a countable non-discrete topological field without non-trivial convergent sequences (cited in SHAKHMATOV[1999]).

Some new interesting results on convergence properties in topological groups were obtained in NOGURA, SHAKHMATOV and TANAKA [1997]. In particular, they established and applied the following curious lemma:

10.16. LEMMA. *A topological group G contains a (closed) topological copy of the countable sequential fan S_ω if and only if G contains a topological copy of the Arens space S_2.*

Observe that the spaces S_ω and S_2 are very different: while S_ω is Fréchet-Urysohn and has only one non-isolated point, the space S_2 is a standard example of a sequential space that is not Fréchet-Urysohn (and S_2 has infinitely many non-isolated points). Note, that the next problem, belonging to Shakhmatov, remains unsolved:

10.17. PROBLEM. Is every countably compact sequential topological group a Fréchet-Urysohn space?

This problem is closely related to the next question which also remains open:

10.18. PROBLEM. Is every countably compact sequential topological group G ω-monolithic, that is, is the closure of arbitrary countable subset of G metrizable? Compact?

Note that the standard example of a non-first countable Fréchet-Urysohn countably compact topological group is the Σ-product of uncountably many copies of the Boolean group $D = \{0, 1\}$. This group G is ω-monolithic; in fact, the closure of any countable subset of it is a metrizable compactum. However, G, obviously, is not Raĭkov complete. This leads us to the next two questions the answers to which are unknown to me:

10.19. PROBLEM. Is there a ZFC-example of a non-metrizable Raĭkov complete Fréchet-Urysohn topological group?

10.20. PROBLEM. Is every ω-monolithic Raĭkov complete Fréchet-Urysohn topological group metrizable?

Though every compact topological group of countable tightness is metrizable (see ARHANGEL'SKII [1987]), one can show by a standard argument that every Abelian compact topological group contains a dense (even a G_δ-dense) countably compact Fréchet-Urysohn topological subgroup. However, the next old question remains open:

10.21. PROBLEM. Is it true that every compact topological group contains a dense (or a G_δ-dense) subspace of countable tightness? A dense Fréchet-Urysohn subspace?

Note that if such subspaces as in the above problem do exist, they might be responsible for the continuity of sequentially continuous homomorphisms of locally compact groups of Ulam non-measurable cardinality into locally compact groups. See in this connection ARHANGEL'SKII [1994b], ARHANGEL'SKII, JUST and PLEBANEK [1996], and COMFORT and REMUS [1994]. Another recent paper with new results on sequentially continuous homomorphisms of topological groups is HUŠEK [1996].

An important event in topological algebra was a solution of K.H. Hoffman's old problem on epimorphisms by USPENSKIJ's [1994]. He has shown that there exist a separable metrizable topological group G and a proper closed subgroup H of G such that any continuous homomorphism f of G into arbitrary topological group K is completely determined by its values on H (that is, if g is any continuous homomorphism of G into K such that $f|H = g|H$, then $f = g$). Clearly, H cannot be an invariant subgroup of G and, hence, H cannot be an Abelian group. In Uspenskij's example, G is the group of all homeomorphisms of a connected compact manifold M onto itself, in the topology of uniform convergence, and H is the stability subgroup at some point of M.

There are several fine new results in the direction of general embedding problem of spaces into topological groups. KATO [1992] has proved that a compact n-dimensional manifold M without boundary can be embedded into an n-dimensional topological group if and only if a group structure compatible with the topology of M can be introduced on M. KIMURA [1992] showed that the figure eight (one-dimensional) cannot be embedded into a 1-dimensional topological group. KULESZA [1993], for every $n > 0$, has constructed a separable metrizable compact n-dimensional space that cannot be embedded into an n-dimensional topological group. On the other hand, it is an old result of BEL'NOV [1978] that every Tychonoff space can be embedded as a closed subspace into a homogeneous space of the same dimension.

SHKARIN [1999] obtained an outstanding result: he proved that there exists an Abelian topological group G with a countable base such that every Abelian topological group with a countable base is topologically isomorphic to a topological subgroup of G. A similar result for the class of all topological groups with a countable base was earlier obtained by V.V. Uspenskij.

Recall that a topological group G is said to be *minimal* if it does not admit a strictly coarser Hausdorff group topology. The class of minimal topological groups is a natural extension of the class of compact groups. USPENSKIJ [2000] has shown that every topological group H can be represented as a topological subgroup of a Raĭkov complete minimal topological group G of the same weight. This should be compared to the fact that only totally bounded topological groups are subgroups of compact topological groups. In his construction Uspenskij relies upon the notion of *Roelke uniformity* on a topological group which is the greatest lower bound of the left and right uniformities of the group. This uniformity is compatible with the topology of the group G ROELKE and DIEROLF [1981]. A topological group is *Roelke precompact* if its Roelke uniformity is precompact USPENSKIJ [2000]. Uspenskij has shown that the minimal group G in his embedding theorem can be selected to be, in addition, Roelke precompact. However, the next problem posed in ARHANGEL'SKII [1987] remains unsolved:

10.22. PROBLEM. Is every topological group topologically isomorphic to a closed subgroup of a minimal topological group?

In Uspenskij's argument, the Roelke compactification (corresponding to Roelke uniformity) of a Roelke precompact topological group plays an essential role. Note that the group G in Uspenskij's construction cannot be made Abelian since, according to a classical result of Iv.Prodanov and L. Stoyanov, every Abelian minimal topological group is totally bounded (see DIKRANJAN, PRODANOV and STOYANOV [1989]). In fact, the minimal group G, in which Uspenskij embeds H, is the group $Is(M)$ of all isometries of a generalized Urysohn metric space onto itself, in the topology of pointwise convergence.

Note, that in the same paper [2000], USPENSKIJ made an important announcement: he withdrew his earlier public claim that every topological group G is a quotient of a minimal topological group. Therefore, the next problem posed by ARHANGEL'SKII [1987] remains open:

10.23. PROBLEM. Is every topological group a quotient of a minimal topological group?

Even a "stronger" question has been asked (see USPENSKIJ [2000]):

10.24. PROBLEM. Is every topological group a retract of a minimal topological group?

Many new interesting results on minimal topological groups were obtained in the last ten or twelve years by D. Dikranjan. A topological group G is called *totally minimal* if every continuous homomorphism of G onto any topological group H is open. It was shown in DIKRANJAN and SHAKHMATOV [1992] that every countably compact totally minimal Abelian topological group G is compact. On the other hand, there exists an ω-bounded minimal zero-dimensional non-compact Abelian group (DIKRANJAN and SHAKHMATOV [1993]). In the light of this example, the next result is especially interesting:

10.25. THEOREM. *Every connected, countably compact, minimal Abelian topological group G of Ulam non-measurable cardinality is compact.*

The assumption in the above theorem that G is Abelian cannot be dropped (DIKRANJAN [1998a]). In [1998b], DIKRANJAN proved that every totally minimal Abelian topological group G is a quotient of zero-dimensional totally minimal Abelian topological group H_G of the same weight. This result should be compared to the well known fact that any quotient group of a zero-dimensional locally compact group is always zero-dimensional. On the other hand, it was established in ARHANGEL'SKII [1981] that every topological group G can be represented as a quotient of a hereditarily paracompact zero-dimensional topological group. TKAČENKO [1991c] proved that every topological group is a quotient of zero-dimensional topological group of the same weight.

M. Megrelishvili (cited in DIKRANJAN [1998a]) showed that a minimal totally disconnected topological group need not be zero-dimensional. However, every pseudocompact totally disconnected topological group admits a weaker zero-dimensional group topology (see DIKRANJAN [1994], [1998a]).

It seems to be open whether there exists a minimal topological group G which is a P-space (that is, G_δ-subsets in G are open). Of course, such a group cannot be Abelian, since every totally bounded topological group which is a P-space has to be finite.

10.26. PROBLEM. Is every minimal topological group a PT-group? Is every minimal topological group Moscow?

Many problems and results on minimal topological groups are discussed in DIKRANJAN and SHAKHMATOV [1993]. Some of these results deeply involve the algebraic structure of minimal topological groups. A separate survey is needed to cover this topic systematically. We refer the reader to DIKRANJAN's excellent survey [1998a] for further results and references, and to MEGRELISHVILI's works [1995], [1998]; he, in particular, has studied G-minimal topological groups and minimal topological rings.

A new method for defining group topologies on Abelian groups was developed by PROTASOV and ZELENYUK [1991a], [1991b]. A filter on an Abelian group G is called a *G-filter* if there exists a (Hausdorff) group topology on G with regards to which ξ converges to the zero element of G. If ξ is a G-filter on G, then there exists the largest group topology on G with respect to which ξ converges to zero. The group G with this topology is denoted by G_ξ. Protasov and Zelenyuk investigated which sequences in an Abelian group G are G-sequences in the sense of the above definition. This helped them to perform several interesting constructions. In particular, they proved that on every infinite Abelian group G there exists a group topology which is sequential but not Fréchet-Urysohn. Protasov and Zelenyuk, using this approach, also introduced an interesting notion of a potentially compact Abelian group.

D. Dikranjan, M.G. Tkačenko, V.V. Tkachuk, A. Tomita, F.J. Trigos-Arrieta, and several other mathematicians obtained a series of results on the following general question: when a topological group is topologically generated by a relatively discrete subspace converging to the neutral element? See in this connection TKAČENKO [1997], DIKRANJAN and TRIGOS-ARRIETA [2000], TOMITA and TRIGOS-ARRIETA [1998], COMFORT, MORRIS, ROBBIE, SVETLICHNY and TKAČENKO [1998], DIKRANJAN, TKAČENKO and TKACHUK [1999], [2000].

The following general problem was considered in ARHANGEL'SKII and BELLA [1993]. Suppose that a topological group G is the remainder in a Hausdorff compactification of a "nice" topological space X. What can we say about G in this situation? For example, it was proved in ARHANGEL'SKII and BELLA [1993] that if a topological group G of countable tightness is the remainder in a Hausdorff compactification of a metrizable space, then G is metrizable and the compactification is an Eberlein compactum.

A closely related question how properties of topological groups are related to properties of remainders in their Hausdorff compactifications was considered in ARHANGEL'SKII [1999b] and [2000d]. In particular, it was established in [2000d] that if G is an extremally disconnected topological group, then $bG \setminus G$ is countably compact, for any Hausdorff compactification bG of G. It was proved in [1999b] that if a topological group G is *paracompact at infinity* (that is, there exists a Hausdorff compactification bG of G such that the remainder $bG \setminus G$ is paracompact), then every closed pseudocompact subspace of G is compact. It was also shown that the Sorgenfrey line is not the remainder of any topological group G in any Hausdorff compactification of G.

10.27. THEOREM (ARHANGEL'SKII [1999b]). *A topological group G with the countable Souslin number is paracompact at infinity if and only if G is a Lindelöf p-space.*

10.28. PROBLEM (ARHANGEL'SKII [1999b]). When is a Lindelöf topological group G metrizable at infinity? Is it true in this case that the Souslin number of G is countable and G is a p-space?

10.29. PROBLEM (ARHANGEL'SKII [1999b]). Is it true that every topological group that is paracompact at infinity is a p-space?

For related open questions and partial results, see ARHANGEL'SKII [1999b]. It was shown in ARHANGEL'SKII and BELLA [2001] that a topological group of cardinality at most ω_1 is metrizable if and only if it is Dieudonné complete at infinity.

References

ALAS, O.T., I.V. PROTASOV, M.G. TKAČENKO, V.V. TKACHUK, R.G. WILSON and I.V. YASHCHENKO

[1998] Almost all submaximal groups are paracompact and σ-discrete, *Fund. Math.* **156**, 241–260.

ALAS, O.T., M.G. TKAČENKO, V.V. TKACHUK and R.G. WILSON

[1999] Connectedness and local connectedness of topological groups and extensions, *Comment. Math. Univ. Carolin.* **40** (4), 735–753.

ARHANGEL'SKII, A.V.

[1967] Groupes topologiques extremalement discontinus, *C. R. Acad. Sci. Paris* **265**, 822–825.

[1981] Classes of topological groups, *Russian Math. Surveys* **36** (3), 151–174.

[1983] Functional tightness, Q-spaces, and τ-embeddings, *Comment. Math. Univ. Carolin.* **24** (1), 105–120.

[1987] Topological homogeneity, topological groups and their continuous images, *Russian Math. Surveys* **42**, 83–131.

[1994a] Bisequential spaces, tightness of products, and metrizability conditions in topological groups, *Trans. Moscow Math. Soc.* **55**, 207–219.

[1994b] On countably compact topologies on compact groups and on dyadic compacta, *Topology Appl.* **57**, 163–181.

[1999a] On a theorem of W.W. Comfort, K.A. Ross, *Comment. Math. Univ. Carolin.* **40**, 133–151.

[1999b] Some connections between properties of topological groups and their remainders, *Moscow Univ. Math. Bull.* **54** (3), 1–6.

[2000a] Moscow spaces, Pestv–Tkačenko problem and C–embeddings, *Comment. Math. Univ. Carolin.* **41** (3), 585–595.

[2000b] On topological and algebraic structure of extremally disconnected semitopological groups, *Comment. Math. Univ. Carolin.* **41** (4), 803–810.

[2000c] Topological groups and C-embeddings, *Topology Appl.* **115** (3), 265–289.

[2000d] Strong disconnectedness properties and remainders in compactifications, *Topology Appl.* **107**, 3–12.

[200?] Moscow spaces and topological groups, *Topology Proc.*, to appear.

ARHANGEL'SKII, A.V. and A. BELLA

[1993] Cardinal invariants in remainders and variations of tightness, *Proc. Amer. Math. Soc.* **119** (3), 947–954.

[2001] On pseudocompactness of remainders of topological groups and some classes of mappings, *Topology Appl.* **111**, 21–33.

ARHANGEL'SKII, A.V. and P.J. COLLINS

[1995] On submaximal spaces, *Topology Appl.* **64** (3), 219–241.

ARHANGEL'SKII, A.V. and M. HUŠEK

[2001] Extensions of topological and semitopological groups and the product operation, *Comment. Math. Univ. Carolin.* **42** (1), 173–186.

ARHANGEL'SKII, A.V., W. JUST and G. PLEBANEK

[1996] Sequential continuity on dyadic compacta and topological groups, *Comment. Math. Univ. Carolin.* **37** (4), 775–790.

ARHANGEL'SKII, A.V., O.G. OKUNEV and V.G. PESTOV

[1989] On free topological groups of metrizable spaces, *Topology Appl.* **33** (1), 63–76.

ARHANGEL'SKII, A.V. and V.I. PONOMAREV
[1984] Fundamentals of General Topology: Problems and Exercises, Reidel.

ARHANGEL'SKII, A.V. and E.A. REZNICHENKO
[2002] Some properties of paratopological groups, preprint.

BANASZCZYK, W.
[1990] Countable products of LCA groups: their closed subgroups, quotients and duality properties, *Colloq. Math.* **59**, 52–57.
[1991] *Additive subgroups of topological vector spaces*, Lecture Notes in Math., vol. 466, Springer Verlag, Berlin-Heidelberg.

BANASZCZYK, W., M.J. CHASCO and E. MARTIN-PEINADOR
[1994] Open subgroups and Pontryagin duality, *Math. Z.* **215**, 195–204.

BEL'NOV, V.K.
[1978] The dimension of topologically homogeneous spaces and free homogeneous spaces, *Soviet Math. Dokl.* **19** (1), 86–89.

BOUZIAD, A.
[1993] The Ellis theorem and continuity in groups, *Topology Appl.* **50**, 73–80.
[1996a] Every Čech-analytic Baire semitopological group is a topological group, *Proc. Amer. Math. Soc.* **24** (3), 953–959.
[1996b] Continuity of separately continuous group actions in p-spaces, *Topology Appl.* **71**, 119–124.

BRUGUERA, M.
[1997] Some properties of locally quasi-convex groups, *Topology Appl.* **77**, 87–94.

BRUGUERA, M., M.J. CHASCO, E. MARTIN-PEINADOR and V. TARIELADZE
[2001] Completeness properties of locally quasi-convex groups, *Topology Appl.* **111**, 81–93.

BURKE, D.K.
[1984] Covering properties, in *Handbook of Set-theoretic Topology*, K. Kunen and J. Vaughan, eds., North-Holland, Amsterdam, 347–422.

CHASCO, M.J.
[1998] Pontryagin duality for metrizable groups, *Arch. Math.* **70**, 22–28.

CHASCO, M.J. and E. MARTIN-PEINADOR
[2001] On strongly reflexive topological groups, *Applied General Topology* **2** (2), 219–226.

CHASCO, M.J., E. MARTIN-PEINADOR and V. TARIELADZE
[1999] On Mackey topology for groups, *Studia Mathematica* **132** (3), 257–284.

CHEN, Y.Q.
[1999] Note on two questions of Arhangel'skii, *Questions and Answers in General Topology* **17**, 91–94.

CHOBAN, M.M.
[1977] Topological structure of subsets of topological groups and their quotients, (Russian), in *Topological Structures and Algebraic Systems*, (Shtiintsa, Kishinev), 117–163.

COLLINS, P.J. and P.M. GARTSIDE
[1997] Mal'tsev spaces, retral spaces and rectifiable diagonals, *Topology Appl.* **77**, 95–103.

COMFORT, W.W.
[1968] On the Hewitt realcompactification of the product space, *Trans. Amer. Math. Soc.* **131**, 107–118.
[1990] Problems on topological groups and other homogeneous spaces, in *Open problems in topology*, J. van Mill, G.M. Reed, eds., Noth-Holland, Amsterdam, 313–348.

COMFORT, W.W., S. HERNÁNDEZ and F.J. TRIGOS-ARRIETA

[1996] Relating a locally compact Abelian group to its Bohr compactification, *Adv. Math.* **120** (3), 322–344.

COMFORT, W.W., K.H. HOFMANN and D. REMUS

[1992] Topological Groups and Semigroups, in *Recent Progress in General Topology*, M. Hušek and J. van Mill, eds., North-Holland, Amsterdam, 239–274.

COMFORT, W.W., O. MASAVEU and H. ZHOU

[1995] Dense subsets of some topological groups, *Ann. New York Acad. Sci.* **767**, 17–27.

COMFORT, W.W. and J. VAN MILL

[1994] Groups with only resolvable group topologies, *Proc. Amer. Math. Soc.* **120** (3), 687–696.

COMFORT, W.W., S.A. MORRIS, D. ROBBIE, S. SVETLICHNY and M. TKAČENKO

[1998] Suitable sets for topological groups, *Topology Appl.* **86**, 25–46.

COMFORT, W.W. and S. NEGREPONTIS

[1966] Extending continuous functions on $X \times Y$ to subsets of $\beta X \times \beta Y$, *Fund. Math.* **59**, 1–12.

COMFORT, W.W. and D. REMUS

[1994] Compact groups of Ulam-measurable cardinality: partial converses to theorems of Arhangel'skii and Varopoulos, *Mathematica Japonica* **39** (2), 203–210.

COMFORT, W.W. and K.A. ROSS

[1966] Pseudocompactness and uniform continuity in topological groups, *Pacific J. Math.* **16**, 483–496.

COMFORT, W.W. and F.J. TRIGOS-ARRIETA

[1991] Remarks on a theorem of Glicksberg, *Ann. of New York Academy of Sciences*, S.Andima, New York, 25–33.

DIKRANJAN, D.

[1994] Zero-dimensionality of some pseudocompact groups, *Proc. Amer. Math. Soc.* **120**, 1299–1308.

[1998a] Recent advances in minimal topological groups, *Topology Appl.* **85**, 53–91.

[1998b] Quotients of zero-dimensional precompact Abelian groups, *Topology Appl.* **86**, 47–62.

DIKRANJAN, D., I. PRODANOV and L. STOYANOV

[1989] *Topological Groups: Characters, Dualities and Minimal Group Topologies*, Marcel Dekker Inc., New York-Basel.

DIKRANJAN, D. and D. SHAKHMATOV

[1992] Compact-like totally dense subgroups of compact groups, *Proc. Amer. Math. Soc.* **114**, 1119–1129.

[1993] Pseudocompact and countably compact Abelian groups: Cartesian products and minimality, *Trans. Amer. Math. Soc.* **335** (2), 775–790.

[1998] *Algebraic structure of the pseudocompact groups*, Memoirs Amer. Math. Soc., vol. 133(633), 1–83.

DIKRANJAN, D. and M.G. TKAČENKO

[2000] Sequential completeness of quotient groups, *Bull. Austral. Math. Soc.* **61**, 129–151.

[2001a] Sequentially complete groups: dimension and minimality, *J. Pure Appl. Algebra* **157**, 215–239.

[2001b] Weakly complete free topological groups, *Topology Appl.* **112**,

DIKRANJAN, D., M.G. TKAČENKO and V.V. TKACHUK

[1999] Some topological groups with and some without suitable sets, *Topology Appl.* **98**, 131–148.

[2000] Topological groups with thin generating sets, *J. Pure Appl. Algebra* **145**, 123–148.

DIKRANJAN, D. and F.J. TRIGOS-ARRIETA

[2000] Suitable sets in some topological groups, *J. Group Theory* **3**, 293–321.

DIKRANJAN, D. and S. WATSON

[2001] A solution to van Douwen's problem on Bohr topologies, *J. Pure Appl. Algebra* **163** (2), 147–158.

DOLECKI, S. and R. PEIRONE

[1992] Topological semigroups of every countable sequential order, in *Recent developments of General Topology and its Applications*, Academie Verlag, 80–84.

VAN DOUWEN, E.K.

[1980] The product of two countably compact topological groups, *Trans. Amer. Math. Soc.* **262**, 417–427.

[1990] The maximal totally bounded group topology on G and the biggest minimal G-space for Abelian groups G, *Topology Appl.* **34**, 69–91.

EDA, K., S. KAMO and H. OHTA

[1993] Abelian groups of continuous functions and their duals, *Topology Appl.* **53**, 131–151.

EDA, K. and H. OHTA

[1985] On Abelian groups of integer-valued continuous functions, their Z-duals and Z-reflexivity, in *Abelian Group Theory*, R. Göbel and E. Walker, eds., Gordon and Breach, London, 241–257.

ELLIS, R.

[1957] A note on the continuity of the inverse, *Proc. Amer. Math. Soc.* **8**, 372–373.

ENGELKING, R.

[1977] *General Topology*, PWN, Warszawa.

FAY, T.H., E.T. ORDMAN and B.V.S. THOMAS

[1979] The free topological group over rationals, *Topology Appl.* **10**, 33–47.

FROLÍK, Z.

[1960] The topological product of two pseudocompact spaces, *Czechoslovak. Math. J.* **10**, 339–349.

[1968] Fixed points of maps of extremally disconnected spaces and complete Boolean Algebras, *Bull. Acad. Polon. Sci., Ser. Math., Astronom., Phys.* **16**, 269–275.

GALINDO, J. and S. HERNÁNDEZ

[1996] On the completion of a MAP group, *Annals New York Academy of Science* **806**, 164–168.

[1997] Transmission of continuity to the Bohr topology, *Topology Appl.* **77**, 161–175.

[1998] On a Theorem of van Douwen, *Extracta Math.* **13** (1), 115–123.

[1999] Pontryagin-van Kampen reflexivity for free Abelian topological groups, *Forum Math.* **11** (4), 399–415.

GARTSIDE, P.M., E.A. REZNICHENKO and O.V. SIPACHEVA

[1997] Mal'tsev and retral spaces, *Topology Appl.* **80**, 115–129.

GILLMAN, L. and M. JERISON

[1960] *Rings of Continuous Functions*, Van Nostrand, Princeton.

GLADDINES, H.
[1995] Countable closed sets that are not a retract of $G^{\#}$, *Topology Appl.* **67**, 81–84.

GLICKSBERG, I.
[1959] Stone–Čech compactifications of products, *Trans. Amer. Math. Soc.* **90**, 369–382.

GUL'KO, A.S.
[1996] Rectifiable spaces, *Topology Appl.* **68** (2), 107–112.

HART, K.P., H. JUNILLA and J. VAN MILL
[1985] A Dowker group, *Comment. Math. Univ. Carolin.* **26**, 799–810.

HART, K.P. and J. VAN MILL
[1991a] Discrete sets and the maximal totally bounded group topology, *J. Pure Appl. Algebra* **70**, 73–80.
[1991b] A countably compact group H such that $H \times H$ is not countably compact, *Trans. Amer. Math. Soc.* **323**, 811–821.

HEATH, R.W.
[1992] Some nonmetric, first countable, cancellative topological semigroups that are generalized metric spaces, *Topology Appl.* **44**, 167–173.

HERNÁNDEZ, C.
[1998] The dimension of an LCA group in its Bohr topology, *Topology Appl.* **86**, 63–67.

HERNÁNDEZ, C. and M.G. TKAČENKO
[1998] Subgroups of $\mathcal{R}$-factorizable groups, *Comment. Math. Univ. Carolin.* **39**, 371–378.

HERNÁNDEZ, S. and S. MACARIO
[2001] Invariance of compactness for the Bohr topology, *Topology Appl.* **111**, 161–173.
[200?] Dual properties in totally bounded Abelian groups, preprint.

HERNÁNDEZ, S., M. SANCHIS and M. TKAČENKO
[2000] Bounded sets in spaces and topological groups, *Topology Appl.* **101**, 21–43.

HERNÁNDEZ, S. and V. USPENSKIJ
[2000] Pontryagin duality for spaces of continuous functions, *Journ. Math. Analysis Appl.* **242**, 135–144.

HINDMAN, N. and D. STRAUSS
[1998] *Algebra in the Stone-Čech Compactification – Theory and Applications*, Walter de Gruyter and Co., Berlin.

HUŠEK, M.
[1970] The Hewitt realcompactification of a product, *Comment. Math. Univ. Carolin.* **11**, 393–395.
[1972] Realcompactness of function spaces and $v(P \times Q)$. *Gen. Topology and Appl.* **2**, 165–179.
[1992] Productivity of properties of topological groups, *Topology Appl.* **44**, 189–196.
[1996] Sequentially continuous homomorphisms on products of topological groups, *Topology Appl.* **70**, 155–165.

ISBELL, J.R.
[1955] Zero-dimensional spaces, *Tohoku Math. J.* **7** (1-2), 3–8.

IVANOVSKIĬ, L.N.
[1958] On a hypothesis of P.S. Aleksandroff, *Dokl. AN SSSR* **123**, 785–786.

KATO, H.
[1992] A note on embeddings of manifolds into topological groups preserving dimensions, *Czechoslovak Math. J.* **42**, 619–622.

KENDEROV, P., I.S. KORTEZOV and W.B. MOORS

[2001] Topological games and topological groups, *Topology Appl.* **109** (2), 157–165.

KIMURA, T.

[1992] Dimensions of topological groups containing the bouquet of two circles, *Proc. Amer. Math. Soc.* **114**, 1109–1113.

KOROVIN, A.V.

[1992] Continuous actions of pseudocompact groups and axioms of topological group, *Comment Math. Univ. Carolin.* **33**, 335–343.

KULESZA, J.

[1993] Spaces which do not embed in topological groups of the same dimension, *Topology Appl.* **50**, 139–145.

KUNEN, K.

[1998] Bohr topologies and partition theorems for vector spaces, *Topology Appl.* **90**, 97–107.

KÜNZI, H.P., S. ROMAGUERA and O.V. SIPACHEVA

[1998] The Doitchinov completion of a regular paratopological group, *Serdica Math.* **24**, 73–88.

LEIDERMAN, A., S.A. MORRIS and V.G. PESTOV

[1997] The free Abelian topological group and the free locally convex space on the unit interval, *J. London Math. Soc.(2)* **56** (3), 529–538.

MALYKHIN, V.I.

[1975] Extremally disconnected and close to them groups, *Dokl. AN SSSR* **220** (1), 27–30.

[1987] Nonpreservation of properties of topological groups on taking their square, *Siberian Math. J.* **28**, 639–645.

MALYKHIN, V.I. and I.V. PROTASOV

[1996] Maximal resolvability of bounded groups, *Topology Appl.* **20**, 1–6.

MARTIN-PEINADOR, E.

[1995] A reflexive admissible topological group must be locally compact, *Proc. Amer. Math. Soc.* **123**, 3563–3566.

MASAVEU, O.

[1995] *Dense subsets of some topological groups*, Ph.D. Thesis, Wesleyan University, Middletown, Connecticut.

MEGRELISHVILI, M. (LEVY)

[1995] Group representations and construction of minimal topological groups, *Topology Appl.* **62**, 1–19.

[1996] Free Topological Groups over (Semi)Group Actions, *Annals of New York Academy of Sciences* **788**, 164–169.

[1998] G-minimal topological groups, in *Abelian Groups, Module Theory, and Topology*, D. Dikranjan and L. Salce, eds., Proceedings in honour of Adalberto's Orsattis 60th birthday, Marcel Dekker, Inc. New York, 289–299.

MORRIS, S. and V. PESTOV

[1993] Open subgroups of free Abelian topological groups, *Math. Proc. Camb. Phil. Soc.* **114**, 439–442.

NOGURA, T., D. SHAKHMATOV and Y. TANAKA

[1993] Metrizability of topological groups having weak topologies with respect to good covers, *Topology Appl.* **54**, 203–212.

[1997] α_4-property versus A-property in topological spaces and groups, *Studia Sci. Math. Hungar.* **33** (4), 351–362.

NYIKOS, P.J.

[1992] Subsets of $^{\omega}\omega$ and the Fréchet-Urysohn property, *Topology Appl.* **48**, 91–62.

OHTA, H.

[1996] Chains of strongly non-reflexive dual groups of integer-valued continuous functions, *Proc. Amer. Math. Soc.* **124** (3), 961–967.

OHTA, H., M. SAKAI and KEN-ICHI TAMANO

[1993] Perfect κ-normality in product spaces, in *Papers on General Topology and Applications, Annals of the New York Academy of Sciences* **704**, 279–289.

OKUNEV, O.G.

[1996] A remark on the tightness of products, *Comment. Math. Univ. Carolin.* **37**, 397–399.

PAPAZYAN, T.

[1991] Extremal topologies on a semigroup, *Topology Appl.* **39** (3), 229–243.

PESTOV, V.G.

[1982] Some properties of free topological groups, *Moscow Univ. Math. Bull.* **37**, 46–49.

[1986] Free Abelian topological groups and the Pontryagin duality, *Vestnik Mosk. Univ., Ser. 1*, **1**, 3–5.

[1995] Free Abelian topological groups and the Pontryagin-van Kampen duality, *Bull. Austral. Math. Soc.* **52** (2), 297–311.

PESTOV, V.G. and M.G. TKAČENKO

[1985] Problem 3.28, in *Unsolved Problems of Topological Algebra*, Academy of Science Moldova, Kishinev, "Shtiinca", p. 18.

PESTOV, V.G. and K. YAMADA

[1999] Free topological groups of metrizable spaces and inductive limits, *Topology Appl.* **98**, 291–301.

PONTRYAGIN, L.S.

[1939] *Continuous Groups*, Moscow. English translation *Topological groups*, Princeton University Press, Princeton 1939.

PROTASOV, I.V.

[1993] Ultrafilters and topologies on groups, *Sib. Mat. Journ.* **34** (5), 163–180.

[1994] Discrete subsets of topological groups, *Mat. Zametki* **55**, 150–151.

[1998a] Absolutely resolvable groups, (Russian), *Ukr. Matem. J.* **48** (3), 383–392.

[1998b] Maximal topologies on groups, *Sib. Mat. Journ.* **39** (6), 1368–1381.

[1998c] Finite groups in βG, *Matematychni Studii* **10**, 17–22.

[1998d] Resolvability of groups, (Russian), *Matematychni Studii* **9**, 130–148.

[1998e] Irresolvable topologies on groups, (Russian), *Ukr. Matem. J.* **50** (12), 1646–1655.

[200?] Remarks on maximal topologies on groups, to appear.

PROTASOV, I.V. and E.G. ZELENYUK

[1991a] Topologies on Abelian groups, *Math USSR Izvestiya* **37** (2), 445–460.

[1991b] Potentially compact Abelian groups, *Math. USSR Sbornik* **69** (1), 299–305.

RAVSKIJ, O.

[2001] Private communication.

REMUS, D. and F.J. TRIGOS-ARRIETA

[1993] Abelian groups which satisfy Pontryagin duality need not respect compactness, *Proc. Amer. Math. Soc.* **117**, 1195–1200.

REZNICHENKO, E.A.
[1994a] Continuity in complete groups, in *Abstracts of Tenth Summer Conference on General Topology and Applications*, Amsterdam, p. 135.
[1994b] Extensions of functions defined on products of pseudocompact spaces and continuity of the inverse in pseudocompact groups, *Topology Appl.* **59**, 233–244.

REZNICHENKO, E.A. and O.V. SIPACHEVA
[1994] On factor mappings on the words of finite length in free topological groups, (Russian), preprint.
[1999] Properties of Fréchet-Urysohn type in topological spaces, groups, and locally convex spaces, *Moscow Univ. Math. Bull.* **54** (3), 33–38.

REZNICHENKO, E.A. and V.V. USPENSKIJ
[1998] Pseudocompact Mal'tsev spaces, *Topology Appl.* **86**, 83–104.

ROELKE, W. and S. DIEROLF
[1981] *Uniform Structures on Topological Groups and their Quotients* McGraw-Hill, New York.

ŠČEPIN, E.V.
[1976] Real-valued functions and canonical sets in Tychonoff products and topological groups, *Russian Math. Surveys* **31**, 19–30.
[1979] On κ-metrizable spaces, *Izv. AN SSSR, Ser. Matem.* **43** (2), 442–478.

SHAKHMATOV, D.B.
[1990] α_i-properties in Fréchet-Urysohn topological groups, *Topology Proc.* **15**, 143–183.
[1993] A note on transfinite inductive dimensions in topological groups, *Topology Appl.* **54**, 213–220.
[1999] A comparative survey of selected results and open problems concerning topological groups, fields, and vector spaces, *Topology Appl.* **91** (1), 51–63.

SHAPIROVSKIJ, B.E.
[1978] Special types of embeddings in Tychonoff cubes. Subspaces of Σ-products and cardinal invariants, in *Coll. Math. Soc. Janos Bolyai, Topology*, vol. 23, Budapest, 1055–1086.

SHIBAKOV, A.
[1996] Sequential group topology on rationals with intermediate sequential order, *Proc. Amer. Math. Soc.* **124**, 2599–2607.
[1998] Sequential topological groups of any sequential order under CH, *Fund. Math.* **155** (1), 79–89.
[1999] Countable Fréchet topological groups under CH, *Topology Appl.* **91**, 119–139.

SHKARIN, S.A.
[1999] On universal Abelian topological groups, (Russian), *Mat. Sb.* **190** (7), 127–144; translation in *Sb. Math.* **190**, 1059–1076.

SIPACHEVA, O.V.
[1991] Compacta with a continuous Mal'tsev operation and retracts of topological groups, *Moscow Univ. Math. Bull.* **46**, 22–24.
[2000] Free topological groups of spaces and their subspaces, *Topology Appl.* **101**, 181–212.

SIROTA, S.
[1969] Products of topological groups and extremal disconnectedness, *Matem. Sb.* **79** (2), 179–192.

TKAČENKO, M.G.

[1983a] On the Souslin property in free topological groups over compact Hausdorff spaces, *Mat. Zametki* **34**, 601–607. English translation in *Math. Notes* **34** (1983).

[1983b] The notion of o-tightness and C-embedded subspaces of products, *Topology Appl.* **15**, 93–98.

[1991a] Factorization theorems for topological groups and their applications, *Topology Appl.* **38**, 21–37.

[1991b] Subgroups, quotient groups and products of $\mathcal{R}$-factorizable groups, *Topology Proc.* **16**, 201–231.

[1991c] On zero-dimensional groups and factorization of homomorphisms with respect to weight and dimension, *Sibirsk. Mat. Zh.* **32** (2), 151–159.

[1994] Free topological groups and inductive limits, *Topology Appl.* **60**, 1–12.

[1997] Generating dense subgroups of topological groups, *Topology Proc.* **22**, 533–582.

[1999] Topological groups for topologists: Part I, *Bol. Soc. Mat. Mexicana* **5**, 735–753.

[2000] Topological groups for topologists: Part II, *Bol. Soc. Mat. Mexicana* **6**, 1–45.

[2002] Topological groups: between compactness and $\aleph_0$-boundedness, this volume.

[200?a] Complete $\aleph_0$-bounded groups need not be $\mathbb{R}$-factorizable, *Comment. Math. Univ. Carolin.*, to appear.

[200?b] Subgroups of Lindelöf P-groups, submitted.

TKACHUK, V.V.

[1991] Methods of the theory of cardinal invariants and the theory of mappings as applied to spaces of functions, *Sibirsk. Math. Journ.* **32**, 93–107.

TODORČEVIĆ, S.

[1993] Some applications of S and L combinatorics, in *The Work of M.E. Rudin*, F.D. Tall, ed., Ann. New York Acad. Sci. **705**, 1301–1367.

TOMITA, A.H.

[1996] On finite powers of countably compact groups, *Comment. Math. Univ. Carolin.* **37** (3), 617–626.

[1997] On infinite products of countably compact groups, in *Proc. 8th Prague Top. Symposium*, P. Simon, ed., Topology Atlas, 362–370.

[1999] A group under MA_ω whose square is countably compact but whose cube is not, *Topology Appl.* **91**, 91–104.

TOMITA, A.H. and F.J. TRIGOS-ARRIETA

[1998] Suitable sets in products of topological groups and in groups equipped with the Bohr topology, in *Abelian Groups, Module Theory, and Topology*, Proc. in Honor of Adalberto Orsatti's 60th Birthday, Lecture Notes in Pure and Appl. Algebra, Marcel Dekker, Inc., pp. 327–342.

TRIGOS-ARRIETA, F.J.

[1991] Continuity, boundedness, connectedness and the Lindelöf property for topological groups, *J. Pure Appl. Algebra* **70**, 199–210.

[1994] Every uncountable Abelian group admits a non-normal group topology, *Proc. Amer. Math. Soc.* **122**, 907–909.

USPENSKIJ, V.V.

[1985] On continuous images of Lindelöf topological groups, *Soviet Math. Dokl.* **32**, 802–806.

[1989a] Topological groups and Dugundji spaces, *Matem. Sb.* **180** (8), 1092–1118.

[1989b] The Mal'tsev operation on countably compact spaces, *Comment. Math. Univ. Carolin.* **30** (2), 395–402.

[1993] The solution of the epimorphism problem for Hausdorff topological groups, *Seminar Sophus Lie* **3**, 69–70.

[1994] The epimorphism problem for Hausdorff topological groups, *Topology Appl.* **57**, 287–294.

[2000] On subgroups of minimal topological groups, *Electronic access:* math.GN/0004119, 1–29.

[2001] The Roelcke compactification of groups of homeomorphisms, *Topology Appl.* **111**, 195–205.

WILLIAMS, S.W.

[1984] Box products, in *Handbook of Set-theoretic Topology*, K. Kunen and J. Vaughan, eds., North-Holland, Amsterdam, 169–200.

YAMADA, K.

[1994] Characterizations of a metrizable space X such that every $A_n(X)$ is a k-space, *Topology Appl.* **49**, 75–94.

[1996] Free Abelian topological groups and k-spaces, *Glas. Math.* **31**, 8–91.

[1997] Tightness of free Abelian topological groups and of finite products of sequential fans, *Topology Proc.* **22**, 363–381.

[1998] Metrizable subspaces of free topological groups on metrizable spaces, *Topology Proc.* **23**, 379–409 (2000).

[200?] Fréchet spaces in free topological groups, preprint.

ZELENYUK, E.G.

[1997] Finite groups in βN are trivial, *Semigroup Forum* **55**, 131–132.

[1998] Resolvability of topological groups, *Ukr. Matem. J.* **50**.

CHAPTER 2

Matrices and Ultrafilters

Joni Baker

Daniel H. Wagner Associates, Hampton, VA 23669, U.S.A.
E-mail: joni@va.wagner.com

Kenneth Kunen

Department of Mathematics, University of Wisconsin, Madison, WI 53706, U.S.A.
E-mail: kunen@math.wisc.edu

Contents

RECENT PROGRESS IN GENERAL TOPOLOGY II
Edited by Miroslav Hušek and Jan van Mill

1. Introduction

In this article, we survey some recent results which produce points with interesting topological properties in the Stone space of a boolean algebra $\mathcal{B}$. Our primary focus is the case where $\mathcal{B} = \mathcal{P}(\kappa)$, a power set algebra; then the points will be in the Čech compactification, $\beta\kappa$ (where κ is discrete). However, these methods also apply to some other complete boolean algebras.

We also present some new results, and we unify all the results under the one umbrella of "hatpoints". In most cases, especially for the new results, we present complete proofs. However, occasionally we refer the reader to the literature when we do not see how to improve on a published proof. In addition, we assume that the reader is already familiar with the Stone Representation Theorem and the theory of Čech compactifications, although we begin by reviewing some basic notation regarding these matters.

If $\mathcal{B}$ is a boolean algebra, we use $\mathrm{st}(\mathcal{B})$ to denote its Stone space. Thus, the elements of $\mathrm{st}(\mathcal{B})$ are the ultrafilters on $\mathcal{B}$, and the clopen sets of $\mathrm{st}(\mathcal{B})$ are all of the form $N_b = \{\mathbf{x} \in \mathrm{st}(\mathcal{B}) : b \in \mathbf{x}\}$, for $b \in \mathcal{B}$. This "st" is a contravariant functor which produces an equivalence between the category of boolean algebras and the category of compact 0-dimensional Hausdorff spaces.

When we work in the category of boolean algebras, the notation $h : \mathcal{B} \to \mathcal{A}$ always implies that h is a homomorphism, and $h : \mathcal{B} \twoheadrightarrow \mathcal{A}$ means that in addition, h is onto. When $h : \mathcal{B} \to \mathcal{A}$, the dual $h^* : \mathrm{st}(\mathcal{A}) \to \mathrm{st}(\mathcal{B})$ (defined by $h^*(\mathbf{x}) = h^{-1}(\mathbf{x})$) will be continuous, and $h : \mathcal{B} \twoheadrightarrow \mathcal{A}$ implies that in addition, $h^* : \mathrm{st}(\mathcal{A}) \hookrightarrow \mathrm{st}(\mathcal{B})$ is 1-1, so that it embeds $\mathrm{st}(\mathcal{A})$ into $\mathrm{st}(\mathcal{B})$.

If $\mathcal{F}$ is a filter on $\mathcal{B}$, with $\mathcal{I}$ its dual ideal, we use both $\mathcal{B}/\mathcal{F}$ and $\mathcal{B}/\mathcal{I}$ to denote the quotient algebra. Then $h^* : \mathrm{st}(\mathcal{B}/\mathcal{F}) \hookrightarrow \mathrm{st}(\mathcal{B})$, where h is the natural surjection. We often identify $\mathrm{st}(\mathcal{B}/\mathcal{F})$ with the subspace of $\mathrm{st}(\mathcal{B})$ consisting of those ultrafilters $\mathbf{x}$ on $\mathcal{B}$ which extend $\mathcal{F}$. We use $a \leq_{\mathcal{I}} b$ or $a \leq_{\mathcal{F}} b$ to mean $a \wedge b' \in \mathcal{I}$.

This is illustrated by our view of $\beta\kappa$:

1.1. DEFINITION. $\mathcal{FR} = \mathcal{FR}(\kappa)$ is the *Fréchet* filter, $\{X \subseteq \kappa : |\kappa \backslash X| < \kappa\}$.

If κ is any infinite cardinal, given the discrete topology, we have $u(\kappa) \subseteq \kappa^* \subseteq \beta\kappa$. $\beta\kappa$ is the space of ultrafilters on κ; that is, $\beta\kappa = \mathrm{st}(\mathcal{P}(\kappa))$. Then, $\kappa^* = \beta\kappa \backslash \kappa = \mathrm{st}(\mathcal{P}(\kappa/fin))$ is the the space of nonprincipal ultrafilters on κ, where *fin* denotes the ideal of finite sets. $u(\kappa) = \mathrm{st}(\mathcal{P}(\kappa)/{<}\kappa) = st(\mathcal{P}(\kappa)/\mathcal{FR}(\kappa))$ is the space of uniform ultrafilters ; that is, $\mathbf{x} \in u(\kappa)$ iff every set in $\mathbf{x}$ has size κ. Here "$<\kappa$" denotes the ideal of sets of size less than κ. For $A, B \in \mathcal{P}(\kappa)$, $A \subseteq^* B$ usually means $A \leq_{\mathcal{FR}(\kappa)} B$.

Our methods construct points in $\mathrm{st}(\mathcal{B})$ which have properties related to "weak P-point":

1.2. DEFINITION. If θ is an infinite cardinal and X is a topological space:

☞ $\mathbf{x} \in X$ is a *P_θ-point* in X iff the intersection of any family of *fewer than* θ neighborhoods of $\mathbf{x}$ is also a neighborhood of $\mathbf{x}$.

☞ $\mathbf{x} \in X$ is a *weak P_θ-point* in X iff $\mathbf{x}$ is not a limit point of any subset of $X \backslash \{\mathbf{x}\}$ of size *less than* θ.

So, a P-point is a P_{ω_1}-point, and a weak P-point is a weak P_{ω_1}-point. Every point is a P_ω-point. In any T_1 space, every P_θ-point is a weak P_θ-point.

We now summarize the results we prove here, and explain how to unify them under one framework.

One specific result is:

1.3. THEOREM. *Let $\mathcal{B}$ be a complete boolean algebra and $\mathcal{I}$ an ideal on $\mathcal{B}$. Assume that $|\mathcal{B}| = 2^\kappa$ and that $\mathcal{B}$ has an antichain $\{a_\xi : \xi < \kappa\}$ such that the a_ξ are distinct and each $a_\xi \notin \mathcal{I}$. Then there is an $\mathbf{x} \in \mathrm{st}(\mathcal{B}/\mathcal{I}) \subseteq \mathrm{st}(\mathcal{B})$ such that $\mathbf{x}$ is a weak P_κ-point in $\mathrm{st}(\mathcal{B})$.*

Applying this to $\mathcal{P}(\kappa)$ and $\mathcal{P}(\kappa)/{<}\kappa$, and noting that κ can be partitioned into κ disjoint sets of size κ, we get:

1.4. COROLLARY. *For every infinite κ, there is an $\mathbf{x} \in u(\kappa)$ which is a weak P_κ-point in $\beta\kappa$.*

In fact, *good* ultrafilters (Definition 2.8) have this property. These were introduced by KEISLER in [1964], who proved that they exist when $2^\kappa = \kappa^+$. A ZFC proof was given by KUNEN in [1972] (see also CHANG and KEISLER [1990], Theorem 6.1.4), and it is a folklore result that these are weak P_κ-points (Lemma 2.13). The fact that one might use good ultrafilters in arbitrary boolean algebras is due to BALCAR and FRANĚK [1982], who proved the following, obtained by setting $\mathcal{I} = \{0\}$:

1.5. COROLLARY. *If $\mathcal{B}$ is a complete boolean algebra such that $|\mathcal{B}| = 2^\kappa$ and $\mathcal{B}$ has an antichain of size κ, then $\mathrm{st}(\mathcal{B})$ contains a weak P_κ-point.*

Note that this article emphasizes *weak* P_κ-points. If $\mathcal{B}$ is complete and of size less than the first measurable cardinal, then no non-isolated point in $\mathrm{st}(\mathcal{B})$ can be a P-point. In fact, the methods discussed here can guarantee that regardless of measurable cardinals, all of the points constructed will not be P-points; see Section 6

The following boolean algebra yields an example of Corollary 1.5:

1.6. DEFINITION. $\mathcal{D}_\kappa$ denotes the regular open algebra of the metric space κ^ω (where κ is discrete).

Alternatively, one may view $\mathcal{D}_\kappa$ as the completion of the forcing order $\mathrm{Fn}(\omega, \kappa)$ (finite partial functions from ω to κ).

1.7. COROLLARY. *There is a dense subset $E \subseteq \mathrm{st}(\mathcal{D}_\kappa)$ such that $|E| = \kappa$ and each point of E is a weak P_κ-point in $\mathrm{st}(\mathcal{D}_\kappa)$.*

□ For each $p \in \mathrm{Fn}(\omega, \kappa)$ let $[p] = \{f \in \kappa^\omega : f \supset p\}$. Then $[p]$ is clopen in κ^ω, so that $[p] \in \mathcal{D}_\kappa$. Note that $N_{[p]}$ is a clopen set in $\mathrm{st}(\mathcal{D}_\kappa)$ which is homeomorphic to all of $\mathrm{st}(\mathcal{D}_\kappa)$, so by Corollary 1.5, we can choose a weak P_κ-point $\mathbf{x}_p \in N_{[p]}$. Let $E = \{\mathbf{x}_p : p \in \mathrm{Fn}(\omega, \kappa)\}$. Then $|E| \leq \mathrm{Fn}(\omega, \kappa) = \kappa$. The $\mathbf{x}_p$ may not all be distinct, but $|E| = \kappa$ because it is dense and all its elements are weak P_κ-points. □

This example also shows that in Corollary 1.5, one cannot strengthen weak P_κ-point to weak P_{κ^+}-point. Likewise, in Corollary 1.4, no $\mathbf{x} \in u(\kappa)$ can be a weak P_{κ^+}-point in $\beta\kappa$, since $\mathbf{x}$ is a limit of the isolated points. However,

1.8. THEOREM. *For every infinite regular κ, there is an $\mathbf{x} \in u(\kappa)$ which is a weak P_{κ^+}-point in $u(\kappa)$.*

This is proved in BAKER and KUNEN [2001], using *mediocre points* (a modification of "good"). The case $\kappa = \omega$ (a weak P-point in $\omega^* = u(\omega)$) is an older result of KUNEN [1980], using *OK points* (another modification of "good"). In fact, "good", "mediocre", and "OK" are all special cases of the notion of *hatpoint* (see Section 2), and we shall derive Theorems 1.3 and 1.8 by one proof, which produces hatpoints in Stone spaces. The exact flavor of hatpoint we get in a specific $\mathrm{st}(\mathcal{B})$ will depend on what kinds of matrices exists in $\mathcal{B}$ (see Section 3). We do not know if Theorem 1.8 holds for singular κ; it would hold if a suitable matrix (described by Lemma 3.9) exists in $\mathcal{P}(\kappa)/{<}\kappa$.

Besides constructing points in $\mathrm{st}(\mathcal{B})$, we shall, by the same proof, construct subsets of $\mathrm{st}(\mathcal{B})$. By way of introduction, consider the following well-known result:

1.9. THEOREM (EFIMOV [1970]). *Let $\mathcal{A}$ be a complete boolean algebra with $|\mathcal{A}| \leq 2^\kappa$. Then $\mathrm{st}(\mathcal{A})$ can be embedded into $u(\kappa)$.*

Equivalently, there is an $h : \mathcal{P}(\kappa)/{<}\kappa \twoheadrightarrow \mathcal{A}$. Since $\mathcal{A}$ is complete, this is easy to prove (as we do in Section 4) using the Sikorski Extension Theorem.

Now, consider the special case where $\mathcal{A}$ is the 2-element algebra and $|\mathrm{st}(\mathcal{A})| = 1$, in which case h can be identified with an ultrafilter, which defines a point in $u(\kappa)$. Of course, Theorem 1.9 is trivial in this case, since it just says that $u(\kappa)$ contains a point – but now we know that in fact it contains a weak P_κ-point, or even a weak P_{κ^+}-point. Thus, we shall investigate improvements of Theorem 1.9 which embed $\mathrm{st}(\mathcal{A})$ as a weak P_θ-*set*; the definition of this notion is part of a general scheme for lifting properties of points to properties of sets:

1.10. DEFINITION. If "❀-point" is some property of points in a space, then $F \subseteq X$ is a ❀-*set* in X iff the *point* F is a ❀-point in the quotient space X/F obtained by collapsing F to a point.

1.11. LEMMA. *F is a weak P_θ-set in X iff no point of F is a limit point of any subset of $X \backslash F$ of size less than θ.*

We shall show:

1.12. THEOREM. *Let $\mathcal{A}$ be a complete boolean algebra with $|\mathcal{A}| \leq 2^\kappa$. Then:*

1. *There is an $F \subseteq u(\kappa)$ such that F is homeomorphic to $\mathrm{st}(\mathcal{A})$ and F is a weak P_κ-set in $\beta\kappa$.*
2. *If κ is regular, then there is an $F \subseteq u(\kappa)$ such that F is homeomorphic to $\mathrm{st}(\mathcal{A})$ and F is a weak P_{κ^+}-set in $u(\kappa)$.*

Note that Theorem 1.8 is the special case of (2) when $\mathcal{A} = \{0, 1\}$. In the case $\kappa = \omega$, (2) is due to SIMON [1985], who produced $\mathrm{st}(\mathcal{A})$ as an OK set in $u(\omega) = \omega^*$. Here, we shall derive Theorems 1.3 and 1.12 as special cases of one result, Theorem 5.6, which produces hat*sets* in Stone spaces.

Now, we can put our results together to obtain dense sets of weak P-points:

1.13. DEFINITION. $D \subseteq X$ is *θ-fuzzy* in X iff $|D| = \theta$, every element of D is a weak P_θ-point in X, and D is dense in itself.

If D is θ-fuzzy, then no element of D is a weak P_{θ^+}-point.

1.14. THEOREM. *For any infinite κ:*

1. *There is an $E \subseteq u(\kappa)$ which is κ-fuzzy in $\beta\kappa$.*
2. *If κ is regular, then there is an $E \subseteq u(\kappa)$ such that E is κ^+-fuzzy in $u(\kappa)$.*

As before, we do not know if the hypothesis that κ is regular can be removed in (2).

Since every point is a P_ω-point, item (1) for $\kappa = \omega$ is immediate from the well-known fact that every non-scattered compact Hausdorff space contains a countable subset which is dense in itself. However (2) has non-trivial content even for $\kappa = \omega$: there is a set of $\aleph_1$ weak P-points in $\omega^* = u(\omega)$ which is dense in itself. We remark that given Theorem 1.8, (2) is easy when $2^\kappa = \kappa^+$, since we can just choose one weak P_{κ^+}-point from each basic clopen set to get a dense set of such points.

PROOF OF THEOREM 1.14.1. Since $|\mathcal{D}_\kappa| = 2^\kappa$, apply Theorem 1.12.1 to let $F \subseteq u(\kappa)$ be a weak P_κ-set in $\beta\kappa$ homeomorphic to $\mathrm{st}(\mathcal{D}_\kappa)$. By Corollary 1.7, let $E \subseteq F$ be κ-fuzzy in F. But then E is also κ-fuzzy in $\beta\kappa$, since no point of F is a limit of any subset of $\beta\kappa \backslash F$ of size less than κ. □

The exact same proof, now applying Theorem 1.12.2 to $\mathcal{D}_{\kappa^+}$, proves Theorem 1.14.2, *provided* that $2^{(\kappa^+)} = 2^\kappa$ (so that $|\mathcal{D}_{\kappa^+}| = 2^\kappa$). If $2^{(\kappa^+)} > 2^\kappa$, then we certainly cannot embed $\mathrm{st}(\mathcal{D}_{\kappa^+})$ into $u(\kappa)$, so we must apply a somewhat more technical version of Theorem 1.12.2 which replaces $\mathcal{A}$ by $\mathcal{A} \cap N$, where N is a κ-closed elementary submodel of the universe with $|N| = 2^\kappa$. This will be discussed in Section 5, where we prove Theorem 1.14.2.

Now, any argument producing weak P_θ-points in $u(\kappa)$ faces the following obstacle: The natural transfinite induction constructs an ultrafilter $\mathbf{x}$ on κ in 2^κ stages. However, $|u(\kappa)| = 2^{2^\kappa}$ by a well-known theorem of POSPÍŠIL [1937] (see Corollary 3.5). Thus, there are too many sequences of ultrafilters to simply diagonalize against them all (e.g., at stage α, make sure that $\mathbf{x}$ is not a limit of the αth sequence). Rather, we define a *base property* of $\mathbf{x}$ which implies weak P_θ-point. Such base properties involve $\mathbf{x}$ and a base for the space $u(\kappa)$. Since the weight of $u(\kappa)$ is only 2^κ, there is a chance of success in building $\mathbf{x}$ in 2^κ steps to have such a property.

The simplest such base property is probably "P-point", and, under CH, W. RUDIN [1956] constructed a P-point in $u(\omega) = \omega^*$ in 2^ω steps. However, for κ regular and strictly between ω and the first measurable cardinal, there cannot be a P-point in $u(\kappa)$ (since it would then be countably complete), and even for $\kappa = \omega$, SHELAH [1982] showed that P-points in $u(\omega)$ cannot be proved to exist in ZFC. If we are looking for ZFC results, we must turn to somewhat more complex base properties, such as Keisler's notion of "good", and modifications thereof, such as "OK" and "mediocre". These are discussed in Section 2.

Even given a base property, it is not necessarily obvious (or even true) that there is an $\mathbf{x} \in u(\kappa)$ with that property. Successful constructions of such $\mathbf{x}$ often proceed with the aid a matrix of sets consisting of 2^κ independent rows. Then, each step in the construction eats up finitely many rows, and the remaining rows provide the necessary freedom to allow the construction to proceed. Matrices are taken up in more detail in Section 3, and their use in Sections 4 and 5.

2. Hatpoints and hatsets

Since we shall frequently be taking finite intersections from a given sequence of sets, the following notation will be useful:

2.1. DEFINITION. Given e_α in a boolean algebra for $\alpha < \theta$, and $s \in [\theta]^{<\omega}$:

$$e_{\boxed{s}} = \bigwedge_{\alpha \in s} e_\alpha \qquad ; \qquad e_{\boxed{\emptyset}} = 1 \ .$$

Of course, this definition applies to sequences of sets, $E_\alpha \in \mathcal{P}(X)$, where $\wedge$ is $\cap$; then $E_{\boxed{\emptyset}} = X$.

2.2. DEFINITION. When $\kappa \leq \theta$, a (θ, κ) *hatfunction* is a function $\widehat{\ } : [\theta]^{<\omega} \to [\kappa]^{<\omega}$.

2.3. DEFINITION. If $\widehat{\ }$ is a (θ, κ) hatfunction, then $\mathbf{x}$ is a $\widehat{\ }$*point* in the space X iff, given neighborhoods U_r $(r \in [\kappa]^{<\omega})$ of $\mathbf{x}$:

✲ There are neighborhoods V_α $(\alpha < \theta)$ of $\mathbf{x}$ such that $V_{\boxed{s}} \subseteq U_{\widehat{s}}$ for each non-empty $s \in [\theta]^{<\omega}$.

Since $\kappa \leq \theta$, we may also regard $\widehat{\ }$ as a (θ, θ) hatfunction; then note that the definition of $\widehat{\ }$point remains unchanged, since only the U_r for $r \in [\kappa]^{<\omega}$ are used. The importance of considering the possibility $\kappa < \theta$ arises in the actual construction of these points. For example, for a suitable $\widehat{\ } : [\kappa^+]^{<\omega} \to [\kappa]^{<\omega}$, every $\widehat{\ }$point is weak P_{κ^+}-point (see Lemma 2.15). In Section 5, we shall construct such points in $u(\kappa)$. The construction works because we construct $\mathbf{x} \in u(\kappa)$ in 2^κ steps and there are only 2^κ possible input sequences $\langle U_r : r \in [\kappa]^{<\omega} \rangle$ of neighborhoods (subsets of κ) to consider, even though $2^{(\kappa^+)}$ may be bigger than 2^κ.

Now, let us consider how the notion of "hatpoint" depends on the hatfunction. First, we point out (Lemma 2.7) that bigger hatfunctions yield stronger hatpoints.

2.4. DEFINITION. A sequence of sets, $\langle U_r : r \in [\kappa]^{<\omega} \rangle$, is *monotone* iff
$r \subseteq p \Rightarrow U_r \supseteq U_p$.

2.5. LEMMA. *In the definition 2.3 of "hatpoint", it is sufficient to verify (✲) only in the case of monotone sequences of neighborhoods,* $\langle U_r : r \in [\kappa]^{<\omega} \rangle$.

□ Replace each U_r by $\bigcap \{U_p : p \subseteq r\}$. □

Note that the hatfunctions form a lattice in the obvious way:

2.6. DEFINITION. Given two (θ, κ) hatfunctions $\widehat{\ }$, $\widetilde{\ }$:

✂ $\widehat{\ } \leq \widetilde{\ }$ iff $\widehat{s} \subseteq \widetilde{s}$ for all $s \in [\theta]^{<\omega}$.
✂ $\widehat{\ } \bigvee \widetilde{\ }$ is the hatfunction $\overline{\ }$ defined by $\overline{s} = \widehat{s} \cup \widetilde{s}$.

2.7. LEMMA. *Given two (θ, κ) hatfunctions $\widehat{\ }$, $\widetilde{\ }$:*

- ✂ *If $\widehat{\ } \le \widetilde{\ }$ then every $\widetilde{\ }$point is a $\widehat{\ }$point.*
- ✂ *If $\mathbf{x}$ is a $(\widehat{\ } \bigvee \widetilde{\ })$point, then $\mathbf{x}$ is both a $\widehat{\ }$point and a $\widetilde{\ }$point.*

□ Apply Lemma 2.5; the result is clear when applied to monotone $\langle U_r : r \in [\kappa]^{<\omega}\rangle$. □

At the bottom of the lattice is the zero hatfunction; more generally, if $\widehat{\ }$ has finite range, then every point is a $\widehat{\ }$point. There is no top hatfunction, but, in the case $\kappa = \theta$, there is a top hatpoint, obtained by using the the identity hatfunction; this is just KEISLER's [1964] notion of good:

2.8. DEFINITION. $\mathbf{x}$ is θ^+-*good* iff $\mathbf{x}$ is a $\widehat{\ }$point, where $\widehat{\ } : [\theta]^{<\omega} \to [\theta]^{<\omega}$ is the identity function.

Keisler defined λ-good to mean θ^+-good for all $\theta < \lambda$.

2.9. LEMMA. *If $\mathbf{x}$ is θ^+-good, then $\mathbf{x}$ is a $\widehat{\ }$point for all (θ, θ) hatfunctions.*

□ Given neighborhoods U_r $(r \in [\theta]^{<\omega})$ of $\mathbf{x}$, let $W_s = U_{\widehat{s}}$. Applying "good", obtain V_α $(\alpha < \theta)$ so that $V_{\boxed{s}} \subseteq W_s$, and hence $V_{\boxed{s}} \subseteq U_{\widehat{s}}$, for each non-empty $s \in [\theta]^{<\omega}$. □

We shall see (Lemma 2.13) that θ^+-good points are weak P_θ-points. Thus, one way to construct an $\mathbf{x} \in u(\kappa)$ which is a weak P_{κ^+}-point would be to make the point κ^{++}-good (i.e., use the identity (κ^+, κ^+) hatfunction). But then we have $2^{(\kappa^+)}$ possible input sequences $\langle U_r : r \in [\kappa^+]^{<\omega}\rangle$ to consider, and it is not clear how to handle them all in only 2^κ steps. It turns out that if κ is regular and $2^\kappa = 2^{(\kappa^+)}$, then there is a κ^{++}-good point in $u(\kappa)$ (see BAKER [2001], Corollary 4.9). However, if κ is regular and uncountable but below the first measurable cardinal, and $2^\kappa = \kappa^+$, then no point in $u(\kappa)$ is κ^{++}-good (see BAKER and KUNEN [2001], Theorem 2.8). Thus, if we want a ZFC result, we need to use a smaller hatfunction, weakening "good" to "mediocre".

2.10. DEFINITION. $\mathbf{x}$ is a κ^+-*mediocre* point iff $\mathbf{x}$ is a $\widehat{\ }$point, where for some $\langle \varphi_\beta : \beta < \kappa^+\rangle$:

- ✈ $\varphi_\beta : \beta \to \kappa$ is 1-1.
- ✈ $\widehat{\ } : [\kappa^+]^{<\omega} \to [\kappa]^{<\omega}$ and $\widehat{\{\alpha, \beta\}} = \{\varphi_\beta(\alpha)\}$ whenever $\alpha < \beta < \kappa^+$.
- ✈ $\widehat{s} = \emptyset$ whenever $|s| \ne 2$.

2.11. LEMMA. $\mathbf{x}$ *is a* κ^+-mediocre *point iff for some* $\langle \varphi_\beta : \beta < \kappa^+\rangle$*:*

- ✈ *$\varphi_\beta : \beta \to \kappa$ is 1-1 and*
- ✈ *Given any κ neighborhoods of $\mathbf{x}$, $\langle U_\xi : \xi < \kappa\rangle$, there are κ^+ neighborhoods of $\mathbf{x}$, $\langle V_\alpha : \alpha < \kappa^+\rangle$, such that $V_\alpha \cap V_\beta \subseteq U_{\varphi_\beta(\alpha)}$ whenever $\alpha < \beta < \kappa^+$.*

We do not know whether this property depends on the particular sequence $\langle \varphi_\beta : \beta < \kappa^+\rangle$ used.

2.12. LEMMA. *If* $\mathbf{x} \in X$ *is a weak* P_κ*-point which is also* κ^+*-mediocre, then* $\mathbf{x}$ *is a weak* P_{κ^+}*-point.*

□ Given $Y = \{y_\xi : \xi \in \kappa\} \subseteq X \backslash \{\mathbf{x}\}$, we wish to show that $\mathbf{x} \notin \overline{Y}$. For $\xi \in \kappa$, let U_ξ be a neighborhood of $\mathbf{x}$ disjoint from $\{y_\eta : \eta \leq \xi\}$. Now, fix neighborhoods V_α $(\alpha \in \kappa^+)$ of $\mathbf{x}$ such that $V_\alpha \cap V_\beta \subseteq U_{\varphi_\beta(\alpha)}$.

Then for some α, $V_\alpha \cap Y = \emptyset$: If not, then we can fix $\xi \in \kappa$ and $E \subseteq \kappa^+$ with $|E| = \kappa^+$ such that $\forall \alpha \in E$ $(y_\xi \in V_\alpha)$. So, if $\alpha < \beta$ and $\alpha, \beta \in E$, then $y_\xi \in V_\alpha \cap V_\beta$, so $\varphi_\beta(\alpha) < \xi$. Now, fixing $\beta \in E$ with $|\beta \cap E| = \kappa$, we contradict the fact that φ_β is 1-1. □

2.13. LEMMA. *If* $\mathbf{x}$ *is* κ^+*-good in the* T_1 *space* X*, then* $\mathbf{x}$ *is a weak* P_κ*-point in* X.

□ Observe that for all $\lambda < \kappa$: $\mathbf{x}$ is λ^{++}-good, and hence (by Lemma 2.9) λ^+-mediocre. Now, use Lemma 2.12, and show, by induction on $\lambda < \kappa$, that $\mathbf{x}$ is a weak P_{λ^+}-point for all $\lambda < \kappa$. □

By Lemmas 2.12 and 2.13, a point which is both κ^+-good and κ^+-mediocre is a weak P_{κ^+}-point. To combine these into one (κ^+, κ) hatfunction, we first pad the domain of the "good" hatfunction from Definition 2.8:

2.14. LEMMA. $\mathbf{x}$ *is* κ^+*-good iff* $\mathbf{x}$ *is a* $\widehat{\ }$*point for the* (κ^+, κ)*-hatfunction defined by:* $\widehat{s} = s \cap \kappa$.

Now, we can apply Lemma 2.7 directly, taking the join of the "mediocre" and the "good" (κ^+, κ)-hatfunctions, to get:

2.15. LEMMA. *For some* (κ^+, κ)*-hatfunction,* $\widehat{\ }$*, every* $\widehat{\ }$*point in every* T_1 *space is a weak* P_{κ^+}*-point.*

Note that we did not discuss hat*sets* in this section, but we do not have to. In view of Definition 1.10, the notion of $F \subseteq X$ being a hatset in X is already defined, and, collapsing F to a point, Lemmas 2.12, 2.13, 2.14, 2.15 apply to obtain various versions of weak P-set from versions of hatset.

The above discussion has been in arbitrary topological spaces. We now see how it applies to Stone spaces. Given $h : \mathcal{B} \twoheadrightarrow \mathcal{A}$ and $h^* : \mathrm{st}(\mathcal{A}) \hookrightarrow \mathrm{st}(\mathcal{B})$, the clopen neighborhoods of $h^*(\mathrm{st}(\mathcal{A}))$ are of the form N_c where $h(c) = 1_{\mathcal{A}}$. Thus,

2.16. LEMMA. *Suppose* $h : \mathcal{B} \twoheadrightarrow \mathcal{A}$ *and* $\widehat{\ }$ *is a* (θ, κ) *hatfunction. Then* $h^*(\mathrm{st}(\mathcal{A}))$ *is a* $\widehat{\ }$*set in* $\mathrm{st}(\mathcal{B})$ *iff: Whenever* $\langle c_r : r \in [\kappa]^{<\omega}\rangle$ *is a sequence of elements of* $\mathcal{B}$ *with each* $h(c_r) = 1_{\mathcal{A}}$*, there are* $d_\alpha \in \mathcal{B}$ *for* $\alpha < \theta$ *such that:*

- *Each* $h(d_\alpha) = 1_{\mathcal{A}}$.
- $d_{\boxed{s}} \leq c_{\widehat{s}}$ *for each nonempty* $s \in [\theta]^{<\omega}$.

When we actually obtain such an h in Section 5, the construction will be easier if we assume that the hatfunction is monotone:

2.17. DEFINITION. A (θ, κ) hatfunction $\widehat{\ }$ is *monotone* iff $\widehat{\emptyset} = \emptyset$ and $s \subseteq t \Rightarrow \widehat{s} \subseteq \widehat{t}$.

2.18. LEMMA. *Given a (θ,κ) hatfunction* $\widehat{\ }$, *there is another* (θ,κ) *hatfunction* $\widetilde{\ }$ *such that the notions of* $\widehat{\ }$*point and* $\widetilde{\ }$*point are equivalent, and such that* $\widetilde{\ }$ is monotone.

□ Let $\widetilde{t} = \bigcup\{\widehat{s} : s \subseteq t\}$ for non-empty t, and $\widetilde{\emptyset} = \emptyset$. □

We conclude this section with some remarks on OK points, although we shall not mention them again in this article. For more, see DOW [1985], KUNEN [1980] and SIMON [1985].

2.19. DEFINITION. $\mathbf{x}$ is *θ-OK* in the space X iff $\mathbf{x}$ is a $\widehat{\ }$*point* for the (θ,ω) hatfunction defined by: $\widehat{s} = \{|s|\}$.

It is easy to see that every point is ω-OK, and that the notion "θ-OK" gets stronger as θ gets bigger.

2.20. LEMMA. *If X is a T_1 space and $\mathbf{x} \in X$ is ω_1-OK, then $\mathbf{x}$ is a weak P-point in X.*

□ Given $Y = \{y_n : n \in \omega\} \subseteq X\backslash\{\mathbf{x}\}$, we wish to show that $\mathbf{x} \notin \overline{Y}$. For $n \in \omega$, let U_n be a neighborhood of $\mathbf{x}$ disjoint from $\{y_m : m \leq n\}$. Now, fix neighborhoods V_α $(\alpha < \omega_1)$ of $\mathbf{x}$ as in the definition (2.3) of $\widehat{\ }$point. So, whenever $\alpha_1, \ldots, \alpha_n$ are distinct, $V_{\alpha_1} \cap \cdots \cap V_{\alpha_n} \subseteq U_n$. Then no y_n can be contained in infinitely many V_α, so some V_α is disjoint from Y. □

However, an ω_2-OK point need not be a weak P_{ω_2}-point. For example, by KUNEN [1980] there are always $2^{\aleph_0}$-OK points in ω^*, but if one adds Cohen reals to a model of CH, one obtains a model of ZFC in which $2^{\aleph_0}$ is arbitrarily large but there are no weak P_{ω_2}-points in ω^*.

2.21. LEMMA. *Every ω_1-mediocre point is ω_1-OK.*

□ Now, we must verify the definition of OK point for a given sequence of neighborhoods of $\mathbf{x}$, $\langle U_n : n \in \omega\rangle$. Thus, we must find neighborhoods V_α $(\alpha < \omega_1)$ of $\mathbf{x}$ so that $V_{\alpha_1} \cap \cdots \cap V_{\alpha_n} \subseteq U_n$ whenever $\alpha_1 < \cdots < \alpha_n$. We may (and shall) assume that $U_0 \supseteq U_1 \supseteq \cdots$.

Applying Lemma 2.11 to $\langle U_{n+2} : n \in \omega\rangle$, fix neighborhoods V_α $(\alpha < \omega_1)$ so that $V_\alpha \cap V_\beta \subseteq U_{\varphi_\beta(\alpha)+2}$ whenever $\alpha < \beta < \kappa^+$. We can also assume that each $V_\alpha \subseteq U_1$, so that we need only consider $n \geq 3$. Then, since φ_{α_n} is 1-1, $j := \max\{\varphi_{\alpha_n}(\alpha_m) : 1 \leq m < n\} \geq n-2$, so $V_{\alpha_1} \cap \cdots \cap V_{\alpha_n} \subseteq U_{j+2} \subseteq U_n$. □

We do not know if the converse of Lemma 2.21 holds.

3. Matrices

We consider here matrices of subsets of κ, and, more generally, of elements of some boolean algebra.

3.1. DEFINITION. A *matrix* in a boolean algebra $\mathcal{B}$ is a sequence $\mathbb{M} = \langle \mathcal{M}^i : i \in I\rangle$ such that each $\mathcal{M}^i \subseteq \mathcal{B}$. Then:

- $\mathbb{M}$ is *independent* with respect to an element $c \in \mathcal{B}$ iff $b_1 \wedge \cdots \wedge b_k \wedge c > 0$ whenever k is finite, each $b_\ell \in \mathcal{M}^{i_\ell}$ $(\ell = 1, \ldots, k)$, and $i_1, \ldots, i_k$ are distinct elements of I.

- ✏ $\mathbb{M}$ is *independent* iff $\mathbb{M}$ is independent with respect to 1.
- ✏ If $\mathcal{F} \subseteq \mathcal{B}$ is a filter, then $\mathbb{M}$ is *independent* with respect to $\mathcal{F}$ iff it is independent with respect to every $c \in \mathcal{F}$.
- ✏ $\mathbb{M}$ is *independent* with respect to an ideal iff it is independent with respect to the dual filter.

Informally, we think of the $\mathcal{M}^i$ as the "rows" of the matrix. Usually, each row conforms to some configuration specified in advance. The most well-known configuration is just a disjoint family:

3.2. DEFINITION. The matrix $\mathbb{M} = \langle \mathcal{M}^i : i \in I \rangle$ is a $|I| \times \kappa$ *disjoint matrix* iff each $\mathcal{M}^i$ is an antichain in $\mathcal{B}$ of size κ.

Thus, a $\theta \times \kappa$ disjoint matrix may be indexed as $\{b^i_\eta : i \in \theta \ \& \ \eta < \kappa\}$, where each $b^i_\eta \wedge b^i_\zeta = 0$ whenever $\eta \neq \zeta$. Independence of the matrix asserts that $b^{i_1}_{\eta_1} \wedge \cdots \wedge b^{i_k}_{\eta_k} > 0$ whenever the $i_1, \ldots, i_k$ are distinct. Independence with respect to a filter $\mathcal{F}$ asserts that in addition, these $b^{i_1}_{\eta_1} \wedge \cdots \wedge b^{i_k}_{\eta_k} \notin \mathcal{I}$, where $\mathcal{I}$ is the dual ideal. When $\theta = 2^\kappa$, $\mathcal{B} = \mathcal{P}(\kappa)$, and $\mathcal{F} = \mathcal{FR}(\kappa)$, the existence of such a matrix is equivalent to the following well-known result on independent functions:

3.3. LEMMA (ENGELKING and KARŁOWICZ [1965]). *For any infinite κ, there are functions $f_i : \kappa \to \kappa$, for $i < 2^\kappa$, which are independent in the sense that whenever k is finite, $i_1, \ldots, i_k < 2^\kappa$ are distinct, and $\eta_1, \ldots, \eta_k < \kappa$ are arbitrary,*

$$|\{\xi < \kappa : f_{i_1}(\xi) = \eta_1 \ \& \ \ldots \ \& \ f_{i_k}(\xi) = \eta_k\}| = \kappa \ .$$

□ Index the functions as $\{f_A : A \subseteq \kappa\}$, and let $f_A : E \to \kappa$, where $E = \{(s,p) : s \in [\kappa]^{<\omega} \ \& \ p : \mathcal{P}(s) \to \kappa\}$; note that $|E| = \kappa$. Define $f_A(s,p) = p(A \cap s)$. Independence is proved by noting that given distinct $A_1, \ldots, A_k \in \mathcal{P}(\kappa)$, there are κ many finite s such that the $A_1 \cap s, \ldots, A_k \cap s$ are distinct, and for these, one can choose p so that each $p(A_\ell \cap s) = \eta_\ell$. □

A direct application of Lemma 3.3, letting $\mathcal{M}^i = \{f_i^{-1}\{\eta\} : \eta < \kappa\}$, yields:

3.4. LEMMA. *In $\mathcal{P}(\kappa)$, there is a $2^\kappa \times \kappa$ disjoint matrix which is independent with respect to $\mathcal{FR}(\kappa)$.*

This improves the earlier theorem of HAUSDORFF [1936] on the existence of 2^κ independent sets, which yields a $2^\kappa \times 2$ disjoint independent matrix. We digress to point out a well-known consequence of this matrix:

3.5. COROLLARY (POSPÍŠIL [1937]). $|u(\kappa)| = 2^{2^\kappa}$.

□ Let $\{A^i_\eta : i \in 2^\kappa \ \& \ \eta < 2\}$ be a $2^\kappa \times 2$ disjoint matrix independent with respect to $\mathcal{FR}(\kappa)$. For each $f : 2^\kappa \to \{0,1\}$, choose an ultrafilter $\mathbf{x}_f \in u(\kappa)$ such that all $A^i_{f(i)} \in \mathbf{x}_f$, and note that these x_f are all distinct. □

Pospíšil's paper does not explicitly mention matrices or Hausdorff's paper, but the construction of the independent matrix is embedded in his proof.

Lemma 3.4 is best possible, in that there cannot be κ^+ disjoint subsets of κ. However, there are κ^+ almost disjoint subsets, and in fact, there is a $2^\kappa \times \kappa^+$ independent disjoint matrix in the boolean algebra $\mathcal{P}(\kappa)/{<}\kappa$ (BAKER [2001], Corollary 4.15; for regular κ, this was due earlier to DOW [1985]).

Lemma 3.4 generalizes to:

3.6. LEMMA. *Let $\mathcal{B}$ be a complete boolean algebra and $\mathcal{I}$ an ideal on $\mathcal{B}$. Assume that $\mathcal{B}$ has an antichain, $\{a_\xi : \xi < \kappa\}$, such that the a_ξ are distinct and each $a_\xi \notin \mathcal{I}$. Then in $\mathcal{B}$ there is a $2^\kappa \times \kappa$ disjoint matrix which is independent with respect to $\mathcal{I}$.*

□ With the f_i as in Lemma 3.3, let $b^i_\eta = \bigvee\{a_\xi : f_i(\xi) = \eta\}$. □

BALCAR and FRANĚK [1982] show that this matrix is sufficient to construct κ^+-good points in $\mathrm{st}(\mathcal{B})$ when $|\mathcal{B}| \le 2^\kappa$, proving Theorem 1.3. However, if we wish to construct hatpoints for a general (θ, κ) hatfunction, we need a more complicated matrix, each row of which is described by the following definition:

3.7. DEFINITION. Let $\widehat{\ }$ be any monotone (θ, κ) hatfunction. If $\mathcal{B}$ is a complete boolean algebra and $\mathcal{G}$ is a filter on $\mathcal{B}$, with dual ideal $\mathcal{J}$: a $\widehat{\ }$*step-family* on $(\mathcal{B}, \mathcal{G})$ is a collection of elements of $\mathcal{B}$ of the form

$$\begin{aligned} \mathcal{M} &= \{e_r : r \in [\kappa]^{<\omega}\} \cup \{a_\alpha : \alpha < \theta\} \\ &\cup \{a_{\boxed{s}} \wedge e_r : s \in [\theta]^{<\omega} \ \&\ r \in [\kappa]^{<\omega} \ \&\ \widehat{s} \subseteq r\} , \end{aligned}$$

where the e_r and a_α satisfy:

S1. $\bigvee\{e_r : r \in [\kappa]^{<\omega}\} = 1$, and $e_r \wedge e_p = 0$ for each distinct $r, p \in [\kappa]^{<\omega}$.
S2. $a_{\boxed{s}} \wedge \bigvee\{e_r : r \not\supseteq \widehat{s}\} \in \mathcal{J}$ for each $s \in [\theta]^{<\omega}$.
S3. $a_{\boxed{s}} \wedge e_r \notin \mathcal{J}$ for each $s \in [\theta]^{<\omega}$ and $r \in [\kappa]^{<\omega}$ such that $\widehat{s} \subseteq r$.

Note that the step-family $\mathcal{M}$ is determined by the e_r and the a_α; we have thrown in the $a_{\boxed{s}} \wedge e_r$ so that independence has the desired meaning. Suppose $\mathbb{M} = \langle \mathcal{M}^i : i \in I\rangle$ is a matrix where each $\mathcal{M}^i$ is a step-family, with corresponding e^i_r and a^i_α. Then independence with respect to c asserts:

$$\left(a^{i_1}_{\boxed{s_1}} \wedge e^{i_1}_{r_1}\right) \wedge \cdots \wedge \left(a^{i_k}_{\boxed{s_k}} \wedge e^{i_k}_{r_k}\right) \wedge c \neq 0 , \qquad (*)$$

whenever k is finite, each $\widehat{s_\ell} \subseteq r_\ell$ ($\ell = 1, \ldots, k$), and $i_1, \ldots, i_k$ are distinct elements of I. It is not necessary to consider the a_α (since $a_{\boxed{s}} \ge a_{\boxed{s}} \wedge e_{\widehat{s}}$) or the e_r (since $e_r \ge a_{\boxed{\emptyset}} \wedge e_r$ and (by monotonicity, Definition 2.17) $\widehat{\emptyset} = \emptyset \subseteq r$).

(S1) says that the e_r, for $r \in [\kappa]^{<\omega}$, form a partition of unity; the $e_r \neq 0$ by (S3) (setting $s = \emptyset$, so $\widehat{s} = \emptyset \subseteq r$). Conversely, if $\theta = \kappa$ and $\widehat{\ }$ is the identity hatfunction, which is what we need for a κ^+-good point, then an antichain yields a matrix of step-families:

3.8. LEMMA. *Let $\mathcal{B}$ be a complete boolean algebra with an ideal $\mathcal{I}$ and dual filter $\mathcal{F}$. Assume that $\mathcal{B}$ has an antichain, $\{a_\xi : \xi < \kappa\}$, such that the a_ξ are distinct and each $a_\xi \notin \mathcal{I}$. Let $\widehat{\ }$ be the identity (κ, κ) hatfunction. Then there is a matrix $\mathbb{M} = \langle \mathcal{M}^i : i \in 2^\kappa \rangle$ in $\mathcal{B}$ such that each $\mathcal{M}^i$ is a $\widehat{\ }$*step-family *on $(\mathcal{B}, \{1\})$, and $\mathbb{M}$ is independent with respect to $\mathcal{F}$.*

□ By Lemma 3.6, let $\mathbb{N} = \langle \mathcal{N}^i : i \in 2^\kappa \rangle$ be a $2^\kappa \times \kappa$ disjoint matrix which is independent with respect to $\mathcal{I}$. Index each row $\mathcal{N}^i$ as $\{e^i_r : r \in [\kappa]^{<\omega}\}$; so, the e^i_r are disjoint as r varies. Since $\mathcal{B}$ is complete, we can expand the e^i_r and assume that $\bigvee\{e^i_r : r \in [\kappa]^{<\omega}\} = 1$, so that (S1) holds. Then, define $a^i_\alpha = \bigvee\{e^i_r : \alpha \in r\}$. Now, $a^i_{\boxed{s}} = \bigvee\{e^i_r : r \supseteq s\}$ for each $s \in [\kappa]^{<\omega}$, so that (S2) holds with $\mathcal{J} = \{0\}$. If $s \subseteq r$, then $a^i_{\boxed{s}} \wedge e^i_r = e^i_r \neq 0$, so (S3) holds and $(*)$ above reduces independence of $\mathbb{M}$ to independence of $\mathbb{N}$. □

The usual constructions of κ^+-good ultrafilters in the literature BALCAR and FRANĚK [1982], CHANG and KEISLER [1990] and KUNEN [1972] work directly from the $2^\kappa \times \kappa$ independent disjoint matrix obtained by Lemmas 3.4 or 3.6, but a step like the above proof is embedded somewhere in the argument. Here, we shall present a general proof which constructs $\widehat{\ }$points in boolean algebras, provided that the correct matrix exists for $\widehat{\ }$; then, Lemma 3.8 yields the correct matrix for the "good" hatfunction whenever $\mathcal{B}$ has an appropriate antichain.

In the most general situation which we shall consider here (see Theorem 5.6), we shall have two filters $\mathcal{G} \subseteq \mathcal{F}$ on the complete boolean algebra $\mathcal{B}$, with a matrix of step-families on $(\mathcal{B}, \mathcal{G})$ which is independent with respect to $\mathcal{F}$. Then $\mathrm{st}(\mathcal{B}/\mathcal{F}) \subseteq \mathrm{st}(\mathcal{B}/\mathcal{G}) \subseteq \mathrm{st}(\mathcal{B})$, and we produce points $\mathbf{x} \in \mathrm{st}(\mathcal{B}/\mathcal{F})$ which are hatpoints in $\mathrm{st}(\mathcal{B}/\mathcal{G})$. For example, with $\mathcal{B} = \mathcal{P}(\kappa)$, $\mathcal{G} = \{1\}$, and $\mathcal{F} = \mathcal{FR}(\kappa)$, Lemma 3.8 will allow us to construct an $\mathbf{x} \in \mathrm{st}(\mathcal{B}/\mathcal{F}) = u(\kappa)$ which is κ^+-good, and hence a weak P_κ-point, in $\mathrm{st}(\mathcal{B}/\mathcal{G}) = \beta\kappa$. $\mathbf{x}$ cannot be a weak P_{κ^+}-point in $\beta\kappa$; if we want $\mathbf{x}$ to be a weak P_{κ^+}-point in $u(\kappa)$, we let $\mathcal{F} = \mathcal{G} = \mathcal{FR}(\kappa)$ and let $\widehat{\ }$ be the appropriate hatfunction (see Lemma 2.15); the same Theorem 5.6 yields a weak $\widehat{\ }$point because the appropriate matrix exists by the following result, which we quote from BAKER and KUNEN [2001]:

3.9. LEMMA. *Assume that κ is regular and $\widehat{\ }$ is any monotone (κ^+, κ) hatfunction. Then there is a matrix $\mathbb{M} = \langle \mathcal{M}^i : i \in 2^\kappa \rangle$ in $\mathcal{P}(\kappa)$ such that each $\mathcal{M}^i$ is a $\widehat{\ }$step-family on $(\mathcal{P}(\kappa), \mathcal{FR}(\kappa))$, and $\mathbb{M}$ is independent with respect to $\mathcal{FR}(\kappa)$.*

We do not know if this lemma is true for singular κ. If it is, then the results proved in this article for regular κ hold also for singular κ, since regularity is not used in the construction of hatpoints or hatsets in Section 5. The construction will use the following consequence of the definition.

3.10. LEMMA. *A step-family as in Definition 3.7 satisfies also:*

S4. $a_{\boxed{s}} \wedge e_r \notin \mathcal{J}$ *iff* $\widehat{s} \subseteq r$.
S5. $a_{\boxed{s}} \leq_{\mathcal{J}} \bigvee\{e_r : r \supseteq \widehat{s}\}$ *for each* $s \in [\theta]^{<\omega}$.

Note that the monotonicity of $\widehat{\ }$ comes in naturally here: If $s \subseteq t$, then $a_{\boxed{s}} \geq a_{\boxed{t}}$; so, if in addition $\widehat{s} \not\subseteq \widehat{t}$, then setting $r = \widehat{t}$ would contradict (S4).

Also, note that monotonicity does *not* imply that $\widehat{\ }$ preserves $\cup$. We might well have $r := \widehat{\{\alpha\}} \cup \widehat{\{\beta\}} \subsetneqq \widehat{\{\alpha, \beta\}}$. Then, modulo $\mathcal{J}$, a_α and a_β meet e_r but $a_\alpha \wedge a_\beta$ does not.

In fact, this does happen with the monotone version of the (κ^+, κ)-hatfunction used for weak P_{κ^+}-points (Lemmas 2.15 and 2.18). Thus, for general step-families, we should not expect the situation occurring in the proof of Lemma 3.8, where each a_α is simply a join of some of the e_r.

In this article, step-families for (θ, κ) hatfunctions will be applied only with $\theta = \kappa$ (using Lemma 3.8) and $\theta = \kappa^+$ (using Lemma 3.9). However, the construction in KUNEN [1980] of $2^{\aleph_0}$-OK points in ω^* using independent linked families could be presented as a construction using an independent matrix of $2^{\aleph_0}$ $\frown$step-families, where $\frown$ is the $(2^{\aleph_0}, \omega)$ hatfunction from Definition 2.19 (made monotone by Lemma 2.18).

4. The Sikorski extension theorem

In this section, we isolate some basic features of the inductive construction of a homomorphism from $\mathcal{B}$ to $\mathcal{A}$. In Section 5, we apply this to the construction of hatpoints and hatsets. Our inductions always follow the pattern:

$$\begin{array}{ccccccccc}
\mathcal{C}_0 & \subseteq & \mathcal{C}_1 & \subseteq \cdots\cdots \subseteq & \mathcal{C}_\mu & \subseteq & \mathcal{C}_{\mu+1} & \subseteq \cdots\cdots \\
\downarrow{\scriptstyle h_0} & & \downarrow{\scriptstyle h_1} & & \downarrow{\scriptstyle h_\mu} & & \downarrow{\scriptstyle h_{\mu+1}} & \\
\mathcal{A}_0 & \subseteq & \mathcal{A}_1 & \subseteq \cdots\cdots \subseteq & \mathcal{A}_\mu & \subseteq & \mathcal{A}_{\mu+1} & \subseteq \cdots\cdots
\end{array}$$

4.1. DEFINITION. If $\mathcal{A}, \mathcal{B}$ are boolean algebras: A λ-*homomorphism sequence* from $\mathcal{B}$ to $\mathcal{A}$ is a sequence $\langle \mathcal{C}_\mu, \mathcal{A}_\mu, h_\mu : \mu < \lambda \rangle$ such that:

P1. $\mathcal{C}_\mu$ is a subalgebra of $\mathcal{B}$, $\mathcal{A}_\mu$ is a subalgebra of $\mathcal{A}$, and $h_\mu : \mathcal{C}_\mu \twoheadrightarrow \mathcal{A}_\mu$.
P2. If $\mu < \nu$ then $\mathcal{A}_\mu \subseteq \mathcal{A}_\nu$, $\mathcal{C}_\mu \subseteq \mathcal{C}_\nu$, and $h_\mu = h_\nu \restriction \mathcal{C}_\mu$.
P3. For limit ν: $\mathcal{C}_\nu = \bigcup_{\mu<\nu} \mathcal{C}_\mu$ and $\mathcal{A}_\nu = \bigcup_{\mu<\nu} \mathcal{A}_\mu$.

Thus, the construction is determined at limit stages. All of the work takes place at successor stages, where, to avoid excessive subscripts, we always display the extension problem as:

$$\begin{array}{ccccc}
\mathcal{C} & \subseteq & \widetilde{\mathcal{C}} & \subseteq & \mathcal{B} \\
\downarrow{\scriptstyle h} & \swarrow{\scriptstyle \widetilde{h}} & & & \\
\mathcal{A} & & & &
\end{array}$$

Here, we are given $h = h_\mu$ and $\mathcal{C} = \mathcal{C}_\mu$, and we have to define $\widetilde{\mathcal{C}} = \mathcal{C}_{\mu+1}$ and $\widetilde{h} = h_{\mu+1}$ to accomplish a desired task (so, $\mathcal{A}_\mu = \operatorname{ran}(h)$ and $\mathcal{A}_{\mu+1} = \operatorname{ran}(\widetilde{h})$). The simplest task is just to make sure that at the end, we get a homomorphism defined on all of $\mathcal{B}$. Then, at a successor stage, we are given some $y \in \mathcal{B} \backslash \mathcal{C}$, and we want to extend to get $y \in \widetilde{\mathcal{C}}$. If $\mathcal{A}$ is complete, this can always be done by Lemma 4.4, which is the main lemma behind the Sikorski Extension Theorem (see SIKORSKI [1969], Theorem 33.1).

4.2. DEFINITION. If $\mathcal{S}$ is any subset of the boolean algebra $\mathcal{B}$, then $((\mathcal{S}))$ is the subalgebra finitely generated by $\mathcal{S}$.

Note that if $\mathcal{S} = \mathcal{C} \cup \{y\}$, where $\mathcal{C}$ is a subalgebra, then the elements of $((\mathcal{S}))$ are all of the form $(y \wedge c_1) \vee (y' \wedge c_2)$, where $c_1, c_2 \in \mathcal{C}$.

4.3. DEFINITION. Suppose that $\mathcal{C}$ is a subalgebra of $\mathcal{B}$ and h is a homomorphism from $\mathcal{C}$ into the complete boolean algebra $\mathcal{A}$. For $y \in \mathcal{B}$, let

$$h^+(y) = \bigwedge\{h(c) : c \in \mathcal{C} \text{ and } c \geq y\} \qquad h^-(y) = \bigvee\{h(c) : c \in \mathcal{C} \text{ and } c \leq y\} .$$

4.4. LEMMA (Sikorski). *With $\mathcal{A}, \mathcal{B}, \mathcal{C}, h, y$ as in Definition 4.3, and z any element of $\mathcal{A}$: $h^-(y) \leq z \leq h^+(y)$ iff there is an extension of h to a homomorphism $\widetilde{h}$ from $((\mathcal{C} \cup \{y\}))$ into $\mathcal{A}$ with $\widetilde{h}(y) = z$. In this case, $\widetilde{h}$ is unique, and is defined by:*

$$\widetilde{h}((y \wedge c_1) \vee (y' \wedge c_2)) = (z \wedge h(c_1)) \vee (z' \wedge h(c_2))$$

for all $c_1, c_2 \in \mathcal{C}$.

For a proof, see SIKORSKI [1969], p 142. Applying this inductively:

4.5. THEOREM (Sikorski Extension Theorem). *Suppose that $\mathcal{C}$ is a subalgebra of $\mathcal{B}$ and h is a homomorphism from $\mathcal{C}$ into the complete boolean algebra $\mathcal{A}$. Then h can be extended to some $\overline{h} : \mathcal{B} \to \mathcal{A}$.*

□ List $\mathcal{B}$ as $\{b_\mu : \mu < \lambda\}$, and construct $\overline{h}$ using a λ-homomorphism sequence. Here, $\mathcal{C}_0 = \mathcal{C}$, $h_0 = h$, and we make sure that $b_\mu \in \mathcal{C}_{\mu+1}$. □

Now, suppose that we want $\overline{h}$ to be onto, so that $\overline{h}^* : \mathrm{st}(\mathcal{A}) \hookrightarrow \mathrm{st}(\mathcal{B})$. In many cases this can be done with the method of independent matrices, obtaining the following lemma, which is the abstract version of Efimov's Theorem 1.9:

4.6. LEMMA. *Suppose that $\mathcal{A}, \mathcal{B}$ are boolean algebras with $\mathcal{A}$ complete and $|\mathcal{A}| \leq \theta$, and suppose that in $\mathcal{B}$, there is a $\theta \times 2$ independent disjoint matrix $\langle \mathcal{M}^i : i < \theta \rangle$. Then there is a homomorphism h from $\mathcal{B}$ onto $\mathcal{A}$.*

□ Here $\mathcal{M}^i = \{c_0^i, c_1^i\}$, where $c_0^i \wedge c_1^i = 0$. We may also assume, expanding the c_0^i, that $c_0^i \vee c_1^i = 1$. Let $\mathcal{C} = ((\bigcup_i \mathcal{M}^i))$. List $\mathcal{A}$ as $\{a_i : i < \theta\}$, and define $h : \mathcal{C} \twoheadrightarrow \mathcal{A}$ by $h(c_0^i) = a_i$. By independence ($\mathcal{C}$ is the free algebra on θ generators; see SIKORSKI [1969], §14), this really defines a homomorphism. Now, extend h to $\mathcal{B}$ by Theorem 4.5. □

PROOF OF THEOREM 1.9. Apply Lemma 4.6, with $\mathcal{B} = \mathcal{P}(\kappa)/{<}\kappa$. The appropriate matrix exists by Lemma 3.4. □

Now, if, in Lemma 4.6, we want $\overline{h}^* : \mathrm{st}(\mathcal{A}) \hookrightarrow \mathrm{st}(\mathcal{B})$ to embed $\mathrm{st}(\mathcal{A})$ into $\mathrm{st}(\mathcal{B})$ as some sort of a hatset, then we must work harder. In these constructions, the use of the matrix is interleaved with the inductive definition of the homomorphism. We take this up in Section 5, and conclude this section with a more elementary remark:

In Lemma 4.4, say we replace $y \in \mathcal{B}$ by a set $\mathcal{Y} \subseteq \mathcal{B}$, and we want to extend h to $\widetilde{h}$ on $((\mathcal{C} \cup \mathcal{Y}))$. Of course, this is possible, by Theorem 4.5, but it is not so simple to describe the possibilities for $\widetilde{h} \restriction \mathcal{Y}$. However, in the following special case, we can send all of $\mathcal{Y}$ to 1:

4.7. LEMMA. *With $\mathcal{A}, \mathcal{B}, \mathcal{C}, h$ as in Lemma 4.4, suppose $\mathcal{Y} \subseteq \mathcal{B}$ is such that $h^+(y_1 \wedge \cdots \wedge y_n) = 1$ for all finite n and all $y_1, \ldots, y_n \in \mathcal{Y}$. Then there is a unique extension of h to a homomorphism $\widetilde{h}$ from $(\!(\mathcal{C} \cup \mathcal{Y})\!)$ into $\mathcal{A}$ such that* $\operatorname{ran} \widetilde{h} = \operatorname{ran} h$ *and $\widetilde{h}(y) = 1$ for all $y \in \mathcal{Y}$.*

□ We just describe $\widetilde{h}$; it is then easy to verify that it works. Fix a $b \in (\!(\mathcal{C} \cup \mathcal{Y})\!)$, and then choose $y_1, \ldots, y_n \in \mathcal{Y}$ such that $b \in (\!(\mathcal{C} \cup \{y_1, \ldots, y_n\})\!)$. Write b in disjunctive normal form as $b = \bigvee_f (c_f \wedge y(f))$, where each $c_f \in \mathcal{C}$, f ranges over all functions $f : \{1, \ldots, n\} \to \{0, 1\}$, and $y(f) = y_1^{f(1)} \wedge \cdots \wedge y_n^{f(n)}$, where $y_\ell^0 = y_\ell$ and $y_\ell^1 = (y_\ell)'$. In particular, if $\vec{0}$ denotes the identically 0 function, then $y(\vec{0}) = y_1 \wedge \cdots \wedge y_n$. Let $\widetilde{h}(b) = h(c_{\vec{0}})$. □

5. Hatsets in Stone spaces

We now take up the problem of embedding $\operatorname{st}(\mathcal{A})$ as a $\frown$set. First, we expand on the general framework outlined in Section 3. We start with complete boolean algebras $\mathcal{A}, \mathcal{B}$, and two filters $\mathcal{G} \subseteq \mathcal{F}$ on $\mathcal{B}$. Then $\operatorname{st}(\mathcal{B}/\mathcal{F}) \subseteq \operatorname{st}(\mathcal{B}/\mathcal{G}) \subseteq \operatorname{st}(\mathcal{B})$, and we wish to embed $\operatorname{st}(\mathcal{A})$ into $\operatorname{st}(\mathcal{B}/\mathcal{F})$ so that it is a hatset in $\operatorname{st}(\mathcal{B}/\mathcal{G})$. This embedding will be via an h^*, where $h : \mathcal{B}/\mathcal{F} \twoheadrightarrow \mathcal{A}$; equivalently, $h : \mathcal{B} \twoheadrightarrow \mathcal{A}$ with $h(\mathcal{F}) = \{1\}$. This will be done using the inductive scheme described in Section 4. Let $\mathcal{I}$ be the ideal dual to $\mathcal{F}$. Now, we start off with $\mathcal{C}_0 = (\!(\mathcal{F})\!) = \mathcal{F} \cup \mathcal{I}$, and $h_0 : \mathcal{C}_0 \twoheadrightarrow \mathcal{A}_0 = \{0, 1\}$ defined by $h_0(\mathcal{F}) = \{1\}$ and $h_0(\mathcal{I}) = \{0\}$. We work with the aid of a matrix which is kept "independent" in the following sense, due to SIMON [1985]:

5.1. DEFINITION. If $\mathbb{M}$ is a matrix in $\mathcal{B}$, $\mathcal{C}$ is a subalgebra of $\mathcal{B}$, and $h : \mathcal{C} \to \mathcal{A}$, then $\mathbb{M}$ is *strongly independent* with respect to h iff $\mathbb{M}$ is independent with respect to every $c \in \mathcal{C}$ such that $h(c) > 0$.

At the start of the induction, strong independence with respect to h_0 means that $\mathbb{M}$ is independent with respect to $\mathcal{F}$, and the appropriate $\mathbb{M}$ will be given by Lemma 3.8 or Lemma 3.9.

Now, we consider a homomorphism sequence (Definition 4.1) augmented by the use of a matrix.

5.2. DEFINITION. Let $\mathbb{M} = \langle \mathcal{M}^i : i \in I \rangle$ be a matrix in $\mathcal{B}$. A *λ-matrix sequence* from $\mathcal{B}$ to $\mathcal{A}$ is a sequence $\langle \mathcal{C}_\mu, \mathcal{A}_\mu, h_\mu, I_\mu : \mu < \lambda \rangle$ such that $\langle \mathcal{C}_\mu, \mathcal{A}_\mu, h_\mu : \mu < \lambda \rangle$ is a homomorphism sequence and:

P4. $I_0 = I$.
P5. If $\mu < \nu$ then $I_\mu \supseteq I_\nu$.
P6. $\mathbb{M} \restriction I_\mu$ is strongly independent with respect to h_μ.
P7. For limit ν: $I_\nu = \bigcap_{\mu < \nu} I_\mu$.

Again, there is no problem at limits, and we use the matrix at successor stages to achieve the goals of the construction. We first modify the extension lemmas, 4.4 and 4.7, to include the matrix:

5.3. LEMMA. *Suppose that $\mathcal{C}$ is a subalgebra of $\mathcal{B}$ and h is a homomorphism from $\mathcal{C}$ into the complete boolean algebra $\mathcal{A}$. Assume that the matrix $\mathbb{M} = \langle \mathcal{M}^i : i \in I \rangle$, is strongly independent with respect to h. Fix any $j \in I$ and let $J = I \backslash \{j\}$. Then:*

1. *If $b \in \mathcal{M}^j$, then $h^+(b) = 1$.*
2. *Assume that $b, e \in \mathcal{M}^j$ and $e \wedge b = 0$. Fix any $z \in \mathcal{A}$. Then $h^+(b) = 1$, $h^-(b) = 0$, and if $\widetilde{h} : ((\mathcal{C} \cup \{b\})) \to \mathcal{A}$ extends h with $h(b) = z$ (as in Lemma 4.4), then $\mathbb{M} \restriction J$ is strongly independent with respect to $\widetilde{h}$.*
3. *Fix $\mathcal{Y} \subseteq \mathcal{B}$. Assume that for all finite n and all $y_1, \ldots, y_n \in \mathcal{Y}$, there are $b \in \mathcal{M}^j$ and $c \in \mathcal{C}$ such that $h(c) = 1$ and $y_1 \wedge \cdots \wedge y_n \geq b \wedge c$. Then $h^+(y_1 \wedge \cdots \wedge y_n) = 1$ for each $y_1, \ldots, y_n \in \mathcal{Y}$, and if $\widetilde{h} : ((\mathcal{C} \cup \mathcal{Y})) \to \mathcal{A}$ with $\widetilde{h}(\mathcal{Y}) = \{1\}$ (as in Lemma 4.7), then $\mathbb{M} \restriction J$ is strongly independent with respect to $\widetilde{h}$.*

□ (1) is clear by strong independence, as is $h^-(b) = 0$ in (2). To prove strong independence of $\mathbb{M} \restriction J$ in (2): Fix distinct $j_1, \ldots, j_k \in J$, and fix $d_\ell \in \mathcal{M}^{j_\ell}$ ($\ell = 1, \ldots, k$). Assume that $\widetilde{h}((b \wedge c_1) \vee (b' \wedge c_2)) > 0$, where $c_1, c_2 \in \mathcal{C}$. We must show that $d_1 \wedge \cdots \wedge d_k \wedge ((b \wedge c_1) \vee (b' \wedge c_2)) > 0$. Since $\widetilde{h}((b \wedge c_1) \vee (b' \wedge c_2)) = (z \wedge h(c_1)) \vee (z' \wedge h(c_2))$, there are two cases:

Case 1. $z \wedge h(c_1) > 0$. Then $h(c_1) > 0$, so $d_1 \wedge \cdots \wedge d_k \wedge b \wedge c_1 > 0$ by strong independence of $\mathbb{M}$ with respect to h.

Case 2. $z' \wedge h(c_2) > 0$. Then $h(c_2) > 0$, so $d_1 \wedge \cdots \wedge d_k \wedge b' \wedge c_2 \geq d_1 \wedge \cdots \wedge d_k \wedge e \wedge$ $\wedge c_2 > 0$.

To verify $h^+(y_1 \wedge \cdots \wedge y_n) = 1$ in (3), fix $z \in \mathcal{C}$ with $z \wedge (y_1 \wedge \cdots \wedge y_n) = 0$; we must show that $h(z) = 0$. Fix b, c as in the hypothesis of (3). Then $b \wedge z \wedge c = 0$ and $b \in \mathcal{M}^j$, so $h(z \wedge c) = 0$ by strong independence. Then, $h(c) = 1$ yields $h(z) = 0$.

To prove strong independence of $\mathbb{M} \restriction J$ in (3): Fix $w \in ((\mathcal{C} \cup \mathcal{Y}))$ with $\widetilde{h}(w) > 0$ and fix distinct $j_1, \ldots, j_k \in I \backslash \{j\}$ and $b_\ell \in \mathcal{M}^{j_\ell}$ ($\ell = 1, \ldots, k$); we must show that $b_1 \wedge \cdots \wedge b_k \wedge w > 0$. But, $\widetilde{h}(w) > 0$ implies that we can find some $y_1, \ldots, y_n \in \mathcal{Y}$ and $z \in \mathcal{C}$ such that $w \geq z \wedge y_1 \wedge \cdots \wedge y_n$ and $h(z) = \widetilde{h}(z \wedge y_1 \wedge \cdots \wedge y_n) > 0$ (see the proof of Lemma 4.7). Fix b, c as in the hypothesis of (3). Then $h(z \wedge c) = h(z) > 0$, and $(z \wedge c) \in \mathcal{C}$, so strong independence of $\mathbb{M}$ with respect to h yields $b_1 \wedge \cdots \wedge b_k \wedge b \wedge z \wedge c > 0$. Now, $w \geq z \wedge b \wedge c$, so $b_1 \wedge \cdots \wedge b_k \wedge w > 0$. □

Lemma 5.3.2 tells us how to make $h : \mathcal{B} \to \mathcal{A}$ *onto* in our inductive construction. Assuming that each row of the matrix contains a pair of disjoint elements, we can, at some stage μ, choose an arbitrary $z \in \mathcal{A}$ and put z into the range of $h_{\mu+1}$, sacrificing one row. However, if y is an arbitrary element of $\mathcal{B}$ (not a matrix element), it is a bit tricky to put y into the domain. We cannot simply quote Lemma 4.4, as there is no guarantee that the matrix will stay strongly independent. However, by the following argument, due to SIMON [1985], we may put y in the domain if we sacrifice $|\operatorname{ran}(h_\mu)|$ rows of the matrix.

5.4. LEMMA. *Let $\mathcal{A}, \mathcal{B}, \mathcal{C}, h, \mathbb{M}$ be as in Lemma 5.3 and fix $y \in \mathcal{B}$. Then there is an extension of h to $\widetilde{h} : ((\mathcal{C} \cup \{y\})) \to \mathcal{A}$ and a $J \subseteq I$ with $|I \backslash J| \leq \max(|\operatorname{ran}(h)|, \aleph_0)$ such that $\mathbb{M} \restriction J$ is strongly independent with respect to $\widetilde{h}$.*

□ Call $d \in \operatorname{ran}(h)$ *bad* iff for some finite k, some $c \in \mathcal{C}$, some distinct $i_1, \ldots, i_k \in I$, and some $b_\ell \in \mathcal{M}^{i_\ell}$ ($\ell = 1, \ldots, k$), we have $h(c) = d$ and $c \wedge y \wedge b_1 \wedge \cdots \wedge b_k = 0$. So, 0 is bad, and the set of bad elements is closed downward in $\operatorname{ran}(h)$. Note that the lemma requires us to set $\widetilde{h}(y) \wedge d = 0$ for each bad d (otherwise, $\widetilde{h}(c \wedge y) = \widetilde{h}(y) \wedge d > 0$ would contradict strong independence).

List all the bad elements as $\{d_\alpha : \alpha < \theta\}$. For each α, choose k_α, c_α, $i_1^\alpha, \ldots, i_{k_\alpha}^\alpha$, and $b_1^\alpha, \ldots, b_{k_\alpha}^\alpha$ as in the definition of "bad". Let $J = I \setminus \{i_\ell^\alpha : \alpha < \theta \;\&\; 1 \le \ell \le k_\alpha\}$. Let $a = (\bigvee_{\alpha<\theta} d_\alpha)' = \bigwedge_{\alpha<\theta}(d_\alpha)' \in \mathcal{A}$. We shall set $\widetilde{h}(y) = a$, applying Lemma 4.4, but first we must verify that $h^-(y) \le a \le h^+(y)$.

To prove $h^-(y) \le a$, it is sufficient to fix $z \le y$ with $z \in \mathcal{C}$ and fix α and show that $h(z) \le (d_\alpha)'$. From $z \le y$ we get $c_\alpha \wedge z \wedge b_1^\alpha \wedge \cdots \wedge b_k^\alpha = 0$, where $k = k_\alpha$. Since $(c_\alpha \wedge z) \in \mathcal{C}$, strong independence yields $h(c_\alpha \wedge z) = 0$, so $h(z) \le (h(c_\alpha))' = (d_\alpha)'$.

To prove $a \le h^+(y)$, we fix $z \ge y$ with $z \in \mathcal{C}$, and show $h(z) \ge a$. From $z \ge y$ we get $z' \wedge y = 0$, so $h(z')$ is bad (and the corresponding k can be 0), so $h(z') = d_\alpha$ for some α, and hence $h(z) = (d_\alpha)' \ge a$.

Now, to verify strong independence, fix $w \in ((\mathcal{C} \cup \{y\}))$ with $\widetilde{h}(w) > 0$, and fix distinct $j_1, \ldots, j_r \in J$ and elements $b_\ell \in \mathcal{M}^{j_\ell}$. We must show that $w \wedge b_1 \wedge \cdots \wedge b_r > 0$. Say $w = (u \wedge y) \vee (v \wedge y')$, where $u, v \in \mathcal{C}$. There are two cases:

Case 1: $\widetilde{h}(u \wedge y) > 0$. That is, $h(u) \wedge a > 0$. Then $h(u)$ is not one of the d_α, so $h(u)$ is not bad, so $u \wedge y \wedge b_1 \wedge \cdots \wedge b_r > 0$.

Case 2: $\widetilde{h}(v \wedge y') > 0$. That is, $h(v) \wedge \bigvee_\alpha d_\alpha > 0$. Fix α with $h(v) \wedge d_\alpha > 0$, so $h(v \wedge c_\alpha) > 0$. By strong independence of $\mathbb{M}$ with respect to h, we have $v \wedge c_\alpha \wedge b_1^\alpha \wedge \cdots \wedge b_k^\alpha \wedge b_1 \wedge \cdots \wedge b_r > 0$, where $k = k_\alpha$. But also $c_\alpha \wedge y \wedge b_1^\alpha \wedge \cdots \wedge b_k^\alpha = 0$, so $y' \ge c_\alpha \wedge b_1^\alpha \wedge \cdots \wedge b_k^\alpha$, and hence $v \wedge y' \wedge b_1 \wedge \cdots \wedge b_r > 0$. □

Finally, while we are building $h : \mathcal{B} \twoheadrightarrow \mathcal{A}$, we must ensure that $h^*(\mathrm{st}(\mathcal{A}))$ becomes a hatset. To do this, we need to obtain the condition in Lemma 2.16, but, bear in mind that $h^*(\mathrm{st}(\mathcal{A}))$ need only be a hatset in $\mathrm{st}(\mathcal{B}/\mathcal{G})$, not in $\mathrm{st}(\mathcal{B})$. So, we phrase the extension lemma as follows:

5.5. LEMMA. *Let $\mathcal{A}, \mathcal{B}, \mathcal{C}, h, \mathbb{M}, j, J$ be as in Lemma 5.3. Let $\frown$ be any monotone (θ, κ) hatfunction. Assume that $\mathcal{B}$ is complete and $\mathcal{G}$ is a filter on $\mathcal{B}$, with dual ideal $\mathcal{J} \subseteq \ker(h)$. Assume that $\mathcal{M}^j$ is a $\frown$step-family on $(\mathcal{B}, \mathcal{G})$, as in Definition 3.7. Let $\langle c_r : r \in [\kappa]^{<\omega}\rangle$ be a sequence of elements of $\mathcal{C}$ with each $h(c_r) = 1_\mathcal{A}$. Then there are $d_\alpha \in \mathcal{B}$ for $\alpha < \theta$ and an extension of h to some $\widetilde{h} : \widetilde{\mathcal{C}} = ((\mathcal{C} \cup \{d_\alpha : \alpha < \theta\})) \to \mathcal{A}$ so that:*

- *Each $\widetilde{h}(d_\alpha) = 1_\mathcal{A}$,*
- *$d_{\boxed{s}} \le_\mathcal{J} c_{\widehat{s}}$ for each nonempty $s \in [\theta]^{<\omega}$,*

$\mathbb{M} \restriction J$ is strongly independent with respect to $\widetilde{h}$, and $\operatorname{ran} \widetilde{h} = \operatorname{ran} h$.

□ We may assume that the sequence $\langle c_r : r \in [\kappa]^{<\omega}\rangle$ is monotone by replacing each c_r by $\bigwedge\{c_p : p \subseteq r\}$, as in the proof of Lemma 2.5. Now, define

$$d_\alpha = a_\alpha \wedge \bigvee\{c_r \wedge e_r : r \in [\kappa]^{<\omega}\} \ .$$

Then, applying S5 (Lemma 3.10), S1, and monotonicity of $\langle c_r : r \in [\kappa]^{<\omega}\rangle$:

$$\begin{aligned} d_{\boxed{s}} &= a_{\boxed{s}} \wedge \bigvee\{c_r \wedge e_r : r \in [\kappa]^{<\omega}\} \\ &\le_\mathcal{J} \bigvee\{e_r : r \supseteq \widehat{s}\} \wedge \bigvee\{c_r \wedge e_r : r \in [\kappa]^{<\omega}\} \\ &= \bigvee\{c_r \wedge e_r : r \supseteq \widehat{s}\} \le \bigvee\{c_r : r \supseteq \widehat{s}\} = c_{\widehat{s}} \ . \end{aligned}$$

Observe that $d_{\boxed{s}} \geq a_{\boxed{s}} \wedge e_{\widehat{s}} \wedge c_{\widehat{s}}$, and that $a_{\boxed{s}} \wedge e_{\widehat{s}} \in \mathcal{M}^j$. We thus get $\widetilde{h}$ by applying Lemma 5.3.3, with $\mathcal{Y} = \{d_\alpha : \alpha < \theta\}$. As in Lemma 4.7, which was used by Lemma 5.3.3, we get $\operatorname{ran} \widetilde{h} = \operatorname{ran} h$. □

We can now put this all together:

5.6. THEOREM. *Let $\mathcal{B}$ be a complete boolean algebra of size 2^κ, with two filters, $\mathcal{G} \subseteq \mathcal{F}$ on $\mathcal{B}$ (so $\mathrm{st}(\mathcal{B}/\mathcal{F}) \subseteq \mathrm{st}(\mathcal{B}/\mathcal{G}) \subseteq \mathrm{st}(\mathcal{B})$). Let $\frown$ be any monotone (θ, κ) hatfunction. Assume that $\mathbb{M} = \langle \mathcal{M}^i : i \in 2^\kappa \rangle$ is a matrix which is independent with respect to $\mathcal{F}$, such that each $\mathcal{M}^i$ is a $\frown$step-family on $(\mathcal{B}, \mathcal{G})$. Then for every complete boolean algebra $\mathcal{A}$ with $|\mathcal{A}| \leq 2^\kappa$, there is an $h : \mathcal{B} \twoheadrightarrow \mathcal{A}$ such that $h(\mathcal{F}) = \{1\}$ and such that $h^*(\mathrm{st}(\mathcal{A})) \subseteq \mathrm{st}(\mathcal{B}/\mathcal{F})$ is a $\frown$set in $\mathrm{st}(\mathcal{B}/\mathcal{G})$.*

□ We follow the pattern described in the beginning of Section 4. We construct h by a 2^κ matrix sequence $\langle \mathcal{C}_\mu, \mathcal{A}_\mu, h_\mu, I_\mu : \mu < 2^\kappa \rangle$. $\mathcal{C}_0 = ((\mathcal{F}))$, $\mathcal{A}_0 = \{0, 1\}$, $h_0(\mathcal{F}) = \{1\}$, and $I_0 = 2^\kappa$.

There are three tasks to be accomplished: making $h^*(\mathrm{st}(\mathcal{A}))$ a hatset, making h onto $\mathcal{A}$, and making h defined on all of $\mathcal{B}$. These are accomplished by Lemmas 5.5, 5.3.2, and 5.4, respectively. When we apply one of these lemmas at step μ, we set $\mathcal{C} = \mathcal{C}_\mu$, $h = h_\mu$, and $I = I_\mu$, and use the appropriate lemma to obtain $\mathcal{C}_{\mu+1} = \widetilde{\mathcal{C}}$, $h_{\mu+1} = \widetilde{h}$, and $I_{\mu+1} = J$. Which one we apply at step μ will depend on μ mod 3.

We inductively assume that $|I \backslash I_\mu| \leq \max(|\mu|, \aleph_0)$, so that in particular I_μ is never empty, which is important when we apply Lemmas 5.5 and 5.3.2. But, since the use of Lemma 5.4 eliminates $|\mathcal{A}_\mu|$ rows of the matrix, we must also assume inductively that $|\mathcal{A}_\mu| \leq \max(|\mu|, \aleph_0)$. We must exercise some caution here; we cannot assume that $|\mathcal{C}_\mu| \leq \max(|\mu|, \aleph_0)$, since as soon as we apply Lemma 5.5, $|\mathcal{C}_\mu|$ will grow to size θ, which might even equal 2^κ. Fortunately, our extension lemmas bound the size of the range.

To ensure that $h^*(\mathrm{st}(\mathcal{A}))$ is a hatset: Before beginning the construction, fix $c^\mu_r \in \mathcal{B}$ for $\mu < 2^\kappa$ and $\mu \equiv 0 \pmod 3$, such that $\langle c^\mu_r : r \in [\kappa]^{<\omega} \rangle$ is a sequence of elements of $\mathcal{B}$ and such that each such sequence is listed cofinally often in 2^κ. At stage μ with $\mu \equiv 0 \pmod 3$: If each $c^\mu_r \in \mathcal{C}_\mu$ and $h_\mu(c^\mu_r) = 1$, we apply Lemma 5.5 at this stage to ensure this instance of hatpoint, so $|I_\mu \backslash I_{\mu+1}| = 1$. If not, then let $\mathcal{C}_{\mu+1} = \mathcal{C}_\mu$, $h_{\mu+1} = h_\mu$, and $I_{\mu+1} = I_\mu$. Note that in either case, $\mathcal{A}_{\mu+1} = \mathcal{A}_\mu$.

To ensure that h is onto: Before beginning the construction, list $\mathcal{A}$ as $\{a_\mu : \mu < 2^\kappa \ \& \ \mu \equiv 1 \pmod 3\}$. Then, at stage μ with $\mu \equiv 1 \pmod 3$, apply Lemma 5.3.2 to make sure that $a_\mu \in \mathcal{A}_{\mu+1}$. Again, $|I_\mu \backslash I_{\mu+1}| = 1$. $\mathcal{A}_{\mu+1}$, which is obtained via Lemma 4.4, is generated from $\mathcal{A}_\mu$ and the one element a_μ, so that $|\mathcal{A}_\mu| \leq \max(|\mu|, \aleph_0)$ implies that $|\mathcal{A}_{\mu+1}| \leq \max(|\mu|, \aleph_0) = \max(|\mu + 1|, \aleph_0)$.

To ensure that h is defined on all of $\mathcal{B}$: Before beginning the construction, list $\mathcal{B}$ as $\{b_\mu : \mu < 2^\kappa \ \& \ \mu \equiv 2 \pmod 3\}$. Then, at stage μ with $\mu \equiv 2 \pmod 3$, apply Lemma 5.4 to make sure that $b_\mu \in \mathcal{C}_{\mu+1}$. Assume, inductively, that $|\mathcal{A}_\mu| \leq \max(|\mu|, \aleph_0)$ and $|I_\mu| \leq \max(|\mu|, \aleph_0)$. Again, $\mathcal{A}_{\mu+1}$ is generated from $\mathcal{A}_\mu$ and one element, so that $|\mathcal{A}_{\mu+1}| \leq \max(|\mu+1|, \aleph_0)$. Also, $|I_\mu \backslash I_{\mu+1}| \leq \max(|\mathcal{A}_\mu|, \aleph_0)$, so $|I \backslash I_{\mu+1}| \leq \max(|\mu+1|, \aleph_0)$. □

In particular, we may apply this to the "good" hatfunction, where the matrix can be constructed directly from a disjoint family (see Lemma 3.8); $\mathcal{G} = \{1\}$ here:

5.7. COROLLARY. *Let $\mathcal{B}$ be a complete boolean algebra of size 2^κ. Let $\mathcal{I}$ be an ideal on $\mathcal{B}$ with dual filter $\mathcal{F}$, and assume that $\mathcal{B}$ has an antichain, $\{a_\xi : \xi < \kappa\}$, such that the a_ξ are distinct and each $a_\xi \notin \mathcal{I}$. Let $\mathcal{A}$ be any complete boolean algebra such that $|\mathcal{A}| \leq 2^\kappa$. Then there is an $h : \mathcal{B} \twoheadrightarrow \mathcal{A}$ such that $h(\mathcal{F}) = \{1\}$ and such that $h^*(\mathrm{st}(\mathcal{A})) \subseteq \mathrm{st}(\mathcal{B}/\mathcal{F})$ is a κ^+-good set (and hence a weak P_κ-set) in $\mathrm{st}(\mathcal{B})$.*

Theorem 1.3 is immediate from this, as is Theorem 1.12.1. To prove Theorem 1.12.2, we apply Theorem 5.6 with $\mathcal{B} = \mathcal{P}(\kappa)$ and $\mathcal{F} = \mathcal{G} = \mathcal{FR}(\kappa)$, where, by Lemma 3.9, we have the correct matrix for every monotone (κ^+, κ) hatfunction. In particular, we can use a hatfunction such that hatsets are weak P_{κ^+}-sets; note that by Lemma 2.18, we can replace every hatfunction by a monotone one. We thus have the following corollary, from which Theorem 1.12.2 follows immediately:

5.8. COROLLARY. *Let κ be any regular cardinal and $\mathcal{A}$ any complete boolean algebra with $|\mathcal{A}| \leq 2^\kappa$. Let $\frown$ be any (κ^+, κ) hatfunction. Then $\mathrm{st}(\mathcal{A})$ can be embedded into $u(\kappa)$ as a $\frown$set.*

As pointed out in the Introduction, we have now proved Theorem 1.14.1. However, we have not yet proved Theorem 1.14.2, except in the case $2^\kappa = 2^{(\kappa^+)}$. Here, we wish to construct $E = \{\mathbf{x}_\delta : \delta < \kappa^+\} \subseteq u(\kappa)$ such that E is κ^+-fuzzy in $u(\kappa)$. We know, by Corollary 1.7, that there is a $D = \{\mathbf{y}_\delta : \delta < \kappa^+\} \subseteq \mathrm{st}(\mathcal{D}_{\kappa^+})$ which is κ^+-fuzzy in $\mathrm{st}(\mathcal{D}_{\kappa^+})$; so each $\mathbf{y}_\delta$ is a weak P_{κ^+}-point, and D is dense in itself. Now, $|\mathcal{D}_{\kappa^+}| = 2^{(\kappa^+)}$, so if $2^\kappa = 2^{(\kappa^+)}$, we can insert $\mathrm{st}(\mathcal{D}_{\kappa^+})$ into $u(\kappa)$ as a weak P_{κ^+}-set. In the general case, we must prove a variant of Theorem 5.6 which applies with $\mathcal{A}$ of arbitrary size.

With any complete $\mathcal{A}$, we can certainly follow all the steps in the proof of Theorem 5.6 and produce a homomorphism $h : \mathcal{B} \to \mathcal{A}$; we just cannot make it onto if $|\mathcal{A}| > 2^\kappa$. If $\widetilde{\mathcal{A}} = \mathrm{ran}(h)$, we have $h : \mathcal{B} \twoheadrightarrow \widetilde{\mathcal{A}}$, and then $h^* : \mathrm{st}(\widetilde{\mathcal{A}}) \hookrightarrow \mathcal{B}$. So, if we have $D = \{\mathbf{y}_\delta : \delta < \kappa^+\} \subseteq \mathrm{st}(\mathcal{A})$, then we will have ultrafilters $\widetilde{\mathbf{y}_\delta} = \mathbf{y}_\delta \cap \widetilde{\mathcal{A}} \in \mathrm{st}(\widetilde{\mathcal{A}})$, and then $\mathbf{x}_\delta = h^*(\widetilde{\mathbf{y}_\delta})$ and $E = \{\mathbf{x}_\delta : \delta < \kappa^+\} \subseteq \mathrm{st}(\mathcal{B})$. Note that $\widetilde{\mathbf{y}_\delta} = i^*(\mathbf{y}_\delta)$, where i is inclusion, $i : \widetilde{\mathcal{A}} \hookrightarrow \mathcal{A}$. If D is dense in itself, then E will be dense in itself by continuity of i^* and h^*, *provided* that the $\mathbf{x}_\delta$ are distinct; that is, $\widetilde{\mathcal{A}}$ must be large enough to distinguish the $\mathbf{y}_\delta$. We can, for $\delta \neq \beta$, fix an element $z_{\delta,\beta} \in \mathbf{y}_\delta \backslash \mathbf{y}_\beta$ (WLOG, the $\mathbf{y}_\delta$ are all distinct). It is easy to modify the proof of Theorem 5.6 to make sure that the κ^+ elements $z_{\delta,\beta}$ all wind up in $\mathrm{ran}\, h$.

Also, the proof of Theorem 5.6 easily makes $h^*(\mathrm{st}(\widetilde{\mathcal{A}}))$ a hatset, and hence a weak P_{κ^+}-set, as in Corollary 5.8, assuming that $\mathcal{B}$ has the correct matrix. Thus, the $\mathbf{x}_\delta \in h^*(\mathrm{st}(\widetilde{\mathcal{A}}))$ will not be limits of sets of size $\leq \kappa$ coming from outside of $h^*(\mathrm{st}(\widetilde{\mathcal{A}}))$. However, they might be limits of such sets coming from inside $h^*(\mathrm{st}(\widetilde{\mathcal{A}}))$, since the $\widetilde{\mathbf{y}_\delta}$ might fail to be weak P_{κ^+}-points in $\mathrm{st}(\widetilde{\mathcal{A}})$ even if the $\mathbf{y}_\delta$ are weak P_{κ^+}-points in $\mathrm{st}(\mathcal{A})$. To make sure that the $\widetilde{\mathbf{y}_\delta}$ are weak P_{κ^+}-points in $\mathrm{st}(\widetilde{\mathcal{A}})$, we generalize Theorem 5.6 to allow a closure operation:

5.9. THEOREM. *Let $\mathcal{B}, \mathcal{G}, \mathcal{F}, \kappa, \frown, \mathbb{M}, \mathcal{A}$ be exactly as in Theorem 5.6,* except *that we do not assume $|\mathcal{A}| \leq 2^\kappa$. Let Φ be a family of 2^κ functions, with $\varphi : \mathcal{A}^\kappa \to \mathcal{A}$ for each $\varphi \in \Phi$. Then there is a subalgebra $\widetilde{\mathcal{A}} \subseteq \mathcal{A}$ and an $h : \mathcal{B} \twoheadrightarrow \widetilde{\mathcal{A}}$ such that $\widetilde{\mathcal{A}}$ is closed under all the $\varphi \in \Phi$, and such that $h(\mathcal{F}) = \{1\}$ and $h^*(\mathrm{st}(\widetilde{\mathcal{A}})) \subseteq \mathrm{st}(\mathcal{B}/\mathcal{F})$ is a $\frown$set in $\mathrm{st}(\mathcal{B}/\mathcal{G})$.*

□ Exactly as for Theorem 5.6, except that to ensure that $\operatorname{ran}(h)$ is closed under Φ: Before beginning the construction, list $2^\kappa \times 2^\kappa$ as $\{(f(\mu), g(\mu)) : \mu < 2^\kappa \ \&\ \mu \equiv 1 \pmod 3\}$, with each $g(\mu) \leq \mu$. At each stage $\nu < 2^\kappa$: list the closure of $\mathcal{A}_\nu$ under Φ as $\{a^\nu_\zeta : \zeta < 2^\kappa\}$. Then, at stage μ with $\mu \equiv 1 \pmod 3$, apply Lemma 5.3.2 to make sure that $a^{g(\mu)}_{f(\mu)} \in \mathcal{A}_{\mu+1}$. □

PROOF OF 1.14.2.. Let the $\mathbf{y}_\delta$ and $z_{\delta,\beta} \in \mathbf{y}_\delta \backslash \mathbf{y}_\beta$ be as described above, with $\mathcal{A} = \mathcal{D}_{\kappa^+}$, $\mathcal{B} = \mathcal{P}(\kappa)$ and $\mathcal{F} = \mathcal{G} = \mathcal{FR}(\kappa)$; $\frown$ is any (κ^+, κ) hatfunction such that all $\frown$sets are weak P_{κ^+}-sets (see Lemma 2.15). Assume that among the functions in Φ are the functions with constant value $z_{\delta,\beta}$; then the $\widetilde{\mathbf{y}_\delta}$ will be distinct.

In addition, we assume that the $\mathbf{y}_\delta$ are weak P_{κ^+}-points in $\operatorname{st}(\mathcal{A})$ *because* they are $\frown$points; in fact, they can be κ^{++}-good (by Corollary 5.7), which implies $\frown$point (by Lemma 2.9). So, given a sequence $\vec{u} = \langle u_r : r \in [\kappa]^{<\omega}\rangle$ of elements of $\mathbf{y}_\delta$, there are elements $v_\alpha = v^\delta_\alpha(\vec{u}) \in \mathbf{y}_\delta$ such that $v_{\boxed{s}} \leq u_{\widehat{s}}$ for each non-empty $s \in [\kappa^+]^{<\omega}$. Now, $\vec{u}$ can be coded as an element of $\mathcal{A}^\kappa$ (since $|[\kappa]^{<\omega}| = \kappa$), so that we may view each v^δ_α as a function from $\mathcal{A}^\kappa \to \mathcal{A}$. Making sure that these functions are all in Φ ensures that the $\widetilde{\mathbf{y}_\delta}$ are also $\frown$points, and hence weak P_{κ^+}-points, in $\operatorname{st}(\widetilde{\mathcal{A}})$. □

We remark that Theorem 5.9 could be viewed in the context of elementary submodels (see DOW [1988]). That is, we can get $\widetilde{\mathcal{A}} = \mathcal{A} \cap N$, where N is a κ-closed elementary submodel of the universe with $|N| = 2^\kappa$. Then, in proving Theorem 1.14.2, we just used closure of N under the Skolem functions needed for the argument to work.

6. Avoiding *P*-points

This article has emphasized *weak* P_κ-points. If κ is a measurable cardinal and $\mathbf{x} \in u(\kappa)$ is a κ-complete ultrafilter, then $\mathbf{x}$ is a P_κ-point in $\beta\kappa$. If $\mathbf{x}$ is a normal ultrafilter, then it is also a P_{κ^+}-point in $u(\kappa)$. Inductive constructions like the ones in Section 5 cannot be guaranteed to generate one of these points. However, they can be guaranteed *not* to generate one of these points. Specific instances of this have been pointed out in the literature. By KUNEN [1980], there are weak P-points in ω^* which are not P-points. By KUNEN [1972], there are good ultrafilters on κ which are countably incomplete, which is important for their use in model theory (see §6.1 of CHANG and KEISLER [1990]). Here, we show that these constructions in CHANG and KEISLER [1990] and KUNEN [1972, 1980] are part of a general procedure which always works.

One way to build a *non-P*-point $\mathbf{x} \in \operatorname{st}(\mathcal{B})$ is to fix some decreasing ω-sequence $b_0 > b_1 > \cdots$ and put all the b_n into $\mathbf{x}$, together with the complement of any element which is below all the b_n. The following lemma lets us integrate this remark into our constructions of weak P_κ-points.

6.1. LEMMA. *Let $\mathcal{B}$ be a boolean algebra with a filter $\mathcal{F}$ and a matrix $\mathbb{M} = \langle \mathcal{M}^i : i \in I\rangle$ which is independent with respect to $\mathcal{F}$. Assume that $J \subseteq I$ and $d_j, e_j \in \mathcal{M}^j$ for $j \in J$ with $d_j \wedge e_j = 0$. Let $\mathcal{F}^\dagger$ be the filter generated by $\mathcal{F}$, all the d_j, and all the $w \in \mathcal{B}$ such that $w' \leq_{\mathcal{F}} d_j$ for infinitely many $j \in J$. Then $\mathcal{F}^\dagger$ is a proper filter and $\mathbb{M} \restriction (I \backslash J)$ is independent with respect to $\mathcal{F}^\dagger$.*

□ We must show that $b_{i_1} \wedge \cdots \wedge b_{i_r} \wedge c > 0$ whenever $i_1, \ldots, i_r$ are distinct elements of $I \backslash J$, each $b_{i_\ell} \in \mathcal{M}^{i_\ell}$ ($\ell = 1, \ldots, r$), and $c \in \mathcal{F}^\dagger$. By definition of $\mathcal{F}^\dagger$, we have $c \geq d_{k_1} \wedge \cdots \wedge d_{k_s} \wedge w_1 \wedge \cdots \wedge w_t$, where $k_1, \ldots, k_s$ are distinct elements of J and each $(w_\ell)' \leq_{\mathcal{F}} d_j$ for infinitely many $j \in J$. Now, choose distinct $j_1, \ldots, j_t \in J \backslash \{k_1, \ldots, k_s\}$ such that each $(w_\ell)' \leq_{\mathcal{F}} d_{j_\ell}$, so that $w_\ell \geq_{\mathcal{F}} e_{k_\ell}$, and then choose $u_\ell \in \mathcal{F}$ such that $w_\ell \geq e_{k_\ell} \wedge u_\ell$. Then $b_{i_1} \wedge \cdots \wedge b_{i_r} \wedge c \geq b_{i_1} \wedge \cdots \wedge b_{i_r} \wedge d_{k_1} \wedge \cdots \wedge d_{k_s} \wedge e_{j_1} \wedge \cdots \wedge e_{j_t} \wedge u_j \wedge \cdots \wedge u_j > 0$ by independence of $\mathbb{M}$ with respect to $\mathcal{F}$. □

So, in Theorems 5.6 and 5.9, where we started with filters $\mathcal{G} \subseteq \mathcal{F}$ on $\mathcal{B}$, we now have $\mathcal{G} \subseteq \mathcal{F} \subseteq \mathcal{F}^\dagger$, and hence $\mathrm{st}(\mathcal{B}/\mathcal{F}^\dagger) \subseteq \mathrm{st}(\mathcal{B}/\mathcal{F}) \subseteq \mathrm{st}(\mathcal{B}/\mathcal{G}) \subseteq \mathrm{st}(\mathcal{B})$. As long as J is infinite in Lemma 6.1, each point of $\mathrm{st}(\mathcal{B}/\mathcal{F}^\dagger)$ will be a non-P-point in the space $\mathrm{st}(\mathcal{B}/\mathcal{F})$, and hence also in the larger spaces $\mathrm{st}(\mathcal{B}/\mathcal{G})$ and $\mathrm{st}(\mathcal{B})$. Here, $I = 2^\kappa$, and as long as $|I \backslash J| = 2^\kappa$, we can replace $\mathcal{F}$ by $\mathcal{F}^\dagger$ in the theorem to get our hatset inside $\mathrm{st}(\mathcal{B}/\mathcal{F}^\dagger)$. Thus, we have:

6.2. THEOREM. *In Theorems 5.6 and 5.9, we can obtain h so that no point of the* $\frown$*set* ($h^*(\mathrm{st}(\mathcal{A}))$ *in 5.6 and* $h^*(\mathrm{st}(\widetilde{\mathcal{A}}))$ *in 5.9) is a P-point in* $\mathrm{st}(\mathcal{B}/\mathcal{F})$.

References

BAKER, J.
[2001] Some Topological Results on Ultrafilters, Ph.D. Thesis, University of Wisconsin.

BAKER, J. and K. KUNEN
[2001] Limits in the uniform ultrafilters, *Trans. Amer. Math. Soc.* **353**, 4083–4093.

BALCAR, B. and F. FRANĚK
[1982] Independent families in complete Boolean algebras, *Trans. Amer. Math. Soc.* **274**, 607–618.

CHANG, C.C. and H.J. KEISLER
[1990] *Model theory*, Third Edition, North-Holland.

DOW, A.
[1985] Good and OK ultrafilters, *Trans. Amer. Math. Soc.* **290**, 145–160.
[1988] An introduction to applications of elementary submodels to topology, *Topology Proc.* **13**, 17–72.

EFIMOV, B.A.
[1970] Extremally disconnected bicompacta and absolutes (Russian), *Trudy Moskov. Mat. Obšč.***23**, 235-276 (English translation: *Trans. Moscow Math. Soc.* **23**, 243–285).

ENGELKING, R. and M. KARŁOWICZ
[1965] Some theorems of set theory and their topological consequences. *Fund. Math.* **57**, 275–285.

HAUSDORFF, F.
[1936] Über zwei Sätze von G. Fichtenholz und L. Kantorovitch, *Studia Math.* **6**, 18–19.

KEISLER, H.J.
[1964] Good ideals in fields of sets, *Ann. of Math.* **79**, 338–359.

KUNEN, K.
[1972] Ultrafilters and independent sets, *Trans. Amer. Math. Soc.* **172**, 299–306.
[1980] Weak P-points in $\mathbb{N}^*$, *Colloq. Math. Soc. János Bolyai* **23**, 741–749.

POSPÍŠIL, B.
[1937] Remark on bicompact spaces, Ann. of Math. (2) **38**, 845–846.

RUDIN, W.
[1956] Homogeneity problems in the theory of Čech compactifications, *Duke Math. J.* **23**, 409–419.

SIKORSKI, R.
[1969] *Boolean Algebras*, Third edition, Springer-Verlag.

SIMON, P.
[1985] Applications of independent linked families, *Colloq. Math. Soc. János Bolyai* **41**, 561–580.

WIMMERS, E.L.
[1982] The Shelah P-point independence theorem, *Israel J. Math.* **43**, 28–48.

CHAPTER 3

Recent Developments in the Topology of Ordered Spaces

Harold R. Bennett

Texas Tech University, Lubbock, TX 79409, U.S.A.
E-mail: graddir@math.ttu.edu

David J. Lutzer

College of William and Mary, Williamsburg, VA 23187-8795, U.S.A.
E-mail: lutzer@math.wm.edu

Contents

RECENT PROGRESS IN GENERAL TOPOLOGY II
Edited by Miroslav Hušek and Jan van Mill

1. Introduction

The last ten years have seen substantial progress in understanding the topology of linearly ordered spaces and their subspaces, the generalized ordered spaces. Some old problems have been solved, and ordered space constructions have been used to solve several problems that were posed in more general settings.

This paper surveys progress in several parts of ordered space theory. Chapter 2 mentions some recent results concerning orderability. Chapter 3 focuses on perfect ordered spaces. Chapter 4 deals with special base properties in ordered spaces. Chapter 5 investigates the role of diagonal and off-diagonal properties in metrization. Chapter 6 discusses Dugundji extension theory in ordered spaces. Chapter 7 briefly mentions Mary Ellen Rudin's recent solution of Nikiel's problem and its consequences for Hahn-Mazurkiewicz theory. Chapter 8 samples recent work on the structure of the Banach space $C(K)$ where K is a compact LOTS and Chapter 9 summarizes recent work on products of ordered spaces.

Recall that a *generalized ordered space* (GO-space) is a Hausdorff space X equipped with a linear order and having a base of order-convex sets. In case the topology of X coincides with the open interval topology of the given linear order, we say that X is a *linearly ordered topological space* (or LOTS). Čech showed that the class of GO-spaces is the same as the class of spaces that can be topologically embedded in some LOTS. (See LUTZER [1971].)

Throughout this paper, we will adopt the convention that all spaces are at least regular and T_1. Of course, the ordered spaces that we will consider have stronger separation – each GO-space is monotonically normal (HEATH, LUTZER and ZENOR [1973]) and hence hereditarily collectionwise normal. We reserve the symbols $\mathbb{N}$, $\mathbb{Z}$, $\mathbb{Q}$, $\mathbb{P}$ and $\mathbb{R}$ for the usual sets of positive integers, all integers, rational numbers, irrational numbers, and real numbers, respectively.

For other related surveys, see TODORČEVIĆ [1984], MAYER and OVERSTEEGEN [1992], and LUTZER [1980].

2. Orderability

The classical orderability problem asks for topological characterizations of spaces whose topology can be given by the usual open interval topology of some linear ordering of the ground set. The survey paper PURISCH [1998] shows that this problem has a long history, going back to the early topological characterizations of the unit interval. The orderability problem for zero-dimensional metric spaces was solved by HERRLICH [1965] and later PURISCH [1977] gave necessary and sufficient conditions for orderability of any metric space. The general orderability problem was solved by VAN DALEN and WATTEL [1973] who proved:

2.1. THEOREM. *A T_1 space X is orderable if and only if X has a subbase $\mathcal{S} = \mathcal{S}_1 \cup \mathcal{S}_2$ where each $\mathcal{S}_i$ is linearly ordered by inclusion and has the property that if $T \in \mathcal{S}_i$ has $T = \bigcap\{S \in \mathcal{S}_i : T \subset S \text{ and } T \neq S\}$, then T also has $T = \bigcup\{S \in \mathcal{S}_i : S \subset T \text{ and} S \neq T\}$.*

More recent work by VAN MILL and WATTEL [1984] significantly sharpened a selection-theoretic orderability theorem that MICHAEL [1951] proved for compact connected spaces.

For any space X, let 2^X be the collection of all closed non-empty subsets of X, topologized using the Vietoris topology , and let $X(2) = \{T \in 2^X : |T| = 2\}$ be topologized as a subspace of 2^X. VAN MILL and WATTEL [1984] showed:

2.2. THEOREM. *For a compact Hausdorff space X, the following are equivalent:*

(a) *X is orderable;*

(b) *there is a continuous function $f : 2^X \to X$ having $f(S) \in S$ for each $S \in 2^X$;*

(c) *there is a continuous function $g : X(2) \to X$ having $g(T) \in T$ for each $T \in X(2)$.*

Functions such as the ones described in parts b) and c) of the previous theorem are called *continuous selections* and *continuous weak selections*, respectively. Theorem 2.2 has been the basis for further research. For example, FUJII and NOGURA [1999] have used selections to characterize compact spaces of ordinals as follows:

2.3. THEOREM. *A compact Hausdorff space is homeomorphic to a compact ordinal space if and only if there is a continuous selection $f : 2^X \to X$ with the property that $f(C)$ is an isolated point of C for each $C \in 2^X$.*

Recently ARTICO, MARCONI, PELANT, ROTTER and TKACHENKO [200?] have proved the following two results:

2.4. THEOREM. *Let X be a space with a continuous weak selection. If X^2 is pseudocompact, then X is countably compact and is a GO-space. In particular, if X is a pseudocompact k-space with a continuous weak selection, then X is a countably compact GO-space, and if X is a countably compact Tychonoff space with a continuous selection, then X is a GO-space.*

2.5. THEOREM. *For a completely regular space X, the following are equivalent:*

(a) *βX is orderable;*

(b) *X is a pseudocompact GO-space;*

(c) *X is countably compact and has a continuous weak selection;*

(d) *X^2 is pseudocompact and X has a continuous weak selection.*

3. Perfect ordered spaces

Recall that a topological space X is *perfect* if each closed subset of X is a G_δ-subset of X. Among GO-spaces, to be perfect is a very strong property.

3.1. THEOREM (LUTZER [1971]). *Any perfect GO-space is hereditarily paracompact and first countable.*

Determining whether or not a given GO-space is perfect is often one of the crucial steps in metrization problems for GO-spaces. (See 4.14 and 5.6 for examples.) The literature contains many generalizations of the property "X is perfect," and it is often useful to know which ones are equivalent to being perfect (in a GO-space). Here is a sample.

3.2. THEOREM (BENNETT, HOSOBUCHI and LUTZER [1999]). *The following properties of a GO-space X are equivalent:*

(a) *X is perfect;*

(b) *each relatively discrete subspace of X is σ-closed-discrete* (FABER [1974])*;*

(c) *for each open set $U \subseteq X$ there are open sets V_n having* $\mathrm{cl}(V_n) \subseteq U$ *and* $U \subseteq \mathrm{cl}(\bigcup\{V_n : n \geq 1\})$ (REED [1971])*;*

(d) *each open subset $U \subseteq X$ contains a dense subset S that is an F_σ-subset of X* (KOČINAC [1986])*;*

(e) *every closed nowhere dense subset of X is a G_δ-subset of X;*

(f) *every regularly closed subset of X is a G_δ-subset of X* (BENNETT *and* LUTZER [1984]).

Before continuing with the main theme of this section, let us pause to mention one interesting generalization of "X is perfect" that does *not* belong in the list given in 3.2. KOČINAC [1983] defined that a topological space is *weakly perfect* if each closed set $F \subseteq X$ contains a set S that is dense in F and is a G_δ-subset of X. It is easy to see that not all GO-spaces are weakly perfect: consider the lexicographically ordered set $X = \mathbb{R} \times [0,1]$ with its usual open interval topology. It is also easy to see that some GO-spaces are weakly perfect but not perfect: consider the usual space of countable ordinals. (In the light of 3.1, the fact that the space of countable ordinals is weakly perfect shows how wide is the gap between weakly perfect and perfect.)

Finding compact spaces that are weakly perfect but not perfect is more of a challenge. The first examples were given by KOČINAC [1983] and involved set theory. A family of ZFC examples was given by HEATH [1989]. In subsequent work BENNETT, HOSOBUCHI and LUTZER [2000] extended Heath's examples and linked the weakly perfect property to certain ideas in classical descriptive set theory.

To see that linkage, recall that a subset $A \subseteq [0,1]$ is *perfectly meager* if for any closed dense-in-itself set $C \subseteq [0,1]$, the set $C \cap A$ is a first category subset of C, where C carries its relative topology. Uncountable perfectly meager sets exist in ZFC (see MILLER [1984]). Next, for any dense subset $A \subseteq [0,1]$, let

$$X(A) = ([0,1] \times \{0\}) \cup (A \times \{-1,1\})$$

with the lexicographic order and usual open interval topology. Then $X(A)$ is always a compact, first-countable LOTS, and we have

3.3. PROPOSITION. *Let $A \subseteq [0,1]$ be dense. Then the following are equivalent:*

(a) *A is a perfectly meager subset of $[0,1]$;*

(b) *$X(A)$ is weakly perfect;*

(c) *$X(A)$ is hereditarily weakly perfect.*

If A is uncountable, then $X(A)$ is not perfect.

Notice that $X(A)$ is weakly perfect if and only if it is hereditarily weakly perfect. That leads to an open question.

3.4. QUESTION. Suppose X is an arbitrary weakly perfect GO-space. Must X be *hereditarily* weakly perfect?

A corollary of 3.3 is a new internal characterization of perfectly meager subsets of $[0, 1]$, namely that a subset $A \subseteq [0, 1]$ is perfectly meager if and only if for each $B \subseteq A$ there is a countable set $C \subseteq B$ that is dense in B and is a G_δ-subset in the relative topology of A. (See GRUENHAGE and LUTZER [2000] for an easier proof.) In addition, various subspaces of $X(A)$ answer questions posed by HEATH [1989] concerning the existence of weakly perfect, non-metrizable, quasi-developable spaces.

As it happens, there is an interesting property that (for GO-spaces) is even better than being perfect, namely the existence of a σ-closed-discrete dense set. Any GO-space with a σ-closed-discrete dense subset must be perfect, but the converse is consistently false: any Souslin space (a non-separable LOTS with countable cellularity) would be a counterexample. In the 1970s, Maurice (see VAN WOUWE [1979]) posed the first of three major problems for perfect GO-spaces, namely:

3.5. QUESTION. Is there a ZFC example of a perfect GO-space that does not have a σ-closed-discrete dense subset?

Although recent work has shed considerable light on what it would take to solve 3.5, Maurice's question remains open today.

A second problem concerning perfect ordered spaces was posed by Heath. PONOMAREV [1967] and BENNETT [1968] independently proved that if there is a Souslin space, then there is a Souslin space with a point-countable base. After Bennett constructed in ZFC a LOTS, now called the "Big Bush" (see 4.5 below for details), that has a point-countable base but not a σ-point-finite base (see BENNETT [1968] [1971]), Heath asked:

3.6. QUESTION. Is there a ZFC example of a perfect GO-space that has a point-countable base and is not metrizable?

Heath's question also remains open. It is linked with Maurice's question 3.5 by

3.7. PROPOSITION (BENNETT and LUTZER [1984]). *If a GO-space has a point-countable base and a σ-closed-discrete dense subset, then it is metrizable.*

Thus, Heath's question boils down to "In ZFC, is there a perfect GO-space with a point-countable base that does not have a σ-closed-discrete dense subset?"

A third question concerning perfect GO-spaces was posed by NYIKOS [1976]. Recall that a topological space is *non-Archimedean* if it has a base that is a tree with respect to inclusion. Nyikos asked:

3.8. QUESTION. In ZFC, is there an example of a perfect non-Archimedean space that is not metrizable?

Question 3.8 is a question about ordered spaces because, as proved in PURISCH [1983], every perfect non-Archimedean space is a LOTS under some ordering. Nyikos' question remains open.

Although the questions of Maurice, Heath, and Nyikos remain open, an important paper by QIAO and TALL [200?] linked them to a generalized Souslin problem by proving:

3.9. THEOREM. *The following statements are equivalent:*

(a) *there is a perfect LOTS that does not have a σ-closed-discrete dense subset;*

(b) *there is a perfectly normal, non-Archimedean space that is not metrizable;*

(c) *there is a LOTS X in which every disjoint collection of convex open sets is σ-discrete, and yet X does not have a σ-closed-discrete dense subset;*

(d) *there is a dense-it-itself LOTS Y that does not have a σ-closed-discrete dense set, and yet each nowhere dense subspace of Y does have a σ-closed-discrete dense subset (in its relative topology).*

The key to the proof of 3.9 is the following lemma of Qiao and Tall:

3.10. LEMMA. *All first-countable GO-spaces contain dense non-Archimedean subspaces.*

In addition, QIAO and TALL [200?] also showed how Heath's question is related to the others by proving:

3.11. THEOREM. *The following statements are equivalent:*

(a) *there is a perfectly normal, non-metrizable, non-Archimedean space having a point-countable base;*

(b) *there is a perfect LOTS that has a point-countable base but does not have a σ-closed-discrete dense subset;*

(c) *there is a LOTS X with a point-countable base and having the property that every pairwise disjoint collection of convex open sets is σ-discrete, and yet X does not have a σ-closed-discrete dense subset;*

(d) *there is a dense-in-itself LOTS Y with a point-countable base that does not have a σ-closed-discrete dense subset, and yet every nowhere dense subspace of Y has a σ-closed-discrete dense subset for its relative topology.*

A family of recent results due to BENNETT, HEATH and LUTZER [2000], BENNETT, LUTZER and PURISCH [1999], and BENNETT and LUTZER [200?a] show how to recognize when a GO-space has a σ-closed-discrete dense set, in terms of relations to more familiar types of spaces. These characterizations may be of use to researchers working on the ZFC questions posed by Maurice, Heath, Nyikos, and Lutzer.

3.12. THEOREM. *The following properties of a perfect GO-space X are equivalent:*

(a) *X has a σ-closed-discrete dense subset;*

(b) *X has a dense metrizable subspace;*

(c) *there is a sequence $\langle \mathcal{G}_n \rangle$ of open covers of X such that for each $p \in X$, the set $\bigcap\{St(p, \mathcal{G}_n) : n \geq 1\}$ has at most two points;*

(d) *there is a sequence $\langle \mathcal{G}_n \rangle$ of open covers of X such that for each $p \in X$, the set $\bigcap\{St(p, \mathcal{G}_n) : n \geq 1\}$ is a separable subspace of X;*

(e) *X is the union of two subspaces, each having a G_δ-diagonal in its relative topology;*

(f) *X is the union of countably many subspaces, each having a G_δ-diagonal in its relative topology;*

(g) *there is a metrizable GO-space Y and a continuous $f : X \to Y$ with $|f^{-1}[y]| \leq 2$ for each $y \in Y$;*

(h) *there is a topological space Z having a G_δ-diagonal and a continuous function $g : X \to Z$ such that $g^{-1}[z]$ is a separable subspace of X for each $z \in Z$;*

(i) *X has a dense subspace $E = \bigcup\{E_n : n \geq 1\}$ where each E_n is a metrizable subspace of X;*

(j) *X is metrically fibered in the sense of* TKACHUK [1994], *i.e., there is a continuous function $f : X \to M$ where M is metrizable and $f^{-1}(m)$ is a metrizable subspace of X for each $m \in M$.*

As a final part of this chapter, we consider the embedding problem for perfect GO-spaces. It is older than the questions of Maurice, Heath, and Nyikos and turns out to be related to them because of work by W-X Shi, as we will explain later.

It has been known since the work of Čech that GO-spaces are precisely the subspaces of linearly ordered topological spaces. Furthermore, there is a canonical construction that produces, for any GO-space X, a LOTS X^* that contains X as a closed subspace and is, in some sense, the smallest LOTS with this property (LUTZER [1971]). Given a GO-space $(X, \tau, <)$, let λ be the usual order topology of the order $<$. Then $\lambda \subseteq \tau$. Let $R = \{p \in X : [p, \to[\in \tau - \lambda\}$ and $L = \{q \in X :]\leftarrow, q] \in \tau - \lambda\}$. Define X^* to be the following lexicographically ordered subset of $X \times \mathbb{Z}$:

$$X^* = (X \times \{0\}) \cup \{(p, n) : p \in R \text{ and } n \leq 0\} \cup \{(q, n) : q \in L \text{ and } n \geq 0\}.$$

Clearly X is homeomorphic to the subspace $X \times \{0\}$ of X^*. It often happens that the GO-space X has a topological property P if and only if the LOTS X^* has property P. For example:

3.13. PROPOSITION (LUTZER [1971]). *Let P be one of the following properties: paracompact, metrizable, Lindelöf, quasi-developable, or has a point-countable base. Then the GO-space X has property P if and only if its LOTS extension X^* has P.*

However, being perfect is one important topological property that cannot be added to the list in 3.13. If S is the Sorgenfrey line, then S is certainly perfect, but a Baire category argument shows that S^* is not perfect. Of course, there is a perfect LOTS that contains S as a subspace, namely the lexicographic product space $\mathbb{R} \times \{0, 1\}$. Examples of that type led to

3.14. QUESTION. Is it true that any perfect GO-space can be topologically embedded in some perfect LOTS? (LUTZER [1971] BENNETT and LUTZER [1990]).

Many related questions have been solved. For example, we know that there are perfect GO-spaces $(X, \tau, <)$ that cannot be embedded:

(a) as a closed subset, or as a G_δ-subset, of any perfect LOTS (LUTZER [1971]);

(b) as a dense subspace of a perfect LOTS whose order extends the given order $<$ of X (SHI [1997] and MIWA and KEMOTO [1993]);

(c) as a dense subspace of a perfect LOTS (SHI, MIWA and GAO [1995]) (with no restriction on the order of the extension).

However, the general question 3.14 remains open today.

We have come to understand that the perfect embedding question 3.14 is closely related to the questions of Maurice, Heath, and Nyikos because SHI [1999b] has proved:

3.15. THEOREM. *If X is a perfect GO-space with a σ-closed-discrete dense set, then there is a perfect LOTS Y that contains X and also has a σ-closed-discrete dense set.*

Shi's theorem links questions 3.14 and 3.5: if there is a model $\mathcal{M}$ of set theory in which every perfect GO-space has a σ-closed-discrete dense set, then in $\mathcal{M}$ every perfect GO-space can be embedded in a perfect LOTS.

It might be tempting to think that one could start with a model of set theory that contains a Souslin space S and modify S by isolating some points and making others "one sided" in the Sorgenfrey sense, in the hope of getting a consistent example of a perfect GO-space that cannot be embedded in a perfect LOTS. That approach cannot work because SHI, MIWA and GAO [1996] showed that if one starts with a perfect LOTS Y and constructs a perfect GO-space X by modifying it as described above, then X will embed in some perfect LOTS Z.

4. Base axioms related to metrizability

As seen in Section 3, topological properties that are distinct in general spaces can coalesce among GO-spaces. The next three theorems illustrate this phenomenon for metrizable spaces, spaces with a σ-disjoint base, and spaces with a point-countable base.

4.1. THEOREM (LUTZER [1971]). *For any GO-space X, the following are equivalent:*

(a) *X is metrizable;*

(b) *X has a σ-locally-countable base;*

(c) *X is developable;*

(d) *X is semi-stratifiable.*

4.2. THEOREM (LUTZER [1971], BENNETT [1971]). *For any GO-space X, the following are equivalent:*

(a) *X has a σ-disjoint base;*

(b) *X has a σ-point-finite base;*

(c) *X is quasi-developable.*

4.3. THEOREM (GRUENHAGE [1992]). *For any GO-space X, the following are equivalent:*

(a) *X has a point-countable base;*

(b) *X has the Collins-Roscoe property "open (G)", i.e., for each $x \in X$ there is a countable collection $\mathcal{B}(x)$ of open neighborhoods of x such that if a sequence x_n converges to x, then the collection $\bigcup\{\mathcal{B}(x_n) : n \geq 1\}$ contains a neighborhood base at the point x.*

Much of recent research on base properties in ordered spaces has focused on two major themes.

- Examine the gaps between the clusters of properties characterized in 4.1, 4.2, and 4.3, i.e., between the properties of metrizability, quasi-developability, and the existence of a point-countable base in GO-spaces;
- Expand the lists of equivalent properties in each of 4.1, 4.2, and 4.3.

We begin by considering the gap between the metrizability cluster described in (4.1) and the properties equivalent to quasi-developability given in (4.2). First of all, there really *is* a gap. The Michael line M is a GO-space example of a quasi-developable space that is not metrizable and we have

4.4. EXAMPLE. The lexicographically ordered space

$$M^* = (\mathbb{R} \times \{0\}) \cup (\mathbb{P} \times \mathbb{Z})$$

is a quasi-developable LOTS that is non-metrizable.

Second, we know exactly what must be added to the properties in 4.2 in order to produce metrizability. One answer comes from the proposition that a quasi-developable GO-space is metrizable if and only if it is perfect. Another is provided by the theorem that a GO-space is metrizable if and only if it is a quasi-developable p-space (in the sense of Arhangel'skii) (BENNETT [1968] [1971]).

The gap between quasi-developability of a GO-space X and the existence of a point-countable base for X is harder to study. Finding *any* ZFC example of a GO-space with a point-countable base and not a σ-point-finite base is complicated.

4.5. EXAMPLE (BENNETT [1971]). The "Big Bush" is a LOTS in ZFC that has a point-countable base but not a σ-disjoint base. One way to describe this space is to view it as a certain lexicographically ordered set. Let X be the set of all functions $f : [0, \omega_1[\to \mathbb{R}$ with the property that for some $\alpha_f < \omega_1$,

$$f(\beta) \in \mathbb{P} \text{ for each } \beta \in [0, \alpha_f[\quad \text{and}$$
$$f(\beta) = f(\alpha_f) \in \mathbb{Q} \text{ for each } \beta \in [\alpha_f, \omega_1[.$$

Endow X with the usual open interval topology of the lexicographic ordering. Then basic neighborhoods of a point $f \in X$ will have the form

$$B(f, n) = \{g \in X : \text{if } \beta < \alpha_f \text{ then } g(\beta) = f(\beta) \text{ and } |g(\alpha_f) - f(\alpha_f)| < 1/n\}.$$

As it happens, the Big Bush is not perfect. So far, as noted in Section 3, obtaining perfect examples of non-metrizable GO-spaces with point-countable bases requires additional set theoretic axioms. If one is willing to accept consistent examples, then one has

4.6. EXAMPLE (BENNETT [1968], PONOMAREV [1967]). If there is a Souslin space (i.e., a non-separable GO-space with countable cellularity) then there is a Souslin space with a point-countable base. Such a Souslin space example is also hereditarily Lindelöf.

As noted in Section 3, Examples 4.5 and 4.6 led Heath to ask whether there is a ZFC example of a perfect non-metrizable LOTS with a point-countable base.

Among GO-spaces, what exactly is the difference between having a point-countable base and the cluster of stronger properties listed in 4.2? In other words, exactly what must one add to the existence of a point-countable base in a GO-space in order to obtain a σ-disjoint base for the space? The answer is a strange property called "Property III" (BENNETT and LUTZER ([1996a], BENNETT, LUTZER and PURISCH [1999]).

4.7. DEFINITION. A topological space X has *Property III* if for each $n \geq 1$ there are subsets U_n and D_n of X satisfying:

(a) each U_n is open in X and D_n is a relatively closed subset of U_n;

(b) each D_n is discrete-in-itself;

(c) if p is a point of an open set G, then for some $n \geq 1$, $p \in U_n$ and $G \cap D_n \neq \emptyset$.

Property III is very weak when considered in the category of all topological spaces. Most generalized metric spaces have Property III (e.g. all semi-stratifiable spaces have Property III) and every topological space X is a closed subset of a topological space $B(X)$ that has Property III. Furthermore, if X is a "nice" space (e.g., regular, normal, has a point-countable base, etc.) then $B(X)$ is nice in the same way. See Section 6 of BENNETT and LUTZER [1996a] for details.

It is clear that if X is hereditarily Lindelöf and has Property III, then X must be separable. Consequently, none of the Souslin space examples in 4.6 can have Property III.

Property III turns out to be exactly what one must add to the existence of a point-countable base (in a GO-space) in order to get quasi-developability.

4.8. THEOREM (BENNETT and LUTZER [1996a, 1996b]). *Let X be a GO-space. Then the following assertions are equivalent:*

(a) *X is quasi-developable;*

(b) *X has a σ-disjoint base;*

(c) *X has a σ-point-finite base;*

(d) *X has a point-countable base and has Property III;*

(e) *X has a $\delta\theta$-base* (AULL [1974]) *and has Property III;*

(f) *X has a point-countable base and a σ-minimal base (see below in this section);*

(g) *X has a point-countable base and has a quasi-G_δ-diagonal* (BENNETT and LUTZER [1996a]).

Theorem 4.8 suggests that a deeper investigation of Property III in GO-spaces is warranted. As a start, we can show that Property III is hereditary among GO-spaces (BENNETT and LUTZER [1996a]) and that Property III implies hereditary paracompactness for a GO-space. It is easy to construct GO-spaces with Property III that are not first-countable. What is more surprising is that there are GO-spaces having Property III that are not first-countable at any point, a fact that shows how far Property III is from the types of properties considered in 4.1, 4.2, and 4.3.

4.9. EXAMPLE. (BENNETT and LUTZER [1996a]) There is a LOTS with Property III that is not first-countable at any point. Let X be the set of sequences $\langle x_i \rangle$ with $x_i \in [0, \omega_1]$ such that for some integer $n = n(\langle x_i \rangle)$ we have $x_i < \omega_1$ for each $i \leq n$ and $x_j = \omega_1$ for all $j > n$. Equip X with the usual open interval topology of the lexicographic order.

In recent years, an idea introduced by HEATH and LINDGREN [1976] has become the basis for understanding the fine structure of the gap between metrizability and quasi-developability in GO-spaces. Heath and Lindgren defined that a collection C is *weakly uniform* if given any $x \neq y$ in X, the collection $\{C \in \mathcal{C} : \{x, y\} \subseteq C\}$ is finite, and they studied topological spaces that have a *weakly uniform base* (WUB) for their topology. They showed that any space with a WUB must have a G_δ-diagonal and concluded:

4.10. PROPOSITION. *Any LOTS with a weakly uniform base must be metrizable.*

They also noted that the Michael line is an example of a non-metrizable GO-space with a WUB, so that 4.10 cannot be extended to be a metrization theorem for GO-spaces in general. However, there is an interesting structure theorem for GO-spaces that have weakly uniform bases, namely:

4.11. THEOREM (BENNETT and LUTZER [1998b]). *For any GO-space $(X, \tau, <)$, the following are equivalent:*

(a) *X has a WUB;*

(b) *X has a G_δ-diagonal and is quasi-developable.*

□ Outline of Proof: The proof of (a) ⇒ (b) that appears in BENNETT and LUTZER [1998b] can be simplified. Suppose that X has a WUB. HEATH and LINDGREN [1976] showed that X must have a G_δ-diagonal. According to a theorem of Przymusinski (see ALSTER [1975]) there is a metrizable topology $\mu \subseteq \tau$ such that $(X, \mu, <)$ is also a GO-space. Consequently there is a set $D = \bigcup\{D_n : n \geq 1\}$ such that each D_n is a closed discrete subset of (X, μ), and hence also of (X, τ), with the property that whenever $a < b$ and $]a, b[\neq \emptyset$, then $D \cap]a, b[\neq \emptyset$. Let $U_0 = \{p \in X : \{p\} \in \tau\}$ and for each $n \geq 1$ let $U_n = X$. The sets U_n and D_n for $n \geq 0$ witness the fact that X has Property III. It is easy to show that any GO-space with a WUB must have a point-countable base, and now the proof that (d) ⇒ (a) in 4.8 shows that X is quasi-developable. Conversely, suppose that the GO-space X has a G_δ-diagonal and is quasi-developable. According to 4.2, X has a σ-disjoint base $\mathcal{B} = \bigcup\{\mathcal{B}_n : n \geq 1\}$. Because X has a G_δ-diagonal, the weaker metric topology μ in Przymusinski's theorem cited above yields a sequence $\langle \mathcal{G}_n \rangle$ of point-finite open covers of X such that $\mathcal{G}_{n+1}$ refines $\mathcal{G}_n$ and $\bigcap\{St(p, \mathcal{G}_m) : n \geq 1\} = \{p\}$ for each $p \in X$. Let $\mathcal{H}_n = \{B \cap G : B \in \mathcal{B}_n, G \in \mathcal{G}_n\}$. Each $\mathcal{H}_n$ is point finite and

$\mathcal{H} = \bigcup\{\mathcal{H}_n : n \geq 1\}$ is a base for X. If p belongs to infinitely many distinct members H_k of $\mathcal{H}$, then point-finiteness of each $\mathcal{H}_n$ forces $H_k \in \mathcal{H}_{n_k}$ for infinitely many distinct n_k so that $\bigcap\{H_k : k \geq 1\} \subseteq \bigcap\{St(p, \mathcal{G}_{n_k} : k \geq 1\} = \{p\}$. Thus, $\mathcal{H}$ is a WUB for the space X. □

The class of GO-spaces with weakly uniform bases lies *strictly* between the class of quasi-developable GO-spaces and the class of metrizable GO-spaces. For example, the Michael line M is non-metrizable and has a weakly uniform base (because it is a quasi-developable GO-space with a G_δ-diagonal). To obtain an example of a quasi-developable GO-space that does not have a weakly uniform base, we use the LOTS extension of the Michael line.

4.12. EXAMPLE. The LOTS M^* described in Example 4.4 is a quasi-developable LOTS that has no weakly uniform base in the light of 4.11 because (being non-metrizable) it cannot have a G_δ-diagonal (see 5.1 (a')).

A recent paper by BALOGH, DAVIS, JUST, SHELAH and SZEPTYCKI [200?] introduced a property that is substantially weaker than having a WUB. A base $\mathcal{B}$ for a space X is a $<\omega$-WUB if, given any infinite set $S \subseteq X$ there is a finite set $F \subseteq S$ such that $\{B \in \mathcal{B} : F \subset B\}$ is finite. This generalizes the notion of an n-WUB, by which we mean a base $\mathcal{B}$ for X with the property that any set with n elements is contained in at most finitely many members of $\mathcal{B}$. Clearly, a weak uniform base is a 2-WUB, and any n-WUB for a space is a $<\omega$-WUB. Examples in BENNETT and LUTZER [1998b] show that none of these implications can be reversed. In addition, we have:

4.13. PROPOSITION. *Any GO-space with a* $<\omega$*-WUB is quasi-developable.*

We do not know how to characterize GO-spaces that have $<\omega$-WUBs.

4.14. QUESTION. For a GO-space X, find a topological property that solves the equation
X is quasi-developable + (?) if and only if X has a $<\omega$-WUB.

[Added in Proof: The property needed in (?) of (4.14) is: there is a sequence $\mathcal{G}_n$ of open covers of X such that for any infinite subset $S \subseteq X$, there is a finite set $F \subseteq S$, a point $p \in F$, and an integer n having $F \not\subseteq \mathrm{St}(p, \mathcal{G}_n)$. See BENNETT and LUTZER [200?b].]

Theorem 4.11 allows us to extend Theorem 4.1, expanding the cluster of generally distinct topological properties that are equivalent to metrizability in GO-spaces. BENNETT and LUTZER [1998b] shows:

4.15. THEOREM. *For any GO-space X, the following are equivalent:*

(a) *X is metrizable;*

(b) *X has a σ-locally countable base;*

(c) *X is developable;*

(d) *X is semi-stratifiable;*

(e) *X has an "open-in-finite" base, i.e., a base $\mathcal{B}$ with the property that $\{B \in \mathcal{B} : U \subseteq B\}$ is finite for every non-empty open set U;*

(f) *X has a sharp base in the sense of* ARHANGELSKII, JUST, REZNICHENKO *and* SZEPTYCKI [2000], *i.e., a base $\mathcal{B}$ with the property that if $\langle B_n \rangle$ is a sequence of distinct members of $\mathcal{B}$ each containing the point p, then the collection $\{\bigcap\{B_j : j \leq n\} : n \geq 1\}$ is a local base at p.*

□ The following argument is shorter than the proof of (4.15) given in BENNETT and LUTZER [1998b]. Every metrizable space satisfies both (e) and (f). To prove that (e) $\Rightarrow$ (a), suppose (e) holds. It is easy to check that X is first-countable. Therefore, X will be quasi-developable if X is the union of countably many quasi-developable subspaces. We may assume that members of the base $\mathcal{B}$ are convex. Let J be the set of all points of X that have either an immediate predecessor or an immediate successor in the given ordering of X. Let J_0 be the set of relatively isolated points of J. Let J_1 be the set of points of $J - J_0$ that have an immediate predecessor in X and let J_2 be the points of $J - J_0$ that have an immediate successor. Clearly J_0 is a quasi-developable subspace of X. One checks that $\{B \cap J_1 : B \in \mathcal{B}\}$ is a WUB for J_1 and concludes from 4.11 that J_1 is also quasi-developable. Analogously, so is J_2. Let $Y = X - J$ and verify that $\{B \cap Y : B \in \mathcal{B}\}$ is a WUB for Y. Because $X = J_0 \cup J_1 \cup J_2 \cup Y$, we see that X is quasi-developable. To complete the proof we show that X is perfect. According to Faber's theorem (see 3.2 (b), above), it is enough to show that every relatively discrete subspace D of X is σ-closed-discrete. Find a collection $\{U(d) : d \in D\}$ of pairwise disjoint open convex sets with $d \in U(d)$ for each $d \in D$. Each set $D_k = \{d \in D : U(d)$ is contained in at most k members of $\mathcal{B}\}$ is closed and discrete, and $D = \bigcup\{D_k : k \geq 1\}$, as required. Thus we have e) $\Rightarrow$ c) $\Rightarrow$ a).

To prove that f) $\Rightarrow$ a), show that any sharp base for X is weakly uniform and once again apply Faber's characterization of perfect GO-spaces to conclude that X is quasi-developable and perfect, whence metrizable. □

For many years, it appeared that one might be able to add another equivalent condition to the list in 4.2. Aull introduced the study of σ-minimal bases in AULL [1974]. A collection $\mathcal{C}$ of subsets of X is *minimal* or irreducible if each $C \in \mathcal{C}$ contains a point $x(C)$ that is not in any other member of $\mathcal{C}$, and a collection that is the union of countably many minimal collections is called σ-minimal. Clearly, any σ-disjoint base is σ-minimal. The first example showing that the converse is not true among GO-spaces appears in BENNETT and BERNEY [1977]:

4.16. EXAMPLE. The lexicographic square $X = [0, 1] \times [0, 1]$ is a compact, non-metrizable LOTS that has a σ-minimal base for its topology, but not a σ-disjoint base. Furthermore, its closed subspace $Y = [0, 1] \times \{0, 1\}$ does not have a σ-minimal base.

Some consequences of the existence of a σ-minimal base for a GO-space are known:

4.17. PROPOSITION (BENNETT and LUTZER [1977]). *Any GO-space with a σ-minimal base is hereditarily paracompact.*

The proof of 4.17 uses stationary set techniques, but is more complicated than usual. Suppose X is a GO-space with a σ-minimal base, and suppose X is not hereditarily paracompact. Then there is a stationary subset S of some uncountable regular cardinal that embeds in X (ENGELKING and LUTZER [1976]). The usual next step would be to say that

S inherits a σ-minimal base, but (as 4.16 shows) that might not be the case. Nevertheless, a more complicated proof still works.

Examples led researchers to suspect that either a GO-space with a σ-minimal base would be quasi-developable, or else that it would contain a certain pathological type of subspace (BENNETT and LUTZER [1990]). This led us to pose two questions:

(a) Is it true that a GO-space X must be quasi-developable provided every subspace of X has a σ-minimal base for its relative topology?

(b) Is it true that a compact LOTS X must be metrizable provided every subspace of X has a σ-minimal base for its relative topology?

Both questions have been answered recently. The first question was answered in BENNETT and LUTZER [1998a] when it was discovered that every subspace of a certain non-metrizable perfect space $E(Y, X)$ has a σ-minimal base for its relative topology. (See 5.5 for a description of $E(Y, X)$.) The second, and harder, question was answered negatively by SHI [1999a] who used a branch space of an Aronszajn tree to construct a non-metrizable compact LOTS X such that every subspace of X has a σ-minimal base for its relative topology.

5. Diagonal and off-diagonal conditions in GO-spaces

There are some striking parallels between the metrization theory for compact Hausdorff spaces and for LOTS. The most basic is:

5.1. THEOREM.

(a) *If X is a compact Hausdorff space having a G_δ-diagonal, then X is metrizable* (ŠNEIDER [1945]).

(a') *Any LOTS with a G_δ-diagonal is metrizable* (LUTZER [1971]).

(b) *A paracompact space that has a G_δ-diagonal and can be p-embedded in a compact Hausdorff space must be metrizable* (BORGES [1966] OKUYAMA [1964]).

(b') *A paracompact GO-space that has a G_δ-diagonal and can be p-embedded in a LOTS must be metrizable* (LUTZER [1971]).

The parallels in 5.1 are not accidental: see LUTZER [1972b] where the following is proved.

5.2. THEOREM. *Suppose X is a p-embedded subspace of a compact Hausdorff space or a p-embedded subspace of a LOTS. Then there is a sequence $\langle \mathcal{B}_n \rangle$ of open bases for X with the property that a collection $\mathcal{L}$ is a local base at a point $p \in X$ whenever*

a) $\bigcap \mathcal{L} = \{p\} = \bigcap \{\mathrm{cl}_X(L) : L \in \mathcal{L}\}$;

b) *$\mathcal{L}$ is a filter base;*

c) *the set $\{n : \mathcal{L} \cap \mathcal{B}_n \neq \emptyset\}$ is infinite.*

Furthermore, any completely regular space with such a sequence of bases and a G_δ-diagonal has a base of countable order so that a space is metrizable if and only if it has a sequence of bases as described above, has a G_δ-diagonal, and is paracompact.

One theme in ordered space research has been to explore how far the parallelism suggested by Theorem 5.1 extends, and that is the focus of this chapter.

One generalization of the notion of a G_δ-diagonal is Hušek's small diagonal property (HUŠEK [1976]). A space X has a *small diagonal* if, given any uncountable subset $T \subseteq X^2 - \Delta(X)$, there is an open set $U \subseteq X^2$ such that $\Delta(X) \subseteq U$ and $|T - U| > \omega$. Some of the metrization theory for spaces with small diagonals is known.

5.3. THEOREM. *Let CH denote the Continuum Hypothesis.*

(a) *Assume CH. Then any compact Hausdorff space with a small diagonal is metrizable* (JUHÁSZ and SZENTMIKLÓSSY [1992])*;*

(b) *In ZFC, any Lindelöf LOTS that has a small diagonal is metrizable* (*van Douwen and Lutzer, announced in* HUŠEK [1977] *and proved in* BENNETT *and* LUTZER [1997b]).

The symmetry suggested by 5.1 is broken, to some degree, in 5.3: if there were a strict parallelism between (a) and (b) in 5.3, then one would not need the Lindelöf hypothesis in 5.3 (b). There is no way to get around that problem because of the following example from BENNETT and LUTZER [1997a].

5.4. EXAMPLE. There is a LOTS with a small diagonal that is not paracompact (and hence not metrizable). The space in question is $S = \{\alpha < \omega_3 : \mathrm{cf}(\alpha) = \omega_2\}$. A result of PURISCH [1977] shows that there is some re-ordering of S under which S is a LOTS. Because S is a stationary subset of ω_3, S is not paracompact, and it is not hard to verify that S has a small diagonal. (This space is Example 6.2 in BENNETT and LUTZER [1997b].)

The history of metrization theory has shown that metrization theorems originally discovered for compact Hausdorff spaces can often be generalized, in the presence of paracompactness, to the progressively larger classes of locally compact Hausdorff spaces, Čech-complete spaces, and finally to the class of p-spaces introduced by Arhangel'skii. The results in Theorem 5.1 (a) and (b) are probably the best known examples. It is natural to wonder whether the same process of generalization would be possible for the Juhász-Szentmiklóssy result in 5.3 (a). The answer is "No" as is shown by:

5.5. EXAMPLE (BENNETT and LUTZER [1998a]). In ZFC there is a paracompact, perfect, Čech-complete LOTS that has a σ-closed-discrete dense subset, weight ω_1, and a small diagonal and yet is not metrizable.

□ The construction of the example in (5.5) begins with a remarkable metric space due to A.H. Stone in STONE [1963]. Stone's metric space is a certain subset $X \subseteq D^\omega$ where D is an uncountable discrete space of cardinality ω_1. The space X has the following properties:

(a) $|X| = \omega_1$;

(b) X is not the union of countably many relatively discrete subspaces;

(c) if S is any countable subset of X, then $\mathrm{cl}_X(S)$ is also countable.

Let Y be the closure of X in D^ω. Then Y is Čech-complete and has weight ω_1 because $w(D^\omega) = \omega_1$. A theorem of HERRLICH [1965] (see also Problem 6.3.2 in ENGELKING [1989]) shows that, with respect to some ordering, Y is a LOTS. Using that ordering, lexicographically order the set $E(Y, X) = (Y \times \{0\}) \cup (X \times \{0, 1\})$. In the open interval topology of the lexicographic ordering, $E(Y, X)$ is a LOTS with a σ-closed-discrete dense subset (whence $E(Y, X)$ is perfect, paracompact, and first-countable), and the natural projection from $E(Y, X)$ onto Y is a perfect mapping. Hence $E(Y, X)$ is Čech-complete. Because the construction of $E(Y, X)$ involves splitting ω_1 points in a LOTS whose weight is ω_1, we know that $w(E(Y, X)) = \omega_1$. The special properties of X yield that the space $E(Y, X)$ is non-metrizable and yet has a small diagonal. □

In addition to showing the limits on possible generalizations of the Juhász-Szentmiklóssy result in 5.3, the space $E(Y, X)$ also answers negatively a question posed by Arhangel'skii and Bella who proved in ARHANGEL'SKII and BELLA [1992] that, assuming CH, any Lindelöf p-space with weight ω_1 and a small diagonal must be metrizable. Then they asked whether, under CH or in ZFC, the Lindelöf hypothesis could be weakened to paracompactness. The space $E(Y, X)$ is a ZFC counterexample to that question. The space $E(Y, X)$ also provides a solution to an old problem about σ-minimal bases, as noted at the end of Section 4, above.

The following question is related to 5.3 (b) and remains open:

5.6. QUESTION. Suppose X is a Lindelöf GO-space with a small diagonal that can be p-embedded in some LOTS. Must X be metrizable?

In the light of Proposition 3.4 of BENNETT and LUTZER [1997a], to prove metrizability of an X with the properties in 5.6, it will be necessary and sufficient to prove that X is perfect.

As it happens, one can prove in ZFC that a countably compact GO-space with a small diagonal must be metrizable, and that contrasts with the situation in more general spaces. GRUENHAGE [200?] and PAVLOV [200?] have proved

5.7. PROPOSITION. *The assertion that any countably compact completely regular space with a small diagonal must be metrizable is consistent with, and independent of, ZFC+CH.*

The metrization results mentioned so far in this section all involve properties of the diagonal. The following theorem of GRUENHAGE [1984] focused attention on the off-diagonal subspace $X^2 - \Delta$ of X^2.

5.8. THEOREM. *A compact Hausdorff space X is metrizable if and only if X is paracompact off of the diagonal (i.e., $X^2 - \Delta$ is paracompact in its relative topology).*

Gruenhage's theorem was generalized in GRUENHAGE and PELANT [1988] to yield a G_δ-diagonal for members of the class of paracompact Σ-spaces that are paracompact off of the diagonal, and led to further investigations of off-diagonal properties by Kombarov and Stepanova who proved:

5.9. THEOREM (KOMBAROV [1989]). *A paracompact Σ-space X has a G_δ-diagonal if and only if there is a rectangular open cover of $X^2 - \Delta$ (i.e. a cover by sets of the form $G \times H$ where G and H are disjoint open subsets of X) that is locally finite in $X^2 - \Delta$.*

5.10. THEOREM (STEPANOVA [1993]). *A paracompact p-space X is metrizable if and only if there is a family of subsets of $X^2 - \Delta$ that is a σ-locally finite (in $X^2 - \Delta$) cover of $X^2 - \Delta$, where each member of the family is a co-zero set in X^2.*

Once again with (5.1) in mind, it is reasonable to ask whether there are off-diagonal metrization theorems for GO-space or LOTS that parallel the results by Gruenhage, Kombarov, and Stepanova, above. This time, the answer is "No," as shown by a single example.

5.11. EXAMPLE (BENNETT and LUTZER [1997b]). Let M^* be the LOTS extension of the Michael line described in 4.4 above. Then M^* is non-metrizable and first-countable, does not have a G_δ-diagonal, and has the properties that $X^2 - \Delta$ is paracompact and admits the kinds of rectangular and co-zero covers described in 5.9 and 5.10.

A subspace of M^* provides a consistent counterexample that answers a question of KOMBAROV [1989]. After proving 5.9 above, Kombarov asked whether a space X must have a G_δ-diagonal provided X is Lindelöf and regular, and $X^2 - \Delta$ has a countable cover by sets of the form $G \times H$, where G and H are disjoint open subsets of X. The next example provides a consistent negative answer.

5.12. EXAMPLE. Assume CH or $\mathbf{b} = \omega_1$. Then there is a Lindelöf LOTS X that does not have a G_δ-diagonal and yet admits a countable cover by sets of the form $G \times H$ where G and H are disjoint open subsets of X.

□ One uses CH or $\mathbf{b} = \omega_1$ to find an uncountable set L of real numbers that is concentrated on the set $\mathbb{Q}$ of rational numbers (i.e., if U is open in $\mathbb{R}$ and contains $\mathbb{Q}$, then $L - U$ is countable). Then the desired space X is the lexicographically ordered LOTS $(\mathbb{Q} \times \{0\}) \cup ((L - \mathbb{Q}) \times \mathbb{Z})$. □

Even though the off-diagonal conditions in 5.8, 5.9, and 5.10 do not yield metrizability or a G_δ-diagonal in GO-spaces, they do yield hereditary paracompactness. One shows that no stationary set in a regular uncountable cardinal can satisfy any of the off-diagonal conditions in 5.8, 5.9, or 5.10, and then a result in ENGELKING and LUTZER [1976] gives hereditary paracompactness in any GO-space with such off-diagonal properties. A stronger result of BALOGH and RUDIN [1992] gives hereditary paracompactness in any monotonically normal space satisfying any one of the off-diagonal conditions of Gruenhage, Kombarov, or Stepanova.

A different kind of off-diagonal property was introduced in STEPANOVA [1994]. She studied the role of a strong form of the Urysohn property in metrization theory. A space X is a *Urysohn space* if for each $(x, y) \in X^2 - \Delta$ there is a continuous, real-valued function $f_{x,y}$ such that $f_{x,y}(x) \neq f_{x,y}(y)$. If the correspondence $(x, y) \to f_{x,y}$ is continuous, where the range space $C_u(X)$ is the set of all continuous, real-valued functions on X with the topology of uniform convergence, then we say that X *has a continuous separating family*. Clearly any metric space (X, d) has a continuous separating family: define $f_{x,y}(z) = d(x, z)$. Stepanova proved:

5.13. THEOREM. *If X is a paracompact p-space, then X is metrizable if and only if X has a continuous separating family.*

The role of continuous separating families in the theory of GO-spaces is not yet clear. Using a stationary set argument, one can show that

5.14. PROPOSITION (BENNETT and LUTZER [2002]). *Any GO-space and any monotonically normal space that has a continuous separating family must be hereditarily paracompact.*

For separable GO-spaces, we understand the role of continuous separating families:

5.15. PROPOSITION. *For any separable GO-space X, and more generally for any GO-space X such that X^2 contains a dense subspace that is either Lindelöf or has countable cellularity, the following are equivalent:*

(a) *X has a G_δ-diagonal;*

(b) *X has a weaker metric topology;*

(c) *X has a continuous separating family.*

Therefore, a separable LOTS with a continuous separating family must be metrizable.

Without separability, little is known. It is easy to see that a LOTS can have a continuous separating family and yet fail to be first-countable. Consider the lexicographic product $([0, \omega_1[\times \mathbb{Z}) \cup \{(\omega_1, 0)\}$. But even if one restricts attention to first-countable GO-spaces, Stepanova's Theorem 5.13 above has no analog for GO-spaces, as can be seen from

5.16. EXAMPLE. (a) The LOTS M^* described in (4.4) above has a σ-disjoint base, is hereditarily paracompact, and has a continuous separating family, but is not metrizable. Under CH or $\mathbf{b} = \omega_1$, there is a Lindelöf LOTS that is not metrizable and yet has a continuous separating family (see 5.12 above).
(b) In ZFC there is a Lindelöf, non-metrizable LOTS that has a σ-disjoint base, is hereditarily paracompact, and has a continuous separating family.

□ For the example mentioned in (b), let $B \subseteq [0, 1]$ be a Bernstein set, i.e., a set such that for each uncountable compact set K, $K \cap B \neq \emptyset$ and $K - B \neq \emptyset$. Such sets exist in ZFC: see OXTOBY [1971]. Let $= (B \times \mathbb{Z}) \cup (C \times \{0\})$, where $C = [0, 1] - B$, and topologize X using the open interval topology of the lexicographic order. It is easy to verify that X is Lindelöf, and has a σ-disjoint base. Hence X is hereditarily paracompact. To see that X has a continuous separating family, suppose $((x, i), (y, j)) \in X^2 - \Delta$. If $x \neq y$ then define $f_{(x,i),(y,j)}(z, k) = |x - z|$. If $x = y$ then $i \neq j$ and $x \in B$ and we let $f_{(x,i),(y,j)}$ be the characteristic function of the set $\{(y, j)\}$. In either case, $f_{(x,i),(y,j)}$ is continuous and separates (x, i) and (y, j). Finally suppose that $((x_n, i_n), (y_n, j_n))$ is a sequence in $X^2 - \Delta$ that converges to $((x_0, i_0), (y_0, j_0)) \in X^2 - \Delta$. A case by case analysis, depending upon which (if any) of the points x_0, y_0 belong to the set B, shows that $\langle f_{(x_n,i_n),(y_n,j_n)} \rangle$ converges uniformly to $f_{(x_0,i_0),(y_0,j_0)}$. Thus, X has a continuous separating family. □

5.17. REMARK. Note that the spaces in 5.16 have uncountable cellularity. It is natural to ask whether separability in 5.15 could be replaced with countable cellularity, i.e., whether a LOTS must be metrizable if it has countable cellularity and has a continuous separating family. Gruenhage has shown that the answer is consistently "No" by showing that if there is a Souslin space, then there is a Souslin space with a continuous separating family. A proof will appear in BENNETT, LUTZER and RUDIN [200?].

A natural question is whether the existence of a continuous separating family in a first-countable GO-space yields special kinds of bases, e.g., a σ-disjoint base or a point-countable base. The Big Bush described in 4.5 provides the necessary counterexample (BENNETT and LUTZER [2002]).

5.18. EXAMPLE. The Big Bush has a continuous separating family and a point-countable base, but does not have a σ-disjoint base. An extension of the Big Bush described in BENNETT and LUTZER [1996b] is a first-countable LOTS that has a continuous separating family and does not have a point-countable base.

5.19. REMARK. Note that in a metric space (X, d), the continuous separating family given by $f_{x,y}(z) = d(x, z)$ really depends only on the parameter x. In a recent paper HALBEISEN and HUNGERBUHLER [200?] proved that a topological space X has a continuous separating family that depends on only one parameter if and only if the space X has a weaker metric topology, and they describe a paracompact space that has a continuous separating family but does not have a one-parameter continuous separating family. In the light of their characterization, the LOTS M^* in (4.4) is an easier example of a paracompact space that has a continuous separating family but does not admit a one-parameter continuous separating family.

The space M^* used above, is a LOTS built on the Michael line. If, instead, one begins with the Sorgenfrey line S, then one obtains the lexicographically ordered LOTS $S^* = \mathbb{R} \times \{n \in \mathbb{Z} : n \leq 0\}$ that is often a useful counterexample in GO-space theory. Whether or not S^* has a continuous separating family may be axiom-sensitive. We have:

5.20. PROPOSITION (BENNETT and LUTZER [2002]). *If there is an uncountable subspace T of the Sorgenfrey line S such that T^2 is a Lindelöf space, then S^* does not have a continuous separating family.*

When does the Sorgenfrey line have a subset T with the properties described in 5.20? MICHAEL [1971] constructed such a subset assuming CH, and BURKE and MOORE [1998] point out that such a T can exist in some models of MA plus not CH, but cannot exist given OCA or PFA. That leads to:

5.21. QUESTION. In ZFC, does S^* have a continuous separating family?

According to BENNETT and LUTZER [2002], a LOTS with a σ-closed-discrete dense subset and a continuous separating family must be metrizable, and the existence of a Souslin line yields a non-metrizable perfect LOTS with a continuous separating family. That raises a question that belongs in the Maurice-Heath-Nyikos family:

5.22. QUESTION. In ZFC, is there a non-metrizable perfect LOTS with a continuous separating family?

We emphasize that 5.22 is a question about LOTS and not a question about GO-spaces, as can be seen from the fact the Sorgenfrey line is a non-metrizable perfect GO-space that is separable and has a continuous separating family.

5.23. QUESTION. In ZFC, is there an example of a GO-space X that has a continuous separating family, but whose LOTS extension X^* does not? (The proof of 5.20 given in BENNETT and LUTZER [2002] shows that the answer is consistently negative.)

6. Dugundji extension theory

For any space X, let $C(X)$ [resp. $C^*(X)$] denote the vector space of continuous [resp. continuous and bounded] real-valued functions on X. As noted in VAN DOUWEN, LUTZER and PRZYMUSINSKI [1977], for any closed subset A of a normal space X, there is a linear function $\Phi : C(A) \to C(X)$ such that $\Phi(f)$ extends f for each $f \in C(A)$. [An analogous assertion holds for bounded functions.] Such a function Φ is called a *linear extender.*

In metric spaces, one can obtain linear extenders that are very well-behaved. DUGUNDJI [1951] proved:

6.1. THEOREM. *If A is a closed subset of a metric space X, then there is a linear extender $\Phi : C(A) \to C(X)$ such that the range of $\Phi(f)$ is contained in the convex hull of the range of f for each $f \in C(A)$.*

Later, BORGES [1966] extended this result to the much larger class of stratifiable spaces. Borrowing terminology from VAN DOUWEN [1975], we will say that the extender in 6.1 is a *ch-extender.* A weaker kind of extender is one for which the range of $\Phi(f)$ is always contained in the *closed* convex hull of the range of f, and such an extender is called a *cch-extender.*

Normal spaces, or even compact Hausdorff spaces, do not always admit cch-extenders (ARENS [1952], MICHAEL [1953]). However GO-spaces do, at least for bounded functions (HEATH and LUTZER [1974]):

6.2. PROPOSITION. *Suppose A is a closed subspace of a generalized ordered space X. Then there is a linear cch-extender from $C^*(A)$ to $C^*(X)$.*

If we consider unbounded functions, then 6.2 can fail.

6.3. EXAMPLE (HEATH and LUTZER [1974]). Let X be the Michael line and let A be the closed subset consisting of all rational numbers. Then there is no linear cch-extender from $C(A)$ to $C(X)$.

In the light of 6.3, HEATH and LUTZER [1974] asked:

6.4. QUESTION. *Suppose A is a closed subset of a perfect LOTS. Is there a linear cch-extender from $C(A)$ to $C(X)$?*

A few years later, VAN DOUWEN [1975] constructed a zero-dimensional separable GO-space having a closed subset that is not a retract and asked whether that space might be a counterexample to 6.4. Recently, GRUENHAGE, HATTORI and OHTA [1998] have proved that van Douwen's space answers 6.4 negatively. The next proposition is a special case of their Theorem 1. It settles questions of HEATH and LUTZER [1974] and of VAN DOUWEN [1975], and ties together several other results in HEATH, LUTZER and ZENOR [1975].

6.5. THEOREM. *Suppose X is a perfect GO-space and that the cardinality of X is non-measurable, and let A be a closed subspace of X. Then the following are equivalent:*

(a) *there is a continuous linear extender from $C(A)$ to $C(X)$ where both function spaces carry the topology of pointwise convergence or both carry the compact-open topology;*

(b) *there is a continuous linear extender from* $C^*(A)$ *to* $C^*(X)$ *where both function spaces carry the topology of pointwise convergence or both carry the compact-open topology;*

(c) *there is a linear cch-extender from* $C(A)$ *to* $C(X)$*;*

(d) *there is a linear ch-extender from* $C(A)$ *to* $C(X)$*;*

(e) *for each space* Y, $A \times Y$ *is* C^**-embedded in* $X \times Y$.

If, in addition, X *is zero-dimensional, then each of the above is equivalent to*

(f) A *is a retract of* X.

Because van Douwen's space is separable and zero-dimensional and has a closed subspace that is not a retract, 6.5 shows that van Douwen's space is a counterexample to 6.2. In addition, GRUENHAGE, HATTORI and OHTA [1998] gave an easier example, namely:

6.6. EXAMPLE. Let X be the lexicographically ordered set $(\mathbb{Q} \times \mathbb{Z}) \cup (\mathbb{P} \times \{-1, 1\})$. With the open interval topology of that order, X is a separable (and hence perfect) zero-dimensional LOTS and its closed subspace $A = \mathbb{P} \times \{-1, 1\}$ is not a retract of X. Hence there is no linear cch-extender from $C(A)$ to $C(X)$.

The space of Example 6.6 gives another answer to a question raised by BORGES [1966]. It is a perfectly paracompact space that does not satisfy the Dugundji extension theorem. (VAN DOUWEN [1975] gave an earlier answer using a different example.)

Finally, GRUENHAGE, HATTORI and OHTA [1998] sharpened the results of HEATH and LUTZER [1974] for perfect GO-spaces by proving:

6.7. PROPOSITION. *Let* A *be a closed* G_δ*-subset of a GO-space* X. *Then there is a linear ch-extender from* $C^*(A)$ *to* $C^*(X)$. *In particular, if* A *is a closed subset of a perfect GO-space* X*, then there is a linear ch-extender from* $C^*(A)$ *to* $C^*(X)$.

7. Rudin's solution of Nikiel's problem, with applications to Hahn-Mazurkiewicz theory

Several authors noticed that compact monotonically normal spaces had remarkable parallels to ordered spaces, and Nikiel asked whether every compact monotonically normal space must be a continuous image of a compact LOTS. Mary Ellen Rudin published three papers that contain the most important and complicated ordered space results in recent years (RUDIN [1998a] [1998b] [200?]):

7.1. THEOREM. *Any compact monotonically normal space is the continuous image of a compact LOTS.*

Rudin's theorem has important consequences for the Hahn-Mazurkiewicz problem that asks for characterizations of topological spaces that are continuous images of some connected compact LOTS. (Compact connected LOTS are often called *arcs*. It is easy to prove that the unit interval is the unique separable arc, and consequently modern Hahn-Mazurkiewicz theory focuses on non-separable arcs.) The most basic result in this area is the original Hahn-Mazurkiewicz theorem that characterized continuous images of *separable arcs* as follows:

7.2. THEOREM. *A topological space X is a continuous image of $[0,1]$ if and only if X is compact, connected, locally connected, and metrizable.*

The systematic study of images of non-separable arcs began with the work of Mardešić in the 1960s. Many theorems in this area begin with the hypothesis that a space X is the continuous image of some compact LOTS and add hypothesis that force X to be the continuous image of some connected, compact LOTS. Rudin's theorem puts such results into a more natural topological context. For example, combining Rudin's theorem with a result of Treybig and Nikiel gives:

7.3. THEOREM. *A space X is the continuous image of a compact connected LOTS if and only if X is compact, connected, locally connected, and monotonically normal.*

For further surveys of the Hahn-Mazurkiewicz problem, see the papers by TREYBIG and WARD [1981], MAYER and OVERSTEEGEN [1992], and NIKIEL, TUNCALI and TYMCHATYN [1993].

8. Applications to Banach spaces

An important problem in Banach space theory asks which Banach spaces have equivalent norms with special properties. For example, a norm $||\cdot||$ on a Banach space is *convex* if $||\frac{x+y}{2}|| < 1$ whenever $||x|| = 1 = ||y||$. A norm is *locally uniformly convex* (LUC) if whenever $||x|| = 1 = ||y_n||$ and $||x + y_n|| \to 2$, then $||x - y_n|| \to 0$, and is called a *Kadec norm* if the weak topology and the norm topology coincide on the norm's unit sphere. An often-studied type of question in Banach space theory is: does a given Banach space have an equivalent norm that is locally uniformly convex (LUC) or is a Kadec norm? It is known that the property of having an equivalent LUC norm is stronger than the property of having an equivalent Kadec norm, and that some Banach spaces have equivalent norms that are LUC or Kadec, while others do not.

Function spaces $C(X)$, where X is a compact Hausdorff space and $C(X)$ carries the sup norm, provide a wide variety of Banach spaces. When X is a compact LOTS, it is possible to study $C(X)$ in great detail, as recent results in JAYNE, NAMIOKA and ROGERS [1995] and HAYDON, JAYNE, NAMIOKA and ROGERS [2000] show. In this section, we present a sample of the results from the second of those papers.

8.1. THEOREM. *Let K be any compact LOTS. Then $C(K)$ has an equivalent Kadec norm and that norm is lower semi-continuous for the pointwise convergence topology on $C(K)$. Furthermore, the norm and pointwise convergence topologies coincide on the unit sphere of the Kadec norm.*

Haydon, Jayne, Namioka, and Rogers then asked for which compact LOTS K would $C(K)$ have an equivalent LUC norm, a property that (as mentioned above) is stronger than having an equivalent Kadec norm. That problem was solved using the ideas of a *dyadic interval system* and a *decreasing interval function* on K. Let J be the collection of all indices $(i_1, \cdots, i_n)$ where $n \geq 1$ and $i_j \in \{0, 1\}$, together with the empty set. A *dyadic interval system* on K is a function from J to the family of all non-empty closed intervals in K where $I(0)$ and $I(1)$ are disjoint closed subintervals of the interval $I(\emptyset)$ and where $I(i_1, \cdots, i_n, 0)$ and $I(i_1, \cdots, i_n, 1)$ are disjoint closed subintervals of $I(i_1, \cdots, i_n)$. By a *decreasing interval function* on K we mean a real valued function ρ defined for each closed non-empty interval in K and having the property that if $J \subseteq I$ are closed intervals, then $\rho(I) \leq \rho(J)$.

8.2. THEOREM. *The following properties of a compact LOTS K are equivalent:*

(a) *$C(K)$ has an equivalent LUC norm that is lower semi-continuous with respect to the pointwise convergence topology on $C(K)$;*

(b) *there is an equivalent strictly convex norm on $C(K)$;*

(c) *there is a bounded decreasing interval function on K that is not constant on any dyadic interval system of K.*

8.3. EXAMPLE. (a) Let α be an ordinal and let K be the lexicographically ordered product $\{0,1\}^\alpha$. Then $C(K)$ has an equivalent LUC norm if and only if α is countable. The same is true if we consider the lexicographic product $L = [0,1]^\alpha$.

(b) The lexicographic product $M = [0,1]^{\omega_1}$ has an equivalent Kadec norm, but not an equivalent LUC norm because it fails to satisfy (8.2-c). This example is considerably more simple than an earlier tree-based construction given by Haydon.

(c) If there is a Souslin space, then there is a compact, connected Souslin space N, and $C(N)$ does not have an equivalent LUC norm.

In HAYDON, JAYNE, NAMIOKA and ROGERS [2000] the authors show that any connected, compact LOTS L is the continuous image of some lexicographic product $[0,1]^\gamma$ under a continuous increasing mapping f (i.e., $x \leq y$ in $[0,1]^\gamma$ implies $f(x) \leq f(y)$ in L), where γ is an appropriately chosen ordinal. Then they prove:

8.4. THEOREM. *Suppose that the compact LOTS K is the continuous image of a closed subset of the lexicographic product $[0,1]^\gamma$, where γ is a countable ordinal. Then $C(K)$ has an equivalent LUC norm.*

It would be interesting to characterize those LOTS that satisfy the hypotheses of the previous theorem.

9. Products of GO-spaces

Between 1940 and 1970, simple GO-spaces proved their utility as counterexamples in product theory. Subspaces of ordinals, the Sorgenfrey line, and the Michael line became standard examples in the product theory of normality, the Lindelöf property, and paracompactness. MICHAEL [1971] showed that subspaces of the Sorgenfrey line and the Michael line can be finely tuned to generate a wide range of important examples.

That same period also saw the discovery of a positive theory for products of certain GO-spaces. Let S be the Sorgenfrey line. HEATH and MICHAEL [1971] showed that S^ω is a perfect space (i.e., closed subsets are G_δ-sets) and LUTZER [1972a] showed that S^ω is hereditarily subparacompact. VAN DOUWEN and PFEFFER [1979] showed that S^n cannot be homeomorphic to T^m for any $m, n \geq 1$, where T is the subspace of S consisting of all irrational numbers, and BURKE and LUTZER [1987] proved that S^n is homeomorphic to S^m if and only if $n = m$. That result is sharpened by BURKE and MOORE [1998] who showed that if X is an uncountable subspace of S, then no power of X can be embedded in a lower power of S. They also characterized subspaces of S that are homeomorphic to S as being those uncountable $X \subseteq S$ that are dense-in-themselves and are both an F_σ-subset and a G_δ-subset of S.

ALSTER [1975] considered the broader class of GO-spaces with G_δ-diagonals and proved, for example, that (hereditary) collectionwise normality is equivalent to (hereditary) paracompactness in finite products of GO-spaces having G_δ-diagonals. He also proved that the Continuum Hypothesis is equivalent to the assertion that $X_1 \times X_2$ is hereditarily subparacompact whenever X_1 and X_2 are Lindelöf GO-spaces with G_δ-diagonals.

More recent investigations have focussed on products of ordinal spaces and their subspaces. (By an *ordinal space* we mean a space $[0, \alpha)$ (where α is an ordinal number) with its usual order topology. Throughout this section, A and B will denote subspaces of an ordinal space.) CONOVER [1972] gave necessary and sufficient conditions for normality of the product of two ordinal spaces. Later KEMOTO and YAJIMA [1992] extended earlier work of SCOTT [1975], showing:

9.1. THEOREM. *Let A and B be subsets of ordinal spaces. Then $A \times B$ is normal if and only if $A \times B$ is orthocompact.*

It is interesting to note that Theorem 9.1 does *not* hold for subspaces of $A \times B$ even when $A = B = [0, \omega_1)$, KEMOTO [1997]. In SCOTT [1977], Scott extended the "normality = orthocompactness" theorem in a different direction, proving it for any finite product of locally compact LOTS.

Some interesting equivalences among normality-related properties of $A \times B$ have been found. Results of KEMOTO, OHTA and TAMANO [1992] and KEMOTO, NOGURA, SMITH and YAJIMA [1996] have been generalized by FLEISSNER [200?a] who proved:

9.2. THEOREM. *Let X be a subspace of the product of finitely many ordinals. The following are equivalent:*

(a) *X is normal;*

(b) *X is normal and strongly zero-dimensional;*

(c) *X is collectionwise normal;*

(d) *every open cover $\mathcal{U}$ of X has an open refinement $\{V(U) : U \in \mathcal{U}\}$ that covers X and has the property that $\mathrm{cl}_X(V(U)) \subseteq U$ for each $U \in \mathcal{U}$.*

The hypothesis of strong zero-dimensionality in 9.2 (b) is not automatic. FLEISSNER, KEMOTO and TERASAWA [200?] prove that $\mathbf{c} = 2^\omega$ is the least cardinal such that $X = [0, \omega] \times [0, \mathbf{c})$ contains a subspace that is not strongly zero-dimensional; in fact X contains a strongly n-dimensional subspace (i.e., a subspace with covering dimension n) for each finite n.

Countable paracompactness is a covering property of every GO-space. In products $A \times B$ of subspaces of ordinal spaces, countable paracompactness is known to be equivalent to the property that for every locally finite closed collection $\mathcal{F}$ in $A \times B$, there is a locally finite open collection $\{U(F) : F \in \mathcal{F}\}$ with $F \subseteq U(F)$ for each $F \in \mathcal{F}$. (See KEMOTO, OHTA and TAMANO [1992].) If $A \times B$ is normal, then it is countably paracompact, but $A \times B$ can be countably paracompact without being normal (the classic example being $[0,\omega_1) \times [0,\omega_1])$. It is an open question whether every countably paracompact subspace of $[0,\omega_1)^2$ is normal. KEMOTO, SMITH and SZEPTYCKI [2000] show that the answer is consistently "yes" but the question remains open in ZFC. One of the few covering properties shared by all products $A \times B$ is hereditary countable metacompactness (KEMOTO and SMITH [1996], FLEISSNER [200?b]).

Some of the equivalences among covering properties in GO-spaces still hold in products $A \times B$ of subspaces of ordinal spaces. For example, combining results of KEMOTO and YAJIMA [1992] with work of FLEISSNER and STANLEY [2001] yields:

9.3. THEOREM. *Let $X = A \times B$ where A and B are subspaces of ordinal spaces. Then the following are equivalent:*

(a) *X is paracompact;*

(b) *X is metacompact;*

(c) *X is subparacompact;*

(d) *X is a D-space, i.e., whenever we have open sets U_x satisfying $x \in U_x$ for each $x \in X$, there is a closed discrete set $D \subseteq X$ such that $\{U_x : x \in D\}$ covers X;*

(e) *no closed subspace of X is homeomorphic to a stationary subspace of an uncountable regular cardinal.*

By way of contrast, metacompactness, paracompactness, and subparacompactness are not equivalent for subspaces of $A \times B$. For example, while metacompact subspaces of $[0,\omega_1)^2$ must be paracompact, there are metacompact subspaces of $[0,\omega_2)^2$ that are not even subparacompact (KEMOTO, TAMANO and YAJIMA [2000]).

In FLEISSNER and STANLEY [2001] Stanley extended earlier work in KEMOTO and YAJIMA [1992], proving:

9.4. THEOREM. *Let X be any subspace of a product of finitely many ordinal spaces. Then the following are equivalent:*

(a) *X is metacompact;*

(b) *X is metaLindelöf;*

(c) *X is a D-space (see 9.3 d);*

(d) *no closed subspace of X is homeomorphic to a stationary subset of a regular uncountable cardinal.*

There is a marked difference between finite and countable products of ordinal spaces. In a recent paper, KEMOTO and SMITH [1997] have shown that the product space $([0,\omega_1))^\omega$ has a subspace that is not countably metacompact, even though every finite power of $[0,\omega_1)$ is hereditarily countably metacompact.

References

ALSTER, K.
[1975] Subparacompactness in Cartesian products of generalized ordered spaces, *Fundamenta Math.* **87**, 7-28.

ARENS, R.
[1952] Extension of functions on fully normal spaces, *Pacific J. Math.* **2**, 11–22.

ARHANGEL'SKII, A. and A. BELLA
[1992] Few observations on topological spaces with small diagonal, *Zbornik radova Filozofkog fakulteta u Nisu* **6** (2), 211–213.

ARHANGEL'SKII, A., JUST, W., REZNICHENKO, E. and P. SZEPTYCKI
[2000] Sharp bases and weakly uniform bases versus point-countable bases, *Topology Appl.* **100**, 39–46.

ARTICO, G., U. MARCONI, J. PELANT, L. ROTTER AND M. TKACHENKO
[200?] Selections and suborderability, preprint.

AULL, C.
[1974] Quasi-developments and $\delta\theta$-bases, *J. London Math. Soc. (2)*, **9**, 192–204.

BALOGH, Z., S. DAVIS, W. JUST, S. SHELAH AND P. SZEPTYCKI
[200?] Strongly almost disjoint sets and weakly uniform bases, to appear.

BALOGH, Z. and M.E. RUDIN
[1992] Monotone normality, *Topology Appl.* **47**, 115–127.

BENNETT, H.
[1968] *On quasi-developable spaces*, Ph.D. dissertation, Arizona State University.
[1971] Point-countability in ordered spaces, *Proc. Amer. Math. Soc.* **28**, 598–606.

BENNETT, H. AND E. BERNEY
[1977] Spaces with σ-minimal bases, *Topology Proc.* **2**, 1–10.

BENNETT, H., R. HEATH AND D. LUTZER
[2000] GO-spaces with σ-closed-discrete dense subspaces, *Proc. Amer. Math. Soc.* **129**, 931–939.

BENNETT, H., M. HOSOBUCHI and D. LUTZER
[1999] A note on perfect generalized ordered spaces, *Rocky Mountain J. Math.* **9**, 1195–1207.
[2000] Weakly perfect generalized ordered spaces, *Houston J. Math.* **26**, 609–627.

BENNETT, H. and D. LUTZER
[1977] Ordered spaces with σ-minimal bases, *Topology Proc.* **2**, 371–382.
[1984] Generalized ordered spaces with capacities, *Pacific J. Math.* **122**, 11–19.
[1990] Problems in perfect ordered spaces, In *Open Problems in Topology*, ed. by J. van Mill and G.M. Reed, North Holland, Amsterdam, pp. 233–237.
[1996a] Point-countability in generalized ordered spaces, *Topology Appl.* **71**, 149–165.
[1996b] A note on property III in generalized ordered spaces, *Topology Proc.* **21**, 15–24.
[1997a] Diagonal conditions in ordered spaces, *Fundamenta Math.* **153**, 99–123.
[1997b] Off diagonal metrization theorems, *Topology Proc.* **22**, 37–58.
[1998a] A metric space of A.H. Stone and an example concerning σ-minimal bases, *Proc. Amer. Math. Soc.* **126**, 2191–2196.
[1998b] Ordered spaces with special bases, *Fundamenta Math.* **158**, 289–299.
[2002] Continuous separating families in ordered spaces and strong base conditions, *Topology Appl.* **119**, 305–314.

[200?a] Metrizably fibered generalized ordered spaces, to appear.
[200?b] Spaces with $< \omega$-weakly uniform bases, to appear.

BENNETT, H., D. LUTZER and S. PURISCH
[1999] On dense subspaces of generalized ordered spaces, *Topology Appl.* **93**, 191–205.

BENNETT, H., D. LUTZER and M.E. RUDIN
[200?] Trees, lines, and branch spaces, to appear.

BORGES, C.
[1966] On stratifiable spaces, *Pacific J. Math.* **17**, 1–25.

BURKE, D. AND D. LUTZER
[1987] On powers of certain lines, *Topology Appl.* **26**, 251–261.

BURKE, D. and J. MOORE
[1998] Subspaces of the Sorgenfrey line, *Topology Appl.* **90**, 57–68.

CONOVER, R.
[1972] Normality and products of linearly ordered spaces, *Gen. Top. Appl.* **2**, 215–25.

VAN DALEN, J. and E. WATTEL
[1973] A topological characterization of ordered spaces, *Gen. Top. Appl.* **3**, 347–354.

VAN DOUWEN, E.
[1975] *Simultaneous Extensions of Continuous Functions*, Ph.D. Thesis, Vrije Universiteit, Amsterdam.

VAN DOUWEN, E., D. LUTZER and T. PRZYMUSINSKI
[1977] Some extensions of the Tietze-Urysohn theorem, *Amer. Math. Monthly* **84**, 435–441.

VAN DOUWEN, E. and W. PFEFFER
[1979] Some properties of the Sorgenfrey line and related spaces, *Pacific J. Math.* **81**, 371–377.

DUGUNDJI, J.
[1951] An extension of Tietze's theorem, *Pacific J. Math.* **1**, 353–367.

ENGELKING, R.
[1989] *General Topology*, Heldermann, Berlin.

ENGELKING, R. and D. LUTZER
[1976] Paracompactness in ordered spaces, *Fundamenta Math.* **94**, 49–58.

FABER, M.J.
[1974] *Metrizability in generalized ordered spaces*, Math. Centre Tracts no. 53, Amsterdam.

FLEISSNER, W.
[200?a] Normal subspaces of products of finitely many ordinals, to appear.
[200?b] Metacompact subspaces of products of ordinals, *Proc. Amer. Math. Soc.*, to appear.

FLEISSNER, W., N. KEMOTO and J. TERASAWA
[200?] Strong zero dimensionality of products of ordinals, to appear.

FLEISSNER, W. and A. STANLEY
[2001] D-spaces, *Topology Appl.* **114**, 261–271.

FUJII, S. and T. NOGURA
[1999] Characterizations of compact ordinal spaces via continuous selections, *Topology Appl.* **91**, 65–69.

GRUENHAGE, G.
[1984] Covering properties of $X^2 - \Delta$ and compact subsets of Σ-products, *Topology Appl.* **28**, 287–304.
[1992] A note on the point-countable base question, *Topology Appl.* **44**, 157–162.
[200?] Spaces having a small diagonal, to appear.

GRUENHAGE, G., Y. HATTORI and H. OHTA

[1998] Dugundji extenders and retracts on generalized ordered spaces, *Fundamenta Math.* **158**, 147–164.

GRUENHAGE, G. AND D. LUTZER

[2000] Baire and Volterra spaces, *Proc. Amer. Math. Soc.* **128**, 3115–3124.

GRUENHAGE, G., T. NOGURA and S. PURISCH

[1991] Normality of $X \times \omega_1$, *Topology Appl.* **39**, 263–275.

GRUENHAGE, G. and J. PELANT

[1988] Analytic spaces and paracompactness of $X^2 - \Delta$, *Topology Appl.* **28**, 11–18.

HALBEISEN, L. and N. HUNGERBUHLER

[200?] On continuously Urysohn and strongly separating spaces, *Topology Appl.*, to appear.

HAYDON, R., J. JAYNE, I. NAMIOKA and C. ROGERS

[2000] Continuous functions on totally ordered spaces that are compact in their order topologies, *J. Funct. Anal.* **178**, 23–63.

HEATH, R.

[1989] On a question of Ljubiša Kočinac, *Publ. Inst. Math. (Beograd) (N.S.)* **46**, 193–195.

HEATH, R. and W. LINDGREN

[1976] Weakly uniform bases, *Houston J. Math.* **2**, 85–90.

HEATH, R. and D. LUTZER

[1974] Dugundji extension theorems for linearly ordered spaces, *Pacific J. Math.* **55**, 419–425.

HEATH, R., D. LUTZER and P. ZENOR

[1973] Monotonically normal spaces, *Trans. Amer. Math. Soc.* **178**, 481–493.

[1975] On continuous extenders, In *Studies in Topology*, ed. by N. Stravrakas and K. Allen, Academic Press, New York, pp. 203–214.

HEATH, R. and E. MICHAEL

[1971] A property of the Sorgenfrey line, *Compositio Math.* **23**, 185–188.

HERRLICH, H.

[1965] Ordnungsfähigkeit total-discontinuierlicher Räume, *Math. Ann.* **159**, 77–80.

HUŠEK, M.

[1976] Continuous mappings on subspaces of products, in *Sympos. Math.* **17**, Academic Press, London, pp. 25–41.

[1977] Topological spaces without κ-accessible diagonals, *Comment. Math. Univ. Carolin.* **18**, 777–788.

JAYNE, J., I. NAMIOKA and C. ROGERS

[1995] Continuous functions on compact totally ordered spaces, *J. Funct. Anal.* **134**, 261–280.

JUHÁSZ, I. and Z. SZENTMIKLÓSSY

[1992] Convergent free sequences in compact spaces, *Proc. Amer. Math. Soc.* **116**, 1153–1160.

KEMOTO, N.

[1993] Normality in products of GO-spaces and cardinals, *Topology Proc.* **18**, 133–142.

[1997] Orthocompact subspaces in products of two ordinals, *Topology Proc.* **22**, 247–263.

KEMOTO, N., T. NOGURA, K. SMITH and Y. YAJIMA

[1996] Normal subspaces in products of two ordinals, *Fundamenta Math.* **151**, 279–297.

KEMOTO, N., H. OHTA and K. TAMANO

[1992] Products of spaces of ordinal numbers, *Topology Appl.* **45**, 245–260.

KEMOTO, N. and K. SMITH

[1996] The product of two ordinals is hereditarily countably metacompact, *Topology Appl.* **74**, 91–96.

[1997] Hereditary countable metacompactness in finite and infinite product spaces of ordinals, *Topology Appl.* **77**, 57–63.

KEMOTO, N., K. SMITH and P. SZEPTYCKI

[2000] Countable paracompactness versus normality in subspaces of ω_1^2, *Topology Appl.* **104**, 141–154.

KEMOTO, N., K. TAMANO and Y. YAJIMA

[2000] Generalized paracompactness of subspaces in products of two ordinals, *Topology Appl.* **104**, 155–168.

KEMOTO, N, and Y. YAJIMA

[1992] Orthocompactness in products, *Tsukuba J. Math.* **16**, 407–422.

KOČINAC, L.

[1983] An example of a new class of spaces, *Mat. Vestnik* **35**, 145–150.

[1986] Some generalizations of perfect normality, *Facta. Univ. Ser. Math. Infor.* **1**, 57–63.

KOMBAROV, A.

[1989] On rectangular covers of $X^2 - \Delta$, *Comment. Math. Univ. Carolinae* **30**, 81–83.

LUTZER, D.

[1971] *On generalized ordered spaces*, Dissertationes Math **89**.

[1972a] Another property of the Sorgenfrey line, *Compositio Math.* **24**, 359–363.

[1972b] On quasi-uniform bases, In *Proc. Univ. of Oklahoma Topology Conference 1972*, ed. by D. Kay, J. Green, L. Rubin, and L. Su, University of Oklahoma, Norman, OK, 104–117.

[1980] Ordered topological spaces, In *Surveys in General Topology*, ed. by G.M. Reed, Academic Press, New York, pp.247–295.

MAYER, J. and L. OVERSTEEGEN

[1992] Continuum Theory, In *Recent Progress in General Topology* ed. by Hušek, M. and J. van Mill, North Holland, Amsterdam, pp. 247–296.

MICHAEL, E.

[1951] Topologies on spaces of subsets, *Trans. Amer. Math. Soc.* **71**, 152–182.

[1953] Some extension theorems for continuous functions, *Pacific J. Math.* **3**, 789–806.

[1971] Paracompactness and the Lindelöf property in finite and countable Cartesian products, *Compositio Math.* **23**, 199–214.

VAN MILL, J. and E. WATTEL

[1984] Orderability from selections: Another solution of the orderability problem, *Fundamenta Math.* **121**, 219–229.

MILLER, A.

[1984] Special subsets of the real line, In *Handbook of Set Theoretic Topology* ed. by K. Kunen and J. Vaughan, Elsevier, New York, pp. 203–223.

MIWA, T., and N. KEMOTO

[1993] Linearly ordered extensions of GO–spaces, *Topology Appl.* **54**, 133–140.

NIKIEL, J., H. TUNCALI and E. TYMCHATYN

[1993] Continuous images of arcs and inverse limit methods, *Memoires Amer. Math. Soc.* **104**, no. 198.

NYIKOS, P.

[1976] A survey of zero-dimensional spaces, In *Topology: Proceedings of the Memphis State University Conference*, ed. by S. Franklin and B. Thomas, Marcel Dekker, New York, pp. 87–114.

OKUYAMA, A.

[1964] On metrizability of M-spaces, *Proc. Japan Acad.* **40**, 176–179.

OXTOBY, J.

[1971] *Measure and Category*, Springer-Verlag, New York.

PAVLOV, O.

[200?] There is a perfect preimage of ω_1 with a small diagonal, preprint.

PONOMAREV, V.

[1967] Metrizability of a finally compact p-space with a point-countable base, *Soviet Math. Doklady* **8**, 765–768.

PURISCH, S.

[1977] The orderability and suborderability of metrizable spaces *Trans. Amer. Math. Soc.* **226**, 59–76.

[1983] Orderability of non-archimedean spaces, *Topology Appl.* **16**, 273–277.

[1998] Orderability of topological spaces, Topology Atlas Invited Contributions, http://at.yorku.ca/z/a/a/b/14.htm

QIAO, Y-Q. and F. TALL

[200?] Perfectly normal non-metrizable non-archimedean spaces are generalized Souslin lines, *Proc. Amer. Math. Soc.*, to appear.

REED, G.M.

[1971] Concerning normality, metrizabilty, and the Souslin property in subspaces of Moore spaces, *Gen. Top. Appl.* **1**, 223–245.

RUDIN, M.E.

[1998a] Compact, separable linearly ordered spaces, *Topology Appl.* **82**, 397–419.

[1998b] Zero-dimensionality and monotone normality, *Topology Appl.* **85**, 319–333.

[200?] Nikiel's conjecture, to appear.

SCOTT, B.

[1975] Toward a product theory for orthocompactness, in *Studies in Topology*, ed. by N. Stavrakas and K. Allen, Academic Press, New York, pp. 517–537.

[1977] Orthocompactness and normality in finite products of locally compact LOTS, in *Set Theoretic Topology* ed. by G. Reed, Academic Press, New York, pp. 339–348.

SHI, W.

[1997] Perfect GO-spaces which have a perfect linearly ordered extension, *Topology Appl.* **81**, 23–33.

[1999a] A non-metrizable compact LOTS each subspace of which has a σ-minimal base, *Proc. Amer. Math. Soc.* **127**, 2783–2791.

[1999b] Extensions of perfect GO-spaces with σ-discrete dense sets, *Proc. Amer. Math. Soc.* **127**, 615–618.

SHI, W., T. MIWA and Y. GAO

[1995] A perfect GO-space which cannot densely embed in any perfect orderable space, *Topology Appl.* **66**, 241–249.

[1996] Any perfect GO-space with the underlying LOTS satisfying local perfectness can embed in a perfect LOTS, *Topology Appl.* **74**, 17–24.

ŠNEIDER, V.

[1945] Continuous mappings of Souslin and Borel sets: metrization theorems, *Dokl. Akad. Nauk SSSR* **50**, 77–79.

STEPANOVA, E.

[1993] Extension of continuous functions and metrizability of paracompact p-spaces, *Mathematical Notes* **53**, 308–314.

[1994] On metrizability of paracompact p-spaces, *Moscow University Mathematics Bulletin* **49**, 41–43.

STONE, A.

[1963] On σ-discreteness and Borel isomorphism, *Amer. J. Math.* **85**, 655–666.

TKACHUK, V.

[1994] A glance at compact spaces which map "nicely" onto the metrizable ones, *Topology Proc.* **19**, 321–334.

TODORČEVIĆ, S.

[1984] Trees and linearly ordered sets, in *Handbook of Set Theoretic Topology*, ed. by K. Kunen and J. Vaughan, North-Holland, Amsterdam, pp. 235–293.

TREYBIG, L. and L. WARD

[1981] The Hahn-Mazurkiewicz problem, in *Topology and Order Structures, Part 1*, ed. by H. Bennett and D. Lutzer, Mathematical Centre Tracts 142, Amsterdam, pp. 95–106.

VAN WOUWE, J.

[1979] *GO-spaces and generalizations of metrizability*, Mathematical Centre Tracts 104, Amsterdam.

CHAPTER 4

Infinite-Dimensional Topology

Jan J. Dijkstra

Divisie der Wiskunde en Informatica, Vrije Universiteit, De Boelelaan 1081a, 1081 HV Amsterdam, The Netherlands

E-mail: dijkstra@cs.vu.nl

Jan van Mill

Divisie der Wiskunde en Informatica, Vrije Universiteit, De Boelelaan 1081a, 1081 HV Amsterdam, The Netherlands

E-mail: vanmill@cs.vu.nl

Contents

RECENT PROGRESS IN GENERAL TOPOLOGY II
Edited by Miroslav Hušek and Jan van Mill

1. Introduction

The aim of this note is to present a survey of the main developments in infinite-dimensional manifold theory that have occurred since we wrote DIJKSTRA and VAN MILL [1992]. Our focus is on topological vector spaces, function spaces, homotopy dense imbeddings, topological classification of semicontinuous functions and hyperspaces.

Infinite-dimensional topology is the creation of R. D. Anderson (see ANDERSON [2002] for some remarks on the early development of infinite-dimensional topology). Several books were written on the subject, or deal with aspects of infinite-dimensional topology. The highlights of infinite-dimensional topology are the theorems of ANDERSON [1966] on the homeomorphy of ℓ^2 and s, of CHAPMAN [1974] on the invariance of Whitehead torsion, of WEST [1977] on the finiteness of homotopy types of compact ANR's and of TORUŃCZYK [1980, 1981] on the topological characterization of manifolds modelled on the Hilbert cube and Hilbert space. A large collection of open problems is WEST's paper [1990]. The subjects that are being touched upon there range from absorbing sets and function spaces to ANR-theory.

2. Definitions and basic theory

We recall the basic ideas that play an important role in infinite-dimensional topology.

A subset A of a space X is called *homotopy dense in* X if there exists a homotopy $H\colon X \times \mathbb{I} \to X$ such that H_0 is the identity and $H(X \times (0,1]) \subset A$. A closed subset F of a space X is called a *Z-set* if $X \setminus F$ is homotopy dense in X. A closed subset F of an ANR X is called a *strong Z-set* if for each open cover $\mathcal{U}$ of X there exists a continuous function $f\colon X \to X$ that is $\mathcal{U}$-close to the identity such that $\mathrm{cl}_X(f(X])) \cap F = \emptyset$. A countable union of (strong) Z-sets is called a *(strong)* σ*Z-set*. A space X that can be written as $X = \bigcup_{i=1}^{\infty} X_i$, where each X_i is a (strong) Z-set in X, is called a *(strong)* σ*Z-space*. An imbedding $f\colon X \to Y$ is called a *Z-imbedding* if $f[X]$ is a Z-set in Y.

It is clear that a Z-set is nowhere dense. It is tempting to think that a 'nice' space, e.g., a vector space, which is meager in itself is in fact a σZ-space. If this were true then some proofs in infinite-dimensional topology would be simpler. However, it is *not* true, as was shown by BANAKH [1999]. His example is the linear span in ℓ^2 of ERDŐS' space from [1940]. It is even absolutely Borel. See also BANAKH, RADUL and ZARICHNYI [1996, Theorem 5.5.19] for details.

We will now recall the definition of an absorber after BESTVINA and MOGILSKI [1986]. Let $\mathcal{C}$ be a topological class that is closed hereditary. In addition, assume that $\mathcal{C}$ is additive: $A \in \mathcal{C}$ whenever A can be written as a union of two closed subsets that are in $\mathcal{C}$. Important examples of such classes are $\mathcal{M}_\alpha$ and $\mathcal{A}_\alpha$, the multiplicative respectively the additive Borel class of level α, $\alpha < \omega_1$. Let $\mathcal{C}_\sigma$ denote the class of spaces that have a countable closed covering consisting of spaces from $\mathcal{C}$. An AR X is called $\mathcal{C}$*-universal* if for every $A \in \mathcal{C}$ there exists a closed imbedding $g\colon A \to X$. An AR X is called *strongly* $\mathcal{C}$*-universal* if for every $A \in \mathcal{C}$ and every map $f\colon A \to X$ that restricts to a Z-imbedding on a closed set $K \subset A$ there exists a Z-imbedding $g\colon A \to X$ that can be chosen arbitrarily close to f with the property $g|K = f|K$. The AR X is called a $\mathcal{C}$*-absorber* if

1. X is a strong σZ-space,
2. $X \in \mathcal{C}_\sigma$,
3. X is strongly $\mathcal{C}$-universal.

Let us call a $\mathcal{C}$-absorber X a *standard $\mathcal{C}$-absorber* if X is a homotopy dense subspace of Hilbert space.

Bestvina and Mogilski proved the Uniqueness Theorem for absorbers:

2.1. THEOREM. *If there exists a standard $\mathcal{C}$-absorber and the spaces X and Y are both $\mathcal{C}$-absorbers then X and Y are homeomorphic.*

If we combine this theorem with Theorem 5.2 of BANAKH [1998] then we get an improved Uniqueness Theorem:

2.2. THEOREM. *If the spaces X and Y are both $\mathcal{C}$-absorbers then X and Y are homeomorphic.*

Bestvina and Mogilski also show that there exists a standard absorber for every Borel class. Let us denote by Ω_α the standard $\mathcal{M}_\alpha$-absorber and by Λ_α the standard $\mathcal{A}_\alpha$-absorber.

The Hilbert cube Q is the product space $\mathbb{I}^\omega$, where $\mathbb{I} = [0,1]$. The pseudointerior and pseudoboundary of Q are the subspaces $s = (0,1)^\omega$ and $B = Q \setminus s$, respectively. According to ANDERSON [1966] s is homeomorphic to the separable Hilbert space ℓ^2. The space B is an important example of an $\mathcal{A}_1$-absorber and B^ω is an example of an $\mathcal{M}_2$-absorber. So B and B^ω are homeomorphic to Λ_1 respectively Ω_2.

Absorbers are generalizations of so-called capsets, which were introduced independently by ANDERSON [19??] and BESSAGA and PEŁCZYŃSKI [1970]. The notion of a capset was a fundamental tool in the early days of infinite-dimensional topology for recognizing topological Hilbert spaces.

3. Topological vector spaces

DUGUNDJI proved in [1951] that every locally convex vector space is an AR. This raised the question whether the local convexity assumption is essential in this result. This was a formidable open problem for several decades. In [1979] DOBROWOLSKI and TORUŃCZYK proved that every separable, infinite-dimensional, complete topological vector space that is an AR is homeomorphic to Hilbert space. This result gave additional importance to finding an answer to the above question. R. Cauty answered it in the negative:

3.1. THEOREM (CAUTY [1994]). *There is a separable, topologically complete vector space that is not an* AR.

Cauty's construction proceeds as follows. The complete example is obtained as a completion of a σ-compact and metrizable vector space E. As basis for the construction of E Cauty considers an infinite-dimensional compact metric space X with the property that it is the cell-like image of a finite-dimensional polyhedron. The existence of such a space follows from DRANISHNIKOV's celebrated construction [1988] of an infinite-dimensional space with cohomological dimension three. Algebraically, E is the free vector space over X. The canonical topology on E is the strongest linear topology that induces on X the original topology. This topology, however, is not metrizable. Cauty constructs a weaker topology τ for E that is metrizable and linear and that has the property that it contains an open set U that does not have the homotopy type of a CW-complex, thereby showing that U is not an ANR and hence (E, τ) cannot be an AR.

An immediate corollary of Theorem 3.1 is:

3.2. COROLLARY. *There exists a separable, topologically complete vector space that is not homeomorphic to any convex subset of a locally convex vector space.*

Bessaga and Dobrowolski proved the following positive result in this direction.

3.3. THEOREM (BESSAGA and DOBROWOLSKI [1977]). *Every locally convex σ-compact metric vector space is homeomorphic to a pre-Hilbert space.*

This result suggested the possibility of simplifying the (difficult) classification problem of incomplete locally convex vector spaces by considering only linear subspaces of Hilbert space. Recall that by the Anderson-Kadec-Toruńczyk Theorem (see TORUŃCZYK [1981]) complete locally convex metric spaces are characterized by their weight. For incomplete spaces, however, Marciszewski found the following obstructions.

3.4. THEOREM (MARCISZEWSKI [1997]). *There exists a separable, normed vector space that is not homeomorphic to any convex subset of Hilbert space.*

3.5. THEOREM (MARCISZEWSKI [1997]). *There exists a separable, locally convex metric vector space that is not homeomorphic to any convex subset of a normed vector space.*

Marciszewski's counterexamples are constructed by transfinite induction and the method of "killing homeomorphisms" that was invented by SIERPIŃSKI [1932]. It is unknown whether there are such examples that are absolute Borel sets.

Even the classification problem for σ-compact pre-Hilbert spaces appears difficult as the following result shows.

3.6. THEOREM (CAUTY [1992]). *There exist a continuum of σ-compact pre-Hilbert spaces such that no two of them have a continuous injection between them.*

Let X be an ANR. It is easy to prove that for every open cover $\mathcal{U}$ of X there exists an open refinement $\mathcal{V}$ of $\mathcal{U}$ such that for every space Y, any two $\mathcal{V}$-close maps $f, g\colon Y \to X$ are $\mathcal{U}$-homotopic. It is a natural problem whether this property of ANR's in fact characterizes the class of all ANR's. This was also a difficult and fundamental problem which remained unanswered for decades. Cauty's example in Theorem 3.1 also solves this problem in the negative. This is because in every vector space close maps can be connected by small homotopies.

To see this, let L be a topological vector space. In addition, let $\mathcal{U}$ be an open cover of L. The function $\lambda\colon L \times L \times \mathbb{I} \to L$ defined by

$$\lambda(x, y, t) = (1 - t) \cdot x + t \cdot y,$$

is defined in terms of the algebraic operations on L and is therefore continuous. For every $x \in L$ pick an element $U_x \in \mathcal{U}$ containing x. Since the function λ is continuous and

$\lambda(\{x\} \times \{x\} \times \mathbb{I}) = \{x\}$, by compactness of $\mathbb{I}$ there exists for every $x \in X$ a neighborhood V_x of x such that $\lambda(V_x \times V_x \times \mathbb{I}) \subseteq U_x$. Put $\mathcal{V} = \{V_x : x \in L\}$. We claim that $\mathcal{V}$ is as required. To this end, let X be a space and let $f, g \colon X \to L$ be continuous $\mathcal{V}$-close functions. Define a homotopy $H \colon X \times \mathbb{I} \to L$ in the obvious way by the formula

$$H(x,t) = (1-t) \cdot f(x) + t \cdot g(x).$$

Then clearly $H_0 = f$ and $H_1 = g$. Fix an arbitrary $x \in X$. Since f and g are $\mathcal{V}$-close, there exists an element $p \in L$ such that $f(x), g(x) \in V_p$. But then

$$\{f(x)\} \times \{g(x)\} \times \mathbb{I} \subseteq V_p \times V_p \times \mathbb{I}$$

from which it follows that $H(x,t) = \lambda(f(x), g(x), t) \in U_p$ for every $t \in \mathbb{I}$. So this indeed proves that f and g are $\mathcal{U}$-homotopic.

The classical (Brouwer)-Schauder-Tychonoff Theorem states that every convex compactum in a locally convex vector space has the fixed point property. Schauder's unsupported claim that this theorem is valid in any metric vector space lead to the formulation of the Schauder Conjecture, which states that every convex compactum in a topological vector space should have the fixed point property. Theorem 3.1 shows that the Schauder Conjecture is a substantially stronger statement than the Schauder-Tychonoff Theorem. Recently, however, Cauty also proved the Schauder conjecture.

3.7. THEOREM (CAUTY [2001]). *Every compact, convex subset of a topological vector space has the fixed point property.*

Cauty's proof is very interesting. For a compact space X, he first considers the space $P(X)$ of probability measures on X with finite support, and let $P_n(X)$ be the subspace of $P(X)$ consisting of those measures whose support has at most n elements. The spaces $P_n(X)$ have a natural compact topology, and the topology on $P(X)$ is just the inductive limit topology induced by the sequence

$$P_1(X) \subset P_2(X) \subset \cdots \subset P_n(X) \subset \cdots;$$

that is, $U \subset P(X)$ is open if and only if $U \cap P_n(X)$ is open in $P_n(X)$ for every n. It is clear that we may identify $P_1(X)$ and X. Cauty proves the following surprising result:

3.8. THEOREM. *Let X be a compact space. Every continuous function $f \colon P(X) \to X$ has a fixed point, i.e., there is an element $x \in X$ such that $f(x) = x$.*

To see that this proves Theorem 3.7, consider a compact convex subset C of some vector space L, and let $f \colon C \to C$ be continuous. It is clear that f can be extended to a continuous function $\bar{f} \colon P(C) \to C$. Hence by Theorem 3.8, f has indeed a fixed point.

For a compact metrizable space X, let $E(X)$ be the free topological vector space over X, and let $\mathcal{T}(X)$ be the collection of all metrizable vector space topologies on $E(X)$ which are finer that the (nonmetrizable) free topology on $E(X)$. Observe that $P(X)$ is homeomorphic to a closed convex subspace of $E(X)$.

If X and Y are compact and $f \colon X \to Y$ is continuous then $\hat{f} \colon P(X) \to P(Y)$ is the natural continuous extension of f.

Observe that no metrizability is assumed in Theorem 3.7. So Cauty first reduces Theorem 3.7 to the metrizable case. Then he proceeds to prove the following result, which is the central element in his construction.

3.9. THEOREM. *Let X be a compact metrizable space. Then there are a compact metrizable space Z and a continuous function $\varphi\colon Z \to X$ such that*

(1) *Z is countable dimensional,*

(2) *If $\tau \in \mathcal{T}(X)$ and $\tau' \in \mathcal{T}(Z)$ are such that*

$$\hat{\varphi}\colon (P(Z), \tau') \to (P(X), \tau)$$

is continuous, then for every τ-open cover $\mathcal{U}$ of $P(X)$ and every countable locally finite simplicial complex N and every continuous function $\xi\colon N \to X$ there is a continuous function $\eta\colon N \to (P(Z), \tau')$ such that $\hat{\varphi} \circ \xi$ is $\mathcal{U}$-close to ξ and $\eta(N) \cup P_2(Z)$ is τ'-compact.

To see that this result implies Theorem 3.8, striving for a contradiction, assume that there are a compact metrizable space X and a continuous function $f\colon P(X) \to X$ without fixed point. Let Z and φ be as in Theorem 3.9 for X. It is not difficult to see that there are topologies $\tau \in \mathcal{T}(X)$ and $\tau' \in \mathcal{T}(Z)$ such that the functions

$$f\colon (P(X), \tau) \to X \quad \text{and} \quad \hat{\varphi}\colon (P(Z), \tau') \to (P(X), \tau)$$

are continuous. There is a τ-open cover $\mathcal{U}$ of $P(X)$ such that

$$U \cap f(U) = \emptyset \tag{1}$$

for every $U \in \mathcal{U}$. Let $\mathcal{V}$ be a τ-open cover of $P(X)$ which is a star-refinement of $\mathcal{U}$. It is not difficult to see that $(P(Z), \tau')$ is countable dimensional, hence it is an AR by a result of GRESHAM [1980]. Since $(P(Z), \tau')$ is separable, there consequently are a countable locally finite simplicial complex N and continuous functions

$$\mu\colon (P(Z), \tau') \to N, \quad \xi\colon N \to (P(Z), \tau')$$

such that

$$\xi \circ \mu\colon (P(Z), \tau') \to (P(Z), \tau')$$

is $\hat{\varphi}^{-1}[\mathcal{V}]$-close to the identity on $P(Z)$. The function $f \circ \hat{\varphi} \circ \xi\colon N \to X$ is continuous. By (ii) of Theorem 3.9 there is a continuous function $\eta\colon N \to (P(Z), \tau')$ such that $\hat{\varphi} \circ \eta$ and $f \circ \hat{\varphi} \circ \xi$ are $\mathcal{V}$-close, while moreover $\eta(N) \cup P_2(Z)$ is τ'-compact. Put $h = \eta \circ \mu$. Then h is a continuous function from $(P(Z), \tau')$ into itself, the range of which has compact closure. Since $(P(Z), \tau')$ is an AR, the function h has a fixed point, say x_0. There is an element V_1 of $\mathcal{V}$ containing the points

$$\hat{\varphi}(x_0) = \hat{\varphi} \circ \eta \circ \mu(x_0), \quad f \circ \hat{\varphi} \circ \xi \circ \mu(x_0).$$

There is also an element V_2 of $\mathcal{V}$ containing the points

$$\hat{\varphi}(x_0), \quad \hat{\varphi} \circ \xi \circ \mu(x_0).$$

Since $\hat{\varphi}(x_0) \in V_1 \cap V_2$ there consequently is an element U_0 of $\mathcal{U}$ which contains the points

$$\hat{\varphi} \circ \xi \circ \mu(x_0), \quad f \circ \hat{\varphi} \circ \xi \circ \mu(x_0).$$

But this contradicts (1).

The most fundamental open problem in this area now seems to be the question whether every *compact* convex subset of a metrizable vector space is an AR.

4. Function spaces

We first consider the function spaces $C_p(X)$, that is the space of all real-valued continuous function on a Tychonoff space X and equipped with the topology of point-wise convergence. Since we are interested in metric spaces we will restrict our attention to spaces X that are countable. The main problem in this field is the topological classification of all such spaces $C_p(X)$ that are Borel. There are many examples of spaces X such that $C_p(X) \in \mathcal{M}_2$, for instance all metric spaces X have this property. In [1985] DIJKSTRA, GRILLIOT, LUTZER and VAN MILL showed that $C_p(X) \in \mathcal{A}_2$ implies that X is discrete. The following result was a major step forward.

4.1. THEOREM (DOBROWOLSKI, MARCISZEWSKI and MOGILSKI [1991]). *If X is a non-discrete countable space with $C_p(X) \in \mathcal{M}_2$ then $C_p(X)$ is an $\mathcal{M}_2$-absorber and hence homeomorphic to Ω_2.*

This result prompted Dobrowolski et al. to conjecture that every $C_p(X)$ that is Borel should be the absorber of the exact Borel class to which it belongs, which would imply by the Uniqueness Theorem that $C_p(X)$ is topologically characterized by its Borel complexity. Further supporting evidence for this conjecture was supplied by the following results.

4.2. THEOREM (CAUTY, DOBROWOLSKI and MARCISZEWSKI [1993]). *If $C_p(X)$ is Borel then it belongs to $\mathcal{M}_\alpha \setminus \mathcal{A}_\alpha$ for some $\alpha \geq 2$, provided that X is not discrete.*

4.3. THEOREM (CAUTY, DOBROWOLSKI and MARCISZEWSKI [1993]). *For each $\alpha \geq 2$ there exists a countable space X_α such that $C_p(X_\alpha)$ is an $\mathcal{M}_\alpha$-absorber and hence homeomorphic to Ω_α.*

A major and surprising break through was Cauty's proof that the conjecture is false.

4.4. THEOREM (CAUTY [1998]). *For each $\alpha > 2$ there exists a countable space Y_α such that $C_p(Y_\alpha) \in \mathcal{M}_\alpha \setminus \mathcal{A}_\alpha$ and yet $C_p(Y_\alpha)$ is not an $\mathcal{M}_\alpha$-absorber.*

In fact, the construction is such that the space $C_p(Y_\alpha)$ does not even contain a closed copy of Λ_2, the $\mathcal{A}_2$-absorber. The spaces Y_α were actually constructed by LUTZER, VAN MILL and POL [1985] to show that there exist spaces $C_p(X)$ of arbitrarily high Borel complexity. We let T_n be the set of functions from $\{0, 1, \ldots, n-1\}$ to $\{0, 1\}$ and define the countable set $T = \bigcup_{n=1}^{\infty} T_n$. If x is an element of the Cantor set 2^ω then $x|n \in T_n$ denotes the restriction of x to the domain $\{0, 1, \ldots, n-1\}$. Let $A_\alpha \subset 2^\omega$ be an element of $\mathcal{M}_\alpha \setminus A_\alpha$ and consider the filter $\mathcal{F}_\alpha$ on T that is generated by the co-finite sets and all sets of the form $\bigcup_{n=1}^{\infty} T_n \setminus \{x|n\}$, where $x \in A_\alpha$. The space Y_α is $T \cap \{\infty\}$ where all points of T are isolated and the neighborhoods of ∞ are the sets $F \cap \{\infty\}$ for $F \in \mathcal{F}_\alpha$. According to CALBRIX [1988] $C_p(Y_\alpha)$ is also in $\mathcal{M}_\alpha \setminus \mathcal{A}_\alpha$.

Let us define

$$s_\alpha = \{f \in \mathbb{R}^T : f|F = 0 \text{ for some } F \in \mathcal{F}_\alpha.\}$$

It is not hard to see that $C_p(Y_\alpha)$ is homeomorphic to a closed subset of $(s_\alpha)^\omega$ so if $C_p(Y_\alpha)$ is $\mathcal{A}_2$-universal then so is $(s_\alpha)^\omega$. But then, according to BANAKH and CAUTY [2000], the pair $((\mathbb{R}^T)^\omega, (s_\alpha)^\omega)$ is $(\mathcal{M}_0, \mathcal{A}_2)$-universal, that is for each compactum K and subset C of K with $C \in \mathcal{A}_2$ there is an imbedding φ of K into $(\mathbb{R}^T)^\omega$ such that $\varphi^{-1}[(s_\alpha)^\omega] = C$.

Let $W(Q, s)$ be the subset of Q^ω consisting of all sequences chosen from Q such that all but finitely many elements are in the pseudo-interior s. It is obvious that $W(Q, s)$ is in $\mathcal{A}_2$. The desired contradiction is obtained by CAUTY via the following lemma, the proof of which occupies essentially the entire paper [1998] and makes extensive use of Homology Theory.

4.5. LEMMA. *For each α there is no continuous function $\varphi\colon Q^\omega \to (\mathbb{R}^T)^\omega$ such that $\varphi^{-1}[(s_\alpha)^\omega] = W(Q, s)$.*

We now turn beyond Borel to the classes of analytic and co-analytic spaces. The following results also exclude a simple answer to the classification problem in these classes.

4.6. THEOREM (MARCISZEWSKI [1993]). *Under $V = L$ there exist countable spaces X and Y such that $C_{\mathrm{p}}(X)$ and $C_{\mathrm{p}}(Y)$ are non-homeomorphic spaces that are both analytic but not co-analytic.*

4.7. THEOREM (MARCISZEWSKI [1993]). *Under $V = L$ there exist countable spaces X and Y such that $C_{\mathrm{p}}(X)$ and $C_{\mathrm{p}}(Y)$ are non-homeomorphic spaces that are both co-analytic but not analytic.*

It follows from results in CAUTY [1998] that both theorems are in fact provable in ZFC.

We now consider the space $\mathcal{C}$ of continuous real-valued functions on the interval $\mathbb{I}$ with the topology of uniform convergence. Let $\mathcal{D}$ and $\mathcal{D}^*$ stand for the subspaces of $\mathcal{C}$ consisting of all differentiable functions respectively all function that are differentiable in at least one point.

4.8. THEOREM (CAUTY [1991]). *$\mathcal{D}$ and $\mathcal{D}^*$ are absorbers for the co-analytic respectively analytic classes.*

5. Homotopy dense imbeddings

The basic theorem concerning dense imbeddings reads as follows.

5.1. THEOREM (BOWERS [1987]). *A separable metric space admits a dense imbedding in Hilbert space if and only if it is nowhere locally compact.*

A space X is said to have the *strong discrete approximation property* (SDAP) if for every sequence of continuous maps $f_1, f_2, \ldots : Q \to X$ and every open cover $\mathcal{U}$ of X there exists another sequence of continuous maps $g_1, g_2, \ldots : Q \to X$ such that each g_i is $\mathcal{U}$-close to f_i and the images of the g_i's form a discrete collection in X. This concept was introduced by TORUŃCZYK [1981] for the purpose of characterizing Hilbert space as the only separable complete metric AR with the SDAP. An imbedding $f\colon X \to Y$ is called *homotopy dense* if $f[X]$ is homotopy dense in Y. The following theorem gives an internal characterization of the homotopy dense subspaces of Hilbert space.

5.2. THEOREM (BANAKH [1996, 1998]). *A separable metric space admits a homotopy dense imbedding in Hilbert space if and only if it is an* AR *with the* SDAP.

A short proof for this theorem can be found in DOBROWOLSKI [1997]. Since BESTVINA and MOGILSKI [1986] have shown that every strong σZ-space has the SDAP an interesting consequence of Theorem 5.2 is that every absorber is imbeddable as a standard absorber and hence Theorem 2.1 improves to Theorem 2.2.

A different approach to homotopy dense imbeddings was taken by CHAPMAN and SIEBENMANN [1976] who introduced the concept of a Z-compactification as the natural infinite-dimensional extension of adding a boundary to a finite-dimensional open manifold, which was the subject of SIEBENMANN's famous thesis [1965]. A *Z-compactification* Y of a (locally compact) space X is a compact metric space containing X such that $Y \setminus X$ is a Z-set in Y. So a locally compact space is Z-compactifiable if and only if it admits a homotopy dense imbedding into some compact space. Model examples are for instance the case that X is the interior of a topological manifold Y or that X is the complement of an endface in the Hilbert cube.

In [1976] CHAPMAN and SIEBENMANN presented criteria for a Hilbert cube manifold X to be Z-compactifiable. Formulated in geometric terms, their result is that X admits a Z-compactification if and only if X is homeomorphic to the product of an inverse mapping telescope with the Hilbert cube. Chapman and Siebenmann were not able to decide whether their characterization can be extended beyond Hilbert cube manifolds to all locally compact ANR's. The existence of that extension depended on an answer to the following question, which was posed in the paper: if $X \times Q$ is Z-compactifiable is X itself Z-compactifiable?

Guilbault answered this question in the negative:

5.3. THEOREM (GUILBAULT [2001]). *There exists a locally compact* 2*-dimensional polyhedron X that is not Z-compactifiable but such that $X \times Q$ has a Z-compactification.*

Surprisingly, the construction of the example X is not complicated. X is the infinite mapping telescope of a direct sequence $S^1 \xrightarrow{\theta} S^1 \xrightarrow{\theta} S^1 \xrightarrow{\theta} \dots$ where θ is a degree 1 map which wraps the circle around itself twice counterclockwise, then once back in the clockwise direction. The fact that $X \times Q$ is Z-compactifiable follows easily from the characterization of Chapman and Siebenmann or by observing that CHAPMAN's characterization of simple homotopy equivalence [1974] implies that $X \times Q$ is homeomorphic to $(S^1 \times Q) \times [0, \infty)$. The proof that X does not admit Z-compactifications, however, is very lengthy and involved.

Although Chapman and Siebenmann's question about Z-compactifications has its origin firmly in Hilbert cube manifold theory, Ferry showed recently that this problem is finite-dimensional rather than infinite-dimensional in nature:

5.4. THEOREM (FERRY [2000]). *If an n-dimensional polyhedron X is such that $X \times Q$ is Z-compactifiable then $X \times \mathbb{I}^{2n+5}$ is also Z-compactifiable.*

6. Topological classification of semicontinuous functions

The primary focus of this research concerns the question whether certain semicontinuous functions of analytic origin that are defined on Hilbert space are topologically indistinguishable. Interesting examples of such functions are the p-norm on the topological Hilbert space $s = \mathbb{R}^{\mathbb{N}}$:

$$|x|_p = \begin{cases} \sqrt[p]{\sum_{n=1}^{\infty} |x_n|^p} & \text{if } p < \infty \\ \sup\{|x_n| : n \in \mathbb{N}\} & \text{if } p = \infty. \end{cases}$$

Because s carries the topology of point-wise convergence these functions are lower semicontinuous but not continuous. In fact, according to VAN MILL and POL [1995] these functions are in a sense universal for all lower semicontinuous functions and they are not even countably continuous, that is their domain cannot be partitioned into countably many sets such that the restrictions are continuous.

If X is a (real) topological vector space endowed with the continuous norms $\|\cdot\|$ and $|\cdot|$, respectively, then there is a norm preserving homeomorphism $f\colon (X, \|\cdot\|) \to (X, |\cdot|)$ defined by $f(\mathbf{0}) = \mathbf{0}$ and

$$f(x) = \frac{\|x\|}{|x|} x$$

if $x \neq \mathbf{0}$. Observe that such a homeomorphism is in general not linear. Consider for example $\mathbb{R}^2$ endowed with the Euclidean norm $\|x\| = \sqrt{x_1^2 + x_2^2}$ and the max norm $|x| = \max\{x_1, x_2\}$. So a norm preserving homeomorphism sends the unit ball $\{(x, y) \in \mathbb{R}^2 : x^2 + y^2 \leq 1\}$ onto the unit brick $[-1, 1]^2$ and consequently changes the shape of a geometric object considerably.

These considerations for continuous norms are not very interesting and the question naturally arises whether something can be said in the case of discontinuous norms. All norms on finite-dimensional vector spaces are continuous, so the question only makes sense within the framework of infinite-dimensional spaces. If X is an infinite-dimensional vector space then it can be endowed with several discontinuous norms. This leads us to consideration of the well-known p-norms from the Banach spaces ℓ^p in combination with the topology of point-wise convergence.

By means of the Bing Shrinking Criterion the authors proved that all the p-norms are topologically indistinguishable:

6.1. THEOREM (DIJKSTRA and VAN MILL [2002]). *For every $p \in (0, \infty)$ there exists a homeomorphism $h\colon s \to s$ such that $|h(x)|_p = |x|_\infty$ for every $x \in s$.*

Sketch of Proof: For $p, q \in (0, \infty)$ it is easy to construct homeomorphisms $H_q^p\colon s \to s$ that are norm preserving, that is $|H_q^p(x)|_p = |x|_q$ for all $x \in s$. Let for $p \in (0, \infty)$ and $q \in (0, \infty]$ the map $H_q^p\colon s \to s$ be defined by the property that for each $x, y \in s$ and $n \in \mathbb{N}$ with $H_q^p(x) = y$ we have

$$y_n = \operatorname{sgn}(x_n) \sqrt[p]{|\xi_n(x)|_q^p - |\xi_{n-1}(x)|_q^p},$$

where $\xi_n(x) = (x_1, x_2, \ldots, x_n, 0, 0, \ldots)$ and $\operatorname{sgn}(x_n)$ is the sign of the number. Note that this definition also works in the case that $q = \infty$. However, H_∞^p is never a homeomorphism but it is a norm-preserving cell-like surjection. The idea of the proof is to take a representative H^1_∞ and to show that this map is shrinkable by homeomorphisms of s that preserve the $|\cdot|_\infty$ norm. Then, according to Bing, H^1_∞ can be approximated by norm-preserving homeomorphisms.

The three figures show the shrinking process in (considerably) simplified form. The first figure shows the unit sphere with respect to the sup norm in the first octant of the first three dimensions with the fibres of the map H^1_∞ indicated by solid lines and a shaded region.

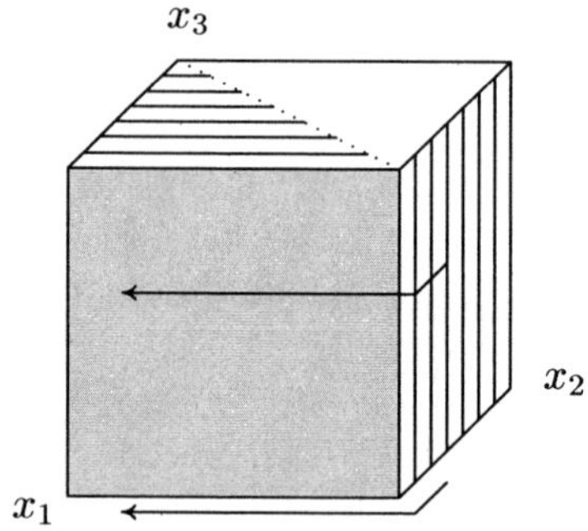

Figure 1

Assuming that the fibres need to be shrunk to a constant size ε the transition from Figure 1 to Figure 2 indicates how the projections of the fibres onto the x_1x_2-plane are shrunk by a rotational move in the x_1x_2-plane that does not involve x_3.

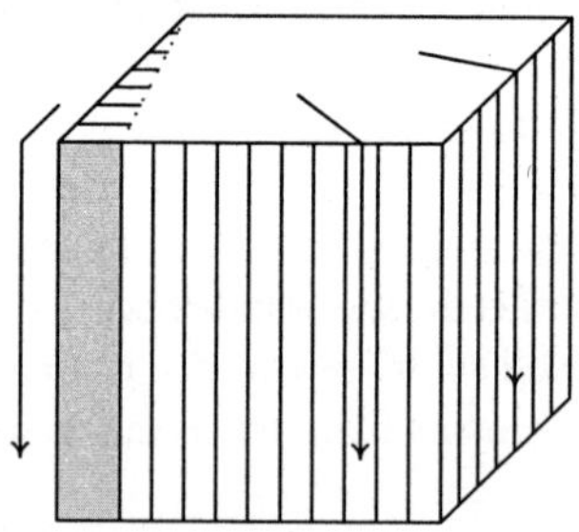

Figure 2

This operation is then followed by a similar move in the planes that contain the x_3-axis, as illustrated by the transition from Figure 2 to Figure 3. The result is that the projections onto the first three dimensions of all fibres have been shrunk to size ε. This process can be continued. If we would be working in the Hilbert cube Q then the process could stop once the length of the n-th coordinate dips below ε. It is not desirable to let the process run through infinitely many coordinates since the norm is not continuous and so limits in general do not preserve norm.

We are, however, working in a highly noncompact space, Hilbert space. This means that we have to work with ε-functions rather than constant ε's. Most of the effort in the paper DIJKSTRA and VAN MILL [2002] goes towards dealing with the tension between this requirement and the rigidity that is caused by the need to preserve the norms of vectors. Also, the use of ε-functions means that the shrinking homeomorphisms are obtained as limits of infinite sequences of homeomorphisms which are kept norm-preserving by making sure that every individual vector is moved only a finite number of times.

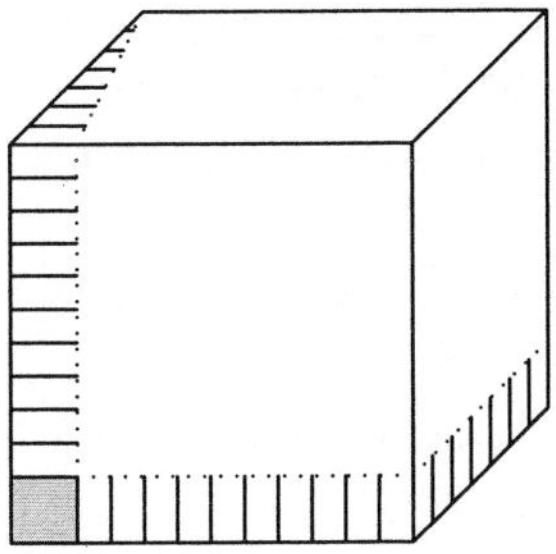

Figure 3

7. Hyperspaces of Peano continua

If X is a compact metric space, then 2^X denotes the hyperspace consisting of all non-empty closed subsets of X, endowed with the Hausdorff metric. $C(X)$ denotes the compact subspace of 2^X consisting of all subcontinua of X. The fundamental theorems are by CURTIS and SCHORI [1978]: 2^X is homeomorphic to Q if and only if X is a non-degenerate Peano continuum and $C(X)$ is homeomorphic to Q if and only if X is a non-degenerate Peano continuum without free arcs. For $k \in \{0, 1, 2, \dots\}$ we let $\mathrm{Dim}_{\geq k}(X)$ denote the subspace consisting of all $\geq k$-dimensional elements of 2^X and we put $\mathrm{Dim}_\infty(X) = \bigcap_{k=0}^\infty \mathrm{Dim}_{\geq k}(X)$.

7.1. THEOREM (DIJKSTRA, VAN MILL and MOGILSKI [1992]). *There exists a homeomorphism α from 2^Q onto Q^ω such that for every $k \in \{0, 1, 2, \dots\}$,*

$$\alpha(\mathrm{Dim}_{\geq k}(Q)) = \underbrace{B \times \cdots \times B}_{k \text{ times}} \times Q \times Q \times \cdots$$

and hence $\mathrm{Dim}_\infty(Q)$ is an $\mathcal{M}_2$-absorber and homeomorphic to B^ω and Ω_2.

The proof of this theorem is based on the technique of absorbing systems, which was introduced in the papers DIJKSTRA, VAN MILL and MOGILSKI [1992] and DIJKSTRA and MOGILSKI [1991]. Subsequently, several authors generalized Theorem 7.1 in different directions. GLADDINES [1992] proved that the theorem remains valid when we consider the sequences $\mathrm{Dim}_{\geq k}(X)$ and $\mathrm{Dim}_{\geq k+1}(X) \cap C(X)$ for X an countable infinite product of Peano continua instead of Q. DOBROWOLSKI and RUBIN [1994a] show that in Theorem 7.1 the covering dimension may be replaced by cohomological dimension. In addition, GLADDINES and VAN MILL [1993a] give an example that shows that the theorem is not valid for all everywhere infinite dimensional Peano continua. The final word on this subject was spoken by Cauty:

7.2. THEOREM (CAUTY [1999]). *A Peano continuum X has the property that every nonempty open subset contains compacta of arbitrarily high finite dimension if and only if here exists a homeomorphism α from 2^X onto Q^ω such that for every $k \in \{0, 1, 2, \ldots\}$,*

$$\alpha(\mathrm{Dim}_{\geq k}(X)) = \underbrace{B \times \cdots \times B}_{k \text{ times}} \times Q \times Q \times \cdots .$$

This result remains valid if we consider $C(X)$ instead of 2^X and also if we replace covering dimension by cohomological dimension.

Gladdines and van Mill have also considered the space $L(X) \subset C(X)$ consisting of all Peano continua in X:

7.3. THEOREM (GLADDINES and VAN MILL [1993b]). *If $n \geq 3$ then $L(\mathbb{I}^n)$ is an $\mathcal{M}_2$-absorber and hence homeomorphic to Ω_2.*

Continuing in this direction Dobrowolski and Rubin found:

7.4. THEOREM (DOBROWOLSKI and RUBIN [1994b]). *If $n \geq 3$ then both $\mathsf{AR}(\mathbb{I}^n)$ and $\mathsf{ANR}(\mathbb{I}^n)$ are $\mathcal{M}_3$-absorbers and hence homeomorphic to Ω_3.*

References

ANDERSON, R.D.

[19??] On sigma-compact subsets of infinite-dimensional manifolds, unpublished manuscript.

[1966] Hilbert space is homeomorphic to the countable infinite product of lines, *Bull. Amer. Math. Soc.* **72**, 515–519.

[2002] The early development of infinite dimensional topology, this volume.

BANAKH, T.

[1998] Characterization of spaces admitting a homotopy dense embedding into a Hilbert manifold, *Topology Appl.* **86**, 123–131.

[1999] Some properties of the linear hull of the Erdős set in ℓ^2, *Bull. Polon. Acad. Sci. Sér. Math. Astronom. Phys.* **47**, 385–392.

BANAKH, T. and R. CAUTY

[2000] Interplay between strongly universal spaces and pairs, *Dissertationes Math.*, vol. 386.

BANAKH, T., T. RADUL and M. ZARICHNYI

[1996] *Absorbing sets in infinite-dimensional manifolds*, Mathematical Studies, vol. 1, VNTL Publishers, Lviv.

BESSAGA, C. and T. DOBROWOLSKI

[1977] Affine and homeomorphic embeddings into l^2, *Proc. Amer. Math. Soc.* **125**, 259–268.

BESSAGA, C. and A. PEŁCZYŃSKI

[1970] The estimated extension theorem homogeneous collections and skeletons, and their application to the topological classification of linear metric spaces and convex sets, *Fund. Math.* **69**, 153–190.

BESTVINA, M. and J. MOGILSKI

[1986] Characterizing certain incomplete infinite-dimensional absolute retracts, *Michigan J. Math.* **33**, 291–313.

BOWERS, P.L.

[1987] Dense embeddings of nowhere locally compact separable metric spaces, *Topology Appl.* **26**, 1–12.

CALBRIX. J.

[1988] Filtres boréliens sur l'ensemble des entiers et espaces des applications continues, *Rev. Roumaine Math. Pures Appl.* **33**, 655–661.

CAUTY, R.

[1991] Charactérisation topologique de l'espace des fonctions dérivables, *Fund. Math.* **138**, 35–58.

[1992] Une famille d'espaces préhilbertiens σ-compacts ayant la puissance du continu, *Bull. Polish Acad. Sci. Math.* **40**, 41–43.

[1994] Un espace métrique linéaire qui n'est pas un rétracte absolu, *Fund. Math.* **146**, 85–99.

[1998] La classe Borélienne ne détermine pas le type topologique de $C_p(X)$, *Serdica Math. J.* **24**, 307–318.

[1999] Suites F_σ-absorbantes en théorie de la dimension, *Fund. Math.* **159**, 115–126.

[2001] Solution du problème de point fixe de Schauder, *Fund. Math.* **170**, 231–246.

CAUTY, R., T. DOBROWOLSKI and W. MARCISZEWSKI

[1993] A contribution to the topological classification of the spaces $C_p(X)$, *Fund. Math.* **142**, 269–301.

CHAPMAN, T.A.

[1974] Topological invariance of Whitehead torsion, *Amer. J. Math.* **96**, 488–497.

CHAPMAN, T.A. and L. C. SIEBENMANN

[1976] Finding a boundary for a Hilbert cube manifold, *Acta. Math.* **137**, 171–208.

CURTIS, D.W. and R.M. SCHORI

[1978] Hyperspaces of polyhedra are Hilbert cubes, *Fund. Math.* **99**, 189–197.

DIJKSTRA, J.J., T. GRILLIOT, D.J. LUTZER and J. VAN MILL

[1985] Function spaces of low Borel complexity, *Proc. Amer. Math. Soc.* **94**, 703–710.

DIJKSTRA, J.J. and J. VAN MILL

[1992] Topological classification of infinite-dimensional spaces with absorbers, in *Recent Progress in General Topology*, (M. Hušek and J. van Mill, eds.), North-Holland Publishing Co., Amsterdam, pp. 145–165.

[2002] Topological equivalence of discontinuous norms, *Israel J. Math.* **128**, 177–196.

DIJKSTRA, J.J., J. VAN MILL and J. MOGILSKI

[1992] The space of infinite-dimensional compacta and other topological copies of $(l^2_f)^\omega$, *Pacific J. Math.* **152**, 255–273.

DIJKSTRA, J.J. and J. MOGILSKI

[1991] The topological product structure of systems of Lebesgue spaces, *Math. Ann.* **290**, 527–543.

DOBROWOLSKI, T.

[1997] Enlarging ANR's with SDAP to l_2-manifolds revisited, *Bull. Polish Acad. Sci. Math.* **45**, 345–348.

DOBROWOLSKI, T., W. MARCISZEWSKI and J. MOGILSKI

[1991] On topological classification of function spaces of low Borel complexity, *Trans. Amer. Math. Soc.* **328**, 307–324.

DOBROWOLSKI, T. and L.R. RUBIN

[1994a] The hyperspaces of infinite-dimensional compacta for covering and cohomological dimension are homeomorphic, *Pacific J. Math.* **164**, 15–39.

[1994b] The space of ANRs in $\mathbf{R}^n$, *Fund. Math.* **146**, 31–58.

DOBROWOLSKI, T. and H. TORUŃCZYK
[1979] On linear metric spaces homeomorphic to l_2 and compact covex sets homeomorphic to Q, *Bull. Polon. Acad. Sci. Sér. Math. Astronom. Phys.* **27**, 883–887.

DRANISHNIKOV, A.N.
[1988] On a problem of P.S. Alexandrov, *Mat. Sb.* **135**, 551–557.

DUGUNDJI, J.
[1951] An extension of Tietze's theorem, *Pac. J. Math.* **1**, 353–367.

ERDŐS, P.
[1940] The dimension of the rational points in Hilbert space, *Annals of Math.* **41**, 734–736.

FERRY, S.
[2000] Stable compactifications of Polyhedra, *Michigan J. Math.* **47**, 287–294.

GLADDINES, H.
[1992] F_σ-absorbing sequences in hyperspaces of compact sets, *Bull. Polish Acad. Sci. Math.* **40**, 175–184.

GLADDINES, H. and J. VAN MILL
[1993a] Hyperspaces of infinite-dimensional compacta, *Compositio Math.* **88**, 143–153.
[1993b] Hyperspaces of Peano continua of Euclidean spaces, *Fund. Math.* **142**, 173–188.

GRESHAM, J.H.
[1980] A class of infinite-dimensional spaces. Part II: an extension theorem and the theory of retracts, *Fund. Math.* **106**, 237–245.

GUILBAULT, C.R.
[2001] A non-Z-compactifiable polyhedron whose product with the Hilbert cube is Z-compactifiable, *Fund. Math.* **168**, 165–197.

LUTZER, D.J., J. VAN MILL and R. POL
[1985] Descriptive complexity of function spaces, *Trans. Amer. Math. Soc.* **291**, 121–128.

MARCISZEWSKI, W.
[1993] On analytic and coanalytic function spaces $C_p(X)$, *Top. Appl.* **50**, 241–248.
[1997] On topological embeddings of linear metric spaces, *Math. Ann.* **308**, 21–30.

VAN MILL, J. and R. POL
[1995] Baire 1 functions which are not countable unions of continuous functions, *Acta Math. Hungar.* **66**, 289–300.

SIEBENMANN, L.C.
[1965] *The obstruction to finding a boundary to an open manifold of dimension greater than five*, Ph.D. thesis, Princeton Univ.

SIERPIŃSKI, W.
[1932] Sur un problème concernant les types de dimensions, *Fund. Math.* **19**, 65–71.

TORUŃCZYK, H.
[1980] On CE-images of the Hilbert cube and characterizations of Q–manifolds, *Fund. Math.* **106**, 31–40.
[1981] Characterizing Hilbert space topology, *Fund. Math.* **111**, 247–262.

WEST, J. E.
[1977] Mapping Hilbert cube manifolds to ANR's: a solution to a conjecture of Borsuk, *Annals of Math.* **106**, 1–18.
[1990] Problems in Infinite-dimensional Topology, in *Open Problems in Topology*, (J. van Mill and G. M. Reed, eds.), North-Holland Publishing Co., Amsterdam, pp. 523–597.

CHAPTER 5

Recent Results in Set-Theoretic Topology

Alan Dow

Mathematics Dept. of UNC Charlotte, 9201 University City Blvd. Charlotte, NC 28223-0001, U.S.A.
E-mail: adow@uncc.edu

Contents

RECENT PROGRESS IN GENERAL TOPOLOGY II
Edited by Miroslav Hušek and Jan van Mill

1. Introduction

This paper has no theme beyond the desire to present very brief excerpts or outlines of a few of the most interesting of the heavily set-theoretic results in topology of the past few years. We are guided most by our own interests. We review some results about $\mathbb{N}^*$ by Farah, about compact spaces of countable tightness by Rabus, Koszmider, and Eisworth. We introduce the reader to club guessing and present a new cardinal inequality result of Arhangel'skii and Buzyakova.

2. Standard tools

Recall that $H(\theta)$ is the set of sets whose transitive closure has cardinality less than θ. A statement of set-theory holds in a model M if the statement with all quantifiers restricted to M holds outright. The sets $H(\theta)$, constitute useful submodels of the universe of sets since each is a transitive model of most of the axioms of ZFC and given a space X and property of interest of X, there is almost surely a sufficiently large θ so that for any statement of interest, if there is an (counter-)example then there is one in $H(\theta)$. In a similar way, set models, such as $H(\theta)$ have *elementary submodels* . These are subsets $M \prec H(\theta)$ with the property that any statement of set-theory using parameters from M holds in M exactly when it holds in $H(\theta)$.

If P is a poset, then a P-name, $\tau \in V^P$, is recursively defined by $\tau \subset P \times V^P$. If G is a generic filter, then $V[G] = \{val_G(\sigma) : \sigma \in V^P\}$ is the forcing extension by P where $val_G(\sigma) = \{val_G(\tau) : (\exists p \in P)(p, \tau) \in \sigma\}$. A filter G is said to be *generic* if $G \cap D \neq \emptyset$ for each dense set $D \subset P$, i.e. for each $p \in P$, there is a $d \in D$ such that $d \leq p$.

A sequence $\{a_\alpha : \alpha \in \omega_1\}$ is a *$\diamondsuit$-sequence* if for each $\alpha \in \omega_1$, $a_\alpha \subset \alpha$, and for all $X \subset \omega_1$, the set $S_X = \{\alpha : X \cap \alpha = a_\alpha\}$ is stationary. Another useful principle is $\clubsuit$ which is the statement obtained by demanding that a_α be cofinal in α for limit α and setting $S_X = \{\alpha : a_\alpha \subset X\}$.

The above is clearly not meant as a suitable introduction to the fundamental tools of this area of topology. There are by now many very adequate references for these basics, perhaps even including an article in the previous edition of this book [1992].

A very interesting result is SHELAH's club guessing, [1987]. This is a $\diamondsuit$-like principle on ω_2 (and larger cardinals) which holds in ZFC. Let S^2_0 denote the set of ordinals in ω_2 which have countable cofinality.

2.1. PROPOSITION. *There is a sequence $\{c_\alpha : \alpha \in S^2_0\}$ where for each $\alpha \in S^2_0$, c_α is cofinal in α and for each cub $C \subset \omega_2$, there is a stationary set $S \subset S^2_0$ such that $c_\alpha \subset C$ for all $\alpha \in S$.*

and the proof is fun:

□ Choose, for every $\alpha \in S^2_0$, an increasing and cofinal sequence $\langle s(\alpha, n) : n \in \omega\rangle$ in α. If this results in a club-guessing sequence then of course we stop. If not then choose a club D_0 that is never guessed, put $C_0 = D'_0$ and define s_0 by $s_0(\alpha, n) = \sup(D_0 \cap s(\alpha, n) + 1)$ if $\alpha \in C_0$ and $s_0(\alpha, n) = s(\alpha, n)$ otherwise.

We continue recursively, defining C_ξ and s_ξ for $\xi < \omega_1$, as follows. At any stage stop if s_ξ defines a club-guessing sequence otherwise choose $D_{\xi+1}$ that is not guessed, set

$C_{\xi+1} = D'_{\xi+1}$ and define $s_{\xi+1}(\alpha, n) = \sup(D_{\xi+1} \cap s_\xi(\alpha, n) + 1)$ if $\alpha \in C_{\xi+1}$ and $s_{\xi+1}(\alpha, n) = s_\xi(\alpha, n)$ otherwise. At limit stages set $D_\xi = \bigcap_{\zeta<\xi} C_\zeta$ and $s_\xi(\alpha, n) = \min_{\zeta<\xi} s_\zeta(\alpha, n)$. Observe that $s_\xi(\alpha, n) \leq s_\zeta(\alpha, n)$ whenever $\zeta < \xi$.

The construction stops before ω_1 because otherwise consider the club set $C = \bigcap_\xi C_\xi$ and take $\alpha \in C$. Because D_ξ was never guessed there will for every ξ be an n such that $s_{\xi+1}(\alpha, n) < s_\xi(\alpha, n)$. But then one n will be chosen uncountably often, yielding an infinite decreasing sequence of ordinals. □

The notion of club guessing has been used in several important applications since its use in pcf theory by SHELAH [1994]. Examples of greatest interest to topologists (and most likely to illustrate its use) are the proof by KOJMAN and SHELAH [1992] that there is no universal linear order of cardinality ω_2 if $\omega_2 < \mathfrak{c}$, and the result of DŽAMONJA [1998] that if $\omega_2 < \mathfrak{c}$ then there is no universal uniform Eberlein compactum of weight ω_2 which complements the result of BELL [2000] that it is independent of ZFC that they exist. This is a strengthening of one direction of Bell's result [2000] that it is independent of ZFC that such spaces exists. Another very pretty application by KOJMAN and SHELAH [2001], is in showing that $\mathfrak{h}(U(\aleph_\omega)) = \omega_1$ (see 5.1 for the definition of $\mathfrak{h}$).

Although the following result does not have independent interest, it does nicely illustrate the use of club-guessing which the reader will see is somewhat different than those from $\diamondsuit$.

Say that a madf $\mathcal{M}$ (of countable sets) on a cardinal κ is embeddable into another, $\mathcal{L}$, if there is an injection of κ so that each member of $\mathcal{M}$ is sent to a subset of a member of $\mathcal{L}$. Of course there are many possible concepts that one might reasonably call embeddability of a madf into another, but our purpose is to simply illustrate the concept of club-guessing.

2.2. PROPOSITION. *There is a family $\{\mathcal{A}_\xi : \xi \in \mathfrak{c}\}$, of pairwise non-embeddable madf's on ω_2.*

□ Fix the club-guessing sequence $\{c_\alpha : \alpha \in S^2_0\}$ as in Proposition 2.1, and let $c_\alpha = \{c^\alpha_n : n \in \omega\}$ be listed in increasing order.

The idea is to construct, for each infinite $X \subset \omega$ a madf $\mathcal{A}_X$ so that for each δ of countable cofinality

(1) for all $a \in \mathcal{A}_X$, $\{n : |a \cap [c^\delta_n, c^\delta_{n+1})| > 1\}$ is either finite or is equal to X, and

(2) there is an $a \in \mathcal{A}_X$ such that $a \cap [c^\delta_n, c^\delta_{n+1}) > 1$ for all $n \in X$.

It is pretty easy to see that $\mathcal{A}_X$ can be constructed. The application of club-guessing is in the proof that if $X \setminus Y$ is non-empty, then $\mathcal{A}_X$ cannot be embedded into $\mathcal{A}_Y$. Indeed, assume that $X \not\subset Y$ and assume that f is an injection from ω_2 into ω_2. There is a cub C on ω_2 so that for each $\delta \in \omega_2$, $f^{-1}([0, \delta)) = \delta$. By the club-guessing principle, there is a $\delta \in S^2_0$ so that $c_\delta \subset C$. Note then that for any cofinal $a \subset \delta$, $f(a)$ is also cofinal in δ. Let a be a member of $\mathcal{A}_X$ with the property that $|a \cap [c^\delta_n, c^\delta_{n+1})| > 1$ exactly for each $n \in X$. Since f is an injection and $c_\delta \subset C$, it follows that $|f(a) \cap [c^\delta_n, c^\delta_{n+1})| > 1$ for all $n \in X$. This, of course, implies that $f(a)$ cannot be contained in any $b \in \mathcal{A}_Y$. □

The proof illustrates the idea of using a subset of ω as a parameter (not merely an index) for the members constructed which controls how the elements interlace with the club-guessing sequence members. This is one of the ideas in the paper KOJMAN and SHELAH [1992] concerning universal linear orders.

3. Linearly Lindelöf spaces

ARHANGEL'SKII and BUZYAKOVA [1999] proved a very nice result about linearly Lindelöf spaces, namely that first countable such spaces have cardinality at most $\mathfrak{c}$. Recall that a space X is linearly Lindelöf if each cover by a chain of open sets has a countable subcover – hence the term linear Lindelöf. This is known to be equivalent to the statement that for each regular uncountable κ, each κ-sized subset of X has a complete accumulation point (i.e. (κ, κ)-compact for each uncountable regular cardinal κ). Most such cardinal inequality results lend themselves naturally to a proof using elementary submodels, or as Arhangel'skii calls it, the radial method. The core to the proof is Lemma 3.4 below, and it certainly does seem resistant to an elementary submodel approach. Theorem 3.2 may be a small improvement of the analogous result in ARHANGEL'SKII and BUZYAKOVA [1999] but otherwise all of these results are in [1999]. We are curious about the possibility of removing the linearly Lindelöf assumption from Lemma 3.4.

3.1. DEFINITION. For a space X, let $Ch(X)$ be defined as the minimum κ such that $\beta X \setminus X$ can be written as the union of at most κ compact sets.

3.2. THEOREM. *Suppose that κ is a cardinal and Y is a subspace of a space X such that $Ch(Y)$ and $|Y|$ are at most κ and, for each $y \in Y$, $\chi(y, X) \leq \kappa$, then Y is a G_κ-set in X.*

□ Fix an elementary submodel M of $H(\theta)$ with $|M| = \kappa$ and Y, X are in M with $\kappa \cup Y \subset M$. To see that Y is a G_κ-set in X we will show that $Y = \bigcap\{U \in M : Y \subset U \text{ and } U \subset X \text{ is open}\}$. Fix a family $\{K_\alpha : \alpha \in \kappa\}$ of compact subsets of $\beta Y \setminus Y$ so that $\beta Y \setminus Y = \bigcup\{K_\alpha : \alpha < \kappa\}$ as per the definition of $Ch(Y)$. By elementarity, we can assume that $\{K_\alpha : \alpha \in \kappa\}$ is in M because some such family is in M. Again elementarity can be applied to deduce that each K_α is a member of M since implicitly $\{K_\alpha : \alpha \in \kappa\}$ represents a function from κ onto the family and with $\kappa \subset M$, the α-th value of the function must also be in M. Suppose that $x \in X \setminus Y$ and let $\mathfrak{F}_x = \{F \subset X : F \text{ is closed and } x \in F \in M\}$. We must show there is an $F \in \mathfrak{F}_x$ such that $F \cap Y = \emptyset$. Assume otherwise and fix any $p \in \beta Y$ such that $p \in \overline{F \cap Y}$ for each member of the filter $\{F \cap Y : F \in \mathfrak{F}_x\}$. Now, $p \notin Y$ since for each $y \in Y$, some neighborhood filter base for y is contained in M and thus there is an open U in M such that $y \in U$ and $x \in F = X \setminus U$. Therefore there is an $\alpha \in \kappa$ such that $p \in K_\alpha$. Since K_α is in M and $Y \subset M$, there is an open cover of Y, $\mathfrak{U}_\alpha$, in M such that for each $U \in \mathfrak{U}_\alpha$, $K_\alpha \cap \overline{U}$ is empty. Of course, since $\mathfrak{U}_\alpha$ is a cover of Y, $F = X \setminus \mathfrak{U}_\alpha$ is a closed set disjoint from Y. However by our assumption $F \notin \mathfrak{F}_x$, hence there is a $U \in \mathfrak{U}_\alpha$ such that $x \in U$, in which case $\overline{U} \in \mathfrak{F}_x$. Of course this contradicts the assumption that $p \in \overline{\overline{U} \cap Y}$. □

3.3. LEMMA. *If X has countable tightness and is (ω_1, ω_1)-compact then for each countably complete filter of closed sets $\mathfrak{F}$ and each $F \in \mathfrak{F}$, there will be a countable set $D \subset F$ such that $\overline{D}$ meets each member of $\mathfrak{F}$.*

□ Inductively choose points $x_\alpha \in F$ and elements $F_\alpha \in \mathfrak{F}$ so that $F_\alpha \cap \overline{\{x_\beta : \beta \in \alpha\}}$ is empty. By always choosing $x_\alpha \in \bigcap\{F_\beta : \beta \leq \alpha\}$ we will be constructing a free sequence. Since X has countable tightness, the sequence $\{x_\alpha : \alpha \in \omega_1\}$ can not have a complete accumulation point, contradicting that X is (ω_1, ω_1)-compact. Therefore, for some α, $D = \{x_\beta : \beta < \alpha\}$ will be as required. □

3.4. LEMMA. *If Y is first countable, separable and linearly Lindelöf, then $Ch(Y) \leq \mathfrak{c}$.*

□ Let $p \in \beta Y$ and fix ρ minimal such that there is a filter base $\{D_\alpha : \alpha < \rho\}$ of separable subsets of Y such that $p \in F_\alpha = \overline{D_\alpha}$ for each α and $Y \cap \bigcap_{\alpha<\rho} F_\alpha$ is empty. We prove that ρ is countable, which, together with $|Y| \leq \mathfrak{c}$, shows that $\beta Y \setminus Y$ is covered by $\mathfrak{c}$ such countable intersections. Since Y is linearly Lindelöf, this ρ must have countable cofinality. Let $\langle \rho_n : n \in \omega \rangle$ be an increasing cofinal sequence in ρ. Let $G_n = \bigcap\{F_\alpha : \alpha < \rho_n\}$. If some closed intersection of neighborhoods of p is disjoint from $G_n \cap Y$ then this family of neighborhoods together with $\{D_\alpha : \alpha < \rho_n\}$ would contradict the minimality of ρ. Therefore, by Lemma 3.3, there is a separable set $E_n \subset G_n \cap Y$ such that $p \in \overline{E_n}$. Now we obviously have that $\bigcap_{n\in\omega} \overline{E_n}$ is disjoint from Y as required. □

We do not know if the assumption of linearly Lindelöf is required in the previous lemma. However, it can be used to give a simple proof of the more general result.

3.5. THEOREM. *If Y is a first countable linearly Lindelöf space of cardinality $\mathfrak{c}$, then $Ch(Y) \leq \mathfrak{c}$.*

□ Fix an elementary submodel M with $Y \cup M^\omega \subset M$ and $|M| = \mathfrak{c}$. We prove that $\beta Y \setminus Y$ is covered by the compact sets $K \subset \beta Y \setminus Y$ such that $K \in M$. Fix any $p \in \beta Y \setminus Y$. Let $\mathcal{F}_p$ denote the collection of zero-sets of Y which are in M and that have p in their closure. If $\mathcal{F}_p$ is not countably complete then, since $M^\omega \subset M$, we have such a K with $p \in K$ as the intersection of countably many closures of members of $\mathcal{F}_p$. If $\mathcal{F}_p$ is not countably complete then, since $M^\omega \subset M$, we have such a K with $p \in K$ as the intersection of countably many closures of members of $\mathcal{F}_p$. If $\mathcal{F}_p$ is countably complete then by Lemma 3.3, there must be a countable $D \subset Y$ such that $p \in \overline{D}$. Of course $D \in M$ and by Lemma 3.4 there is a collection $\mathcal{K} \in M$ of compact subsets of $\overline{D} \cap (\beta Y \setminus Y)$ with $|\mathcal{K}| \leq \mathfrak{c}$. Of course $D \in M$ and by Lemma 3.4 there is a collection $\mathcal{K} \in M$ of compact sets covering $\overline{D} \cap (\beta Y \setminus Y)$ with $|\mathcal{K}| \leq \mathfrak{c}$. Since $\mathcal{K} \subset M$ there is some $K \in \mathcal{K} \cap M$ with $p \in K$ as required. □

3.6. THEOREM. *If X is first countable and linearly Lindelöf, then X has cardinality at most $\mathfrak{c}$.*

□ Fix an increasing sequence $\{M_\alpha : \alpha \in \omega_1\}$ of elementary submodels, each of cardinality $\mathfrak{c}$, such that for each α, $\{M_\beta : \beta < \alpha\} \cup [M_\alpha]^\omega$ are contained in M_α. In addition, we assume that X and its topology are members of M_0, hence it suffices to prove that M_{ω_1} contains X. Let for $\alpha \in \omega_1$, $\mathcal{U}_\alpha$ equal the collection of all open $W \subset X$ such that $X_\alpha = X \cap M_\alpha$ is a subset of W. Note that $\mathcal{U}_\alpha$ is an element of $M_{\alpha+1}$ and by Lemma 3.2 and 3.5, $X_\alpha = \bigcap \mathcal{U}_\alpha$. Let $x \in X \setminus \bigcup_{\alpha<\omega_1} X_\alpha$ and $W_\alpha \in \mathcal{U}_\alpha$ so that $x \notin W_\alpha$. The sequence $\{W_\alpha : \alpha < \beta\}$ is a member of M_β for each β and so the open cover, $\{W_\alpha : \alpha < \omega_1\}$, of $\bigcup_{\alpha<\omega_1} X_\alpha$ has no countable subcover, contradicting that this closed set should be ω_1-Lindelöf. □

4. Stone-Čech compactification of $\mathbb{N}$

Our first two main results in this section are from FARAH's new book [2000].

4.1. DEFINITION. An ideal $J \subset P(\mathbb{N})$ is *ccc over fin* if there do not exist uncountably many almost disjoint members of $P(\mathbb{N}) \setminus J$. Dually, we will say that a closed set $K \subset \mathbb{N}^*$ is ccc over fin, if for every disjoint open family of subsets of $\mathbb{N}^*$, only countably many will meet K; equivalently, the ideal J, consisting of those $A \subset N$ such that $A^* \cap K = \emptyset$, is ccc over fin.

The next theorem is a special case for the ideal of finite sets of much more general results contained in FARAH [2000]. In his terminology, the function $\varphi_h : P(\mathbb{N}) \to P(\mathbb{N})$ defined by $\varphi(B) = h^{-1}(B)$ is an almost lifting of the homomorphism F. Both of the following two results are remarkable and have powerful applications. These results continue the investigation begun by SHELAH [1982], JUST [1989a], SHELAH and STEPRANS [1988], and VELIČKOVIĆ [1992]. Of course it started with Shelah's result that it was consistent that there are no non-trivial autohomeomorphisms of $\mathbb{N}^*$.

4.2. THEOREM (OCA and MA). *If F is a homomorphism from $P(\mathbb{N})/fin$ onto itself, then there is an h from a subset A of ω into ω so that $F(B) = h^{-1}(B)$ for all B in some ccc over fin ideal J.*

4.3. THEOREM (OCA and MA). *If a subset of $\mathbb{N}^*$ is a continuous image of $\mathbb{N}^*$, then it is equal to the disjoint union of a clopen set and a nowhere dense set.*

This first sample application is proven by DOW and HART in [2000] and the second is an older result of JUST [1989] but it relates to a very interesting problem that is still open. One of the tricks to most applications of 4.2 is to show that there is a member of the ideal on which it is interesting to know that F has a lifting. In an unrelated vein, many readers will also be interested to know that J.T. MOORE [200?] has shown that the conjunction with another version of OCA implies that $\mathfrak{c} = \omega_2$.

4.4. THEOREM (OCA and MA). $\mathbb{N}^*$ *does not map onto the Stone space of the measure algebra.*

4.5. THEOREM (OCA and MA). *The only P-sets of $\mathbb{N}^*$ which are homeomorphic to $\mathbb{N}^*$ are the clopen sets.*

□ Suppose that $K \subset \mathbb{N}^*$ is a P-set and that $CO(K)$ is isomorphic to $P(\mathbb{N})/fin$. By theorem 4.3, we may assume that K is nowhere dense (i.e. it has a relative clopen subset which is nowhere dense in $\mathbb{N}^*$). Setting $F(A) = A^* \cap K$ for $A \subset \omega$, defines a homomorphism onto $CO(K)$. Thus there is an ideal $\mathcal{I}$ which is ccc over fin such that F is induced by h^{-1} for each $B \in \mathcal{I}$. Now K is homeomorphic to $\mathbb{N}^*$, hence there is an uncountable family $\{B_\alpha : \alpha \in \omega_1\}$ of infinite subsets of $\mathbb{N}$ such that $B^*_\alpha \cap K$ are pairwise disjoint. Since K is a P-set and, for each $\beta < \alpha$, $(B_\alpha \cap B_\gamma)^*$ is disjoint from K, there is an infinite $C_\alpha \subset B_\alpha$ such that $C_\alpha \cap B_\beta$ is finite for each $\beta < \alpha$, and $C^*_\alpha \cap K \neq \emptyset$. Thus the C_α's are almost disjoint, hence there is some α such that C_α is in $\mathcal{I}$, i.e. h induces the homomorphism. However this is impossible as can be seen as follows. Let $B = C_\alpha \cap h(A)$ where A is given as in Theorem 4.2. Since $C^*_\alpha \cap K$ is not empty, it follows that B is infinite. However as K is nowhere dense, B has an infinite subset B_1 such that $B^*_1 \cap K$ is empty. It follows then that $F(B_1) = \emptyset \neq^* h^{-1}(B_1)$. □

If D is any countable discrete subset of $\mathbb{N}^*$, then $\overline{D}$ is homeomorphic to $\beta\mathbb{N}$. Therefore, of course, there is a copy of $\mathbb{N}^*$, namely $\overline{D} \setminus D$ which is ccc over fin but which is not itself ccc. In addition, there is a function F from $P(\mathbb{N})/fin$ onto $P(D)/fin$ induced by $F(B) = \{d \in D : B \in d\}$. It is easily seen, just as in the case of the nowhere dense

P-set, that there is no function $h : D \to \mathbb{N}$ which induces F; the ideal J of Theorem 4.2 would be $\{B \subset \mathbb{N} : \overline{B} \cap D = \emptyset\}$

If a P-set is ccc over fin, then it is ccc, and as we see above, if K is contained in a ccc set then K is also ccc over fin. I do not know if the following is a theorem of ZFC. We use Kunen's notion of an ω_1-OK point (see the article by Baker and Kunen in this volume) and such points are not the limit of any ccc subset.

4.6. PROPOSITION. *It follows from* $\diamondsuit$ *that there is a closed set* K *of* $\mathbb{N}^*$ *which is ccc over fin but which is not contained in a ccc subset of* $\mathbb{N}^*$.

□ Construct a sequence of $\mathfrak{c}$-OK points $\{x_\alpha : \alpha \in \omega_1\}$ in $\mathbb{N}^*$ and infinite subsets, $b(\alpha, n)$, of $\mathbb{N}$ as follows. Let $\{a_\beta : \beta \in \omega_1\}$ be an enumeration of $[\omega]^\omega$ and let $\{S_\alpha : \alpha \in \omega_1$ and $\lim(\alpha)\}$ be a $\clubsuit$-sequence ($\diamondsuit$ is equivalent to $CH + \clubsuit$). For each limit β and $\beta < \alpha < \omega_1$:

1. there is a sequence $\langle \gamma_n : n \in \omega \rangle$ increasing cofinal in β such that $b(\beta, n) \in x_{\gamma_n}$
2. if there is $\langle \zeta_n : n \in \omega \rangle$ increasing cofinal in β such that $S_\beta \supset \{\zeta_n : n \in \omega\}$ and $a_{\zeta_n} \in x_{\gamma_n}$, then $b(\beta, n) = a_{\zeta_n}$,
3. x_α is in the closure of $\bigcup\{b(\beta, n)^* : n \in \omega\}$.

At stage α, first check if α is a limit. If condition (ii) holds choose $b(\alpha, n)$ as required, if it does not let $b(\alpha, n) \in x_{\gamma_n}$ for any sequence γ_n cofinal in α. It is easy to check by induction that the family of cozero sets $\{C_\beta : \lim(\beta)$ and $\beta \le \alpha\}$, where $C_\beta = \bigcup\{b(\beta, n)^* : n \in \omega\}$ has the finite intersection property. It is also easy to see that there are uncountably many ω_1-OK points in $\bigcap\{\overline{C_\beta} : \lim(\beta)$ and $\beta \le \alpha\}$ so we may choose x_α in this intersection distinct from x_β for each $\beta < \alpha$.

To see that $K = \overline{\{x_\alpha : \alpha \in \omega_1\}}$ is ccc over fin assume that there is an uncountable $I \subset \omega_1$ such that the family $\{a_\gamma : \gamma \in I\}$ are almost disjoint and that for each $\gamma \in I$ there is a ζ_γ such that $y_\gamma \in x_{\zeta_\gamma}$. There is a cub $C \subset \omega_1$ such that for each $\delta \in C$ and each $\gamma \in \delta$, $\zeta_\gamma \in \delta$. In addition, we may assume that $I \cap \delta$ is cofinal in δ for each $\delta \in C$. Therefore there is a $\beta \in C$ such that $S_\beta \subset I$. By our construction, $\{b(\beta, n) : n \in \omega\}$ is a subset of $\{a_\gamma : \gamma \in S_\beta\}$ and for all $\zeta \in \omega_1 \setminus \beta$ and $a \in x_\zeta$, we have arranged that $a \cap b(\beta, n)$ is infinite for some n. Of course this implies that I is countable. In addition, since each x_α is an ω_1-OK point, no x_α is a limit point of a ccc subspace of $\mathbb{N}^*$, hence $\{x_\alpha : \alpha \in \omega_1\}$ is not contained in a ccc subset of $\mathbb{N}^*$. □

A subset $K \subset \mathbb{N}^*$ which has the form $\overline{D} \setminus D$ for some countable discrete $D \subset \beta\mathbb{N}$ is a trivially occurring copy of $\mathbb{N}^*$. It is unknown if there is a copy of $\mathbb{N}^*$ in $\mathbb{N}^*$ which is not trivially occurring. It is interesting to note that each copy of $\mathbb{N}^*$ in $2^{\mathfrak{c}}$ is trivially occurring. It may be interesting to note the following.

Consider the case that X is the free union of a family $\{X_n : n \in \omega\}$ where each X_n is a copy of the Cantor set (any compact ccc space with no isolated points would be interesting). A filter $\mathcal{F}$ on a family $\{X_n : n \in \omega\}$ is called *nice* if for each $F \in \mathcal{F}$, there are only finitely many n such that $F \cap X_n$ is empty. Let us say that a nice filter is *maximal* if for each sequence $\{W_n : n \in \omega\}$ of clopen sets with $W_n \subset X_n$, there is an $F \in \mathcal{F}$, such that for each n, either $F \cap W_n$ is empty or $W_n \subset F$. A filter on a space X is *remote* if for each nowhere dense set the filter has a member completely separated from it.

4.7. PROPOSITION. *If there is a remote maximal nice filter on X, then there is a copy of $\mathbb{N}^*$ which is not trivially occurring.*

It seems at least possible that MA implies there is a maximal remote nice filter on X.

5. Distributivity of $\mathbb{N}^* \times \mathbb{N}^*$

Another result we report on is the following result of SHELAH and SPINAS [2000]. They prove that $\mathfrak{h}(\mathbb{N}^* \times \mathbb{N}^*) = \omega_1$ in the Mathias model (which is known to satisfy $\mathfrak{h}(\mathbb{N}^*) = \omega_2$).

5.1. DEFINITION. The cardinal number $\mathfrak{h}(X)$ is the minimum number of dense open subsets of X whose intersection has empty interior. The number $\mathfrak{h}(\mathbb{N}^*)$ is usually denoted by $\mathfrak{h}$ and is equal to the distributivity degree of $P(\mathbb{N})/\text{fin}$.

In his MR review (MR2001f:03095), A. Blass writes:

> "This result follows fairly easily from two (difficult) propositions that are of considerable interest in their own right."

We were interested to see if the approach of DOW [1998] could also be used, that is, to determine a combinatorial property of a tree π-base which will guarantee that the forcing will not diagonalize any branches (new or old). The combinatorial property used by DOW in [1998] for the reals was remote filter. Just as Blass says, the result follows easily once we know how to construct such a tree. We leave that aspect of the construction to the interested reader, we satisfy ourselves with re-proving that, if CH is assumed, then there is a tree π-base for $\mathbb{N}^* \times \mathbb{N}^*$ which is not diagonalized by the Mathias iteration.

5.2. THEOREM. *In the model obtained by adding ω_2-Mathias reals to a model of CH, $\mathfrak{h}(\mathbb{N}^* \times \mathbb{N}^*) = \omega_1 < \mathfrak{h} = \omega_2$.*

5.3. DEFINITION. The Mathias poset $\mathbb{M}$ consists of pairs (a, A) where $a \in [\omega]^{<\omega}$ and $A \in [\omega \setminus (\max a)]^\omega$. The ordering is defined by $(a, A) < (b, B)$ if $b \subset a$ and $a \subset b \cup B$. In the iteration, let $\dot{g}_\alpha$ denote the enumeration of the α-th Mathias real.

The Mathias real $\mathfrak{m}$, given by a generic filter $G \subset \mathbb{M}$, equals to $\bigcup\{a : (\exists A)((a, A) \in G)\}$. The strictly increasing enumerating function $e_\mathfrak{m}$ of $\mathfrak{m}$, eventually dominates all the ground model functions $f \in \omega^\omega$. It is useful to notice that a condition (a, A) precisely determines the values of $e_\mathfrak{m}(i)$ for each $i < |a|$ and leaves available exactly the values in A for $e_\mathfrak{m}(|a|)$. In the iteration, let $\dot{g}_\alpha$ denote the enumerating function of the α-th Mathias real.

The Mathias poset $\mathbb{M}$ has two important structural properties. The first is a consequence of Ramsey-like properties of $[\omega]^\omega$. It is called the *Prikry property* and it says that given $(a, A) \in M$ and some formula ϕ there is an infinite subset B of A such that (a, B) decides ϕ, i.e., either $(a, B) \Vdash \phi$ or $(a, B) \Vdash \neg\phi$. In practice this is often used in the following form: given a condition (a, A), a name $\dot{x}$ and a finite set F (from the ground model) then there is an infinite subset B of A such that either $(a, B) \Vdash (\dot{x} \notin F)$ or there is an $f \in F$ with $(a, B) \Vdash (\dot{x} = f)$.

Even though the counting function of the Mathias real is dominating we do have some control on the new reals: if $f \in {}^\omega\omega$ and if $\dot{x}$ is a name for a function such that $p \Vdash$

$(\forall n)(\dot{x}(n) < f(n))$ then there are a condition $q \leq p$ and a sequence $\langle F_n : n \in \omega\rangle$ of finite sets such that $p \Vdash (\dot{x}(n) \in F_n)$ and $|F_n| \leq 2^n$ for all n. This property is called the *Laver property* [1982] and in it the function $n \mapsto 2^n$ can be replaced by any increasing unbounded function like $\log n$.

Now the next result is a consequence of the Laver property holding over the ground model $V[G_0]$ where $G_0 = G \cap P_1$ for a P_{ω_2}-generic filter G and P_1 would be isomorphic to $\mathbb{M}$. For convenience, let us treat the iteration as $\mathbb{M} * P_{\omega_2}$.

5.4. PROPOSITION. *If $\{\dot{x}_n : n \in \omega\}$ is a $\mathbb{M} * P_{\omega_2}$-name of a subset of ω such that for each n, $1 \Vdash \dot{g}_0(n) < \dot{x}_n$, and $p \in \mathbb{M} * P_{\omega_2}$ is given, then there is an $q < p$ and a sequence $\{\dot{y}(n,i) : n \in \omega, i < 2^n\}$ of $\mathbb{M}$-names such that $q \Vdash \{\dot{x}_m : m \leq n\} \cap [g_0(n), g_0(n+1)] \subset \{\dot{y}(n,i) : i < 2^n\} \subset [g_0(n), g_0(n+1)]$.*

The next result is a standard approach using Mathias forcing and repeatedly applying the Prikry property so as to thin out the set B until the desired A is found after an induction of length ω.

5.5. PROPOSITION. *Let $\{\dot{y}(n,i) : n \in \omega, i < 2^n\}$ be a sequence of $\mathbb{M}$-names as in Proposition 5.4 and let $(b, B) \in \mathbb{M}$, then there is an $A = \{a_n : n \in \omega\} \subset B$ such that*

$$(\forall m \leq n)(\forall i < 2^m)(\forall k \leq n)(\forall e \subset a_n + 1) \text{ either } (e, \{a_\ell : n < \ell \in \omega\}) \Vdash k = \dot{y}(m,i) \text{ or } (e, \{a_\ell : n < \ell \in \omega\}) \Vdash k \neq \dot{y}(m,i)$$

and, for each $m \leq n$, $i < 2^m$ and $e \subset a_n + 1$ if there is $A' \subset A$ and k such that $(e, A') \Vdash k = \dot{y}(m,i)$, then $k < a_{n+1}$.

Once we have the name $\{\dot{x}_n : n \in \omega\}$ and the condition (b, A) in the right form as given by the previous two results we can present the desired combinatorial property of the elements of the desired π-basis for the product $\mathbb{N}^* \times \mathbb{N}^*$.

5.6. PROPOSITION. *Assume that E, F are disjoint infinite subsets of ω such that for each $e \in E$ and $f \in F$, $|A \cap ((e,f) \cup (f,e))| > 2$ then (b, A) forces that one of the following two sets, $\{n : E \cap \{\dot{y}(2n,i) : i < 2^{2n}\} \neq \emptyset\}$ or $\{n : F \cap \{\dot{y}(2n+1,i) : i < 2^{2n+1}\} \neq \emptyset\}$ is finite.*

□ Given any $(b_1, A_1) \leq (b, A)$ choose $A_2 \subset A_1$ so that either $a \in A_2$ implies $\min((E \cup F \cup A) \setminus a+1) \in E$ or $a \in A_2$ implies $\min((E \cup F \cup A) \setminus a+1) \in F$. Assume the former and let $n > \max(b_1)$ and $(b_2, A_3) < (b_1, A_2)$ such that $(b_2, A_3) \Vdash \dot{y}(2n,i) = e \in E$ and $\dot{y}(2n+1,j) = f \in F$.

By the assumption on (E, F), let $a_1 < a_2 < a_3 < f < a_4 < a_5 < a_6$ where $a_i \in A$ and $E \cap (a_1, a_6) = \emptyset$. Note then that $A_2 \cap [a_1, a_5]$ is empty. Since $(b_2, A_2) \Vdash y(n,j) = f$, it follows that $|b_2 \cap f| \geq n$ (since $1 \Vdash \dot{g}_0(n) \leq \dot{y}(n,j)$) and, since $b_2 \cap f \subset a_1$, it follows that $n \leq a_1$.

By the construction of A it follows that $(b_2 \cap a_4, A \setminus a_4 + 1)$ forces that $\dot{y}(n,j) = f$. Again since $b_2 \setminus b_1 \subset A_2$, and the properties of A, it follows that $(b_2 \cap a_2, A_2 \setminus a_2 + 1)$ forces that $\dot{y}(n,j) = f$ which is supposed to imply that $f < a_3$ - a contradiction. □

One can complete the proof that $\mathfrak{h}(\mathbb{N}^* \times \mathbb{N}^*) = \omega_1$ by constructing (along with careful bookkeeping) a tree π-base for $\mathbb{N}^* \times \mathbb{N}^*$ so that for each name $\{\dot{y}(n,i) : n \in \omega, i < 2^n\}$ and (b, A) as above, there is a maximal antichain of the tree consisting of pairs (E, F) as in Proposition 5.6. Nonetheless, certainly Proposition 5.6 can be used to show that (assuming CH) there is a tree π-base for $\mathbb{N}^* \times \mathbb{N}^*$ which is not diagonalized by the iteration of Mathias forcing as follows. Fix an enumeration in order type ω_1 of all the combinations of conditions $(b, A) \in \mathbb{M}$ and $\mathbb{M}$-names $\{\{\dot{y}(n,i) : i < 2^n\} : n \in \omega\}$ as in Proposition 5.4. The tree π-base would consist of pairs (E, F) at level α which mod finite satisfy the condition of Proposition 5.6 with respect to all the names with index less than α. Now consider any Mathias name $\{\dot{x}_n : n \in \omega\}$ of a subset of N with the property that the pair $(\{\dot{x}_{2n} : n \in \omega\}, \{\dot{x}_{2n+1} : n \in \omega\})$ diagonalizes the tree. By passing to a subset we may assume that $g_0(n) < x_n$ for all n. Obtain the $\mathbb{M}$-names $\{\{\dot{y}(n,i) : i < 2^n\} : n \in \omega\}$ and (b, A) from Propositions 5.4 and 5.5 for this name. Some extension (b', A') of (b, A) will have to force that $(\{\dot{x}_{2n} : n \in \omega\}, \{\dot{x}_{2n+1} : n \in \omega\})$ is mod finite below some (E, F) from the tree on a level so as to satisfy 5.6 with respect to (a, A) and $\{\{\dot{y}(n,i) : i < 2^n\} : n \in \omega\}$. But we have a contradiction from Proposition 5.6 since we will get that one of E or F is almost disjoint from $\{\dot{x}_n : n \in \omega\}$ since this latter set is forced to be contained in the union of $\{\{\dot{y}(n,i) : i < 2^n\} : n \in \omega\}$.

6. Countable tightness in compact spaces

The following two results were proven by the author and represent a strongly held interest in the topic of the section. It is a pleasure to report on brilliant improvements by RABUS [1996], KOSZMIDER [1999] and EISWORTH [2001]. In the case of Eisworth (and Theorem 6.10), the result we include is building on an earlier paper by EISWORTH and ROITMAN [1999]. Each of these are major results and we can do little more than to outline the main ideas of the constructions.

This first result was also proven independently by van Douwen.

6.1. THEOREM (DOW [1980]). *Under CH, each initially ω_1-compact space of countable tightness is compact.*

6.2. THEOREM (DOW [1988]). *It is consistent with $MA(\omega_1)$ that every compact space of countable tightness contains a point of countable character.*

Rabus proved the following.

6.3. THEOREM. *It is consistent that there is an initially ω_1-compact space of countable tightness which is not compact.*

We will be working with special Boolean subalgebras of $\mathcal{P}(\omega_2)$ and the relationships between them. For each $L \subset \omega_2$ say that a Boolean algebra B is L-minimal if there is $\{a_x : x \in L\} \subseteq B$ with the following properties:

(1) B is generated by $\{a_x : x \in L\}$ with $x \in a_x \subset [0, x] \subset \omega_2$,

(2) if $x < y$, then $a_x \cap a_y \in B_x$, where B_x is the subalgebra of B generated by $\{a_z : z \leq x\}$.

The space we seek will be the Stone space of a particular ω_2-minimal Boolean algebra. This Boolean algebra will be constructed by forcing with a poset of finite minimal Boolean algebras as above with a particular order relation. It is easy to describe the structure of the Stone spaces of such algebras. Let B be a ω_2-minimal algebra generated by $\{a_\alpha : \alpha \in \omega_2\}$ and let X be its Stone space. The underlying set is the ordinal $\omega_2 + 1$ and the neighborhood base for the point $\alpha \in \omega_2$ is simply given by $\{a_\alpha - \bigcup_{\beta \in s} a_\beta : s \in [\alpha]^{<\omega}\}$, similarly, the neighborhood base for ω_2 is obtained by treating a_{ω_2} as 1.

Suppose that B is a K-minimal Boolean algebra, $K \in [\omega_2]^{<\omega}$, generated by $\{a_x : x \in K\}$. Of course B is atomic with atoms $b_x = a_x - \bigcup\{a_y : y < x\}$ and possibly $b_K = \omega_2 \setminus \bigcup\{a_x : x \in K\}$ if it is non-empty. When we speak of the atoms of B we will actually be ignoring b_K and be most interested in $\{b_x : x \in K\}$.

Let $L \subseteq K$. We say that L generates a subalgebra of B if $\{a_x : x \in L\}$ generates an L-minimal algebra. This is equivalent to say that for $x < y$ in L, the intersection $a_x \cap a_y$ is in the algebra generated by $\{a_v : v \leq x,\ v \in L\}$. Note that if L is an initial part of K, then L generates a subalgebra of B.

Suppose now that K' is another finite subset of the same size as K. Let $\Delta = K \cap K'$ and let B' be a K'-minimal algebra generated by $\{a'_y : y \in K'\}$. Suppose that the structures (K, Δ) and (K', Δ) are isomorphic, i.e, Δ has the same position in both K and K'. Suppose also that the isomorphism between (K, Δ) and (K', Δ) extends to the isomorphism between B and B'. Finally, assume that Δ generates a subalgebra of B, (thus also B').

Let $L = K \cup K'$, and define the minimal amalgamation of B and B' to be an L-minimal Boolean algebra C generated by $\{c_x : x \in L\}$ such that C restricted to K is isomorphic to B and C restricted to K' is isomorphic to B'. Moreover, the amalgamation is minimal in a sense that the intersection, $c_x \cap c_y$ where $x \in K - \Delta$, $y \in K' - \Delta$, is as small as possible. In particular if $\Delta = \emptyset$, then $c_x \cap c_y = \emptyset$.

For the sake of readability in this definition, we may assume that $B \subset \mathcal{P}(K)$, hence b_x is simply x and similarly for $B' \subset \mathcal{P}(K')$ although it is not necessary. The new algebra will be a subalgebra of $\mathcal{P}(L)$. We define some auxiliary sets. For $x \in K - \Delta$ let $D_x = \{x\}$, for $z \in \Delta$ let $D_z = a'_z - \bigcup\{a'_v : v \in \Delta, v < z\}\}$ and $D'_z = a_z - \bigcup\{a_v : v \in \Delta, v < z\}$. For $y \in K' - \Delta$ let $D_y = \{y\}$. Then for $x \in K \setminus \Delta$ we get $c_x = a_x \cup \bigcup\{D_\beta : \beta \in a_x \cap \Delta\}$ and similarly for $x' \in K' \setminus \Delta$. For $z \in \Delta$, we have $c_z = \bigcup\{D_y \cup D'_y : y \in a_\alpha \cap \Delta\}$.

The poset of finite minimal Boolean algebras on ω_2 when ordered by simple embeddability is not ccc. The idea introduced next is a device for extracting a subposet which will be ccc. This idea originates with BAUMGARTNER and SHELAH[1987]

A Δ *function* is a function $f : [\omega_2]^2 \to [\omega_2]^{\leq\omega}$ with the following properties:

(1) $f\{x, y\} \subseteq \min\{x, y\} + 1$ and $\min\{x, y\} \in f\{x, y\}$.

(2) For every uncountable family D of finite subsets of ω_2 there are distinct $a, b \in D$ such that whenever $x \in a - b$, $y \in b - a$ and $z \in a \cap b$:

   (a) if $x, y > z$, then $z \in f\{x, y\}$,

   (b) if $y > z$, then $f\{x, z\} \subseteq f\{x, y\}$,

   (c) if $x > z$, then $f\{y, z\} \subseteq f\{x, y\}$.

It has been shown by BAUMGARTNER and SHELAH in [1987], that a Δ function can be forced by a σ-closed ω_2-cc poset P. In addition, Todorčević has shown that such a function exists whenever there is a ρ-function as in TODORČEVIĆ [1987] and BEKKALI [1991], hence the non-existence of a Δ function implies there are large cardinals.

We now present the ccc poset Q which forces an ω_2-minimal Boolean algebra A generated by $\{a_\alpha : \alpha \in \omega_2\}$. A pair (B, L) is a condition in Q if $L = \{x_1, \ldots, x_k\}$ is a subset of ω_2, and B is a L-minimal Boolean algebra generated by some $\{c_x = B(x) : x \in L\}$. We will abuse notation and assume that B denotes the algebra as well as the function which selects the generators. For the most part we will assume that the generators are clear from the context. If $x, y \in L$ the element $c_x \cap c_y$ is in the Boolean algebra generated by $\{c_z : z \leq \min\{x, y\}$ and $z \in f\{x, y\}\}$.

A condition (B', L') extends (B, L) if $L \subseteq L'$ and L generates a subalgebra of B' which is isomorphic to B.

6.4. LEMMA. *The forcing Q has the ccc.*

□ Let $\{(B_\alpha, L_\alpha) : \alpha \in \omega_1\}$ be an uncountable subset of Q. By thinning out we can assume that $\{L_\alpha : \alpha \in \omega_1\}$ form a Δ-system with the root Δ; for every $\alpha \neq \beta$, the structures (L_α, Δ) and (L_β, Δ) are isomorphic and the isomorphism lifts to the isomorphism between B_α and B_β. Let $A \in [\omega_2]^\omega$ be such that A is closed under f and $\Delta \subseteq A$. Let α, β, $\alpha \neq \beta$ be such that $L_\alpha \cap A = \Delta$ and $L_\beta \cap A = \Delta$ and L_α, L_β satisfy conditions $(a), (b), (c)$ of a Δ-function. Note that it follows that Δ generates subalgebras of both B_α and B_β.

Let $L = L_\alpha \cup L_\beta$ and let C be the minimal amalgamation of B_α and B_β. Suppose that C is generated by $\{c_x : x \in L\}$. We have to prove that (C, L) is a condition in Q. To see this it is enough to prove that if $x \in L_\alpha - \Delta$, $y \in L_\beta - \Delta$, $x < y$, then the intersection $c_x \cap c_y$ is in the algebra generated by $\{c_v : v \leq \min\{x, y\}, v \in L \cap f\{x, y\}\}$. Assume e.g. that $x < y$. For $z \in \Delta$, $z < x$, the intersection $c_x \cap c_z$ is in the algebra generated by $\{c_v : v \leq z, v \in L_\alpha \cap f\{z, x\}\}$, and also $c_y \cap c_z$ is in the algebra generated by $\{c_v : v \leq z$, $v \in L_\beta \cap f\{z, y\}\}$. Note that we have that $f\{x, z\} \cup f\{y, z\} \subseteq f\{x, y\}$. Therefore the above intersections are in the algebra generated by $\{c_v : v \leq x, v \in L \cap f\{x, y\}\}$. Similarly we show that for $w \in \Delta$, $x < w < y$ we have $c_x \cap (c_w - \bigcup\{c_v : v < w, v \in \Delta\})$ is in the algebra generated by $\{c_v : v \leq x, v \in L \cap f\{x, y\}\}$. This follows from the fact that $f\{x, v\} \subseteq f\{x, y\}$ for $v \in \Delta$, $v < y$. □

Note that Q forces an ω_2-minimal algebra A. If G is Q-generic, then for each $\alpha \in \omega_2$, $a_\alpha = \{\beta \leq \alpha : (\exists (B, L) \in G)\ (\beta, \alpha \in L$ and $B(\beta) \in B(\alpha))\}$. The desired space is ω_2 with the topology induced by the Stone space of B, i.e. $S(B) \setminus \omega_2$. It is reasonably straightforward to show that $S(B)$ has countable tightness and that ω_2 is not the unique limit of any ω_1-sequence in ω_2 (see RABUS [1996]). To complete the proof of initial ω_1-compactness we must show that every countably infinite set has an accumulation point in ω_2, which is equivalent to showing that no ω-sequence in ω_2 converges to ω_2.

The following result is the most important new idea which distinguishes it from the algebra in BAUMGARTNER and SHELAH [1987]. As we will see, the proof (lifted liberally from RABUS [1996]) is very involved. The idea is roughly as follows. A condition q is chosen that decides a clopen neighborhood of a sufficiently large γ which is disjoint from a countable sequence. The condition q is *reflected* to an isomorphic condition below γ which in turn is extended to absorb a member of the sequence. The key step is the invention of a way to extend this new condition in such a way to guarantee that q is compatible with an extension that allows a_γ to be maximal (rather than the usual minimal).

6.5. LEMMA. *There are no ω-sequences in X converging to ω_2.*

Its proof depends on the following property of the Δ-function f.

6.6. LEMMA. *Let $\gamma \in \omega_2$, $cf(\gamma) = \omega_1$. Let $B \in [\gamma]^\omega$, $E \in [\omega_2 - \gamma]^{<\omega}$, $E = \{x_1, \ldots, x_k\}$. Then there is $A \in [\gamma]^\omega$ containing B and $E' \in [\gamma - \delta]^k$, $E' = \{x'_1, \ldots, x'_k\}$ and $F \in [[\gamma]^2]^k$ such that $F = \{\{y_i^1, y_i^2\} : i \leq k\}$ and letting $D = E' \cup \bigcup F$ we have:*

(1) *For all $i, j \leq k$, $\sup(A) < x'_i < y_j^1 < y_j^2$.*

(2) *For $i < j$, $y_i^2 < y_j^1$.*

(3) *If $z \in D \cup A$ and $y \in A$, then $f\{z, y\} \subseteq A$.*

(4) *For $i \leq k$, if z is any element of D with a subscript i, then $f\{x_i, y\} = f\{z, y\}$ for every $y \in A$.*

(5) *For $i \leq k$, for every $v \in D$, $(A \cup \{z : z \in D, z < v\}) \subseteq f\{x_i, v\}$.*

(6) *For every $v, w \in D$, if $v < w$, then $(A \cup \{z : z \in D, z < v\}) \subseteq f\{v, w\}$.*

We also recall the following from RABUS [1996].

6.7. LEMMA. *Let $A \subseteq \omega_2$ be closed under f. Suppose that $q_1 = (B_1, L_1)$ and $q_2 = (B_2, L_2)$ are two conditions in Q such that $L_1 \subseteq A$, $D = L_2 - A$ is such that if $y \in A$ and $z \in D$ then $y < z$ and $A \cup D$ is closed under f. Suppose that a condition $q = (B, L)$ extends both q_1 and q_2. Then the restriction of q to $A \cup D$ is a condition in Q extending q_1 and q_2.*

PROOF OF LEMMA 6.5. Suppose that $\{x_n : n \in \omega\} \subseteq \omega_2$ is a sequence converging to ω_2. We work in the model V^P. For $n \in \omega$ let A_n be a maximal antichain in Q that determines x_n, i.e., for each $u \in A_n$ there is some $\alpha_n \in L(u)$ such that $u \Vdash x_n = \alpha_n$. Let $\gamma \in \omega_2$, $\mathrm{cf}(\gamma) = \omega_1$ be such that for every n, for every $u \in A_n$, $L(u) \subseteq \gamma$.

Since a_γ is a clopen set avoiding the point ω_2, there is some condition $q \in Q$ and $m \in \omega$ such that $q \Vdash x_n \notin a_\gamma$ for every $n > m$. Let $q = (B(q), L(q))$ and let $B(q)$ be generated by $\{a_x(q) : x \in L(q)\}$ with the atoms $\{d_x(q) : x \in L(q)\}$. We can assume that $\gamma \in L(q)$. Then $L(q) = L' \cup E$, where $L' = L(q) \cap \gamma$, $E = L(q) - L'$. Let $\{x_1, \ldots, x_k\}$ be an enumeration of E in the increasing order. Note that $x_1 = \gamma$. Let $A \in [\gamma]^\omega$ be such that A contains $L(q) \cap \gamma$ and *suppp*(u) for every $u \in A_n$, $n \in \omega$ and let E', F be as in Lemma 6.6.

Let us now give the idea of the rest of the proof. Our intention is, of course, to find a condition $r \leq q$ and $n > m$ such that $r \Vdash x_n \in a_\gamma$. To do this we first find a condition s, such that $L(s) = L' \cup E'$, and $B(s)$ is isomorphic to $B(q)$ via bijection from E to E'.

Let U be a name for $\bigcup\{a_x : x \in L(s)\}$. If $p \in Q$ is any condition such that $L(s) \subseteq L(p)$, then by $U(p)$ we denote $\bigcup\{a_x(p) : x \in L(s)\}$. In particular $p \Vdash U(p) = U$. Next we find an auxiliary condition $t \leq s$ such that $L(t) = L(s) \cup \bigcup F$, $B(t)$ restricted to $L(s)$ is isomorphic to $B(s)$. Since t forces that U is a clopen set disjoint from $\{\omega_2\}$ and $\{x_n : n \in \omega\}$ converges to the point ω_2, we can find a condition $t' \leq t$ and $n > m$ such that $t' \Vdash x_n \notin U$. We can assume that t' extends some condition $u \in A_n$, so there is some $\alpha_n \in L(t')$ such that $t' \Vdash \alpha_n = x_n$. By Lemma 6.7, since $L(u) \subseteq A$, we can assume that

$L(t') - L(t) \subseteq A$. Let W be a name for $\bigcup\{a_x(t') : x \in L(t')\}$. For any condition p such that $L(t') \subseteq L(p)$, let $W(p) = \bigcup\{a_x(p) : x \in L(t')\}$.

Finally we define $r \leq t'$. Our intention is to define r such that r extends q and $W(r) - U(r) \subseteq a_\gamma$. Then $r \Vdash x_n \in a_\gamma$, a contradiction.

To finish the proof we show that s and t are conditions in Q, then we define r and show that it has required properties.

Recall that $L(s) = L' \cup E'$ and the bijection between $L(s)$ and $L(q)$, constant on L', lifts to an isomorphism of $B(s)$ with $B(q)$. Hence for $y < x$ in $L(s)$, if $y, x \in L'$, then $d_y(s) \in a_x(s)$ iff $d_y(q) \in a_x(q)$. If $y \in L'$, $x \in E'$, then $x = x'_i$ for some $i \leq k$ and $d_y(s) \in a_{x'_i}(s)$ iff $d_y(q) \in a_{x_i}(q)$. Finally, if $y, x \in E'$, then $y = x'_j$, $x = x'_i$ and $d_{x'_j}(s) \in a_{x'_i}(s)$ iff $d_{x_j}(q) \in a_{x_i}(q)$.

We check that s is a condition in Q. Let $y < x$ in $L(s)$. We have to show that $a_x(s) \cap a_y(s)$ is in the algebra generated by $\{a_v(s) : v \leq y,\ v \in L(s) \cap f\{x, y\}\}$. If $x, y \in L'$ we have nothing to do since the isomorphism is constant on L'. Assume that $y \in L'$, $x = x'_i$. Then $a_y(s) \cap a_{x'_i}(s)$ has the same representation by $\{a_v(s) : v \leq y\}$ as $a_y(q) \cap a_{x_i}(q)$ by $\{a_v(q) : v \leq y\}$. Moreover, since q is a condition the intersection $a_y(q) \cap a_{x_i}(q)$ is in the algebra generated by $\{a_v(q) : v \in L(q) \cap f\{x_i, y\}\}$. By Lemma 6.6(4), we have $f\{x_i, y\} = f\{x'_i, y\}$, hence the intersection $a_y(s) \cap a_{x'_i}(s)$ is in the algebra generated by $\{a_v(s) : v \in L(s) \cap f\{x'_i, y\}\}$.

If $y = x'_j$, $x = x'_i$ with $j < i$, then by Lemma 6.6(6), $\{v : v \in L(s),\ v \leq y\} \subseteq f\{x'_i, x'_j\}$, hence $a_{x'_j}(s) \cap a_{x'_i}(s)$ is in the algebra generated by $\{a_v(s) : v \in L(s) \cap f\{x'_i, x'_j\}\}$ and we are done.

Now we define t, extending s. Let $L(t) = L(s) \cup \bigcup F$. We define $B(t)$ such that $B(t)$ restricted to $L(s)$ is isomorphic to $B(s)$. We define $a_{y_i^1}(t)$ and $a_{y_i^2}(t)$ for $i = 1, \ldots, k$ such that $a_{y_i^\mu}(t) \cap a_{y_j^\nu}(t) = U(t)$ for every $(\mu, i) \neq (\nu, j)$. Recall that $U(t) = \bigcup_{v \in L(s)} a_v(t)$. To check that t is a condition note that if $(\mu, i) \neq (\nu, j)$, then by Lemma 6.6(6), $L(s) \subseteq f\{y_i^\mu, y_j^\nu\}$.

Recall that $t' \leq t$ is a condition such that $L(t') - L(t) \subseteq A \cap \gamma$. Finally we define r. Put $L(r) = L(t') \cup E$. The algebra $B(r)$ restricted to $L(t')$ is isomorphic to $B(t')$. We have to define $a_{x_i}(r)$ for $i = 1, \ldots, k$. Recall that $W(r) = \bigcup\{a_x(r) : x \in L(t')\}$. For $x \in L(q)$ we define auxiliary sets D_x as follows. If $x \in L'$, then put $D_x = a_x(r) - \bigcup\{a_y(r) : y < x,\ y \in L'\}$. Assume $x \in E$, i.e., $x = x_i$ for some i. Assume first that $i = 1$. Put $D_{x_1} = (a_{x'_1}(r) - \bigcup\{a_y(r) : y < x'_1,\ y \in L(s)\}) \cup (W(r) - U(r)) \cup \{d_{x_1}(r)\}$. For $i > 1$ define $D_{x_i}(r) = (a_{x'_i}(r) - \bigcup\{a_y(r) : y < x'_i,\ y \in L(s)\}) \cup \{d_{x_i}(r)\}$.

Now define $a_{x_i}(r)$ for $i \leq k$. Each $a_{x_i}(r)$ is a union of some sets D_x, $x \in L(q)$. We put $D_x \subseteq a_{x_i}(r)$ if $d_x(q) \in a_{x_i}(q)$, otherwise we put $D_x \cap a_{x_i}(r) = \emptyset$. Note that this is well defined since the sets D_x are pairwise disjoint.

CLAIM. For $x_i \in E$, $a_{x_i}(r) \cap U(r) = a_{x'_i}(r)$. Moreover $a_{x_i}(r) \cap W(r) = a_{x_i}(r) \cap U(r)$ if $d_{x_1}(q) \notin a_{x_i}(q)$ and $a_{x_i}(r) \cap W(r) = (a_{x_i}(r) \cap U(r)) \cup (W(r) - U(r))$ if $d_{x_1}(q) \in a_{x_i}(q)$.

Proof. Note that $a_{x_i}(r) \cap U(r)$ is a union of sets of the form $D'_x = a_x(r) - \bigcup\{a_y(r) : y < x,\ y \in L(s)\}$, for $x \in L(s)$. Since $B(r)$ restricted to $L(s)$ is isomorphic to $B(s)$ and the bijection between E' and E lifts to the isomorphism between $B(s)$ and $B(q)$, it follows that $a_{x'_i}(r)$ is also a union of sets D'_x for $x \in L(s)$. Moreover $D'_x \subseteq a_{x'_i}(r)$ if and only if $D'_x \subseteq a_{x_i}(r) \cap U(r)$. The second part is obvious by the definition of $a_{x_i}(r)$.

Consider now $B(r)$ restricted to $L(q)$. For $x \in L(q)$ the set $a_x(r) - \bigcup\{a_y(r) : y < x, y \in L(q)\}$ is equal to D_x. By the definition, $D_y \subseteq a_x(r)$ if $d_y(q) \in a_x(q)$, and $D_y \cap a_x(r) = \emptyset$ otherwise. Hence the restriction of $B(r)$ to $L(q)$ is isomorphic to $B(q)$.

Finally we have to show that r is a condition in Q, i.e., we have to show that if $y < x$ in $L(r)$, then $a_x(r) \cap a_y(r)$ is in the algebra generated by $\{a_v(r) : v \in L(s) \cap f\{x, y\}\}$. Since $B(r)$ restricted to $L(t')$ is isomorphic to $B(t')$ we can assume that $x \in E$, i.e., $x = x_i$ for some $i \leq k$.

Assume first that $y \in L(t') - (E' \cup \bigcup F)$. By the definition, since $a_y(r) \subseteq W(r)$ for $r \in L(t')$ we have $a_{x_i}(r) \cap a_y(r)$ is equal to either $a_{x'_i}(r) \cap a_y(r)$ or $(a_{x'_i}(r) \cap a_y(r)) \cup (a_y(r) - U(r))$. Recall that $U(r) = a_{y_i^1}(r) \cap a_{y_i^2}(r)$. Hence $a_y(r) - U(r) = a_y(r) - ((a_y(r) \cap a_{y_i^1}(r)) \cap (a_y(r) \cap a_{y_i^2}(r)))$. Moreover, since $(L(t') - (E' \cup \bigcup F)) \subseteq A$, it follows by Lemma 6.6(4) that the sets $f\{y, x_i\}$, $f\{y, x'_i\}$, $f\{y, y_i^1\}$, $f\{y, y_i^2\}$ are equal and we are done.

Assume that $y \in E'$. Then $y = x'_j$ for some $j \leq k$ and then, since $a_{x'_j}(r) \subseteq U(r)$, we have $a_{x_i}(r) \cap a_{x'_j}(r) = a_{x'_i}(r) \cap a_{x'_j}(r)$. But, by Lemma 6.6 (5), we have $\{v : v \in L(t'), v \leq x'_j\} \subseteq f\{x_i, x'_j\}$ and of course $a_{x'_i}(r) \cap a_{x'_j}(r)$ is in the algebra generated by $\{a_v(r) : v \in L(t'), v \leq x'_j\}$.

Finally, if $y \in \bigcup F$, say $y = y_j^1$, then $a_{x_i}(r) \cap a_{y_j^1}(r)$ is equal either to $a_{x'_i}(r)$ or $a_{x'_i}(r) \cup (a_{y_j^1}(r) - U(r))$. In the first case note that $x'_i \in f\{x_i, y_j^1\}$. In the second case recall that $U(r) = \bigcup\{a_v(r) : v \in L(s)\}$ and $L(s) = L' \cup E'$. Hence $a_{y_j^1}(r) - U(r) = a_{y_j^1}(r) - \bigcup\{a_{y_j^1}(r) \cap a_v(r) : v \in L(s)\}$. By Lemma 6.6 (5) it follows that $L(s) \subseteq f\{y_j^1, x_i\}$. Hence r is a condition in Q. This finishes the proof of the lemma. □

This next result is again an impressive forcing construction by KOSZMIDER [1999]. Spaces of countable tightness in which there are no points of first countability have been constructed from ◇ (FEDORCHUK [1975]) and consistent with any cardinal arithmetic (MALYHIN [1987]). However this is the first example, in any model, of a first countable space with such a continuous image.

6.8. THEOREM. *It is consistent* (*with MA*) *that there is a first countable space which maps onto a space* (*of countable tightness*) *in which there are no points of countable character.*

The paper KOSZMIDER [1999] explores many applications of the following forcing

$$P(A, u) = \{(p_{-1}, p_1) \in [A]^2 : (p_{-1} \cup p_1) \notin u, p_{-1} \cap p_1 = \emptyset\}$$

where A is a Boolean algebra and u is an ultrafilter on A. The effect of the forcing is to add a generic element g so that in $A \cup \{g\}$ the only ultrafilter which is split is u. There is complete symmetry, but for definiteness, we take $g = \bigcup\{p_1 : (\exists p_{-1})(p_{-1}, p_1) \in G\}$.

These forcings are not, in general, ccc; in fact, if they are, the Stone space has countable tightness. If $P(A, u)$ is ccc, then

$$\forall a \notin u \; |A \restriction a| \leq \omega$$

and $t(A) = \omega$. Moreover if A is a Boolean algebra such that for each ultrafilter u of A there is a subalgebra $A_u \subset A$ such that u is generated from A_u and $P(A_u, A_u \cap u)$ is ccc, then $t(A)$ is countable. The plan is to construct such an A so that the character of each ultrafilter u is uncountable.

Slightly more is required to get the first countable preimage but this is the most interesting part. The construction actually ensures that the countable finite support product, $P^\omega(A_u, A_u \cap u)$, is ccc and of cardinality ω_1. If, for each n, a_n is the generic element added by the n-th coordinate of the product then consider the algebra generated by $A' = A \cup \{a_n : n \in \omega\}$. Every ultrafilter of A' which extends the filter u has countable character and every other ultrafilter v of A remains an ultrafilter in A'. The genericity over the product $P^\omega(A_u, A_u \cap u)$ guarantees that for distinct ultrafilters $v_1, v_2 \neq u$ on A_u, there will be an n such that $a_n \in v_1$ and $a_n \notin v_2$ and that $\{a_n : n \in \omega\}$ forms an independent family. That is, the closed set K in the Stone space of A' consisting of the ultrafilters extending u is a copy of the Cantor set, and for each such $k \in K$, there's at most one $v \in S(A_u) \setminus \{u\}$ such that for all n, $a_n \in k \cap v$ or $-a_n \in k \cap v$. Using Martin's Axiom, we so extend A for each ultrafilter u.

A careful structure is needed for the algebra A which will use complete branching subtrees of $\{-1, 1\}^{<\omega_1}$, T. A T-algebra is a Boolean algebra with a set of generators $\{a_t : t \in T\}$ such that for each $t \in T$

1. $\{a_s : s < t\}$ is a filter and a_t is minimal for (A_t, u_t) where A_t is the subalgebra generated by $\{a_s : s < t\}$ and u_t is the filter generated by $\{a_s : s < t\}$;
2. $a_{t^\frown -1} = -a_{t^\frown 1}$.

6.9. LEMMA. *If u is any ultrafilter on A, there is a maximal branch b of T such that $u = u_b$ is generated by $\{a_t : t \in b\}$. Conversely, for each such maximal branch b, u_b is an ultrafilter.*

Notice that the minimum character is at least the cofinality of a maximal branch.

By induction on $\alpha < \omega_2$, define

1. a finite support iteration of forcings P_α of length α such that $P_\alpha \restriction \beta = P_\beta$ for each $\beta < \alpha$;
2. a P_α-name for a tree $\dot{T}_\alpha$ such that $\dot{T}_\beta \prec \dot{T}_\alpha$ for each $\beta < \alpha$;
3. a P_α-name $\dot{A}_\alpha = \langle \dot{a}_t : t \in \dot{T}_\alpha \rangle$ for a $\dot{T}_\alpha$-algebra such that $\dot{A}_\beta \prec \dot{A}_\alpha$ for each $\beta < \alpha$.

$P_{\beta+1} = P_\beta * (\dot{P}^\beta * \dot{Q}^\beta)$ where $\dot{P}^\beta$ is $\Pi\{P^\omega(A_{\dot{b}}, u_{\dot{b}}) : \dot{b}$ is a maximal branch of $\dot{T}_\beta\}$ and, of course $P^\omega(A, u)$ is simply the finite support product of countably many copies of $P(A, u)$.

If we want MA to hold we must let $\dot{Q}^\beta$ be a suitably chosen (Souslin free (see DEVLIN [1978])) ccc poset. For our purposes we are satisfied to skip this part; the Souslin free is necessary to maintain control over when new uncountable branches are added to the $\dot{T}_\alpha$. Note though, that we are including $\dot{b}$ which are uncountable maximal branches of $\dot{T}_\beta$ and so we will have the necessary generics for the first countable extension.

We refer the reader to KOSZMIDER [1999] for the proof that P_{ω_2} is ccc although we do outline some of the main ideas. The most problematic is to analyze the product of those $P(A_b, u_b)$ where b is an uncountable branch. Again, we are only giving a very superficial overview which will give the flavor.

If $\mathrm{cf}(\alpha) = \omega_1$, then for each $\dot{b}$, a new uncountable branch of $\dot{T}_\alpha$, there is a function $f_b : \omega_1 \to \alpha$ (cofinal) such that for each $\xi \in \omega_1$, we have

$$b \restriction \xi \in V^{P_{f(\xi)}} \wedge b \restriction \xi \text{ is maximal}$$

(or of course $\dot{b}$ might be a subset of T_β but the forcing ensures that such $\dot{b}$ first occur in V^{P_β}).

If the branches in the proof of ccc come from a fixed V^{P_β}, then the corresponding $P(A,u)$ have become σ-centered hence cause no difficulty. So we assume that we have those functions f and we work in the forcing extension by P_α.

Fix a maximal antichain X included in the product. There is a non-decreasing function f from ω_1 into β so that for each $\xi \in \omega_1$, $X \restriction \xi$, defined as

$$X \cap [P^n(A_{b_1 \restriction \xi}, u_{b_1 \restriction \xi}) \times \ldots \times P^n(A_{b_k \restriction \xi}, u_{b_k \restriction \xi})]\,,$$

we have $X \restriction \xi \in V^{P_{f(\xi)}}$ and for each ξ in a cub C,

$$X \restriction \xi \text{ is a maximal antichain in } P^n(A_{b_1 \restriction \xi}, u_{b_1 \restriction \xi}) \times \cdots \times P^n(A_{b_k \restriction \xi}, u_{b_k \restriction \xi})\,.$$

Of course we can find a $\xi \in C$ such that $f_{b_i}(\xi) = f(\xi) = \theta$ for each i and so that all the branches b_i have split. Then by induction on η for $\theta \le \eta < \beta$ one can prove that $X \restriction \xi$ is a maximal antichain in $P^n(B_1^\eta, \langle u_{b_1 \restriction \xi}\rangle^{Fi}) \times \cdots \times P^n(B_k^\eta, \langle u_{b_k \restriction \xi}\rangle^{Fi})$ where $B_i^\eta = \langle \{a_t : t \in T_\eta, b_i \restriction \xi \subset t\}\rangle$ and $\langle u\rangle^{Fi}$ denotes the filter generated by u.

Once this has been shown, closer examination shows that

$$[P^n(A_{b_1}, u_{b_1}) \times \ldots \times P^n(A_{b_k}, u_{b_k})]$$

is a subposet of

$$P^n(B_1^\eta, \langle u_{b_1 \restriction \xi}\rangle^{Fi}) \times \cdots \times P^n(B_k^\eta, \langle u_{b_k \restriction \xi}\rangle^{Fi})$$

hence the countable set $X \restriction \xi$ will be dense in the smaller poset.

Obviously a lot of details have been omitted but let us include for illustration one of the interesting ideas for carrying out the induction mentioned above. If we set $A_0 = A_{b \restriction \xi}$ and $F = u_{b \restriction \xi}$ for any one of the b_i's and assume that X (for $X \restriction \xi$) is a maximal antichain in $P(A_0, F)$. Further assume that B is a Boolean algebra containing A_0 such that X remains maximal in $P(B, \langle F\rangle^{Fi})$, then we show that X remains maximal in $P(B', \langle F\rangle^{Fi})$ where B' is generated by $B \cup \{g\}$ where g (like some a_t) is the generic added by $P(B,u)$ for some ultrafilter $u \supset F$ on B. Note that F is an ultrafilter on A_0 but not on B.

The general element of $P(B', \langle F\langle^{Fi})$ has the form $r = ((g \cap e_1) \cup (f_1 - g)\,,\ (g \cap e_2) \cup (f_2 - g))$ for elements $e_1, f_1, e_2, f_2 \in B$. Now this is taking place in the extension by g, so that some $p \in P(B,u)$ forces that $r \in P(B', \langle F\rangle^{Fi})$ is incompatible with each member of X. Since $r_{-1} \cap r_1$ is empty, it follows that neither $e_1 \cap e_2$ nor $f_1 \cap f_2$ can be in u. In addition, $e_1 \cap f_1$ and $e_2 \cap f_2$ may be assumed not to be in u since otherwise both r_{-1} and r_1 can be shown to be in B. Therefore we may assume that $p \Vdash r = ((g \cap e_1) \cup f_1,$ $e_2 \cup (f_2 - g))$. Set $p^* = ((p_1 \cap e_1) \cup f_1\ ,\ e_2 \cup (p_{-1} \cap f_2))$ and check that r^* is in $P(B, \langle F\rangle^{Fi})$. For example, $f_1 \cap (p_{-1} \cap f_2)$ is empty since $p \Vdash p_{-1} \cap f_2 \subset f_2 - g$. Choose an $x = (x_{-1}, x_1) \in X$ which is compatible with r^* and set $p^* = (p_{-1} \cup (e_1 \cap x_1)\ ,\ p_1 \cup (f_2 \cap x_{-1})) \in P(B,u)$. The fact that r^* is compatible with x ensures $p_{-1}^* \cap p_1^* = \emptyset$ and $x \in P(B, \langle F\rangle^{Fi})$ ensures that $x_{-1} \cup x_1$ is disjoint from a member of F.

Finally, we check that p^* forces that x is compatible with r. All that needs to be checked is that x_{-1} is disjoint from r_1 and x_1 is disjoint from r_{-1}. We check the first case.

$$x_{-1} \cap (e_2 \cup (f_1 - g)) \subset (x_{-1} \cap e_2) \cup ((x_{-1} \cap f_2) - g) .$$

Now $x_{-1} \cap e_2$ is empty because x is compatible with r^* while the second term is empty since $p^* \Vdash x_{-1} \cap f_2 \subset g$.

Finally we turn to the result of Eisworth. This result uses the technique originating with Jensen's proof that CH is consistent with there being no Souslin trees, or perhaps more accurately with SHELAH's general machinery for iterating certain proper forcings without adding reals [1982]. EISWORTH and ROITMAN in [1999] also attribute the paper of AVRAHAM and TODORČEVIĆ [1997] as a major influence. The chief consideration is to establish the extra conditions on the posets which will permit them to be iterated without adding reals. We refer the reader to either EISWORTH [2001] or EISWORTH and ROITMAN [1999] for the innovation that expanded this class. We are content with examining the more basic properties of the single stage poset in this case.

6.10. THEOREM. *It is consistent with CH that any locally compact countably compact first countable space which has compactification of countable tightness is already compact.*

In fact, if $\mathcal{U}$ is any maximal free filter of closed subsets of a space X as in the theorem, then there is a poset P which does not add reals and which forces there to be a copy of ω_1 in the space X which meets every member of $\mathcal{U}$. Note that by Lemma 3.3, we may assume that each member of $\mathcal{U}$ contains a separable member of $\mathcal{U}$. In the following, *large* means meets every member of $\mathcal{U}$, while *small* will mean not large.

Given a neighborhood assignment f on a large subset, dom f, of X define $Ban(f) = \{y \in X : y \in f(x) \text{ for a small set of } x\text{'s } \}$. Using that X is first countable and that $\mathcal{U}$ is countably complete, one can show that $Ban(f)$ is closed. Also, using the fact that $\mathcal{U}$ has a base of separable members, one can show that $Ban(f)$ is small. Such f will be *promises* in the conditions of the poset which is promising that the elements of the new copy of ω_1 will not be coming from $Ban(f)$; in fact quite a bit more is promised.

6.11. DEFINITION. Define a notion of forcing $P = P_X$ by putting p into P if and only if $p = ([p], \Phi_p)$ where

1. $[p]$ is a countable closed subset of X;
2. Φ_p is a countable collection of promises.

A condition q extends p if $[q] \supset [p]$, $\Phi_q \supset \Phi_p$, for each $f \in \Phi_p$, the set

$$Y(f,q,p) = \{x \in \text{dom} f : [q] \setminus [p] \subset f(x)\}$$

is large and $f \restriction Y(f,q,p) \in \Phi_q$.

One of the key steps to the proof is that given a dense set D of P, and a condition p, let

$$Bad(p,D) = \{x \in X : x \text{ has a neighborhood } U_x \text{ such that there is no } q \leq p \text{ with } q \in D \text{ and } [q] \setminus [p] \subset U_x\}$$

then $Bad(p,D)$ is small.

To see this, assume that $Bad(p, D)$ is large. Let $f(x) = U_x$ for $x \in Bad(p, D)$ where U_x is as in the definition of $Bad(p, D)$. It is trivial that $([p], \Phi_p \cup \{f\}) < p$ and so has an extension $q \in D$. By the definition of $<$, $Y(f, q, p)$ is large

We finish by showing that P is proper and does not add reals. More specifically, given a countable $M \prec H(\lambda)$ with $X, \mathcal{U}$ and P in M and $p \in M \cap P$, there is a $q < p$ which such that $q \in D$ for each dense open $D \in M$. Let $\{D_n : n \in \omega\}$ list all dense open subsets of P that are members of M. The following inductive construction is really the heart of the matter (after suitably defining the poset of course). Build $p_{n+1} \leq p_n$ by induction with $p_0 = p$ and $p_{n+1} \in M \cap D_n$ in such a way to ensure there is a lower bound. We must ensure that for each $m \in \omega$ and each promise $f \in \Phi_m$,

$$Y = \{y : \bigcup_{n\in\omega} [p_n] \setminus [p_m] \subset f(y)\} \text{ is large}$$

If so, we'll set $\Phi_q = \bigcup_{n\in\omega} \Phi_{p_n} \cup \{f \restriction Y : f \in \Phi_{p_n} n \in \omega\}$. Moreover to ensure that the set $\bigcup\{[p] : p \in G\}$ for a generic G is homeomorphic to ω_1, we have to ensure that there is a point x such that $\{x\} \cup \bigcup_{n\in\omega} [p_n]$ is compact.

Let $\{x_n : n \in \omega\}$ be chosen so that it converges to some point x and for each $U \in \mathcal{U} \cap M$, $\{x_n : n \in \omega\} \setminus U$ is finite. Check that if $f \in M$ is a promise, then $x \notin Banf$. In defining the sequence p_n we also define a function $g \in \omega^\omega$. We also enumerate all the promises in M because without loss the dense open sets of M guarantee they will be in some Φ_m. We can assume that promise f_n comes from Φ_{p_m} for some $m \leq n$. Also let $U_n(x)$ enumerate a neighborhood base at x.

Set $B = Y(f_n, p_n, p_m) = \{y \in \operatorname{dom} f_n : [p_n] \setminus [p_m] \subset f_n(y)\}$. By definition of extension, B is large and $f_n \restriction B \in \Phi_{p_n}$. Since x is not banned by $f_n \restriction B \in M$ we can find a neighborhood $V \subset U_n(x)$ of x such that $K(V, f_n) = \{y \in B : V \subset f_n(y)\}$ is large. For ease of notation, assume that $V = U_n(x)$. Since $Bad(p_n, D_n) \cap \{x_n : n \in \omega\}$ is finite, we can choose $g(n)$ so that $x_{g(n)} \in V$ and there is an open neighborhood $V_{g(n)} \subset V$ of $x_{g(n)}$ which is contained in $V \setminus Bad(p_n, D_n)$. Now $V_{g(n)}$ is a neighborhood of $x_{g(n)} \notin B(p_n, D_n)$ so in M we can find $p_{n+1} \leq p_n$, in D_n such that $[p_{n+1}] \setminus [p_n] \subset V_{g(n)} \subset V$.

Let $B' = \{y \in \operatorname{dom} f_n : [p_n] \setminus [p_m] \subset f(y)\}$. Since $[q] \setminus [p_n] \subset V$, it follows that $[q] \setminus [p_m] \subset f(y)$ for enough y.

Finally, to ensure that the new set is uncountable one proves it as follows.

6.12. PROPOSITION. *If A is large, then there is a dense set of conditions q with $[q] \cap A \neq \emptyset$.*

□ Fix any condition p and recall that $Ban(f)$ is small for each $f \in \Phi_p$. Therefore there is a $U \in \mathcal{U}$ such that $U \cap Ban(f)$ is empty for each $f \in \Phi_p$. Since A is large, there is an $a \in A \cap U$ and, of course, $Y(f, p) = \{x \in \operatorname{dom}(f) : a \in f(x)\}$ is large for each $f \in \Phi_p$. Set $[q] = [p] \cup \{a\}$ and $\Phi_q = \Phi_p \cup \{f \restriction Y(f, p) : f \in \Phi_p\}$. Then $q < p$ as required. □

References

ABRAHAM, U. and S. TODORČEVIĆ

[1997] Partition properties of ω_1 compatible with CH, *Fund. Math.* **152** (2), 165–181.

ARHANGEL'SKII, A.V. and R.Z. BUZYAKOVA

[1999] On linearly Lindelöf and strongly discretely Lindelöf spaces, *Proc. Amer. Math. Soc.* **127** (8), 2449–2458.

BAUMGARTNER, J.E.
[1983] Iterated forcing, in *Surveys in set theory*, Cambridge Univ. Press, Cambridge, pp. 1–59.

BAUMGARTNER, J.E. and S. SHELAH
[1987] Remarks on superatomic Boolean algebras, *Ann. Pure Appl. Logic* **33** (2), 109–129.

BEKKALI, M.
[1991] *Topics in set theory*, Springer-Verlag, Berlin. Lebesgue measurability, large cardinals, forcing axioms, rho-functions, Notes on lectures by Stevo Todorčević.

BELL, M.
[2000] Universal uniform Eberlein compact spaces, *Proc. Amer. Math. Soc.* **128** (7), 2191–2197.

DEVLIN, K.J.
[1978] $\aleph_1$-trees, *Ann. Math. Logic* **13** (3), 267–330.

DOW, A.
[1980] Absolute C-embedding of spaces with countable character and pseudocharacter conditions. *Canad. J. Math.*, **32** (4), 945–956.
[1988] An introduction to applications of elementary submodels to topology. *Topology Proceedings*, **13**, 17–72.
[1998] The regular open algebra of $\beta\mathbb{R} \setminus \mathbb{R}$ is not equal to the completion of $\mathcal{P}(\omega)/\text{fin}$, *Fund. Math.* **157** (1), 33–41.

DOW, A. and K.P. HART
[2000] The measure algebra does not always embed, *Fund. Math.* **163** (2), 163–176.

DŽAMONJA, M.
[1998] On uniform Eberlein compacta and c-algebras, in *Proceedings of the 1998 Topology and Dynamics Conference* (Fairfax, VA), vol. 23, pp. 143–150.

EISWORTH, T.
[2001] CH and first countable, countably compact spaces, *Topology Appl.* **109** (1), 55–73.

EISWORTH, T. and J. ROITMAN
[1999] ch with no Ostaszewski spaces, *Trans. Amer. Math. Soc.* **351** (7), 2675–2693.

FARAH, I.
[2000] *Analytic quotients: theory of liftings for quotients over analytic ideals on the integers*, Mem. Amer. Math. Soc. vol. 148 (702), xvi+177.

FEDORČUK, V.V.
[1975] The cardinality of hereditarily separable bicompacta, *Dokl. Akad. Nauk SSSR* **222** (2), 302–305.

HUŠEK, M. and J. VAN MILL, EDITORS
[1992] *Recent progress in general topology*, North-Holland Publishing Co., Amsterdam.

JUST, W.
[1989] Nowhere dense P-subsets of ω, *Proc. Amer. Math. Soc.* **106** (4), 1145–1146.
[1989a] The space $(\omega^*)^{n+1}$ is not always a continuous image of $(\omega^*)^n$, *Fund. Math.* **132** (1), 59–72.

KOJMAN, M. and S. SHELAH
[1992] Nonexistence of universal orders in many cardinals, *J. Symbolic Logic* **57** (3), 875–891.
[2001] Fallen cardinals, *Ann. Pure Appl. Logic* **109** (1-2,Dedicated to Petr Vopěnka), 117–129.

KOSZMIDER, P.
[1999] Forcing minimal extensions of Boolean algebras, *Trans. Amer. Math. Soc.* **351** (8), 3073–3117.

MALYKHIN, V.I.
[1987] A Fréchet-Uryson compact set without points of countable character, *Mat. Zametki* **41** (3), 365–376, 457.

MOORE, J.T.
[200?] Open colorings, the continuum, and the second uncountable cardinal, *Proceedings of the American Mathematical Society*, to appear.

RABUS, M.
[1996] An ω_2-minimal Boolean algebra, *Trans. Amer. Math. Soc.* **348** (8), 3235–3244.

SHELAH, S.
[1982] *Proper forcing*, Springer-Verlag, Berlin.
[1987] Universal classes, in *Classification Theory* (Chicago, IL, 1985), Springer, Berlin, pp. 264–418.
[1994] *Cardinal Arithmetic*, The Clarendon Press Oxford University Press, New York. Oxford Science Publications.

SHELAH, S. and O. SPINAS
[2000] The distributivity numbers of $\mathcal{P}(\omega)/\text{fin}$ and its square, *Trans. Amer. Math. Soc.* **352** (5), 2023–2047 (electronic).

SHELAH, S. and J. STEPRĀNS
[1988] PFA implies all automorphisms are trivial, *Proc. Amer. Math. Soc.* **104** (4), 1220–1225.

TODORČEVIĆ, S.
[1987] Partitioning pairs of countable ordinals, *Acta Math.* **159** (3-4), 261–294.

VELIČKOVIĆ, B.
[1992] Applications of the open coloring axiom, in *Set Theory of the Continuum* (Berkeley, CA, 1989), Springer, New York, pp. 137–154.

CHAPTER 6

Topics in Topological Dynamics, 1991 to 2001

Eli Glasner

Department of Mathematics, Tel Aviv University, Tel Aviv, Israel
E-mail: glasner@math.tau.ac.il

Contents

RECENT PROGRESS IN GENERAL TOPOLOGY II
Edited by Miroslav Hušek and Jan van Mill

1. Introduction

As in most areas of mathematical research the extent and boundaries of Topological Dynamics are quite flexible and in its broadest scope it can include almost every aspect of Dynamics. Traditionally though there is a rough division of Dynamics into three parts: Measurable Dynamics (or Ergodic Theory), Topological Dynamics and Differentiable Dynamics. This division can serve (mostly by elimination) as means of deciding which subject really belongs to topological dynamics, but even then we are left with a great number of topics which may qualify as legitimate subjects for a review such as the present one. My choice of topics is, to some extent, arbitrary and should not necessarily be considered as representing the most important developments in the area during the last ten years. Of course importance was one factor which influenced my choice but personal interest, taste, and knowledge played equal roles in choosing the subjects for this review. In fact the word "review" here is misleading and I prefer to think of the following three sections as being three colloquium lectures addressed to the members of, say, the mathematics department at my university. Throughout this article a dynamical system, with no further qualifications, is a pair (X,T) where X is a compact metric space and $T : X \to X$ a self-homeomorphism.

2. Orbit equivalence of Cantor minimal dynamical systems

1. Orbit and strong orbit equivalence

In 1995 appeared a paper GIORDANO, PUTNAM and SKAU [1995]. The beautiful results which were obtained in this work were all the more surprising because the authors made a substantial use of C^*-algebras as well as homological algebra techniques in proving their main results, even though these can be stated in purely topological terms. It was shown later (see GLASNER and WEISS [1995]) how at least the dependence of the proofs on C^*-algebra theory can be eliminated, but the question whether homological algebra is really necessary for some parts of the proof is still open. The ground for the work of Giordano Putnam and Skau was prepared by the fundamental paper HERMAN, PUTNAM and SKAU [1992]. In turn these papers use ideas of earlier works from VERSHIK [1981], [1981]. In this section I shall describe two of the main theorems of GIORDANO, PUTNAM and SKAU [1995] and shall then try to demonstrate their power and generality by means of some examples and further developments.

The objects we are dealing with here are *Cantor minimal systems* (CM systems). These are dynamical systems (X,T) where X is the Cantor set and $T : X \to X$ is a self homeomorphism with the property that each orbit $\mathcal{O}_T(x) = \{T^n x : n \in \mathbb{Z}\}$ is dense in X. Two such systems (X,T) and (X,S) are *isomorphic* (or *conjugate*) if there exists a homeomorphism $F : X \to X$ that intertwines the T and S actions; i.e. $FTx = SFx$ for every $x \in X$. The classification of CM systems up to isomorphism is a formidable task which occupies researchers in topological dynamics since the beginning of this theory. Moreover in some (precise) sense a complete such classification is impossible.

Two CM systems are called *orbit equivalent* (OE) if we only require $F(\mathcal{O}_T(x)) = \mathcal{O}_S(Fx)$ for every $x \in X$. Equivalently: there are functions $n : X \to \mathbb{Z}$ and $m : X \to \mathbb{Z}$ such that for every $x \in X$ $F(Tx) = S^{n(x)}(Fx)$ and $F(T^{m(x)}) = S(Fx)$. An old result of M. Boyle implies that the requirement that, say, the function $n(x)$ be continuous already implies that the two systems are *flip conjugate*; i.e. (X,S) is isomorphic either to (X,T) or to (X,T^{-1}). However, if we require that both $n(x)$ and $m(x)$ have at most one point of discontinuity we get the new and, as it turns out, useful notion of *strong orbit equivalence* (SOE). The two results of HERMAN, PUTNAM and SKAU [1995] alluded to above give us complete classification of CM systems up to OE and SOE respectively. These classifications are achieved via the notion of dimension group which we now proceed to define.

A countable abelian group G is a *dimension group* if it is: (i) Partially ordered; i.e. it is equipped with a translation invariant partial order such that the positive cone $G^+ = \{x \in G : x \geq 0\}$ generates G as group. Thus the pair (G, G^+) with $G^+ + G^+ \subset G^+$, $G^+ - G^+ = G$ and $G^+ \cap (-G^+) = \{0\}$, determines the partial order on G. (ii) G is *unperforated*; i.e. $nx \geq 0$ for some $x \in G$ and $n \in \mathbb{N}$ implies $x \geq 0$ (this implies that G is torsion-free). (iii) G satisfies the *Riesz interpolation property*; i.e. given $x_1, x_2, y_1, y_2 \in G$ such that $x_i \leq y_j$ for all i, j, there exists $z \in G$ such that $x_i \leq z \leq y_j$ for all i, j.

The last condition is usually not easy to verify and there are more practicable definitions of dimension groups, but this is the most economical one.

Let G be a dimension group, an element $u \in G^+$ is called an *order-unit* if for every $x \in G^+$ there is an $n \in \mathbb{N}$ with $nu \geq x$. G is called *simple* if every non-zero element of G^+ is an order-unit. Let G be a non-trivial simple dimension group and fix an order-unit $u \in G^+$. A function $p : G \to \mathbb{R}$ is a *state* if p is positive (i.e. $p(G^+) \subset \mathbb{R}^+$) and $p(u) = 1$. When equipped with the topology of pointwise convergence the collection of states $S_u(G)$ is a Choquet simplex. Moreover the order on G is determined by the cone:

$$G^+ = \{x \in G : p(x) > 0 \text{ for all } p \in S_u(G)\} \cup \{0\}.$$

An element $x \in G$ is an *infinitesimal* if $p(x) = 0$ for all $p \in S_u(G)$ (it is easy to see that this notion does not depend on the choice of an order-unit). The collection of infinitesimal elements forms a subgroup of G denoted by Inf (G).

Next define the functor

$$K^0 : \{\text{Minimal Cantor systems}\} \to \{\text{Simple dimension groups with a distinguished order-unit}\},$$

which will implement the classification of CM systems up to OE and SOE. Let (X,T) be a CM system, and let $Z(X,\mathbb{Z})$ denote the (countable) collection of continuous functions $f : X \to \mathbb{Z}$. An element $f \in Z(X,\mathbb{Z})$ is a *coboundary* if there exists $g \in Z(X,\mathbb{Z})$ such that $f = g \circ T - g$. We let $B(X,T)$ denote the collection of coboundaries in $Z(X,\mathbb{Z})$. Clearly $Z(X,\mathbb{Z})$ forms a group under pointwise addition and $B(X,T)$ is a subgroup. Set

$$K^0(X,T) = Z(X,\mathbb{Z})/B(X,T), \qquad K^0(X,T)^+ = \{f + B(X,T) : f \geq 0\}, \quad \text{and} \quad \mathbf{1} = \mathbf{1}_X + B(X,T),$$

the *first cohomology group* of the dynamical system (X,T). We have the following:

2.1. PROPOSITION. *Let (X,T) be a CM system and let $M_T(X)$ denote the compact convex set of T-invariant probability measures on X (with respect to the weak* topology).*

1. $G = (K^0(X,T), K^0(X,T)^+, \mathbf{1})$ *is a simple dimension group with distinguished order-unit.*
2. *For each* $\mu \in M_T(X)$ *the map* $p_\mu : f \mapsto \int_X f(x)\, d\mu(x)$ *defines a state* $p_\mu \in S_1(G)$, *and the map* $\mu \mapsto p_\mu$ *is an affine homeomorphism of* $M_T(X)$ *onto* $S_1(G)$.
3. $\mathrm{Inf}\,(G) = \{f + B(X,T) : \int f\, d\mu = 0, \forall \mu \in M_T(X)\}$.

We now have all the notions needed for the statement of the classification theorems.

2.2. THEOREM (SOE). *Let* (X,T) *and* (Y,S) *be Cantor minimal systems; the following conditions are equivalent:*

1. (X,T) *and* (Y,S) *are strong orbit equivalent.*
2. $K^0(X,T) \cong K^0(Y,S)$ *as ordered dimension groups with order-unit.*

2.3. THEOREM (OE). *Let* (X,T) *and* (Y,S) *be Cantor minimal systems; the following conditions are equivalent:*

1. (X,T) *and* (Y,S) *are orbit equivalent.*
2. $K^0(X,T)/Inf(X,T) \cong K^0(Y,S)/Inf(Y,S)$ *are isomorphic as simple ordered dimension groups with order-unit.*
3. *There exists a homeomorphism* $F : X \to Y$ *such that the induced map* $F_* : M_T(X) \to M_S(Y)$ *is an affine homeomorphism.*

2. Kakutani-Rohlin towers and Bratteli-Vershik diagrams

The key to understanding the ideas of the proofs of these theorems is a construction called the *Bratteli-Vershik* representation of a CM dynamical system. In turn these representations are defined by means of *Kakutani-Rohlin (KR) towers.* If U is a non-empty clopen subset of the CM system (X,T) such that $TU \cap U = \emptyset$ then, by minimality, there exists a positive integer N such that the collection $U, TU, \ldots, T^N U$ covers X. We now define the *first return time* function $r_U(x) = \min\{0 \le j \le N : T^j x \in U\}$ and consider the partition of U into finitely many ($\le N$) disjoint clopen subsets $U_j = \{x \in U : r_U(x) = j\}$. This defines a corresponding partition of X into the clopen sets $\mathcal{B} = \{T^i U_j : 0 \le j \le N, 0 \le i < j, j \in \mathrm{range}\, r_U\}$. The latter is the KR *tower with basis* U and the subsets $\mathbf{c}_j = \{U_j, TU_j, \ldots, T^{j-1}U\}$ are called the *columns* of $\mathcal{B}$. If $x \in U_j$ then we say that the set $\{x, Tx, \ldots, T^{j-1}x\}$ is a *fiber* of the corresponding column. If $\mathcal{P}$ is any finite partition of X into clopen sets one can refine the partition $\{U_j\}$ of U in such a way that the resulting KR tower $\mathcal{B}$ is finer than $\mathcal{P}$.

It is always possible to construct a refining sequence $\mathcal{B}_n$ of KR towers such that (i) $U^n \searrow \{x_0\}$, where U^n is the basis of the tower $\mathcal{B}_n$, and x_0 is some point in X, (ii) the collection $\bigcup \mathcal{B}_n$ generates the topology. The fact that for each n, U^n is a subset of U^{n+1}, implies that a fiber of an $n+1$ column is composed of various fibers of order n. We then define the corresponding *Bratteli-Vershik diagram* to be the graph $\Gamma = (V, E)$, where V — the collection of vertices of Γ — consists of the columns of the KR towers $\mathcal{B}_n$, and a column $\mathbf{c}_i^n$ of $\mathcal{B}_n$ is connected by an edge to a column $\mathbf{c}_j^{n+1}$ in $\mathcal{B}_{n+1}$ —

an element of E, the collection of edges of Γ, denoted by $(\mathbf{c}_i^n, \mathbf{c}_j^{n+1}, \ell)$ — whenever a fiber of $\mathbf{c}_i^n$ appears in the column $\mathbf{c}_j^{n+1}$ at height ℓ. Multiple appearances define multiple edges. Additional structure is obtained when we define a partial order on the set of edges E as follows. $(\mathbf{c}_i^n, \mathbf{c}_j^{n+1}, \ell) \prec (\mathbf{c}_{i'}^{n'}, \mathbf{c}_{j'}^{n'+1}, \ell')$ iff $n = n', j = j'$, and $\ell \le \ell'$. With a Bratteli-Vershik diagram of this sort one can associate a dimension group. This time we use the characterization of dimension groups as a direct limit of *simplicial* dimension groups. These are just the groups $\mathbb{Z}^k$ with the usual partial order given by the cone $(\mathbb{Z}^k)^+ = \{(a_1, \dots, a_k) : a_i \ge 0\}$. More explicitly, in the directed system:

$$\mathbb{Z} \xrightarrow{M_1} \mathbb{Z}^{k(1)} \xrightarrow{M_2} \mathbb{Z}^{k(2)} \longrightarrow \cdots \qquad D = \lim_{\rightarrow} \mathbb{Z}^{k(n)},$$

the canonical generators of $\mathbb{Z}^{k(n)}$ correspond to the columns of $\mathcal{B}_n$ and the non-negative integer matrix M_n describes the number of connecting edges between the vertices of the n-th and $n+1$-st levels of the Bratteli-Vershik diagram.

It is shown in HERMAN, PUTNAM and SKAU [1992] that the dimension group D so obtained is isomorphic to $K^0(X,T)$. And conversely that with an ordered Bratteli-Vershik diagram with certain additional requirements, one can associate a Cantor set X — the *Bratteli compactum* — and a self homeomorphism T — the *Vershik* transformation — such that the resulting dynamical system (X,T) is a CM system. The great flexibility in the construction of KR towers, or equivalently, in the description of a CM system as a Vershik map on a Bratteli compactum, is the main tool used in the proofs of Theorems 2.2 and 2.3.

Another crucial ingredient of the proofs is yet another characterization of noncyclic (i.e. not equal to $\mathbb{Z}$) simple dimension groups with a trivial group of infinitesimals (see EFFROS [1981]).

Suppose that Q is a Choquet simplex, H is a countable dense subgroup of the linear topological vector space $A(Q)$ of continuous affine real-valued functions on Q, with the partial order given by: $f >> g$ iff $f(x) > g(x), \forall x \in Q$, so that $A(Q)^+ = \{f \in A(Q) : f >> 0\} \cup \{0\}$. Suppose further that $\theta : G \to H$ is a homomorphism of a torsion-free abelian group G onto H and set $G^+ = \{g \in G : \theta(g) \in A(Q)^+\}$. Then G is a simple dimension group in which $\ker \theta$ consists of the infinitesimal elements. Moreover, if G is a noncyclic simple dimension group with order unit u and trivial group of infinitesimals, then the canonical embedding $\theta : G \to A(S_u(G))$ defined by $\theta(g)(p) = p(g), g \in G$, $p \in S_u(G)$, is an isomorphism of dimension groups.

I shall not elaborate further on the proofs of these theorems, but instead shall demonstrate their power and scope by examining a few applications and some examples.

3. *Examples and applications*

A well known theorem of Oxtoby and Ulam asserts that, a full (i.e. positive on non-empty open sets) non-atomic probability measures μ on a Euclidean ball B with $\mu(\partial B) = 0$ is the image under a homeomorphism of normalized Lebesgue measure on B. Consider two non-atomic probability measures on the Cantor set X, say μ and ν; when are they equivalent in the sense that there exists a self homeomorphism $\phi : X \to X$ such that $\phi_* \mu = \nu$? From the outset it is clear that the countable collection of real numbers $S(\mu) = \{\mu(U) : U \text{ is a clopen subset of } X\}$ is an invariant for this equivalence relation.

In a recent work AKIN [1999], the author introduces the following definitions. A measure μ satisfies the *subset condition* if:

> whenever U and V are clopen subsets of X with $\mu(U) \leq \mu(V)$ then, there exists a clopen subset U_1 of X such that $U_1 \subset V$, and $\mu(U_1) = \mu(U)$.

Call a probability measure μ on X *good* if it is (i) non-atomic, (ii) full, and (iii) satisfies the subset condition. Akin then proves the following result.

2.4. PROPOSITION (AKIN). *Let μ and ν be two good probability measures on a Cantor set X, then μ and ν are equivalent iff $S(\mu) = S(\nu)$.*

Here is a sketch of a proof. It is not hard to see that the subset condition already implies the following seemingly stronger property $(*)$:

> Given a clopen partition $\mathcal{A} = \{A_1, A_2, \ldots, A_k\}$ of X with $\mu(A_i) = p_i$ and a probability vector $q = (q_1, \ldots, q_m)$, with $q_j \in S(\mu), j = 1, 2, \ldots, m$, which refines $p = (p_1, \ldots, p_k)$ (i.e. there is a partition $\{1, 2, \ldots, k\} = J_1 \cup \cdots \cup J_n$ such that for each $1 \leq i \leq k$, $p_i = \sum_{j \in J_i} q_j$), there exists a clopen partition $\mathcal{B} = \{B_1, B_2, \ldots, B_m\}$ of X which refines $\mathcal{A}$ and for which $\mu(B_j) = q_j, j = 1, 2, \ldots, m$.

Given μ and ν good measures on Cantor sets X and Y, with $S(\mu) = S(\nu)$, it is now easy to construct — using property $(*)$ and going back and forth between X and Y — two sequences of refining finite clopen partitions $\mathcal{B}_n$ and $\mathcal{C}_n$ on (X, μ) and (Y, ν) respectively with the following properties. (i) $\lim_{n\to\infty} \sup\{\text{diam}\,(A) : A \in \mathcal{B}_n\} = 0$, (ii) a similar condition for $\mathcal{C}_n$ and (iii) for each n the probability vectors which correspond to $\mathcal{B}_n$ and $\mathcal{C}_n$ are equal. This data then suffices to define a homeomorphism F from X to Y which carries μ onto ν.

Using similar ideas and a basic construction from HERMAN, PUTNAM and SKAU [1992] it is now possible to prove the following.

2.5. PROPOSITION. *Let μ be a good probability measure on the Cantor set X, then there exists a homeomorphism $T : X \to X$ such that the system (X, T) is minimal and such that $M_T(X) = \{\mu\}$; i.e. the Cantor minimal system (X, μ, T) is* uniquely ergodic.

□ The subset condition also implies that the set $S(\mu)$ is "group like" in the sense that the group G generated by $S(\mu)$ is a dense subgroup of $\mathbb{R}$ such that $S(\mu) = G \cap [0, 1]$. Now, as we have seen above, every countable dense subgroup of $\mathbb{R}$ is a simple dimension group with a trivial subgroup of infinitesimals and with a unique state. By Theorem 6.2 of HERMAN, PUTNAM and SKAU [1992] there exists a uniquely ergodic CM dynamical system (Y, ν, S) with $K^0(Y, S) \cong G$ as ordered simple dimension groups with order-unit. In particular then $S(\mu) = S(\nu)$. By Lemma 2.5 of GLASNER and WEISS [1995] the measure ν satisfies the subset condition and by Akin's result there exits a homeomorphism $F : X \to Y$ with $F_*(\mu) = \nu$. Defining $T : X \to X$ by $T = F^{-1} \circ S \circ F$, we now see that (X, μ, T) is the required uniquely ergodic CM system. □

It now follows that if μ and ν are good measures on a Cantor set X with $S(\mu) = S(\nu)$, then there exist self homeomorphisms T and S of X such that (X, T) and (Y, S) are uniquely ergodic CM systems with dimension groups whose infinitesimal subgroups are trivial. Theorem 2.2 implies that (X, T) and (Y, S) are strong orbit equivalent.

Compare this with the following corollary of Theorem 2.3 (Corollary 1 in HERMAN, PUTNAM and SKAU [1995]), where we are given two uniquely ergodic CM systems (X, T, μ) and (Y, S, ν) but without the triviality assumption on the infinitesimal subgroups.

2.6. COROLLARY. *Let (X, T, μ) and (Y, S, ν) be uniquely ergodic CM systems, then the following conditions are equivalent.*

1. $S(\mu) = S(\nu)$.
2. *There exists a homeomorphism $F : X \to Y$ with $F_*(\mu) = \nu$.*
3. *The two systems are orbit equivalent.*

We conclude this section with the following remarkable theorems, due to ORMES [1997], which simultaneously generalize the theorems of JEWETT [1970] and KRIEGER [1970], about the realization of an arbitrary ergodic measure preserving system as a uniquely ergodic CM system, and a theorem of DOWNAROWICZ [1991] which, given any Choquet simplex Q, provides a CM system (X, T) with $M_T(X)$ affinelly homeomorphic with Q.

2.7. THEOREM (ORMES).

1. *Let $(\Omega, \mathcal{B}, \nu, S)$ be an ergodic, non-atomic, probability measure preserving, dynamical system. Let (X, T) be a Cantor minimal system such that whenever* $\exp(2\pi i/p)$ *is a (topological) eigenvalue of (X, T) for some $p \in \mathbb{N}$ it is also a (measurable) eigenvalue of $(\Omega, \mathcal{B}, \nu, S)$. Let μ be any element of the set of extreme points of $M_T(X)$. Then, there exists a homeomorphism $T' : X \to X$ such that* (i) *T and T' are strong orbit equivalent,* (ii) *$(\Omega, \mathcal{B}, \nu, S)$ and $(X, \mathcal{X}, \mu, T')$ are isomorphic as measure preserving dynamical systems.*
2. *Let $(\Omega, \mathcal{B}, \nu, S)$ be an ergodic, non-atomic, probability measure preserving, dynamical system. Let (X, T) be a Cantor minimal system and μ any element of the set of extreme points of $M_T(X)$. Then, there exists a homeomorphism $T' : X \to X$ such that* (i) *T and T' are orbit equivalent,* (ii) *$(\Omega, \mathcal{B}, \nu, S)$ and $(X, \mathcal{X}, \mu, T')$ are isomorphic as measure preserving dynamical systems.*
3. *Let $(\Omega, \mathcal{B}, \nu, S)$ be an ergodic, non-atomic, probability measure preserving dynamical system. Let Q be any Choquet simplex and q an extreme point of Q. Then there exists a CM system (X, T) and an affine homeomorphism $\phi : Q \to M_T(X)$ such that, with $\mu = \phi(q)$, $(\Omega, \mathcal{B}, \nu, S)$ and $(X, \mathcal{X}, \mu, T)$ are isomorphic as measure preserving dynamical systems.*

3. Williams' conjecture

1. Introduction

For an integer $\ell \geq 2$, let $\mathfrak{L} = \{1, 2, \ldots, \ell\}$ and set $\Omega = \Omega(\ell) = \mathfrak{L}^{\mathbb{Z}}$. Let $\sigma : \Omega \to \Omega$ be the shift transformation: $(\sigma x)_n = x_{n+1}$. If $X \subset \Omega$ is a closed σ-invariant subset then the pair (X, σ) is called a *symbolic dynamical system* or sometimes a *subshift*. The branch of Dynamics that studies such systems is called symbolic dynamics. (Sometimes also countable alphabets are considered.) Of special interest are the *subshifts of finite type* (SFT). A symbolic system $X \subset \Omega$ as above is a SFT if there exists a finite list of finite words $w_1, w_2, \ldots, w_k$ on the alphabet $\mathfrak{L}$ such that X is exactly the set of bi-infinite words (i.e. elements of Ω) in which these words do not appear.

The history of symbolic dynamics goes back to J. Hadamard who used coding in his investigation of smooth dynamical systems. GOTTSCHALK and HEDLUND, whose book [1955] can be viewed as the foundation of modern theory of topological dynamics, devoted a chapter to symbolic dynamics.

The notion of *Markov partitions* is an important tool in the study of smooth dynamical systems by means of SFTs. The first example of a Markov partition is due to Adler and Weiss and the notion was developed later by Sinai, Bowen and many other authors, who applied such partitions to the study of Anosov and axiom A diffeomorphisms.

Roughly speaking, given a smooth dynamical system (M, ϕ) — where M is a compact manifold, ϕ a self diffeomorphism — and an "attractor" $Y \subset M$, a Markov partition is a finite partition of M, say $\{A_1, A_2, \ldots, A_\ell\}$, such that there exists a SFT, $X \subset \Omega(\ell)$ and a continuous onto map (homomorphism) $\pi : X \to Y$ satisfying (i) $\pi(\sigma x) = \phi\pi(x)$, $\forall x \in X$, (ii) there exists a ϕ-invariant dense G_δ subset $Y_0 \subset Y$ with card $(\pi^{-1}(y)) = 1$ for $y \in Y_0$ and (iii) for $y \in Y_0$ we have for $x = \pi^{-1}(y)$, $x_n = j$ iff $\phi^n y \in A_j$.

In another direction SFTs can be used successfully in the study of abstract ergodic theory. See for example DENKER, GRILLENBERGER and SIGMUND [1976] for a proof of Ornstein's isomorphism theorem where the main tool is a method by which a general symbolic dynamical system with an invariant ergodic measure is being approximated by a sequence of mixing SFTs equipped with their naturally defined invariant measure of maximal entropy.

Of course for many years now the subject of SFT is a central area of research on its own. The recent, negative, solution of the famous and long standing Williams' conjecture by KIM and ROUSH [1999], drew the attention of many mathematicians to this interesting and intricate subject. In the present section I shall survey some of the ideas involved in the construction of the counterexample.

2. *The shift equivalence problem for SFTs*

A homomorphism $\pi : (X, \sigma) \to (Y, \sigma)$ of symbolic systems $X \subset \Omega(\ell), Y \subset \Omega(r)$ is called a *block code*. The reason for this terminology is the observation (due to Curtis-Hedlund and Lyndon) that such a homomorphism is necessarily of the form

$$\pi(x)_i = F(x_{i-n}, \ldots, x_{i+n}), \qquad x \in X, i \in \mathbb{Z}$$

for a fixed n and $F : \{1, 2, \ldots, \ell\}^{2n+1} \to \{1, 2, \ldots, r\}$. When there exists such a code π which is a bijection of X onto Y we say that the systems are *isomorphic* or *conjugate* and write $(X, \sigma) \cong (Y, \sigma)$. We say that the symbolic systems (X, σ) and (Y, σ) are *eventually conjugate* if there exists some n_0 such that for every $n \geq n_0$ the dynamical systems (X, σ^n) and (Y, σ^n) are isomorphic. Finally we say that the symbolic systems (X, σ) and (Y, σ) are *almost conjugate* if there exists a third symbolic system (Z, σ) and two block codes $\phi : Z \to X$ and $\psi : Z \to Y$ such that on some dense G_δ subsets $X_0 \subset X$ and $Y_0 \subset Y$ we have card $\phi^{-1}(x) = 1$ and card $\psi^{-1}(y) = 1$ for all $x \in X_0, y \in Y_0$. Again it is easy to see that almost conjugacy is an equivalence relation.

As usual a central theme of SFT theory is the classification up to conjugacy of these systems. A meaningful classification of general dynamical systems up to isomorphism is not possible. In order to get successful classification results one needs to either restrict the class of dynamical systems one wishes to classify, or replace isomorphism by a less stringent equivalence relation (or do both). Perhaps the best known such classification is obtained by Ornstein's isomorphism theorem which I already mentioned: two Bernoulli measure-preserving dynamical systems are isomorphic iff they have equal entropy. It turns out that an analogous theorem exists for SFT systems. The following theorem is (a special case of) a theorem of ADLER and MARCUS [1979]. Recall that for a symbolic system (X, σ) the *topological entropy* can be defined as

$$h(X) = \lim_{n\to\infty} \frac{1}{n} \log N_n(X),$$

where $N_n(X)$ is the cardinality of the set of words of length n occurring in elements of X.

3.1. THEOREM. *Two mixing SFTs (X, σ) and (Y, σ) are almost conjugate iff $h(X) = h(Y)$.*

In order to describe the current situation of the classification problem for SFTs we first make the following observations.

An efficient and useful way of describing a SFT (X, σ) is by means of a square matrix A whose entries are in $\mathbb{Z}^+$, the set of non-negative integers. This is done as follows. Given $A = (a_{ij})$ we consider a finite directed graph (V, E) with exactly a_{ij} edges (elements of E) connecting the vertex $i \in V$ to the vertex $j \in V$. If we label the elements of E by the set $\mathfrak{L}$, where $\ell = \operatorname{card} E$, then a point $x \in \Omega(\ell)$ is an element of X iff the sequence $x = (\ldots, x_{-1}, x_0, x_1, \ldots)$ represents a continuous bi-infinite path on the graph. Of course infinitely many matrices can represent the same SFT. We write (X_A, σ_A) for the SFT corresponding to A. It is not hard to see that (for a nondegenerate matrix A; i.e. having no zero rows and no zero columns) the system (X_A, σ_A) is transitive (i.e. having a dense forward orbit) iff A is an irreducible matrix, and it is mixing iff A is primitive. (A is irreducible if for every entry (i, j) there exists $k > 0$ such that $A^k_{ij} > 0$ and primitive if there exists $k > 0$ such that every entry of A^k is strictly positive.) In the sequel we shall only deal with mixing SFTs. In some sense, which I shall not make precise, the study of the general SFT can be reduced to the study of mixing SFTs.

We say that two square matrices A and B with entries in $\mathbb{Z}^+$ (not necessarily of the same dimension), are *elementary strong shift equivalent* if there are matrices R, S with entries in $\mathbb{Z}^+$ such that $A = RS$ and $B = SR$. We write this relation as $(R, S) : A \to B$. A and B are *strong shift equivalent* (SSE) if they are linked by a finite chain of elementary strong shift equivalences. A and B are *shift equivalent* (SE) if there are matrices R, S with $\mathbb{Z}^+$ entries and a positive integer q such that

$$A^q = RS, \quad B^q = SR, \quad \text{and} \quad AR = RB, \quad BS = SA.$$

It will be convenient to have versions of these equivalence relations over a distinguished subset Λ of $\mathbb{Z}$, by requiring that all matrices in the definitions above will have their coefficients in Λ. The useful choices for Λ are $\{0, 1\}, \mathbb{Z}^+$ or $\mathbb{Z}$.

The following theorem (due to WILLIAMS [1973] and KIM and ROUSH [1979]) is our starting point.

3.2. THEOREM. *Two SFTs (X_A, σ_A) and (X_B, σ_B) are conjugate (eventually conjugate) iff the matrices A and B are SSE (SE) respectively.*

The outstanding problem (usually referred to as Williams's conjecture), that remained open for over twenty years, was whether these two notions of isomorphism of SFTs are in fact the same. The theorem above reduces the question to an equivalent question about matrices and the latter was finally resolved, in the negative, by Kim and Roush who produced counter examples (i.e. two matrices A and B which are SE but not SSE) first for reducible matrices and finally in KIM and ROUSH [1999] in the irreducible, in fact primitive (mixing), case. It was shown by WILLIAMS [1992] that SE and SSE are the same over $\mathbb{Z}$ and thus Williams' conjecture was refuted by Kim and Roush by showing that:

> There exist two primitive $\mathbb{Z}^+$ square matrices A and B which are SSE over $\mathbb{Z}$ but not over $\mathbb{Z}^+$.

The interest in Williams' conjecture stems from the fact that SE for matrices over $\mathbb{Z}^+$ is much easier to handle and understand than SSE over $\mathbb{Z}^+$. An affirmation of the conjecture would have been a great simplification of the classification theory. However the main goal, was and remains, finding good (preferably a complete set of) invariants for conjugacy of SFTs. The importance of the efforts made by many mathematicians to solve Williams' problem — as well as the importance of KIM and ROUSH [1999] itself, which in turn, is based on these efforts — lies in the development of a machinery aimed at that goal rather than in the confutation of the conjecture. We shall have a glimpse of some such tools in the rest of this section where I shall outline a construction of a counter example to Williams' conjecture as described in WAGONER [1999] and KIM and ROUSH [1999] (see also KIM, ROUSH and [1999]).

3. *The CW complex $SSE(\mathbb{Z})$*

The definition of SSE leads naturally to the idea that a chain of elementary strong shift equivalences is a kind of path between the matrices A and B. This idea motivated the work of J. B. Wagoner, who in a series of papers developed the algebraic topology of the strong shift equivalence spaces (CW complexes) $SSE(\Lambda)$. To be specific we describe the space $SSE(\mathbb{Z}^+)$. The vertices of $SSE(\mathbb{Z}^+)$ are finite, square matrices A with nonnegative integer entries. The edges are elementary strong shift equivalences $(R, S) : A \to B$ over $\mathbb{Z}^+$. The 2-cells are triangles

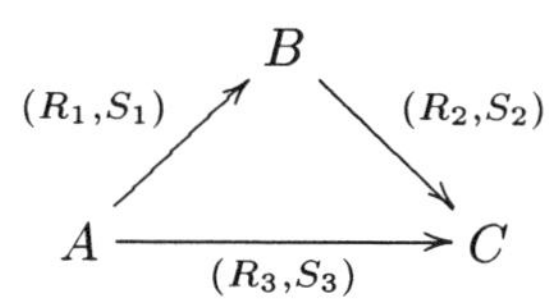

where the *triangle identities* hold: $R_1R_2 = R_3, R_2S_3 = S_1, S_3R_1 = S_2$. Higher dimensional cells are defined similarly.

If we let $c(R, S) : (X_A, \sigma_A) \to (X_B, \sigma_B)$ be the conjugacy arising from the elementary strong shift equivalence $(R, S) : A \to B$ (getting c(R,S) well defined requires some care), then the triangle identities yield the relation

$$c(R_1, S_1)c(R_2, S_2) = c(S_3, S_3). \tag{6.1}$$

It follows from the definitions and the paper WILLIAMS [1973] that

$$\begin{aligned}\pi_0(SSE(\{0,1\})) &\cong \text{strong shift equivalence classes over } \{0,1\}\\ \pi_0(SSE(Z^+)) &\cong \text{strong shift equivalence classes over } \mathbb{Z}^+\\ \pi_0(SSE(\mathbb{Z})) &\cong \text{strong shift equivalence classes over } \mathbb{Z}.\end{aligned}$$

4. The group Aut (σ_A)

In order to understand the structure of the group Aut (σ_A) of automorphisms of the SFT (X_A, σ_A) we should consider the idea of the proof of Theorem 3.2 and recall how an elementary strong shift equivalence $A = RS, B = SR$ over $\mathbb{Z}^+$ determines an elementary conjugacy c from σ_A to σ_B. View A and B as adjacency matrices of directed graphs with disjoint vertex sets, and view R and S as adjacency matrices for sets of edges between these vertex sets. According to the defining equations, for any vertices i, j in the graph G_A the number of A-edges a from i to j equals the number of (two-edge) RS paths rs from i to j (here the initial vertex of the edge r is i, the terminal vertex of the edge s is j, and the terminal vertex of r, in G_B, equals the initial vertex of s). So we may choose a bijection α of A-edges and RS-paths respecting initial and terminal vertices. Similarly we may choose a bijection β of B-edges and SR-paths respecting initial and terminal vertices. Now the conjugacy c is defined by the composition

$$\begin{aligned}x = \dots x_{-1}x_0x_1\dots &\mapsto \dots(r_{-1}s_{-1})(r_0s_0)(r_1s_1)\dots\\ &\mapsto \dots(s_{-1}r_0)(s_0r_1)(s_1r_2)\dots\\ &\mapsto \dots y_0y_1y_2\dots = y\end{aligned}$$

where the first and third bijections are defined by coordinatewise application of the bijections α and β, and the middle bijection just shifts parentheses.

If we apply this procedure to each edge in a loop in $SSE(\mathbb{Z}^+)$ based at A and then take the composition of the corresponding conjugacies we obtain an element of Aut (σ_A). Of course there is a lot of freedom in the choice of the bijections α and β and we can get rid of this ambiguity if we consider the normal subgroup of simple automorphisms Simp $(\sigma_A) \subset$ Aut (σ_A).

An *elementary simple* automorphism arises from an automorphism of the graph G_A (or G_P for another matrix P with $(X_P, \sigma_P) \cong (X_A, \sigma_A)$) that permutes the edges and keeps the vertices fixed. By definition Simp (σ_A) is the normal subgroup generated by the elementary simple automorphisms.

5. The dimension group G_A

The next ingredient is the "dimension group homomorphism". Let A be a square $n \times n$ matrix over $\mathbb{Z}^+$ and consider the "stationary" dimension group G_A defined by the direct limit

$$\mathbb{Z}^n \xrightarrow{A} \mathbb{Z}^n \xrightarrow{A} \mathbb{Z}^n \longrightarrow \cdots \qquad G_A = \varinjlim \mathbb{Z}^n.$$

We denote by s_A the automorphism of G_A induced by A (see section 1 of this review for a discussion of dimension groups). When A is primitive with $\det A = \pm 1$ we find that G_A is the simplicial group $\mathbb{Z}^n$ and $s_A = A$.

The map $A \mapsto G_A$ is an important invariant of SSE. A theorem of Krieger asserts that *matrices A and B are shift equivalent over $\mathbb{Z}^+$ iff (G_A, G_A^+, s_A) is isomorphic to (G_B, G_B^+, s_B)*.

We let Aut (s_A) denote the group of automorphisms of the dimension pair (G_A, s_A). By definition, this consists of those automorphisms α of the group G_A which commute with the isomorphism s_A (we do not require α to preserve the nonnegative set G^+). When $\det(A) = \pm 1$ we have a very concrete description of Aut (s_A), namely it consists of the elements of $GL_n(\mathbb{Z})$ which commute with A.

Using the relations (6.1), which are derived from the triangular identities, WAGONER establishes in [1999] the following correspondences.

3.3. THEOREM. *There are canonical isomorphisms*

$$\pi_1(SSE(\{0,1\}), A) \cong \text{Aut}\,(\sigma_A)$$
$$\pi_1(SSE(\mathbb{Z}^+), A) \cong \text{Aut}\,(\sigma_A)/\text{Simp}\,(\sigma_A)$$
$$\pi_1(SSE(\mathbb{Z}), A) \cong \text{Aut}\,(s_A).$$

6. The dimension representation δ_A

An important representation of Aut (σ_A) is the dimension group homomorphism δ_A : Aut $(\sigma_A) \to$ Aut (s_A) which was first defined by Krieger. In view of Theorem 3.3, it also can be defined using the homomorphism of fundamental groups

$$\delta_A : \pi_1(SSE(\{0,1\}), A) \to \pi_1(SSE(\mathbb{Z}^+), A) \to \pi_1(SSE(\mathbb{Z}), A).$$

7. The Δ-strategy

Let $\mathcal{S} = SSE_m(\mathbb{Z}^+)$ denote the union of those components of $SSE(\mathbb{Z}^+)$ containing vertices A satisfying

$$\text{Trace}\,(A) = \cdots = \text{Trace}\,(A^m) = 0.$$

Let $\pi_1(SSE(\mathbb{Z}), \mathcal{S})$ denote homotopy classes of paths in $SSE(\mathbb{Z})$ with endpoints in $\mathcal{S}$. Suppose we have a function

$$\Delta : \pi_1(SSE(\mathbb{Z}), \mathcal{S}) \to G,$$

where G is an abelian group with the following properties:

$$\Delta(\alpha * \beta) = \Delta(\alpha) + \Delta(\beta), \qquad \Delta(\alpha^{-1}) = -\Delta(\alpha), \tag{6.2}$$
$$\Delta(\alpha) = 0, \qquad \text{whenever } \alpha \text{ lies in } \mathcal{S} \tag{6.3}$$
$$\Delta(\alpha) = 0, \qquad \text{whenever } \alpha \text{ is in } \pi_1(SSE(\mathbb{Z}), A). \tag{6.4}$$

Let A and B be vertices in $\mathcal{S}$, and choose a path α from A to B in $SSE(\mathbb{Z})$. If β is another path from A to B, we have

$$\Delta(\alpha) = \Delta(\beta) + \Delta(\alpha * \beta^{-1}). \tag{6.5}$$

If now primitive matrices A and B can be found in $\mathcal{S}$ such that there exists a path β from A to B in $\pi_1(SSE(\mathbb{Z}), \mathcal{S})$ for which $\Delta(\beta) \neq 0$, then these matrices will provide a counterexample to Williams' conjecture.

In fact, β will testify to the fact that A and B are SSE over $\mathbb{Z}$. On the other hand if A and B are also SSE over $\mathbb{Z}^+$ then this would imply that there is a path α from A to B in $SSE(\mathbb{Z}^+)$. Now $\mathcal{S}$ being a union of components of $SSE(\mathbb{Z}^+)$, it follows that α lies in $\mathcal{S}$ and thus $\Delta(\alpha) = 0$ by (6.3). However, as $\alpha * \beta^{-1} \in \pi_1(SSE(\mathbb{Z}), A)$, we also have $\Delta(\alpha * \beta^{-1}) = 0$ by (6.4), hence $0 = \Delta(\alpha) = \Delta(\beta)$, by (6.5). Thus no such α exists and A and B are not SSE over $\mathbb{Z}^+$.

8. *Sign-gyration-compatibility-condition*

So far we have considered the abstract functorial setup which is used in the construction of the counterexample. The sign-gyration-compatibility-condition is the combinatorial key to the construction.

In [1987] Boyle and Krieger defined the sign and gyration number homomorphisms:

$$OS_m : \text{Aut}\,(\sigma_A) \to \mathbb{Z}/2\mathbb{Z}, \quad \text{for } m \geq 1$$
$$GY_m : \text{Aut}\,(\sigma_A) \to \mathbb{Z}/m\mathbb{Z}, \quad \text{for } m \geq 2.$$

$OS_m(\alpha)$ is the sign of the permutation induced by α on the orbits of length m, and $GY_m(\alpha)$ is the average measure of how α moves orbits of length m parallel to themselves. To define $GY_m(\alpha)$, list the orbits of length m and choose a point b_i on the i'th orbit. Write $\sigma_A(b_i) = \sigma_A^{r_i}(b_j)$. Then

$$GY_m(\alpha) = \sum_i r_i \qquad (\text{mod } m).$$

This is independent of the choice of base points b_i.

Now define the *sign-gyration-compatibility-condition homomorphism*

$$SGCC_m = GY_m + \sum_{i>0} OS_{m/2^i},$$

where $OS_{m/2^i} = 0$ if $m/2^i$ is not integral and where $Z/2\mathbb{Z}$ is identified with the subgroup $\{0, m/2\}$ of $\mathbb{Z}/m\mathbb{Z}$ when m is even. For example, when $m = 2$ we have $SGCC2 = GY_2 + OS_1$.

Kim, Roush and Wagoner [1992] have shown that the homomorphism $SGCC_m$: $\text{Aut}\,(\sigma_A) \to \mathbb{Z}/m\mathbb{Z}$ vanishes on the kernel of the dimension representation δ_A, the group $\text{Inert}\,(\sigma_A) \subset \text{Aut}\,(\sigma_A)$. This is the content of their factorization theorem which developed gradually from the work of Boyle and Krieger [1987] and through the work of several other authors.

9. *The KRW factorization theorem*

We present the dimension representation $\delta_A : \text{Aut}\,(\sigma_A) \to \text{Aut}\,(s_A)$ as $\delta_A : \text{Aut}\,(\sigma_A) \to \pi_1(SSE(\mathbb{Z}), A)$, as explained above. The homomorphism $SGCC_m$ defined on paths of edges in $SSE(\mathbb{Z})$ depends only on the homotopy class of a path and therefore induces a homomorphism from $\pi_1(SSE(\mathbb{Z}), A)$ into $\mathbb{Z}/m\mathbb{Z}$. We regard this as a map, which we denote by $sgcc_m$, from Aut (s_A) into $\mathbb{Z}/m\mathbb{Z}$.

3.4. THEOREM (The factorization theorem). *There is a commutative diagram*

$$\begin{array}{ccccc} \text{Aut}\,(\sigma_A) & \xrightarrow{\delta_A} & \text{Aut}\,(s_A) & & \cong \text{Aut}\,(\sigma_A)/\text{Inert}\,(\sigma_A) \\ & \searrow^{SGCC_m} & \downarrow{\scriptstyle sgcc_m} & & \\ & & \mathbb{Z}/m\mathbb{Z} & & \end{array}$$

In particular we have the following explicit formulas for $sgcc_2$. If $(R, S) : M \to N$ is an edge in $SSE(\mathbb{Z})$, then $sgcc_2(R, S)$ in $\mathbb{Z}/2\mathbb{Z}$ is given by

$$sgcc_2(R,S) = \sum_{i<j,l<k} R_{ik}S_{ki}R_{jl}S_{lj} + \sum_{i<j,l\leq k} R_{ik}S_{kj}R_{jl}S_{li} + \frac{1}{2}\sum_{i,k} R_{ik}(R_{ik}-1)S_{ki}^2. \tag{6.6}$$

Let A and B be vertices in $SSE_2(\mathbb{Z}^+)$. Consider a path γ from A to B in $SSE(\mathbb{Z})$. Write γ as a concatenation

$$\gamma = \prod_{k=1}^{m} \gamma(R_k, S_k)^{\epsilon_k},$$

where $\epsilon_k = +1$ if $(R_k, S_k) : A_{k-1} \to A_k$ and $\epsilon_k = -1$ if $(R_k, S_k) : A_k \to A_{k-1}$.

The formula

$$sgcc_2(\gamma) = \sum_{k=1}^{m} \epsilon_k \, sgcc_2(R_k, S_k), \qquad (\text{mod } 2)$$

defines the function $sgcc_2 : \pi_1(SSE(\mathbb{Z}), SSE_2(\mathbb{Z}^+)) \to \mathbb{Z}/2\mathbb{Z}$.

Now the crux of the proof is the fact that $sgcc_2$ can serve as the sought for Δ function.

3.5. THEOREM. *The function $sgcc_2$ satisfies the conditions* (6.2) - (6.5).

See WAGONER [1999] for details on the history of these results which represent the culmination of ten years' work by several authors.

10. The counterexample

Using some guess-work, a computer, and some luck as well, the following matrices were found. Let M be the 4×4 matrix

$$M = \begin{pmatrix} 0 & 0 & 1 & 1 \\ 1 & 0 & 0 & 0 \\ 0 & 1 & 0 & 0 \\ 0 & 0 & 1 & 0 \end{pmatrix}$$

with characteristic polynomial $t^4 - t - 1$. It satisfies the equation $(M - I)(M^4 + M^3 + M^2) = M$. Let $E = M - I, F = M^4 + M^3 + M^2$. Next extend E and F to matrices of the form

$$R = \begin{pmatrix} E & 0 \\ 0 & I \end{pmatrix}, \qquad S = \begin{pmatrix} F & X \\ Y & Z \end{pmatrix},$$

with

$$X = \begin{pmatrix} 2 & 0 & 0 \\ 2 & 0 & 0 \\ 2 & 0 & 0 \\ 2 & 0 & 0 \end{pmatrix}, \quad Y = \begin{pmatrix} 0 & 0 & 0 & 0 \\ 1 & 1 & 2 & 1 \\ 1 & 1 & 2 & 1 \end{pmatrix}, \quad Z = \begin{pmatrix} 0 & 0 & 0 \\ 3 & 0 & 0 \\ 0 & 1 & 0 \end{pmatrix}.$$

Now set $A = RS, B = SR$. We have:

$$A = RS = \begin{pmatrix} 0 & 0 & 1 & 1 & 2 & 0 & 0 \\ 1 & 0 & 0 & 0 & 0 & 0 & 0 \\ 0 & 1 & 0 & 0 & 0 & 0 & 0 \\ 0 & 0 & 1 & 0 & 0 & 0 & 0 \\ 0 & 0 & 0 & 0 & 0 & 0 & 1 \\ 1 & 1 & 2 & 1 & 3 & 0 & 0 \\ 1 & 1 & 2 & 1 & 0 & 1 & 0 \end{pmatrix}, \qquad B = SR = \begin{pmatrix} 0 & 0 & 1 & 1 & 2 & 0 & 0 \\ 1 & 0 & 0 & 0 & 2 & 0 & 0 \\ 0 & 1 & 0 & 0 & 2 & 0 & 0 \\ 0 & 0 & 1 & 0 & 2 & 0 & 0 \\ 0 & 0 & 0 & 0 & 0 & 0 & 1 \\ 0 & 1 & 0 & 0 & 3 & 0 & 0 \\ 0 & 1 & 0 & 0 & 0 & 1 & 0 \end{pmatrix}$$

Finally let β be the path $(R, S) : A \to B$.

With $\Delta = sgcc_2$ and $G = \mathbb{Z}/2\mathbb{Z}$ the following claims are established:

1. A and B are primitive, nonnegative integral matrices with $\det(A) = \det(B) = \pm 1$ and $\mathrm{Trace}\,(A) = \mathrm{Trace}\,(B) = \mathrm{Trace}\,(A^2) = \mathrm{Trace}\,(B^2) = 0$.
2. $\Delta(\beta) \neq 0$.
3. $\Delta(\alpha) = 0$ for all α in $\pi_1(SSE(\mathbb{Z}), A) \cong \mathrm{Aut}\,(s_A)$.

In view of Subsection 3.7 this completes the construction.

4. Mean dimension

1. Introduction

"Of all the theorems of analysis situs, the most important is that which we express by saying that space has three dimensions". Thus said Poincaré in 1912, the last year of his life. The next year came Brouwer's fundamental work where he introduced dimension as a topological invariant with the property $\dim \mathbb{R}^n = n$. The year 1922 yielded the works of Menger and Urysohn who independently created dimension theory. Already with the book HAUSDORFF [1927] and, with even greater vigor, in the advent of fractals, real valued dimension functions have become a central theme in several areas of mathematics.

In particular in the theory of dynamical systems the strong connection between the fundamental notions of entropy and Hausdorff dimension was realized. Perhaps the first to point out this connection was FURSTENBERG in his seminal work [1967]. Today this direction is a recognized branch of dynamics, see e.g. the book PESIN [1997].

Recently there appeared two papers LINDENSTRAUSS and WEISS [2000] and LINDENSTRAUSS [1999] where, following a suggestion of M. Gromov, a revolutionary new "mean dimension" theory is developed. In many ways this theory is a topological dynamics' parallel of classical dimension theory. It should be realized however that this new invariant is designed to deal with complicated, infinite dimensional systems, since finite dimensional as well as finite entropy systems have mean dimension zero. It was in order to handle this kind of large systems that Gromov introduced mean dimension and I refer the reader to his work GROMOV [1999] for an elaboration of these ideas and for applications.

In this section I shall survey the main results of LINDENSTRAUSS and WEISS [2000] and LINDENSTRAUSS [1999].

2. *Mean and metric mean dimension of a dynamical system*

In this subsection we shall introduce the notions of "mean dimension" and "metric mean dimension". As we shall see they are intimately connected with the classical notions of "dimension" and "entropy" respectively.

We begin by recalling the definition of topological dimension. Let X be a (compact metric) topological space. For finite open covers α and β of X set

$$\operatorname{ord}(\beta) = \max_{x \in X} \Big(\sum_{U \in \beta} \mathbf{1}_U(x) \Big) - 1, \quad \text{and}$$

$$D(\alpha) = \min_{\beta \succ \alpha} \operatorname{ord}(\beta),$$

where β runs over the finite open covers that refine α. Recall that β refines α ($\beta \succ \alpha$) if every $B \in \beta$ is contained in some $A \in \alpha$. Now the dimension of X is defined by

$$\dim X = \sup_{\alpha} D(\alpha), \tag{6.7}$$

where α runs over all finite open covers of X.

The following lemma is proved in LINDENSTRAUSS and WEISS [2000] in order to prove the subadditivity of D, that is, the property

$$D(\alpha \vee \beta) \leq D(\alpha) + D(\beta). \tag{6.8}$$

Besides doing that, the lemma also hints at several later developments of the theory.

4.1. LEMMA. *If α is a finite open cover of X, then $D(\alpha) \leq k$ iff there is a k-dimensional simplicial complex K, a continuous function $f : X \to K$ and an open cover β of K such that $f^{-1}(\beta) \succ \alpha$.*

Using this lemma and the fact that $\dim(K \times L) \leq \dim K + \dim L$ the subadditivity (6.8) is easy to deduce.

Let now (X, T) be a dynamical system. For a finite open cover α and $n \in \mathbb{N}$ we let $\alpha_0^n = \alpha \vee T^{-1}\alpha \vee \cdots \vee T^{-n}\alpha$.

The *mean dimension* of (X, T) is defined as

$$\operatorname{mdim}(X, T) = \sup_{\alpha} \lim_{n \to \infty} \frac{1}{n} D(\alpha_0^n),$$

(the limit exists by the subadditivity of D).

We immediately note that — as for each n, $D(\alpha_0^n) \leq \dim X$ — the mean dimension of any finite dimensional dynamical system, vanishes.

Next recall Bowen's definition of the topological entropy of (X,T) . Fix a compatible metric d on X. A subset A of X is called (n,ϵ,d)-*spanning* if for every $x \in X$ there is $y \in A$ such that $\max_{0 \leq k < n} d(T^k x, T^k y) < \epsilon$. Set

$$sp_n(\epsilon) = \min\{\text{card}\, A : A \text{ is } (n,\epsilon,d)\text{-spanning}\},$$

$$sp(\epsilon) = \limsup_{n\to\infty} \frac{1}{n} \log sp_n(\epsilon),$$

and finally

$$h_{\text{top}}(X,T) = \lim_{\epsilon\to 0} sp(\epsilon). \tag{6.9}$$

One shows that $h_{\text{top}}(X,T)$, the *topological entropy* of (X,T) , does not depend on the metric d.

The *metric mean dimension* of (X,T) with respect to the metric d is defined by

$$\text{mdim}_M(X,T,d) = \lim_{\epsilon\to 0} \frac{sp(\epsilon)}{|\log(\epsilon)|}.$$

Comparing this definition with (6.9) we see that the metric mean dimension measures how fast the terms that approximate the entropy $sp(\epsilon)$ tend to infinity as $\epsilon \to 0$. Again we conclude that $\text{mdim}_M(X,T,d) = 0$ whenever $h_{\text{top}}(X,T) < \infty$.

Unlike $h_{\text{top}}(X,T)$ the value $\text{mdim}_M(X,T,d)$ does in general depend on the metric d and in order to get a topological invariant we set

$$\text{mdim}_M(X,T) = \inf_d \text{mdim}_M(X,T,d),$$

where d runs over all metrics d compatible with the topology on X.

As we shall see in the sequel the class of systems (X,T) with $\text{mdim}(X,T) = 0$ is an interesting class whose study may unify some of the facts known about diverse classes of systems such as those with finite entropy, those having finite dimensional phase space, or those admitting at most countably many ergodic invariant measures. However, the main thrust of the theory should be sought for in the domain of systems with positive (or infinite) mean dimension — a domain which so far received little attention. In LINDENSTRAUSS and WEISS [2000] the authors assert (but provide no detailed proof) that for every $0 \leq t \leq \infty$ there exists a minimal system with $\text{mdim}(X,T) = t$.

Since many of the results in LINDENSTRAUSS [1999] are proved for dynamical systems that admit an infinite minimal factor, it will be convenient to give these systems a provisional name; let us say that such a system is an IMF system. The first assertion of the next theorem is from LINDENSTRAUSS and WEISS [2000], the second from LINDENSTRAUSS [1999].

4.2. THEOREM. *Let (X,T) be a dynamical system.*

1. *For every compatible metric d on X we have*

$$\text{mdim}(X,T) \leq \text{mdim}_M(X,T,d),$$

hence $\text{mdim}(X,T) \leq \text{mdim}_M(X,T)$.

2. *If* (X,T) *is in IMF then* $\text{mdim}\,(X,T) = \text{mdim}_M(X,T)$ *and moreover there exists a compatible metric* d' *such that* $\text{mdim}\,(X,T) = \text{mdim}_M(X,T,d')$.

Note the analogy between this theorem and that of PONTRJAGIN and SCHNIRELMANN [1932].

4.3. THEOREM. *For a compact metric space* X,

$$\dim X = \inf_d \liminf_{\epsilon\to 0} \frac{sp_0(\epsilon)}{|\log(\epsilon)|},$$

where d *runs over all metrics* d *compatible with the topology on* X.

There is another classical theorem behind the equality part of Theorem 4.2. Denote by $\dim_H(X,d)$ the Hausdorff dimension of the metric separable space (X,d). Then (i) $\dim(X) \le \dim_H(X,d)$, and (ii) there is a compatible metric d' such that $\dim(X) = \dim_H(X,d')$. In fact the proof of this statement (see HUREWICZ and WALLMAN [1941], Chapter VII) serves as a model for the proof given in LINDENSTRAUSS [1999] to Theorem 4.2.2.

3. The embedding theorem

A famous theorem (due to Menger and Nöbeling, see HUREWICZ and WALLMAN [1941], page 56) asserts that every compact metric space of dimension $\le n$ is homeomorphic to a subset of the $2n+1$-dimensional cube $[0,1]^{2n+1}$. In fact the proof shows that the set of topological embeddings in the space $C(X,[0,1]^{2n+1})$ with the topology of uniform convergence forms a dense G_δ set.

Given a dynamical system (X,T) and a continuous function $f : X \to K$, with, say $K = [0,1]^n$, set

$$I_f : X \to K^{\mathbb{Z}}, \qquad I_f(x) = (\ldots, f(T^{-1}x), f(x), f(T^1x), \ldots).$$

It is easy to check that $I_f : (X,T) \to (K^{\mathbb{Z}},\sigma)$ is a homomorphism of dynamical systems. The first assertion of the next theorem is from LINDENSTRAUSS and WEISS [2000], the second from LINDENSTRAUSS [1999].

4.4. THEOREM. *Let* (X,T) *be a dynamical system.*

1. *A necessary condition for* (X,T) *to be embeddable in the shift dynamical system* $(([0,1]^d)^{\mathbb{Z}},\sigma)$ *is that* $\text{mdim}\,(X,T) \le d$.

2. *If* (X,T) *is IMF and* $\text{mdim}\,(X,T) < \frac{1}{36}d$ *then for a dense* G_δ *subset of functions* $f \in C(X,[0,1]^d)$, *the map* I_f *is an embedding.*

Compare this theorem with Krieger's finite generator theorem which states that an ergodic probability measure preserving dynamical system $(X,\mathcal{X},\mu,T)$ is embeddable in $(\{1,\ldots,d\}^{\mathbb{Z}},\sigma)$ when $h_\mu(T) < \log d$.

4. The small boundary property

The definition (6.7) is the "Lebesgue cover" definition of topological dimension. The "inductive" definition, in Menger's formulation, is as follows:

a) the empty set has dimension -1,
b) the dimension of a space is the least integer n for which every point has arbitrarily small neighborhoods whose boundaries have dimension less than n.

A basic theorem of dimension theory asserts that for all values of n the two definitions are equivalent. This fact is rather trivial for $n = 0$, nonetheless mean dimension theory has, a far from trivial, analogous statement for mean dimension zero and the proof again is modelled on the dimension theory proof (HUREWICZ and WALLMAN [1941], Theorem V.5).

The relevant notion here is that of the "small boundary property". For any set E, the *orbit capacity* of E, denoted by $\mathrm{ocap}\,(E)$ (a notion due to SHUB and WEISS [1991]), is the number $\lim_{n\to\infty} \sup_{x\in E} \sum_{i=0}^{n-1} \mathbf{1}_E(T^i x)$, which for closed sets E is the supremum of $\mu(E)$, taken over all T-invariant probability measures μ on X. A subset E has the *small boundary property* (SBP) if $\mathrm{ocap}\,(E) = 0$.

Again the first assertion of the next theorem is from LINDENSTRAUSS and WEISS [2000], the second from LINDENSTRAUSS [1999].

4.5. THEOREM. *Let (X,T) be a dynamical system.*

1. *If there is a basis for the topology of X consisting of sets with the SBP, then* $\mathrm{mdim}\,(X,T) = 0$.
2. *If (X,T) is IMF and* $\mathrm{mdim}\,(X,T) = 0$ *then there is a basis for the topology of X consisting of sets with the SBP.*

5. Applications

In this subsection I shall list some applications and corollaries which are obtained in LINDENSTRAUSS and WEISS [2000] and LINDENSTRAUSS [1999] from the main results discussed above.

• If a system (X,T) has only countably many ergodic T-invariant measures then one has $\mathrm{mdim}\,(X,T) = 0$. In particular this is the case when (X,T) is uniquely ergodic.

• Any (X,T) with at most countably many ergodic T-invariant measures, and in particular any uniquely ergodic system, can be embedded in $([0,1]^{\mathbb{Z}}, \sigma)$. The same is true for minimal (X,T) with finite dimensional phase space X (a theorem due to Jaworski, see AUSLANDER [1988]) as well as any (X,T) with finite topological entropy.

• Not every minimal (X,T) can be embedded in $([0,1]^{\mathbb{Z}}, \sigma)$. This result in LINDENSTRAUSS and WEISS [2000] answers a long standing open problem.

• Ornstein and Sinai's theory of Bernoulli systems implies the existence — for a measure preserving system $(X, \mathcal{X}, \mu, T)$ with positive entropy $0 < h \le \infty$ and any $0 < t \le h$ — of a factor with entropy exactly t. Naturally the question arises whether the same property holds for topological dynamical systems. It was partially answered first by SHUB and WEISS [1991] and then by LINDENSTRAUSS [1995]. In particular in the latter paper Lindenstrauss has shown that the full shift system $([0,1]^{\mathbb{Z}}, \sigma)$, which has infinite entropy, does not admit any nontrivial finite entropy factor. Moreover he also constructed a minimal system with the same property. For the class of IMF systems a rather complete answer is given in LINDENSTRAUSS [1999]. For (X,T) in IMF there exists a unique maximal factor (Y,T) with mean dimension zero. Two points x and y in X can be distinguished by low (or finite) entropy factors iff their images in Y are distinct. In particular for (X,T) in IMF with $\mathrm{mdim}\,(X,T) = 0$ and $h_{\mathrm{top}}(X,T) = h > 0$, the range of the entropy values obtained by nontrivial factors of (X,T) includes the interval $(0,h]$.

• A system (X,T) in IMF has mean dimension zero iff (X,T) is the inverse limit of a directed system $\{(X_i,T)\}_{i\in\mathcal{I}}$ of factors with finite entropy.

6. Some words about proofs

The works LINDENSTRAUSS and WEISS [2000] and LINDENSTRAUSS [1999] extend results and techniques of SHUB and WEISS [1991] and LINDENSTRAUSS [1995]. Both papers do not require much machinery beyond the standard methods of topological dynamics and classical dimension theory. A crucial new feature in LINDENSTRAUSS [1999] is the following Rohlin-type lemma.

If (X,T) is a nontrivial minimal system then, for any $\delta > 0$, there is a continuous function $n : X \to \mathbb{R}$ such that the set $E = \{x \in X : n(Tx) \neq n(x) + 1\}$ has $\mathrm{ocap}\,(E) < \delta$.

This lemma and Baire's category theorem are the main ingredients in the proofs of the IMF parts of Theorems 4.2, 4.4 and 4.5.

References

ADLER, R. and B. MARCUS

[1979] *Topological entropy and equivalence of dynamical systems*, Mem. Amer. Math. Soc. **219**.

AKIN, E.

[1999] Good measures on Cantor space, preprint.

AUSLANDER, J.

[1988] *Minimal Flows and their Extensions*, Mathematics Studies 153, Notas de Matemática.

BOYLE, M. and W. KRIEGER

[1987] Periodic points and automorphisms of the shift, *Trans. Amer. Math. Soc.* **302**, 125–149.

DENKER, M., C. GRILLENBERGER and K. SIGMUND

[1976] *Ergodic theory on compact spaces*, Lecture Notes in Math. vol. 527, Springer-Verlag.

DOWNAROWICZ, T.

[1991] The Choquet simplex of invariant measures for minimal flows, *Israel J. of Math.* **74**, 241–256.

EFFROS, E.G.

[1981] *Dimensions and C^*-algebras*, CBMS regional Conf. Series in Math. vol. 46, Amer. Math. Soc. , Providence, R. I.

FURSTENBERG, H.

[1967] Disjointness in ergodic theory, minimal sets, and a problem in Diophantine approximation, *Math. System Theory* **1**, 1–49.

GIORDANO, T., I.F. PUTNAM and C.F. SKAU

[1995] Topological orbit equivalence and C^*-crossed products, *J. reine angew. Math.* **469**, 51–111.

GLASNER, E. and B. WEISS

[1995] Weak orbit equivalence of Cantor minimal systems, *Internat. J. Math.* **6**, 559–579.

GOTTSCHALK, W. H. and G.A. HEDLUND

[1955] *Topological Dynamics*, AMS Colloquium Publications vol 36.

GROMOV, M.

[1999] Topological invariants of dynamical systems and spaces of holomorphic maps. I, *Math. Phys. Anal. Geom.* **2**, 323-415.

HAUSDORFF, F.

[1927] *Mengenlehre* (zweite Auflage), Berlin-Leipzig.

HERMAN, R.H., I.F. PUTNAM and C.F. SKAU

[1992] Ordered Bratteli diagrams, dimension groups and topological dynamics, *Internat. J. Math.* **3**, 827–864.

HUREWICZ, W. and H. WALLMAN

[1941] *Dimension Theory*, Princeton Univ. Press, Princeton, N. J.

JEWETT, R.I.

[1970] The prevalence of uniquely ergodic systems, *J. Math. Mech.* **19**, 717–729.

KIM, K.H. and F.W. ROUSH

[1979] Some results on decidability of shift equivalence, *J. Combinatorics, Info. Sys. Sci.* **4**, 123–146.

[1999] The Williams conjecture is false for irreducible subshifts, *Annals of Math.* **149**, 545–558.

KIM, K.H., F.W. ROUSH and J. WAGONER

[1992] Automorphisms of the dimension group and gyration numbers, *JAMS* **5**, 191–212.

[1999] The shift equivalence problem, *Math. Intelligencer* **21**, 18–29.

KRIEGER, W.

[1970] On unique ergodicity, in:*Proc. Sixth Berkeley Simposium on Math. Stat. and Prob.*, 327–346.

LIND, D. and B. MARCUS

[1995] *An introduction to symbolic dynamics and coding*, Cambridge University Press.

LINDENSTRAUSS, E.

[1995] Lowering topological entropy, *J. d'Analyse Math.* **67**, 231–267.

[1999] Mean dimension, small entropy factors and an embedding theorem, *Publ. Math. IHES* **89**, 227–262

LINDENSTRAUSS, E. and B. WEISS

[2000] On mean dimension, *Israel. J. of Math.* **115**, 1–24.

ORMES, N.S.

[1997] Strong orbit realization for minimal homeomorphisms, *J. Anal. Math.* **71**, 103–133.

PESIN, Y.B.

[1997] *Dimension theory in dynamical systems*, Chicago Lecture in Mathematics, The University of Chicago Press, Chicago and London.

PONTRJAGIN, L. and L. SCHNIRELMANN

[1932] Sur une propriété métric de la dimension, *Annals of Math.* II, Ser. **33**, 156–162.

SHUB, M. and B. WEISS

[1991] Can one always lower topological entropy?, *Ergod. Th. and Dynam. Sys.* **11**, 535–546.

VERSHIK, A.M.

[1981] A theorem on periodical Markov approximation in ergodic theory, in: *Ergodic theory and related topics, Math. Res.*, vol. 12, Akademie-Verlag, Berlin (Vitte, 1981), 195–206.

[1981] Uniform algebraic approximation of shift and multiplication operators, *Soviet Math. Dokl.* **24**, 97–100.

WAGONER, J.B.

[1999] Strong shift equivalence theory and the shift equivalence problem, *Bull. Amer. Math. Soc.* **36**, 271–296.

WILLIAMS, R.F.

[1973] Classification of subshifts of finite type, *Annals of Math.* **98**, 120–153; Errata, ibid. **99**, (1974), 380–381.

[1992] Strong shift equivalence of matrices in $GL(2, Z)$, *Contemp. Math.* **135**, 445–451.

CHAPTER 7

Banach Spaces of Continuous Functions on Compact Spaces

Gilles Godefroy

Equipe d'analyse, Université Paris 6, Case 186, 4, Place Jussien, 75252 Paris cedex 05, France
E-mail: gig@ccr.jussieu.fr

Contents

RECENT PROGRESS IN GENERAL TOPOLOGY II
Edited by Miroslav Hušek and Jan van Mill

1. Introduction

This survey is an expanded version of a lecture I gave in the 9th Prague Topological Symposium in August 2001. I am glad to express my warmest thanks to the organizers of this Symposium, and in particular to Professor M. Hušek, for their kind invitation.

The interaction between Banach spaces and general topology is a huge field of functional analysis, and it would be a formidable task to sketch what has been done in this domain in the last ten years. We refer to NEGREPONTIS [1984] and MERCOURAKIS and NEGREPONTIS [1992] for important surveys of this field. The fundamental book BENYAMINI and LINDENSTRAUSS [2000] contains a wealth of updated information on this topic (and many others). Several chapters of the Handbook of Banach spaces (volumes 1 and 2) are also devoted to topological matters (see e.g. GODEFROY [2001a] and ZIZLER [2002]). The book FABIAN [1997] contains an abundant information on topological properties of non-separable Banach spaces from the point of view of differentiability. The books HABALA, HÁJEK and ZIZLER [1996] and FABIAN, HABALA, HÁJEK, MONTESINOS, PELANT and ZIZLER [2001] provide excellent introductions to Banach space theory.

The present work is by no means an exhaustive review of the topics covered by the title. It deals almost exclusively with recent work, where recent means "less than ten years old". Even under this restriction, choices had to be made and they were greatly influenced by the author's own taste (and limited knowledge).

$\mathcal{C}(K)$ In this note, I will focus on Banach spaces of continuous functions from a compact space K to the real line, equipped with the norm of uniform convergence, or sometimes with some equivalent norm. Such a space is denoted. Our purpose in studying this class is twofold: on one hand, it is sometimes possible to construct compact spaces with prescribed properties, and thus $\mathcal{C}(K)$ spaces frequently provide examples of Banach spaces with specific features, mainly in the non-separable case (which corresponds of course to non-metrizable compact spaces). On the other hand, the properties of the Banach space $\mathcal{C}(K)$ reflect quite faithfully those of the compact space K. The simplest result in this direction is the Banach-Stone theorem stating that the spaces $\mathcal{C}(K)$ and $\mathcal{C}(L)$ are isometric if and only if the compact spaces K and L are homeomorphic, but there is a number of more elaborate statements which belong to the isomorphic and even to the nonlinear theory where "common features" can be transferred from the Banach spaces to the compact sets. We will meet such results below.

A significative progress has recently been achieved on the understanding of some families of $\mathcal{C}(K)$ spaces, although many natural questions remain open. It is clear anyway that this class shares the (inexhaustible) complexity and richness of the concept of compactness. Let us now outline the contents of this survey.

In Section 2 below, we examine the linear classification of $\mathcal{C}(K)$ spaces. The separable case is completely elucidated for some time now, and we just recall the results. In the non-separable case, it matters a lot to know if we can "split" the space into separable subspaces, or if we can't. The spaces which can be split in separable pieces are of course easier to handle since several constructions by transfinite induction can be completed for such spaces, which are not available in the irreducible case. Such spaces can now be characterized. More precisely, Banach spaces which admit a long sequence of projections for every equivalent norm can be characterized through their dual unit ball. The existence of transfinite biorthogonal systems, which brings out a quite surprising situation, will also be considered.

Renormings provide a bridge between linear and nonlinear theory, and quite naturally Section 3 is devoted to renormings. Banach spaces $\mathcal{C}(K)$ are equipped with the natural (uniform) norm defined by

$$\|f\|_\infty = sup\{|f(x)|;\ x \in K\}$$

but this norm fails to enjoy any special property of convexity or smoothness; this is indeed clear since any Banach space is isometric to a subspace of a $\mathcal{C}(K)$ space (take for K the dual unit ball). It is therefore quite natural to wonder when it is possible to equip a $\mathcal{C}(K)$ space with an equivalent norm with good properties of convexity and/or smoothness. This is now fully understood in the separable case, but very far from clear in the non-separable case, despite spectacular recent progress due in particular to HAYDON [1999], and natural questions are still open. The complexity of the situation still allows some general theorems: a weak uniformity in the smoothness of $\mathcal{C}(K)$ provides indeed a strong information on the compact set K.

The final Section 4 deals with the nonlinear theory. We now consider a Banach space, stripped of its linear structure, as a metric space or as a uniform space. When two spaces are isomorphic as metric spaces, through a bi-Lipschitz map, are they linearly isomorphic? The answer is no in full generality (and $\mathcal{C}(K)$ spaces provide natural counterexamples) but no separable counterexamples are known so far, and few positive results are available even in the separable situation. The uniform situation, where isomorphisms are bi-uniform bijective maps, is still more difficult. A Banach space version of the Cantor derivation provides however an useful invariant.

This survey contains a number of open questions and problems. Some of them are simply included in the comments (but then it is clearly said that they are, to the best of my knowledge, open questions), others are explicitly stated.

2. Linear classification of *C*(*K*) spaces

The linear classification of $\mathcal{C}(K)$ spaces, when K is a metrizable compact set, is the following: all $\mathcal{C}(K)$ spaces, where K is metrizable and uncountable (equivalently, metrizable and not scattered), are isomorphic (MILUTIN [1966]). We now turn to the scattered case. Let us define the Cantor-Bendixon derivation δ as follows: given a compact set K, its derivative $\delta(K)$ is obtained by removing from K all its isolated points. This operation can of course be transfinitely iterated, if we define derivatives of order λ, when λ is a limit ordinal, as the intersection of the previously defined derivatives. When K is a countable compact set, we denote $\beta(K)$ the smallest ordinal α such that the Cantor-Bendixon derivative of K of order ω^α is empty. Cardinality shows that $\beta(K)$ exists and is a countable ordinal, while compactness implies that $\beta(K)$ is necessarily a successor ordinal. Now it turns out that if K and L are two countable compact sets, then the spaces $\mathcal{C}(K)$ and $\mathcal{C}(L)$ are isomorphic if and only if $\beta(K) = \beta(L)$ (BESSAGA and PEŁCZYŃSKI [1960]).

We now consider non-separable $\mathcal{C}(K)$ spaces. The "splitting" property we mention in the introduction is made precise in the following definition:

2.1. DEFINITION. Let X be a Banach space. We denote by μ the density character $dens(X)$ of X, that is, the smallest possible cardinal of a dense subset of X. A *projectional resolution of identity* (in short, P.R.I.) on X is a collection $\{P_\alpha;\ \omega_0 \le \alpha \le \mu\}$ of projections of norm one from X to X which satisfy for all α the following conditions:

(i) $P_\alpha P_\beta = P_\beta P_\alpha = P_\alpha$ if $\omega_0 \le \alpha \le \beta \le \mu$.

(ii) $dens(P_\alpha(X)) \le |\alpha|$, where $|\alpha|$ denotes the cardinality of α.

(iii) $\bigcup\{P_{\beta+1}(X);\ \beta < \alpha\}$ is norm dense in $P_\alpha(X)$.

(iv) $P_\mu = Id_X$.

This concept has been defined in the seminal paper LINDENSTRAUSS [1965]. Then AMIR and LINDENSTRAUSS [1968] show the existence of such a P.R.I. on every weakly compactly generated Banach space, that is, on every Banach space X such that there exists a weakly compact subset W of X which spans a dense linear subspace of X (see DEVILLE, GODEFROY and ZIZLER [1993], Chapter VI, and ZIZLER [2002]). In terms of $\mathcal{C}(K)$ spaces, this means that if K is an *Eberlein compact* set, that is, if K is homeomorphic to a weakly compact subset of some Banach space, then $\mathcal{C}(K)$ admits a P.R.I.

This result has been subsequently extended by several authors (see TALAGRAND [1979], VASAK [1981], GUL'KO [1979], JOHN and ZIZLER [1974], FABIAN and GODEFROY [1988], ORIHUELA and VALDIVIA [1989],...). A relevant notion is that of a Corson compact (see CORSON [1959]) and its generalization, which we recall now. Let I be a set, and let $\Sigma(I)$ denote the set of all functions $x(i) \in \mathbf{R}^I$ such that $x(i) = 0$ except for a countable number of i's. We equip the set $\Sigma(I)$ with the topology of pointwise convergence on I. With this notation, one has:

2.2. DEFINITION. A compact set K is a *Corson compact* if it is homeomorphic to a compact subset of $\Sigma(I)$, for some set I. A compact set K is a *Valdivia compact* if it is homeomorphic to a subset L of $\mathbf{R}^I$ such that $L \cap \Sigma(I)$ is dense in L.

A simple saturation argument (ARGYROS, MERCOURAKIS and NEGREPONTIS [1988], VALDIVIA [1990], see DEVILLE, GODEFROY and ZIZLER [1993], Lemma VI.7.5), provides a long sequence of continuous retractions on a Valdivia compact set, from which derives the existence of a P.R.I. on $\mathcal{C}(K)$ spaces when K is Valdivia, with the additional property that the projections P_α are continuous for the topology of pointwise convergence on K. In particular, when K is a Corson compact set, then $\mathcal{C}(K)$ has a P.R.I. We refer to KALENDA [2000a] for a comprehensive survey on Valdivia compact sets and their use in topology and Banach space theory.

It follows from VALDIVIA [1991] that if X is a Banach space such that the dual unit ball B_{X^*} (equipped of course with the weak* topology) is a Corson compact set, then the space X has a P.R.I. for every equivalent norm. A remarkable recent result of O. Kalenda, extending some preliminary results from FABIAN, GODEFROY and ZIZLER [1999], states the converse. It thus shows that the class of Corson compact sets provides the right isomorphic notion when the existence of P.R.I. is concerned. Indeed, one has by KALENDA [2000b], Theorem 1:

2.3. THEOREM. *Let X be a Banach space. Then the following assertions are equivalent:*

(i) *B_{X^*} is a Corson compact set.*

(ii) *for every equivalent norm $|\,.\,|$ on X, the unit ball of the dual space $(X^*, |\,.\,|^*)$ is a Valdivia compact set.*

If the density character of X is equal to ω_1, these conditions are also equivalent to:

(iii) *For every equivalent norm $|\,.\,|$ on X, the space $(X, |\,.\,|)$ has a P.R.I.*

It follows for instance from this result that the space $l_1(\omega_1)$ has an equivalent norm with no P.R.I. It follows from ARGYROS, MERCOURAKIS and NEGREPONTIS [1988] that if one assumes the continuum hypothesis, there is a Corson compact space K and an equivalent norm $|\,.\,|$ such that $(\mathcal{C}(K), |\,.\,|)$ has no P.R.I. Indeed, a compact set K is a Corson compact with property (M) (this means that the support of every Radon measure is separable) if and only if the unit ball of the space $\mathcal{M}(K) = \mathcal{C}(K)^*$ is a Corson compact, KALENDA [2000a], Theorem 5.4, and under the continuum hypothesis there exists (ARGYROS, MERCOURAKIS and NEGREPONTIS [1988]) a Corson compact set of weight ω_1 failing property (M). Note however that if we assume Martin's axiom and the negation of the continuum hypothesis, then every Corson compact with the countable chain condition is metrizable (see ARGYROS, MERCOURAKIS and NEGREPONTIS [1982], COMFORT and NEGREPONTIS [1982]) and thus every Corson compact has Property (M).

Since the property of having a Corson dual unit ball is clearly stable under isomorphism, it follows from the above that if $\mathcal{C}(K)$ is isomorphic to $\mathcal{C}(L)$ and K is a Corson compact with property (M), then L also is a Corson compact with (M). It is not known whether one can dispense in full generality with assuming (M) in the above statement; in other words, is the property "K Corson" determined by the space $\mathcal{C}(K)$ in (ZFC)?

We recall that a compact set is called a *Radon-Nikodym compact* if it is homeomorphic to a weak* compact subset of a dual space X^* with the Radon-Nikodym property (see BOURGIN [1983] for a survey of this notion). Scattered compact sets are Radon-Nikodym since the spaces $l_1(\Gamma)$ are dual spaces with the Radon-Nikodym property. Since reflexive spaces are in particular dual spaces with the Radon-Nikodym property, we have by DAVIS, FIGIEL, JOHNSON and PEŁCZYŃSKI [1974] that every Eberlein compact is a Radon-Nikodym compact. Moreover, AMIR and LINDENSTRAUSS [1968] show in particular that Eberlein compact sets are Corson. There is a nice converse to these implications: a compact set is Eberlein if and only if it is Radon-Nikodym and Corson (ORIHUELA, SCHACHERMAYER and VALDIVIA [1991], STEGALL [1990]). We refer to FABIAN and WHITFIELD [1994] for a proof of this theorem relying on projectional resolutions.

It is not known whether a continuous image of a Radon-Nikodym compact set is Radon-Nikodym, and not even whether the class is stable under (non disjoint) union of two sets. We refer to ARVANITAKIS [2001] for very interesting partial results on the continuous image problem.

We now turn to biorthogonal systems and Markushevich bases, according to the following definitions:

2.4. DEFINITION. Let X be a Banach space. A *biorthogonal system* is a subset $\{(x_i, x_i^*);\ i \in I\}$ of $X \times X^*$ such that $x_i^*(x_j) = 0$ if $j \neq i$ and $x_i^*(x_i) = 1$. If moreover, we have the following two properties:

(i) $\{x_i; i \in I\}$ spans a dense linear subspace of X.

(ii) if $x_i^*(x) = 0$ for every $i \in I$, then $x = 0$,

then the biorthogonal system (in short, b.s.) is called a *Markushevich basis* (in short, M-basis).

Although a biorthogonal system is a subset of $X \times X^*$, we will sometimes say that X (or X^*) contains a biorthogonal system of a given cardinality, or that X has an M-basis.

Every separable Banach space has an M-basis such that $\|x_i\| < 1+\epsilon$ and $\|x_i^*\| < 1+\epsilon$ for all i (OVSEPIAN and PEŁCZYŃSKI [1975], PEŁCZYŃSKI [1976]), with $\epsilon > 0$ arbitrary. It seems to be open whether this result still holds with $\epsilon = 0$. We refer to PLICHKO [1986] and TERENZI [1998] for important results on biorthogonal systems in separable spaces.

When X is not separable, it is natural to look for uncountable b. s. which "testify" of the non-separability. When X is isomorphic to a subspace of $l_\infty(\mathbf{N})$ which is analytic as a subset of $\mathbf{R}^{\mathbf{N}}$ (that is, when X is a *representable* Banach space), then if X is non-separable it contains a b. s. of cardinality $\mathbf{c}$ (continuum) (GODEFROY and TALAGRAND [1982]); in particular, when K is a separable non-metrizable Rosenthal compact (see GODEFROY [1980]), then the space $\mathcal{C}(K)$ contains a b. s. with $\mathbf{c}$ elements. The construction is a modification of Stegall's proof (STEGALL [1975]) showing that if X is separable and X^* is not, then X^* contains a b. s. which is weak* homeomorphic to the Cantor set. It is considerably harder to construct M-bases, and in fact there are non-separable spaces X which are such that their only subspaces which have an M-basis are the separable ones. Actually, if K is a separable and non-metrizable Rosenthal compact, the space $X = \mathcal{C}(K)$ has this property (see SUAREZ [1985] and FINET and GODEFROY [1989]).

If one assumes the continuum hypothesis, there exists by a theorem of K. Kunen (see NEGREPONTIS [1984]) a $\mathcal{C}(K)$ space, with K a separable, scattered and non metrizable compact set, which satisfies the following property: if A is any uncountable subset of $\mathcal{C}(K)$, there exists $a \in A$ such that $a \in \overline{conv}(A\backslash\{a\})$. It clearly follows that this $\mathcal{C}(K)$ space constructed by Kunen is a non-separable Banach space which contains no uncountable biorthogonal system; it follows for instance that any closed subspace of the Kunen space is a *countable* intersection of closed hyperplanes FINET and GODEFROY [1989]. This space provides an important result in renorming theory (see Section 3 below).

On the other hand, if one assumes Martin's axiom and the negation of continuum hypothesis, the situation gets drastically different: indeed, a recent deep result due to S. Todorčević (relying in part on TODORČEVIĆ [1993]) asserts then that if K is totally disconnected and non-metrizable, then $\mathcal{C}(K)$ contains an uncountable biorthogonal system.

We conclude this section by quoting a recent result due to MARCISZEWSKI [2001], which will be used in Section 4. If E is a set, we denote by $[E]^{\leq n}$ the set of subsets of E of cardinality at most n. This set is clearly a closed subset of the power set 2^E of E and thus it is compact in the product topology. Its closed subsets can be characterized as follows MARCISZEWSKI [2001]:

2.5. THEOREM. *Let K be a compact set. The following assertions are equivalent:*

(i) *K is homeomorphic to a subset of $[E]^{\leq n}$ for some set E and some integer n.*

(ii) *The Banach space $\mathcal{C}(K)$ is isomorphic to the space $c_0(\Gamma)$ for some set Γ.*

(iii) *The Banach space $\mathcal{C}(K)$ is isomorphic to a subspace of the space $c_0(\Gamma)$ for some set Γ.*

3. Renormings of *C*(*K*) spaces

Unless the compact set K reduces to one point, the natural norm of a $C(K)$ space fails to have any special property of convexity or smoothness: it is nor strictly convex neither differentiable, even in the weaker (Gâteaux) sense. The question arises whether nicer norms can be constructed. We refer to DEVILLE, GODEFROY and ZIZLER [1993] for renorming theory as it was in 1992, to HAYDON [1999] for a fundamental contribution to the subject, to GODEFROY [2001a] for a recent survey on renormings of Banach spaces, and to ZIZLER [2002] for the corresponding work on non-separable Banach spaces. Let us recall some definitions.

3.1. DEFINITION. Let X be a Banach space, and $\|\ .\ \|$ be an equivalent norm on X. We denote $S_X = \{x \in X;\ \|x\| = 1\}$. This norm is:

(i) *locally uniformly rotund* (in short, LUR) if $\|x\| = \|x_n\| = 1$ and $lim\ \|x + x_n\| = 2$ implies that $lim\ \|x - x_n\| = 0$.

(ii) *Gâteaux smooth* (in short, G-smooth) if for every $x \in S_X$ and $y \in S_X$,

$$lim_{t\to 0} \frac{\|x + ty\| - 1}{t} = f_x(y) \tag{7.1}$$

exists.

(iii) *Fréchet smooth* (in short, F-smooth) if for every $x \in S_X$, the above limit is uniform in $y \in S_X$.

(iv) *uniformly Gâteaux smooth* (in short, UG-smooth) if for every $y \in S_X$, the above limit is uniform in $x \in S_X$.

It is easily seen that the functional f_x defined by (7.1) belongs to X^*, and that a norm is F-smooth if and only if the map $x \to f_x$ is norm-to-norm continuous from $X\backslash\{0\}$ to X^*, that is, if and only if the norm is C^1. It is then possible to iterate this process and to consider C^2 or C^∞ norms.

The metrizable case is now fully understood, and the results can be summarized as follows.

3.2. PROPOSITION. *let K be a metrizable compact set. The space $C(K)$ has an equivalent norm which is LUR and UG. Moreover, the following assertions are equivalent:*

(i) *K is countable.*

(ii) *the space $C(K)$ has an equivalent F-smooth norm.*

(iii) *the space $C(K)$ has an equivalent C^∞ smooth norm.*

Indeed, any separable Banach space X has an equivalent norm which is LUR and UG (see DEVILLE, GODEFROY and ZIZLER [1993], Theorem II.7.1), and it has a F-smooth norm if and only if X^* is separable. Finally, it was shown in HAYDON [1992] that when K is countable, then the space $C(K)$ has an equivalent C^∞ smooth norm. This result has been considerably extended in HÁJEK [2001], HÁJEK [200?] and then in HÁJEK and HAYDON [200?], where it is shown in full generality that when there is a dual LUR norm on $C(K)^*$, then $C(K)$ has an equivalent C^∞-smooth equivalent norm and C^∞-smooth partitions of unity.

Renorming non-separable $C(K)$ spaces is delicate, and many connections occur with topological properties of K. We recall that a topological space is a Baire space is every countable intersection of dense open sets is dense. The following definition is motivated by the pioneering work NAMIOKA [1974].

3.3. DEFINITION. A compact set K has the *Namioka property* if for any Baire topological space E and any separately continuous function $f : E \times K \to \mathbf{R}$, there is a dense G_δ subset Ω of E such that f is jointly continuous at every point of $\Omega \times K$.

The separately continuous function f induces a continuous map $\phi : E \to (\mathcal{C}(K), \tau_p)$, where τ_p denotes the topology of pointwise convergence on K. The Namioka property amounts to assert that this map is continuous from E to $\mathcal{C}(K)$ equipped with the norm topology at every point of a dense G_δ subset of E.

For any τ_p- lower semi-continuous (in short, τ_p- l.s.c) equivalent norm $\| \,.\, \|$ on $\mathcal{C}(K)$, the map $\|\phi(.)\|$ is l.s.c. on the Baire space E and it is therefore continuous at every point of a dense G_δ subset Ω_0. It is then easy to show that if there is an equivalent τ_p-l.s.c. norm on $\mathcal{C}(K)$ which is such that the τ_p and the norm topologies coincide on the unit sphere, then K has the Namioka property. This is the case, in particular, when there is an equivalent τ_p-l.s.c. and LUR norm. Note that when K is scattered, the τ_p topology coincides with the weak topology on bounded subsets of $\mathcal{C}(K)$ and thus every equivalent norm is τ_p-l.s.c.

This implies for instance that any continuous image of a Valdivia compact set (see Definition 2.2) has the Namioka property, any scattered compact set K such that $K^{(\omega_1)} = \emptyset$ has the Namioka property (see DEVILLE, GODEFROY and ZIZLER [1993], Theorem VII.7.5), and that the two-arrows space has the Namioka property (JAYNE, NAMIOKA and ROGERS [1992]). On the other hand, there are scattered compact sets which fail to have the Namioka property, HAYDON [1995].

Using the existence of a coanalytic subset of $\mathbf{R}$ with cardinality $\mathbf{c}$ and containing no perfect subsets (which is independent of ZFC), there is (by NAMIOKA and POL [1992]) a scattered compact set K with the Namioka property such that the space $\mathcal{C}(K)$ has no equivalent norm such that the weak and norm topologies coincide on the unit sphere. It seems to be an open question whether such an example exists in ZFC.

The Namioka property leads to the following open problem:

3.4. PROBLEM. Does there exist a Baire space E, a compact set K, and a separately continuous function $f : E \times K \to \mathbf{R}$ with no point of joint continuity?

It seems hopeless to characterize compact spaces K such that $\mathcal{C}(K)$ has an equivalent LUR norm, or an equivalent G-smooth or F-smooth norm. In special classes however, this can be done, such as for trees in HAYDON [1999]. We recall that following the definition given in HAYDON [1999], a *tree* is a partially ordered set $(T, \leq)$ such that for any $t \in T$, the set $S_t = \{s \in T;\ s \leq t\}$ is well-ordered by $\leq$. The tree T is equipped with the coarsest topology such that all the sets S_t are open and closed, and it is assumed that the tree is Hausdorff in the sense of FREMLIN [1984], which means that this topology is Hausdorff. Trees are scattered locally compact spaces, and HAYDON [1990], [1999] deal with $\mathcal{C}_0(T)$ spaces, with T a tree. The full tree T of height ω_1, that is, $T = \cup_{\alpha \in \omega_1} \omega_1^\alpha$ is such that $\mathcal{C}_0(T)$ has no G-smooth nor strictly convex equivalent norm HAYDON [1990]. This example answered negatively the problem whether every Asplund space has an equivalent F-smooth norm. A weaker notion, however, is whether every Asplund space has an equivalent norm with the Mazur intersection property. As shown in JIMENEZ SEVILLA and MORENO

[1997], the Kunen space provides a negative answer, under the continuum hypothesis, but no example is known in ZFC, and in view of the recent contribution due to S. Todorčević (a refinement of TODORČEVIĆ [1993]) it seems to be a non trivial problem to dispense with axioms.

It is interesting to notice that when T is a tree, an LUR equivalent norm exists on $\mathcal{C}_0(T)$ if and only if there is a F-smooth norm on that space, HAYDON [1999]. The article HAYDON [1999] gives examples of $\mathcal{C}_0(T)$ spaces which have a strictly convex dual norm (and thus a G-smooth norm) but no strictly convex equivalent norm. However, the following question is still open.

3.5. PROBLEM. Let X be a Banach space which has an equivalent F-smooth norm. Does there exist an equivalent LUR norm on X?

The most general positive result in this direction is a very recent theorem of HAYDON [200?], which states in full generality that when there is a norm $\|\,.\,\|$ on X whose dual norm is LUR (this is a stronger condition than $\|\,.\,\|$ F-smooth), then there is an equivalent LUR norm on X. A motivation for this result is that using VANDERWERFF [1992] (see DEVILLE, GODEFROY and ZIZLER [1993], Theorem VIII.3.12), it follows that when X has a norm whose dual norm is LUR, then it has $\mathcal{C}^1$-smooth partitions of unity; that is, continuous functions defined on X can be uniformly approximated by $\mathcal{C}^1$-smooth functions.

We refer to MOLTO, ORIHUELA, TROYANSKI and VALDIVIA [1999], MOLTO, ORIHUELA, TROYANSKI and VALDIVIA [2000], JAYNE, NAMIOKA and ROGERS [1992], NAMIOKA and POL [1992], CASCALES, MANJABACAS and VERA [1998], RAJA [1999] and related works of these authors for the use of fragmentability techniques in renorming theory and in topology of (usually non separable) Banach spaces. The article HAYDON, JAYNE, NAMIOKA and ROGERS [2000] addresses the question of LUR equivalent norms on $\mathcal{C}(K)$ spaces, where K is a totally ordered compact space.

It is not known whether every Banach space which has an equivalent F-smooth norm has $\mathcal{C}^1$-smooth partitions of unity. Several partial positive results are available, and one might formulate the optimistic conjecture that it is true in full generality. A positive answer to Problem 3.5 would be a (big) step in this direction. Note that the case of $\mathcal{C}(K)$ spaces is positively solved; indeed for $1 \leq k \leq \infty$, a $\mathcal{C}(K)$ space on which there is a $\mathcal{C}^k$ smooth bump function has $\mathcal{C}^k$ smooth partitions of unity, HÁJEK and HAYDON [200?].

Unexpectedly, the case of UG norms leads to a satisfactory structure theorem, which gives a partial positive answer to LINDENSTRAUSS' question, [1972], Problem 9, asking whether "smooth" spaces are subspaces of weakly compactly generated spaces. We recall that a compact set K is *uniformly Eberlein* if it is homeomorphic to a weakly compact subset of a Hilbert space, or equivalently to a weakly compact subset of a super-reflexive space, BENYAMINI and STARBIRD [1976]. With this notation, we have (FABIAN, GODEFROY and ZIZLER [2001]):

3.6. THEOREM. *Let X be a Banach space. There exists an equivalent UG-smooth norm on X if and only if the dual unit ball $(B_{X^*}, weak*)$ equipped with the weak* topology is a uniformly Eberlein compact set.*

Outline of proof: an important step in the proof consists into showing that the space X is a $K_{\sigma\delta}$ subset of its bidual X^{**} equipped with its weak* topology. It then follows

(TALAGRAND [1979], VASAK [1981]) that it admits a P.R.I. (see Definition 2.1), and this decomposition into separable pieces allows to use the results of FABIAN, HÁJEK and ZIZLER [1997]. We only state and prove here a topological lemma (see FABIAN, GODEFROY and ZIZLER [2001], Lemma 1):

3.7. LEMMA. *Let X be a Banach space with an equivalent uniformly Gâteaux smooth norm. then X is a $K_{\sigma\delta}$ subset of (X^{**}, w^*).*

□ Assume that $\| \,.\, \|$ is any equivalent norm on X. Pick any $G \in X^{**} \backslash X$. Let $H = G^{-1}(0)$ be the subspace of X^* consisting of the elements which vanish on G. The space H is a norming subspace of X^*, that is, there is $\delta > 0$ such that for all $x \in X$,

$$sup\{|f(x)|;\ f \in H,\ \|f\| \leq 1\} \geq \delta \|x\|.$$

Indeed we have an isometric identification $H^* = X^{**}/H^{\perp}$, and it follows that for all $x \in X$ one has

$$sup\{|f(x)|;\ f \in H,\ \|f\| \leq 1\} = inf\{\|x - \lambda G\|;\ \lambda \in \mathbf{R}\}$$

and the claim follows.

For any equivalent norm $\| \,.\, \|$ on X, we define for all $n, p \in \mathbf{N}$ the subsets $S_{n,p}(\| \,.\, \|)$ in X as follows:

$$S_{n,p}(\| \,.\, \|) = \{x \in X;\ |f(x) - g(x)| \leq 1/p\, if\ \|f\| \leq 1, \|g\| \leq 1\ and\ \|f + g\| > 2 - 2/n\}.$$

We now assume that $\| \,.\, \|$ an equivalent uniformly Gâteaux smooth norm on X. Then an easy duality argument shows that the dual norm has the so-called W^*UR property (see DEVILLE, GODEFROY and ZIZLER [1993], Theorem II.6.7), and thus for any $p \in \mathbf{N}$, one has

$$\bigcup_{n \geq 1} S_{n,p}(\| \,.\, \|) = X\,. \tag{7.2}$$

We will show that

$$X = \bigcap_{p \geq 1} \bigcup_{n \geq 1} \overline{(S_{n,p}(\| \,.\, \|)}^{*}.$$

It follows from (7.2) that it suffices to find, for any $G \in X^{**} \backslash X$, some $p \in \mathbf{N}$ such that

$$G \notin \bigcup_{n \geq 1} \overline{S_{n,p}(\| \,.\, \|)}^{*}. \tag{7.3}$$

We set as before $H = G^{-1}(0)$, and we define an equivalent norm q on X by the formula

$$q(x) = sup\{|f(x)|;\ f \in H,\ \|f\| \leq 1\}.$$

We claim that

$$S_{n,p}(\| \,.\, \|) \subset S_{n,p}(q).$$

To prove this claim, we observe that the bipolar theorem shows that the dual unit ball B_{q^*} satisfies

$$B_{q^*} = \{f \in X^*;\ q^*(f) \leq 1\} = \overline{\{f \in H,\ \|f\| \leq 1\}}^{*}. \tag{7.4}$$

Therefore if $q^*(f) \leq 1$, $q^*(g) \leq 1$ and $q^*(f+g) > 2 - 2/n$, there are nets (f_α) and (g_α) in H, weak* convergent to f and g, such that $\|f_\alpha\| \leq 1$ and $\|g_\alpha\| \leq 1$ for all α. Since the norm q^* is weak*-l.s.c., one has $q^*(f_\alpha + g_\alpha) > 2 - 2/n$ when α is large enough, and thus $\|f_\alpha + g_\alpha\| > 2 - 2/n$ since q^* and $\|\ .\ \|$ coincide on H. If now $t \in S_{n,p}(\|\ .\ \|)$, we have $|f_\alpha(t) - g_\alpha(t)| \leq 1/p$ for α large enough and it follows that $t \in S_{n,p}(q)$. This shows our claim.

To prove (7.3), it therefore suffices to show that for $p \in \mathbf{N}$ large enough, one has

$$G \notin \bigcup_{n \geq 1} \overline{S_{n,p}(q)}^{*}.$$

Fix two integers n and p with $p > 1/q^{**}(G)$ and set for simplicity $S_{n,p}(q) = S$. We will show that $G \notin \overline{S}^{*}$. Pick $f \in B_{q^*}$ such that $G(f) > 1/p$, and let $x \in X$ be such that $q(x) \leq 1$ and $f(x) > 1 - 1/n$. By (7.4), there is $g \in H$ with $q^*(g) \leq 1$ and $g(x) > 1 - 1/n$. We have then

$$q^*(f+g) \geq (f+g)(x) > 2 - 2/n.$$

For all $z \in S$, one has $(f-g)(z) \leq 1/p$. On the other hand, $G(f-g) = G(f) > 1/p$ and thus $G \notin \overline{S}^{*}$.

This concludes the proof of Lemma 3.7. and the outline of proof of Theorem 3.6. □

It is not known whether there exists a Banach space X which is a Borel subset of (X^{**}, w^*) without actually being a $K_{\sigma\delta}$ in that space; this question goes back to TALAGRAND [1979]. Note that when X is weakly compactly generated (in short, w.c.g.), there is a weakly compact subset W of X such that $(\cup_{n \geq 1} nW)$ is dense in X, and then we have

$$X = \bigcap_{p \geq 1} \bigcup_{n \geq 1} (nW + \frac{1}{p} B_{X^{**}})$$

therefore any w.c.g. space is a $K_{\sigma\delta}$ of its weak* bidual, and moreover we can write

$$X = \bigcap_{p \geq 1} \bigcup_{n \geq 1} K_{n,p}$$

with $K_{n,p} \subset X + \epsilon(p) B_{X^{**}}$ for all n and $\epsilon(p) \to 0$ when $p \to \infty$. This property easily goes to subspaces of w.c.g. spaces, and conversely it has been very recently shown that it characterizes subspaces of w.c.g. spaces, FABIAN, MONTESINOS and ZIZLER [2001]. This provides an alternative proof of the fact that the class of Eberlein compact sets is stable under continuous image, BENYAMINI, RUDIN and WAGE [1977]. With the proof of Lemma 3.7, it also shows that a space which has an equivalent UG-smooth norm is a subspace of a w.c.g. Banach space.

In fact, more is true and a Banach space X has an equivalent UG-smooth norm if and only if X is a subspace of a space Y such that there is an operator with dense range from some Hilbert space $l_2(\Gamma)$ into Y, FABIAN, GODEFROY and ZIZLER [2001]. This last result is optimal and we refer to FABIAN, GODEFROY, HÁJEK and ZIZLER [200?] for more about this, and for the related notion of strongly UG-smooth norm.

In the case of $\mathcal{C}(K)$ spaces, the result reads: K is uniformly Eberlein compact if and only if $\mathcal{C}(K)$ has an equivalent UG smooth norm (FABIAN, GODEFROY and ZIZLER [2001], Theorem 2). An immediate corollary is that the class of uniformly Eberlein compact sets is stable under continuous image.

4. Nonlinear classification of *C(K)* spaces

Banach spaces are in particular metric spaces, and we can decide to forget their linear structure and to allow nonlinear isomorphisms between them. If we do so, the sentence "isomorphism between X and Y" can mean "homeomorphism between the topological spaces X and Y", or "bi-uniform homeomorphism between the uniform spaces X and Y", or "bi-Lipschitz homeomorphism between the metric spaces X and Y". We refer to the recent and authoritative book BENYAMINI and LINDENSTRAUSS [2000] where the links between this theory and the deepest parts of functional analysis are displayed.

Thanks to the theorems of KADETS [1967] and TORUŃCZYK [1981], two Banach spaces are homeomorphic if and only if they have the same density character, hence the trivial necessary condition for two spaces to be homeomorphic turns out to be sufficient. The situation is much less clear when the uniform structure or the metric structure are considered, even for special classes of Banach spaces such as C(K) spaces. Indeed, the following question is open even within the class of C(K) spaces with K countable.

4.1. PROBLEM. Let X and Y be two separable Banach spaces, such that there exists a bi-Lipschitz homeomorphism between X and Y. Does it follow that X and Y are linearly isomorphic?

For the simplest possible C(K) spaces, namely those which are isomorphic to $c_0(\mathbf{N})$, Problem 4.1 has a positive answer, GODEFROY, KALTON and LANCIEN [2000]. It turns out that a special property of the norm, which we now define, is useful in this context.

4.2. DEFINITION. Let X be a separable Banach space. The norm of X is said to be *Lipschitz weak-star Kadec-Klee* (in short, LKK^*) if there exists $c \in (0, 1]$ such that its dual norm satisfies the following property: for any $x^* \in X^*$ and any weak* null sequence $(x_n^*)_{n\geq 1}$ in X^* ($x_n^* \xrightarrow{w^*} 0$),

$$\limsup \|x^* + x_n^*\| \geq \|x^*\| + c \limsup \|x_n^*\|.$$

This property is in fact an asymptotic smoothness property of the norm of X (in the sense of MILMAN [1971]) which is expressed as a convexity property of the dual norm. Its importance lies in the fact that it provides a characterization of subspaces of $c_0(\mathbf{N})$: indeed a separable space is isomorphic to a subspace of $c_0(\mathbf{N})$ if and only if it has an equivalent LKK^* norm, GODEFROY, KALTON and LANCIEN [2000], Theorem 2.4. This result somehow states that $c_0(\mathbf{N})$ is smoother than any other space; visualizing it as a cube with no vertices gives some intuition of what is going on. We can now sketch a proof of the main result of GODEFROY, KALTON and LANCIEN [2000]:

4.3. THEOREM. *The class of all Banach spaces that are linearly isomorphic to a subspace of $c_0(\mathbf{N})$ is stable under Lipschitz isomorphisms.*

Outline of proof: The following general topological lemma, called Gorelik's principle, serves as a substitute to the lack of weak continuity for nonlinear maps between Banach spaces.

4.4. LEMMA. *Let E and X be two Banach spaces and U be a homeomorphism from E onto X with uniformly continuous inverse. Let b and d be two positive constants and let E_0 be a subspace of finite codimension of E. If*

$$d > \omega(U^{-1}, b) = sup\{\|U^{-1}(x) - U^{-1}(y)\|;\ \|x - y\| \leq b\}$$

then there exists a compact subset K of X such that

$$bB_X \subset K + U(2dB_{E_0}).$$

□ The following claim is due to GORELIK [1994]. It relies on an application of Schauder's fixed point theorem.

Claim:*For every $\varepsilon > 0$ and $d > 0$, there exists a compact subset A of dB_E such that, whenever Φ is a continuous map from A to E satisfying $\|\Phi(a) - a\| < (1 - \varepsilon)d$ for any a in A, then $\Phi(A) \cap E_0 \neq \emptyset$.*

Now, fix $\varepsilon > 0$ such that $d(1 - \varepsilon) > \omega(U^{-1}, b)$. Let $K = -U(A)$, where A is the compact set obtained in the Claim. Consider now $x \in bB_X$ and the map Φ from A to E defined by $\Phi(a) = U^{-1}(x + Ua)$. It is clear that for any $a \in A$, $\|\Phi(a) - a\| < (1 - \varepsilon)d$. Then, it follows from the Claim that there exists $a \in A$ so that $U^{-1}(x + Ua) \in 2dB_{E_0}$. This concludes the proof of Lemma 4.4. □

□ We can now proceed to prove Theorem 4.3. Let U be a Lipschitz isomorphism from a subspace E of c_0 onto the Banach space X. We need to build an equivalent LKK^* norm on X. This norm will be defined as follows. For x^* in X^*, set:

$$|||x^*||| = \sup\left\{\frac{|x^*(Ue - Ue')|}{\|e - e'\|};\ (e, e') \in E \times E,\ e \neq e'\right\}.$$

Since U and U^{-1} are Lipschitz maps, $|||\ |||$ is an equivalent norm on X^*. It is clearly weak* lower semicontinuous and therefore is the dual norm of an equivalent norm on X that we will also denote $|||\ |||$.

Consider $\varepsilon > 0$, $x^* \in X^*$ and $(x_k^*)_{k\geq 1} \subset X^*$ such that $x_k^* \xrightarrow{w^*} 0$ and $\|x_k^*\| \geq \varepsilon > 0$ for all $k \geq 1$. Fix $\delta > 0$ and then e and e' in E so that

$$\frac{x^*(Ue - Ue')}{\|e - e'\|} > (1 - \delta)|||x^*|||.$$

By using translations in order to modify U, we may as well assume that $e = -e'$ and $Ue = -Ue'$. Since E is a subspace of c_0, it admits a finite codimensional subspace E_0 such that

$$\forall f \in \|e\| B_{E_0}, \quad \|e + f\| \vee \|e - f\| \leq (1 + \delta)\|e\|. \tag{7.5}$$

Let C be the Lipschitz constant of U^{-1}. By Lemma 4.4, for every $b < \frac{\|e\|}{2C}$ there is a compact subset K of X such that $bB_X \subset K + U(\|e\| B_{E_0})$. Since (x_k^*) converges uniformly to 0 on any compact subset of X, we can construct a sequence $(f_k) \subset \|e\| B_{E_0}$ such that:

$$\liminf x_k^*(-Uf_k) \geq \frac{\varepsilon \|e\|}{2C}.$$

We deduce from (7.5) that $x^*(Uf_k + Ue) \leq (1 + \delta)\|e\| \, |||x^*|||$ and therefore $x^*(Uf_k) \leq 2\delta \|e\| \, |||x^*|||$. Using again the fact that $x_k^* \xrightarrow{w^*} 0$, we get that:

$$\liminf (x^* + x_k^*)(Ue - Uf_k) \geq (1 - 3\delta)\|e\| \, |||x^*||| + \frac{\varepsilon \|e\|}{2C}.$$

Since δ is arbitrary, by using the definition of $|||\ |||$ and (2.1), we obtain

$$\liminf |||x^* + x_k^*||| \geq |||x^*||| + \frac{\varepsilon}{4C}.$$

This proves that $|||\ |||$ is LKK^*, and concludes the proof of Theorem 4.3. □

It follows from Theorem 4.3 and classical results from Banach space theory (due to HEINRICH and MANKIEWICZ [1982] and JOHNSON and ZIPPIN [1972]) that a space which is Lipschitz-isomorphic to $c_0(\mathbf{N})$ is in fact linearly isomorphic to that space, and quantitative versions of this result are also available (GODEFROY, KALTON and LANCIEN [2000]). However, the proof does not provide any constructive way of obtaining a linear isomorphism once a Lipschitz isomorphism is given. It is not known whether this result on $c_0(\mathbf{N})$, or equivalently on $C(K)$ spaces when $\beta(K) = 1$, extends to all countable spaces. That is, the following question is open: let K be a countable compact set, and X be a Banach space which is Lipschitz-isomorphic to $\mathcal{C}(K)$; is X linearly isomorphic to $\mathcal{C}(K)$? We refer to DUTRIEUX [2001a], DUTRIEUX [2001b] for partial results along these lines. Note that when L and K are metrizable compact spaces, and $\mathcal{C}(K)$ and $\mathcal{C}(L)$ are Lipschitz-isomorphic with Lipschitz constants close enough to 1, then K and L are homeomorphic (JAROSZ [1989]) and thus a nonlinear version of the Amir-Cambern theorem (AMIR [1965], CAMBERN [1966]) holds true; see also LOVBLOM [1986] for a related result.

Theorem 4.3 says that if a Banach space X is Lipschitz-isomorphic to a *subspace* of $c_0(\mathbf{N})$, then it is linearly isomorphic to a subspace of $c_0(\mathbf{N})$. However AHARONI [1974] showed that any separable Banach space is Lipschitz-isomorphic to a *subset* of $c_0(\mathbf{N})$. It is not known whether $c_0(\mathbf{N})$ is the smallest space with this property; in other words, if a Banach space Y contains a subset which is Lipschitz-isomorphic to $c_0(\mathbf{N})$, does it contain a linear copy of $c_0(\mathbf{N})$? Also, this result does not extend to a non-separable frame: it has been shown by PELANT, who also gave some quantitative improvements of Aharoni's result in [1994], that the space $C_0(\omega_1)$ is not even uniformly homeomorphic to a subset of a $c_0(\Gamma)$ space, PELANT [2001].

We now consider the non-separable theory, which looks quite different. It turns out that it is relevant to know whether the spaces involved are w.c.g. since in that case they decompose into separable pieces and are easier to deal with. But the property of being w.c.g. is not stable under Lipschitz-isomorphisms since they are not weakly continuous, and following AHARONI and LINDENSTRAUSS [1978] and DEVILLE, GODEFROY and ZIZLER [1990] this provides simple examples of Lipschitz-isomorphic spaces which are not linearly isomorphic. More precisely, we obtain from a combination of Theorems 4.7 and 4.8 from GODEFROY, KALTON and LANCIEN [2000]:

4.5. THEOREM. *Let K be a compact space. Then:*

(i) *The Cantor-Bendixon derivative of order ω_0 of K is empty if and only if the Banach space $\mathcal{C}(K)$ is Lipschitz-isomorphic to a $c_0(\Gamma)$ space.*

(ii) *If the density character of $\mathcal{C}(K)$ is ω_1, then $\mathcal{C}(K)$ is linearly isomorphic to $c_0(\omega_1)$ if and only if K is an Eberlein compact and its Cantor-Bendixon derivative of order ω_0 is empty.*

Since the $c_0(\Gamma)$ spaces are w.c.g., it is clear that if $\mathcal{C}(K)$ is linearly isomorphic to $c_0(\Gamma)$ then K is Eberlein. Therefore by the above assertion (i), any non-Eberlein compact set L which has some finite derivative empty provides an example of two Banach spaces (namely, $\mathcal{C}(L)$ and $c_0(\Gamma)$) which are Lipschitz but not linearly isomorphic. Here is a simple example:

4.6. EXAMPLE. Let 2^{ω_0} be the Cantor set equipped with its natural topology. For any finite sequence $s \in 2^{<\omega_0}$, let V_s be the corresponding clopen subset of 2^{ω_0}. We consider the subset L of the space of first Baire class functions on 2^{ω_0} consisting of the characteristic functions of the sets V_s, the characteristic functions of the points of 2^{ω_0}, and the function which is identically 0. It is easily seen that L is compact for the topology of pointwise convergence on 2^{ω_0} (L is therefore a Rosenthal compact), that the third derivative of L is empty, and that L is not Eberlein since it is separable but not metrizable.

The density restriction which is attached to the condition (ii) can partly be released, since by checking the arguments in GODEFROY, KALTON and LANCIEN [2000], one may extend (ii) to the case where the density character is less than ω_{ω_0}, that is, the ω_0-th cardinal. But a drastic change occurs at this level, as shown by BELL and MARCISZEWSKI [2001]. Indeed, they prove the existence of an Eberlein compact set L of cardinal ω_{ω_0} with third derivative empty, such that L is not homeomorphic to a subset of $[E]^{\leq n}$ for some set E and some integer n, and thus by Theorem 2.5 above the space $\mathcal{C}(L)$ is not linearly isomorphic to $c_0(\Gamma)$. This provides the first example of a pair of w.c.g. spaces which are Lipschitz but not linearly isomorphic. In this same paper BELL and MARCISZEWSKI [2001], it is also shown that an Eberlein compact set K such that $K^{(\omega_0+1)} = \emptyset$ is uniformly Eberlein, and $(\omega_0 + 1)$ is optimal for this result.

This unexpected behavior of the singular cardinal ω_{ω_0} is related to the failure of the non separable version of Sobczyk's theorem at this level: indeed, the spaces $c_0(\Gamma)$ are complemented in w.c.g. superspaces if the cardinality of Γ is less than ω_{ω_0}, but it fails for this cardinal, ARGYROS, CASTILLO, GRANERO, JIMENEZ and MORENO [2001].

We now change our notion of "isomorphisms" and consider bi-uniform homeomorphisms between two Banach spaces X and Y. When such a map exists, we say that X

and Y are uniformly homeomorphic. There exist uniformly homeomorphic separable Banach spaces X and Y with X reflexive and Y^* non-separable RIBE [1984]; note that such spaces cannot be Lipschitz-isomorphic since having a separable dual is invariant under Lipschitz-isomorphisms, BATES, JOHNSON, LINDENSTRAUSS, PREISS and SCHECHTMAN [1999]. We refer to VAN MILL, PELANT and POL [2001] and related work for results on uniform homeomorphisms between $\mathcal{C}(K)$ spaces equipped with the pointwise convergence topology. A classical reference on such spaces is ARHANGEL'SKII [1989].

It is interesting to notice that although the separability of the dual is not stable under uniform homeomorphisms, it does work if the dual is "very separable". Let us try to give a meaning to this sentence. For this, we recall the definition of the Szlenk index and the Szlenk derivation. Suppose X is a separable infinite-dimensional Banach space and K is a weak* compact subset of X^*. If $\epsilon > 0$ we let $\mathcal{V}$ be the set of all weak* open subsets V of X^* such that $diam(V \cap K) \leq \epsilon$ and the ϵ interior of K is $\iota_\epsilon K = K \setminus \cup\{V : V \in \mathcal{V}\}$. We then define $\iota_\epsilon^\alpha K$ for any ordinal α by $\iota_\epsilon^{\alpha+1} K = \iota_\epsilon \iota_\epsilon^\alpha K$ and $\iota_\epsilon^\alpha K = \cap_{\beta<\alpha} \iota_\epsilon^\beta K$ if α is a limit ordinal.

We denote by B_{X^*} the closed unit ball of X^*. We then define $Sz(X, \epsilon)$ (or $Sz(\epsilon)$ if no confusion can arise) to be the least countable ordinal α so that $\iota_\epsilon^\alpha B_{X^*} = \emptyset$, if such an ordinal exists. Otherwise we will put $Sz(X, \epsilon) = \omega_1$. The *Szlenk index* is defined by $Sz(X) = \sup_{\epsilon>0} Sz(X, \epsilon)$. It follows from Baire's theorem that $Sz(X) < \omega_1$ if and only if X^* is separable.

Note that $Sz(X, \epsilon) \geq \epsilon^{-1}$ if $\epsilon > 0$, and compactness requires that $Sz(X, \epsilon)$ is not a limit ordinal. Thus $Sz(X) = \omega_0$ is equivalent to $Sz(X, \epsilon) < \omega_0$ for every $\epsilon > 0$, where ω_0 denotes the first limit ordinal.

The Szlenk index measures "how close" the weak* and norm topologies on the dual unit ball are to each other. We refer to GODEFROY [2001b] for a survey of this notion. We are mainly interested here in the case when the Szlenk index is as small as possible, namely when $Sz(X) = \omega_0$. In this case, the quantity $Sz(X, \epsilon)$ is a function of ϵ which takes values into $\mathbf{N}$ and we may investigate its quantitative behavior. One way to do that is to refine the approach by allowing the removal of slices of different sizes: Following KNAUST, ODELL and SCHLUMPRECHT [1999], we will say that X admits a *summable Szlenk index* if there exists a constant C so that $\sum_{i=1}^n \epsilon_i \leq C$ whenever $\iota_{\epsilon_1} \ldots \iota_{\epsilon_n} B_{X^*} \neq \emptyset$. It is clear that when X has a summable Szlenk index, then $Sz(X) = \omega_0$ and moreover $Sz(X, \epsilon) \leq C\epsilon^{-1}$ for some constant C.

When the norm of X is LKK^* (see Definition 4.2), it is easily seen that the Szlenk index is summable, and thus the Szlenk index of subspaces of $c_0(\mathbf{N})$ is summable. The converse is not true, since there is a reflexive space which satisfies this condition, namely Tsirelson's space (KNAUST, ODELL and SCHLUMPRECHT [1999]), and no infinite dimensional subspace of $c_0(\mathbf{N})$ is reflexive. Hence, summability of the Szlenk index does not suffice for obtaining an equivalent LKK^* norm. However, it *nearly* suffices: more precisely, we have by GODEFROY, KALTON and LANCIEN [2001], Theorem 4.10, that a separable Banach space X has a summable Szlenk index if and only if for any function $f : (0, 1) \to (0, 1)$ which satisfies $\lim_{\tau \to 0} \tau^{-1} f(\tau) = 0$, there is an equivalent norm $|\cdot|$ on X and a constant $c > 0$ so that if $0 < \tau < 1$ and $x^*, x_n^* \in X$ satisfy $|x^*| = 1$, $|x_n^*| = \tau$ and $\lim_{n\to\infty} x_n^* = 0$ weak* then

$$\liminf_{n\to\infty} |x^* + x_n^*| \geq 1 + cf(\tau).$$

Note that a norm is LKK^* if and only if it satisfies this condition with $f(\tau) = \tau$.

The proof of Theorem 4.3 shows that the existence of a LKK^* norm is invariant under Lipschitz isomorphisms. The above considerations suggest that the quantitative behavior of the Szlenk index should be invariant under such isomorphisms. It is indeed so, and in fact more is true, as shown in GODEFROY, KALTON and LANCIEN [2001]:

4.7. THEOREM. *Let X and Y be two separable and uniformly homeomorphic Banach spaces. Then $Sz(X) = \omega_0$ if and only if $Sz(Y) = \omega_0$, and $Sz(X)$ is summable if and only if $Sz(Y)$ is summable.*

However, it is still unknown whether the space $c_0(\mathbf{N})$ is determined by its uniform structure.

4.8. PROBLEM. Let X be a Banach space which is uniformly homeomorphic to $c_0(\mathbf{N})$. Is the space X linearly isomorphic to $c_0(\mathbf{N})$?

Les us collect some comments about this question. Its answer is positive when X is Lipschitz isomorphic to $c_0(\mathbf{N})$, and also when X is isomorphic to a complemented subspace of a $C(K)$ space JOHNSON, LINDENSTRAUSS and SCHECHTMAN [1996], Corollary 3.2. It follows from Theorem 4.7 and HEINRICH and MANKIEWICZ [1982] that if X is uniformly homeomorphic to $c_0(\mathbf{N})$, then X is an isomorphic predual of $l_1(\mathbf{N})$ with a summable Szlenk index. It is shown in ALSPACH [2000] that there exist isomorphic preduals B of $l_1(\mathbf{N})$ with $Sz(B) = \omega_0$ which are not isomorphic to $c_0(\mathbf{N})$, but it follows from HAYDON [2000] that the Szlenk index of these spaces is not summable.

It is not known whether an isomorphic predual of $l_1(\mathbf{N})$ with summable Szlenk index is isomorphic to $c_0(\mathbf{N})$. If it is so, then of course Problem 4.8 has a positive answer. The possibility remains that $l_1(\mathbf{N})$ could have an isomorphic predual sharing many features of $c_0(\mathbf{N})$ without being isomorphic to that space.

References

AHARONI, I.

[1974] Every separable metric space is Lipschitz equivalent to a subset of c_0, *Israel J. Math* **19**, 284–291.

AHARONI, I. and J. LINDENSTRAUSS

[1978] Uniform equivalence between Banach spaces, *Bull. Amer. Math. Soc* **84**, 281–283.

ALSPACH, D.

[2000] The dual of the Bourgain-Delbaen space, *Israel J. Math* **117**, 239–259.

AMIR, D.

[1965] On isomorphisms of continuous function spaces, *Israel J. Math* **3**, 205–210.

AMIR, D. and J. LINDENSTRAUSS

[1968] The structure of weakly compact sets in Banach spaces, *Annals of Math* **88**, 35–44.

ARGYROS, S., J. CASTILLO, A. GRANERO, M. JIMENEZ and J. MORENO

[2001] On Sobczyk's theorem, preprint.

ARGYROS, S., G. GODEFROY and H. P. ROSENTHAL

[2002] Descriptive set theory and Banach spaces, in *Handbook of Banach Spaces, vol. 2*, North Holland, Amsterdam, to appear.

ARGYROS, S., S. MERCOURAKIS and S. NEGREPONTIS

[1982] Analytic properties of Corson-compact spaces, *Proc. Fifth Prague Topological Symposium*, August 1981, Heldermann-Verlag, Berlin, 12–23.

[1988] Functional-analytic properties of Corson-compact spaces, *Studia Math* **89**, 197–229.

ARHANGEL'SKII, A.V.

[1989] *Topological Function Spaces*, Moscow State Univ. 1989 (in Russian); English translation: Kluwer Acad. Publ. 1992.

ARVANITAKIS, A.D.

[2001] Some remarks on Radon-Nikodym compact spaces, preprint.

BATES, S., W.B. JOHNSON, J. LINDENSTRAUSS, D. PREISS and G. SCHECHTMAN

[1999] Affine approximation of Lipschitz functions and nonlinear quotients, *Geom. Funct. Anal.* **6**, 1092–1127.

BELL, M. and W. MARCISZEWSKI

[2001] On scattered Eberlein compact spaces, preprint.

BENYAMINI, Y. and J. LINDENSTRAUSS

[2000] *Geometric Nonlinear Functional Analysis*, volume 1, Amer. Math. Soc., Colloquium Publications 48.

BENYAMINI, Y., M.E. RUDIN and M. WAGE

[1977] Continuous images of weakly compact subsets of Banach spaces, *Pacific J. Math.* **70**, 309–327.

BENYAMINI, Y. and T. STARBIRD

[1976] Embedding weakly compact sets into Hilbert spaces, *Israel J. Math.* **23**, 137–141.

BESSAGA, C. and A. PEŁCZYŃSKI

[1960] Spaces of continuous functions (IV), *Studia Math.* **19**, 53–62.

BOURGIN, R.D.

[1983] *Geometric Aspects of Convex Sets with the Radon-Nikodym Property*, Lecture Notes in Math. 993, Springer–Verlag .

CAMBERN, M.

[1966] A generalized Banach-Stone theorem, *Proc. Amer. Math. Soc.* **17**, 396–400.

CASCALES, B., G. MANJABACAS and G. VERA

[1998] Fragmentability and compactness in $\mathcal{C}(K)$ spaces, , *Studia Math.* **131** (1), 73–87.

COMFORT, W.W. and S. NEGREPONTIS

[1982] *Chain Conditions in Topology*, Cambridge Tracts in Math., Cambridge Univ. Press.

CORSON, H. H.

[1959] Normality in subsets of product spaces, *Amer. J. Math.* **81**, 785–796.

DAVIS, W., T. FIGIEL, W.B. JOHNSON and A. PEŁCZYŃSKI

[1974] Factoring weakly compact operators, *J. Funct. Anal.* **17**, 311–327.

DEVILLE, R., G. GODEFROY and V. ZIZLER

[1990] The three space problem for smooth partitions of unity and $\mathcal{C}(K)$ spaces, *Math. Ann.* **288**, 613–625.

[1993] *Smoothness and Renormings in Banach Spaces*, Pitman Monographs and Surveys 64, Longman Ed.

DUTRIEUX, Y.

[2001a] Quotients of c_0 and Lipschitz-homeomorphisms, *Houston J. Math.* **27** (3), 593–599.

[2001b] Lipschitz-quotients and the Kunen-Martin theorem, *Comment. Math. Univ. Carolinae* **42** (4), 641–648.

FABIAN, M.

[1997] *Gâteaux Differentiability of Convex Functions and Topology, Weak Asplund Spaces*, Canadian Math. Soc. Monographs, Wiley-Interscience.

FABIAN, M. and G. GODEFROY

[1988] The dual of every Asplund space admits a projectional resolution of the identity, *Studia Math.* **91**, 141–151.

FABIAN, M., G. GODEFROY, P. HÁJEK and V. ZIZLER

[200?] Hilbert generated spaces, *J. of Funct. Anal.*, to appear.

FABIAN, M., G. GODEFROY and V. ZIZLER

[1999] A note on Asplund generated Banach spaces, *Bull. Polish Acad. Sciences. Math* **47** (3), 221–230.

[2001] The structure of uniformly Gâteaux smooth Banach spaces, *Israel J. Math.* **124**, 243–252.

FABIAN, M., P. HABALA, P. HÁJEK, V. MONTESINOS, J. PELANT and V. ZIZLER

[2001] *Functional Analysis and Infinite-dimensional Geometry*, CMS Books in Mathematics 8, Springer-Verlag, New York.

FABIAN, M., P. HÁJEK and V. ZIZLER

[1997] Uniform Eberlein compacta and uniformly Gâteaux smooth norms, *Serdica Math. J.* **23**, 351–362.

FABIAN, M., V. MONTESINOS and V. ZIZLER

[2001] An inner characterization of subspaces of weakly compactly generated Banach spaces, preprint.

FABIAN, M. and J.H.M. WHITFIELD

[1994] On equivalent characterizations of weakly compactly generated Banach spaces, *Rocky Mountain J. Math.* **24**, 1363–1378.

FINET, C. and G. GODEFROY

[1989] Biorthogonal systems and big quotient spaces, in *Proceedings of a research workshop*, Bor Luh-Lin ed., *Contemporary Math* vol. 85, 87–110.

FREMLIN, D.

[1984] *Consequences of Martin's Axiom*, Cambridge Univ. Press.

GODEFROY, G.

[1980] Compacts de Rosenthal, *Pacific J. Math.* **91** (2), 293–306.

[2001a] Renormings of Banach spaces, in *Handbook of the Geometry of Banach spaces*, vol 1, Elsevier, 781–835.

[2001b] The Szlenk index and its applications, in: *General Topology in Banach Spaces*, T. Banakh ed., Nova Sci. Publ., Huntington, New York, 71–79.

GODEFROY, G., N.J. KALTON and G. LANCIEN

[2000] Subspaces of $c_0(\mathbf{N})$ and Lipschitz isomorphisms, *Geom. Funct. Anal.* **10**, 798–820.

[2001] Szlenk indices and uniform homeomorphisms, *Trans. Amer. Math. Soc.* **353** (10), 3895–3918.

GODEFROY, G. and M. TALAGRAND

[1982] Espaces de Banach representables, *Israel J. Math.* **41** (4), 321–329.

GORELIK, E.
[1994] The uniform non-equivalence of L_p and l_p, *Israel J. Math.* **87**, 1–8.

GUL'KO, S.P.
[1979] On the structure of spaces of continuous functions and their complete paracompactness, *Russian Math. Surveys* **34** (6), 36–44.

HABALA, P., P. HÁJEK and V. ZIZLER
[1996] *Introduction to Banach spaces*, volumes I and II, Matfyzpress, Prague.

HÁJEK, P.
[2001] Smooth partitions of unity on certain $\mathcal{C}(K)$ spaces, preprint.
[200?] Smooth norms on certain $\mathcal{C}(K)$ spaces, *Proc. Amer. Math. Soc.*, to appear.

HÁJEK, P. and R. HAYDON
[200?] Smooth renormings of $\mathcal{C}(K)$ spaces, in preparation.

HAYDON, R.
[1990] A counterexample to several questions about scattered compact spaces, *Bull. London Math. Soc.* **22**, 261–268.
[1992] Normes indéfiniment différentiables sur certains espaces de Banach, *Note aux C. R. A. S. Paris* **315**, 1175–1178.
[1995] Baire trees, bad norms and the Namioka property, *Mathematika* **42** (1), 30–42.
[1999] Trees in renorming theory, *Proc. London Math. Soc.* **78**, 541–584.
[2000] Subspaces of the Bourgain-Delbaen space, *Studia Math.* **139**, 275–293.
[200?] On locally uniformly rotund renormings, in preparation.

HAYDON, R., J. JAYNE, I. NAMIOKA and A. ROGERS
[2000] Continuous functions on totally ordered spaces that are compact in their order topology, *J. Funct. Anal.* **178** (1), 23–63.

HEINRICH, S. and P. MANKIEWICZ
[1982] Applications of ultrapowers to the uniform and Lipschitz classification of Banach spaces, *Studia Math.* **73**, 225–251.

JAROSZ, K.
[1989] Nonlinear generalization of the Banach Stone theorem, *Studia Math.* **93**, 97–107.

JAYNE, J., I. NAMIOKA and A. ROGERS
[1992] σ-fragmentable Banach spaces, *Mathematika* **39**, 197–215.

JIMENEZ SEVILLA, M. and J. P. MORENO
[1997] Renorming Banach spaces with the Mazur intersection property, *J. of Funct. Anal.* **144**, 486–504.

JOHN, K. and V. ZIZLER
[1974] Smoothness and its equivalents in weakly compactly generated Banach spaces, *Journal of Funct. Anal.* **15**, 161–166.

JOHNSON, W.B., J. LINDENSTRAUSS and G. SCHECHTMAN
[1996] Banach spaces determined by their uniform structure, *Geom. Funct. Anal.* **6**, 430–470.

JOHNSON, W.B. and M. ZIPPIN
[1972] On subspaces and quotients of $(\Sigma G_n)_{l_p}$ and $(\Sigma G_n)_{c_0}$, *Israel J. Math.* **13**, 311–316.

KADETS, M.I.
[1967] A proof of the topological equivalence of separable Banach spaces,(Russian), *Funkcional Anal. Prilozen.* **1**, 53–62.

KALENDA, O.

[2000a] Valdivia compact spaces in topology and Banach space theory, *Extracta Math.* **15** (1), 1–85.

[2000b] Valdivia compacta and equivalent norms, *Studia Math.* **138** (2), 179–191.

KNAUST, H., E. ODELL and T. SCHLUMPRECHT

[1999] On asymptotic structure, the Szlenk index and UKK properties in Banach spaces, *Positivity* **3**, 173–199.

LINDENSTRAUSS, J.

[1965] On reflexive spaces having the metric approximation property, *Israel J. Math.* **3**, 199–204.

[1972] Weakly compact sets, their topological properties and the Banach spaces they generate, in *Annals of Math. Studies* 69, Princeton University Press, 235–276.

LOVBLOM, G.M.

[1986] Isometries and almost isometries between spaces of continuous functions, *Israel J. Math.* **56**, 143–159.

MARCISZEWSKI, W.

[2001] On Banach spaces $\mathcal{C}(K)$ isomorphic to $c_0(\Gamma)$, preprint.

MERCOURAKIS, S. and S. NEGREPONTIS

[1992] Banach spaces and topology II, in *Recent Progress in General Topology*, M.Hušek and J.van Mill, eds., Noth-Holland, Amsterdam, 493–536.

VAN MILL, J., J. PELANT and R. POL

[2001] Note on function spaces with the topology of pointwise convergence, preprint

MILMAN, V.

[1971] Geometric theory of Banach spaces II. Geometry of the unit ball,(Russian), *Uspehi Mat. Nauk* **26** (6), 73–149 ; English Translation: Russian Math. Surveys **26** (6), 79–163.

MILUTIN, A.

[1966] Isomorphisms of the spaces of continuous functions over compact sets of the cardinality of the continuum, (Russian), *Teor. Funkcii Funkcional. Anal. Prilozen.* **2**, 150–156.

MOLTO, A., J. ORIHUELA, S. TROYANSKI and M. VALDIVIA

[1999] On weakly locally uniformly rotund Banach spaces, *J. Funct. Anal.* **163** (2), 252–271.

[2000] Kadec and Krein-Milman properties, *Note aux C. R. A. S. Paris* **331** (6), 459–464.

NAMIOKA, I.

[1974] Separate continuity and joint continuity, *Pacific J. Math.* **51**, 515–531.

NAMIOKA, I. and R. POL

[1992] Mapping of Baire spaces into function spaces and Kadec renorming, *Israel J. Math.* **78** (1), 1–20.

NEGREPONTIS, S.

[1984] Banach spaces and topology, in *Handbook in set-theoretic topology*, K.Kunen and J.Vaughan, eds., North-Holland, Amsterdam, 1045–1142.

ORIHUELA, J., W. SCHACHERMAYER and M. VALDIVIA

[1991] Every Radon-Nikodym Corson compact is Eberlein compact, *Studia Math.* **98**, 157–174.

ORIHUELA, J. and M. VALDIVIA

[1989] Projective generators and resolution of identity in Banach spaces, *Rev. Math. Univ. Compl. Madrid*, 179–199.

OVSEPIAN, R.I. and A. PEŁCZYŃSKI

[1975] The existence in every separable Banach space of a fundamental total and bounded biorthogonal sequence and related constructions of uniformly bounded orthonormal systems in L^2, *Studia Math.* **54**, 149–159.

PELANT, J.

[1994] Embeddings into c_0, *Topology Appl.* **57**, 259–269.

[2001] Complexity of uniform covers in function spaces, preprint.

PEŁCZYŃSKI, A.

[1976] All separable Banach spaces admit for every $\epsilon > 0$ fundamental and total biorthogonal sequences bounded by $1 + \epsilon$, *Studia Math.* **55**, 295–304.

PLICHKO, A.

[1986] On bounded biorthogonal systems in some function spaces, *Studia Math.* **84**, 25–37.

RAJA, M.

[1999] Locally uniformly rotund norms, *Mathematika* **46** (2), 343–358.

RIBE, M.

[1984] Existence of separable uniformly homeomorphic nonisomorphic Banach spaces, *Israel J. Math.* **48**, 139–147.

STEGALL, C.

[1975] The Radon-Nikodym property in conjugate Banach spaces, *Trans. Amer. Math. Soc.* **206**, 213–223.

[1990] More facts about conjugate Banach spaces with the Radon-Nikodym property, *Acta Univ. Carolinae Math. Phys.* **31**, 107–117.

SUAREZ, G.A.

[1985] Some uncountable structures and the Choquet-Edgar property in non-separable Banach spaces, in *Proceedings on the 10th Spanish-Portuguese Conf. in Math.*, vol. III, Murcia, 397–406.

TALAGRAND, M.

[1979] Espaces de Banach faiblement K-analytiques, *Annals of Math.* **110**, 407–438.

TERENZI, P.

[1998] A positive anwser to the basis problem, *Israel J. Math.* **104**, 51–124.

TODORČEVIĆ, S.

[1993] Irredundant sets in Boolean Algebras, *Trans. Amer. Math. Soc.* **339**, 35–44.

TORUŃCZYK, H.

[1981] Characterizing Hilbert space topology, *Fund. Math.* **111**, 247–262.

VALDIVIA, M.

[1990] Projective resolutions of identity in $\mathcal{C}(K)$ spaces, *Arch. Math.* **54**, 493–498.

[1991] Simultaneous resolutions of the identity operator in normed spaces, *Collect. Math.* **42**, 265–285.

VANDERWERFF, J.

[1992] Smooth approximations in Banach spaces, *Proc. Amer. Math. Soc.* **115**, 113–120.

VASAK, L.

[1981] On one generalization of weakly compactly generated Banach spaces, *Studia Math.* **70**, 11–19.

ZIZLER, V.

[2002] Nonseparable Banach spaces, in *Handbook on Banach Spaces*, vol.2, to appear.

CHAPTER 8

Metrizable Spaces and Generalizations

Gary Gruenhage[1]

Department of Mathematics, Auburn University, Auburn, Alabama 36830, USA
E-mail: garyg@auburn.edu

Contents

[1] Partially supported by NSF DMS-0072269

RECENT PROGRESS IN GENERAL TOPOLOGY II
Edited by Miroslav Hušek and Jan van Mill

1. Introduction

Classes of "generalized" metric or metrizable spaces are those which possess some of the useful structure of metrizable spaces. They have had many applications in the theory of topological groups, in function space theory, dimension theory, and other areas. Even some applications in theoretical computer science are appearing–see, e.g., the article by G.M. Reed in this volume.

In GRUENHAGE [1992], we discussed research activity in generalized metrizable spaces and metrization which occurred primarily over the seven years previous to the 1991 Prague Topological Symposium. Here we discuss activity in the ten years since that time. Of course, there were too many results to include everything, so this article is a quite imperfect selection of them, reflecting to some extent the author's interests, as well as his lack of expertise in certain areas. The article is divided into sections on topics where most of the recent activity has occurred. In the final section, we discuss a variety of open problems.

There are a number of sources for more basic information about the concepts discussed here, e.g., GRUENHAGE [1984] in the Handbook of Set-theoretic Topology, and several articles in the book Topics in Topology (MORITA and NAGATA [1989]).

Unless otherwise stated, all spaces are assumed to be regular and T_1.

2. Metrics, metrizable spaces, and mappings

In our previous article, not much about metrizable spaces or metrics themselves were discussed, but let us mention here a few results in this area that have a set-theoretic topology flavor.

First, an outstanding result in the dimension theory of metrizable spaces was obtained by MRÓWKA [1997] [2000]. Mrówka constructed a metrizable space M (he denoted it by $\nu\mu_0$) with ind $M = 0$, such that if the set-theoretic axiom denoted by $S(\aleph_0)$ is assumed, then any completion of M contains an interval (hence Ind $M \geq 1$) , and further every completion of M^2, which of course also has small inductive dimension 0, contains the square, and hence Ind $M^2 \geq 2$. M^2 is the first known metrizable space in which the gap between the small and large inductive dimensions is at least 2. KULESZA [200?] extended this to show that every completion of M^n contains an n-cube, and hence the gap between these dimensions can be arbitrarily large.

The only rub with these fascinating examples is that they are far from being ZFC examples. The space M is constructed in ZFC, and M *fails* to be an example under the continuum hypothesis! Furthermore, the axiom $S(\aleph_0)$ under which M is an example is very strong. DOUGHERTY [1997] showed $S(\aleph_0)$ is consistent modulo large cardinals and has large cardinal strength. More specifically, its consistency follows from the existence of the Erdös cardinal $E(\omega_1 + \omega)$ and implies the consistency of $E(\omega)$.

Some interesting mapping theory questions of E. Michael were answered. A continuous map $f : X \to Y$ is *compact covering* (resp., *countable-compact covering*) if every compact (resp., compact countable) subset L of Y is the image of some compact subset K of X, and is *inductively perfect* if there is some $X' \subset X$ such that the restricted map $f{\restriction}X'$ is a perfect map of X' onto Y. MICHAEL [1981] asks the following questions, which were repeated (in Problems 392 and 393) in his article in the book Open Problems in Topology:

Suppose X and Y are separable metric spaces, and $f : X \to Y$ continuous.

(a) Suppose f is compact covering. Must f be inductively perfect if either (i) Y is countable, or (ii) each $f^{-1}(y)$ is compact?

(b) If each $f^{-1}(y)$ is compact, and f is countable-compact covering, must f be compact covering?

DEBS and SAINT RAYMOND ([1996] and [1997]) give counterexamples to (a)(ii) and (b), respectively[2] (contradicting Theorem 2.4 in JUST and WICKE [1994] and Theorem 0.2 in CHO and JUST [1994]). On the other hand, the answer to (a)(i) is positive (JUST and WICKE [1994]), even if the condition on Y is generalized to σ-compact (OSTROVSKII [1994]).

Of course, (a) is a special version of the more general question of when compact covering maps between separable metric spaces must be inductively perfect, and it turns out that this can be the case if X is "nice" in a descriptive set-theoretical sense. For example, it is known to be the case if X is Polish, CHRISTENSEN [1973], SAINT RAYMOND [1971-1973]. Under Analytic Determinacy, it holds if X is absolutely Borel (DEBS and SAINT RAYMOND [1996]), but under $V = L$, there is a counterexample where X is an F_σ-subset of the irrationals (DEBS and SAINT RAYMOND [1999]).

Another question on compact covering maps, due to Michael and Nagami and also appearing in Open Problems in Topology, was answered by H. CHEN [1999]. Chen constructed a Hausdorff space Y which is the image of a metrizable space under a quotient map with separable fibers, such that Y is not a compact covering image of any metrizable space. His space Y is not regular; he asks if there can be a regular example.

A space Y is called a *connectification* of a space X if X is dense in Y and Y is connected. It is easy to see that if X has a compact open subset, then X has no Hausdorff connectification. There seem to be no other obvious general conditions which preclude spaces from having "nice" connectifications. WATSON and WILSON [1993] gave the first systematic study of when spaces have a Hausdorff connectification. Included in this work, they show that every metrizable nowhere locally compact space has a Hausdorff connectification. ALAS, TKACHUK, TKAČENKO and WILSON [1996] then showed that every separable metrizable space with no compact open sets has a metrizable connectifaction, and asked if this is true in the non-separable case as well. This question was answered in the negative by GRUENHAGE, KULESZA and LE DONNE [1998], who gave a construction (due primarily to Kulesza) of a metrizable space with no compact open sets which does not have a metrizable, or even perfectly normal, connectification. It is also proven there that nowhere locally compact metrizable spaces do have metrizable connectifications. Whether or not every metrizable space with no compact open sets has a Tychonoff connectification remains an open question.

Now we present a sampling of results about metrics with special properties. *Ultrametric spaces*, also called *non-Archimedean* metric spaces, are metric spaces with a distance d such that $d(x,z) \le max\{d(x,y), d(y,z)\}$. The metrizable spaces which admit such a metric are exactly those having covering dimension 0. They have a long history and have found many diverse applications. Here we mention recent results on universal (in the sense of isometry) ultrametric spaces. Any universal space for ultrametric spaces of cardi-

[2] This implies a negative answer to Michael's question on triquotient maps mentioned in MICHAEL [1981]

nality two must have cardinality continuum. A. LEMIN and V. LEMIN [2000] constructed for every infinite cardinal number τ, a universal ultrametric space LW_τ with weight τ^ω. The Lemin's asked if their result could be improved for cardinals $\tau > \mathfrak{c}$. This was answered by VAUGHAN [1999], who showed that there is a subspace LW'_τ of LW_τ which is universal for ultrametric spaces of cardinality τ, and assuming the singular cardinal hypothesis, has weight τ whenever $\tau \geq \mathfrak{c}$ (and in ZFC has weight τ for an unbounded set of cardinals). It is unknown if this can be done in ZFC; if so, apparently a different example would be needed.

NAGATA [1983] showed that every metric space has a metric d such that, for each $\epsilon > 0$, the collection $\mathcal{B}_d(\epsilon)$ of all ϵ-balls with respect to d is closure-preserving. His method shows that for separable metric spaces $\mathcal{B}_d(\epsilon)$ can be made finite. In [1999] he asked if any metrizable space admits a metric d such that $\mathcal{B}_d(\epsilon)$ is locally finite. GRUENHAGE and BALOGH [200?] gave a negative answer by showing that the class of metrizable spaces which admit a metric d such that $\mathcal{B}_d(\epsilon)$ is locally finite for all $\epsilon > 0$ is exactly the class of strongly metrizable spaces, i.e., those spaces which embed in $\kappa \times [0,1]^\omega$ for some cardinal κ, where κ carries the discrete topology.

Nagata has also asked if every metrizable space admits a metric d such that X has a σ-locally finite (σ-discrete) base consisting of open d-balls. HATTORI [1986] has shown that the answer to the σ-locally finite question is positive; the σ-discrete question is still unsettled.

"Midset" metric properties have been studied by several authors. The midset between points x and y is the set of all z such that $d(z,x) = d(z,y)$. HATTORI and OHTA [1993] showed that a separable metric space X is homeomorphic to a subspace of the real line iff there exists a metric d for X such that the cardinality of each midset is at most one, and for each x there are at most two points the same distance from x. A metrizable space X is said to have the *unique midset property (UMP)* if there is a metric d on X such that each midset has exactly one point. ITO, OHTA and ONO [1999] showed that discrete spaces with the UMP are exactly the ones of cardinality $\leq \mathfrak{c}$ other than 2 or 4. They also showed that the countable power of any discrete space of size $\leq \mathfrak{c}$ has the UMP; hence, the Cantor set and the irrationals have the UMP. But the question of Hattori and Ohta, whether any separable metrizable space having the UMP must be homeomorphic to a subset of the real line, remains open.

3. Networks

Recall that $\mathcal{F}$ is a *network* for a space X if $x \in U$, where U is open, implies $x \in F \subset U$ for some $F \in \mathcal{F}$. A σ-space is a space with a σ-discrete network. Spaces with a countable network are exactly the continuous images of separable metric spaces, and are sometimes called *cosmic spaces*.

DELISTATHIS and WATSON [2000] made an important advance in the dimension theory of general spaces by constructing, under CH, a cosmic space X (in fact, X is the union of countably many separable metrizable subspaces) in which $\mathit{dim}\, X \neq \mathit{Ind}\, X$. Of course, all three of the standard dimensions agree for separable metrizable spaces; this shows that they may differ for their continuous images, and answers a question of Arhangel'skii. It was also known that the dimensions agree for paracompact Hausdorff spaces which are μ-spaces, i.e., embeddable in the countable product of paracompact spaces which are F_σ-metrizable (= the union of countably many closed metrizable subspaces). So this is also a consistent example of a cosmic space which is not a μ-space. TAMANO [2001] later obtained a ZFC example of a cosmic non-μ-space; and subsequently TAMANO and TODORČEVIĆ [2001] obtained rather natural examples by showing that for separable metric spaces X, $C_p(X)$ is not a μ-space if X is not σ-compact. See Section 5 for the relevance of these examples to the "stratifiable implies M_1" problem.

I am a little embarrassed to mention that the definition of Σ-spaces (a generalization of σ-spaces) given in my article in the Handbook of Set-theoretic Topology is incorrect, as was pointed out by TAMANO [1997]. I had said that X is a Σ-space if there are a cover $\mathcal{C}$ by closed countably compact sets and a σ-discrete collection $\mathcal{F}$ of subsets of X such that, for any $C \in \mathcal{C}$ and open U with $C \subset U$, there is $F \in \mathcal{F}$ with $C \subset F \subset U$. I should have replaced "σ-discrete" with "σ-locally finite", and required members of $\mathcal{F}$ to be closed. For the class of σ-spaces, i.e., where $\mathcal{C}$ can be taken to be the collection of singletons, these differences can be ignored. However, Tamano showed that they can't be ignored here by obtaining an example which satisfies my definition, but is not a Σ-space. It is apparently not known if my definition would have been equivalent to the original had I required the members of $\mathcal{F}$ to be closed (but keeping "σ-discrete" in place of "σ-locally finite").

Cosmic spaces, which are exactly the Lindelöf σ-spaces, are properly contained in the class of Lindelöf Σ-spaces, which can be characterized as the continuous images of perfect pre-images of separable metric spaces. One motivation for studying Lindelöf Σ-spaces comes from Banach space theory. If X is Eberlein compact, then $C_p(X)$ is Lindelöf Σ. Indeed, the class of Gul'ko compacta is precisely the class of compact spaces which have Lindelöf Σ function spaces and is an important generalization of the class of Eberlein compacta.

Several questions of Arhangel'skii concerning Lindelöf Σ-spaces were answered. Let $C_{p,1}(X) = C_p(X)$, and $C_{p,n+1}(X) = C_p(C_{p,n}(X))$. OKUNEV [1993] showed that if X and $C_p(X)$ are Lindelöf Σ, then so is $C_{p,n}(X)$ for all $n > 0$; hence a compact space X is Gul'ko compact iff $C_{p,n}(X)$ is Lindelöf Σ for some $n \in \omega\setminus\{0\}$ iff $C_{p,n}(X)$ is Lindelöf Σ for all $n \in \omega\setminus\{0\}$. TKACHUK [2000] shows further that there are exactly four possibilities for which $C_{p,n}(X)$'s are Lindelöf Σ: either this holds for no n, for all n, or for exactly all even n, or exactly all odd n. He also shows that if ω_1 is a caliber of X (equivalently, $C_p(X)$ has a small diagonal[3]), or [2001] if X has countable spread, then $C_p(X)$ Lindelöf Σ implies X is cosmic. (Arhangel'skii had obtained these results consistently.) OKUNEV and TKACHUK [2001] answered another question of Arhangel'skii by showing that the aforementioned countable spread result fails if this condition is weakened to $p(X) = \omega$, i.e., every point-finite open collection in X is countable. It is not known if $C_p(X)$ Lindelöf Σ and ω_1 a caliber for $C_p(X)$ implies X cosmic.

Arhangel'skii (see YASHCHENKO [1994]) also asked the following question about network properties in $C_p(X)$, which is still open: Does $C_p(X)$ a σ-space imply that X and $C_p(X)$ are cosmic?

GRUENHAGE [200?] partially answered another question of Arhangel'skii by showing that under CH a Lindelöf Σ-space with a small diagonal is cosmic. See Problem 10 in the

[3] A space Y has a *small diagonal* if any uncountable subset Z of $Y^2\setminus\Delta$ contains an uncountable Z' such that $\overline{Z'} \cap \Delta = \emptyset$

problems section for other results on small diagonals.

A stronger network notion is that of a *k-network* for a space X, i.e., a collection $\mathcal{F}$ of subsets of X such that, whenever K is compact and U is an open set containing K, then $K \subset \bigcup\mathcal{F}' \subset U$ for some finite $\mathcal{F}' \subset \mathcal{F}$. k-networks have been useful, among other things, in the study of certain kinds of images of metrizable spaces (e.g., see my earlier surveys GRUENHAGE [1984] and [1992]). In the last ten years, many results and examples concerning k-networks that are point-countable, star-countable, compact-countable, and so forth have been obtained. Rather than attempt to summarize them here, we refer the interested reader to the excellent and very complete surveys of Y. TANAKA [1994] [2001].

4. Monotone normality

The definition of *monotonically normal*, due to Heath, Lutzer, and Zenor, is probably what you would guess if asked to define "normal in a monotone way". It means that one can assign to each pair (H, K) of disjoint closed sets an open set $U(H, K)$ with $H \subset U(H, K) \subset \overline{U(H, K)} \subset X \setminus K$, so that $H \subset H'$ and $K \supset K'$ implies $U(H, K) \subset U(H', K')$. Every metrizable space and every linearly ordered space is monotonically normal.

Surely the most exciting recent development in this area is the proof of RUDIN [2001] that the compact monotonically normal spaces are precisely the continuous images of compact ordered spaces. This answered a question of J. Nikiel. By an earlier result of Nikiel and (independently) Treybig, it also implies the following non-metric analogue of the Hahn-Mazurkiewicz theorem: X is a continuous image of a connected ordered compact space iff X is compact, connected, locally connected, and monotonically normal.

The earlier work of WILLIAMS and ZHOU [1991] [1998] on the structure of compact monotonically normal spaces, which was discussed to some extent in HUŠEK and VAN MILL [1992], has continued to play a role, in particular, the so-called "Williams-Zhou" trees. The idea of these trees is part of the difficult proof of Rudin's result above, and the trees are used by GARTSIDE [1997] in his thorough study of cardinal invariants of monotonically normal spaces.

Another result of RUDIN [1996] answered a question of Purisch; she constructed a locally compact monotonically normal space which has no monotonically normal compactification.

Some interesting results regarding products were obtained. PURISCH and RUDIN [1997] showed that if X and Y are monotonically normal, and Y is countable, then $X \times Y$ is normal. They construct an example demonstrating that the monotone normality assumption on Y is necessary.

NYIKOS [1999] studied monotone normality in trees with the interval topology. He shows that a tree is monotonically normal iff it is the topological sum of convex chains of the tree (hence of ordinal spaces); this generalizes a result proven by K. P. Hart for $\aleph_1$-trees.

A few more results about monotonically normal spaces are mentioned in the next two sections, since they are related to the classes discussed there. Also, we refer the reader to COLLINS [1996] for an excellent survey of monotone normality up to 1996.

5. Stratifiable and related spaces

CEDER [1961] defined the class of M_1 spaces to be those spaces which have a σ-closure preserving base. He also defined M_2 and M_3-spaces, now known to be equal and usually known as *stratifiable* spaces. A nice characterization of stratifiable spaces is that they are exactly the *monotonically perfectly normal* spaces; i.e., to each closed set H one can assign a sequence $U_n(H)$ of open sets satisfying $H = \bigcap_{n<\omega} U_n = \bigcap_{n<\omega} \overline{U_n}$ so that $H \subset H'$ implies $U_n(H) \subset U_n(H')$. They are also exactly the class of monotonically normal σ-spaces.

The main question about stratifiable spaces remains open: are they the same as M_1-spaces, i.e., do they have a σ-closure-preserving base? There have been a number of interesting advances and related partial results. MIZOKAMI and SHIMANE [2000a] improved Ito's much earlier result that first-countable stratifiable spaces are M_1 by proving that stratifiable k-spaces (or, equivalently in this context, sequential spaces are M_1. Their result is a bit more general than that, as follows:

5.1. THEOREM. *Suppose X is stratifiable and has the following property:*

(∗) *Whenever U is open and $x \in \overline{U}\backslash U$, there exists a closure-preserving collection $\mathcal{F}$ of closed subsets of X that is a net at x and $\overline{F \cap U} = F$ for each $F \in \mathcal{F}$.*

Then X is an M_1-space.

It is easy to see that Fréchet spaces satisfy (∗): let $\mathcal{F}$ be the collection of tails of a sequence in U converging to x (along with the limit x itself). Mizokami and Shimane show that sequential stratifiable spaces satisfy (∗).[4] The quite lengthy proof of their main theorem involves using (∗) to build closure-preserving collections of various sorts until finally all can be put together to get the sought-after σ-closure-preserving base.

Recently, this result has been improved by MIZOKAMI, SHIMANE and KITAMURA [200?], who showed that Theorem 5.1 holds with (∗) restricted to dense open sets U.

We mentioned the class of μ-spaces in the section on networks. The examples of Delistathis and Watson, and Tamano, mentioned there are the first known examples of paracompact σ-spaces which are not μ-spaces. This is relevant here because stratifiable spaces are paracompact σ-spaces, and it is known that all stratifiable μ-spaces are M_1, but it is not known whether all stratifiable spaces are μ-spaces.

Gartside and others have looked, with some success, for "natural" classes of spaces that are stratifiable. First, in a negative direction, GARTSIDE [1998] proved that a compact κ-metrizable (in the sense of SHCHEPIN [1980]; one may use Dugundji compacta here instead) space with a dense monotonically normal subspace is metrizable. Since Tychonoff cubes are κ-metrizable, and any linear subspace of a product of lines can be re-embedded densely in a product of lines, and hence densely in a compact κ-metrizable space, it follows that (i) $C_p(X)$ is monotonically normal (or stratifiable, or metrizable) iff X is countable, and (ii) a Banach space in its weak topology is monotonically normal (or stratifiable, or metrizable) iff it is finite dimensional. Results (i) and (ii) answer questions of Arhangel'skii and Wheeler, respectively. In a more positive direction, SHKARIN [1999] shows that the locally convex direct sum of stratifiable locally convex spaces, as well as

[4] In fact, one can use an induction on the sequential order of a point x with respect to U to show more generally that regular sequential spaces with points G_δ satisfy (∗)

the strict inductive limit of a sequence of metrizable locally convex spaces, is stratifiable (see ROBERTSON and ROBERTSON [1964] for definitions of the functional analytic terms used here).

Gartside's result that $C_p(X)$ is stratifiable if and only if X is countable was proven independently by both YASHCHENKO [1994] and SAKAI [1995]. On the other hand, GARTSIDE and REZNICHENKO [2000] show that for $C_k(X)$, the space of continuous real-valued functions on X with the compact-open topology, the situation is quite different:

5.2. THEOREM. *Let X be a Polish space (i.e., complete separable metric). Then $C_k(X)$ is stratifiable.*

This result provides us with a very natural class of stratifiable spaces. It is interesting that the proof gives no clue as to whether or not these function spaces are in general M_1. In particular, it is not known if $C_k(\mathbb{P})$, where $\mathbb{P}$ is the space of irrationals, is M_1. In unpublished results, Gruenhage and Balogh have shown that there is no σ-closure-preserving base consisting of finite unions of standard basic open sets, and K. Tamano showed that standard basic open sets cannot witness another base property known to imply that the space is a μ-space (and hence M_1).

It is also not known if $C_k(X)$ for other separable metric spaces X are stratifiable; deciding whether or not $C_k(\mathbb{Q})$ is stratifiable is probably key here.

Besides function spaces, spaces of subsets with the Vietoris topology were studied by several researchers. It follows easily from classical results that the hyperspace of all (non-empty) closed sets is monotonically normal, stratifiable, or cosmic iff X is compact metric. FISHER, GARTSIDE, MIZOKAMI and SHIMANE [1997] show that the space $\mathcal{F}(X)$ of all finite subsets of X is monotonically normal iff X^2 is monotonically normal iff (by Gartside's result mentioned in the next section) X^n is monotonically normal for all $n < \omega$, and thereby obtain as a corollary the result of MIZOKAMI and KOIWA [1987] that $\mathcal{F}(X)$ is stratifiable iff X is. Furthermore, they show that the space $\mathcal{K}(X)$ of all compact subsets of X is stratifiable if it is monotonically normal and every non-empty open set in X contains an infinite compact set.

GUO and SAKAI [1993] showed that if X is a connected CW-complex, then the space of compact (resp. compact connected) subsets of X is an absolute retract (AR) for the class of stratifiable spaces. CAUTY, GUO and SAKAI [1995] showed that the space of non-empty finite subsets of X is an absolute neighborhood retract (resp., AR) for stratifiable spaces iff X is stratifiable and 2-hyper-locally connected (resp., and connected). In the negative direction, CAUTY [1998] obtained a result implying that none of the classical characterizations of ANR's for metrizable spaces extend in general to the class of stratifiable spaces.

A long and difficult argument of SIPACHEVA [1993] shows that the free abelian group of a stratifiable space is stratifiable. Arhangel'skii's question whether the same result holds in the non-abelian case remains unsolved. In any case, Sipacheva's result shows that any stratifiable space can be embedded as a closed subspace of a stratifiable abelian group.

KUBIAK [1993] characterized monotonically normal spaces in terms of the insertion of a continuous function between upper and lower semi-continuous functions. Specifically, X is monotonically normal iff X has the *monotone insertion property*, i.e., for every pair (f, g) of real-valued functions on X with f upper semi-continuous and g lower semi-continuous, and $f(x) \leq g(x)$ for all $x \in X$, one can assign a continuous function $\lambda(f, g)$

on X with $f(x) \le \lambda(f,g)(x) \le g(x)$ for all x, and $f \le f', g \le g'$ implies $\lambda(f,g)(x) \le \lambda(f',g')(x)$ for all x.

LANE, NYIKOS and PAN [200?] showed that also requiring $\lambda(f,g)(x)$ to be strictly between $f(x)$ and $g(x)$ for all x with $f(x) < g(x)$ characterizes stratifiability; GOOD and STARES [2000] show that stratifiability is also characterized by requiring this strict betweenness only for those pairs (f,g) with $f(x) < g(x)$ for all $x \in X$. [5]

6. Some higher cardinal generalizations

Some interesting work has been done on higher cardinal generalizations of metrizable and stratifiable spaces, and related classes. A space X is *non-archimedean* if it admits a base which is a tree of open sets under reverse inclusion. If each level of such a tree base covers X, and the height of the tree has uncountable cofinality ω_μ (i.e., $\mu > 0$), then X is *ω_μ-metrizable*. A space X with topology $\mathcal{T}$ is said to be *stratifiable over the cardinal* ω_μ if for each open set U, U can be written as the increasing union of open subsets $U(\alpha)$, $\alpha < \omega_\mu$, whose closures are also contained in U, such that $U \subset U'$ implies $U(\alpha) \subset U'(\alpha)$ for each $\alpha < \omega_\mu$. If X is stratifiable over some cardinal, X is called *linearly stratifiable*, and is *ω_μ-stratifiable* if ω_μ is the least cardinal over which it is stratifiable. If X is ω_μ-stratifiable and each point has a totally ordered local base, X is *ω_μ-Nagata*. ω_μ-metrizability implies ω_μ-Nagata implies ω_μ-stratifiable.

See VAUGHAN [200?] for a nice discussion of the above classes of spaces, along with other characterizations of them and examples illustrating their differences. He shows there that ω_μ-Nagata spaces are ultraparacompact (for $\mu > 0$ of course); it is apparently not known if the same is true for ω_μ-stratifiable spaces. He also shows that, unlike ω_μ-metrizable spaces, ω_μ-Nagata spaces need not have an ortho-base (i.e., a base $\mathcal{B}$ such that for any $\mathcal{B}' \subset \mathcal{B}$, either $\cap\mathcal{B}'$ is open, or $\cap\mathcal{B}'$ is a single point and $\mathcal{B}'$ is a base at that point). Vaughan's original argument [1972] that linearly stratifiable spaces are paracompact turned out to have a gap, which is filled here, though it was earlier fixed in a slightly different way by HARRIS [1991] (who was the one who noticed the gap).

An interesting result of STARES and VAUGHAN [1996] shows that the space 2^{ω_1} with the countable box topology is an ω_1-metrizable topological group without the Dugundji extension property. This demonstrates that certain results claimed by van Douwen and Borges, namely, that the Dugundji extension property holds for ω_μ-metrizable and linearly stratifiable spaces, respectively, are false.

A non-archimedean space can equivalently be described as a space with a *rank 1 base* $\mathcal{B}$; i.e., $B, B' \in \mathcal{B}$ and $B \cap B' \neq \emptyset$ implies either $B \subset B'$ or $B' \subset B$. X is *proto-metrizable* if X has a *rank 1 pair-base* $\mathcal{B}$, i.e, $\mathcal{B} = \{B = (B_1, B_2) : B \in \mathcal{B}\}$ such that (i) B_1 is open; (ii) $B_1 \subset B_2$; (iii) $x \in U$, U open, implies $x \in B_1 \subset B_2 \subset U$ for some $B \in \mathcal{B}$; and (iv)$B, B' \in \mathcal{B}$ and $B_1 \cap B_1' \neq \emptyset$ implies either $B_1 \subset B_2'$ or $B_1' \subset B_2$.

All of the classes of spaces discussed in this section are hereditarily paracompact and monotonically normal. GARTSIDE [1999] proved that X^2 monotonically normal implies

[5] The above insertion results should be compared with classical results of Katětov and Tong, Michael, and Dowker, asserting that the existence of at least *one* continuous function between pairs as above characterizes normal spaces, perfectly normal spaces, and normal and countably paracompact spaces, respectively.

X^n is monotonically normal and hereditarily paracompact for all $n < \omega$, and under some further assumptions is linearly stratifiable. On the other hand, he constructs a non-linearly stratifiable topological group every finite power of which is monotonically normal. CAMMAROTO [1994] obtains a related result: X is ω_μ-stratifiable iff $X \times Y$ is monotonically normal for every ω_μ-metrizable Y.

GARTSIDE and MOODY [1993] characterized proto-metrizable spaces as "monotonically paracompact" spaces, and JUNNILA and KÜNZI [1993] characterized them as those spaces which are both monotonically normal and "monotonically orthocompact". Here, a space X is *monotonically paracompact* if one can assign to each open cover $\mathcal{U}$ a star-refinement $\mu(\mathcal{U})$ such that $\mathcal{V}$ refines $\mathcal{U}$ implies $\mu(\mathcal{V})$ refines $\mu(\mathcal{U})$; monotonic orthocompactness, which we will not define here, has a bit more complicated definition in terms of "transitive neighbornets". We should mention that Gartside and Moody also characterize ω_μ-metrizable topological groups.

A class of spaces related to the above classes is the class of *elastic* spaces, due to H.TAMANO and VAUGHAN [1971]. GARTSIDE and MOODY [1992] showed that every elastic space has the "well-ordered F" property (called "well-ordered point-network" in GRUENHAGE [1992]); recall that well-ordered F spaces are also monotonically normal and hereditarily paracompact. Later [1997a], they showed that every proto-metrizable space is elastic, and in [1997b] they obtained a counterexample to a long-standing conjecture of Tamano and Vaughan by constructing an example of a perfect image of an elastic space which is not elastic.

7. Moore and developable spaces

Recall that a space X is *developable* if there is a sequence $\mathcal{G}_n$, $n < \omega$, of open covers such that, for every $x \in X$, $\{st(x, \mathcal{G}_n) : n \in \omega\}$ is a base at x. A *Moore space* is a regular developable space.

SHAKHMATOV, TALL and WATSON [1996] showed that it is consistent for there to be a normal Moore space which is not submetrizable (i.e., it has no weaker metrizable topology). TALL [1994] shows that there are also models of set theory in which there is a non-metrizable normal Moore space and a non-submetrizable countably paracompact Moore space, yet every normal Moore space is submetrizable.

There have been other results related to countable paracompactness in Moore spaces. KNIGHT [1993] used a complicated forcing argument to obtain (consistently) a subset Δ of the real line which is a Δ-set but not a Q-set, where $A \subset \mathbb{R}$ is a *Δ-set* if for every decreasing sequence of subsets A_n, $n < \omega$, of A with empty intersection, there are open subset $U_n \supset A_n$ such that $\bigcap_{n \in \omega} U_n = \emptyset$. It follows that it is consistent that there is a separable countably paracompact non-normal Moore space (namely, the tangent disk space over Δ).

A space is *δ-normal* if every pair of disjoint closed sets, one of which is a regular G_δ-set, can be separated by disjoint open sets. GOOD and TREE [1994], answer a question of Nyikos by constructing a δ-normal Moore space Y which is not countably paracompact (recall that normal Moore spaces must be countably paracompact). It is also the case that $Y \times [0, 1]$ fails to be δ-normal.

LABERGE [1999] answered a question of Tall by constructing a consistent example of a normal Moore space in which every open cover of cardinality $\aleph_1$ has a locally finite open refinement, but not every open cover of cardinality ω_2 does. (Tall did not include the Moore property in the statement of his question.) The continuum hypothesis holds in the model, and the example has the property that every countable-to-one pre-image of the space is normal.

A *π-base* for a space X is a collection $\mathcal{P}$ of non-empty open sets such that every non-empty open set in X contain a member of $\mathcal{P}$. G. M. Reed had shown long ago that a Moore space has a σ-discrete π-base iff it can be densely embedded in a developable T_2-space with the Baire property, and asked if "developable T_2-space" could be replaced with "Moore space". This question appears as Problem 303 in the book Open Problems in Topology. D. FEARNLEY [1999] answered the question in the negative by constructing a Moore space with a σ-discrete π-base which cannot be densely embedded in any Moore space with the Baire property.

GARTSIDE, GOOD, KNIGHT and MOHAMAD [2001] have answered a couple of questions of P. Nyikos by constructing a quasi-developable (defined like developable, except that the $\mathcal{G}_n$'s need not be covers) manifold with a G_δ-diagonal which is not developable, and a consistent example which is also countably metacompact. TREE and WATSON [1993] answer questions of Reed by constructing two non-metrizable Moore manifolds, one of which is pseudonormal, while the other, done under CH, is pseudocompact.

"Property (a)" is a property which had its origins in the theory of countably compact spaces and appears close to normality. X has *property (a)* if for every open cover $\mathcal{U}$ of X and every dense subset D of X, there is a closed discrete subset F of D such that $st(F,\mathcal{U}) = X$. RUDIN, STARES and VAUGHAN [1997] proved that monotonically normal spaces have property (a). MATVEEV [1997] showed that separable Moore spaces having property (a) are metrizable. However, this does not carry over to the non-separable case, as JUST, MATVEEV and SZEPTYCKI [2000] show that in ZFC there is a non-metric Moore space having property (a).

8. Bases with certain order properties

There has been some interesting work done on weakly uniform bases and related properties. Recall that a base $\mathcal{B}$ is *weakly uniform* if the intersection of any infinite subcollection of $\mathcal{B}$ is either empty or a singleton. Also, a base $\mathcal{B}$ is *uniform* (resp, *sharp*) if every infinite subfamily is a local base (resp., subbase) at each point of its intersection. Sharp bases were introduced by ALLECHE, ARHANGEL'SKII and CALBRIX [2000]. Note that if a base is sharp, it is also weakly uniform. A weak uniform base could reasonably be (and now is) also called 2-*in-finite*, since any *two* distinct points are in only *finitely* many members of the base. Similarly, a point-countable base might be called "1-in-countable". This leads to the notions of n-in-finite, ω-in-countable, and so forth.

Motivating some of the results here is an old question of HEATH and LINDGREN [1976]: Does every first-countable space with a weakly uniform base have a (possibly different) point-countable base? In other words, does the existence of a 2-in-finite base along with first-countability imply the existence of a 1-in-countable base? Put in those terms, it shouldn't be too surprising that some interesting combinatorics get involved. Old partial results of DAVIS, REED and WAGE [1976] say that there is a counterexample under

MA$_{\omega_2}$, though the answer is positive in ZFC if there are not more than $\aleph_1$-many isolated points.

Here are some interesting new results. ARHANGEL'SKII, JUST, REZNICHENKO and SZEPTYCKI [2000] show that a space with a sharp base has a point-countable base, generalizing the corresponding known (and easy) result for uniform bases. They also show, under CH, that every first-countable space with a weakly uniform base and no more than $\aleph_\omega$-many isolated points has a point-countable base. BALOGH, DAVIS, JUST, SHELAH and SZEPTYCKI [2000] obtain a stronger (topological) result, which in particular eliminates any condition on the number of isolated points, by introducing the axiom CECA, which is equivalent to GCH plus a bit of $\Box_\lambda$ for singular λ and thus follows from $V = L$. They show that the following holds under CECA: X has a point-countable base if it is first-countable and has a base $\mathcal{B}$ such that, for every infinite subset A of X, some finite subset of A is included in only finitely many members of $\mathcal{B}$. Note that the stated base condition is weaker than n-in-finite for any fixed n.

In the paper of ALLECHE ET AL above, an example of a non-developable space with a sharp base is given. In ARHANGEL'SKII ET AL it is asked whether a pseudocompact space with a sharp base must be metrizable (this is known to be the case if "sharp" is strengthened to "uniform"). GOOD, KNIGHT and MOHAMAD [200?] answer this in the negative; their counterexample has the additional property that its product with the unit interval fails to have a sharp base, which answers a question of ALLECHE ET AL.

BALOGH and GRUENHAGE [2001] generalize the classical result that compact spaces with a point-countable base are metrizable, by showing that compact spaces with an ω-in-countable base are metrizable. They also show that the corresponding statement for countably compact spaces is independent of ZFC. Generalizing results of Peregudov, and Burke and Davis, they show that a locally compact space is metrizable if it has an n-in-countable base, or, provided it has no isolated points, if it has a $\mathfrak{c}$-in-countable base.

BALOGH, BENNETT, BURKE, GRUENHAGE, LUTZER and MASHBURN [2000] study the notion of an *open-in-finite (OIF)* base $\mathcal{B}$, i.e., every non-empty open set is contained in at most finitely many members of $\mathcal{B}$. They show that a base $\mathcal{B}$ is uniform iff the restriction of $\mathcal{B}$ to any subspace Y is an OIF base for Y. They also show, among other things, that every space is an open perfect image as well as a closed subset of a space with an OIF base, and give an example of a space with a point-countable base and an OIF base which is not quasi-developable.

BENNETT and LUTZER [1998a] study several of these base properties in ordered spaces. They show that a linearly ordered space has a point-countable base iff it has an ω-in-countable base. They also show that a generalized ordered space has a weak uniform base iff it is quasi-developable with a G_δ-diagonal, and is metrizable iff it has an OIF base iff it has a sharp base.

There have been some results on products involving point-countable bases. ZHU [1993] showed that if X is a metalindelöf Morita P-space (see Section 9 for the definition), and Y has a point-countable base, then $X \times Y$ is metalindelöf. ALSTER and GRUENHAGE [1995] showed that the same holds if X is a paracompact monotonically normal space, part of the motivation here being the corollary that there can be no monotonically normal counterexample in ZFC to Michael's question whether the product of a Lindelöf space with the irrationals must be Lindelöf. (A Lindelöf version of the Michael line is a monotonically normal counterexample which exists under CH, or more generally, under $\mathfrak{b} = \omega_1$.) Zhu also showed $X \times Y$ is metalindelöf in case X and Y are metalindelöf and Y is a strong Σ-space.

BALOGH ([200?a], [200?b]) obtained some interesting reflection theorems regarding point-countable bases. He showed that for spaces of density not greater than $\aleph_1$, if every subspace of cardinality ω_1 has a point-countable base, then so does the whole space. A very interesting question is whether or not this can be consistently true, for first-countable spaces, without the density restriction (it is known to be consistently false). Balogh also proved that under the so-called Axiom R of Fleissner, a locally compact space is metrizable if every subspace of cardinality ω_1 has a point-countable base. (Compare with DOW's ZFC theorem [1988]that compact space is metrizable if every subset of cardinality $\aleph_1$ is metrizable.)

9. Normality in products

A most exciting development here is the solution by Larson and Todorčević of the following long-standing problem of KATĚTOV [1948]: If X^2 is compact and hereditarily normal, must X be metrizable? Katětov had shown the answer to be positive if X^3 is hereditarily normal. Nyikos discovered a counterexample under MA_{ω_1}, and Gruenhage under CH. The question remained whether or not there is a consistent positive answer.

The difficulty in the problem lay in the fact that certain consequences of CH simultaneously with certain consequences of MA_{ω_1} appeared to be needed to solve it. In particular, it is (in effect) shown in GRUENHAGE and NYIKOS [1993] that a model with no Q-sets and no $S-$ or $L-$ subspaces of first countable compacta would be a model in which the answer to Katětov's question is affirmative. LARSON and TODORČEVIĆ [200?] succeeded in constructing such a model. To get their model, they take a model M in which there is a "coherent" Souslin tree S such that MA_{ω_1} holds for all posets P such that $P \times S$ is *ccc*. (Such models were already known to exist.) They prove that in the model obtained from M by forcing with S, there are no Q-sets (in fact, there are no Q-sets in any model obtained by forcing with a Souslin tree), and that also (here is the most difficult part of the argument) a certain partition relation holds which implies that the hereditary Lindelöf property and the hereditary separable property are equivalent in subspaces of first countable compacta.

It is not difficult to observe that in a model with no Q-sets and no S- or L- subspaces of first countable compacta, e.g., the Larson-Todorčević model, any compact space X such that $X^2 \setminus \Delta$ is perfectly normal must be metrizable; this answers a question mentioned in GRUENHAGE and NYIKOS [1993]. However, Przymusinski's problem whether there are ZFC examples of non-metrizable compact X and Y such that $X \times Y$ is perfectly normal is still open.

Another old problem involving normality in products was solved by Z. Balogh. Morita long ago characterized the spaces X such that $X \times Y$ is normal for every metric space Y; such X are now called (Morita) P-spaces. Morita also stated the following three problems:

(1) Must X be discrete if $X \times Y$ is normal for any normal space Y?

(2) Must X be metrizable if $X \times Y$ is normal for any normal P-space Y?

(3) Must X be metrizable and σ-locally compact if $X \times Y$ is normal for any countably paracompact normal space Y?

In GRUENHAGE [1992], we reported that M. Atsuji and M.E. Rudin had answered (1) affirmatively in ZFC, and that K. Chiba, T. Przymusinski, and Rudin had answered (2) and (3) affirmatively assuming $V = L$. Now, BALOGH [2001] has used extensions of the idea of his recent ZFC Dowker space construction to give affirmative answers to (2) and (3) in ZFC.

Motivated by normality of product questions, JUNNILA and YAJIMA [1998] introduce some new classes of spaces defined by special networks. One such class is the class of *LF-netted* spaces, which are spaces X having a σ-locally finite network $\mathcal{F}$ such that, for every closed $H \subset X$, the collection $\{F \in \mathcal{F} : F \cap H \neq \emptyset\}$ is locally finite at each point of $X \setminus H$. They show that if X is LF-netted and Y normal and countably paracompact, then $X \times Y$ is normal iff countably paracompact. Since metric spaces are LF-netted, this extends classical results of Morita, and Rudin and Starbird. Since MIZOKAMI and SHIMANE [2000b] show that every *Lašnev space*, i.e., closed image of a metric space, is LF-netted, it also generalizes a result of Hoshina. It is not known if every stratifiable space, or stratifiable μ-space, is LF-netted, though stratifiable F_σ-metrizable spaces are.[6]

Next we mention an interesting result on normality in Σ-products, where X is a *Σ-product* of the the spaces X_α $\alpha < \kappa$, if there are $p_\alpha \in X_\alpha$ such that

$$X = \{\vec{x} \in \Pi_{\alpha<\kappa} X_\alpha : |\{\alpha : x_\alpha \neq p_\alpha\}| \leq \omega\}.$$

Recall that Gul'ko and Rudin independently proved that any Σ-product of metrizable spaces is normal. There is by now a whole theory of normality in Σ-products inspired by this result. One of the more interesting open problems, due to Kodama and mentioned in my earlier survey GRUENHAGE [1992], is whether or not the Gul'ko-Rudin theorem holds for the wider class of Lašnev spaces. There is now a consistent negative example, essentially due to P. Koszmider and appearing in EDA, GRUENHAGE, KOSZMIDER, TAMANO and TODORČEVIĆ [1995]. Sequential fans are key in this result. The *sequential fan with κ-many spines*, denoted by $S(\kappa)$, is the (Lašnev) space obtained from the topological sum of κ-many convergent sequences by identifying their limit points. Since the space ω_1 of countable ordinals appears as a closed subspace in any uncountable Σ-product of non-trivial T_1-spaces, the product $S(\omega_2)^2 \times \omega_1$ is a closed subspace of a Σ-product of Lašnev spaces. Koszmider constructs a model in which this product is non-normal. On the other hand, he shows that this product *is* normal under MA_{ω_2} and the negation of Chang's Conjecture. It is still open if there are ZFC examples of non-normal Σ-products of Lašnev spaces. In particular, it is not known if $S(2^{\mathfrak{c}})^2 \times \omega_1$ is non-normal in ZFC.

10. Sums of metrizable subspaces

TKACHUK [1994] defines a property P to be *weakly n-additive* if X has P whenever X^n is the union of at most n subspaces having property P. He showed many properties–most of them local properties such as first-countability and local compactness–are weakly n-additive for finite n. He also showed that metrizability is weakly n-additive for finite n in the class of compact or *ccc* spaces. He asked if metrizability is weakly n-additive in general. BALOGH, GRUENHAGE and TKACHUK [1998] show that the answer is positive

[6] Mizokami recently announced that stratifiable μ-spaces are LF-netted.

for pseudocompact spaces, but negative in general. Any "ladder space" X on the countable ordinals is a counterexample; indeed, X^n is the union of two metrizable subspaces for any finite n. Independently, OHTA and YAMADA [1998] obtained another example.

Tkachuk's investigation was motivated by an old question of Arhangel'skii. To put it in the above terminology, it asks whether metrizability is weakly ω-additive, i.e., must X be metrizable if X^ω is the union of countably many metrizable subspaces? Tkačenko had shown that the answer is positive for separable or countably compact spaces, and Tkachuk showed it to be the case if X^ω is the union of finitely many metrizable subspaces. However, GRUENHAGE [1997] answered the question in the negative by showing that any ladder system space on ω_1 in which the set of non-isolated points is G_δ (such spaces can easily be constructed in ZFC) is a counterexample.

In another direction, ISMAIL and SZYMANSKI [1995], [1996], [2001] have a series of papers in which they investigate the *metrizability number* $m(X)$ of a space X; $m(X)$ is the least cardinal κ such that X is the union of κ-many metrizable subspaces. They obtain nice structure theorems for locally compact X when $m(X)$ is finite; e.g., in this case X must have a dense open metrizable subspace whose complement has smaller metrizability number. From their structure theorems, they easily conclude that, in the finite case, the metrizability number of a locally compact space cannot be raised by a perfect mapping. It is not known if this remains true for infinite metrizability numbers, even if the domain is compact (in particular, the case $m(X) = \omega$ is unsettled).

We take the opportunity to mention here an old unsolved problem on metrizability number due to VAN DOUWEN, LUTZER, PELANT and REED [1980]: Is $m(X) \leq \mathfrak{c}$ whenever X has a point-countable base?

11. Open problems

In this section we give a selection of several open problems. In some cases, there are recent partial results to be mentioned that didn't conveniently fit into previous categories. That is one purpose; another is to show the rich variety of interesting questions that remain, and thereby, I hope, stimulate further research in generalized metric spaces and metrization. Some of these questions have appeared in my previous article [1992], and in one case earlier in this article. One is completely new. None are due to me originally.

1. *Are stratifiable spaces M_1?*

This question of Ceder from 1961, the oldest one on my list, was discussed in detail in Section 5; in particular, $C_k(\omega^\omega)$ was suggested as a potential counterexample.

2. *Is it consistent that there are no symmetrizable L-spaces?*

This is an old problem of Arhangel'skii dating back to 1966. Recall that a function $d : X \times X \to [0, \infty)$ is a *symmetric* on X if $d(x, y) = d(y, x)$ and $d(x, y) = 0 \iff x = y$. Then a space X is *symmetrizable* if there is a symmetric d on X such that U is open in X iff for each $x \in U$, there is some $\epsilon > 0$ such that the ϵ-ball $B(x, \epsilon)$ about x is contained in U. (Note: As d need not satisfy the triangle inequality, $B(x, \epsilon)$ itself need not be open.) SHAKHMATOV [1992] obtained a consistent example of a symmetrizable L-space by forcing, but no ZFC example is known.

Next we list three problems which would take finding a certain Dowker space (or prove such a Dowker space cannot exist) to answer. In each case, without further assumptions on the space, there are no consistent theorems or counterexamples. I.e, the solution, for all we know, could go either way, in ZFC!

3 (a) *Is there a symmetrizable Dowker space?*

(b) *Suppose X is normal, and the union of countably many open metrizable subspaces. Must X be metrizable?*

(c) *Is every normal space with a σ-disjoint base paracompact?*

Problem 3(a) is due to S. Davis. A positive answer would imply a negative answer to an old question of E. Michael: Must every point of a symmetrizable space be G_δ? 3(b) is one of Mike Reed's favorite problems. In unpublished work, Reed has shown the answer is positive for spaces of weight less than $\mathfrak{b}$, and has constructed a regular non-developable space which is the increasing union of open metrizable subspaces. And 3(c) is one of Mary Ellen Rudin's favorites. A counterexample to 3(b) is easily seen to be a counterexample to (c) too.

4. (THE POINT-COUNTABLE BASE PROBLEM.) *Does a space X have a point-countable base iff X has a countable open point-network?*

This problem is due to Collins, Reed, and Roscoe. The property "countable open point-network", also called "open(G)", means that one can assign to each $x \in X$ a countable open base $\mathcal{B}(x)$ for x such that, whenever a sequence of points x_n converges to x, then $\bigcup_{n\in\omega} \mathcal{B}(x_n)$ contains a base at x. It is easily seen that a space with a point-countable base $\mathcal{B}$ has a countable open point-network (let $\mathcal{B}(x) = \{B \in \mathcal{B} : x \in B\}$). It is known that the answer to Question 4 is "yes" for spaces of density $\aleph_1$, so a positive answer (necessarily consistent) to the reflection problem mentioned in Section 8 would give a consistent positive answer to this one. See COLLINS, REED and ROSCOE [1990] for more insight and partial results related to this problem.

5. *If every $\aleph_1$-sized subspace of a first-countable space X is metrizable, must X be metrizable?*

This reflection problem, a version of a problem due to P. Hamburger, was also mentioned in GRUENHAGE [1992], where further information can be found. Except for Balogh's related results on point-countable base reflection (see Section 8), I know of no progress since then.

6. *Is Arhangel'skii's class MOBI preserved by perfect mappings?*

Recall that MOBI is the smallest class of spaces containing all metrizable spaces, and closed under open compact images. The above problem is the only one still open of those mentioned in ARHANGEL'SKII's classic survey [1966]. However, it is still open at least in part because some other fundamental questions about MOBI remain unsolved, especially whether or not there is some positive integer n such that every space X in MOBI is "n^{th}-generation", i.e., there is a metrizable space M and a sequence $f_1, f_2, ... f_n$ of open compact mappings such that $X = f_n \circ f_{n-1} ... \circ f_1(M)$. Indeed, it is possible, perhaps likely, that such an n, if it exists, can be 2, as this is the case in a very natural way for every known example. See my earlier survey [1992] for more information.

Part of the motivation for MOBI was the search for "nice" classes of spaces which generalized metrizable ones; part of the definition of "nice" included preservation under various standard topological constructions. Now we state a new question of this sort, asked of the author in a recent private communication by E. Michael.

7. *Is there a class of spaces (and if so, describe it) which:*

(i) *contains all metrizable spaces;*

(ii) *is closed under the taking of closed subspaces, closed images, perfect pre-images, and countable products; and*

(iii) *is contained in the class of paracompact spaces?*

If one only asked for preservation under perfect mappings, then the class of paracompact p-spaces (i.e., the class of all perfect pre-images of metrizable spaces) would satisfy all the conditions. The required class of course must contain all paracompact p-spaces, but also Lašnev spaces (=closed images of metric spaces). Both Lašnev and paracompact p-spaces are subclasses of the class of paracompact Σ-spaces, which satisfies all conditions except preservation under closed maps. The somewhat wider class of paracompact $\Sigma^\#$-spaces[7] is closed under closed maps, and it would work if paracompactness were countably productive in the class of $\Sigma^\#$-spaces. But that is an unsolved problem! In fact, it is not known if X, Y paracompact $\Sigma^\#$ implies $X \times Y$ is paracompact (it is $\Sigma^\#$).

Another approach to the question might be to consider the smallest class $\mathcal{M}^\#$ containing all metrizable spaces and closed under the operations mentioned in (ii). Then $\mathcal{M}^\#$ is contained in the class of $\Sigma^\#$ spaces. The question becomes: Is every member of $\mathcal{M}^\#$ paracompact? If the answer is affirmative, one would also like an internal characterization of $\mathcal{M}^\#$. I don't know if there is a paracompact $\Sigma^\#$-space which is not in $\mathcal{M}^\#$.

8. *Is there in ZFC a non-metrizable perfectly normal non-archimedean space?*

Recall that X is *non-archimedean* if it has a base which is a tree of open sets under reverse inclusion. A Souslin tree implies a consistent counterexample (namely, the "branch space" of a Souslin tree). QIAO and TALL [200?] proved that this problem, originally due to Nyikos, is actually equivalent to the following problem of Maurice: "Does every perfect (= closed sets are G_δ) linearly ordered space have a σ-discrete dense set?" See Lutzer's article in this volume for more information.

There is a more general question, which is due to Mike Reed and came out of research from the '60's and '70's on dense metrizable or dense Moore subspaces, which is also unsolved:

9. *Is there in ZFC a regular perfect first-countable space with no σ-discrete dense subset?*

This question seems to be unsolved even without the "first-countable" assumption. Note that L-spaces do not have σ-discrete dense sets. But it's consistent that there are no first-countable L-spaces, and may be consistent that there are no L-spaces in general, though this is still unsettled.

10. *Is there a non-metrizable compact space with a small diagonal?*

This is an old problem of Hušek. As was reported in my previous survey, results of Hušek, Dow, and Juhász and Szentmiklóssy show that the answer is "no" under CH or if Cohen reals are added to a model of CH. GRUENHAGE [200?] answered questions of Zhou and Shakhmatov by showing that the same question for countably compact spaces is independent of ZFC. A deep result of EISWORTH and NYIKOS [200?] about the consistency with CH of countably compact non-compact first-countable spaces containing a copy of the countable ordinals (which does not have a small diagonal) implies the positive result.

[7] $\Sigma^\#$-spaces are defined like Σ-spaces were in Section 3, except that the collection $\mathcal{F}$ is assumed to be only σ-closure-preserving instead of σ-locally finite.

PAVLOV [200?], answered one of my questions by showing that, under $\diamondsuit^*$, there is a perfect pre-image of ω_1 with a small diagonal; together with my positive result above, this showed independence of the countably compact question with $ZFC + CH$. Pavlov also has shown that the negation of CH implies that there is a Lindelöf space with a small diagonal but no G_δ-diagonal. Gruenhage also showed that there are consistent examples of hereditarily Lindelöf, consistent with CH examples of Lindelöf, and ZFC examples of locally compact spaces having a small diagonal but no G_δ-diagonal.

We should also mention that ARHANGEL'SKII and BELLA [1992] showed that CH implies that every Lindelöf p-space (i.e., every perfect pre-image of a separable metrizable space) with a small diagonal is metrizable, and BENNETT and LUTZER [1998b] answered one of their questions by obtaining a ZFC example of a paracompact p-space (i.e., a perfect pre-image of a metrizable space) with a small diagonal which is not metrizable. See Lutzer's article in this volume for more details.

References

ALAS, O., M. TKAČENKO, V. TKACHUK and R. WILSON
[1996] Connectifying some spaces, *Topology Appl.* **71**, 203–215.

ALLECHE, B., A.V. ARHANGEL'SKII and J. CALBRIX
[2000] Weak developments and metrization, Special issue in honor of Howard H. Wicke, *Topology Appl.* **100** (1), 23–38.

ALSTER, K. and G. GRUENHAGE
[1995] Products of Lindelöf spaces and GO-spaces, *Topology Appl.* **64** (1), 23–36.

ARHANGEL'SKII, A.V.
[1966] Mappings and spaces, *Uspehi Mat. Nauk* **21** (4),133–184 (Russian); translated as *Russian Math. Surveys* **21** (4), 115–162.

ARHANGEL'SKII, A.V. and A. BELLA
[1992] Few observations on topological spaces with small diagonal, *Zbornik radova Filozofkog fakultete u Nisu* **6** (2), 211–213.

ARHANGEL'SKII, A.V., W. JUST, E.A. REZNICHENKO and P.J. SZEPTYCKI
[2000] Sharp bases and weakly uniform bases versus point-countable bases, Special issue in honor of Howard H. Wicke, *Topology Appl.* **100** (1), 39–46.

BALOGH, Z.
[2001] Nonshrinking open covers and K. Morita's duality conjectures, *Topology Appl.* **115**, 333–341.
[200?a] Reflecting point-countable families, *Proc. Amer. Math. Soc.*, to appear.
[200?b] Locally nice spaces and Axiom R, preprint.

BALOGH, Z., H. BENNETT, D. BURKE, G. GRUENHAGE, D. LUTZER and J. MASHBURN
[2000] OIF spaces, *Questions Answers Gen. Topology* **18**, 129–141.

BALOGH, Z. T., S.W. DAVIS, W. JUST, S. SHELAH and P. J. SZEPTYCKI
[2000] Strongly almost disjoint sets and weakly uniform bases, *Trans. Amer. Math. Soc.* **352**, 4971–4987.

BALOGH, Z. and G. GRUENHAGE

[2001] Base multiplicity in compact and generalized compact spaces, *Topology Appl.* **115**, 139–151.

[200?] When the collection of ϵ-balls is locally finite, *Topology Appl.*, to appear.

BALOGH, Z., G. GRUENHAGE and V.V. TKACHUK

[1998] Additivity of metrizability and related properties, Proceedings of the International Conference on Set-theoretic Topology and its Applications, Part 2 (Matsuyama, 1994), *Topology Appl.* **84**, 91–103.

BENNETT, H. and D. LUTZER

[1998a] Ordered spaces with special bases, *Fund. Math.* **158** (3), 289–299.

[1998b] A metric space of A.H. Stone and an example concering σ-minimal bases, *Proc. Amer. Math. Soc.* **126**, 2191–2196.

CAMMAROTO, F.

[1994] A note on ω_μ-stratifiable spaces, *Atti Sem. Mat. Fis. Univ. Modena* **42** (1), 65–68.

CAUTY, R.

[1998] Sur les rétractes absolus P_n-valués de dimension finie [On finite-dimensional P_n-valued absolute retracts], *Fund. Math.* **158** (3), 241–248.

CAUTY, R., B. GUO and K. SAKAI

[1995] The hyperspace of finite subsets of a stratifiable space, *Fund. Math.* **147** (1), 1–9.

CEDER, J.

[1961] Some generalizations of metric spaces, *Pacific J. Math.* **11**,105–125.

CHEN, H.

[1999] Weak neighborhoods and Michael-Nagami's question, *Houston J. Math.* **25**, 297–309.

CHO, M. and W. JUST

[1994] Countable-compact-covering maps and compact-covering maps, *Topology Appl.* **58**, 127–143.

CHRISTENSEN, J.P.R.

[1973] Necessary and sufficient conditions for measurability of certain sets of closed sets, *Math. Annal.* **200**, 189–193.

COLLINS, P.J.

[1996] Monotone normality, *Topology Appl.* **74**, 179–198.

COLLINS, P.J., G.M. REED and A.W. ROSCOE

[1990] The Point-countable Base Problem, in VAN MILL and REED [1990], pp. 237–250.

DAVIS, S.W., G.M. REED and M.L. WAGE

[1976] Further results on weakly uniform bases, *Houston J. Math.* **2** (1), 57–63.

DEBS, G. and J. SAINT RAYMOND

[1996] Compact covering and game determinacy, *Topology Appl.* **68**, 153–185.

[1997] Compact-covering-properties of finite-to-one mappings, *Topology Appl.* **81**, 55–84.

[1999] Cofinal Σ_1^1 and Π_1^1 subsets of ω^ω, *Fund. Math.* **159**, 161–193.

DELISTATHIS, G. and S. WATSON

[2000] A regular space with a countable network and different dimensions, *Trans. Amer. Math. Soc.* **352**, 4095–4111.

DOUGHERTY, R.

[1997] Narrow coverings of ω-ary product spaces, *Ann. Pure Appl. Logic* **88**, 47–91.

VAN DOUWEN, E.K., D.J. LUTZER, J. PELANT and G.M. REED

[1980] On unions of metrizable spaces, *Canad. J. Math.* **32**, 76–85.

Dow, A.

[1988] An empty class of nonmetric spaces, *Proc. Amer. Math. Soc.* **104**, 999–1001.

Eda, K., G. Gruenhage, P. Koszmider, K. Tamano and S. Todorčević

[1995] Sequential fans in topology, *Topology Appl.* **67** (3), 189–220.

Eisworth, T. and P.J. Nyikos

[200?] On first-countable closed preimages of ω_1, preprint.

Fearnley, D.L.

[1999] A Moore space with a σ-discrete π-base which cannot be densely embedded in any Moore space with the Baire property, *Proc. Amer. Math. Soc.* **127** (10), 3095–3100.

Fisher, S., P. Gartside, T. Mizokami and N. Shimane

[1997] Near metric properties of hyperspaces, *Topology Proc.* **22** (Spring), 197–211.

Gartside, P.M.

[1997] Cardinal invariants of monotonically normal spaces, *Topology Appl.* **77**, 303–314.

[1998] Nonstratifiability of topological vector spaces, *Topology Appl.* **86** (2), 133–140.

[1999] Monotone normality in products, *Topology Appl.* **91** (3), 181–195.

Gartside, P.M., C. Good, R.W. Knight and A.M. Mohamad

[2001] Quasi-developable manifolds, *Topology Appl.* **111** (3), 207–215.

Gartside, P.M. and P.J. Moody

[1992] A note on proto-metrisable spaces, *Topology Appl.* **52** (1), 1–9.

[1993] Well-ordered (F) spaces, *Topology Proc.* **17**, 111–130.

[1997a] Proto-metrisable spaces are elastic, *Topology Appl.* **79** (1), 49–62.

[1997b] Elastic spaces may snap under perfect maps, *Topology Appl.* **79** (1), 63–74.

Gartside, P. and E.A. Reznichenko

[2000] Near metric properties of function spaces, *Fund. Math.* **164** (2), 97–114.

Good, C., R.W. Knight and A.M. Mohamad

[200?] Metrization and sharp bases, preprint.

Good, C. and I. Stares

[2000] Monotone insertion of continuous functions, *Topology Appl.* **108** (1), 91–104.

Good, C. and I.J. Tree

[1994] On δ-normality, *Topology Appl.* **56** (2), 117–127.

Gruenhage, G.

[1984] Generalized metric spaces, in Kunen and Vaughan [1984], pp. 423–501.

[1992] Generalized metric spaces and metrization, in Hušek and van Mill [1992], pp. 239–274.

[1997] A non-metrizable space whose countable power is σ-metrizable, *Proc. Amer. Math. Soc.* **125** (6), 1881–1883.

[200?] Spaces having a small diagonal, *Topology Appl.*, to appear.

Gruenhage, G., J. Kulesza and A. Le Donne

[1998] Connectifications of metrizable spaces, Special volume in memory of Kiiti Morita, *Topology Appl.* **82**, 171–179.

Gruenhage, G. and P.J. Nyikos

[1993] Normality in X^2 for compact X, *Trans. Amer. Math. Soc.* **340** (2), 563–586.

Guo, B. and K. Sakai

[1993] Hyperspaces of CW-complexes, *Fund. Math.* **143** (1), 23–40.

Harris, M.J.

[1991] Ph.D. dissertation, University of Pittsburgh.

HATTORI, Y.

[1986] On special metrics characterizing topological properties, *Fund. Math.* **126**, 133–145.

HATTORI, Y. and H. OHTA

[1993] A metric characterization of a subspace of the real line, *Topology Proc.* **18**, 75–87.

HEATH, R.W. and W.F. LINDGREN

[1976] Weakly uniform bases, *Houston J. Math.* **2** (1), 85–90.

HUŠEK, M. and J. VAN MILL, editors

[1992] *Recent Progress in General Topology*, North Holland, Amsterdam.

ISMAIL, M. and A. SZYMANSKI

[1995] On the metrizability number and related invariants of spaces, *Topology Appl.* **63** (1), 69–77.

[1996] On the metrizability number and related invariants of spaces. II, *Topology Appl.* **71** (2), 179–191.

[2001] On locally compact Hausdorff spaces with finite metrizability number, *Topology Appl.* **114** (3), 285–293.

ITO, M., H. OHTA and J. ONO

[1999] A graph-theoretic approach to the unique midset property of metric spaces, *J. London Math. Soc.* **60**, 353–365.

JUNNILA, H.J.K. and H.P. KÜNZI

[1993] Ortho-bases and monotonic properties, *Proc. Amer. Math. Soc.* **119** (4), 1335–1345.

JUNNILA, H.J.K. and Y. YAJIMA

[1998] Normality and countable paracompactness of products with σ-spaces having special nets, *Topology Appl.* **85**, 375–394.

JUST, W., M.V. MATVEEV and P.J. SZEPTYCKI

[2000] Some results on property (a), Special issue in honor of Howard H. Wicke, *Topology Appl.* **100** (1), 67–83.

JUST, W. and H. WICKE

[1994] Some conditions under which tri-quotient or compact-covering maps are inductively perfect, *Topology Appl.* **55**, 289–305.

KATĚTOV, M.

[1948] Complete normality of Cartesian products, *Fund. Math.* **35**, 271–274.

KNIGHT, R.W.

[1993] Δ-sets, *Trans. Amer. Math. Soc.* **339** (1), 45–60.

KUBIAK, T.

[1993] Monotone insertion of continuous functions, *Questions Answers Gen. Topology* **11** (1), 51–59.

KULESZA, J.

[200?] Some new properties of Mrówka's space $\mu\nu_0$, *Proc. Amer. Math. Soc.*, to appear.

KUNEN, K. and J. VAUGHAN, editors

[1984] *Handbook of Set-theoretic Topology*, North Holland, Amsterdam.

LABERGE, T.

[1999] A space with normal countable-to-one regular preimages, *Topology Appl.* **91** (3), 227–232.

LANE, E., P. NYIKOS and C. PAN

[200?] Monotone insertion properties of stratifiable spaces, to appear.

LARSON, P. and S. TODORČEVIĆ

[200?] Katětov's Problem, *Trans. Amer. Math. Soc.*, to appear.

LEMIN, A.J. and V.A. LEMIN

[2000] On a universal ultrametric space, *Topology Appl.* **103**, 339–345.

MATVEEV, M.V.

[1997] Some questions on property (a), *Questions Answers Gen. Topology* **15** (2), 103–111.

MICHAEL, E.

[1981] Inductively perfect maps and tri-quotient maps, *Proc. Amer. Math. Soc.* **82**, 115–119.

VAN MILL, J. and G.M. REED, editors

[1990] *Open Problems in Topology*, North Holland, Amsterdam.

MIZOKAMI, T. and K. KOIWA

[1987] On hyperspaces of compact and finite subsets, *Bull. Joetsu Univ. Ed.* **6** (3), 1–14.

MIZOKAMI, T. and N. SHIMANE

[2000a] On the M_3 versus M_1 problem, *Topology Appl.* **105** (1), 1–13.

[2000b] A Lasnev space is LF-netted, *Houston J. Math.* **26**, 537–542.

MIZOKAMI, T., N. SHIMANE and Y. KITAMURA

[200?] A characterization of a certain subclass of M_1-spaces, to appear.

MORITA, K. and J. NAGATA, editors

[1989] *Topics in General Topology*, North Holland, Amsterdam.

MRÓWKA, S.

[1997] Small inductive dimension of completions of metric spaces, *Proc. Amer. Math. Soc.* **125**, 1545–1554.

[2000] Small inductive dimension of completions of metric spaces, II, *Proc. Amer. Math. Soc.* **128**, 1247–1256.

NAGATA, J.

[1983] *Modern Dimension Theory*, Heldermann Verlag, Berlin.

[1999] Remarks on metrizability and generalized metric spaces, *Topology Appl.* **91**, 71–77.

NYIKOS, P.J.

[1999] Metrizability, monotone normality, and other strong properties in trees. II, Iberoamerican Conference on Topology and its Applications (Morelia, 1997), *Topology Appl.* **98**, 269–290.

OHTA, H. and K. YAMADA

[1998] Simple examples showing that various topological properties are not finitely additive in the sense of V. V. Tkachuk, *Math. Japon.* **47** (1), 93–95.

OKUNEV, O.G.

[1993] On Lindelöf Σ-spaces of continuous functions in the pointwise topology, *Topology Appl.* **49**, 149–166.

OKUNEV, O.G. and V.V. TKACHUK

[2001] Lindelöf Σ property in $C_p(X)$ and $p(C_p(X)) = \omega$ do not imply countable network weight in X, *Acta. Math. Hungar.* **90**, 119–132.

OSTROVSKII, A.V.

[1994] New class of maps connected with compact-covering maps, *Vestnik Moskov. Univ.* **4**, 24–27.

PAVLOV, O.

[200?] There is a perfect pre-image of ω_1 with a small diagonal, to appear.

PURISCH, S. and M.E. RUDIN

[1997] Products with linear and countable type factors, *Proc. Amer. Math. Soc.* **125** (6), 1823–1830.

QIA, Y-Q. and F. TALL

[200?] Perfectly normal non-metrizable non-archimedean spaces are generalized Souslin lines, *Proc. Amer. Math. Soc.*, to appear.

ROBERTSON, R.A. and W. ROBERTSON

[1964] *Topological Vector Spaces*, Cambridge University Press.

RUDIN, M.E.

[1996] Monotone normality and compactness, Proceedings of the International Conference on Set-theoretic Topology and its Applications (Matsuyama, 1994), *Topology Appl.* **74** (1-3), 199–205.

[2001] Nikiel's conjecture, *Topology Appl.* **116** (3), 305–331.

RUDIN, M.E., I.S. STARES and J.E. VAUGHAN

[1997] From countable compactness to absolute countable compactness, *Proc. Amer. Math. Soc.* **125** (3), 927–934.

SAINT RAYMOND, J.

[1971-1973] Charactérisation d'espaces polonais, *Seminaire d'Initiation à l'Analyse, 11-12ème anées* **5**.

SAKAI, M.

[1995] Embeddings of k-metrizable spaces into function spaces, *Topology Appl.* **65**, 155–165.

ŠČEPIN, E.V.

[1980] On κ-metrizable spaces, *Math. USSR Izvestija* **14**, 407–440, Russian original: *Izv. Akad. Nauk. SSSR, ser. Math.* **43** (2), (1979), 3–62.

SHAKHMATOV, D.B.

[1992] Final compactness and separability in regular symmetrizable spaces, *J. Soviet Math.* **60** (6), 1796–1815.

SHAKHMATOV, D., F. TALL and S. WATSON

[1996] A normal Moore space which is not sub-metrizable, preprint posted on Topology Atlas.

SHKARIN, S.A.

[1999] On stratifiable locally convex spaces, *Russ. J. Math. Phys.* **6** (4), 435–460.

SIPACHEVA, O.

[1993] Stratifiability of free abelian topological groups, *Topology Proc.* **18**, 271–311.

STARES, I. and J. VAUGHAN

[1996] The Dugundji extension property can fail in ω_μ-metrizable spaces, *Fund. Math.* **150**, 11–16.

TALL, F.

[1994] σ-centred forcing and reflection of (sub)metrizability, *Proc. Amer. Math. Soc.* **121** (1), 299–306.

TAMANO, H. and J. VAUGHAN

[1971] Paracompactness and elastic spaces, *Proc. Amer. Math. Soc.* **28**, 299–303.

TAMANO, K.

[1997] Definitions of Σ-spaces, *Topology Proc.* **22** (Summer), 529–532.

[2001] A cosmic space which is not a μ-space, *Topology Appl.* **115**, 259–263.

TAMANO, K. and S. TODORČEVIĆ

[2001] $C_p(\omega^\omega)$ is not a μ-space, Note of September, 2001.

TANAKA, Y.

[1994] Theory of k-networks, *Questions Answers Gen. Topology* **12**, 139–164.

[2001] Theory of k-networks, II, *Questions Answers Gen. Topology* **19**, 27–46.

TKACHUK, V.V.

[1994] Finite and countable additivity of topological properties in nice spaces, *Trans. Amer. Math. Soc.* **341** (2), 585–601.

[2000] Behaviour of the Lindelöf Σ-property in iterated function spaces, *Topology Appl.* **107**, 297–305.

[2001] Lindelöf Σ-property of $C_p(X)$ together with countable spread of X implies X is cosmic, *New Zealand J. Math.* **30**, 93–101.

TREE, I.J. and W.S. WATSON

[1993] Two Moore manifolds, *Topology Appl.* **51** (1), 27–39.

VAUGHAN, J.

[1972] Linearly stratifiable spaces, *Pacific J. Math.* **43**, 253–266.

[1999] Universal ultrametric spaces of smallest weight, *Topology Proc.*, **24** (Summer), 611–619.

[200?] Examples of linearly stratifiable spaces, to appear.

WATSON, S. and R. WILSON

[1993] Embeddings in connected spaces, *Houston J. Math.* **19**, 469–481.

WILLIAMS, S.W. and H. ZHOU

[1991] Strong versions of normality, in General topology and applications (Staten Island, NY, 1989), *Lecture Notes in Pure and Appl. Math.* **134** Dekker, New York, pp. 379–389.

[1998] Order-like structure of monotonically normal spacesm, *Comment. Math. Univ. Carolin.* **39** (1), 207–217.

YASHCHENKO, I.V.

[1994] On the monotone normality of function spaces, (Russian), *Vestnik Moskov. Univ. Ser. I Mat. Mekh.* (3), 95–96; translation in *Moscow Univ. Math. Bull.* **49** (3), 62–63 (1995).

ZHU, P.

[1993] The products of meta-Lindelöf spaces, *Topology Proc.* **18**, 221–229.

CHAPTER 9

Recent Progress in the Topological Theory of Semigroups and the Algebra of βS

Neil Hindman[1]

Department of Mathematics, Howard University, Washington, DC 20059, USA
E-mail: nhindman@aol.com, nhindman@howard.edu

Dona Strauss

Department of Pure Mathematics, Hull University, Hull HU67RX, UK
E-mail: d.strauss@maths.hull.ac.uk

Contents

[1] This author acknowledges support received from the National Science Foundation (USA) via grant DMS-0070593.

RECENT PROGRESS IN GENERAL TOPOLOGY II
Edited by Miroslav Hušek and Jan van Mill

1. Introduction

Throughout this article, we shall assume that all hypothesized topological spaces are Hausdorff.

Let S be a semigroup which is also a topological space. S is said to be a *topological semigroup* if the operation $\cdot : S \times S \to S$ is continuous. Given $x \in S$, define $\lambda_x : S \to S$ and $\rho_x : S \to S$ by $\lambda_x(y) = x \cdot y$ and $\rho_x(y) = y \cdot x$. If one only assumes that each λ_x is continuous and each ρ_x is continuous, then S is a *semitopological* semigroup. If one only assumes that each ρ_x is continuous, then S is a *right topological* semigroup. (Some authors call this a *left topological* semigroup because multiplication is continuous in the left variable.)

From our point of view, probably the most fundamental theorem about right topological semigroups is the following.

1.1. THEOREM. *Let S be a compact right topological semigroup. Then S has a smallest two sided ideal $K(S)$. Further $K(S)$ is the union of all of the minimal left ideals of S and is also the union of all of the minimal right ideals of S. Given any minimal left ideal L of S and any minimal right ideal R of S, $L \cap R$ is a maximal subgroup of S. Also, any two minimal left ideals of S are isomorphic, any two minimal right ideals of S are isomorphic, and any two maximal subgroups of $K(S)$ are isomorphic.*

Theorem 1.1 was established for finite semigroups by SUSCHKEWITSCH [1928], for topological semigroups by WALLACE [1955], and for right topological semigroups by RUPPERT [1973]. A crucial contribution to the result for right topological semigroups was the proof by ELLIS [1969] that any compact right topological semigroup has an idempotent.

Classic (and neo-classic) references are the books by CLIFFORD and PRESTON [1961] on the algebraic theory of semigroups, by HOFMANN and MOSTERT [1996] on compact topological semigroups, by RUPPERT [1984] on semitopological semigroups, and by BERGLUND, JUNGHENN and MILNES [1989] on right topological semigroups.

Suppose that S is both a semigroup and a topological space. A *semigroup compactification of S* is a pair (ϕ, T) such that T is a compact right topological semigroup, $\phi : S \to T$ is a continuous homomorphism, $\phi[S]$ is dense in T and $\lambda_{\phi(s)} : T \to T$ is continuous for every $s \in S$. (In this case, we may simply call T a semigroup compactification of S. Note that a semigroup compactification need not be a topological compactification, because ϕ is not required to be an embedding.)

Let $\mathcal{P}$ be a property of semigroups which are topological spaces. A semigroup compactification (ϕ, T) of S is said to be the *universal $\mathcal{P}$-semigroup compactification of S* if T has property $\mathcal{P}$ and if, for every semigroup compactification (ϕ', T') of S for which T' has property $\mathcal{P}$, there is a continuous homomorphism $\theta : T \to T'$ such that $\phi' = \theta \circ \phi$.

We shall discuss the *weakly almost periodic compactification* wS of S and the *$\mathcal{LMC}$ compactification* $S^{\mathcal{LMC}}$ of S. We define (η, wS) to be the universal $\mathcal{P}$-semigroup compactification of S, where $\mathcal{P}$ denotes the property of being a semitopological semigroup. A bounded continuous function $f : S \to \mathbb{C}$ is *weakly almost periodic* if and only if there is a continuous $\gamma : wS \to \mathbb{C}$ such that $\gamma \circ \eta = f$.

We define $S^{\mathcal{LMC}}$ to be the universal $\mathcal{P}$-semigroup compactification of S, where $\mathcal{P}$ denotes the property of being a right topological semigroup.

It has been known for some time that if S is a discrete semigroup, the operation on S can be uniquely extended to the Stone-Čech compactification βS of S so that βS becomes a semigroup compactification of S, and in fact $\beta S = S^{\mathcal{LMC}}$. See the notes to Chapter 4 of HINDMAN and STRAUSS [1998b] for a discussion of the origins of this fact.

We shall also mention the *uniform compactification* uG of a topological group G. We define this in terms of the right uniform structure on G, which has the sets of the form $\{(x, y) \in G \times G : xy^{-1} \in U\}$, where U denotes a neighborhood of the identity in G, as a base for the vicinities. This compactification has the property that a continuous bounded real-valued function defined on G has a continuous extension to uG if and only if it is uniformly continuous. It is a semigroup compactification of G in which G is embedded. In the case in which G is locally compact, $uG = G^{\mathcal{LMC}}$.

The semigroup βS plays a significant role in topological dynamics. Whenever a discrete semigroup S acts on a compact topological space X, the enveloping semigroup (defined as the closure in ${}^{X}X$ of the functions corresponding to elements of S), is a semigroup compactification of S and therefore a quotient of βS. For this reason, some of the concepts related to the algebra of βS originated in topological dynamics. Several of these are described in Section 5 below.

Because the points of βS can be viewed as ultrafilters on S one obtains built in applications to the branch of combinatorics known as *Ramsey Theory*. That is, as soon as one knows that there is an ultrafilter on S which is contained in some set $\mathcal{G}$ of "good" subsets of S, one automatically has a corresponding Ramsey Theoretic result, namely that whenever S is divided into finitely many parts, one of these parts is a member of $\mathcal{G}$.

In this paper we propose to survey progress in the areas mentioned above in the last decade, i.e. since the publication of HUŠEK and VAN MILL [1992]. Section 2 of COMFORT, HOFMANN and REMUS [1992] dealt primarily with topological semigroups, with brief mention of results in semitopological semigroups, right topological semigroups, and the algebra of βS. In Section 2 of this paper we shall only mention a few recent results of which we are aware from the theory of topological semigroups. This light treatment is dictated by two facts. Most importantly, neither of the authors is an expert in the theory of topological semigroups. Secondly, a thorough treatment of progress during the last decade of the theory of topological semigroups would consume much more space than is allocated for this paper.

In 1998 our book HINDMAN and STRAUSS [1998b] was published. Sections 3 through 5 of this paper will survey results in subjects covered in that book, and will concentrate on progress since the manuscript went to the publisher. Section 3 will deal with results in the theory of right topological semigroups. Section 4 will present recent progress in the algebra of βS. And Section 5 will survey recent progress in the applications of the algebra of βS to Ramsey Theory.

In Section 6 we deal with a subject, the Stone-Čech compactification of *partial semigroups*, that has only recently emerged as an area of productive research, both in terms of abstract algebra and in terms of combinatorial applications.

2. Topological and semitopological semigroups

For reasons mentioned in the introduction, we are unable to give a substantive review of recent progress in the theory of topological semigroups. The excellent article HOFMANN [2000] discusses the theory beginning in antiquity (meaning in this case the 19th century) and continues through results as recent as 1998, with an emphasis on the Lie theory of semigroups. See also the volume HOFMANN and MISLOVE [1996], which has several relevant articles, and the survey HOFMANN and LAWSON [1996].

In a now classic result, ELLIS [1957] showed that any semitopological semigroup which is locally compact and is algebraically a group is in fact a topological group. BOUZIAD [1993] describes a class $\mathcal{C}$ of Baire spaces and shows that if G is a left topological group which acts on a space X in a separately continuous fashion and if G and X both belong to the class $\mathcal{C}$, then the action is jointly continuous.

It is reasonably easy to see that if S is an infinite discrete cancellative semigroup, then βS contains at least $2^{\mathfrak{c}}$ idempotents. (See HINDMAN and STRAUSS [1998b, Section 6.3].) In the case of wS, even for S discrete, the situation is not so simple. Using techniques of harmonic analysis, BROWN and MORAN [1972] established in 1972 that $w\mathbb{Z}$ has $2^{\mathfrak{c}}$ idempotents. The proof of this result was simplified by an elementary (but still complicated) exhibition of specific weakly almost periodic functions on $\mathbb{Z}$ by RUPPERT [1991]. BORDBAR [1998], gave a much simpler construction of enough weakly almost periodic functions on $\mathbb{Z}$ to guarantee the existence of $2^{\mathfrak{c}}$ idempotents in $w\mathbb{Z}$.

Bordbar's construction used the base -2 expansion of an arbitrary integer. It is a simple, but not so well known, fact that for any $p \in \mathbb{N}$ with $p \geq 2$, any $x \in \mathbb{Z}$ has a unique expansion to the base $-p$ using only the digits $\{0, 1, 2, \ldots, p-1\}$. This expansion has the virtue that, so long as the supports of x and y are disjoint, there is no borrowing and no carrying when x and y are added. We shall have occasion to refer to another use of this representation in Section 4.

BERGLUND [1980] asked whether the set of idempotents in any compact monothetic semitopological semigroup must be closed. (A semigroup S with topology is *monothetic* provided there is some $x \in S$ for which $\{x^n : n \in \mathbb{N}\}$ is dense.) BORDBAR and PYM [2000] used the base -2 expansion of integers to show that the set of idempotents in $w\mathbb{N}$ is not closed, thereby answering Berglund's question. They also showed that the set of idempotents in $w\mathbb{Z}$ is not closed. Independently, BOUZIAD, LEMAŃCZYK and MENTZEN [2001] also answered Berglund's question by constructing a class of compact semitopological semigroups, each containing a dense topological group which is monothetic (as a semigroup), in which the set of idempotents is not closed. Notice that because of the universal extension property of $w\mathbb{N}$ and $w\mathbb{Z}$, this latter result implies that the set of idempotents in $w\mathbb{N}$ is not closed and that the set of idempotents in $w\mathbb{Z}$ is not closed.

BORDBAR and PYM [1998] investigated the structure of wG, where G is the direct sum of countably many finite groups. In any semigroup there is a natural ordering of the idempotents according to which one has $e \leq f$ if and only if $e = ef = fe$. Bordbar and Pym established that not only does wG have $2^{\mathfrak{c}}$ idempotents, it in fact has an antichain consisting of $2^{\mathfrak{c}}$ idempotents. They showed further that under the continuum hypothesis, there is also a chain of $2^{\mathfrak{c}}$ idempotents.

Notice that in the definition of the weak almost periodic compactification, the continuous homomorphism from S into wS is not required to be an embedding. Of course, if S is not a semitopological semigroup, then it could not possibly be an embedding. In the following remarkable result, M. Megrelishvili established that it can be very far from being an embedding, even when S is not only a topological semigroup, but in fact a topological group.

2.1. THEOREM (MEGRELISHVILI). *Let G be the set of all orientation preserving self homeomorphisms of the interval $[0, 1]$ with the compact-open topology. Then G is a topological group and all weakly almost periodic functions on G are constant. Consequently $|wG| = 1$.*

□ MEGRELISHVILI [2001, Theorem 3.1] □

If G is a locally compact group, the homomorphism mapping G into wG is an embedding. However, wG need not be much larger than G. RUPPERT [1984, Theorem 6.3] has shown that, if G is a simple non-compact Lie group, then wG is the one-point compactification of G. It was recently shown by FERRI [2001] that wG is large if G is an IN group (i.e. a group in which the identity has a compact neighborhood invariant under conjugation). More precisely, let κ be the cardinal denoting the smallest number of compact subsets of G required to cover G. Assuming that G is non-compact, wG has at least 2^{2^κ} points. If G is a non-compact SIN group (i.e. a group in which the identity has a basis of compact neighborhoods invariant under conjugation), S. Ferri showed that $uG \setminus G$ has a dense open subset W of cardinality 2^{2^κ} with the following property: for every $w \in W$, $\{w\} = \phi^{-1}[\{\phi(w\}]$, where $\phi : uG \to wG$ denotes the natural homomorphism. This extends a result due to RUPPERT [1973], who had previously proved this fact for a discrete group G.

3. Right (or left) topological semigroups

As we mentioned in the introduction, if S is a discrete semigroup, then βS is in a natural way a right topological semigroup.

If S is a right topological semigroup, its topological center $\Lambda(S)$ is the set of points $s \in S$ for which $\lambda_s : S \to S$ is continuous. In the case of discrete commutative S, it is easy to see that the topological center and the algebraic center of βS coincide by a simple consideration of the functions λ_x and ρ_x. (See HINDMAN and STRAUSS [1998b, Theorem 4.24].) If S is weakly left cancellative (meaning that for all $u, v \in S$, $\{x \in S : ux = v\}$ is finite), then the algebraic center of βS is equal to the algebraic center of S, and the algebraic center of $S^* = \beta S \setminus S$ is empty (See HINDMAN and STRAUSS [1998b, Theorem 6.54].)

If $q \in S^*$, the question of the continuity of λ_q restricted to S^* is not straightforward. The following is an old result of E. van Douwen. (The date on the paper is 1991, but the result was established in 1979.)

3.1. THEOREM (VAN DOUWEN). *Let S be a countable cancellative semigroup.*

(a) *There is a dense subset D of S^* such that for all $p \in D$ and all $q \in S^*$, the restriction of λ_q to S^* is discontinuous at p.*

(b) *There is a P-point in $\mathbb{N}^*$ if and only if there is a dense subset E of S^* such that for all $p \in E$ and all $q \in S^*$, the restriction of the operation $\cdot$ to $S^* \times S^*$ is continuous at (q, p).*

□ These conclusions follow from Theorems 9.7 and 9.8 of VAN DOUWEN [1991] respectively. □

The following theorem about joint continuity was proved by PROTASOV.

3.2. THEOREM (PROTASOV [1996]). *If G is a countable discrete abelian group, with only a finite number of elements of order 2, then there is no point in $G^* \times G^*$ at which the operation $\cdot$ from $\beta G \times \beta G$ to βG is continuous.*

□ PROTASOV[1996, Theorem 4.1] □

In the same paper, PROTASOV [1996, Example 4.4] showed that, if G denotes a discrete abelian group for which $|G|$ is Ulam measurable, then $G^* \times G^*$ does contain a point at which the operation $\cdot$ from $\beta G \times \beta G$ to βG is continuous. ZELENYUK [1996b] showed that Martin's Axiom implies that the same statement holds if G is a countable Boolean group. We do not know whether examples of this kind of joint continuity can be constructed in ZFC.

We shall continue with the discussion of continuity in G^* momentarily. However, in this discussion we shall use the notion of strongly summable ultrafilters, which we pause now to introduce. An ultrafilter on a semigroup $(S, +)$ is said to be *strongly summable* if it has a base of sets of the form $FS(\langle x_n \rangle_{n=1}^{\infty})$, where $FS(\langle x_n \rangle_{n=1}^{\infty}) = \{\sum_{n \in F} x_n : F$ is a finite nonempty subset of $\mathbb{N}\}$. BLASS and HINDMAN [1987] showed that Martin's Axiom implies the existence of strongly summable ultrafilters on $\mathbb{N}$, but that their existence cannot be established in ZFC. This result was extended from $\mathbb{N}$ to arbitrary countable abelian groups by HINDMAN, PROTASOV and STRAUSS [1998a]. If p is a strongly summable ultrafilter of a certain kind on a countable abelian group G, it has a remarkable algebraic property. The equation $x + y = p$ can only hold with $x, y \in G^*$ if $x = a + p$ and $y = -a + p$ for some $a \in G$. The existence of ultrafilters p with this property follows from Martin's Axiom. This extends to many non-commutative groups. If G is any countable group which can be embedded algebraically in a compact topological group, MA guarantees the existence of ultrafilters p on G with the property that whenever $xy = p$, with $x, y \in G^*$, one must have that $x = pa^{-1}$ and $y = ap$ for some $a \in G$.

Strongly summable ultrafilters on a countable Boolean group are particularly interesting, because they can be used to define topologies for which the group is an extremally disconnected non-discrete topological group. This construction is due to MALYKHIN [1975]. It is not known whether extremally disconnected non-discrete topological groups can be defined in ZFC.

Suppose that G is a countable discrete group. For each $p \in \beta G$, let λ_p^* denote the restriction of λ_p to G^*. It is easy to see that, if q is a P-point in G^*, then λ_p^* is continuous at q for every $p \in G^*$. Conversely, PROTASOV [200?] has announced that if G can be algebraically embedded in a compact topological group and if $q \in G^*$ has the property that λ_p^* is continuous at q for every $p \in G^*$, then q is a P-point in G^*; also if $p \in G^*$ is idempotent, the continuity of λ_p^* at p implies the existence of a P-point in ω^*. So the existence of an idempotent p with the property that λ_p^* is continuous at p cannot be established in ZFC. However, as we have just observed, if G is a countable abelian group, then Martin's Axiom implies that there is an idempotent $p \in G^*$ which is strongly summable. If G is Boolean and countable and $p \in G^*$ is strongly summable, it is easy to show that λ_p^*

is continuous at p. We do not know whether it is consistent with ZFC that there exists an infinite discrete group G with the property that λ_p^* is discontinuous at q for all $p, q \in G^*$.

Recall that, if G is a locally compact group, then $G^{\mathcal{LMC}} = uG$, and if S is discrete, then $S^{\mathcal{LMC}} = \beta S$. Since in general G is embedded in uG we may pretend that $G \subseteq uG$ (just as we pretend that $S \subseteq \beta S$) and one may then let $G^* = uG \backslash G$. I. Protasov and J. Pym proved that the topological center of G^* is empty for any locally compact topological group G. They also obtained the following generalization of Theorem 3.1(a).

3.3. THEOREM (PROTASOV and PYM). *Let G be a locally compact, noncompact, σ-compact topological group. There is a dense subset D of G^* such that for all $p \in D$ and all $q \in G^*$, the restriction of λ_q to G^* is discontinuous at p.*

□ PROTASOV and PYM [2001, Theorem 1]. □

Recall from Theorem 1.1 that any compact right topological semigroup S has a smallest two sided ideal which is the union of all minimal left ideals, and each minimal left ideal is the union of pairwise isomorphic groups. Further, given a minimal left ideal L of S and a point $x \in L$, $L = Sx = \rho_x[S]$ so L is compact, and thus closed.

3.4. THEOREM (LAU, MILNES and PYM). *Let G be a locally compact noncompact topological group and let L be a minimal left ideal of uG. Then L is not a group.*

□ LAU, MILNES and PYM [1999]. □

In the process of proving Theorem 3.4, LAU, MILNES and PYM establish for "nearly all groups" the stronger result that no maximal subgroup of the smallest ideal can be closed.

The following result is a *local structure theorem* for uG, when G is a locally compact topological group.

3.5. THEOREM (LAU, MEDGHALCHI and PYM). *Let G be a locally compact topological group and let U be an open symmetric neighborhood of the identity with $c\ell_G(U)$ compact. Let $X \subseteq G$ be maximal with respect to the property that $\{Ux : x \in X\}$ is a disjoint family. Then $\overline{X} = c\ell_{uG}(X)$ is homeomorphic with βX. Also, for each open neighborhood V of the identity with $c\ell_G(V) \subseteq U$, the subspace $V\overline{X}$ is open in uG and homeomorphic with $V \times \beta X$.*

Moreover, given any $\mu \in uG$ one may choose an open symmetric neighborhood of the identity with $c\ell_G(U)$ compact and $X \subseteq G$ maximal with respect to the property that $\{Ux : x \in X\}$ is a disjoint family such that $\mu \in \overline{X}$.

□ LAU, MEDGHALCHI and PYM [1993, Theorem 2.10] and PYM [1999]. □

PYM [1999] used Theorem 3.5 to provide a short proof of a theorem of W. Veech, namely that if G is a locally compact group, $s \in G$, and s is not the identity of G, then for all $\mu \in uG$, $s\mu \neq \mu$ (VEECH [1977, Theorem 2.2.1]).

M. Filali and J. Pym have recently extended some results known to hold for βS (for a discrete semigroup S) to $uG = G^{\mathcal{LMC}}$ for a locally compact group G.

3.6. THEOREM (FILALI). *Let G be a locally compact noncompact abelian topological group. Then the set of points in G^* which are right cancellable in uG has dense interior in G^*. If, in addition, G is countable, then for each $x \in G^*$, $\{y \in G^* : (G^* + y) \cap (G^* + x) \neq \emptyset\}$ is nowhere dense in G^*.*

□ FILALI [1997, Corollary 1] □

In FILALI and PYM [200?] this result was extended and the commutativity assumption was eliminated.

3.7. THEOREM (FILALI and PYM). *Let G be a locally compact noncompact topological group. Then the set of points in G^* which are right cancellable in uG has dense interior in G^*. If κ is the cardinal denoting the smallest number of compact subsets of G required to cover G, then $G^{\mathcal{LMC}}$ has 2^{2^κ} minimal left ideals.*

□ FILALI and PYM [200?, Theorem 1 and Corollary 3] □

S. Ferri and one of the authors have obtained results of this kind for a class of topological groups larger than the class of locally compact groups, in which case one need not have $uG = G^{\mathcal{LMC}}$.

3.8. THEOREM (FERRI and STRAUSS). *Let G be a topological group. For each neighborhood U of the identity in G, let κ_U be the cardinal denoting the smallest number of sets of the form Uy, where $y \in G$, required to cover G, and let $\kappa = \sup\{\kappa_U :$ U is a neighborhood of the identity in $G\}$. If κ is infinite and there is a neighborhood U of the identity in G for which G cannot be covered by fewer than κ sets of the form xUy with $x, y \in G$, then there are at least 2^{2^κ} points in G^* which are right cancellable in uG and at least 2^{2^κ} minimal left ideals in uG.*

□ FERRI and STRAUSS [2001, Theorem 1.3] □

Observe that the hypotheses of Theorem 3.8 are satisfied if G is a topological group which is not totally bounded and is either locally compact or separable. It is an open problem whether there exists a topological group G, which is not totally bounded, for which uG has precisely one minimal left ideal.

In collaboration with I. Protasov, we described a method for obtaining topologies on a semigroup S that are completely determined by the algebra of S and make S into a *left* topological semigroup by using idempotents in the *right* topological compactification βS. (Of course, if one takes βS to be left topological, the resulting topologies are right topological.)

3.9. THEOREM (HINDMAN, PROTASOV and STRAUSS). *Let S be a cancellative semigroup. For any idempotent $p \in \beta S$, let $\mathcal{T}_p = \{V \subseteq S :$ for all $x \in V$, $V \in xp\}$ and let $\mathcal{V}_p = \{\rho_p^{-1}[U] \cap S : U$ is open in $\beta S\}$. Then for each idempotent $p \in \beta S$, $\mathcal{V}_p$ and $\mathcal{T}_p$ are Hausdorff topologies on S making S into a left topological semigroup. If $|S| = \kappa$, then there are 2^{2^κ} noncomparable topologies of the form $\mathcal{V}_p$. One always has that $\mathcal{V}_p \subseteq \mathcal{T}_p$ and the inclusion is proper unless p has the property that $\{q \in \beta S : q \cdot p = p\} = \{p\}$. If S is a group, the property that $\{q \in \beta S : q \cdot p = p\} = \{p\}$ guarantees that $\mathcal{V}_p = \mathcal{T}_p$.*

□ HINDMAN, PROTASOV and STRAUSS [1998b, Theorems 3.4, 3.6, 4.1, 4.2, and 5.1 and Corollary 3.13]. □

An idempotent $p \in \beta S$ such that $\{q \in \beta S : q \cdot p = p\} = \{p\}$ is said to be *strongly right maximal*. They are certainly rare birds, but it is a result of I. Protasov that their existence can be established in ZFC. (See HINDMAN and STRAUSS [1998b, Theorem 9.10].) If S is an infinite group and p a strongly right maximal idempotent in βS, then $\mathcal{V}_p = \mathcal{T}_p$ and this topology on S is extremally disconnected and maximal subject to having no isolated points. (See HINDMAN and STRAUSS [1998b, Corollary 9.17].) This fact answers an old question posed by E. van Douwen: is it possible in ZFC to define a regular homogeneous topology on $\mathbb{Z}$ which is maximal subject to having no isolated points?

PROTASOV has obtained results about ω-resolvability by using the algebra of the Stone-Čech compactification. He showed that any non-discrete left topological group G, which is not of first category, is ω-resolvable; i.e. it can be partitioned into infinitely many disjoint dense subsets, PROTASOV [2001a].

In HINDMAN and STRAUSS [1995d] we investigated topological properties of certain algebraically defined subsets of βS, where S denoted a countable commutative discrete semigroup. In any compact right topological semigroup, all minimal left ideals are homeomorphic as well as isomorphic. However, we showed that, if the minimal left ideals of βS are infinite, then the minimal right ideals of βS belong to $2^{\mathfrak{c}}$ different homeomorphism classes. The same statement is true for the maximal groups contained in any minimal left ideal of βS. If, in addition, S is cancellative, then the sets of the form $S + e$, where e denotes an idempotent in S^*, also belong to $2^{\mathfrak{c}}$ homeomorphism classes. We also showed that, if e and e' are idempotents in $\beta\mathbb{N}$, with e' being non-minimal, then there is no continuous surjective homomorphism from $\beta\mathbb{N} + e$ onto $\beta\mathbb{N} + e'$, apart from the identity.

4. Algebra of βS

Let us begin with a little history about a difficult and annoying open problem which has attracted some significant attention. In 1979, E. van Douwen asked (in VAN DOUWEN [1991], published much later) whether there are topological and algebraic copies of the right topological semigroup $(\beta\mathbb{N}, +)$ contained in $\mathbb{N}^* = \beta\mathbb{N}\backslash\mathbb{N}$. This question was answered in STRAUSS [1992a], where it was in fact established that if φ is a continuous homomorphism from $\beta\mathbb{N}$ to $\mathbb{N}^*$, then $\varphi[\beta\mathbb{N}]$ is finite. The problem to which we refer is whether one can have such a continuous homomorphism with $|\varphi[\beta\mathbb{N}]| > 1$. We conjecture that one cannot.

Another old and difficult problem in the algebra of $\beta\mathbb{N}$ was solved by ZELENYUK [1996a] who showed that there are no nontrivial finite groups contained in $\mathbb{N}^*$. (See HINDMAN and STRAUSS [1998b, Section 7.1] for a presentation of this proof.)

PROTASOV has generalized Zelenyuk's Theorem by characterizing the subgroups of βG, where G denotes a countable discrete group.

4.1. THEOREM (PROTASOV). *If G is a countable discrete group, every finite subgroup of G^* has the form Hp, where H is a finite subgroup of G and p an idempotent in G^* which commutes with all the elements of H.*

□ PROTASOV [1998]. □

Using Zelenyuk's Theorem, it is not hard to show that there is a nontrivial continuous homomorphism from $\beta\mathbb{N}$ to $\mathbb{N}^*$ if and only if there exist distinct p and q in $\mathbb{N}^*$ such that $p+p = q = q+q = q+p = p+q$. (See HINDMAN and STRAUSS [1998b, Corollary 10.20].)

The question of which finite semigroups can exist in $\mathbb{N}^*$ has implications for a large class of semigroups of the form βS. It is not hard to prove that any finite semigroup in $\mathbb{N}^*$ is contained in $\mathbb{H} = \bigcap_{n\in\mathbb{N}} c\ell_{\beta\mathbb{N}}(2^n\mathbb{N})$. Now if S is any infinite discrete semigroup which is right cancellative and weakly left cancellative, S^* contains copies of $\mathbb{H}$. (See HINDMAN and STRAUSS [1998b, Theorem 6.32].) Thus a finite semigroup which occurs in $\mathbb{N}^*$ also occurs in S^*, if S is any semigroup of this kind.

In collaboration with I. Protasov and J. Pym, one of us established a technical lemma that has several corollaries relating to continuous homomorphisms. We combine a few of these in the following.

4.2. THEOREM (PROTASOV, PYM and STRAUSS). *Let G be a countable discrete group.*

(a) *If S is a cancellative discrete semigroup, then any continuous injective homomorphism from βS to βG is the extension of an injective homomorphism from S to G.*

(b) *If S is a countable discrete semigroup and $\varphi : \beta S \to G^*$ is a continuous homomorphism, then every element of $\varphi[S]$ has finite order.*

(c) *If $\varphi : \beta\mathbb{N} \to G^*$ is a continuous homomorphism, then $\varphi[\beta\mathbb{N}]$ is finite and $\varphi[\mathbb{N}^*]$ is a finite group.*

(d) *If C is a compact subsemigroup of G^*, then every element of the topological center of C has finite order.*

□ PROTASOV, PYM and STRAUSS [2000, Theorems 6.5, 6.6 and Corollaries 6.7, 6.8]. □

The conjecture above can be stated equivalently by saying that $\mathbb{N}^*$ contains no elements of finite order, other than idempotents. This conjecture has implications about the nature of possible continuous homomorphisms from βS into $\mathbb{N}^*$, where S is any countable semigroup at all. It follows from Theorem 4.2(b) that, if this conjecture is true, then any continuous homomorphism from βS into $\mathbb{N}^*$ must map all the elements of S to idempotents.

DAVENPORT, HINDMAN, LEADER and STRAUSS [2000] showed that the existence of the two element subsemigroup of $\mathbb{N}^*$ mentioned above implies the existence of a three element semigroup $\{p, q, r\}$ where $p + p = q = q + q = q + p = p + q$, $r + r = r$, $p = p + r = r + p$, and $q = q + r = r + q$. We also showed that if there is a nontrivial continuous homomorphism from $\beta\mathbb{N}$ into $\mathbb{N}^*$, then there is a subset A of $\mathbb{N}$ with the property that, whenever A is finitely colored, there must exist a sequence $\langle x_n \rangle_{n=1}^{\infty}$ in $\mathbb{N}\backslash A$ such that $\{\sum_{t\in F} x_t : F \in \mathcal{P}_f(\mathbb{N}) \text{and } |F| \geq 2\}$ is a monochrome subset of A. (When we refer to a "k-coloring" of a set X we mean a function $\phi : X \to \{1, 2, \ldots, k\}$. The assertion that a set B is "monochrome" is the assertion that ϕ is constant on B.)

Finite subsemigroups of $\mathbb{N}^*$ of any size do exist, for trivial reasons. Any minimal right or left ideal of $\beta\mathbb{N}$ contains $2^{\mathfrak{c}}$ idempotents and if e and f are idempotents in the same minimal left (respectively right) ideal then $e + f = e$ (respectively $e + f = f$). It was shown some time ago in BERGLUND and HINDMAN [1992] that there are idempotents in the smallest ideal of $\beta\mathbb{N}$ whose sum is not idempotent. (Idempotents in the smallest ideal are *minimal* idempotents.) This raised the question of whether there are any minimal idempotents whose sum is again idempotent but not equal to either of them. This question has recently been answered affirmatively by ZELENYUK [2001] in a grand fashion.

4.3. DEFINITION. A semigroup S is an *absolute coretract* if and only if for any continuous homomorphism f from a compact Hausdorff right topological semigroup T onto a compact Hausdorff right topological semigroup containing S algebraically there exists a homomorphism $G : S \to T$ such that $f \circ g$ is the identity on S.

There is a copy of any absolute coretract in $\beta\mathbb{N}$. ZELENYUK [2001] produced a class of countable semigroups of idempotents, showed that each of them is an absolute coretract, and showed that any finite semigroup of idempotents which is an absolute coretract is a member of this class. The self contained proof of the following special case of Zelenyuk's result can be found in HINDMAN [2001].

4.4. THEOREM (ZELENYUK [2001]). *There exist $p \in \mathbb{H}$ and $\{\alpha_{11}, \alpha_{12}, \alpha_{21}, \alpha_{22}\} \subseteq K(\mathbb{H}) = K(\beta\mathbb{N}) \cap \mathbb{H}$ such that the listed elements are all distinct and the operation $+$ satisfies*

$+$	p	α_{11}	α_{12}	α_{21}	α_{22}
p	p	α_{11}	α_{12}	α_{21}	α_{22}
α_{11}	α_{12}	α_{11}	α_{12}	α_{11}	α_{12}
α_{12}	α_{12}	α_{11}	α_{12}	α_{11}	α_{12}
α_{21}	α_{22}	α_{21}	α_{22}	α_{21}	α_{22}
α_{22}	α_{22}	α_{21}	α_{22}	α_{21}	α_{22}

In particular, α_{11}, α_{22}, and α_{12} are idempotents in $K(\beta\mathbb{N})$ and $\alpha_{11} + \alpha_{22} = \alpha_{12}$.

Some recent results deal with the ability to solve certain equations in βS. An element e of βS satisfying the equation $xe = x$ for all $x \in \beta S$ is a *right identity* for βS. Recall that for any ultrafilter p, the *norm* of p, $||p|| = \min\{|A| : A \in p\}$. J. Baker, A. Lau, and J. Pym recently obtained the following result, which implies that if βS has a two sided identity e, then $e \in S$.

4.5. THEOREM (BAKER, LAU and PYM). *Let S be a discrete semigroup, let $e \in \beta S \setminus S$ be a right identity for βS, and let $\kappa = ||e||$. Then βS has 2^{2^κ} right identities.*

□ BAKER, LAU and PYM [1999, Theorem 6]. □

HINDMAN, MALEKI and STRAUSS [2000] showed that for any distinct positive integers a and b, if $(S, +)$ is any commutative cancellative semigroup, and the equation $n \cdot s = n \cdot t$ has at most finitely many solutions with $s, t \in S$ and $n = ab|a - b|$, then the equation $u + a \cdot p = v + b \cdot p$ has no solutions with $u, v \in \beta S$ and $p \in \beta S \setminus S$. (Note for example that $2 \cdot p$ is the continuous extension of the function $s \mapsto 2 \cdot s$ to βS applied at p and it is usually not true that $2 \cdot p = p + p$.) We also showed that if S can be embedded in the circle group $\mathbb{T}$, then the equation $a \cdot p + u = b \cdot p + v$ has no solutions with $u, v \in \beta S$ and $p \in \beta S \setminus S$.

ADAMS [2001] has shown that the above statements hold if S is a countable commutative group and a and b are distinct elements of $\mathbb{Z} \setminus \{0\}$.

We mentioned above a Ramsey Theoretic consequence of the (unknown) existence of a nontrivial continuous homomorphism from $\beta\mathbb{N}$ to $\mathbb{N}^*$. In Section 5 we shall present several Ramsey Theoretic results that have been obtained recently using the algebraic structure of βS. The relationship between combinatorics and topological algebra goes both ways. Recently, in collaboration with I. Leader, we established a Ramsey Theoretic result which had the following as a corollary. We shall not attempt to explain the Ramsey Theoretic result of which it is a corollary, but remark that the proof used the idea of expansion of numbers to negative bases which we mentioned in Section 2.

4.6. THEOREM (HINDMAN, LEADER and STRAUSS). *Let $n, m \in \mathbb{N}$ and let $a_1, a_2, \ldots, a_n$, $b_1, b_2, \ldots, b_m \in \mathbb{Z}\backslash\{0\}$ such that $a_i \neq a_{i+1}$ and $b_j \neq b_{j+1}$ for all $i \in \{1, 2, \ldots, n-1\}$ and $j \in \{1, 2, \ldots, m-1\}$. If $p + p = p \in \beta\mathbb{N}$ and $a_1 \cdot p + a_2 \cdot p + \ldots + a_n \cdot p = b_1 \cdot p + b_2 \cdot p + \ldots + b_m \cdot p$, then $\langle a_1, a_2, \ldots, a_n\rangle = \langle b_1, b_2, \ldots, b_m\rangle$.*

□ HINDMAN, LEADER and STRAUSS [200?b, Corollary 4.2]. □

It is an open question whether the assumption that $p = p + p$ in Theorem 4.6 can be replaced by the weaker assumption that $p \in \mathbb{N}^*$.

The choice to make βS a right topological semigroup rather than a left topological semigroup is an arbitrary one. Let us denote by $\odot$ the operation on βS making βS a left topological semigroup with S contained in its topological center (in this case, $\{p \in \beta S : \rho_p$ is continuous$\}$). One might suspect that results for $(\beta S, \odot)$ and $(\beta S, \cdot)$ would be simply left-right switches of each other. If S is commutative, this is correct because for any $p, q \in \beta S$, $p \odot q = q \cdot p$. In particular a subset of βS is a subsemigroup under one operation if and only if it is a subsemigroup under the other and the smallest ideals $K(\beta S, \cdot)$ and $K(\beta S, \odot)$ are identical. It has been known since 1994 that both conclusions can fail given sufficient noncommutativity of S. EL-MABHOUH, PYM and STRAUSS [1994a] showed that if S is the free semigroup on countably many generators, then there is a subsemigroup H of $(\beta S, \cdot)$ with the property that given any $p, q \in H$, $p \odot q \notin H$. And it was shown by ANTHONY [1994a] that if S is the free semigroup or free group on two generators, then $K(\beta S, \cdot)\backslash c\ell K(\beta S, \odot) \neq \emptyset$. On the other hand, it was also shown in ANTHONY [1994a] that for any semigroup S whatever, $K(\beta S, \cdot) \cap c\ell K(\beta S, \odot) \neq \emptyset$. It was recently shown by BURNS [2001] that if S is the free semigroup or free group on two generators, then $K(\beta S, \cdot) \cap K(\beta S, \odot) = \emptyset$. In fact the following much stronger result was established in the same paper.

4.7. THEOREM. *Let S be the free semigroup on two generators. If $p \in c\ell K(\beta S, \cdot) \cap c\ell K(\beta S, \odot)$, then p is right cancellable in $(\beta S, \cdot)$ or left cancellable in $(\beta S, \odot)$.*

□ [2001, Theorem 3.13]. □

ADAMS [2001] has proved the corresponding theorem for the free group on two generators.

5. Applications to Ramsey Theory

We were first led to study the algebra of βS because of the very simple proof given in 1975 by F. Galvin and S. Glazer of the Finite Sums Theorem (whose proofs had previously been very complicated). See the notes to Chapter 5 of HINDMAN and STRAUSS [1998b] for details of the discovery of this proof. Over a quarter of a century later, new applications of the algebra of βS to Ramsey Theory continue to be discovered.

One of the classic results of Ramsey Theory is the Hales-Jewett Theorem (HALES and JEWETT [1963]). Given an alphabet A, a *variable word* over A is a word over the alphabet $A \cup \{v\}$ in which v actually occurs (where v is a "variable" not in A). Given a variable word w and $a \in A$, $w(a)$ has its obvious meaning, namely the replacement of all occurrences of v by a. The Hales-Jewett Theorem says that whenever A is a finite alphabet, $r \in \mathbb{N}$, and the set of finite words over A are r-colored, there is a variable word w

over A such that $\{w(a) : a \in A\}$ is monochrome. For a simple algebraic proof of the Hales-Jewett Theorem see HINDMAN and STRAUSS [1998b, Section 14.2].

Notice that one can color words based on what their leftmost and rightmost letters are. Consequently, the variable word guaranteed by the Hales-Jewett Theorem cannot be a *left variable* word (i.e., one whose leftmost letter is v) or a *right variable word*. However, in collaboration with R. McCutcheon, one of us obtained the following theorem which extends previous generalizations of the Hales-Jewett Theorem due to CARLSON [1988] and to CARLSON and SIMPSON [1984]. The proof of Theorem 5.1 uses in an intricate fashion the structure of the smallest ideal $K(\beta S)$. The products that are "obviously forbidden" are those beginning with a left variable word, ending with a right variable word, or having a right variable word immediately followed by a left variable word. The latter is forbidden because one may count the number of occurrences of a 1 followed immediately by a 2 and divide $\mathcal{W}_{k+1}$ according to whether this count is even or odd. (See HINDMAN and MCCUTCHEON [200?b, Theorem 2.10].) In an expression of the form $\prod_{n\in F} x_n$, the terms occur in the order of increasing indices.

5.1. THEOREM (HINDMAN and MCCUTCHEON). *Let $\mathcal{W}_k$ be the free semigroup on the alphabet $\{1, 2, \ldots, k\}$. Let $\mathcal{W}_k$ and $\mathcal{W}_{k+1}\backslash\mathcal{W}_k$ be finitely colored. There exists a sequence $\langle w_n\rangle_{n=1}^{\infty}$ of variable words over $\mathcal{W}_k$ such that*

(1) *for each $n \in \mathbb{N}$, if $n \equiv 1 \pmod 3$, then w_n is a right variable word;*

(2) *for each $n \in \mathbb{N}$, if $n \equiv 0 \pmod 3$, then w_n is a left variable word; and*

(3) *all products of the form $\prod_{n\in F} w_n\big(f(n)\big)$ that lie in $\mathcal{W}_k$ are monochrome and all of those that lie in $\mathcal{W}_{k+1}$ are monochrome, except for those that are obviously forbidden.*

□ HINDMAN and MCCUTCHEON [200?a, Theorem 2.9]. □

DEUBER, HINDMAN, GUNDERSON and STRAUSS [1997] obtained results in graph theory which depended on properties of idempotents in βS. A. Hajnal had asked whether, for every triangle-free graph on $\mathbb{N}$, there is a sequence $\langle x_n\rangle_{n=1}^{\infty}$ in $\mathbb{N}$ for which $FS\langle x_n\rangle_{n=1}^{\infty}$ is an independent set. We showed that the answer is "no". However, we showed that for every K_m-free graph G on a semigroup S, there exists a sequence $\langle x_n\rangle_{n=1}^{\infty}$ in S such that $\{\prod_{n\in F} x_n, \prod_{n\in H} x_n\} \notin E(G)$ whenever F and H are disjoint nonempty finite subsets of $\mathbb{N}$. We also showed that, for every $K_{m,m}$-free graph on a cancellative semigroup S, there exists a sequence $\langle x_n\rangle_{n=1}^{\infty}$ for which $FP(\langle x_n\rangle_{n=1}^{\infty})$ is independent, where $FP(\langle x_n\rangle_{n=1}^{\infty}) = \{\prod_{n\in F} x_n : F \text{ is a finite nonempty subset of } \mathbb{N}\}$.

In another purely combinatorial result whose proof relies heavily on facts about idempotents in βS, HINDMAN and STRAUSS [200?b] have shown, extending (and using) a result of GUNDERSON, LEADER, PRÖMEL and RÖDL [2001], that given any $m \in \mathbb{N}$ and any graph on $\mathbb{N}$ which does not include a complete graph on m vertices, there is a sequence of arithmetic progressions of all lengths such that there are not edges within or between the progressions nor between certain specified sums of the terms of those progressions.

As mentioned in the introduction, the relationship between the algebra of βS and topological dynamics has always been strong. Several notions from topological dynamics are important in describing the algebraic structure of βS. For example given an ultrafilter p

on S, $p \in c\ell K(\beta S)$ if and only if every member of p is *piecewise syndetic*. Another notion, originally defined in terms of topological dynamics, is *central*. A central set is quite simply characterized as one which is a member of a minimal idempotent in βS. Central sets are guaranteed to have substantial combinatorial structure. For example, the chosen monochrome sets in Theorem 5.1 above can both be chosen to be central (in $\mathcal{W}_k$ and in $\mathcal{W}_{k+1}$ respectively). Two other notions of largeness that originated in topological dynamics, namely *syndetic* and *thick* have simple characterizations in terms of βS. A set A is thick if and only if $\overline{A}$ contains a left ideal of βS, while A is syndetic if and only if $\overline{A}$ meets every left ideal of βS.

Let $u, v \in \mathbb{N} \cup \{\omega\}$. A $u \times v$ matrix with rational entries and only finitely many nonzero entries in each row is *image partition regular* provided that whenever $\mathbb{N}$ is finitely colored, there exists $\vec{x} \in \mathbb{N}^v$ such that the entries of $A\vec{x}$ are monochrome. Such a matrix is *kernel partition regular* provided that whenever $\mathbb{N}$ is finitely colored, there exists $\vec{x} \in \mathbb{N}^v$ such that $A\vec{x} = \vec{0}$ and the entries of $\vec{x}$ are monochrome. A computable characterization of finite kernel partition regular matrices was found by RADO [1933] and several characterizations of finite image partition regular matrices were found by HINDMAN and LEADER [1993].

For finite matrices which are either image partition regular or kernel partition regular, one may always choose the color class in which solutions are found to be a central set. It was shown by DEUBER, HINDMAN, LEADER and LEFMANN [1995] that this need not hold for infinite image partition regular matrices. HINDMAN, LEADER and STRAUSS investigated infinite matrices with entries from $\mathbb{Z}$ which satisfied the requirement that images could be found in any central set, which we call *centrally image partition regular*. We defined the compressed form of a finite vector with entries in $\mathbb{Z} \setminus \{0\}$ to be the vector obtained from the given one by deleting every entry equal to its predecessor. Let A be any matrix with entries from $\mathbb{Z}$ with finitely many nonzero entries in each row and no row equal to $\vec{0}$. Assume that the rows of A have the same compressed form with positive last entry and for some $s \in \mathbb{Z} \setminus \{0\}$, each row of A has a sum of terms equal to s. By using extensively the algebraic properties of $\beta\mathbb{N}$, we showed that, for every central subset C of $\mathbb{N}$, there is an infinite increasing sequence $\langle x_n \rangle_{n=1}^{\infty}$ in $\mathbb{N}$ with the property that $\sum_{i=1}^{\infty} a_i \cdot x_i \in C$ for every row $\vec{a}$ of A. This implies the following new result in Ramsey Theory.

5.2. THEOREM (HINDMAN, LEADER and STRAUSS). *Let $\mathcal{E}$ denote the set of all finite vectors of the form $\langle a_1, a_2, \ldots, a_m \rangle$ where each $a_i \in \mathbb{Z} \setminus \{0\}$, $a_m > 0$ and $a_1 + a_2 + \ldots + a_m \neq 0$. Let a finite coloring of $\mathbb{N}$ be given. For each $\epsilon = \langle a_1, a_2, \ldots, a_m \rangle \in \mathcal{E}$, there is an infinite increasing sequence $\langle x_n(\epsilon) \rangle_{n=1}^{\infty}$ in $\mathbb{N}$ such that, if $Y_\epsilon = \{a_1 x_{n_1}(\epsilon) + a_2 x_{n_2}(\epsilon) + \ldots + a_m x_{n_m}(\epsilon) : n_1 < n_2 < \ldots < n_m\}$, then $\bigcup_{\epsilon \in \mathcal{E}} Y_\epsilon$ is monochrome. Furthermore, the sequences $\langle x_n(\epsilon) \rangle_{n=1}^{\infty}$ can be chosen so that the sets Y_ϵ are pairwise disjoint.*

□ HINDMAN, LEADER and STRAUSS [200?a, Corollary 3.8]. □

HINDMAN and STRAUSS [200?a] provide, again using the algebra of $\beta\mathbb{N}$ as well as some elementary combinatorics, ways of producing new centrally image partition regular matrices from old ones.

FURSTENBERG and GLASNER [1998] showed, in an extension of van der Waerden's Theorem, that whenever B is a piecewise syndetic subset of $\mathbb{Z}$ and $l \in \mathbb{N}$, then the set of length l arithmetic progressions in B is not only nonempty, but is in fact piecewise syndetic in the set of all arithmetic progressions. Using some simple facts about the algebra of βS, BERGELSON and HINDMAN [2001, Theorem 3.7] generalized this result by showing that for a large number of notions of largeness (including "piecewise syndetic", "central", and "thick"), if S is a semigroup, $l \in \mathbb{N}$, E is a subsemigroup of S^l with $\{(a, a, \ldots, a) : a \in S\} \subseteq E$, I is an ideal of E, and B is a large subset of S, then $B^l \cap I$ is a large subset of I.

In a similar vein, HINDMAN, LEADER and STRAUSS [2002, Theorem 4.5] showed for the same notions of largeness mentioned above, that if $u, v \in \mathbb{N}$, A is a $u \times v$ matrix with entries from $\mathbb{Q}$, $I = \{A\vec{x} : \vec{x} \in \mathbb{N}^v\} \cap \mathbb{N}^u$, $\vec{1} \in I$, and C is a large subset of $\mathbb{N}$, then $I \cap C^u$ is a large subset of I.

It is a simple fact that if $A \subseteq \mathbb{N}$ has positive upper density, then $A - A = \{x - y : x, y \in A \text{ and } y < x\}$ meets $FS(\langle x_n \rangle_{n=1}^\infty)$ for every sequence $\langle x_n \rangle_{n=1}^\infty$ in $\mathbb{N}$. BERGELSON, HINDMAN and MCCUTCHEON [1998] investigated the relationship between "left" and "right" versions of *syndetic*, *thick*, and *piecewise syndetic*, in an arbitrary semigroup S. (The "right" versions are the usual notions. The "left" versions correspond to the left topological structure on βS.) They then investigated the conditions under which AA^{-1} or $A^{-1}A$ can be guaranteed to meet $FP(\langle x_n \rangle_{n=1}^\infty)$ for every sequence $\langle x_n \rangle_{n=1}^\infty$ in S, where $AA^{-1} = \{x \in S : (\exists y \in A)(xy \in A)\}$ and $A^{-1}A = \{x \in S : (\exists y \in A)(yx \in A)\}$.

6. Partial semigroups

The study of algebraic operations defined for only some members of $S \times S$ has a long history. (See the book EVSEEF and LJAPIN [1997].) Its relationship to algebra in the Stone-Čech compactification is of much more recent origin. In 1987 PYM [1987] introduced the concept of an "oid". He showed that the oid structure of $\mathbb{N}$, in which the sum of two numbers is defined as usual but only when they have disjoint binary supports, already induces all of the semigroup structure of the set $\mathbb{H} = \bigcap_{n \in \mathbb{N}} c\ell_{\beta\mathbb{N}}(2^n\mathbb{N})$. This approach was extended in BERGELSON, BLASS and HINDMAN [1994].

6.1. DEFINITION. A *partial semigroup* is a pair $(S, \cdot)$ where S is a set and there is some set $D \subseteq S \times S$ such that $\cdot : D \to S$ and the operation is associative where it is defined (in the sense that for any $x, y, z \in S$, if either of $(x \cdot y) \cdot z$ or $x \cdot (y \cdot z)$ is defined, then so is the other and they are equal). Given $x \in S$, $\varphi(x) = \{y \in S : (x, y) \in D\}$. The partial semigroup $(S, \cdot)$ is *adequate* if and only if for every finite nonempty set $F \subseteq S$, $\bigcap_{x \in F} \varphi(x) \neq \emptyset$. If S is adequate, then $\delta S = \bigcap_{x \in S} \overline{\varphi(x)}$.

Notice that the requirement that S be adequate is exactly what is needed to have $\delta S \neq \emptyset$. From our point of view, the most important thing about adequate partial semigroups is that δS is a (compact right topological) *semigroup*, with all of the structure known for such objects. In BERGELSON, BLASS and HINDMAN [1994] several Ramsey Theoretic results related to the Hales-Jewett Theorem were obtained.

In 1992, W. Gowers established a Ramsey Theoretic result as a tool to solve a problem about Banach spaces. While he did not state it this way, his result is naturally stated in terms of partial semigroups. Let $k \in \mathbb{N}$ and let $Y = \{f : f : \mathbb{N} \to \{0, 1, \ldots, k\}$ and $\{x \in \mathbb{N} : f(x) \neq 0\}$ is finite$\}$. Given $f \in Y$, let $\text{supp}(f) = \{x \in \mathbb{N} : f(x) \neq 0\}$ and for $f, g \in Y$, define $f + g$ pointwise, but only when $\text{supp}(f) \cap \text{supp}(g) = \emptyset$. Then $(Y, +)$ is an adequate partial semigroup. Let $Y_k = \{f \in Y : \max(f[\mathbb{N}]) = k\}$. Define $\sigma : Y \to Y$

by

$$\sigma(f)(x) = \begin{cases} f(x) - 1 & \text{if } f(x) > 0 \\ 0 & \text{if } f(x) = 0 . \end{cases}$$

Notice that σ is a partial semigroup homomorphism in the sense that it holds $\sigma(f+g) = \sigma(f) + \sigma(g)$ whenever $f + g$ is defined.

6.2. THEOREM (GOWERS). *Let k, Y, Y_k and σ be as defined above, let $r \in \mathbb{N}$, and let $Y = \bigcup_{i=1}^{r} C_i$. Then there exist $i \in \{1, 2, \ldots, r\}$ and a sequence $\langle f_n \rangle_{n=1}^{\infty}$ in Y_k such that* $\text{supp}(f_n) \cap \text{supp}(f_m) = \emptyset$ *for all $m, n \in \mathbb{N}$ and $\{\sum_{n \in F} \sigma^{t(n)}(f_n) : F \in \mathcal{P}_f(\mathbb{N})$, $t : F \to \{0, 1, \ldots, k-1\}$, and $t^{-1}[\{0\}] \neq \emptyset\} \subseteq C_i$.*

□ GOWERS [1992, Theorem 1]. □

FARAH, HINDMAN and MCLEOD derived a simultaneous generalization of Theorem 6.2 and one of the results of BERGELSON, BLASS and HINDMAN [1994]. This generalization is quite complicated to state in its entirety, but we shall describe a reasonably simple corollary.

6.3. THEOREM (FARAH, HINDMAN and MCLEOD). *Let S, T, and R be the free semigroups with identity e on the alphabets $\{a, b, c\}$, $\{a, b\}$, and $\{a\}$ respectively. Given $x, y, z \in \{a, b, c, e\}$, let f_{xyz} be the endomorphism of S determined by $f(a) = x$, $f(b) = y$, and $f(c) = z$. For every $r \in \mathbb{N}$ and every partition $S = \bigcup_{j=1}^{r} C_j$ there exist an infinite $\langle x_n \rangle_{n=1}^{\infty}$ in $S \setminus T$ and $\gamma\colon \{a, b, c\} \to \{1, 2, \ldots, r\}$ such that if $\sigma \in \{f_{eab}, f_{aeb}, f_{aab}\}$ and $\mathcal{F} = \{f_{abc}, f_{abb}, f_{aba}, f_{abe}, \sigma\} \cup \{f_{xyz} | x, y, z \in \{a, e\}\}$, then we have*

$$\{\textstyle\prod_{n \in F} g_n(x_n) : F \in \mathcal{P}_f(\mathbb{N}),\ \textit{and for each } n \in F,\ g_n \in \mathcal{F}\} \cap (S \setminus T) \subsetneq C_{\gamma(a)}$$

$$\{\textstyle\prod_{n \in F} g_n(x_n) : F \in \mathcal{P}_f(\mathbb{N}),\ \textit{and for each } n \in F,\ g_n \in \mathcal{F}\} \cap (T \setminus R) \subsetneq C_{\gamma(b)}$$

$$\{\textstyle\prod_{n \in F} g_n(x_n) : F \in \mathcal{P}_f(\mathbb{N}),\ \textit{and for each } n \in F,\ g_n \in \mathcal{F}\} \cap R \setminus \{e\}) \subsetneq C_{\gamma(c)} .$$

□ FARAH, HINDMAN and MCLEOD [200?, Corollary 3.14]. □

As we have previously mentioned, several dynamical notions of largeness in a semigroup S, including "syndetic", "thick", and "piecewise syndetic" have simple characterizations in terms of the algebra of βS. These notions (for a discrete semigroup) also have simple combinatorial characterizations. For example, a subset A of S is syndetic if and only if there is a finite nonempty subset H of S such that $S = \bigcup_{t \in H} t^{-1}A$, where $t^{-1}A = \{s \in S : ts \in A\}$. Each of these notions has a completely obvious analogue for partial semigroups in terms of the algebra of δS. (So that, for example, a subset A of the partial semigroup S is syndetic if and only if for every left ideal of δS, $\overline{A} \cap S \neq \emptyset$.) There are also natural, though somewhat less obvious, analogues of the combinatorial characterizations. For example A is č-syndetic if and only if there exists finite nonempty $H \subseteq S$ such that $\bigcap_{t \in H} \varphi(t) \subseteq \bigcup_{t \in H} t^{-1}A$.

MCLEOD [2001] and [200?], showed that for each of these (and other) notions of largeness, the natural algebraic and the natural combinatorial versions (the ones preceded by č) need not be equivalent. She also showed that in each case one of the notions implies the others. (For example "syndetic" implies "č-syndetic", while "č-thick" implies "thick".)

A *VIP system* is a polynomial type generalization of the notion of an IP system, i.e., a set of finite sums. HINDMAN and MCCUTCHEON [2001], extended the notion of VIP system to commutative partial semigroups and obtained an analogue of the Central Sets Theorem for these systems which extends the polynomial Hales-Jewett Theorem of BERGELSON and LEIBMAN [1996]. Several Ramsey Theoretic consequences, including the Central Sets Theorem itself, were then derived from these results.

References

ADAMS, P.

[2001] *Topics in the algebra of βS*, Ph.D. Dissertation, Hull University.

ANTHONY, P.

[1994a] *Ideals in the Stone-Čech compactification of noncommutative semigroups*, Ph.D. Dissertation , Howard University.

[1994b] The smallest ideals in the two natural products on βS, *Semigroup Forum* **48**, 363–367.

BAKER, J., N. HINDMAN and J. PYM

[1992a] κ-topologies for right topological semigroups, *Proc. Amer. Math. Soc.* **115**, 251–256.

[1992b] Elements of finite order in Stone-Čech compactifications, *Proc. Edinburgh Math. Soc.* **36**, 49–54.

BAKER, J., A. LAU and J. PYM

[1999] Identities in Stone-Čech compactifications of semigroups, *Semigroup Forum* **59**, 415–417.

BALCAR, B. and F. FRANEK

[1997] Structural properties of universal minimal dynamical systems for discrete semigroups, *Trans. Amer. Math. Soc.* **349**, 1697–1724.

BENINGSFIELD, K.

[1999] Cancellation and embedding theorems for compact uniquely divisible semigroups, *Semigroup Forum* **58**, 336–347.

BERGELSON, V., A. BLASS and N. HINDMAN

[1994] Partition theorems for spaces of variable words, *Proc. London Math. Soc.* **68**, 449–476.

BERGELSON, V., W. DEUBER and N. HINDMAN

[1992] Rado's Theorem for finite fields, in Proceedings of the Conference on Sets, Graphs, and Numbers, Budapest, 1991, *Colloq. Math. Soc. János Bolyai* **60**, (1992), 77–88.

BERGELSON, V., W. DEUBER, N. HINDMAN and H. LEFMANN

[1994] Rado's Theorem for commutative rings, *J. Comb. Theory (Series A)* **66**, 68–92.

BERGELSON, V. and N. HINDMAN

[1992a] Ramsey Theory in non-commutative semigroups, *Trans. Amer. Math. Soc.* **330**, 433–446.

[1992b] Some topological semicommutative van der Waerden type theorems and their combinatorial consequences, *J. London Math. Soc.* **45**, 385–403.

[1993] Additive and multiplicative Ramsey Theorems in $\mathbb{N}$ – some elementary results, *Comb. Prob. and Comp.* **2**, 221–241.

* In addition to items cited in the paper, this list of references includes relevant articles of which we are aware that were published since 1992. Many of these are not discussed here because they were covered in HINDMAN and STRAUSS [1998b].

BERGELSON, V. and N. HINDMAN

[1994] IP*-sets and central sets, *Combinatorica* **14**, 269–277.

[1996] IP* sets in product spaces, in Papers on General Topology and Applications. S. Andima et. al. eds. *Annals of the New York Academy of Sciences*, **806**, 28–41. [1]

[2001] Partition regular structures contained in large sets are abundant, *J. Comb. Theory (Series A)* **93**, 18–36.

BERGELSON, V., N. HINDMAN and B. KRA

[1996] Iterated spectra of numbers — elementary, dynamical, and algebraic approaches, *Trans. Amer. Math. Soc.* **348**, 893–912.

BERGELSON, V., N. HINDMAN and I. LEADER

[1996] Sets partition regular for n equations need not solve $n + 1$, *Proc. London Math. Soc.* **73**, 481–500.

[1999] Additive and multiplicative Ramsey Theory in the reals and the rationals, *J. Comb. Theory (Series A)* **85**, 41–68.

BERGELSON, V., N. HINDMAN and R. MCCUTCHEON

[1998] Notions of size and combinatorial properties of quotient sets in semigroups, *Topology Proceedings* **23**, 23–60.

BERGELSON, V., N. HINDMAN and B. WEISS

[1997] All-sums sets in $(0, 1]$ – category and measure, *Mathematika* **44**, 61–87.

BERGELSON, V. and A. LEIBMAN

[1996] Polynomial extensions of van der Waerden's and Szemerédi's theorems, *Journal Amer. Math. Soc.*, **9**, 725–753.

BERGLUND, J.

[1980] Problems about semitopological semigroups, *Semigroup Forum* **14**, 373–383.

BERGLUND, J. and N. HINDMAN

[1992] Sums of idempotents in $\beta\mathbb{N}$, *Semigroup Forum* **44**, 107–111.

BERGLUND, J., H. JUNGHENN and P. MILNES

[1989] *Analysis on Semigroups*, Wiley, N.Y.

BLASS, A.

[1993] Ultrafilters: where topological dynamics = algebra = combinatorics, *Topology Proceedings* **18**, 33–56.

BLASS A. and N. HINDMAN

[1987] On strongly summable ultrafilters and union ultrafilters, *Trans. Amer. Math. Soc.* **304**, 83–99.

BLÜMLINGER, M.

[1996] Lévy group action and invariant measures on $\beta\mathbb{N}$, *Trans. Amer. Math. Soc.* **348**, 5087–5111.

BORDBAR, B.

[1998] Weakly almost periodic functions on $\mathbb{N}$ with a negative base, *J. London Math. Soc.* **57**, 706–720.

[1] All of the items in this list of references that include Hindman as an author and have a publication date of 1995 or later are currently available in dvi and pdf forms at `http://members.aol.com/nhindman/` except for items HINDMAN and STRAUSS [1995d] and [1998b].

BORDBAR, B. and J. PYM

[1998] The weakly almost periodic compactification of a direct sum of finite groups, *Math. Proc. Cambr. Phil. Soc.* **124**, 421–449.

[2000] The set of idempotents in the weakly almost periodic compactification of the integers is not closed, *Trans. Amer. Math. Soc.* **352**, 823–842.

BOUZIAD, A.

[1992] Continuité d'une action d'un semilattis compact, *Semigroup Forum* **44**, 79–86.

[1993] The Ellis theorem and continuity in groups, *Topology and its Applications* **50**, 73–80.

BOUZIAD, A., M. LEMAŃCZYK, AND M. MENTZEN

[2001] A compact monothetic semitopological semigroup whose set of idempotents is not closed, *Semigroup Forum* **62**, 98–102.

BROWN, G. and W. MORAN

[1972] The idempotent semigroup of a compact monothetic semigroup, *Proc. Royal Irish Acad., Section A*, **72**, 17–33.

BUDAK, T.

[1993] Compactifying topologised semigroups, *Semigroup Forum* **46**, 128–129.

BUDAK, T., N. ISIK and J. PYM

[1994] Subsemigroups of Stone-Čech compactifications, *Math. Proc. Cambr. Phil. Soc.* **116**, 99–118.

BURNS, S.

[2000] *The existence of disjoint smallest ideals in the left continuous and right continuous structures in the Stone-Čech compactification of a semigroup*, Ph.D. Dissertation, Howard University.

[2001] The existence of disjoint smallest ideals in the two natural products on βS, *Semigroup Forum* **63**, 191–201.

CARLSON, T.

[1988] Some unifying principles in Ramsey Theory, *Discrete Math.* **68**, 117–169.

CARLSON, T. and S. SIMPSON

[1984] A dual form of Ramsey's Theorem, *Advances in Math.* **53**, 265–290.

CLIFFORD, A. and G. PRESTON

[1961] *The Algebraic Theory of Semigroups*, American Mathematical Society, Providence.

COMFORT, W., K. HOFMANN and D. REMUS

[1992] Topological groups and semigroups, in HUŠEK and VAN MILL [1992], pages 59–144.

DAVENPORT, D., N. HINDMAN, I. LEADER and D. STRAUSS

[2000] Continuous homomorphisms on $\beta\mathbb{N}$ and Ramsey Theory, *New York J. Math.* **6**, 73–86.

DEUBER, W., D. GUNDERSON, N. HINDMAN and D. STRAUSS

[1997] Independent finite sums for K_m-free graphs, *J. Comb. Theory (Series A)* **78**, 171–198.

DEUBER, W., N. HINDMAN, I. LEADER and H. LEFMANN

[1995] Infinite partition regular matrices, *Combinatorica* **15**, 333–355.

VAN DOUWEN, E.

[1991] The Čech-Stone compactification of a discrete groupoid, *Topology and its Applications*, **39**, 43–60.

ELLIS, R.

[1957] Locally compact transformation groups, *Duke Math. J.* **24**, 119–125.

[1969] *Lectures on Topological Dynamics*, Benjamin, New York.

EL-MABHOUH, A., J. PYM and D. STRAUSS

[1994a] On the two natural products in a Stone-Čech compactification, *Semigroup Forum* **48**, 255–257.

[1994b] Subsemigroups of $\beta\mathbb{N}$, *Topology and its Applications* **60**, 87–100.

EVSEEV A. and E. LJAPIN

[1997] *The Theory of Partial Algebraic Operations*, Kluwer Academic Publishers, Dordrecht.

FARAH, I., N. HINDMAN and J. MCLEOD

[200?] Partition theorems for layered partial semigroups, *J. Comb. Theory (Series A)*, to appear.

FERRI S.

[2001] *A study of some universal semigroup compactifications*, Ph.D. Dissertation, Hull University.

FERRI S. and D. STRAUSS

[2001] Ideals, idempotents and right cancelable elements in the uniform compactification, *Semigroup Forum* **63**, 449–456.

FILALI, M.

[1996a] Right cancellation in βS and UG, *Semigroup Forum* **52**, 381–388.

[1996b] Weak p-points and cancellation in βS, in Papers on General Topology and Applications. S. Andima et. al. eds. *Annals of the New York Academy of Sciences*, **806**, 130–139.

[1997] On some semigroup compactifications, *Topology Proc.* **22**, 111–123.

[1999] On the semigroup βS, *Semigroup Forum* **58**, 241–247.

FILALI, M. and J. PYM

[200?] Right cancellation in the $\mathcal{LUC}$-compactification of a locally compact group, *Bull. London Math. Soc.*, to appear.

FURSTENBERG, H. and E. GLASNER

[1998] Subset dynamics and van der Waerden's Theorem, *Canad. J. Math.* **32**, 197–203.

GARCÍA-FERREIRA, S.

[1993] Three orderings on $\beta(\omega)\backslash\omega$, *Topology and its Applications* **50**, 199–216.

[1994] Comfort types of ultrafilters, *Proc. Amer. Math. Soc.* **120** (1994), 1251–1260.

GARCÍA-FERREIRA, S., N. HINDMAN AND D. STRAUSS

[1999] Orderings of the Stone-Čech remainder of a discrete semigroup, *Topology and its Applications* **97**, 127–148.

GLASNER, E.

[1998] On minimal actions of Polish groups, *Topology and its Applications* **85**, 119–125.

GOWERS, W.

[1992] Lipschitz functions on classical spaces, *European J. Combinatorics* **13**, 141–151.

GUNDERSON, D., I. LEADER, H. PRÖMEL, AND V. RÖDL

[2001] Independent arithmetic progressions in clique-free graphs on the natural numbers, *J. Comb. Theory (Series A)* **93**, 1–17.

HALES, A. and R. JEWETT

[1963] Regularity and positional games, *Trans. Amer. Math. Soc.* **106**, 222–229.

HILGERT, J. and K. NEEB

[1993] *Lie Semigroups and their Applications*, Springer-Verlag, Berlin.

HINDMAN, N.

[1993] The topological-algebraic system $(\beta\mathbb{N}, +, \cdot)$, in Papers on General Topology and Applications, S. Andima et. al. eds. *Annals of the New York Academy of Sciences*, **704**, 155–163.

[1995] Recent results on the algebraic structure of βS, in Papers on General Topology and Applications. S. Andima et. al. eds. *Annals of the New York Academy of Sciences*, **767**, 73–84.

[1996] Algebra in βS and its applications to Ramsey Theory, *Math. Japonica* **44**, 581–625.

[2001] Problems and new results in the algebra of βS and Ramsey Theory, in *Unsolved problems on mathematics for the* 21st *century*, J. Abe and S. Tanaka (eds.). IOS Press, Amsterdam.

HINDMAN, N., J. LAWSON and A. LISAN

[1994] Separating points of $\beta\mathbb{N}$ by minimal flows, *Canadian J. Math.* **46**, 758–771.

HINDMAN, N. and I. LEADER

[1993] Image partition regularity of matrices, *Comb. Prob. and Comp.* **2**, 437–463.

[1999] The semigroup of ultrafilters near 0, *Semigroup Forum* **59**, 33–55.

HINDMAN, N., I. LEADER and D. STRAUSS

[2002] Image partition regular matrices – bounded solutions and preservations of largeness, *Discrete Math.* **242**, 115–144.

[200?a] Infinite partition regular matrices – solutions in central sets, *Trans. Amer. Math. Soc.*, to appear.

[200?b] Separating Milliken-Taylor Systems with negative entries, preprint.

HINDMAN, N. and H. LEFMANN

[1993] Partition regularity of $(\mathcal{M}, \mathcal{P}, \mathcal{C})$-systems, *J. Comb. Theory (Series A)* **64**, 1–9.

[1996] Canonical partition relations for (m,p,c)-systems, *Discrete Math.* **162**, 151–174.

HINDMAN, N. and A. LISAN

[1994] Points very close to the smallest ideal of βS, *Semigroup Forum* **49**, 137–141.

HINDMAN, N., A. MALEKI and D. STRAUSS

[1996] Central sets and their combinatorial characterization, *J. Comb. Theory (Series A)* **74**, 188–208.

[2000] Linear equations in the Stone-Čech compactification of $\mathbb{N}$, *Integers* **0**, #A02, 1–20.

HINDMAN, N. and R. MCCUTCHEON

[1999] Weak VIP systems in commutative semigroups, *Topology Proceedings* **24**, 199–201.

[2001] VIP systems in partial semigroups, *Discrete Math.* **240**, 45–70.

[200?a] One sided sdeals and Carlson's theorem, *Proc. Amer. Math. Soc.* **130**, 2559–2567.

[200?b] Partition theorems for left and right variable words, preprint.

HINDMAN, N., J. VAN MILL and P. SIMON

[1992] Increasing chains of ideals and orbit closures in $\beta\mathbb{Z}$, *Proc. Amer. Math. Soc.* **114**, 1167–1172.

HINDMAN, N., I. PROTASOV and D. STRAUSS

[1998a] Strongly summable ultrafilters on abelian groups, *Matem. Studii* **10**, 121–132.

[1998b] Topologies on S determined by idempotents in βS, *Topology Proceedings* **23**, 155–190.

HINDMAN, N. and D. STRAUSS

[1994] Cancellation in the Stone-Čech compactification of a discrete semigroup, *Proc. Edinburgh Math. Soc.* **37**, 379–397.

[1995a] Nearly prime subsemigroups of $\beta\mathbb{N}$, *Semigroup Forum* **51**, 299–318.

[1995b] Topological and algebraic copies of $\mathbb{N}^*$ in $\mathbb{N}^*$, *New York J. Math.* **1**, 111–119.

[1995c] Chains of idemptotents in $\beta\mathbb{N}$, *Proc. Amer. Math. Soc.* **123**, 3881–3888.

[1995d] Algebraic and topological equivalences in the Stone-Čech compactification of a discrete semigroup, *Topology and its Applications* **66**, 185–198.

[1996a] Prime properties of the smallest ideal of $\beta\mathbb{N}$, *Semigroup Forum* **52**, 357–364.

[1996b] Compact subsemigroups of $(\beta\mathbb{N}, +)$ containing the idempotents, *Proc. Edinburgh Math. Soc.* **39**, 291–307.

[1998a] An algebraic proof of Deuber's Theorem, *Comb. Prob. and Comp.* **7**, 167–180.

[1998b] *Algebra in the Stone-Čech Compactification*, Walter de Gruyter & Co., Berlin.

[200?a] Infinite partition regular matrices, II – extending the finite results, *Topology Proceedings*, to appear.

[200?b] Independent sums of arithmetic progressions in K_m-free graphs, *Ars. Combinatorica*, to appear.

HINDMAN, N. and W. WOAN

[1993] Central sets in semigroups and partition regularity of systems of linear equations, *Mathematika* **40**, 169–186.

HOFMANN, K.

[1994] Semigroups and Hilbert's Fifth Problem, *Mathematica Slovaca* **44**, 365–377.

[1995] From a topological theory of semigroups to a geometric one, *Semigroup Forum* **50**, 123–134.

[2000] A history of topological and analytical semigroups – a personal view, *Semigroup Forum* **61**, 1–25.

HOFMANN, K. and J. LAWSON

[1996] Linearly ordered semigroups: historical origins and A. H. Clifford's influence, *London Math. Soc. Lecture Note Series* **231**, 15–39.

HOFMANN, K. and M. MISLOVE, EDITORS

[1996] *Semigroup Theory and its Applications*, Cambridge Univ. Press, Cambridge.

HOFMANN, K. and P. MOSTERT

[1996] *Elements of Compact Semigroups*, Charles E. Merrill, Columbus.

HUŠEK, M. and J. VAN MILL, EDITORS

[1992] *Recent Progress in General Topology*, Elsevier, Amsterdam.

KÜNZI, H., J. MARÍN and S. ROMAGUERA

[2001] Quasi-uniformities on topological semigroups and bicompletion, *Semigroup Forum* **62**, 403–422.

LAU, A., A. MEDGHALCHI and J. PYM

[1993] On the spectrum of $L^\infty(G)$, *J. London Math. Soc.* **48**, 152–166.

LAU, A., P. MILNES and J. PYM

[1997] Locally compact groups, invariant means and the centres of compactifications, *J. London Math. Soc.*, **56**, 77–90.

[1999] On the structure of minimal left ideals in the largest compactification of a locally compact group, *J. London Math. Soc.* **59**, 133–152.

LAWSON, J.

[1992] Historical links to a Lie theory of semigroups, *J. of Lie Theory* **2**, 263–278.

[1996] The earliest semigroup paper, *Semigroup Forum* **52**, 55–60.

MALEKI, A. and D. STRAUSS

[1996] Homomorphisms, ideals and commutativity in the Stone-Čech compactification of a discrete semigroup, *Topology and its Applications* **71**, 47–61.

MALYKHIN, V.

[1975] Extremally disconnected and similar groups, *Soviet Math. Dokl.* **16**, 21–25.

MCLEOD, J.

[2001] *Notions of size in partial semigroups*, Ph.D Dissertation, Howard University.

[200?] Some notions of size in partial semigroups, *Topology Proceedings*, to appear.

MEGRELISHVILI, M.

[2001] Every semitopological semigroup compactification of the group $H_+[0,1]$ is trivial, *Semigroup Forum* **63**, 357–370.

PROTASOV, I.

[1993] Ultrafilters and topologies on groups (Russian), *Sibirsk. Math. J.* **34**, 163–180.

[1996] Points of joint continuity of a semigroup of ultrafilters of an abelian group, *Math. Sbornik* **187**, 131–140.

[1997] *Combinatorics of numbers*, VTNL, Ukraine.

[1998] Finite groups in βG, *Matem. Studii* **10**, 17–22.

[2001a] Resolvability of left topological groups, *Voprosy Algebry, Izv. Gomel. University* **17**, 72–78.

[2001b] Extremal toplogies on groups, *Matem. Stud.* **15**, 9–22.

[200?] Continuity in G^*, *Manuscript.*

PROTASOV, I. and J. PYM

[2001] Continuity of multiplication in the largest compactification of a locally compact group, *Bull. London Math. Soc.* **33**, 279–282.

PROTASOV I., J. PYM and D. STRAUSS

[2000] A lemma on extending functions into F-spaces and homomorphisms between Stone-Čech remainders, *Topology and its Applications* **105**, 209–229.

PYM, J.

[1987] Semigroup structure in Stone-Čech compactifications, *J. London Math. Soc.* **36**, 421–428.

[1999] A note on $G^{\mathcal{LUC}}$ and Veech's Theorem, *Semigroup Forum* **59**, 171–174.

RADO, R.

[1933] Studien zur Kombinatorik, *Math. Zeit.* **36**, 242–280.

RUPPERT, W.

[1973] Rechtstopologische Halbgruppen, *J. Reine Angew. Math.* **261**, 123–133.

[1984] *Compact Semitopological Semigroups: an Intrinsic Theory*, Springer-Verlag, Berlin.

[1991] On signed α-adic expansions and weakly almost periodic functions, *Proc. London Math. Soc.* **63**, 620–656.

SHI, H. and H. YANG

[1996] Nonmetrizable topological dynamical characterization of central sets, *Fundamenta Math.* **150**, 1–9.

SMITH, G.

[1994] *Partition regularity of sums of products of natural numbers*, Ph.D. Dissertation, Howard University.

[1995] Partitions and (m and n) sums of products, *J. Comb. Theory (Series A)* **72**, 77–94.

STRAUSS, D.

[1992a] $\mathbb{N}^*$ does not contain an algebraic and topological copy of $\beta\mathbb{N}$, *J. London Math. Soc.* **46**, 463–470.

[1992b] Semigroup structures on $\beta\mathbb{N}$, *Semigroup Forum* **41**, 238–244.

[1994] Ideals and commutativity in $\beta\mathbb{N}$, *Topology and its Applications* **60**, 281–293.

SUSCHKEWITSCH, A.

[1928] Über die endlichen Gruppen ohne das Gesetz der eindeutigen Umkehrbarkeit, *Math. Annalen* **99**, 30–50.

TERRY, E.

[1997] *Finite sums and products in Ramsey Theory*, Ph.D. Dissertation, Howard University.

UMOH, H.

[1995] Maximal density ideal of βS, *Math. Japonica* **42**, 245–247.

VEECH, V.

[1977] Topological dynamics, *Bull. Amer. Math. Soc.* **83**, 775–830.

WALLACE, A.

[1955] The structure of topological semigroups, *Bull. Amer. Math. Soc.* **23**, 95–112.

ZELENYUK, Y.

[1996a] Finite groups in $\beta\mathbb{N}$ are trivial (Russian), *Ukranian National Academy of Sciences Institute of Mathematics* **96** (3), 12pp.

[1996b] Topological groups with finite semigroups of ultrafilters (Russian), *Matematychni Studii* **6**, 41–52.

[2001] On subsemigroups of $\beta\mathbb{N}$ and absolute retracts, *Semigroup Forum* **63**, 457–465.

CHAPTER 10

Recent Progress in Hyperspace Topologies

L'ubica Holá

Institute of Mathematics, Slovak Academy of Sciences, Štefánikova 49, 814 73 Bratislava, Slovakia
E-mail: hola@mat.savba.sk

Jan Pelant

Institute of Mathematics, Academy of Sciences of the Czech Republic
Žitná 25, 115 67 Praha 1, Czech Republic
E-mail: pelant@math.cas.cz

Contents

RECENT PROGRESS IN GENERAL TOPOLOGY II
Edited by Miroslav Hušek and Jan van Mill

This paper will present some recent results in the theory of hyperspace topologies, which are important (from our point of view) for further development in this field as well as some older results which are missing in the book BEER [1993a] which is now regarded as a basic reference for topologies on hyperspaces.

We are fully aware of the unavoidable fact that there are many important papers dealing with hyperspaces from various points of view which will not be mentioned in this survey. We omit completely the continuum theory as well as infinite-dimensional aspects of theory of hyperspaces. Some relevant information on the latter subject could be found in VAN MILL [1989], [2001] and also in the papers by DIJKSTRA and VAN MILL and by POL and TORUŃCZYK in this volume; there are interesting papers in this area dealing with the relatively new notion of Wijsman topology (to be defined below), e.g. KUBIŚ [200?] or KUBIŚ, SAKAI and YAGUCHI [200?]. Finally, an exposition of results connected with "exponential" spaces (i.e. topological spaces homeomorphic to their own hyperspace), given by TODORČEVIĆ [1997, Chapter IV], is very informative. Also the selection theory is related to our topic; a very recent and comprehensive survey is presented by REPOVŠ and SEMENOV in this volume.

We will concentrate mainly on "classical" hyperspace topologies and we have to emphasize once more that there are many important and interesting papers dealing with so called (proximal) hit-and-miss topologies as well as weak topologies generated by gap and excess functionals which we could not include here because of a lack of space.
Both authors gratefully acknowledge the support of the Slovak-Czech Grant 180-15. The second author gratefully acknowledges the support of the grant GA ČR 201/00/1466.

1. Preliminaries

We refer to BEER [1993a], ENGELKING [1989], GRUENHAGE [1984] and KECHRIS [1995] for basic notions.

For a set X and an integer $n \in \omega$, $[X]^n$ denotes a collection of all subsets of X of cardinality n and $Fin(X) = \bigcup_{n\in\omega}[X]^n$.
Let $CL(X)$ ($K(X)$) denote the family of all nonempty closed (compact) sets of a T_2 topological space (X, τ) and let 2^X stand for all closed subsets of X. Notice that some authors denote 2^X as $CL_\emptyset(X)$.

Historically, there have been two topologies of particular importance: the Vietoris topology and the Hausdorff metric topology, as considered by MICHAEL [1951] in his fundamental article on hyperspaces.

The Fell topology can be regarded also as a classical one, it has found numerous applications in different fields of mathematics (see e.g. MATHERON [1975], ATTOUCH [1984]). A further classical and very important notion in hyperspace theory is that of Kuratowski - Painlevé convergence of sets.

1.1. DEFINITION. Given a net $\{A_\lambda : \lambda \in \Lambda\}$ in $CL(X)$, we define its lim-sup $Ls\{A_\lambda : \lambda \in \Lambda\}$ to be the set of all points $x \in X$ such that for every open neighborhood U of x the set $\{\lambda \in \Lambda : A_\lambda \cap U \neq \emptyset\}$ is cofinal in Λ. We define its lim-inf $Li\{A_\lambda : \lambda \in \Lambda\}$ to be the set of all points $x \in X$ such that for every open neighborhood U of x the set $\{\lambda \in \Lambda : A_\lambda \cap U \neq \emptyset\}$ is residual in Λ.

1.2. DEFINITION. We say that a net $\{A_\lambda : \lambda \in \Lambda\}$ *Kuratowski - Painlevé converges* to $A \subset X$ if $A = LiA_\lambda = LsA_\lambda$, and that a net $\{A_\lambda : \lambda \in \Lambda\}$ *upper Kuratowski - Painlevé converges* to $A \subset X$ if $LsA_\lambda \subset A$.

To describe the topologies mentioned above, we need to introduce some notation. For $U \subset X$ put

$$U^+ = \{A \in CL(X) : A \subset U\} \text{ and } U^- = \{A \in CL(X) : A \cap U \neq \emptyset\}.$$

Subbase elements of the *Vietoris* (*locally finite*) *topology* **V** (**LF**) on $CL(X)$ are of the form U^+ with $U \in \tau$ and $\bigcap_{U \in \mathcal{U}} U^-$ with $\mathcal{U} \subset \tau$ finite (locally finite). Thus, a base for the Vietoris topology consists of sets of the form

$$\langle U_0, \ldots, U_n \rangle = \Big(\bigcup_{k \le n} U_k\Big)^+ \cap \bigcap_{k \le n} U_k^-$$

where $U_0, \ldots, U_n \in \tau$ and $n \in \omega$.

We will recall later an obvious:

1.3. REMARK. If Y is a dense subset of X then $Fin(Y)$ is a dense subset of $(CL(X), \mathbf{V})$.

The *Fell topology* **F** on $CL(X)$ is coarser than **V**. **F** has a subbase whose elements are of the form U^- and V^+, where $U \in \tau$ and V has a compact complement in X.

1.4. DEFINITION. Let (X, d) be a metric space. The *Hausdorff* (extended) *metric* H_d is defined as

$$H_d(A, B) = \max\{(e_d(A, B), e_d(B, A))\},$$

where

$$e_d(A, B) = \sup\{d(a, B) : a \in A\}.$$

The topology $\mathbf{H}_d$ generated by H_d on $CL(X)$ is called the *Hausdorff metric topology.*

For metrizable spaces, the supremum of all Hausdorff metric topologies corresponding to topologically equivalent metrics is the locally finite topology **LF** defined above (see BEER, HIMMELBERG, PRIKRY and VAN VLECK [1987]; cfr. MARJANOVIC [1966], NAIMPALLY and SHARMA [1988]).

For the next definition, we use a widely accepted convention that $d(x, \emptyset) = +\infty$ for each point $x \in X$ (see also the subsection 6.3 below).

1.5. DEFINITION. For a metric space (X, d), the *Wijsman topology* $\mathbf{W}_d$ on $CL(X)$ (or on 2^X) is the weak topology determined by the family $\{d(x, \cdot) : x \in X\}$.

Recall that this convergence for sequences of convex sets in Euclidean spaces was introduced and investigated by WIJSMAN [1966].

So a net $\{A_\lambda : \lambda \in \Lambda\}$ converges in $\mathbf{W}_d$ to $A \subset X$ iff the distance functions $\{d(\cdot, A_\lambda) : \lambda \in \Lambda\}$ converge to $d(\cdot, A)$ in $C_p(X)$. To apply various techniques, developed within metrizable spaces, for an investigation of Wijsman topologies, it is quite crucial to realize the following

1.6. REMARK. $\mathbf{W}_d$ is metrizable iff (X, d) is separable.

The lower Vietoris topology $\mathbf{V}^-$ on $CL(X)$ is generated by all subcollections of the form G^-, where G is an open subset of X; similarly the upper Vietoris topology $\mathbf{V}^+$ is generated by all G^+, where G is open. The supremum $\mathbf{V}^+ \vee \mathbf{V}^-$ is the Vietoris topology $\mathbf{V}$.

The co-compact topology $\mathbf{C}$ is generated by all G^+, where G is the complement of a compact subset of X. The supremum $\mathbf{C} \vee \mathbf{V}^-$ is the Fell topology $\mathbf{F}$.

For every space X, there exists the strongest topology $\tau_{\mathbf{UK}}$ on 2^X such that upper Kuratowski - Painlevé convergence forces $\tau_{\mathbf{UK}}$-convergence. This topology is usually called the upper Kuratowski topology and is often referred to as a topologization of the upper Kuratowski - Painlevé convergence.

DOLECKI, GRECO and LECHICKI [1995] pointed out that the upper Kuratowski topology $\tau_{\mathbf{UK}}$ on the complete lattice 2^X is homeomorphic to the so-called Scott topology studied in SCOTT [1972].

Similarly, for every space X there exists the strongest topology $\tau_{\mathbf{Z}}$ on 2^X such that Kuratowski - Painlevé convergence forces $\tau_{\mathbf{Z}}$-convergence. Since $\tau_{\mathbf{Z}}$ is a topologization of the Kuratowski - Painlevé convergence, in view of the obvious analogy with the upper Kuratowski topology, the topology $\tau_{\mathbf{Z}}$ should have been naturally called the Kuratowski topology, but it is usually referred to as the convergence topology on 2^X. It was introduced by FLACHSMEYER [1964], rediscovered one year later by EFFROS [1965] and subsequently extensively studied by CHRISTENSEN [1974].

The supremum $\tau_{\mathbf{UK}} \vee \mathbf{V}^-$ of the upper Kuratowski topology $\tau_{\mathbf{UK}}$ and the lower Vietoris topology $\mathbf{V}^-$ is called the Kuratowski topology, and will be denoted by $\tau_{\mathbf{K}}$.

2. Cardinal invariants of hyperspaces

Some cardinal functions on hyperspaces - and in particular the local character - have been extensively studied, especially for the Vietoris topology, by ČOBAN [1971], MALYKHIN [1972], MIZOKAMI [1976], HOLÁ and LEVI [1997], DI MAIO and HOLÁ [1995].

For instance, (MALYKHIN [1972]) proved that, for the hyperspace of a normal space endowed with the Vietoris topology, all the following properties are equivalent:

- first countability
- being Fréchet
- sequentiality
- countable tightness.

ČOBAN [1971] characterized the first countability of the Vietoris topology on $CL(X)$ for a normal space X, see Theorem 2.3 below.

2.1. DEFINITION. Given a cardinal function $f(X)$, denote by $hf(X)$ the hereditary version of $f(X)$, i.e.:

$$hf(X) = \sup\{f(Y) : Y \subset X\};$$

Define also the open hereditary version of $f(X)$ as

$$h_o f(X) = \sup\{f(Y) : Y \text{ open in } X\}.$$

1. *Tightness, character and pseudocharacter*

It is easy to verify (see HOLÁ and LEVI [1997], COSTANTINI, HOLÁ and VITOLO [200?a]) that the character of the Vietoris (Fell, locally finite) topology can be calculated as the product of characters of the corresponding upper and lower parts. So it is useful to study the tightness and character of upper and lower parts of these topologies.

To describe the character of the co-compact and the Fell topologies, we define the compact cofinality, $ck(X)$, as the least cardinality of an infinite cofinal family in $K(X)$ (with respect to inclusion), so that $ck(X) \leq \omega$ means that X is hemicompact. The Lindelöf number $L(X)$ is bounded by $ck(X)$. For example, if X is the space Q of rational numbers then $L(X) = \omega$ but $ck(X) = \mathfrak{d} \geq \omega_1$ (see VAN DOUWEN [1984]). On the other hand, if X is locally compact then $ck(X) = L(X)$.

2.2. THEOREM (COSTANTINI, HOLÁ and VITOLO [200?a], MIZOKAMI [1976], HOU [1998]).

1. $\chi(CL(X), \mathbf{V}^-) = hd(X) \cdot \chi(X)$;
2. $\chi(CL(X), \mathbf{C}) = h_o ck(X)$;
3. $\chi(CL(X), \mathbf{V}^+) = \sup\{\chi(A, X) : A \in CL(X)\}$;
4. $\chi(CL(X), \mathbf{F}) = hd(X) \cdot \chi(X) \cdot h_o ck(X)$;
5. $\chi(CL(X), \mathbf{V}) = hd(X) \cdot \sup\{\chi(A, X) : A \in CL(X)\}$.

Some estimates for the character of the locally finite topology can be found in BELLA [1998] and DI MAIO, HOLÁ and PELANT [2001]. From the above theorem we have that the Fell topology is first countable if and only if X is first countable, hereditarily separable and each open set is hemicompact (see also HOLÁ and LEVI [1997], HOU [1998], BEER [1993b]). Note that there exist spaces, that are not first countable, whose all open sets are hemicompact (see ENGELKING [1989, Ex. 1.6.19]). So the first countability of the co-compact topology does not imply the first countability of the base space.

We mention now a result on the first countability of the Vietoris topology, improving ČOBAN [1971].

It was proved in HOLÁ and LEVI [1997] that the first countability of $(CL(X), \mathbf{V}^+)$ implies that X is normal and the derived set X' of X is countably compact. A natural example from DI MAIO, HOLÁ and PELANT [2001] shows that the sequentiality of $(CL(X), \mathbf{V}^+)$ is not enough to have X' countably compact. However, if X is a regular space then the Fréchetness of $(CL(X), \mathbf{V}^+)$ implies that X' is countably compact (see DI MAIO, HOLÁ and PELANT [2001]).

We have the following results concerning the first countability of the Vietoris topology and the locally finite one.

2.3. THEOREM (for X normal ČOBAN [1971], for X Hausdorff HOLÁ and LEVI [1997]). *The first countability of $(CL(X), \mathbf{V})$ is equivalent to the following conditions:*

(i) *X is perfectly normal;*

(ii) *the set of isolated points of X is countable;*

(iii) *X' is countably compact, hereditarily separable and $\chi(X', X) \leq \aleph_0$.*

In metrizable spaces, there is the following interesting characterization of the first countability using a notion of UC spaces (or Atsuji spaces, in recognition of his paper ATSUJI [1958]).

A metric space X is called UC space or Atsuji space if each continuous function on the space X with values in an arbitrary metric space is uniformly continuous. RAINWATER [1959] proved that a metrizable space X admits a compatible UC metric if and only if X' is compact.

2.4. THEOREM (DI MAIO and HOLÁ [1995], HOLÁ and LEVI [1997]). *Let X be a metrizable space. Then*

(i) $(CL(X), \mathbf{V}^+)$ *is first countable if and only if X is a topologically UC space* (i.e., X admits a compatible UC metric)*;*

(ii) $(CL(X), \mathbf{V})$ *is first countable if and only if X is a separable topologically UC space.*

2.5. THEOREM (DI MAIO, HOLÁ and PELANT [2001]). *The first countability of the space $(CL(X), \mathbf{LF})$ is equivalent to the following conditions:*

(i) *X' is countably compact and hereditarily separable;*

(ii) *$\chi(A, X) \leq \aleph_0$ for every $A \in CL(X)$;*

(iii) *X is normal.*

Notice that Theorem 2.5 improves results from NAIMPALLY and SHARMA [1988].
To consider tightness of hyperspace topologies we start with a generalization of MALYKHIN [1972].

2.6. PROPOSITION (COSTANTINI, HOLÁ and VITOLO[200?b]; the countable case MALYKHIN [1972], DI MAIO, HOLÁ and MECCARIELLO [200?]). *Let X be a normal space. Then $t(CL(X), \mathbf{V}^+) = \chi(CL(X), \mathbf{V}^+)$ and also $t(CL(X), \mathbf{V}) = \chi(CL(X), \mathbf{V})$.*

□ We prove only that $t(CL(X), \mathbf{V}) = \chi(CL(X), \mathbf{V})$ ($\mathbf{V}^+$-case is trivial).

Let $\kappa = t(CL(X), \mathbf{V})$. Let $A \in CL(X)$. Denote by

$$\mathcal{G} = \{\overline{U} : A \subset U,\ U \text{ open}\} \text{ and } \mathcal{F} = \{B : B \subset A, B \text{ finite }\}.$$

The normality of X implies that $A \in \mathbf{V}$-closure of $\mathcal{G}$. It is easy to verify that A belongs also to $\mathbf{V}$-closure of $\mathcal{F}$ and that $\chi(X) \leq \kappa$. Thus there are $\mathcal{G}' \subset \mathcal{G}$ and $\mathcal{F}' \subset \mathcal{F}$ such that $|\mathcal{G}'| \leq \kappa$, $|\mathcal{F}'| \leq \kappa$, $A \in \mathbf{V}$-closure of $\mathcal{G}'$ and also $A \in \mathbf{V}$-closure of $\mathcal{F}'$.

Put $L = \cup\{F : F \in \mathcal{F}'\}$. Then also $|L| \leq \kappa$. For every $x \in X$ let $\mathcal{B}(x)$ denote a base of open neighborhoods of x. The family

$$\{U^+ \cap \bigcap_{s \in S} B(s)^- : \overline{U} \in \mathcal{G}',\ S \subset L,\ S \text{ finite},\ B(s) \in \mathcal{B}(s)\}$$

is a local base of A in $(CL(X), \mathbf{V})$. Thus $\chi(CL(X), \mathbf{V}) \leq \kappa$. □

The tightness of the lower Vietoris topology is described using the subspace $X^{[n]}$, $n \in \omega$, of the product X^n consisting of all the n-tuples $(x_1, ..., x_n)$ where $x_i \neq x_j$ for $i \neq j$ and by $\mathcal{E}_n(X) = [X]^n$, i.e. the collection of all subsets of X of cardinality n.

2.7. THEOREM (COSTANTINI, HOLÁ and VITOLO [200?a]). *Let X be a Hausdorff topological space. Then $t(CL(X), \mathbf{V}^-) = hd(X) \cdot \sup_{m \in N} t(X^{[m]})$.*

□ Let $\kappa = t(CL(X), \mathbf{V}^-)$, $\lambda = hd(X)$ and $\mu = \sup_{m \in N} \mu_n$, where $\mu_n = t(X^{[n]})$ for each natural number n. First observe that we have

$$(*) \qquad \sup\{t(F, (CL(X), \mathbf{V}^-)) : F \in \mathcal{E}_n(X)\} = \mu_n.$$

Indeed, it is seen easily that any finite set $F = \{x_1, ..., x_n\} \subset X$ of cardinality n belongs to the $\mathbf{V}^-$-closure of $\mathcal{A} \subset CL(X)$ if and only if $(x_1, ..., x_n) \in \overline{\bigcup_{C \in \mathcal{A}} C^{[n]}}$, hence $t(F, (CL(X), \mathbf{V}^-)) \leq t(X^{[n]})$.

On the other hand $t(F, (CL(X), \mathbf{V}^-)) \geq t(F, (\mathcal{E}_n(X), \mathbf{V}^-))$. There is a local homeomorphism of $X^{[n]}$ onto $(\mathcal{E}_n(X), V^- \restriction \mathcal{E}_n(X))$ (see e.g. COSTANTINI, HOLÁ and VITOLO [200?a]). Thus we get $t(\mathcal{E}_n(X), V^-) = \mu_n$.

Also, it follows from $(*)$ that $\kappa \geq \mu$. Since $t(CL(X), V^-) \geq hd(X)$, we have $\kappa \geq \lambda \cdot \mu$.

Now let $C \in CL(X)$ and let D be a dense subset of C, with $|D| \leq \lambda$. Denote by $\mathcal{H}$ the $\mathbf{V}^-$-closure of some $\mathcal{G} \subset CL(X)$. Since $C \in \mathcal{H}$ if and only if $F \in \mathcal{H}$ for every finite subset F of D, and $t(F, (CL(X), V^-)) \leq \mu$ by $(*)$, we immediately conclude that $t(C, (CL(X), V^-)) \leq \lambda \cdot \mu$, and the result follows as the choice of C was arbitrary. □

We pass now to the tightness of the co-compact topology. We show first that it coincides with the open hereditary version of the k-Lindelöf number. We give then suitable estimates of it using other cardinal functions; such estimates will make it often easier to compute tightness of the co-compact topology.

We recall that a network for X is a collection $\mathcal{N}$ of subsets of X, such that for every $x \in X$ and every open subset Ω of X with $x \in \Omega$ there exists $N \in \mathcal{N}$ with $x \in N \subset \Omega$; the netweight, $nw(X)$, is the least cardinality of a network for X. Such a notion may be generalized in the following way (see also MCCOY and NTANTU[1988]): given any two collections $\mathcal{M}, \mathcal{N}$ of subsets of X, we say that $\mathcal{N}$ is a network for $\mathcal{M}$ if for every $M \in \mathcal{M}$ and every open subset Ω of X with $M \subset \Omega$ there exists $N \in \mathcal{N}$ with $M \subset N \subset \Omega$. The netweight of $\mathcal{M}$ in X, $nw(\mathcal{M}, X)$, is the least cardinality of a network for $\mathcal{M}$.

In particular we will say that a collection $\mathcal{N}$ of subsets of X is k-network for X (see MCCOY and NTANTU[1988]) if it is a network for $K(X)$; and we will define the k-netweight, $knw(X)$, to be $nw(K(X), X)$. It is clear that every k-network is also a network, and that if $\mathcal{B}$ is a base then the collection $\mathcal{N}$ of all finite unions of members of $\mathcal{B}$ is a k-network. Hence $nw(X) \leq knw(X) \leq w(X)$.

Both the previous inequalities may be strict (see COSTANTINI, HOLÁ and VITOLO [200?a]). Notice that the k-netweight is hereditary (as well as the netweight) and that for a p-space X we have $nw(X) = knw(X) = w(X)$ (cfr. GRUENHAGE [1984, Definition 3.15 and Theorem 4.2]).

We also introduce a new cardinal function $cknw(X)$, the cofinal k-netweight of X defined by:

$$cknw(X) = \min\{nw(\mathcal{H}, X) : \mathcal{H} \text{ is cofinal in } K(X)\}.$$

It is apparent from the definition that $cknw(X) \leq ck(X)$ and $cknw(X) \leq knw(X)$. Both these inequalities may be strict (see COSTANTINI, HOLÁ and VITOLO [200?a]).

Another cardinal function, which will play a crucial role, is the k-Lindelöf number $kL(X)$. Like the k-netweight, it belongs to a broad family of the "k-modifications" of previously defined notions (see also the concept of a weak k- development in ALLECHE,

ARHANGEL'SKII and CALBRIX [2000]). The earliest reference about $kL(X)$, we are aware of, is MCCOY and NTANTU[1988].

We will say that a collection $\mathcal{U}$ of (open) subsets of a space X is a (open) *k-cover* of X, if every compact subset of X is included in some member of $\mathcal{U}$. The k-Lindelöf number, $kL(X)$, is the least infinite cardinal κ such that every open k-cover $\mathcal{U}$ of X has a subcollection $\mathcal{V}$ with $\mathcal{V} \leq \kappa$, which is still a k-cover.

The following results are from COSTANTINI, HOLÁ and VITOLO [200?a].

2.8. PROPOSITION. *Let X be a Hausdorff topological space. Then*

$$L(X) \leq kL(X) \leq cknw(X).$$

2.9. THEOREM. *Let X be a Hausdorff topological space. Then* $t(CL(X), \mathbf{C}) = h_o kL(X)$.

To show further estimates for the tightness of the co-compact topology the following notion will be useful.

Let μ be a cardinal. A complete μ-system for a space X (FROLÍK [1960]) is a family $\{\mathcal{U}_k\}_{k\in\mu}$ of open covers such that if $\mathcal{W}$ is a collection of open sets with the finite intersection property and $\mathcal{W} \cap \mathcal{U}_k \neq \emptyset$ for each $k \in \mu$, then $\cap_{W\in\mathcal{W}} \overline{W} \neq \emptyset$.

2.10. DEFINITION. We call the *Čech number* of X the least infinite cardinal μ such that X has a complete μ-system, and denote it by $C(X)$. The family Γ of all open covers of X is a complete $|\Gamma|$-system, hence

$$C(X) \leq 2^{w(X)^{hL(X)}}.$$

If $C(X) = \omega$, we say that X is *quasi Čech-complete*; thus X is Čech-complete if and only if it is quasi Čech-complete and completely regular.

2.11. PROPOSITION. *If X is a regular space, then* $cknw(X) \leq C(X) \cdot L(X)$.

2.12. COROLLARY. *If X is a Hausdorff topological space, then* $hL(X) \leq t(CL(X), C) \leq h_o cknw(X)$.

2.13. COROLLARY. *If X is a regular space, then* $t(CL(X), \mathbf{C}) \leq hL(X) \cdot C(X)$.

□ Use Corollary 2.12 and the open hereditary version of the inequality stated in Proposition 2.11, taking into account that $h_o C(X) = C(X)$ for a regular space. □

As a consequence, if X is a regular quasi Čech-complete space then $t(CL(X), \mathbf{C}) = hL(X)$.

Observe further that in a locally compact space X, since $ck(X) = L(X)$, we have $h_o ck(X) = hL(X)$. Thus we have $\chi(CL(X), \mathbf{C}) = t(CL(X), \mathbf{C})$.

Concerning the tightness of the Fell topology we have the following result:

2.14. PROPOSITION. *Let X be a Hausdorff topological space. Then*
$t(CL(X), \mathbf{C}) \cdot t(CL(X), \mathbf{V}^-) = hd(X) \cdot \sup_{m\in N} t(X^{[m]}) \cdot h_o kL(X) \leq t(CL(X), \mathbf{F}) \leq hd(X) \cdot \min\{\chi(X) \cdot h_o kL(X), h_o ck(X) \cdot \sup_{m\in N} t(X^{[m]})\}$.

□ For the first inequality, we can use a similar argument as for the character. For the second inequality, we use the fact that $t(S, \tau' \vee \tau'') \leq \chi(S, \tau') \cdot t(S, \tau'')$ for two topologies τ', τ'' on the same set S, and the above results. □

It is not known whether the first inequality in Proposition 2.14 may be strict.
If X is metrizable, then the above proposition gives

$$t(CL(X),\mathbf{F}) = t(CL(X),\mathbf{V}^-)\cdot t(CL(X),\mathbf{C}) = w(X).$$

Let us conclude with another estimate from COSTANTINI, HOLÁ and VITOLO [200?a] for the tightness of the Fell topology, which turns out to be independent on that provided by Proposition 2.14 :

2.15. THEOREM. $t(2^X,\mathbf{F}) \le h_o cknw(X)\cdot \sup_{m\in\omega} hd(X^{[m]})$.

2.16. COROLLARY. $t(2^X,\mathbf{F}) \le h_o cknw(X)\cdot nw(X) \le knw(X)$.

Results concerning pseudocharacter of hyperspace topologies may be found in BELLA [1998], MIZOKAMI [1976], COSTANTINI, HOLÁ and VITOLO[200?b]. To mention some of them, we need the following notions.

Let $A \in CL(X)$. The pseudocharacter of X at A (ARHANGEL'SKII [1992]) is the minimal cardinality of a family $\mathcal{U}$ of open sets in X such that $\cap\mathcal{U} = A$; this cardinal number is denoted by $\psi(A,X)$.

We say that $\mathcal{F}$ is a *compact cover* in X if $\mathcal{F}\subset K(X)$ and $\mathcal{F}$ is a cover of X. We define

$$k(X) = min\{|\mathcal{F}| : \mathcal{F} \text{ is an infinite compact cover in } X\}.$$

By $\pi w(X)$ we mean the π-weight of X.

2.17. THEOREM (COSTANTINI, HOLÁ and VITOLO[200?b]). *Let X be a Hausdorff topological space. Then $\psi(CL(X),V) = \sup\{\psi(A,X) : A\in CL(X)\}\cdot h\pi w(X)$.*

2.18. THEOREM (COSTANTINI, HOLÁ and VITOLO[200?b]). *Let X be a Hausdorff topological space. Then $\psi(CL(X),F) = h_o k(X)\cdot h\pi w(X)$.*

2. *Cellularity and density*

It is known that $d(X) = d(CL(X),V)$ for any topological space X, see Remark 1.3
FEDORCHUK and TODORČEVIĆ [1997] proved the following result:

2.19. THEOREM. *For every compact space X we have*

$$c(CL(X),V) = \sup\{c(X^n) : n\in\omega\}.$$

Since $c(B) = c(X)$ for any dense subset B of X and because of Remark 1.3, the assertion of the above Theorem holds also for every Tychonoff space.
There are ZFC-examples of compact spaces X such that $c(X) < c(X\times X)$ (TODORČEVIĆ [1986]), thus it may happen that $c(X) < c(CL(X),V)$.
The cellularity of the locally finite topology was estimated by BELLA [1998]:

2.20. THEOREM. $c(CL(X),\mathbf{LF}) \le 2^{c(X)}$.

Note that taking as X a discrete space we obtain $c(CL(X),\mathbf{LF}) = 2^{c(X)} = 2^{|X|}$, while taking as X the Alexandroff compactification of a discrete space we obtain $c(CL(X),\mathbf{LF}) = c(X) = |X|$. The density of the hyperspace of a metric space equipped with the Hausdorff metric topology $\mathbf{H}_d$ and the locally finite topology $\mathbf{LF}$ was studied in BARBATI

and COSTANTINI [1997]; some estimates for the density of the locally finite topology **LF** for any topological space can be found also in BELLA [1998] and DI MAIO, HOLÁ and PELANT [2001]. To study the density of the hyperspace of a metric space, the notions of a totally bounded space in the generalized sense and a compact space in the generalized sense were introduced in BARBATI and COSTANTINI [1997].
A metric space (X, d) is said to be *totally bounded in the generalized sense* (briefly *GTB*) if for every $\varepsilon > 0$ there exists an ε-dense subset $N \subset X$ with $|N| < d(X)$.
Clearly, the cofinality of $d(X)$ of any GTB metric space X is countable.
A topological space X is *compact in the generalized sense* (briefly *GK*) if for every open cover $\mathcal{U}$ of X, there exists a subcover $\mathcal{H}$ such that $|\mathcal{H}| < d(X)$.

2.21. THEOREM (BARBATI and COSTANTINI [1997]). *If (X, d) is a metric space, then*

- $d(CL(X), \mathbf{H}_d) = 2^{<d(X)}$ *if X is GTB*
- $d(CL(X), \mathbf{H}_d) = 2^{d(X)}$ *if X is not GTB.*

2.22. THEOREM (BARBATI and COSTANTINI [1997]). *If X is a metric space, then*

- $d(CL(X), \mathbf{LF}) = 2^{<d(X)}$ *if X is GK*
- $d(CL(X), \mathbf{LF}) = 2^{d(X)}$ *if X is not GK.*

BARBATI and COSTANTINI [1997] noticed that a metrizable space X is not GK if and only if there is a closed discrete subset of cardinality equal to $d(X)$.

3. Consonance

A topological space X is said to be *consonant* if the co-compact topology $\mathbf{C}$ on $CL(X)$ coincides with the upper Kuratowski topology $\tau_{\mathbf{UK}}$; in the opposite case X is called *dissonant.*

The class of consonant spaces was introduced by DOLECKI, GRECO and LECHICKI [1995] and has been studied recently rather intensively. DOLECKI, GRECO and LECHICKI [1995] proved that every Čech-complete space is consonant.

We call a space X *hyperconsonant* if the Fell topology and the convergence topology $\tau_{\mathbf{Z}}$ coincide (ARAB and CALBRIX[1997]). The property $\mathbf{F} = \tau_{\mathbf{Z}}$ is called hyperconsonance. FLACHSMEYER [1964] showed that $\tau_{\mathbf{Z}} = \mathbf{F}$ for every locally compact space X. Topsoe asked when does the equation $\tau_{\mathbf{Z}} = \mathbf{F}$ hold and conjectured that the above result of Flachsmeyer could be reversed for separable metric spaces (Topsoe's question was cited in CHRISTENSEN [1974]). FREMLIN [1974] disproved Topsoe's conjecture and answered his question completely for metric spaces. This result of Fremlin remained unpublished, which partially explains why Topsoe's conjecture was disproved once more in SPAHN [1980].

Recently ARAB and CALBRIX [1994] proved the following extension of Fremlin's result.

3.1. THEOREM (for X metrizable FREMLIN [1974]). *Let X be a first countable, locally paracompact, T_3 space. Then X is hyperconsonant if and only if X possesses at most one point without a compact neighborhood.*

Of course, every hyperconsonant space is consonant. The space of irrationals is an example of consonant space, which is not hyperconsonant. ARAB and CALBRIX [1997] and ALLECHE and CALBRIX [1999] obtained that the property of consonance is equivalent to that of hyperconsonance for a class of spaces similar to that introduced in Theorem 3.1.

3.2. THEOREM (ALLECHE and CALBRIX [1999]). *If X is a Hausdorff topological space such that at most one point of X is without compact neighborhood, then X is consonant if and only if it is hyperconsonant.*

Studying the notion of consonance, DOLECKI, GRECO and LECHICKI [1995] called a collection $\mathcal{P}$ of subsets of a topological space X a *compact* family on X if the following two conditions are satisfied:

(1) If $A \in \mathcal{P}$ and G is an open subset such that $A \subset G$, then $G \in \mathcal{P}$;

(2) If $\mathcal{H}$ is a collection of open subsets of X such that $\cup\mathcal{H} \in \mathcal{P}$, then there exists a finite subcollection $\mathcal{H}' \subset \mathcal{H}$ such that $\cup\mathcal{H}' \in \mathcal{P}$.

They gave the following remarkable characterization: a collection $\mathcal{F}$ of closed subsets of a topological space X is an open subset for the topology $\tau_{\mathbf{UK}}$ if and only if the family $\mathcal{F}_C = \{X \setminus F : F \in \mathcal{F}\}$ is a compact family on X. This characterization allows them to obtain many results, and in particular: every regular quasi Čech-complete space is consonant.

A family $\mathcal{G}$ of open subsets of a topological space (X, τ) is compactly generated (see DOLECKI, GRECO and LECHICKI [1995]) if there exists $\mathcal{K} \subset K(X)$ such that

$$\mathcal{G} = \{G \in \tau : \text{there is } K \in \mathcal{K}, K \subset G\}.$$

A compactly generated family $\mathcal{G}$ is a compact family. Furthermore a family $\mathcal{G}$ of open subsets of X is a compactly generated if and only if $\mathcal{G}_C$ is **C**-open. Thus to see that X is not consonant, it suffices to find a compact family $\mathcal{G}$ of open subsets of X which is not compactly generated, that is there exists $G_0 \in \mathcal{G}$ such that for every compact subset $K \subset G_0$, $\{G \in \tau : K \subset G\} \setminus \mathcal{G} \neq \emptyset$.

It was proved in DOLECKI, GRECO and LECHICKI [1995] that consonance is not preserved by G_δ-sets, quotient maps and finite products. However consonance is hereditary with respect to closed and open subspaces.

Answering a question from DOLECKI, GRECO and LECHICKI [1995], NOGURA and SHAKHMATOV [1996] first proved that there exist metrizable dissonant spaces. They proved also that every locally Čech-complete space is consonant as well as every regular (locally) k_ω-space. (Recall that k_ω-spaces are those spaces whose topology is determined by countably many compact subsets.) Answering a question of NOGURA and SHAKHMATOV [1996], ALLECHE and CALBRIX [1999] showed that every Hausdorff locally k_ω-space is consonant.

ALLECHE and CALBRIX [1999] used the concept of Radon measure to study consonance and dissonance; they established a criterion of dissonance, later generalized by BOUZIAD [1996].

By a measure on a space X, we mean a finite measure defined on the Borel σ-algebra $\mathcal{B}(X)$ of X. A measure μ is a τ-additive measure on X if for every family $\mathcal{H}$ of open subsets of X and for every $\epsilon > 0$, there exists $\mathcal{H}' \subset \mathcal{H}$, $\mathcal{H}'$ finite, and $\mu(\cup\mathcal{H}') > \mu(\cup\mathcal{H}) - \epsilon$. Every measure in a hereditarily Lindelöf space is τ-additive. A measure μ on X is called a Radon measure if for each $B \in \mathcal{B}(X)$, we have

$$\mu(B) = \sup\{\mu(K) : K \in K(X), K \subset B\}.$$

Recall that a space X is called a Radon space (respectively a pre-Radon space) if each measure (respectively τ-additive measure) on X is Radon. Every Polish space is a Radon space and every Čech-complete space is a pre-Radon space.

3.3. THEOREM (BOUZIAD [1996]). *Every Hausdorff consonant space is pre-Radon.*

□ Let X be such a space and μ a finite τ-additive measure. Since μ is finite, to establish that μ is Radon it suffices to show that every open set $U \subset X$ is μ-Radon; i.e. $\mu(U) = \sup\{\mu(K); K \subset U, K \in K(X)\}$. Let $U_0 \subset X$ be an open set and $\epsilon > 0$ such that $\mu(U_0) > \epsilon$. We want to find a compact set $K \subset U_0$ such that $\mu(K) \geq \epsilon$. Let $\mathcal{H}$ be the family of all open sets $O \subset X$ such that $\mu(U) > \epsilon$. Since μ is τ-additive, the family $\mathcal{H}$ is compact. Hence there exists a compact set $K \subset U_0$, such that $U \in \mathcal{H}$ for every open set $U \subset X$, which contains K. Since the space X is Hausdorff we have $K = \cap\{\overline{U} : K \subset U,$ U open$\}$; hence, since μ is τ-additive, we obtain $\mu(K) \geq \epsilon$. □

3.4. COROLLARY (BOUZIAD [1996]). *The Sorgenfrey line S is a dissonant space.*

□ Recall that the Sorgenfrey line S is defined on the set R of reals and it is endowed with the topology τ_S generated by the collection $\{[a, b) : a < b; a, b \in R\}$.

It is well known that S is a hereditarily Lindelöf space such that every compact subset is a countable set. Furthermore the Borel σ-algebra in R for the Sorgenfrey topology coincides with that for the usual topology τ_u. Take a diffused probability measure μ in (R, τ_u). We have that $\sup\{\mu(K) : K \in K(R, \tau_S)\} = 0 < 1 = \mu(R)$. That is, S is not a pre-Radon space, and hence it is dissonant. □

Corollary 3.4 answers negatively a question of DOLECKI, GRECO and LECHICKI [1995]. Note that COSTANTINI, HOLÁ and VITOLO [200?a] obtained this result using only topological arguments.

BOUZIAD [1996] examined also the relationship between consonance and Prohorov's property. The topological space X is called a Prohorov space if for every compact $\mathcal{M} \subset \mathcal{P}(X)$ and every $\epsilon > 0$ there is a compact set $K \subset X$ such that $\mu(K) \geq 1 - \epsilon$ for every $\mu \in \mathcal{M}$ ($\mathcal{P}(X)$ denotes the space of all probability Radon measures on the σ-algebra of Borel sets in X).

3.5. THEOREM (BOUZIAD [1996]). *Let X be a completely regular consonant space. Then X is a Prohorov space.*

The above Theorem has two important consequences:

3.6. THEOREM (BOUZIAD [1996]). *Let X be a metrizable separable co-analytic space. The following are equivalent:*

1. *X is consonant:*
2. *X is Prohorov;*
3. *X is a Polish space.*

□ (1) $\Rightarrow$ (2) is Theorem 3.5.
(2) $\Rightarrow$ (3) is a result of PREISS [1973]. □

It follows from the above Theorem that the space Q of rationals is a dissonant space. The dissonance of rationals follows also from COSTANTINI and WATSON [1998] as they proved that every metrizable separable zero-dimensional space X without isolated points, such that every compact subset of X is scattered, is dissonant.

3.7. THEOREM (BOUZIAD [1996]). *Let X be a regular first countable consonant space. Then X is hereditarily Baire.*

□ Let X be such a space. Suppose that X contains a closed non-Baire subspace. It follows from a result of DEBS [1988] that X contains a closed subspace F which is homeomorphic to the space Q of rational numbers. Thus it follows from Theorem 3.6 that F is not consonant which is a contradiction. □

However there are dissonant hereditarily Baire separable metrizable spaces as was proved in ALLECHE and CALBRIX [1999] and COSTANTINI and WATSON [1998].

3.8. THEOREM (ALLECHE and CALBRIX [1999]). *If X is a regular hereditarily Lindelöf space and μ is a Radon measure on X, then every non-μ-measurable subspace A of X is a dissonant space.*

3.9. COROLLARY. *Every Bernstein space is a dissonant, hereditarily Baire, separable metrizable space.*

(A topological space X is called a Bernstein space if it is a subspace of a nonvoid Polish space X with no isolated point, such that if K is an uncountable compact subset of X, then $B \cap K \neq \emptyset \neq (X \setminus B) \cap K$.))
An interesting problem, posed by NOGURA and SHAKHMATOV [1996, Problem 11.4], is to find a non-completely metrizable consonant space. This is not possible in the realm of separable co-analytic spaces (Theorem 3.6) and the question is not decidable within the analytic spaces:

3.10. THEOREM (BOUZIAD [1999]). *The statement "all analytic metric consonant spaces are completely metrizable" is consistent and independent of the usual axioms of set theory.*

□ The reasoning reminds that used in VAN MILL, PELANT and POL [1996, Remark 5.2]. Let C be the Cantor set. By MARTIN and SOLOVAY [1970], it is consistent with MA+$\neg$CH that for every $A \subset C$ of the cardinality $\aleph_1$, a set $C \setminus A$ is analytic, but then $C \setminus A$ cannot be Čech-complete as $\omega < |A| < \mathfrak{c}$. However BOUZIAD proves that $C \setminus A$ is consonant.

On the other hand, KANOVEĬ and OSTROVSKIĬ [1981] presented a model of the set theory in which every analytic metrizable space, which is not completely metrizable, contains a closed homeomorphic copy of the rationals Q. Since Q is not consonant and consonance is hereditary with respect to closed subsets, every analytic metrizable consonant space is completely metrizable in this model. □

BOUZIAD [1993] proved that the hyperspace $K(X)$ of all nonempty compact subsets of a metrizable consonant space X, endowed with the Vietoris topology $\mathbf{V}$, is hereditarily Baire. Comparing this result with the above problem of Nogura and Shakhmatov, the following question of Bouziad becomes very natural: Does there exist a ZFC example of a non-completely metrizable space X such that $(K(X), \mathbf{V})$ is hereditarily Baire?
BOUZIAD, HOLÁ and ZSILINSZKY [2001] provided an affirmative answer to this question, making use of a ZFC construction of SAINT-RAYMOND [1994] of a non-completely metrizable space, each separable closed subspace of which is completely metrizable.
In fact, the following result is proved in BOUZIAD, HOLÁ and ZSILINSZKY [2001]:

3.11. PROPOSITION. *Let X be a completely regular space such that all compact subsets of X are separable and of countable character. If the separable closed subspaces of X are consonant then $(K(X), \mathbf{V})$ is hereditarily Baire.*

A natural question arises in this context: does there exist a non-consonant metrizable space X such that all separable closed subsets of X are completely metrizable?
BOUZIAD, HOLÁ and ZSILINSZKY [2001] constructed such a space under CH (in fact, they used an L-space from KUNEN [1981]):

3.12. THEOREM (BOUZIAD, HOLÁ and ZSILINSZKY [2001]). *Under CH, there exists a metrizable non-consonant space, each separable closed subspace of which is completely metrizable.*

BOUZIAD [1999] proved that continuous open surjections, defined on a consonant space, are compact-covering which gives a generalization of the classical theorem of Pasynkov stated for Čech-complete spaces. His results are based on a characterization of consonance using a special property of lower semicontinuous set-valued maps.
Following NOGURA and SHAKHMATOV [1996], if $H(X)$ is a sublattice of $CL(X)$, the space X is called H-trivial if $\tau_{UK} \restriction H(X) = \mathbf{C} \restriction H(X)$. Spaces that are CL-trivial are precisely the consonant spaces, called uK-trivial in NOGURA and SHAKHMATOV [1996]. The study of K-trivial and Fin-trivial spaces (recall that $Fin(X)$ stands for the set of all finite subsets of X) is initiated in NOGURA and SHAKHMATOV [1996].
BOUZIAD [200?] proved that if $C_k(X)$ is a Baire space or more generally if X has "the moving off property" of Gruenhage and Ma, then X is K-trivial. (A collection $\mathcal{K} \subset K(X)$ is called a moving off collection if, for any compact set $L \subset X$, there exist some $K \in \mathcal{K}$ disjoint from L. Following GRUENHAGE and MA [1997], we say that X has a *moving off property* provided every moving off collection of nonempty compact sets contains an infinite subcollection which has a discrete open expansion in X.)
If X is countable, then $C_p(X)$ is Baire if and only if X is Fin-trivial and all compact subsets of X are finite.
As for consonant spaces, it is shown in BOUZIAD [200?] that every regular K-trivial space is Prohorov and that this result remains true for any regular Fin-trivial space in which all compact subsets are scattered.
Examples of K-trivial non-consonant spaces, of Fin-trivial K-nontrivial spaces and of countably compact Prohorov Fin-nontrivial spaces, are described in BOUZIAD [200?]. In particular, it is shown that all (generalized) Fréchet-Urysohn fans are K-trivial, answering a question of NOGURA and SHAKHMATOV [1996].

4. Generalized metric properties of hyperspaces

In this part we will mention some results concerning monolithicity, developability, paracompactness and hereditary normality for hyperspaces.

1. Monolithicity and monotone normality

We start with the definition of a monolithic space given by ARHANGEL'SKII [1976]:

4.1. DEFINITION. Let X be a topological space and let κ be an infinite cardinal. X is called *κ-monolithic* if for every subset A of X such that $|A| \le \kappa$ we have $nw(\overline{A}) \le \kappa$, where $nw(A)$ is the netweight of A. X is called *monolithic*, if it is κ-monolithic for all cardinals κ.

Monolithicity is hereditary and ω-productive. Of course, for compact Hausdorff spaces, we can use weight instead of netweight in the above definition as $w(\overline{A}) = nw(\overline{A})$ in this case. Examples of monolithic spaces include: all metric spaces and all spaces of countable netweight.

ARHANGEL'SKII [1987] wondered when $(CL(X), V)$ is monolithic and whether X must be metrizable when its hyperspace is monolithic. BELL [1996] obtained the following results on these questions:

4.2. THEOREM (BELL [1996]). *Let X be a T_1-space. If $(CL(X), V)$ is monolithic then X is monolithic, hereditarily Lindelöf and compact.*

4.3. COROLLARY (MA_{ω_1}). *Let X be a compact Hausdorff space. $(CL(X), \mathbf{V})$ is monolithic if and only if X is metrizable.*

□ $\mathrm{MA}(\omega_1)$ implies that all first countable, ccc, compact Hausdorff spaces are separable (see JUHÁSZ [1979]), hence X has a countable weight and therefore is metrizable. □

4.4. THEOREM (BELL [1996]). *Let X be a compact orderable space. Then $(CL(X), \mathbf{V})$ is monolithic if and only if X is monolithic and hereditarily Lindelöf.*

Currently the main examples of non-metrizable, monolithic and HL compact Hausdorff spaces are a Souslin continuum and Kunen's compact L-spaces.
BELL proved that under the negation of the Souslin hypothesis, a Souslin continuum would be an example of a non-metrizable HL compact space with a monolithic hyperspace. BELL [1996] asked whether Kunen's compact L space, constructed under CH, is another example with these properties.

BRANDSMA and VAN MILL [1998b] showed that this need not be the case, modifying Kunen's construction to ensure that its hyperspace is non-monolithic. This shows also that the necessary conditions, given by Theorem 4.2, are not sufficient in general. Note that Kunen's L-space is indeed monolithic, as every separable subspace has a countable base, and its weight is ω_1.

For monotonically normal spaces, Theorem 4.4 was generalized by BRANDSMA and VAN MILL [1998a]. (Remind that a topological space X is called monotonically normal if X is T_1 and there exists for every pair x and U, where $x \in U$ and U is an open subset of X, an open set $\mu(x, U)$ with $x \in \mu(x, U) \subset U$ and the following two properties hold:

(a) if $U \subset V$, then $\mu(x, U) \subset \mu(x, V)$

(b) $\mu(x, X \setminus \{y\}) \cap \mu(y, X \setminus \{x\}) = \emptyset$.

It is well-known that monotonically normal spaces are hereditarily collectionwise normal and that every stratifiable and every generalized ordered space is monotonically normal.)

4.5. THEOREM (BRANDSMA and VAN MILL [1998a]). *Let X be a monotonically normal space. Then $(CL(X), \mathbf{V})$ is monolithic if and only if X is monolithic, compact and hereditarily Lindelöf.*

BRANDSMA and VAN MILL [1998a] proved also:

4.6. THEOREM. *Let X be a monotonically normal space. $(CL(X), \mathbf{V})$ is $\aleph_0$-monolithic if and only if X is $\aleph_0$-monolithic, compact and hereditarily Lindelöf.*

4.7. THEOREM. *The following are equivalent:*

1. *There is a nonmetric, monotonically normal space with an $\aleph_0$-monolithic hyperspace;*
2. *There is a Souslin line ($\neg$ SH).*

□ If (1) holds, then we know that X is hereditarily Lindelöf, compact and $\aleph_0$-monolithic. So X is not separable. This implies (2) by a result of WILLIAMS and ZHOU [1991].

For the other direction, if there is a Souslin line, by standard techniques we can make it compact, and such that every separable space is second-countable. Then Theorem 4.6 implies that its hyperspace is also $\aleph_0$-monolithic. □

The following result, proved in BRANDSMA and VAN MILL [1998a], is a nice preservation result for spaces with monolithic hyperspaces. This will have as a corollary the fact that there are (consistent) non-monotonically normal spaces with a monolithic hyperspace. Up to now all spaces with monolithic hyperspaces were monotonically normal and it was shown in BRANDSMA and VAN MILL [1998b] that Kunen's compact L-space from CH has a non-monolithic hyperspace. This latter space cannot be monotonically normal as it carries a nonseparable Radon measure (BRANDSMA and VAN MILL [1998a]).

4.8. THEOREM. *Let X be a compact space such that $(CL(X), \mathbf{V})$ is κ-monolithic, and let Y be a compact space such that $w(Y) \leq \kappa$. Then $(CL(X \times Y), \mathbf{V})$ is κ-monolithic.*

4.9. COROLLARY. *Assuming $\neg$ SH, there exists a non-monotonically normal space with an $\aleph_0$-monolithic hyperspace.*

2. *When does metrizability occur?*

FEDORCHUK's [1990] theorem, cited in HERRLICH and HUŠEK [1992, Theorem 4.20], gives as a special case that a compact Hausdorff space X is metrizable if and only if $(CL(X), \mathbf{V})$ is hereditarily normal. As normality of $(CL(X), \mathbf{V})$ implies compactness of X, this special case is closely related to the classical result due to KATĚTOV [1948] stating that a compact space X is metrizable iff $X \times X \times X$ is hereditarily normal. FEDORCHUK's proof is rather sketchy but a detailed proof for hyperspaces is given in BRANDSMA [1998].

So we arrive to the result, proved directly in BRANDSMA and VAN MILL [1996]:

4.10. THEOREM. *The hyperspace $(CL(X), \mathbf{V})$ is monotonically normal if and only if X is compact metrizable.*

For similar reasons, $(CL(X), \mathbf{V})$ is stratifiable if and only if X is compact metrizable and the same remains true if X is Hausdorff space and $(CL(X), \mathbf{V})$ is cosmic (i.e. it has a countable network).

HOLÁ, PELANT and ZSILINSZKY [200?] stated explicitly that if X is a Hausdorff space and $(CL(X), \mathbf{V})$ is a developable, or just a σ-space (i.e. it has a σ-discrete network), then $(CL(X), \mathbf{V})$ is metrizable, too.

We will present here proofs of the above results, using the following interesting fact discovered by POPOV [1978, Example 5]. A slightly different proof is taken from ISMAIL, PLEWIK and SZYMANSKI [2000].

4.11. THEOREM. *The space $[\omega]^\omega$ $(= \{A : A \subset \omega, |A| = \aleph_0\})$, equipped with the Vietoris topology induced from $CL(\omega)$, contains a subspace homeomorphic to the Sorgenfrey line S.*

□ Let Q denote the set of rational numbers with the discrete topology. Then $CL(Q)$ is homeomorphic to $CL(\omega)$. Let $X = \{C \in CL(\omega) : C \text{ is a cut }\}$. Recall that a proper subset C of Q is a cut if C has no largest element and for each $p \in C$, $(-\infty, p] \cap Q \subset C$. Also, if C and D are cuts and C is a proper subset of D, then we write $C < D$.

Let us show that the subspace X of $[\omega]^\omega$ is homeomorphic to the Sorgenfrey line S whose basic neighborhoods face to the left.

Let C and D be cuts such that $C < D$, and let E be a cut from $(C, D]$. Then for any $q \in E \setminus C$, take a Vietoris open set $W_q = E^+ \cap (\{q\})^-$. Then $W_q \cap X \subset (C, D]$. This shows that $(C, D]$ is open in X.

Conversely, let $W = A^+ \cap (\{a_1\})^- \cap \ldots \cap (\{a_n\})^- \cap X$, where $a_i \in A$ for every i and let $D \in W$. Let $a_{\bar{n}}$ be the largest element from $\{a_i : i = 1, 2, \ldots n\}$. Let $C = \{r \in Q : r < a_{\bar{n}}\}$. Then $C < D$ and $(C, D] \subset W$. This shows that W is open in the topology generated by sets of the form $(C, D]$. □

Recall that the Sorgenfrey line S is not a σ-space. Because developable regular spaces are σ-spaces (see GRUENHAGE [1984]), the hereditarity of these properties implies

4.12. COROLLARY. *$(CL(\omega), \mathbf{V})$ is not a σ-space. In particular, $(CL(\omega), \mathbf{V})$ is not developable.*

Recall that the metrizability number of a space X (ISMAIL, PLEWIK and SZYMANSKI [2000]), denoted by $m(X)$, is the smallest cardinal κ such that X can be represented as a union of κ many metrizable subspaces.

4.13. COROLLARY (ISMAIL, PLEWIK and SZYMANSKI [2000]). $m(CL(\omega), \mathbf{V}) = 2^{\aleph_0}$.

□ Let X be a subspace of $[\omega]^\omega$ which is homeomorphic to the Sorgenfrey line. Then X is hereditarily separable and the netweight of X is $2^{\aleph_0}$. Hence $m(X) = 2^{\aleph_0}$. □

4.14. THEOREM (HOLÁ, PELANT and ZSILINSZKY[200?]). *Let X be a Hausdorff topological space. The following are equivalent:*

1. $(CL(X), \mathbf{V})$ *has a σ-discrete network;*

2. $(CL(X), \mathbf{V})$ *is developable;*
3. $(CL(X), \mathbf{V})$ *is Moore;*
4. $(CL(X), \mathbf{V})$ *is metrizable;*
5. X *is compact and metrizable.*

□ Let us show first that (1) ⇒ (5). Under (1), X must be a countably compact space with a σ-discrete network; i.e., X must have a countable network. Thus X must be Lindelöf; i.e. X must be compact with a countable base.
(2) ⇒ (3) By Theorem 2.3, even the first countability of $(CL(X), \mathbf{V})$ implies the normality of X which gives the regularity of $(CL(X), \mathbf{V})$; and regular developable spaces are Moore spaces.
(3) ⇒ (5) If $(CL(X), \mathbf{V})$ is Moore, then X is a normal Moore space (regularity of $(CL(X), V)$ implies normality of X and developability is hereditary). □

4.15. THEOREM (ISMAIL, PLEWIK and SZYMANSKI [2000]). *Let X be a Hausdorff topological space such that $m(CL(X), \mathbf{V}) \leq \aleph_0$. Then X (hence $(CL(X), \mathbf{V})$) is compact and metrizable.*

HOLÁ, PELANT and ZSILINSZKY[200?] proved also that the developability of the Fell, locally finite and bounded Vietoris topology is equivalent to the metrizability.

By VELICHKO [1975], compactness, paracompactness, Lindelöfness and normality are equivalent for the Vietoris topology on $CL(X)$.
We can conclude from the above results that the hyperspace of closed sets equipped with the Vietoris topology is "too large" to possess interesting generalized metric properties without being compact and metrizable.
The situation is different for the Fell topology: normality, Lindelöfness and paracompactness are equivalent and these properties do not force the compactness of the underlying space as proved in HOLÁ, LEVI and PELANT [1999]:

4.16. THEOREM. *Let X be a Hausdorff topological space. The following are equivalent:*

1. X *is locally compact and Lindelöf;*
2. $(CL(X), \mathbf{F})$ *is σ-compact and regular;*
3. $(CL(X), \mathbf{F})$ *is Lindelöf;*
4. $(CL(X), \mathbf{F})$ *is paracompact;*
5. $(CL(X), \mathbf{F})$ *is normal.*

3. $K(X)$ and $Fin(X)$ with the Vietoris topology.

In contrast with the case for $CL(X)$, properties of $K(X)$ and $Fin(X)$, equipped with the Vietoris topology are much closer to those of X.
The problem of knowing whether $K(X)$ or $Fin(X)$ has a topological property $\mathcal{P}$, when the space X has it, has been studied for several years. The hereditarity of $K(X)$ and $Fin(X)$ with respect to many properties, such as metrizability and most separation axioms, is known already from MICHAEL [1951]. Recently, MIZOKAMI [1995], [1996] and

[1997] and FISHER, GARTSIDE, MIZOKAMI and SHIMANE [1997] studied this problem for some generalized metric spaces: Moore spaces, $w\Delta$-spaces, σ-spaces, spaces with G_δ-diagonal, Lasnev spaces, monotonically normal spaces, stratifiable spaces and others. Moore spaces are distinguished among generalized metrizable spaces for their nice behavior:

4.17. THEOREM (MICHAEL [1951]; MIZOKAMI [1995], [1997]). *A topological space X is Moore iff $K(X)$ is Moore iff $Fin(X)$ is Moore.*

For other classes of spaces, there are rather results showing that corresponding properties are not preserved, at least simultaneously, by $K(X)$ and $Fin(X)$. Sample results follow:

4.18. EXAMPLE (BORGES [1980], MIZOKAMI [1996]). There exists a stratifiable space X such that $K(X)$ is not a Σ-space.

4.19. THEOREM (MIZOKAMI and KOIWA [1987]). *X is stratifiable iff $Fin(X)$ is stratifiable.*

4.20. THEOREM (FISHER, GARTSIDE, MIZOKAMI and SHIMANE [1997]). *Let $K(X)$ be monotonically normal.*

1. *If X contains an infinite compact set then X is stratifiable.*
2. *If each non-empty open subset of X contains an infinite compact set then $K(X)$ is stratifiable.*

Hereditarity of weak developability and weak k-developability of the hyperspaces $Fin(X)$ and $K(X)$ endowed with the Vietoris topology was studied by ALLECHE [2001], who proved that a Tychonoff space X is weakly developable if and only if $Fin(X)$ is a weakly developable space and that $K(X)$ is weakly developable for a Tychonoff weakly k-developable X.

5. Completeness properties of hyperspaces

In this part we will mention mainly results concerning the Wijsman topology, which is one of the most important hyperspace topologies; at the end of this section we mention also results concerning completeness properties of the Vietoris topology on the hyperspace of compact sets.
The Wijsman topology has many interesting applications to analysis, measure theory and descriptive set theory (BEER [1993a], KECHRIS [1995]).
If (X, d) is a separable and complete metric space, then the Wijsman topology $\mathbf{W}_d$ on $CL(X)$ is Polish, i.e. separable and completely metrizable (see BEER [1991]).

Polish spaces are essential in descriptive set theory and its applications to measurable multifunctions and probability theory. Thus the problem of finding Polish topologies for the hyperspace of a Polish space, as considered by BEER [1991], is really of great importance.

5.1. REMARK. EFFROS [1965] already provided the effective result in this direction. He established that if X is a Polish space, then considering any totally bounded compatible metric ρ on X, the Hausdorff topology H_ρ relative to this metric on the hyperspace of X is Polish. Recall that the Hausdorff topology $\mathbf{H}_\rho$ and the Wijsman topology $\mathbf{W}_\rho$ coincide for a totally bounded metric space (X, ρ) (BEER [1993a]). Therefore the article of Effros is, in fact, the first to endow the hyperspace of a Polish space with a (suitable) Polish Wijsman topology.

COSTANTINI [1995] proved the following theorem:

5.2. THEOREM. *For every metric space (X, d), the space $(CL(X), \mathbf{W}_d)$ is Polish if and only if X is Polish.*

If the metric space (X, d) is not separable, then $(CL(X), \mathbf{W}_d)$ is neither separable nor metrizable, as each of these properties is in fact equivalent to the separability of the base space (see BEER [1993a]). However, it is worth wondering whether the complete metrizability of (X, d), or at least its actual completeness, can imply some suitable form of completeness for $(CL(X), \mathbf{W}_d)$.

ZSILINSZKY [1998b] proved the following result:

5.3. THEOREM. *Let X be a completely metrizable space. Then $(CL(X), \mathbf{W}_d)$ is a strong Choquet space for any compatible metric d on X.*

In presence of separability, this yields Costantini's result, and in general case one still concludes that the Wijsman hyperspaces are Baire. Zsilinszky asked if $(CL(X), \mathbf{W}_d)$ is hereditarily Baire, provided X is completely metrizable.

CHABER and POL [200?] provided an answer to the question of Zsilinszky. They proved the following two interesting results:

5.4. THEOREM. *Let X be a separable completely metrizable space, let d, e be metrics generating the topology of X, and let $\mathbf{W}_d, \mathbf{W}_e$ be the corresponding Wijsman topologies on the hyperspace $CL(X)$. If $\mathcal{A} \subset CL(X)$ is a countable set without $\mathbf{W}_d$-isolated points, then $\mathcal{A}$ is not closed with respect to $\mathbf{W}_e$.*

Notice that Theorem 5.4 implies Theorem 5.2 by the classical result due to Hurewicz.

Letting in Theorem 5.4, $d = e$, we see that for completely metrizable separable X, the space $(CL(X), \mathbf{W}_d)$ contains no closed copy of the rationals. Since $\mathbf{W}_d$ is a subfamily of the Effros Borel structure on $CL(X)$, the Wijsman hyperspace is absolutely Borel, and in effect, by Hurewicz's theorem, completely metrizable, cf. KECHRIS [1995, Theorem 12.6 and Corollary 21.21].

5.5. THEOREM (CHABER and POL [200?]). *Let X be a metrizable space such that the set of points in X without any compact neighborhood has weight $2^{\aleph_0}$. Then for any metric d generating the topology of X, $N^{2^{\aleph_0}}$ embeds as a closed subspace in $(CL(X), \mathbf{W}_d)$. In particular, the Wijsman hyperspace contains a closed copy of the rationals.*

□ Let $\mathcal{D}(X)$ be the family of metrics on X generating the topology. Let $d \in \mathcal{D}(X)$ and let $d(z, A)$ denote the distance function given by d. The assumptions about X yield $r > 0$ and $S \subset X$ of cardinality $2^{\aleph_0}$ such that

(1) $d(s, t) > r$ for $s, t \in S$, $s \neq t$, and no $s \in S$ has a compact neighborhood.

For $s \in S$, we set:

(2) $B_s = \{x : d(s,x) < r/4\}$, $E_s = \{x : d(s,x) \leq r/5\}$.

The neighborhood E_s of s is not compact, hence it contains a countable closed discrete set M_s. We write:

(3) $M_s = \{a_{s,n} : n \in N\} \subset E_s$, $M = \bigcup_{s \in S} M_s$, where $a_{s,n} \neq a_{s,m}$ for $n \neq m$.

Let N^S be the space of functions $u : S \to N$ with the pointwise topology; i.e. topologically - the Tychonoff product $N^{2^{\aleph_0}}$. We shall define $F : N^S \to CL(X)$ by the formula:

(4) $F(u) = \{a_{s,u(s)} : s \in S\} \cup (X \setminus \bigcup_{s \in S} B_s)$.

Then $\mathcal{F} = F(N^S)$ is closed in $(CL(X), \mathbf{W}_d)$ and $F : N^S \to \mathcal{F}$ is a homeomorphism, where $\mathcal{F}$ is considered with the relative Wijsman topology.
Finally, the space of rationals Q embeds onto a closed subspace of N^S (VAN DOUWEN [1984]). Therefore, the hyperspace $(CL(X), W_d)$ contains a closed copy of Q. □

Theorem 5.5 gives also an alternative justification of COSTANTINI'S result [1998], that the Wijsman hyperspace of the complete metric space may not be Čech-complete.

5.6. REMARK. COSTANTINI [1998] demonstrated that the Borel structure of the Wijsman hyperspace of a non-separable completely metrizable space X depends on the choice of a metric inducing the topology of X. The reasoning in the proof of Theorem 5.5 can also be used to that effect. To that end, let us consider the discrete space X of cardinality $\aleph_1$. Let e be the discrete metric on X, and let d be a metric on X generating the discrete topology, such that one can find in the metric space (X, d) sets with the properties (1), (2) and (3). Then $N^{\aleph_1}$ embeds onto a closed subset H of $(CL(X), \mathbf{W}_d)$. On the other hand, $(CL(X), \mathbf{W}_e)$ can be identified with the Cantor Cube $\{0,1\}^{\aleph_1}$ without the point having all coordinates zero. Since H is a Baire space without any dense Čech-complete subspace, H cannot be embedded as a Borel set into any compact space. In effect, H is closed with respect to $\mathbf{W}_d$ but not Borel with respect to $\mathbf{W}_e$.

A topological space X is called analytic if it is Hausdorff and there exists a continuous mapping from a Polish space onto X.

In a similar vein as COSTANTINI [1995], BARBATI [1993] proved that the Wijsman hyperspace of an analytic metrizable space X is analytic, for any choice of a compatible metric d on X.

We present a proof of this fact as was done by ALLECHE, AMARA and ARAB [200?], using a theorem due to FROLÍK [1970], which gives a powerful sufficient condition for a metrizable space to be analytic via measurable mappings from a Polish space.

5.7. THEOREM (BARBATI [1993], ALLECHE, AMARA and ARAB [200?]). *Let X be a metrizable analytic space. Then, for every compatible metric d, $(CL(X), \mathbf{W}_d)$ is a metrizable analytic space.*

□ Let P be a Polish space, and $f : P \to X$ a continuous mapping from P onto X. Take any compatible metric d_p on P. The space $(CL(P), \mathbf{W}_{d_p})$ is Polish (see Theorem 5.2). Consider the mapping $\zeta_f : CL(P) \to CL(X)$ defined as $\zeta_f(F) = \overline{f(F)}$. It is easy to verify that ζ_f is onto and $(\mathcal{E}(P) - \mathcal{E}(X))$-measurable, where $\mathcal{E}(Z)$ is the Effros σ-algebra on $CL(Z)$; i.e. the smallest σ-algebra on $CL(Z)$ generated by the collection $\{V^- : V$ is open in $Z\}$. It is known (BEER [1993a]) that if (Z, e) is a separable metric space, then $\mathcal{E}(Z)$ is the Borel σ-algebra generated by W_e. By Frolík's theorem, $(CL(X), \mathbf{W}_d)$ is analytic. □

5.8. REMARK. We will confront some facts about separable metric spaces with Remark 5.6. If (X, d) is a separable metric space then the Borel σ-algebra $\mathcal{B}(\mathbf{W}_d)$ does not depend on the choice of an equivalent metric. Really, it is quite straightforward to see that both the identity map Id : $(CL(X), \mathbf{W}_d) \to (CL(X), \mathbf{W}_e)$ and its inverse Id^{-1} map open sets to G_δ-sets for any two equivalent metrics d and e. Theorem 5.7 and Theorem 5.2 mean that a separable $\big(CL(X), \mathbf{W}_d\big)$ is analytic or even absolutely G_δ iff X has the same property. It would be nice to know something more about the preservation of higher absolute classes. However, possible affirmative results are actually very limited. SAINT RAYMOND [1978] proved a deep fact, that the Effros σ-algebra $\mathcal{E}(X)$ on $CL(X)$ is standard iff X is a union of a Polish space and a σ-compact space. It is connected to our topic because of Hess' theorem (BEER [1993a, Theorem 6.5.14]), which says that the Effros σ-algebra on $CL(X)$ coincides with the Borel σ-algebra $\mathcal{B}(\mathbf{W}_d)$ for each separable metric space (X, d). So using SAINT RAYMOND [1978], we obtain that if $Z = Q \times N^N$, where Q is the space of rationals and N^N is the space of irrationals, then for any equivalent metric d on Z, $(CL(Z), \mathbf{W}_d)$ is analytic by Theorem 5.7 but it does not belong to any absolute Borel class.

Concerning completeness properties of further topologies, let us just point out that Baireness of $(CL(X), \mathbf{V})$ was studied by MCCOY [1975]; some sufficient conditions for Baireness and α-favorability of the Fell topology are given in HOLÁ and ZSILINSZKY [2001]- they extend and complement results of ZSILINSZKY [1996], [1998a].
In the last part of this section, we will mention results on completeness properties of the hyperspace of compact sets equipped with the Vietoris topology.

It is well known (KURATOWSKI [1966]) that $(K(X), \mathbf{V})$ is completely metrizable if and only if X is completely metrizable, and that $(K(X), \mathbf{V})$ is Čech-complete if and only if X is Čech-complete (ZENOR [1970], ČOBAN [1971]).
So we have also that $(K(X), \mathbf{V})$ is Polish if and only if X is Polish.
HOLÁ and ZSILINSZKY [200?] proved that the same is true also for sieve completeness. Sieve complete spaces (MICHAEL [1977]) are continuous open images of Čech-complete spaces. In particular, Čech-complete spaces are sieve complete. On the other hand, paracompact sieve complete spaces are Čech-complete (MICHAEL [1977]).

5.9. THEOREM (HOLÁ and ZSILINSZKY [200?]). *The following are equivalent:*

1. $(K(X), \mathbf{V})$ *is sieve complete;*
2. X *is sieve complete.*

□ Only (2) $\Rightarrow$ (1) needs some explanation: let $f : Z \to X$ be an open continuous mapping from a Čech-complete space Z onto X. Define $F : (K(Z), \mathbf{V}) \to (K(X), \mathbf{V})$ as $F(K) = f(K)$ for each $K \in K(X)$. Then F is continuous, and since f is compact-covering (ENGELKING [1989, Problem 5.5.11(e)], F is onto.

Also, F is an open mapping, since if $\mathbf{U} = \langle U_0, ..., U_n \rangle \in \mathbf{V(Z)}$, then $F(\mathbf{U}) = \langle f(U_0), ..., f(U_n) \rangle = \mathbf{G}$. Indeed, clearly $F(\mathbf{U}) \subset \mathbf{G}$. On the other hand, if $K \in \mathbf{G}$, we can find some $x_i \in K \cap f(U_i)$ and a corresponding $z_i \in U_i$ with $f(z_i) = x_i$ for each $i \leq n$. Now, $U = \bigcup_{i \leq n} U_i \subset Z$ is Čech-complete and $f \restriction U$ is a continuous open mapping of U into $\bigcup_{i \leq n} f(U_i)$, and hence compact-covering. This means that for some Z-compact $L_0 \subset U$, $f(L_0) = K$ and therefore $L = L \cup \{z_0, ..., z_n\} \in \mathbf{U}$ and $F(L) = K$. Finally, since $(K(Z), \mathbf{V})$ is Čech-complete, we see that $(K(X), \mathbf{V})$ is sieve complete. □

Note that hereditary Baireness, which is another completeness property, behaves differently: hereditary Baireness of X is only necessary but not sufficient for hereditary Baireness of $(K(X), \mathbf{V})$.

5.10. REMARK. In fact, since for every $n \in N$ the natural mapping $S_n : X^n \to \mathcal{F}_n(X)$ ($\mathcal{F}_n(X) = \{F \in CL(X) : |F| \leq n\}$) is perfect, $\mathcal{F}_n(X)$ is a closed subspace of $(K(X), \mathbf{V})$, and a regular space which is a perfect preimage of a hereditarily Baire metric space is itself hereditarily Baire BOUZIAD [1997]. The hereditary Baireness of $(K(X), \mathbf{V})$ for a metrizable X implies the hereditary Baireness of X^n for every $n \in N$. Thus if we take the hereditarily Baire (separable) metric space X from AARTS and LUTZER [1973] with a non-hereditarily Baire square X^2, then $(K(X), \mathbf{V})$ is not hereditarily Baire. Some sufficient conditions for hereditary Baireness of $(K(X), \mathbf{V})$ are given in Proposition 3.11.

CHRISTENSEN [1974] and SAINT RAYMOND [1973] proved the following remarkable

5.11. THEOREM. *For a metrizable space Y, the following properties are equivalent:*

1. *Y is Polish.*
2. *There is a Polish space X and a map $F : K(X) \to K(Y)$ such that*
   i) *F is monotone, i.e. $(A, B \in K(X), A \subset B) \Rightarrow F(A) \subset F(B)$.*
   ii) *$F(K(X))$ is cofinal in $K(Y)$.*
3. *$(K(Y), \mathbf{V})$ is analytic.*

Let us mention that properties of the map F in *(2)* of Theorem 5.11 are closely related to Tukey functions which were examined e.g. in FREMLIN [1991a] and [1991b] with many results on the descriptive structure of separable metrizable spaces.
We mention interesting generalizations of Theorem 5.11. We shall need:

5.12. DEFINITION. Let X be a topological space. X is said to be a *q-space* if every point from X has a sequence $\{U_n\}_{n \in \omega}$ of neighborhoods such that: if $x_n \in U_n$ for each $n \in \omega$, then $\{x_n\}$ has a cluster point in X.

5.13. DEFINITION.

(a) A subset $A \subset X$ is said to be *bounded* if each locally finite family (in X) has a finite restriction to A.

(b) X is said to be *μ-complete* if every closed and bounded subset of X is compact.

Recall that each first-countable space is a q-space and that both paracompactness and realcompactness imply μ-completeness. For a completely regular space X, a subset A of X is bounded, as defined above, iff all real-valued functions defined on X are bounded on A.

5.14. THEOREM (DUBE and VALOV [2001]). *Let Y be a regular μ-complete q-space. $\mathcal{P}$ denotes either Čech-completeness or local compactness. The following conditions are equivalent:*

(a) *Y is a Lindelöf space with the property $\mathcal{P}$.*

(b) *There is a Lindelöf space $X \in \mathcal{P}$ and a monotone map $F : K(X) \to K(Y)$ such that $F(K(X))$ is cofinal in $K(Y)$.*

(c) *There is a separable metrizable space $M \in \mathcal{P}$ and a monotone map $F : K(M) \to K(Y)$ such that $F(K(M))$ is cofinal in $K(Y)$.*

Using continuity, BOUZIAD and CALBRIX [1996] proved earlier:

5.15. THEOREM. *A regular q-space Y is Lindelöf Čech-complete iff there is a continuous map f from a Lindelöf Čech-complete space X into $(K(Y), \mathbf{V}^+)$ such that $f(X)$ is cofinal in $K(Y)$.*

6. Compactness in hyperspaces

1. Around countable compactness

It is well-known that the compactness of the whole hyperspace $CL(X)$, equipped with some of the topologies mentioned above, is equivalent to the compactness of X.

Concerning countable compactness, the situation is different. Countable compactness of the Hausdorff metric and Wijsman topologies on $CL(X)$ is equivalent to their compactness since they are defined for a metric space X.

It was proved in HOLÁ and KÜNZI [1998] that $(CL(X), \mathbf{F})$ is countably compact if and only if X is countably compact.

There is no satisfactory characterization of countable compactness of the Vietoris topology on $CL(X)$ in the literature. As far as we know the only papers in which we can find some results concerning this subject are Keesling's articles KEESLING [1970a], [1970b] and GINSBURG [1975a], [1975b] and NATSHEH [1998].

Of course, it is easy to see (by using of the same argumentation as in Remark 5.10) that the countable compactness of $(CL(X), \mathbf{V})$ implies that all finite powers of X must be countably compact. GINSBURG [1975a] provided an example of a completely regular space X, all finite powers of which are countably compact, such that $(CL(X), \mathbf{V})$ fails to be countably compact.

KEESLING [1970a] showed that $(CL(X), \mathbf{V})$ can be countably compact and noncompact, as he presented some sufficient conditions for countable compactness of the Vietoris topology which fail to force it to be compact. But these conditions - X is a T_1 normal ω-bounded space - give simultaneously also ω-boundedness of the Vietoris topology.

A natural question arises whether countable compactness and ω-boundedness are different for the Vietoris topology on $CL(X)$. It has been shown in HOLÁ and KÜNZI [1998] that this is the case.

The situation for $(K(X), \mathbf{V})$ is different; countable compactness and ω-boundedness coincide in $(K(X), \mathbf{V})$ (MILOVANCEVIC [1985]). In fact, they are equivalent to the condition that each σ-compact set in X has a compact closure. It is readily seen that for spaces X, where every $K \in K(X)$ is separable, ω-boundedness of X is equivalent to ω-boundedness of $(K(X), \mathbf{V})$; the space $X = \Sigma(0)$ – the Σ-product of $[0,1]^{\aleph_1}$ – is an example of an ω-bounded space, for which $(K(X), \mathbf{V})$ is not ω-bounded.

2. *Relative compactness for the Vietoris topology*

The problem when a given subspace of a hyperspace is compact or relatively compact turns out to be quite different and, of course, more complicated than characterizing the compactness of the whole hyperspace. COSTANTINI, LEVI and PELANT [200?] deal with this subject.
The results concerning the (relative) compactness for subsets of the Vietoris hyperspace were obtained independently and almost simultaneously (in a different form, and using other techniques) by O'BRIEN and WATSON [1998]. The results about (relative) compactness for subsets of the Wijsman hyperspace have been used in a slightly wider context by SIANESI [1999].

6.1. THEOREM (COSTANTINI, LEVI and PELANT [200?]). *Let X be a regular space and $\mathcal{K}$ a subset of $CL(X)$. Then $\mathcal{K}$ is compact with respect to the Vietoris topology if and only if it is closed and satisfies the following condition:*

$$(*) \quad \begin{array}{l} \forall C \in CL(X) : \forall \mathcal{U} \text{ open cover of } C \text{ in } X: \\ \exists \mathcal{F} \text{ finite subcollection of } \mathcal{U}: C^- \cap \mathcal{K} \subset \bigcup_{F \in \mathcal{F}} F^-. \end{array}$$

A subset Y of a topological (T_2)-space X is said to be *weakly relatively compact*, if every net in Y has a subnet converging to a point of X (see O'BRIEN and WATSON [1998, Definition 4]). Y is said to be *strongly relatively compact* if $\overline{Y}$ is compact in X.

6.2. THEOREM (COSTANTINI, LEVI and PELANT [200?]). *Let X be a regular space and $\mathcal{K}$ a subset of $CL(X)$. Then $\mathcal{K}$ is weakly relatively compact in $(CL(X), \mathbf{V})$ if and only if it satisfies condition $(*)$ of Theorem 6.1. Moreover, if X is normal, then condition $(*)$ is equivalent to the strong relative compactness of $\mathcal{K}$.*

3. *Relative compactness for the Wijsman topology*

We are going to mention briefly the Wijsman topology. In contrast with the Vietoris topology, the empty set $\emptyset$ need not be isolated in $(2^X, \mathbf{W}_d)$. The following facts may be found in LECHICKI and LEVI [1987]:

6.3. THEOREM. *Let (X, d) be a metric space.*

- $(CL(X), \mathbf{W}_d)$ *is compact* $\Longleftrightarrow$ (X, d) *is compact;*
- $(2^X, \mathbf{W}_d)$ *is compact* $\Longleftrightarrow$ *each closed ball in* (X, d) *is compact.*

We should recall that the standard convention defining $\text{dist}(x, \emptyset) = +\infty$ for all $x \in X$ is used here to define $\mathbf{W}_d$ on 2^X. One could meet also different convention using $\text{dist}(x, \emptyset) = \sup\{d(x, y) \ : \ y \in X\}$ for all $x \in X$. An analysis of differences between these two conventions may be found e.g. in COSTANTINI, LEVI and PELANT [200?], where the Wijsman topology on 2^X defined with the latter convention is denoted as $\widetilde{\mathbf{W}}_d$.

6.4. THEOREM (COSTANTINI, LEVI and PELANT [200?]).

Let (X, d) be a metric space and $\mathcal{K}$ a subset of 2^X. Then $\mathcal{K}$ is compact with respect to the topology induced by $\mathbf{W}_d$ if and only if it is $\mathbf{W}_d$-closed in 2^X, and satisfies the following condition (where $S_d(x, r)$ denotes an open ball $\{y \in X : d(x, y) < r\}$):

$$(\Delta) \qquad \begin{array}{l} \forall x \in X\, \forall r > 0\, \forall \varepsilon > 0\, \forall \mathcal{U} \text{ open cover of } S_d(x, r+\varepsilon) \text{ in } X: \\ \exists \mathcal{F} \text{ finite subcollection of } \mathcal{U}: \{C \in \mathcal{K} : d(x, C) \le r\} \subseteq \bigcup_{F \in \mathcal{F}} F^- \end{array}$$

Clearly, the weak relative compactness and the strong relative compactness coincide for completely regular spaces, and $(2^X, \mathbf{W}_d)$ is one of them.

6.5. THEOREM (COSTANTINI, LEVI and PELANT [200?]). *Let (X, d) be a metric space and $\mathcal{K}$ a subset of 2^X. Then $\mathcal{K}$ is relatively compact in $(2^X, \mathbf{W}_d)$ if and only if it satisfies condition (Δ) of Theorem 6.4.*

4. Final remark

At the end of this survey, we would like to mention at least some important references concerning hit-and-miss topologies and weak topologies generated by gap and excess functionals on hyperspaces: BEER and LUCCHETTI [1993], HOLÁ and LUCCHETTI [1996], LOWEN and SIOEN [1996], [1998], NAIMPALLY [200?], ZSILINSZKY [1996], [1998a] and references therein.

References

AARTS, J.M. and D.J. LUTZER
[1973] The product of totally nonmeager spaces, *Proc. Amer. Math. Soc.* **38**, 198–200.

ALLECHE, B.
[2001] Weakly developable and weakly k-developable spaces and the Vietoris topology, *Topology Appl.* **111**, 3–19.

ALLECHE, B., A. AMARA and M. ARAB
[200?] The Wijsman topology of an analytic space is analytic, preprint.

ALLECHE, B., A.V. ARHANGEL'SKII and J. CALBRIX
[2000] Weak developments and metrization, *Topology Appl.* **100**, 23–38.

ALLECHE, B. and J. CALBRIX
[1995] Topology superieure de Kuratowski et mesure de Radon, preprint.
[1999] On the coincidence of the upper Kuratowski topology with the cocompact topology, *Topology Appl.* **93**, 207–218.

ARAB, M. and J. CALBRIX
[1994] Sur la topologie de Fell et la topologie de la convergence, *C.R. Acad. Sci. Paris Ser. I* **318**, 549–552.
[1997] Comparison of some classical topologies on hyperspaces within the framework of point spaces, *Set-Valued Anal.* **5**, 47–55.

ARHANGEL'SKII, A.V.
[1976] On some topological spaces that occur in functional analysis, *Russian Math. Surveys* **31**, 14–30.
[1987] Topological homogenity, topological groups and their continuous images, *Russian Math. Surveys* **42**, 83–131.
[1992] *Topological Function Spaces*, Kluwer Academic Publishers.

ATSUJI, M.
[1958] Uniform continuity of continuous functions of metric spaces, *Pacific J. Math.* **8**, 11–16.

ATTOUCH, H.
[1984] *Variational Convergence for Functions and Operators*, Pitman, Boston,

BARBATI, A.
[1993] The hyperspace of an analytic metrizable space is analytic, *Rendiconti dell' Istituto di Matematica dell'a Universita di Trieste* **XXV**, 15–21.

BARBATI, A. and C. COSTANTINI
[1997] On the density of the hyperspace of a metric space, *Comment. Math. Univ. Carolinae* **38**, 349–360.

BEER, G.
[1991] A Polish topology for the closed subsets of a Polish space, *Proc. Amer. Math. Soc* **113**, 1123–1133.
[1993a] *Topologies on Closed and Closed Convex Sets*, Kluwer, Dordrecht
[1993b] On the Fell Topology, *Set-Valued Anal.* **1**, 69–80.

BEER, G.A., C.J. HIMMELBERG, K. PRIKRY and F. S. VAN VLECK
[1987] The locally finite topology on 2^X, *Proc. Amer. Math. Soc.* **101**, 168–172.

BEER, G.A. and R. LUCCHETTI
[1993] Weak topologies on the closed subsets of a metrizable space, *Trans. Amer. Math. Soc.* **335**, 805–822.

BELL, M.
[1996] The hyperspace of a compact space I, *Top. Appl.* **72**, 39–46.

BELLA, A.
[1998] Some cardinality properties of a hyperspace with the locally finite topology, *Proc. Amer. Math. Soc.* **104**, 1274-1278.

BORGES, C.R.
[1980] Normality of hyperspaces, *Math. Japonica* **25**, 507–510.

BOUZIAD, A.
[1993] A note on consonance of G_δ-sets, *Top. Appl.* **87**, 53–61.
[1996] Borel measures in consonant spaces, *Top. Appl.* **70**, 125–132.
[1997] Preimage d'espaces hereditairement de Baire, *Fund. Math.* **153**, 191–197.
[1999] Consonance and topological completeness in analytic spaces, *Proc. Amer. Math. Soc.* **127**, 3733–3737.
[200?] Coincidence of the upper Kuratowski topology with the co-compact topology on compact sets, and the Prohorov property, *Top. Appl.*, to appear.
[200?] H-trivial spaces, preprint.

BOUZIAD, A. and J. CALBRIX
[1996] Čech-complete spaces and the upper topology, *Top. Appl.* **71**, 119–124.

BOUZIAD, A., Ľ. HOLÁ and L. ZSILINSZKY
[2001] On hereditary Baireness of the Vietoris topology, *Topology Appl.* **115**, 247–258.

BRANDSMA, H.

[1998] *Monolithic hyperspaces*, PhD. Thesis, Vrije Universiteit, Amsterdam.

BRANDSMA, H. and J. VAN MILL

[1996] Monotonically normal hyperspaces are metrizable, Report WS-469, Vrije Universiteit, Amsterdam.

[1997] Every Kunen-like L-space has a non-monolithic hyperspace, *Topology Proc.* **22**, 15–24.

[1998a] Monotone normality, measures and hyperspace, *Topology Appl.* **85**, 287–298.

[1998b] A compact HL space need not have a monolithic hyperspace, *Proc. Amer. Math. Soc.* **126**, 3407–3411.

CHABER, J.

[1976] Conditions which imply compactness in countably compact spaces, *Bull. Acad. Pol. Sci. Ser. Math.* **24**, 993–998.

CHABER, J. and R. POL

[200?] Note on the Wijsman hyperspaces of completely metrizable spaces, preprint.

CHRISTENSEN, J.P.R.

[1974] *Topology and Borel Structure*, North-Holland, Amsterdam.

ČOBAN, M.

[1971] Note sur topologie exponentielle, *Fundam. Math.* **71**, 27–42.

COSTANTINI, C.

[1995] Every Wijsman topology relative to a Polish space is Polish, *Proc. Amer. Math. Soc.* **123**, 2569–2574.

[1998] On the hyperspace of a non-separable metric space, *Proc. Amer. Math. Soc.* **126**, 3393–3396.

COSTANTINI, C., Ľ. HOLÁ and P. VITOLO

[200?a] Tightness and character of hyperspace topologies, preprint.

[200?b] preprint.

COSTANTINI, C., S. LEVI and J. PELANT

[200?] Compactnes and local compactness in hyperspaces, *Topology Appl.*, to appear.

COSTANTINI, C. and S. WATSON

[1998] On the dissonance of some metrizable spaces, *Topology Appl.* **84**, 259–268.

DEBS, G.

[1988] Espaces hereditairement de Baire, *Fund. Math.* **129**, 199–206.

DI CONCILIO, A.

[1992] Uniform properties and hyperspace topologies for α-uniformities, *Top. Appl.* **44**, 115–123.

DI MAIO, G. and Ľ. HOLÁ

[1995] On hit–and–miss topologies, *Rend. Acc. Sc. fis. mat. Napoli* **LXII**, 103–124.

DI MAIO, G., Ľ. HOLÁ and E. MECCARIELLO

[200?] Some compactness properties in hyperspaces: A new approach, preprint.

DI MAIO, G., Ľ. HOLÁ and J. PELANT

[2001] Properties related to the first countability of hyperspace topologies, *Questions and Answers in General Topology* **19**, 139–157.

DOLECKI, S., G.H. GRECO and A. LECHICKI

[1995] When do the upper Kuratowski topology(homeomorphicaly, Scott topology) and the co-compact topology coincide?, *Trans. Amer. Math. Soc.* **8** , 2869–2884.

VAN DOUWEN, E.

[1984] The integers in topology, in *Handbook of Set-Theoretic Topology* (K. Kunen and J.E. Vaughan, eds.) North Holland, Amsterdam, 116–167.

DUBE, T. and V. VALOV

[2001] Generalized tri-quotient maps and Čech-completeness, *Comment. Math. Univ. Carolin.* **42**, 187–194.

EFFROS, E.

[1965] Convergence of closed subsets in a topological space, *Proc. Amer. Math. Soc.* **16**, 929–931.

ENGELKING, R.

[1989] *General Topology*, Helderman, Berlin,

FEDORCHUK, V.V.

[1990] On some geometric properties of functors, *Rend. Circ. Mat. Palermo Suppl.* **24**, 73–78.

FEDORCHUK, V.V. and S. TODORČEVIĆ

[1997] Cellularity of covariant functors, *Topology Appl.* **76**, 125–150.

FISCHER, S., P. GARTSIDE, T. MIZOKAMI and N. SHIMANE

[1997] Near metric properties of hyperspace, *Topology Proc.* **22**, 197–211.

FLACHSMEYER, J.

[1964] Verschiedene Topologisierungen im Raum der abgeschLossenen Mengen, *Math. Nachrichten* **26**, 321–337.

FREMLIN, D.

[1974] Topologies on spaces of subsets, unpublished manuscript.

[1991a] Families of compact sets and Tukey's ordering, *Atti Sem. Mat. Fis.-Univ. Modena* **39**, 29–50.

[1991b] The partially ordered sets of measure theory and Tukey's ordering, *Note Mat.* **11**, 177–214.

FROLÍK, Z.

[1960] Generalizations of the G_δ-property of complete metric spaces, *Czechoslovak Mathematical Journal* **10 (85)**, 359-378

[1970] A measurable map with analytic domain and metrizable range is quotient, *Bull. Amer. Math. Soc.* **76**, 1112–1117.

GINSBURG, J.

[1975a] Some results on the countable compactness and pseudocompactness of hyperspaces, *Can. J. Math.* **XXVII**, 1392–1399.

[1975b] On the Stone-Cech compactification of the space of closed sets, *Trans. Amer. Math. Soc.* **215**, 301–311.

GRUENHAGE, G.

[1984] Generalized metric spaces, in *Handbook of Set-Theoretic topology*, edited by K. Kunen and J. Vaughan, North Holland, Amsterdam, 423–501.

GRUENHAGE, G. and D.K. MA

[1997] Baireness of $C_k(X)$ for locally compact X, *Topology Appl.* **80**, 131–139.

GULDEN, S. L., W.M. FLEISCHMAN and J.H. WESTON

[1970] Linearly Ordered Topological Spaces, *Proc. Amer. Math. Soc.* **24**, 760–766.

HERRLICH, H. and M. HUŠEK

[1992] Categorical Topology, in *Recent Progress in General Topology*, edited by M. Hušek and J. van Mill, North-Holland, 370–403.

HOLÁ, Ľ. and H.P. KÜNZI

[1998] Properties related to compactness in hyperspaces, *Topology Proc.* **23**, 191–205.

HOLÁ, Ľ. and S. LEVI

[1997] Decomposition Properties of Hyperspace Topologies, *Set-Valued Anal.* **5**, 309–321.

HOLÁ, Ľ., S. LEVI and J. PELANT

[1999] Normality and paracompactnees of the Fell topology, *Proc. Amer. Math. Soc.* **127**, 2193–2197.

HOLÁ, Ľ. and R. LUCCHETTI

[1996] Polishness of weak topologies generated by gap and excess functionals, *Journal of Convex Analysis* **3**, 283–294.

HOLÁ, Ľ., J. PELANT and L. ZSILINSZKY

[200?] Developable hyperspaces are metrizable, preprint.

HOLÁ, Ľ. and L. ZSILINSZKY

[2001] Completeness properties of the generalized compact-open topology on partial functions with closed domains, *Topology Appl.* **110**, 303–321.

[200?] Completeness properties of the Vietoris topology on partial maps with compact domains, preprint.

HOU, J.C.

[1998] Character and tightness of hyperspaces with the Fell topology, *Top. Appl.* **84**, 199–206.

ISMAIL, M., SZ. PLEWIK and A. SZYMANSKI

[2000] On subspaces of Exp(N), *Rendiconti Del Circolo Matematico Di Palermo* **XLIX**, 397–414.

JUHÁSZ, I.

[1979] *Cardinal Functions in Topology*, Mathematical Centre Tracts 34, Mathematical Centrum, Amsterdam.

KANOVEĬ, V.G. and A.V. OSTROVSKIĬ

[1981] On non-Borel F_{II}-sets, *Soviet. Math. Dokl.* **24**, 386–389.

KATĚTOV, M.

[1948] Complete normality of Cartesian products, *Fund. Math.* **36**, 271–274.

KECHRIS, A.S.

[1995] *Classical Descriptive Set Theory*, Springer-Verlag, New York,

KEESLING, J.

[1970a] Normality and properties related to compactness in hyperspaces, *Proc. Amer. Math. Soc.* **24**, 760–766.

[1970b] On the equivalence of normality and compactness in hyperspaces, *Pacific J. Math.* **33**, 657–667.

KUBIŚ, W.

[200?] Completely metrizable absolute neighborhood retracts, preprint.

KUBIŚ, W., K. SAKAI and M. YAGUCHI

[200?] Hyperspaces of separable Banach spaces with the Wijsman topology, preprint.

KUNEN, K.

[1981] A compact L-space under CH, *Topology Appl.* **12**, 283–287.

KURATOWSKI, K.

[1966] *Topology I*, Academic Press, New York.

LECHICKI, A. and S. LEVI

[1987] Wijsman convergence in the hyperspace of a metric space, *Boll. Un. Mat. Ital. B* (7) **1**, 439–451.

LOWEN, R. and M. SIOEN

[1996] The Wijsman and Attouch-Wets topologies on hyperspaces revisited, *Topology Appl.* **70**, 179–197.

[1998] Proximal hyperspace topologies revisited, *Set-Valued Analysis* **6**, 1–19.

MALYKHIN, V.I.

[1972] On tightness and Suslin number in exp X and in a product of spaces, *Soviet Mathematics - Doklady* **13** (2), 496–499.

MARJANOVIC, M.

[1966] Topologies on collections of closed subsets, *Publ. inst. Math. (Beograd)* **20**, 125–130.

MARTIN, D.A. and R.M. SOLOVAY

[1970] Internal Cohen extension, *Ann. Math. Logic* **2**, 143–178.

MATHERON, G.

[1975] *Random Sets and Integral Geometry*, Wiley, New York.

MCCOY, R.A.

[1975] Baire spaces and hyperspaces, *Pacific J. Math.* **58**, 133–142.

MCCOY, R.A. and I. NTANTU

[1988] *Topological properties of spaces of continuous functions*, Lecture Notes in Math. 1315, Springer-Verlag, Berlin.

MICHAEL, E.

[1951] Topologies on spaces of subsets, *Trans. Amer. Math. Soc.* **71**, 152–182.

[1977] Complete spaces and tri-quotient maps, *Illinois J. Math.* **21**, 716–733.

VAN MILL, J.

[1989] *Infinite Dimensional Topology: Prerequisities and Introduction*, North-Holland, Amsterdam.

[2001] *The Infinite-dimensional Topology of Function Spaces*, North-Holland, Amsterdam.

VAN MILL, J., J. PELANT and R. POL

[1996] Selections that characterize topological completeness, *Fund. Math.* **149**, 127–141.

MILOVANCEVIC, D.

[1985] A property between compact and strongly countably compact, *Publ. de L'Inst. Math.* **38**, 193–201.

MIZOKAMI, T.

[1976] Cardinal functions on hyperspaces, *Colloq. Math* **41**, 201–205.

[1995] Hyperspaces of a Moore space and a d-paracompact space, *Glas. Mat.* **30**, 69–72.

[1996] On hyperspaces of spaces around Moore spaces, *Houston Jour. of Mathematics* **22**, 297–306.

[1997] On hyperspaces of generalized metric spaces, *Topology Appl.*, **76**, 169–173.

MIZOKAMI, T. and K. KOIWA

[1987] On hyperspaces of compact and finite subsets, *Bull. Joetsu Univ. of Educ.* **6**, 1–14.

NAIMPALLY, S.

[200?] All hypertopologies are hit–and–miss, preprint.

NAIMPALLY, S. and P.L. SHARMA

[1988] Fine uniformity and the locally finite hyperspace topology, *Proc. Amer. Math. Soc.* **103**, 641–646.

NATSHEH, M.A.

[1998] On pseudocompactness of hyperspaces, *Questions and Answers in General Topology* **16**, 67–71.

NOGURA, T. and D. SHAKHMATOV

[1996] When does the Fell topology on a hyperspace of closed sets coincide with the meet of the upper Kuratowski and the lower Vietoris topologies?, *Topology Appl.* **70**, 213–243.

O'BRIEN, G.L. and S. WATSON

[1998] Relative compactness for capacities, measures, upper semicontinuous functions and closed sets, *J. Theoret. Probab.* **11**, 577–588.

POPOV, V.

[1978] On the subspaces of exp X, in *Colloquia Mathematica Societatis Janos Bolyai, Topology*, Budapest, vol. 23, 977–984.

PREISS, D.

[1973] Metrics in which Prohorov's theorem is not valid, *Z. Wahrsch. verw. Geb.* **27**, 109–116.

RAINWATER, J.

[1959] Spaces whose finest uniformity is metric, *Pacific J. Math* **9**, 567–570.

SAINT RAYMOND, J.

[1973] Caracterisation d'espaces Polonais, *Sem. Choquet (Initiation Anal.)* **5**, 1–10

[1978] La structure borélienne d'Effros est-elle standard? *Fund. Math.* **100**, 201–210.

[1994] Peut-on caracteriser les espaces metrisables par leurs sous-espaces fermes separables, *Publ. Math. Univ. Pierre et Marie Curie* **107**, 6 pp.

SCOTT, D.

[1972] Continuous lattices, *Lecture Notes in Math.* **274**, 97–136.

SIANESI, F.

[1999] Relative compactness for hyperspaces, *Serdica Math. Jour.* **25**, 11–18.

SPAHN, B.S.

[1980] On topologies generating the Effros Borel structure and on the Effros measurability of the boundary operation, *Colloq. Math.* **43**, 251–260.

TODORČEVIĆ, S.

[1986] Remarks on cellularity in products, *Compositio Math.* **57**, 357–372.

[1997] *Topics in Topology*, Lecture Notes in Math. 1652, Springer-Verlag.

VELICHKO, N.V.

[1975] On the space of closed subsets, *Sibirsk. Math. Z.* **16**, 627–629.

WIJSMAN, R.

[1966] Convergence of sequences of convex sets, cones and functions, II, *Trans. Amer. Math. Soc.* **123**, 32–45.

WILLIAMS, S.W. and H. ZHOU

[1991] Strong versions of normality, in *Gen. Topology Appl., Lecture Notes in Pure and Appl. Math.*, Dekker, New York, vol. 134, 379–389.

ZENOR, P.

[1970] On the completeness of the space of compact subsets, *Proc. Amer. Math. Soc.* **26**, 190–192.

ZSILINSZKY, L.

[1996] Baire spaces and hyperspace topologies, *Proc. Amer. Math. Soc.* **124**, 2575–2584.

[1998a] Topological games and hyperspace topologies, *Set-Valued Analysis* **6**, 187–207.

[1998b] Polishness of the Wijsman topology revisited, *Proc. Amer. Math. Soc.* **126**, 3763–3765.

CHAPTER 11

Some Topics in Geometric Topology

Kazuhiro Kawamura

Institute of Mathematics, University of Tsukuba, Tsukuba, Ibaraki 305-8571, Japan
E-mail: kawamura@math.tsukuba.ac.jp

Contents

RECENT PROGRESS IN GENERAL TOPOLOGY II
Edited by Miroslav Hušek and Jan van Mill

1. Introduction

The present article discusses some topics on geometric topology for the last decade. The choice of the materials is rather arbitrary and many important topics and results will not be discussed or mentioned. Also the results stated are not necessarily the most updated at the time of this writing. Actually excellent survey articles are already available on geometric topology (see DAVERMAN and SHER[2002] for example) and the main intention of the author is to give a brief introduction to some topics in this research area. No completeness of the references is claimed. The author hopes, in spite of this insufficiency, the article might be of some use.

The author would like to express his sincere thanks to Professor D. Repovš for the information on some references on Section 2.

Throughout the present article, a compactum means a compact metric space and a connected compactum is called a continuum.

2. Generalized manifolds and the recognition problem of topological manifolds

A finite dimensional locally compact separable metric ANR X is called an *n-dimensional generalized manifold* if $\mathrm{H}_*(X, X \setminus \{x\}) \cong \mathrm{H}_*(\mathbf{R}^n, \mathbf{R}^n \setminus \{0\})$ for each $x \in X$. It is a fundamental problem to detect the class of topological manifolds among the class of generalized manifolds. An answer is given by the following theorem due to Edwards and Quinn.

2.1. THEOREM (EDWARDS [1980], QUINN [1987]). *Let n be an integer with $n \geq 5$. An n-dimensional generalized manifold X is a topological manifold if and only if $i(X) = 1$ and X has the disjoint disks property.*

Here $i(X)$ is the *Quinn's local index* $\in 1 + 8\mathbf{Z}$ defined for every generalized manifold of dimension at least 4. It has a feature that, for an $n(\geq 4)$-dimensional generalized manifold X, $i(X) = 1$ if and only if there exists a cell-like map, called a resolution of X, $f : M \to X$ of a topological n-manifold M onto X. The *disjoint disks property* (abbreviated to the DDP) refers to the following property: each pair of maps $\alpha, \beta : D \to X$ of the 2-dimensional disk D to X is approximated arbitrarily closely by maps $\alpha', \beta' : D \to X$ such that $\mathrm{Im}(\alpha') \cap \mathrm{Im}(\beta') = \emptyset$. In Edwards' theorem above, it is proved that the hypotheses $n \geq 5$ and the DDP imply that the cell-like map $f : M \to X$ of a manifold M is approximated arbitrarily closely by homeomorphisms and in particular, $M \approx X$ (see also DAVERMAN [1986]).

It was a fundamental open problem as to whether there exists a non-resolvable generalized manifold, that is, a generalized manifold X with $i(X)$ not equal to 1. The following outstanding theorem solved the problem.

2.2. THEOREM (BRYANT, FERRY, MIO and WEINBERGER [1996], see also PEDERSEN, QUIN and RANICKI [200?]). *For each $n \geq 6$ and for each $k \in 1 + 8\mathbf{Z}$, there exists a generalized manifold X with $i(X) = k$.*

Also it has been shown that every homology manifold of dimension at least 6 is a cell-like image of a generalized manifold of the same dimension with the DDP (BRYANT, FERRY, MIO and WEINBERGER [200?]).

2.3. REMARK. (1) Every generalized manifold of dimension at most 2 is a topological manifold.
(2) If every generalized 3-manifold is resolvable by a 3-manifold, then the Poincaré Conjecture is true.
(3) The general position properties for dimension 3 that correspond to the DDP have been studied in DAVERMAN and REPOVŠ [1989], DAVERMAN and REPOVŠ [1992].
(4) An important ingredient in the proof of Edwards' Theorem is a 1-LCC shrinking theorem. A complete analogue of the theorem in dimension 4 has not been known yet and a version is given by BESTVINA, DAVERMAN, VENEMA and WALSH [2001].

Some results on the detection of generalized manifolds have been obtained in BRYANT [1987] and DYDAK and WALSH [1987].

The following two conjectures should be stated in this context.

2.4. CONJECTURE (BING and BORSUK [1965]). *If a finite dimensional locally compact separable ANR X is topologically homogenous, then X is a topological manifold.*

A space X is said to be *topologically homogenous* if, for each pair of points $x, y \in X$, there exists a homeomorphism $h : X \to X$ such that $h(x) = y$. The above conjecture is true if $\dim X \leq 2$ (BING and BORSUK [1965]). Jakobsche JAKOBSCHE [1980] showed that the validity of the above conjecture for dimension 3 implies the validity of the Poincaré conjecture.

2.5. CONJECTURE. *If a generalized manifold X has the DDP, then X is topologically homogenous.*

The validity of the above conjecture together with the existence of a non- resolvable generalized manifold provides a counterexample to the Bing-Borsuk conjecture for higher dimension.

Attempts have been made to build an analogous theory for generalized manifolds with the DDP to the one of topological manifolds. See a survey article BRYANT [2001], and also REPOVŠ [1992], REPOVŠ [1994].

Application to Metric geometry:

The Edwards-Quinn Theorem has applications in metric geometry which we shall briefly discuss below.

1. Busemann G-space conjecture

The precise definition of the Busemann G-space (BUSEMANN [1955]) is not given here. Here we just mention that it is a locally compact separable metric space (X, d) with the following features:

(i) for each pair of two points $x, y \in X$, there exists an isometry (called *geodesics*) $\alpha : [0, d(x, y)] \to X$ such that $\alpha(0) = x$ and $\alpha(d(x, y)) = y$.

(ii) every geodesics is locally extendable in a unique way.

2.6. CONJECTURE (BUSEMANN). *Every finite dimensional G-space is a topological manifold.*

This has been known to be true for dimension at most 3 (BUSEMANN [1955], KRAKUS [1968]). In THURSTON [1996], it is shown that every finite dimensional G-space is a generalized manifold and every 4-dimensional G-space admits a resolution. Applying a 4-dimensional analogue of a Daverman's shrinking criterion (DAVERMAN [1981]), the resolution is shown to be approximated arbitrarily closely by homeomorphisms, verifying the above conjecture for dimension 4 (THURSTON [1996]).

2. Gromov-Hausdorff convergence of compacta

For a compact metric spaces $(X, d_X), (Y, d_Y)$, the *Gromov-Hausdorff distance* is defined by

$$\begin{aligned} d_{GH}(X,Y) &= \inf\{\rho_H(i(X), j(Y))| \\ &\qquad (Z,\rho) \text{ is a compact metric space,} \\ &\qquad i : X \to Z, j : Y \to Z \text{ are isometric embeddings }\}, \end{aligned}$$

where ρ_H is the Hausdorff distance induced by the metric ρ. This defines a metric on $\mathcal{CM}$, the isometry classes of compact metric spaces. The metric plays a basic role in the collapsing theory of Riemannian manifolds and some finiteness or precompactness theorems. See FERRY [1994], FERRY [1998], GROVE, PETERSEN and WU [1990] etc. For example, if a sequence (X_i, d_i) of closed Riemannian manifolds converges to a finite dimensional compact metric space X in the Gromov-Hausdorff convergence and if the local contractibility of (X_i, d_i)'s is chosen to be uniform (in an appropriate sense), then the limit X is a generalized manifold. For some results on metric geometry from topological view point, see, for ANCEL and GUILBAULT [1997], FERRY and OKUN [1995], MOORE [1995], WU [1999] etc. See section 4 as well.

3. Cohomological dimension theory

For a paracompact topological space X and an abelian group G, we say that the *cohomological dimension of X with respect to G is at most n*, denoted by $\mathrm{c} - \dim_G X \leq n$, if $\check{\mathrm{H}}^{n+1}(X, A; G) = 0$ for each closed subset A of X, where $\check{\mathrm{H}}^*(X, A; G)$ is the Čech cohomology of (X, A) with the coefficient group G. If there is no such n, then we write $\mathrm{c} - \dim_G X = \infty$. It is well known that $\mathrm{c} - \dim_G X \leq n$ if and only if the Eilenberg-MacLane complex $K(G, n)$ is an absolute extensor of X, that is, each map $A \to K(G, n)$ defined on each closed subset A of X extends to a map $X \to K(G, n)$.

After the breakthrough due to Dranishnikov (see DRANISHNIKOV [1988a]), extensive research has been made on cohomological dimension theory, its generalization and applications. Excellent survey articles DYDAK [2002], DRANISHNIKOV [200?b] are already available (see also KOYAMA [2001]) and here we give a brief discussion on the subject.

1. Resolution Theorem

The following theorem due to Edwards-Walsh has been playing the fundamental role in geometric topology, in that it describes an interplay between cohomological dimension theory and the theory of cell-like maps.

3.1. THEOREM (WALSH [1981]). *Let X be a compactum. Then* $\mathrm{c}-\dim_{\mathbf{Z}}X \leq n$ *if and only if there exists a cell-like map $f : Z \to X$ of a compact metric space Z with* $\dim Z \leq n$.

The existence of an infinite (covering) dimensional compactum of cohomological dimension at most 2, DRANISHNIKOV [1988a], DYDAK and WALSH [1993b] (see LEVIN [2001] for a generalization of the construction), implies the following.

3.2. COROLLARY. *There exist a cell-like map $S^5 \to X$ of the 5-dimensional sphere S^5 onto an infinite dimensional compactum X*

It is shown in KOZLOWSKI and WALSH [1983], WALSH [1983] that every cell-like image of each 3-manifold is finite dimensional. The remaining open case is the one for 4-manifolds.

3.3. PROBLEM. Let $f : M^4 \to X$ be a cell-like map of a topological 4-manifold M onto a compact metric space X. Does X have a finite covering dimension?

A criterion for the finite dimensionality of such X is given in MITCHELL, REPOVŠ and ŠČEPIN [1992].

3.4. REMARK. WATANABE [1995] showed that the integral cohomological dimension is equal to the covering dimension for each approximate movable space.

A key of the proof of the Edwards-Walsh Theorem WALSH [1981] is the construction, now called the Edwards-Walsh construction. This is explicitly formulated in DYDAK and WALSH [1993a]. In the construction, for a given abelian group G, an integer n and a simplicial complex L, there corresponds a combinatorial map $\pi_L : EW_G(L) \to L$ of a CW complex $EW_G(L)$ with certain properties. See the above paper for the precise formulation. A sufficient condition on the possibility of the Edwards-Walsh construction is given by DYDAK and WALSH [1993a] and KOYAMA and YOKOI [200?]. Concerning the necessity of the condition given in these papers, see YOKOI [2000]. These results, with additional arguments, provide some variations of the Edwards-Walsh Theorem for various coefficients. A simple but weak form of the results is stated below. A map $f : Z \to X$ between compacta is said to be G-acyclic, G being an abelian group, if each fiber of f has the trivial Čech cohomology with coefficient G.

3.5. THEOREM (KOYAMA and YOKOI [200?]). *Let G be one of the Bockstein groups* (see DYDAK [2002] for the definition). *For each compact metric space X with* $\mathrm{c}-\dim_G X \leq n$, *then there exists a G-acyclic map $f : Z \to X$ such that* $\dim Z \leq n+1$.

3.6. PROBLEM (KOYAMA and YOKOI [200?]). . Does the above theorem hold for arbitrary abelian group? Is it possible to choose above Z so that $\mathrm{c}-\dim_G Z \leq n$?

In these results, no further topological properties of X and its resolution $f : Z \to X$ other than (cohomological) dimension are under consideration. It is then natural to ask as to whether we may resolve a compactum X with a given cohomological dimension and with some additional property by a compactum with that property. In this direction, Ferry proved the following.

3.7. THEOREM (FERRY [1998]). *If a locally n-connected compactum X has the integral cohomological dimension at most n, then there exists a cell-like map $f : P \to X$ of a compact polyhedron P with* $\dim P \leq 2n + 1$.

2. *The existence of universal spaces*

Let $\mathcal{C}$ be a class of spaces and a space X is said to be *universal* for the class $\mathcal{C}$ if $X \in \mathcal{C}$ and every space $Y \in \mathcal{C}$ is topologically embedded in X. For example the n-dimensional universal Menger compactum is universal for the class of all separable metric space of dimension at most n. In the context of cohomological dimension theory, the following has been obtained in CHIGOGIDZE and VALOV [2001]. See also DYDAK and MOGILSKI [1994], OLSZEWSKI [1995b], ZARICHNYI [1997], ZARICHNYI [1997], LEVIN [200?], etc.

3.8. THEOREM. *Let G be a countable abelian group. Then the class of all metrizable spaces of* $\mathrm{c}-\dim_G \leq n$ *with given weight has a universal space.*

The following conjecture has been made by several people and seems to be still open at the time of this writing.

3.9. CONJECTURE. *For each $n \geq 2$, there does not exist a universal space for the class of all compacta of integral cohomological dimension at most n.*

The above problem is closely related to the cohomological-dimension-preserving compactification problem. An explicit formulation of this connection is given in CHIGOGIDZE [1999]. It is known in DYDAK and WALSH [1991] that there exists a metrizable space which has no (Hausdorff) compactification of the same cohomological dimension (see also DYDAK [1993]). On the other hand, the cohomological-dimension-preserving completions have been known to exist for several cases, OLSZEWSKI [1995a], LEVIN [200?].

Actually the subject should be discussed in the context of extension dimension and extension type, DRANISHNIKOV and DYDAK [1996], DYDAK [2002], DYDAK [200?], DRANISHNIKOV [1998], LEVIN [200?], MIYATA [1996]. Theory of extension dimension has been investigated by many authors, but is out of scope of this article. See also DYDAK [1996], ŠČEPIN [1998]. Cohomological dimension theory of Tychonoff spaces is studied in CHIGOGIDZE [1997].

4. Compactifications in geometric topology

In this section we discuss two compactifications of locally compact separable metrizable spaces: *$\mathcal{Z}$-compactifications* and *Higson-Roe compactifications*. These compactifications are playing significant roles in the recent development of geometric topology and related areas.

1. *$\mathcal{Z}$-compactifications*

Let X be a locally compact separable ANR. A metric compactification $\overline{X}$ of X is called a *$\mathcal{Z}$-compactification* of X if $\overline{X} \setminus X$ is a Z-set. The remainder $\overline{X} \setminus X$ is then called a *$\mathcal{Z}$-boundary* of X and is sometimes denoted by $\partial_{\mathcal{Z}} X$. It follows automatically that $\overline{X}$

is also an ANR. A typical example is a manifold compactification $\overline{M}$ of an open manifold M (the manifold boundary $\partial\overline{M} = \overline{M} \setminus M$ is a Z-set). Siebenmann's thesis SIEBENMANN [1965] gives a necessary and sufficient condition for an open manifold of dimension at least 6 to have a manifold compactification. In addition to this typical example, $\mathcal{Z}$-compactifications appear naturally in several contexts.

HILBERT CUBE MANIFOLD THEORY. Here the Hilbert cube, i.e., the countable product of closed intervals, is denoted by Q. CHAPMAN and SIEBENMANN [1976] gave a necessary and sufficient condition for a Hilbert cube manifold X to have a $\mathcal{Z}$-compactification $\overline{X}$. By FERRY [2000] or the characterization theorem of Toruńczyk, $\overline{X}$ is a Hilbert cube manifold.

Noticing that $K \times Q$ is a Hilbert cube manifold for each locally compact polyhedron, the following question was asked by Chapman (cf. VAN MILL and REED [1990], QM8).

4.1. QUESTION. Let X be a locally compact polyhedron and suppose that $X \times Q$ admits a $\mathcal{Z}$-compactification. Then does X itself admits a $\mathcal{Z}$-compactification?

The comment after [1990, QM8] was corrected as follows.

4.2. THEOREM (GUILBAULT [2001]). *There exists a 2-dimensional locally compact polyhedron X such that $X \times Q$ admits a $\mathcal{Z}$-compactification and X itself does not admit a $\mathcal{Z}$-compactification.*

A stable compactification theorem is proved in FERRY [2000]:

4.3. THEOREM. *Let X be an n-dimensional locally compact polyhedron such that $X \times Q$ admits a $\mathcal{Z}$-compactification. Then $X \times [0,1]^{2n+5}$ admits a $\mathcal{Z}$-compactification.*

$\mathcal{Z}$-COMPACTIFICATIONS OF OPEN MANIFOLDS. The following theorem illustrates that the concept of $\mathcal{Z}$-compactifications serves as a convenient framework for the study of open manifolds which do not admit manifold compactifications.

4.4. THEOREM (ANCEL and GUILBAULT [1999]). *Let M and N be open $n(\geq 5)$-manifolds which admit $\mathcal{Z}$-compactifications with homeomorphic $\mathcal{Z}$-boundaries $\partial_{\mathcal{Z}} M \approx \partial_{\mathcal{Z}} N$. Then*

$M \cup_{\partial_{\mathcal{Z}} M = \partial_{\mathcal{Z}} N} N$ is an n-manifold.

The higher dimensional Poincaré conjecture with the above implies:

4.5. COROLLARY (ANCEL and GUILBAULT [1999]). *Let M and N be open contractible $n(\geq 5)$-manifolds which admit $\mathcal{Z}$-compactifications with homeomorphic $\mathcal{Z}$-boundaries $\partial_{\mathcal{Z}} M \approx \partial_{\mathcal{Z}} N$. Then $M \cup_{\partial_{\mathcal{Z}} M = \partial_{\mathcal{Z}} N} N$ is homeomorphic to the n-sphere.*

The above corollary may be used to construct involutions on S^n with exotic fixed point sets.

Contractible open manifolds which admit $\mathcal{Z}$-compactifications are determined by their $\mathcal{Z}$-boundaries.

4.6. THEOREM (ANCEL and GUILBAULT [1999]). *Let M^n and N^n be contractible open manifolds of dimension $n \geq 5$ such that $\partial_Z M^n \approx \partial_Z N^n$. Then $M^n \approx N^n$.*

The above provides a partial answer to the following conjecture. A CW-complex P is said to be *aspherical* if $\pi_i(P) = 0$ for each $i \geq 2$, or equivalently, its universal cover is contractible.

4.7. CONJECTURE (weak BOREL). *If P and Q are closed aspherical manifolds with isomorphic fundamental groups, then their universal covers are homeomorphic.*

Corollary 2.1 of ANCEL and GUILBAULT [1999] states that the above conjecture is valid, if $\dim P = \dim Q \geq 5$ and $\pi_1(P) \cong \pi_1(Q)$ is a CAT(0) or word-hyperbolic group (see the next section for the definitions). Original Borel conjecture, still being open, is stated as follows.

4.8. CONJECTURE (BOREL). *If P are Q are closed aspherical manifolds with isomorphic fundamental groups, then P and Q are homeomorphic.*

WORD-HYPERBOLIC GROUPS AND CAT(0)-GROUPS. Word metrics of finitely generated groups enable us to apply geometric idea to the study of groups. One of the origins of this idea is due to GROMOV [1981], [1988]. Intensive research has been made and several connections with geometric topology have been recognized. We give here a brief discussion on the aspect of the theory.

Let Γ be a finitely generated group with a finite set S of generators. We assume that $S^{-1} := \{s^{-1} | s \in S\} = S$. The *length* of an element $\gamma \in \Gamma$ with respect to S is defined by

$$\|\gamma\|_S = \min\{n | \text{there exists } s_1, \ldots, s_n \in S \text{ such that } \gamma = s_1 \cdots s_n\}.$$

The *word metric* d_S of Γ is defined by $d_S(\alpha, \beta) = \|\alpha^{-1}\beta\|_S$, for $\alpha, \beta \in \Gamma$.

It is easy to see the above metric has the following two properties.

(1) d_S is invariant under left multiplications, that is, $d_S(\gamma\alpha, \gamma\beta) = d_S(\alpha, \beta)$ for all $\alpha, \beta, \gamma \in \Gamma$.

(2) If S_1 and S_2 are two sets of generators of Γ, then there exists a constant $L \geq 1$ such that $\frac{1}{L} \cdot d_{S_2} \leq d_{S_1} \leq L \cdot d_{S_2}$.

These indicate that the properties of Γ which are to be studied should be invariant under left multiplications and should be Lipschitz invariant.

The *Cayley graph* $K(\Gamma, S)$ is a graph defined as follows: the set of vertices $= \Gamma$; two vertices $\gamma_1, \gamma_2 \in \Gamma$ are joined by an edge if and only if $\gamma_1^{-1}\gamma_2 \in S$. Since S is a finite set, $K(\Gamma, S)$ is a locally finite graph. The word metric d_S is naturally extended to the metric $\overline{d}_S$ on $K(\Gamma, S)$ by declaring that each edge has the length 1. Each left multiplication of an element of Γ naturally extends to a simplicial $\overline{d}_S$-isometry of $K(\Gamma, S)$.

A finitely generated group Γ is said to be *word hyperbolic* if the Cayley graph $K(\Gamma, S)$ for some (or equivalently any) finite set of generators S is hyperbolic in the following sense. A metric d of a locally compact separable metrizable space X is said to be *proper* if each d-bounded subset of X has the compact closure.

4.9. DEFINITION. Let (X,d) be a proper metric space and $\delta \geq 0$. Fix a base point $x_0 \in X$ and for each pair of points $x, y \in X$, define:

$$(x|y)_{x_0} := \tfrac{1}{2}(d(x,x_0) + d(y,x_0) - d(x,y)).$$

The space (X,d) is said to be *δ-hyperbolic* if

$$(x|y)_{x_0} \geq \min((x,z)_{x_0}, (y,z)_{x_0}) - \delta$$

for all $x, y, z \in X$. If (X,d) is δ-hyperbolic for some $\delta \geq 0$, then we simply say that (X,d) is *hyperbolic*.

4.10. REMARK. The hyperbolicity does not depend on the choice of the point x_0 (the constant δ would depend on x_0) and also is invariant under quasi-isometries. Here a map $\varphi : (X,d_X) \to (Y,d_Y)$ is called a *quasi-isometry* if there exist constants $\lambda > 1$ and $C, D > 0$ such that

$$\tfrac{1}{\lambda} d_X(x,y) - C \leq d_Y(\varphi(x),\varphi(y)) \leq \lambda d_X(x,y) + C \text{ and}$$
$$\varphi(X) \text{ is } D\text{-dense in } (Y,d_Y).$$

This invariance guarantees that the word hyperbolicity of a group does not depend on the choice of finite sets of generators.

Also the above has an important consequence. Suppose that a finitely generated group Γ acts properly discontinuously on a proper metric space (X,d) as isometries so that X/Γ is compact. Fix a point $x_0 \in X$ and take a finite set of generators S of Γ such that $S^{-1} = S$. We define a map $e : \Gamma \to X$ by $e(\gamma) = \gamma \cdot x_0, \gamma \in \Gamma$. Then the map e is a quasi-isometry with respect to d_S and d (cf. COORNAERT, DELZANT and PAPADOPOULOS [1990]).

Hence we have

4.11. PROPOSITION. *Let Γ be a finitely generated group acting properly discontinuously on a hyperbolic metric space (X,d) as isometries such that X/Γ is compact. Then Γ is a word hyperbolic group.*

Typical examples of hyperbolic metric spaces are:
the hyperbolic space $\mathbf{H}^n = \{(x_1,\ldots,x_n) | x_n > 0\}$ (endowed with the Riemannian metric $ds^2 = \frac{\sum_{i=1}^n (dx_i)^2}{x_n^2}$),
(infinite) trees with the path length metrics (where the constant $\delta = 0$), etc.

The above two examples and Proposition 5.7 imply that finitely generated free groups and the fundamental groups of closed orientable surfaces of genus at least 2 are word hyperbolic.

For a word hyperbolic group Γ (with a finite set of generators) and a sufficiently large $d > 0$, we define a simplicial complex $P_d(\Gamma)$, called the *Rips complex* as follows: the set of vertices is Γ and vertices $x_1,\ldots,x_n$ span a simplex if and only if $\operatorname{diam}_{d_S}\{x_1,\ldots,x_n\} \leq d$. It is known that $P_d(\Gamma)$ is contractible when d is sufficiently large and it also admits a $\mathcal{Z}$-compactification $\overline{P_d(\Gamma)} = P_d(\Gamma) \cup \partial\Gamma$. The homeomorphism type of the remainder $\partial\Gamma$ does not depend on the choice of S or d and is called the *boundary* of Γ. For example, if Γ is a free group of finite rank, then $\partial\Gamma$ is homeomorphic to the Cantor set. If Γ is the fundamental group of a closed orientable surface of genus at least 2, then $\partial\Gamma$ is homeomorphic to the simple closed curve.

Several good expository articles are already available. See for example COORNAERT, DELZANT and PAPADOPOULOS [1990], GHYS, HAEFLIGER and VERJOVSKY [1993], GHYS and DE LA HARPE [1990] etc.

The topology of $\partial\Gamma$ is rather complicated in general and it is natural to study the connection with the topology of $\partial\Gamma$ and the group theoretic property of Γ. The following amazing theorem describes such a connection.

4.12. THEOREM (BESTVINA and MESS [1991]). *Let R be a commutative ring with unity.*

(1) $\mathrm{c} - \dim_R \partial\Gamma = \mathrm{cd}_R\Gamma - 1$ *if* $\mathrm{cd}_R\Gamma < \infty$.

(2) *If Γ is torsion free, then* $\dim \partial\Gamma = \mathrm{cd}\Gamma - 1$.

*Where $\mathrm{cd}_R\Gamma$ denotes the cohomological dimension of the group Γ with respect to R and $\mathrm{cd}\Gamma = \mathrm{cd}_{\mathbf{Z}}\Gamma$ (*BROWN [1982]*).*

A homological complexity is described in BESTVINA [1996], SWENSON [1999b] in a more general setting. Specializing their result to word hyperbolic groups, it is stated as follows.

4.13. THEOREM (BESTVINA [1996], SWENSON [1999b]). *For "almost all" points z of the boundary of a word hyperbolic group Γ, one of the following statements holds for Steenrod homology with the countable coefficient field* $\mathbf{L}$.

(1) $\mathrm{H}_q(\partial\Gamma) \to \mathrm{H}_q(\partial\Gamma, \partial\Gamma \setminus \{z\})$ *is an isomorphism and* $\dim_{\mathbf{L}} \mathrm{H}_q(\partial\Gamma) < \infty$.

(2) $\mathrm{H}_q(\partial\Gamma, \partial\Gamma \setminus \{z\})$ *is uncountable.*

On the other hand, the boundaries of many word hyperbolic groups are locally connected.

4.14. THEOREM (BESTVINA and MESS [1991], BOWDITCH [1999], LEVITT [1998], SWARUP [1996], SWENSON [1999a]). *For each one-ended word hyperbolic group Γ, $\partial\Gamma$ is locally connected.*

4.15. DEFINITION. A group Γ acting on a compactum Z is called a *convergence group* if the following condition is satisfied: for each sequence $\{\gamma_i\}$ of elements of Γ, there exist a subsequence $\{\gamma_{i_k}\}$ and two distinct points $p, q \in Z$ such that
for each neighborhood U of q, there exist an integer $N > 0$ and a neighborhood V of p satisfying: $k > N \Rightarrow \gamma_{i_k}(Z \setminus V) \subset U$.

The above condition is equivalent to the following: the natural action of Γ on $\mathrm{F}_3(Z)$, the space of all mutually distinct triple points of Z, induced by the one on Z is properly discontinuous. Further this provides a topological characterization of the word hyperbolic groups BOWDITCH [1998]. Dynamics of the action of a word hyperbolic group is represented by a quotient of a symbolic dynamics as is shown in COORNAERT and PAPADOPOULOS [1993].

4.16. REMARK. The $\mathcal{Z}$-compactifications arising in this context often admit group actions and one of the axioms described in BESTVINA [1996] takes into account of these actions.

As is seen in the typical examples mentioned above, one of the models of word hyperbolic groups is the class of finitely generated groups acting properly discontinuously on the hyperbolic space $\mathbf{H}^n$ as isometries in such a way that the orbit spaces are compact. CAT(0)-groups are groups acting on "non-positively curved" spaces in a similar way to the above. The notion of non-positively curved spaces is defined in the following way, being motivated by the Toponogov comparison theorem CHEEGER and EBIN [1973]. It is a metric analogue of Hadamard manifolds, simply connected Riemannian manifolds with sectional curvature ≤ 0 everywhere. A geodesic segment joining a point p with another point q of a proper metric space (X, d) (= an isometric image of $[0, d(p, q)]$ with the end points p and q) is denoted by $\overline{pq}$.

4.17. DEFINITION. A proper metric space (X, d) is called a CAT(0)-space if it satisfies the following condition: for each triple x, y, z of points of X, take a triangle $\tilde{x}, \tilde{y}, \tilde{z} \in \mathbf{R}^2$ such that $||\tilde{x} - \tilde{y}|| = d(x, y)$, $||\tilde{y} - \tilde{z}|| = d(y, z)$ and $||\tilde{z} - \tilde{x}|| = d(z, x)$. For each point $w \in \overline{xy}$, take the unique point $\tilde{w} \in \overline{\tilde{x}\tilde{y}}$ such that $d(x, w) = ||\tilde{x} - \tilde{w}||$. Then the inequality

$$d(z, w) \leq ||\tilde{z} - \tilde{w}||$$

holds.

A group Γ is called a CAT(0)-group if it acts properly discontinuously on a CAT(0)-space (X, d) as isometries such that X/Γ is compact.

For a Hadamard manifold M and a fixed point $x_0 \in M$, the visual boundary M_∞ is defined as the set of all geodesic rays emanating from x_0 with the uniform convergence topology on compact sets. By the same way, the boundary ∂X of a CAT(0)-space (X, d) is defined as follows.

4.18. DEFINITION. Let (X, d) be a CAT(0)-space and fix some point x_0. Let $\partial X = \{\alpha : [0, \infty) \to X | \alpha \text{ is a geodesic ray with } \alpha(0) = x_0\}$. The set ∂X endowed with the uniform convergence topology on compact sets is called the *ideal boundary* of X. The topological type of ∂X is known to be independent on the choice of x_0.

For a CAT(0)-group Γ acting properly discontinuously on a CAT(0)-space (X, d) such that X/Γ is compact, ∂X is denoted by $\partial\Gamma$ and is called a *boundary* of Γ.

4.19. REMARK. If Γ is a word-hyperbolic group, its boundary in the previous sense is homeomorphic to the boundary defined above, being applied to a hyperbolic-and-CAT(0) metric space on which Γ acts properly discontinuously so that the orbit space is compact.

The visual boundaries of Hadamard n-manifolds are always homeomorphic to the $(n-1)$-sphere. On the other hand, boundaries of CAT(0)-spaces may be rather complicated in general as are illustrated in the following results.

4.20. THEOREM.

(1) *There exist two isomorphic CAT(0)-groups Γ_1 and Γ_2 such that $\partial\Gamma_1$ and $\partial\Gamma_2$ are not homeomorphic* (CROKE *and* KLEINER [2000]). *In other words, the topology of the boundary of a CAT(0)-group does depend on the CAT(0)-space on which the group acts.*

(2) *There exist CAT(0)-groups with non locally connected boundaries* (MIHALIK *and* RUANE [1999], [2001]).

(3) *There exist CAT(0)-groups with the boundary homeomorphic to the Menger universal compacta, the Sierpiński curve, and the Pontrygin surface respectively* (BENAKLI [1992], DRANISHNIKOV [200?a], DRANISHNIKOV [1999]).

In Introduction of MIHALIK and RUANE [1999], it is mentioned that R. Geoghegan pointed out that a result of ONTANEDA [1996] can be used to show that the shape types of the boundaries of isomorphic CAT(0)-groups are the same, so the shape theory seems to be an appropriate framework for the study of boundaries of CAT(0)-groups.

An important class of CAT(0)-groups is the class of finitely generated Coxeter groups. A *Coxeter group* is a group Γ with a presentation $\Gamma = < s_1, \ldots, s_n | (s_i s_j)^{m_{ij}} = 1 >$, where

$m_{ii} = 1, m_{ij} = m_{ji}$ and $m_{ij} \geq 2$ for each distinct pair i, j,

denoted by V here (this is an ad hoc notation and is not standard).

Given a presentation V above, there exists a canonical way to define a simplicial complex $\Sigma(\Gamma, V)$ which is endowed with a CAT(0)-metric and Γ acts on $\Sigma(\Gamma, V)$ with finite isotropy groups such that $\Sigma(\Gamma, V)/\Gamma$ is compact, DAVIS [2002], MOUSSONG [1988]. Hence Γ is a CAT(0)-group and $\partial\Gamma = \partial\Sigma(\Gamma, V)$. The following conjecture is still open at the time of this writing.

4.21. CONJECTURE (Rigidity). *Let Γ_1 and Γ_2 be two isomorphic Coxeter groups (with possibly different presentations). Then $\partial\Gamma_1$ and $\partial\Gamma_2$ are homeomorphic.*

The topology of the boundaries of Coxeter groups is a subject of further study.

4.22. THEOREM.

(1) *The Bestvina-Mess formula on cohomological dimensions of word hyperbolic groups holds as well for Coxeter groups* (DRANISHNIKOV [1997a]).

(2) *The Čech cohomology group of the boundary is computed from a given presentation of a Coxeter group* (DAVIS [1998], HOSAKA [200?]).

(3) *The boundary of each one ended Coxeter group has the shape type of a locally connected continuum* (MIHALIK [1996]).

2. *Higson-Roe compactification*

The Higson-Roe compactification plays an important role in coarse geometry and asymptotic topology. We shall not discuss the background materials here and interested readers should consult ROE [1993], DRANISHNIKOV [2000a], etc. In what follows, $C^*(X)$ denotes the ring of all bounded real-valued continuous functions on a space X with the supremum norm.

4.23. DEFINITION. Let (X, d) be a proper metric space and let $C^*_d(X)$ be the closed subring of $C^*(X)$ consisting of all continuous real-valued functions which satisfy the following condition $(*)_d$:

for each $r > 0$ and for each $\epsilon > 0$, there exists a compact subset $K_{r,\epsilon}$ of X such that, if $x \in X \setminus K_{r,\epsilon}$, then $\operatorname{diam} f(B_d(x, r)) < \epsilon$, where $B_d(x, r)$ denotes the closed r-ball centered at x with radius r.

The compactification associated with $C^*_d(X)$ is denoted by $\overline{X}^d$ and is called the *Higson-Roe* (or *Higson*) *compactification* of X. The remainder $\overline{X}^d \setminus X$ is called the *Higson-Roe* (or *Higson*) *corona* and denoted by $\nu_d X$.

The topology of the compactification is rather complicated. For example, $\beta\omega$ is embedded in $\overline{X}^d$ for each proper metric space (KEESLING [1994]). Moreover,

4.24. THEOREM.

(1) $\check{\mathrm{H}}^k(\overline{\mathbf{R}^n}^d; \mathbf{Q})$ *is not trivial for each* $k = 1, \ldots, n$ (KEESLING [1994] for $n = 1$ and DRANISHNIKOV *and* FERRY [1997] for $n \geq 2$). *Here* $\mathbf{R}^n$ *has the standard Euclidean metric.*

(2) $\check{\mathrm{H}}^2(\overline{\mathbf{H}^n}^d) = \check{\mathrm{H}}^4(\overline{\mathbf{H}^n}^d) = \check{\mathrm{H}}^8(\overline{\mathbf{H}^n}^d) = 0$ *for each* n, *and* $\check{\mathrm{H}}^n(\overline{\mathbf{H}^n}^d)$ *has at most 2-torsion for every even* n (DRANISHNIKOV *and* FERRY [1997]).

The dimension of the Higson-Roe corona is estimated in terms of the asymptotic dimension.

4.25. DEFINITION. Let (X, d) be a proper metric space. We say that the *asymptotic dimension* of (X, d) is at most n, denoted by $\operatorname{asdim}(X, d) \leq n$ if, for each $R > 0$, there exists a uniformly bounded family $\mu_1, \ldots, \mu_{n+1}$ of subsets of X such that

(1) $\bigcup_{i=1}^{n+1} \mu_i$ is a covering of X and

(2) for each pair E, F of distinct elements of μ_i, we have $d(E, F) > R$.

4.26. THEOREM (DRANISHNIKOV, KEESLING and USPENSKIJ [1998], DRANISHNIKOV [2000a]). *For each proper metric space* (X, d) *we have an inequality* $\dim \nu_d X \leq \operatorname{asdim} X$. *If* $\operatorname{asdim} X < \infty$, *then* $\dim \nu_d X = \operatorname{asdim} X$. *In particular, the Higson corona of* $\mathbf{R}^n$ *with the standard metric has dimension* n.

5. Approximate fibrators

A proper map $p : E \to B$ between locally compact separable metric ANR's is called an *approximate fibration* if it has the following approximate lifting property: for each metric space X, for each pair of maps $H : X \times [0, 1] \to B$ and $h : X \times \{0\} \to E$ such that $p \cdot h = H|X \times \{0\}$ and for each open cover $\mathcal{U}$ of B, there exists a map $\overline{H} : X \times [0, 1] \to E$ such that $\overline{H}|X \times \{0\} = h$ and $p \cdot \overline{H}$ and H are $\mathcal{U}$-close, CORAM and DUVALL [1977]. The notion of approximate fibrations provides an appropriate bundle theory for maps between ANR's and detecting approximate fibrations is a fundamental problem. In this direction, R.J. Daverman initiated the study of approximate fibrators.

5.1. DEFINITION. An n-dimensional closed manifold N^n is called a *codimension k fibrator* (resp. a *codimension k orientable fibrator*) if each map $p : M^{n+k} \to B$ defined on an $(n + k)$-manifold (resp. an orientable $(n + k)$-manifold) M^{n+k} with each fiber $p^{-1}(b)$ being shape equivalent to N is an approximate fibration.

So the basic problem is to recognize codimension k fibrators. By definition, every codimension k fibrator is a codimension k orientable fibrator.

In what follows, we will confine ourselves to the problem of detecting codimension 2 (orientable) fibrator. In order to state known results, we need some terminologies. A closed orientable manifold N is said to be *hopfian* if each degree one map $N \to N$ is a homotopy equivalence.

5.2. DEFINITION. Let G be a finitely generated group.

(1) G is said to be *hopfian* if every epimorphism $G \to G$ is an isomorphism.

(2) G is said to be *cohopfian* if every monomorphism $G \to G$ is an isomorphism.

(3) G is said to be *hyperhopfian* if every homomorphism $f : G \to G$ such that $f(G)$ is a normal subgroup of G and $G/f(G)$ is cyclic is an isomorphism.

5.3. REMARK.

(1) By definition, every hyperhopfian group is hopfian.

(2) Every torsion free word hyperbolic group is hopfian (SELA [1999]).

(3) The fundamental group of each compact hyperbolic manifold and compact surface (with or without boundary) are hyperhopfian (cf. DAVERMAN [1993a], [200?]).

(4) If a hopfian group is presented by s generators and t relations with $s > t + 1$, then it is hyperhopfian (DAVERMAN [1993a]).

The following problem posed by H. Hopf is still open.

5.4. PROBLEM. If a closed manifold N has the hopfian fundamental group, is N hopfian?

Some partial solution has been obtained in DAVERMAN [1993a].
First we state some results to recognize codimension 2 *orientable* fibrators.

5.5. THEOREM. *Let N be a closed manifold. Then each of the following conditions implies that N is a codimension 2 orientable fibrator.*

(1) *N is hopfian and:*

(1.1) $\pi_1(N)$ *is hyperhopfian,* DAVERMAN [1993a], *or*

(1.2) $\pi_1(N)$ *is hopfian and, either* $\chi(N)$ *is not zero or* $\mathrm{H}_1(N) \cong \mathbf{Z}_{2^t}$ *for some t,* DAVERMAN [1993a], DAVERMAN *and* KIM [200?].

(2) *N is orientable and:*

(2.1) $\pi_1(N)$ *is finite or hopfian,* $\mathrm{H}_1(N)$ *is finite, N is aspherical and N admits no maps of degree d provided* $\mathrm{H}_1(N)$ *contains a cyclic subgroup of order d,* DAVERMAN [1991b], *or*

(2.2) $\pi_1(N)$ *is finite and* $\mathrm{H}_1(N) \cong$ [*the finite direct sum of* $\mathbf{Z}_2$], DAVERMAN *and* KIM [200?].

It is an open problem as to whether every hopfian manifold with the hyperhopfian fundamental group is a codimension 2 fibrator.

Next results provide criteria for codimension 2 fibrators. Clearly they serve as criteria for codimension 2 orientable fibrators as well.

5.6. THEOREM. *Let N be a closed manifold. Each of the following conditions implies that N is a codimension 2 fibrator.*

(1) *N is hopfian,* $\pi_1(N)$ *is hopfian and either* $\chi(N)$ *is nonzero or* $\mathrm{H}_1(N) \cong \mathbf{Z}_2$, IM *and* KIM [1999].

(2) $\pi_1(N)$ *is finite and:*

(2.1) *$\chi(N)$ is nonzero,* CHINEN [1998], *or*

(2.2) *$\pi_1(N)$ is the direct product of finitely many $\mathbf{Z}_{2^r}$ for some r,* CHINEN [2000], *or*

(2.3) *$\pi_1(N)$ is an abelian 2-group,* DAVERMAN *and* KIM [200?].

The hypothesis "an abelian 2-group" in (2.3) above cannot be replaced by "an abelian p-group" for an odd prime p in general, DAVERMAN [1999a]. When a given manifold N is represented by a connected sum or a product of some manifolds, further results are known, IM [1995], IM and KIM [2000], KIM [2000].

Approximate fibrators in other codimensions, in PL category and in low dimensional manifolds have been studied in DAVERMAN [1991a], [1991b], [1993b], [1995a], [1995b], [1999b], etc.

6. Some other topics

1. *Characterization of Nöbeling spaces*

Since the outstanding work due to BESTVINA [1988], it is widely recognized that Menger manifold theory is a finite dimensional analogue of Hilbert cube manifold theory (see CHIGOGIDZE [1996], CHIGOGIDZE, KAWAMURA and TYMCHATYN [1995], KAWAMURA [2000]). The next natural step is to establish a finite dimensional analogue of l_2- manifold theory and the following was a central question in this direction.

6.1. QUESTION. Let X be a Polish space with the following properties:

(1) $\dim X = n$, X is locally $(n-1)$-connected and $(n-1)$-connected, and

(2) for each Polish space Z with $\dim Z \leq n$, every map $f : Z \to X$ is approximated arbitrarily closely by closed embeddings.

Is then X homeomorphic to the n-dimensional Nöbeling space
$N_n^{2n+1} = \{(x_i) \in \mathbf{R}^{2n+1} |$ at most n coordinates x_i's are rational$\}$?

Recently AGEEV [200?] announced the affirmative answer to the above question.

2. *General position properties of compacta in the Euclidean spaces*

General position has played one of the central roles in PL topology and several attempts have been made to establish analogous results for general compacta (not necessarily subpolyhedra) in the Euclidean spaces. One of the recent results can be stated as follows.

6.2. THEOREM (DRANISHNIKOV [2000b]). *Let $f : X \to \mathbf{R}^n$ and $g : Y \to \mathbf{R}^n$ be maps of compacta and assume that* $\max(\dim X, \dim Y) < n - 2$. *Then f and g are approximated arbitrarily closely by maps $f' : X \to \mathbf{R}^n$ and $g' : Y \to \mathbf{R}^n$ such that* $\dim(f'(X) \cap g'(Y)) \leq \dim(X \times Y) - n$.

In what follows, we briefly discuss the solution of Chogoshvili's conjecture. In attempting to generalize a classical theorem of P.S. Alexandroff, G. Chogoshvili announced the following statement, the argument for which turned out to be incomplete.

6.3. CONJECTURE (CHOGOSHVILI). *Let X be a compactum in $\mathbf{R}^n$. If* $\dim X \geq k$*, then there exist an $(n-k)$-dimensional affine subspace L of $\mathbf{R}^n$ and an $\epsilon > 0$ such that, for each map $f : X \to \mathbf{R}^n$ with $d(f, id) < \epsilon$, $f(X)$ intersects with L* (when this happens, we say that *"X intersects stably with L"*).

Chogoshvili's original statement was that the above holds for every subspace of $\mathbf{R}^n$ and a noncompact counterexample has been given by K.A. Sitnikov. If X is (for example) a 2-dimensional ANR, then the above conjecture was verified in DOBROWOLSKI, LEVIN and RUBIN [1997]. However a counterexample in DRANISHNIKOV [1997b] (for $k = 2$, $n = 4$) shows that the conjecture is not valid in general. See LEVIN and STERNFELD [1998] for a related result.

There are plenty of works on the embeddings of compacta in Euclidean spaces. See REPOVŠ and SKOPENKOV [1999], REPOVŠ, SKOPENKOV and ŠČEPIN [1996], AKHMET'EV [2000], etc.

3. Dimension raising actions of compact zero dimensional groups on compacta

Dimension raising actions of compact zero dimensional groups on compacta, especially those of the p-adic integers, have been investigated for many authors in the connection with the Hilbert-Smith conjecture (cf. YANG [1960]).

The group of p-adic integers A_p is defined by:

$$A_p = \lim_{\leftarrow}(\mathbf{Z}/p\mathbf{Z} \leftarrow \mathbf{Z}/p^2\mathbf{Z} \leftarrow \cdots \leftarrow \mathbf{Z}/p^n\mathbf{Z} \leftarrow \mathbf{Z}/p^{n+1}\mathbf{Z} \leftarrow \cdots)$$

where the bonding maps are the canonical projections. As a topological space, A_p is homeomorphic to the Cantor set.

Several dimension raising actions of A_p onto non-manifolds are known (KOLMOGOROV [1937], RAYMOND and WILLIAMS [1963], BESTVINA and EDWARDS [1987]). Also every universal Menger compactum admits a free action of every compact zero-dimensional group actions, DRANISHNIKOV [1988b], MAYER and STARK [200?], SAKAI [1994], IWAMOTO [1996], AGEEV and REPOVŠ [2000], etc. Also in AGEEV [1993], Ageev pointed out a connection of the existence of the dimension lowering equivariant maps between Menger compacta with the Hilbert-Smith conjecture. Also LEE [1984] contains related materials. A dimension raising action of a compact zero-dimensional torsion group on a compactum is constructed in DRANISHNIKOV and WEST [1997].

4. Fundamental groups and singular homology groups of non-CW complexes

The class of spaces which have the homotopy type of CW-complexes serves as the fundamental ground on which algebraic topology is highly developed. The study of spaces without the homotopy type of CW-complexes has relied on shape theoretic approach, or in other word, Čech type approach. On the other hand, some attempts recently revived to study fundamental groups, homotopy groups or singular (co)homology groups of spaces which do not have the homotopy type of CW-complexes. Such study necessarily need to handle "infinitary words" and infinitely generated groups. EDA [1992a] and recently CANNON and CONNER [2000a] developed some technique to handle these objects (see CANNON and CONNER [2000b] as well). As an application, the singular homology group of the Hawaiian earring was computed in EDA and KAWAMURA [2000b]. On the fundamental group of general compacta, the following theorem shows, for example, that the group $\mathbf{Q}$ of the rational numbers cannot be the fundamental group of a compactum, while the result of EDA and KAWAMURA [2000b] states that it is contained in the first singular homology of the Hawaiian earring.

6.4. THEOREM (SHELAH [1988], PAWLIKOWSKI [1998]). *The fundamental group of a compact metric space is either finitely generated or uncountable.*

Also in ZASTROW [1998] and CANNON, CONNER and ZASTROW [200?], it is shown that every subspace of $\mathbf{R}^2$ is a $K(\pi, 1)$ space, and for each one-dimensional metric space X, $\pi_1(X)$ is isomorphic to a subgroup of the shape group $\check{\pi}_1(X)$ (EDA and KAWAMURA [1998]). However very little is known about singular homology groups and homotopy groups of higher dimensional spaces (BARRATT and MILNOR [1962], EDA [1991], [1992b], EDA and KAWAMURA [2000a], KARIMOV and REPOVŠ [2001], REPOVŠ and SKOPENKOV [2001], etc).

References

AGEEV, S.M.

[1993] Classifying spaces for free actions and the Hilbert-Smith conjecture, *Russian Acad. Sci. Sbornik Math.* **75**, 137–144.

[200?] Axiomatic method of partitions in the theory of Menger and Nöbeling spaces, preprint.

AGEEV, S.M. and D. REPOVŠ

[2000] A new construction of semi-free actions on Menger manifolds, *Proc. Amer. Math. Soc.* **129**, 1551–1562.

AKHMET'EV, P.M.

[2000] Embedding of compacta, stable homotopy groups of spheres and singularity theory, *Russian Math. Surveys.* **55** (3), 405–462.

ANCEL, F.D. and C.R. GUILBAULT

[1997] Interiors of compact contractible N-mafnifolds are hyperbolic (N $\geq$ 5), *Pacific J. Math.* **45**, 1–32.

[1999] $\mathcal{Z}$-compactifications of open manifolds, *Topology* **38**, 1265–1280.

BARRATT, M.G. and J. MILNOR

[1962] An example of anomalous singular theory, *Proc. Amer. Math. Soc.* **13**, 293–297.

BENAKLI, N.

[1992] *Polyhdres hyperboliques passage du lacal an global*, PhD thesis, Univ. de Paris-Sud, Centre d'Orsay.

BESTVINA, M.

[1988] *Characterizing k-dimensional Universal Menger Compactum*, Mem. Amer. Math. Soc. vol. 71 (380).

[1996] Local homology properties of boundaries of groups, *Michigan Math. J.* **43**, 123–139.

BESTVINA, M., R.J. DAVERMAN, G.A. VENEMA and J.J. WALSH

[2001] A 4-dimensional 1-lcc shrinking theorem, *Top. Appl.* **110**, 3–20.

BESTVINA, M. and R.D. EDWARDS

[1987] Some remarks on the Hilbert-Smith conjecture, mimeographical note by D. Garity.

Bestvina, M. and G. Mess
[1991] The boundary of negatively curved groups, *J. Amer. Math. Soc.* **4**, 469–481.

Bing, R.H. and K. Borsuk
[1965] Some remarks concerning topologically homogeneous spaces, *Ann. Math.* **81**, 100–111.

Bowditch, B.H.
[1998] A topological characterization of hyperbolic groups, *J. Amer. Math. Soc.* **11**, 643–667.
[1999] *Treelike Structures Arising from Continua and Convergence Groups*, Mem. Amer. Math. Soc. vol. 139 (662).

Brechner, B.
[1984] Prime ends and group actions, unpublished preprint.

Brown, K.S.
[1982] *Cohomology of Groups*, Graduate Texts in Math., vol. 87, Springer, Berlin.

Bryant, J.L.
[1987] Homogeneous ENR's, *Top. Appl.* **27**, 301–306.
[2001] A survey of recent results on generalized manifolds, *Top. Appl.* **113**, 13–22.

Bryant, J., S. Ferry, W. Mio and S. Weinberger
[1996] Topology of homology manifolds, *Ann. Math.* **143**, 435–467.
[200?] Dissingularizing homology manifolds, preprint.

Busemann, H.
[1955] *The Geometry of Geodesics*, Monograph in Pure and Applied Math., vol. 6, Adademic Press.

Cannon, J. and G. Conner
[2000a] The combinatorical structure of the Hawaiian earring, *Top. Appl.* **106**, 225–271.
[2000b] The big fundamental group, big Hawaiian earring and the big free groups, *Top. Appl.* **106**, 273–291.

Cannon, J., G. Conner and A. Zastrow
[200?] One-dimensional sets and planar sets are aspherical, preprint.

Chapman, T.A. and L.C. Siebenmann
[1976] Finding a boundary for a Hilbert cube manifold, *Acta Math.* **137**, 171–208.

Cheeger, J. and D.G. Ebin
[1973] *Comparison Theorem in Riemannian Geometry*, North Holland Math. Lib., vol. 9, North Holland, Amsterdam.

Chigogidze, A.
[1996] *Inverse Spectra*, North-Holland Math. Lib., vol. 53, North Holland, Amsterdam.
[1997] Cohomological dimension of Tychonoff spaces, *Top. Appl.* **79**, 197–228.
[1999] Compactifications and universal spaces in extension theory, *Proc. Amer. Math. Soc.* **128**, 2187–2190.

Chigogidze, A., K. Kawamura and E.D. Tymchatyn
[1995] Menger manifolds, in *Continua with Houston Problem Book*, H. Cook, W.T. Ingram, K.T. Kuperberg, A. Lelek and P. Minc, eds., Marcel Dekker, pp. 37–88.

Chigogidze, A. and V. Valov
[2001] Universal metric spaces and extension dimension, *Top. Appl.* **113**, 23–28.

Chinen, N.
[1998] Manifolds with nonzero Euler characteristic and codimension 2 fibrators, *Top. Appl.* **86**, 151–167.
[2000] Finite groups and approximate fibrations, *Top. Appl.* **102**, 59–88.

COORNAERT, M., T. DELZANT and A. PAPADOPOULOS
[1990] *Géométrie et thèorie des groupes*, Lecture Notes in Math., vol. 1441, Springer, Berlin.

COORNAERT, M. and A. PAPADOPOULOS
[1993] *Symbolic dynamics and hyperbolic groups*, Lecture Notes in Math., vol. 1539, Springer, Berlin.

CORAM, D.S. and JR. F. DUVALL
[1977] Approximate fibrations, *Rocky Mt. J. Math.* **7**, 275–288.

CROKE, C.B. and V. KLEINER
[2000] Spaces with nonpositive curvature and their boundaries, *Topology* **39**, 549–556.

DAVERMAN, R.J.
[1981] Detecting the disjoint disks property, *Pacific J. Math.* **93**, 277–298.
[1986] *Decomposition of Manifolds*, Pure and applied mathematics, vol. 124, Academic Press.
[1991a] 3-manifolds with geometric structure and approximate fibrations, *Indiana Univ. Math. J.* **40**, 1451–1469.
[1991b] Manifolds with finite first homology as condimension 2 fibrators, *Proc. Amer. Math. Soc.* **113**, 471–477.
[1993a] Hyperhopfian groups and approximate fibrations, *Compositio Math.* **86**, 159–176.
[1993b] The PL fibrators among aspherical geometric 3-manifolds, *Michigan Math. J.* **8**, 571–585.
[1995a] Complex projective spaces as PL fibrators, *Top. Appl.* **66**, 67–74.
[1995b] Manifolds that induce approximate fibrations in the PL category, *Top. Appl.* **66**, 267–297.
[1999a] Codimension 2 nonfibrators with finite fundamental groups, *Proc. Amer. Math. Soc.* **127**, 881–888.
[1999b] Real projective spaces are nonfibrators, *Top. Appl.* **94**, 61–66.
[200?] Hyperbolic groups are hyperhopfian, preprint.

DAVERMAN, R.J. and Y. KIM
[200?] 2-groups and approximate fibrations, preprint.

DAVERMAN, R.J. and D. REPOVŠ
[1989] A new 3-dimensional shrinking theorem, *Trans. Amer. Math. Soc.* **315**, 219-230.
[1992] General position properties that characterize 3-manifolds, *Canad. J. Math.* **44**, 234–251.

DAVERMAN, R.J. and R.B. SHER, (EDITORS)
[2002] Handbook of Geometric Topology, Elsevier Sci. B.V., Amsterdam.

DAVIS, M.W.
[1998] The cohomology of a Coxeter group with group ring coefficients, *Duke Math. J.* **91**, 297–314.
[2002] Nonpositive curevature and reflection, in *Handbook of Geometric Topology*, R.J. Daverman and R.B. Sher, eds., Elsevier Sci. B.V., Amsterdam, pp. 373–422.

DOBROWOLSKI, T., M. LEVIN and L. RUBIN
[1997] Certain 2-stable embeddings, *Top. Appl.* **80**, 81–90.

DRANISHNIKOV, A.N.
[1988a] Homological dimension theory, *Russian Math. Surveys* **43** (4), 11–83.
[1988b] On free actions of zero-dimensional compact groups, *Izv. Akad. Nauk USSR* **32**, 217–232.
[1997a] On the virtual cohomological dimensions of Coxeter groups, *Proc. Amer. Math. Soc.* **125**, 1885–1891.
[1997b] On Chogoshvili's conjecture, *Proc. Amer. Math. Soc.* **125**, 2155–2160.

[1998] Extension theory for maps of compact spaces, *Russian Math. Surveys* **53** (5), 929–935.
[1999] Boundaries of Coxeter groups and simplicial complexes with given links, *J. Pure Appl. Alg.* **137**, 139–151.
[2000a] Asymptotic topology, *Russian Math. Surveys* **55** (6), 1085–1129.
[2000b] On the dimension of the product of two compacta and the dimension of their intersection in general position in Euclidean space, *Trans. Amer. Math. Soc.* **352**, 5599–5618.
[200?a] Boundaries and cohomological dimension of Coxeter groups, preprint.
[200?b] Basic elements of the cohomological dimension theory of compact metric spaces, preprint.

DRANISHNIKOV, A.N. and J. DYDAK
[1996] Extension dimension and extension types, in *Proc. Steklov Inst. Math.* **212**, 55–88.

DRANISHNIKOV, A.N. and S. FERRY
[1997] On the Higson-Roe corona, *Russian Math. Surveys* **52** (5), 1017–1028.

DRANISHNIKOV, A.N., J. KEESLING and V.V. USPENSKIJ
[1998] On the Higson corona of uniformly contractible spaces, *Topology* **37**, 791–803.

DRANISHNIKOV, A.N. and J. WEST
[1997] Compact group actions that raise dimension to infinity, *Top. Appl.* **80**, 101–114.

DYDAK, J.
[1993] Compactifications and cohomological dimension, *Top. Appl.* **50**, 1–10.
[1996] Extension theory: The interface between set-theoretic and algebraic topology, *Top. Appl.* **20**, 1–34.
[2002] Cohomological dimension theory, in *Handbook of Geometric Topology*, R.J. Daverman and R.B. Sher, eds., Elsevier Sci. B.V., Amsterdam, pp. 423–470.
[200?] Geometry and algebra in dimension theory, preprint.

DYDAK, J. and J. MOGILSKI
[1994] Universal cell-like maps, *Proc. Amer. Math. Soc.* **122**, 943–948.

DYDAK, J. and J.J. WALSH
[1987] Sheaves that are locally constant with applications to homology manifolds, in *Geometric topology and Shape theory*, S. Mardešić and J. Segal, eds., Lecture Notes in Math., vol. 1283, Springer, Berlin, pp. 65–87.
[1991] Spaces without dimension preserving compactifications, *Proc. Amer. Math. Soc.* **113**, 1155–1162.
[1993a] Complexes that arise in cohomological dimension theory: A unified approach, *J. London Math. Soc.* **48**, 329–347.
[1993b] Infinite dimensional compacta having cohomological dimension two: An application of the Sullivan conjecture, *Topology* **32**, 93–104.

EDA, K.
[1991] The first integral singular homology groups of one point unions, *Quart. J. Math. Oxford* **42**, 443–456.
[1992a] Free σ-products and noncommutative slender groups, *J. Algebra* **148**, 243–263.
[1992b] A locally simply connected space and fundamental groups of one point unions of cones, *Proc. Amer. Math. Soc.* **116**, 239–249.

EDA, K. and K. KAWAMURA
[1998] The fundamental groups of 1-dimensional spaces, *Top. Appl.* **87**, 163–172.
[2000a] Homotopy and homology groups of the n-dimensional Hawaiian earring, *Fund. Math.* **165**, 17–28.
[2000b] The singular homology of the Hawaiian earring, *J. London Math. Soc.* **62**, 305–310.

EDWARDS, R.D.

[1980] The topology of manifolds and cell-like maps, in *Proc. Internat. Cong. Mathematicians*, O. Lehto ed., Helsinki, Acad. Sci. Fenn., Helsinki, pp. 111–127.

FERRY, S.C.

[1994] Topological finiteness theorem in Gromov-Hausdorff space, *Duke Math. J.* **74**, 95–106.

[1998] Limits of polyhedra in Gromov-Hausdorff convergence, *Topology* **37**, 1325–1338.

[2000] Stable compactifications of polyhedra, *Michigan Math. J.* **47**, 287–294.

FERRY, S.C. and B.L. OKUN

[1995] Approximating topological metrics by Riemannian metrics, *Proc. Amer. Math. Soc.* **123**, 1865–1872.

GHYS, E., A. HAEFLIGER and A. VERJOVSKY

[1993] *Groups Theory from a Geometrical View Point*, World Scientific, Singapore.

GHYS, E. and P. DE LA HARPE

[1990] *Sur les Groupes Hyperboliques d'après Mikhael Gromov*, Progress in Math., vol. 83, Birkhäuser, Basel.

GROMOV, M.

[1981] Groups of polynomical growth and expanding maps, *Pub. Math. I.H.E.S.* **53**, 53–73.

[1988] Hyperbolic groups, in *Essays in Group Theory*, S.M. Gersten editor, M.S.R.I. Pub., vol. 8.

GROVE, K., P. PETERSEN and J.-Y. WU

[1990] Geometric finiteness theorems via controlled topology, correction, *Inv. Math.* **99, 104**, 205–213, 222–223.

GUILBAULT, C.R.

[2001] A non-$\mathcal{Z}$-compactifiable polyhedron whose product with the Hilbert cube is $\mathcal{Z}$-compactifiable, *Fund. Math.* **168**, 165–197.

HOSAKA, T.

[200?] On the cohomology of the Coxeter groups, *J. Pure Appl.*, to appear.

IM, Y.H.

[1995] Products of surface that induce approximate fibrations, *Houston J. Math.* **21**, 339–348.

IM, Y.H. and Y. KIM

[1999] Hopfian and strongly hopfian manifolds, *Fund. Math.* **159**, 127–134.

[2000] Necessary and sufficient conditions for s-hopfian manifolds to be codimension-2 fibrators, *Proc. Amer. Math. Soc.* **129**, 2135–2140.

IWAMOTO, Y.

[1996] Fixed points of transformation groups of Menger manifolds, their pseudo-interiors and their pseudo-boundaries, *Top. App.* **68**, 267–283.

JAKOBSCHE, W.

[1980] The Bing-Borsuk conjecture is stronger than the Poincaré conjecture, *Fund. Math.* **106**, 127–134.

KARIMOV, U.H. and D. REPOVŠ

[2001] On embeddability of contractible k-dimensional compacta into $\mathbb{R}^{2n}$, *Top. Appl.* **113**, 81–85.

KAWAMURA, K.

[2000] A survey on Menger manifold theory – Update, *Top. Appl.* **101**, 83–91.

KEESLING, J.

[1994] The one-dimensional Čech cohomology of the Higson compactification and its corona, *Topology Proc.* **19**, 129–148.

KIM, Y.

[2000] Connected sums of manifolds which induce approximate fibrations, *Proc. Amer. Math. Soc.* **128**, 1497–1506.

KOLMOGOROV, A.N.

[1937] Über offene Abbildüngen, *Ann. Math.* **38**, 488–498.

KOYAMA, A.

[2001] Recent developement of cohomological dimension theory - Edwards-Walsh resolution and its applications - (in Japanese), *Sugaku* **53**, 1349–1372.

KOYAMA, A. and K. YOKOI

[200?] Cohomological dimension and acyclic resolutions, *Top. Appl.*, to appear

KOZLOWSKI, G. and J.J. WALSH

[1983] Cell-like mappings on 3-manifolds, *Topology* **22**, 147–151.

KRAKUS, B.

[1968] Any 3-dimensional G-space is a manifold, *Bull. Acad. Pol. Sci.* **16**, 737–740.

LEE, J.S.

[1984] Almost periodic homeomorphisms and p-adic transformation groups on compact 3-manifolds, *Proc. Amer. Math. Soc.* **121**, 267–273.

LEVIN, M.

[2001] Constructing compacta of different extensional dimensions, *Cand. Math. Bull.* **44**, 80–86.

[200?] On extensional dimension of metrizable spaces, preprint.

LEVIN, M. and Y. STERNFELD

[1998] Atomic maps and the Chogoshvili-Pontrjagin claim, *Trans. Amer. Math. Soc.* **350**, 4623–4632.

LEVITT, G.

[1998] Non-nesting actions on real trees, *Bull. London Math. Soc.* **30**, 46–54.

MALEŠIČ, J.

[1997] Hilbert-Smith conjecture for actions by Hölder maps, *Russian Math. Surveys* **52** (2), 407–408.

MAYER, J.C. and C.W. STARK

[200?] Dimension-raising p-adic actions on Menger manifolds, preprint.

MIHALIK, M.

[1996] Semistability of Artin and Coxeter groups, *J. Pure Appl Alg.* **111**, 205–211.

MIHALIK, M. and K. RUANE

[1999] CAT(0) groups with non-locally connected boundary, *J. London Math. Soc.* **60**, 757–770.

[2001] CAT(0) HNN- extensions with non-locally connected boundary, *Top. Appl.* **110**, 83–98.

VAN MILL, J. and G.M. REED, EDITORS

[1990] *Open Problems in Topology*, North Holland, Amsterdam.

MITCHELL, W.J.R., D. REPOVŠ and E.V. ŠČEPIN

[1992] On 1-cycles and the finite dimensionality of homology 4-manifolds, *Topology* **31**, 605–623.

MIYATA, T.
[1996] Extension type and homology, *Bull. Pol. Acad. Sci. Math.* **44**, 1–4.

MOORE, T.E.
[1995] Gromov-Hausdorff convengence to nonmanifolds, *J. Geom. Anal.* **5**, 411–418.

MOUSSONG, G.
[1988] *Hyperbolic Coxeter groups*, PhD thesis, Ohio State Univ.

OLSZEWSKI, W.
[1995a] Completion theorem for cohomological dimensions, *Proc. Amer. Math. Soc.* **123**, 2261–2264.
[1995b] Universal separable metric spaces of given cohomological dimension, *Top. Appl.* **61**, 293–299.

ONTANEDA, P.
[1996] Local extendability of CAT(0)-groups, preprint.

PAWLIKOWSKI, J.
[1998] The fundamental groups of a compact metric space, *Proc. Amer. Math. Soc.* **126**, 3083–3087.

PEDERSEN, E., F. QUINN and A. RANICKI
[200?] Controlled surgery with trivial local fundamental groups, preprint.

QUINN, F.
[1987] An obstruction to the resolution of homology manifolds, *Michigan Math. J.* **34**, 285–291.

RAYMOND, F. and R.F. WILLIAMS
[1963] Examples of p-adic transformation groups, *Ann. Math.* **78**, 92–106.

REPOVŠ, D.
[1992] Detection of higher dimensional topological manifolds among topological spaces, Seminari di geometria, *Giornate de topologia e geometria delle varietà*, 113–143.
[1994] The recognition problem for topological manifolds: A survey, *Kodai Math. J.* **17**, 538–548.

REPOVŠ, D. and E.V. ŠČEPIN
[1997] A proof of the Hilbert-Smith conjecture for actions by Lipschitz maps, *Math. Ann.* **308**, 361–364.

REPOVŠ, D. and A.B. SKOPENKOV
[1999] New results on embedding of polyhedra and manifolds in Euclidean spaces, *Russian Math. Survey* **54** (6), 1149–1196.
[2001] On contractible n-dimensional compacta, non-embeddable into $\mathbf{R}^{2n}$, *Proc. Amer. Math. Soc.* **129**, 627–628.

REPOVŠ, D., A.B. SKOPENKOV and E.V. ŠČEPIN
[1996] C^1-homogeneous compacta in $\mathbf{R}^n$ are C^1-submanifolds in $\mathbf{R}^n$, *Proc. Amer. Math. Soc.* **124**, 1219–1226.

ROE, J.
[1993] *Coarse cohomology and index theory on complete Riemannian manifolds*, Mem. Amer. Math. Soc., vol. 104(497), A.M.S.

SAKAI, K.
[1994] Free actions of zero-dimensional compact groups on Menger manifolds, *Proc. Amer. Math. Soc.* **122**, 647–648.

ŠČEPIN, E.V.

[1998] Arithmetic of dimension theory, *Russian Math. Surveys* **53** (5), 975–1069.

SELA, Z.

[1999] Endomorphism of hyperbolic groups I; the Hopf property, *Topology* **38**, 301–321.

SHELAH, S.

[1988] Can the fundamental (homotopy) group of a space be the rationals? *Proc. Amer. Math. Soc.* **103**, 627–632.

SIEBENMANN, L.C.

[1965] *The obstruction to finding a boundary for an open manifolds of dimension greater than five*, PhD thesis, Princeton Univ.

SWARUP, G.A.

[1996] On the cup point conjecture, *Electron. Res. Announc. Amer. Math. Soc.* **2**, MR 97f, 20048.

SWENSON, E.L.

[1999a] A cut point theorem for CAT(0) groups, *J. Diff. Geom.* **53**, 327–358.

[1999b] On axion H, *Michigan Math. J.* **46**, 3–11.

THURSTON, P.

[1996] 4-dimensional Busemann G-spaces are 4-manifolds, *J. Diff. Geom.* **6**, 245–270.

WALSH, J.J.

[1981] Dimension, cohomological dimension and cell-like mappings, in *Shape theory and geometric topology*, S. Mardesič and J. Segal, eds. Lecture Notes in Math. vol. 870, Springer, Berlin, pp. 105–118.

[1983] The finitie dimensionalily of integral homology 3-manifols, *Proc. Amer. Math. Soc.* **88**, 154–156.

WATANABE, T.

[1995] A note on cohomological dimension of approximate movable spaces, *Proc. Amer. Math. Soc.* **123**, 2883–2885.

WU, J.-Y.

[1999] A generalization of a theorem of Edwards, *Proc. Amer. Math. Soc.* **127**, 3119–3123.

YANG, C. T.

[1960] p-Adic transformation groups, *Michigan Math. J.* **7**, 201–218.

YOKOI, K.

[2000] Edwards-walsh resolutions of complexes and abelian groups, *Bull. Austr. Math. Soc.* **62**, 407–416.

ZARICHNYI, M.

[1997] Absorbing sets in the Hilbert cube related to cohomological dimension, *Top. Appl.* **80**, 195–200.

[200?] Universal spaces and absolute extensors for integral cohomological dimension, preprint.

ZASTROW, A.

[1998] *Planar sets are aspherical*, Habilitationsschrift, Bochum.

CHAPTER 12

Quasi-Uniform Spaces in the Year 2001

Hans-Peter A. Künzi

Department of Mathematics and Applied Mathematics, University of Cape Town, Rondebosch 7701
South Africa
E-mail: kunzi@maths.uct.ac.za

Contents

RECENT PROGRESS IN GENERAL TOPOLOGY II
Edited by Miroslav Hušek and Jan van Mill

1. Introduction

In the last two decades most of the research conducted in the area of uniform structures dealt either with topological groups, or so-called quasi-uniformities or quasi-pseudometrics, that is, uniformities (resp. pseudometrics) that do not necessarily satisfy the symmetry condition. Since the work related to topological groups is discussed in another chapter of this book, we shall concentrate on the second topic in this article.

The two monographs of MURDESHWAR and NAIMPALLY [1966] and FLETCHER and LINDGREN [1982] contain many basic facts about the category **Quu** of quasi-uniform spaces and quasi-uniformly continuous maps. In the following we shall assume that the reader is familiar with these two books and mainly discuss results proved after the publication of FLETCHER and LINDGREN [1982].

Recent studies in the area of quasi-uniform spaces often dealt with applications of asymmetric structures to problems in Theoretical Computer Science and Topological Algebra. Much progress can be reported from the areas of quasi-uniform function spaces and hyperspaces. Remarkable contributions were also made concerning extensions and completions of quasi-uniform spaces; more work however still seems to be needed to understand better the connections between the various suggested completions.

Special attention will be given in this note to some new developments in the field that are not yet covered in the article of KÜNZI [2001]. The latter article provides a comprehensive recent survey (without proofs) of the area of quasi-uniform spaces. We refer the reader to it for many results that had to be omitted here because of lack of space.

Further interesting information on quasi-uniform structures can be found in the survey articles of BRÜMMER [1999], CSÁSZÁR [1999], DEÁK [1993b], KOFNER [1980], KOPPERMAN [1995] and KÜNZI [1993, 1995a, 1998]. A modern treatment of the theory of uniform spaces was presented in the book of HOWES [1995]. Other interesting facts about the history of uniform structures were discussed in the article of BENTLEY, HERRLICH and HUŠEK [1998].

In order to fix the terminology let us recall a few concepts.

1.1. DEFINITION. A *quasi-uniformity* on a set X is a filter $\mathcal{U}$ on $X \times X$ such that each member of $\mathcal{U}$ is reflexive and for each $U \in \mathcal{U}$ there is $V \in \mathcal{U}$ such that $V^2 = V \circ V \subseteq U$. Here $V^2 = \{(x, z) \in X \times X : \text{there is } y \in X \text{ such that } (x, y) \in V \text{ and } (y, z) \in V\}$.

The filter $\mathcal{U}^{-1} = \{U^{-1} : U \in \mathcal{U}\}$ where $U^{-1} = \{(x, y) \in X \times X : (y, x) \in U\}$ is called the *conjugate quasi-uniformity* of $\mathcal{U}$. A quasi-uniformity $\mathcal{U}$ is a *uniformity* provided that $\mathcal{U} = \mathcal{U}^{-1}$.

A map $f : (X,\mathcal{U}) \to (Y, \mathcal{V})$ between two quasi-uniform spaces $(X,\mathcal{U})$ and $(Y, \mathcal{V})$ is called *quasi-uniformly continuous* if for each $V \in \mathcal{V}$ there exists $U \in \mathcal{U}$ such that $(f \times f)U \subseteq V$.

If $\mathcal{U}_1$ and $\mathcal{U}_2$ are two quasi-uniformities on X such that $\mathcal{U}_1 \subseteq \mathcal{U}_2$, then $\mathcal{U}_1$ is called *coarser* than $\mathcal{U}_2$. The coarsest uniformity $\mathcal{U}^s$ finer than a quasi-uniformity $\mathcal{U}$ is generated by the subbase $\mathcal{U} \cup \mathcal{U}^{-1}$.

A quasi-uniformity $\mathcal{U}$ determines a topology $\tau(\mathcal{U})$ given by the neighborhood filters $\mathcal{U}(x) = \{U(x) : U \in \mathcal{U}\}$ $(x \in X)$ where $U(x) = \{y \in X : (x, y) \in U\}$. Each quasi-uniformly continuous map between quasi-uniform spaces is continuous.

Every topological space (X, τ) is *quasi-uniformizable*, that is, its topology coincides with the topology induced by the *Pervin quasi-uniformity* of X generated by the subbase $\{[(X \setminus G) \times X] \cup [G \times G] : G \in \tau\}$, and it admits a finest compatible quasi-uniformity called its *fine quasi-uniformity*.

A quasi-uniformity $\mathcal{U}$ on a set X induces the bitopological space (that is, bispace) $(X, \tau(\mathcal{U}), \tau(\mathcal{U}^{-1}))$. Exactly the pairwise completely regular bispaces can be obtained in this way (see e.g. KÜNZI [2001]).

In the following, $[0, \infty)$ will denote the set of nonnegative reals.

1.2. DEFINITION. A *quasi-pseudometric* on a set X is a function $d : X \times X \to [0, \infty)$ that satisfies $d(x, x) = 0$ and $d(x, z) \leq d(x, y) + d(y, z)$ whenever $x, y, z \in X$.

Then d^{-1} defined by $d^{-1}(x, y) = d(y, x)$ whenever $x, y \in X$ is the *conjugate quasi-pseudometric* of d. A quasi-pseudometric d on X is called a *quasi-metric* if $x, y \in X$ and $d(x, y) = 0$ imply that $x = y$; it is called *non-archimedean* whenever the inequality $d(x, z) \leq \max\{d(x, y), d(y, z)\}$ holds for $x, y, z \in X$.

It should be noted that many interesting questions about asymmetric distances originate in combinatorics (see e.g. DEZA and PANTELEEVA [2000]).

Each quasi-pseudometric d on a set X generates a quasi-uniformity $\mathcal{U}_d$ having the base $\{\{(x, y) : d(x, y) < \epsilon\} : \epsilon > 0\}$. For each quasi-uniformity $\mathcal{U}$ possessing a countable base there is a quasi-pseudometric d such that $\mathcal{U} = \mathcal{U}_d$ (compare FLETCHER and LINDGREN [1982]). Hence each quasi-uniformity can be represented as the supremum of a family (= gauge) of quasi-pseudometric quasi-uniformities.

Other approaches (also discussed in KÜNZI [2001]) to the concept of a quasi-uniformity are known, e.g. via pair-covers, or via quasi-pseudometrics taking values in ordered structures more general than the reals (see e.g. GOLAN [1998] and HEITZIG [200?] for such investigations related to Kopperman's idea of a *continuity space*, which was dealt with e.g. in FLAGG [1997] and FLAGG and KOPPERMAN [1997]).

Recently FERREIRA and PICADO [200?] pointed out that (quasi-)uniform structures can also be described by Galois connections in two different ways, namely in terms of polarities and in terms of axialities and that Galois connections rather than entourages are the root of the FLETCHER, HUNSAKER and LINDGREN [1995] approach to frame quasi-uniformities.

CLEMENTINO and HOFMANN [200?] described quasi-uniform spaces as reflexive and transitive lax algebras for the identity monad in a convenient 2-category. In the spirit of Lawvere, SCHMITT [200?] applied the language of enriched categories to quasi-uniform spaces.

Many computer scientists developed by means of quasi-uniformities a theory that combines common features (like completions and fixed-point theorems) of the theories of metric spaces and partial orders (see e.g. HITZLER and SEDA [1999] and SEDA [1997]). BRATTKA [200?a, 200?b] showed that the most important function spaces and hyperspaces of computable analysis can be considered recursive quasi-(pseudo)metric spaces.

LOWEN and WINDELS [1998] introduced a notion of an approach quasi-uniformity which uniformizes the concept of an approach space and quantifies the concept of a (quasi-)uniform space.

1.3. DEFINITION. A quasi-uniform space $(X,\mathcal{U})$ is called *precompact* (resp. *totally bounded*) if for each $U \in \mathcal{U}$, $\{U(x) : x \in X\}$ has a finite subcover (resp. $\mathcal{U}^s$ is precompact).

The concept of a totally bounded quasi-uniform space is (in the categorical sense) equivalent to the notion of a quasi-proximity space (see e.g. FLETCHER and LINDGREN [1982]); precompact quasi-uniform spaces need not be totally bounded.

Each quasi-uniformity $\mathcal{U}$ contains a finest totally bounded quasi-uniformity $\mathcal{U}_\omega$ coarser than $\mathcal{U}$; furthermore $\tau(\mathcal{U}_\omega) = \tau(\mathcal{U})$. For a quasi-uniformity $\mathcal{U}$ on X, the set of all quasi-uniformities $\mathcal{V}$ on X such that $\mathcal{V}_\omega = \mathcal{U}_\omega$ is called the *quasi-proximity class* of $\mathcal{U}$. For an arbitrary topological space X the Pervin quasi-uniformity is the finest compatible totally bounded quasi-uniformity of X.

1.4. DEFINITION. A quasi-uniformity is called *transitive* if it has a base consisting of transitive relations.

The following construction is due to Fletcher (compare e.g. FLETCHER and LINDGREN [1982]):

Let (X,τ) be a topological space and let $\mathcal{A}$ be a collection of interior-preserving open covers $\mathcal{C}$ of X such that $\bigcup\mathcal{A}$ is a subbase for τ. For any $\mathcal{C}$ set $U_\mathcal{C} = \bigcup_{x\in X}(\{x\} \times \bigcap\{D : x \in D \in \mathcal{C}\})$. Then the collection $\{U_\mathcal{C} : \mathcal{C} \in \mathcal{A}\}$ is a subbase for a compatible transitive quasi-uniformity $\mathcal{U}_\mathcal{A}$ on X.

If $\mathcal{A}$ is the collection of all finite (resp. point-finite, locally finite, interior-preserving) open covers of X, then $\mathcal{U}_\mathcal{A}$ is the Pervin (resp. point-finite, locally finite, fine transitive) quasi-uniformity of X. As is shown in FLETCHER and LINDGREN [1982], the semicontinuous quasi-uniformity can also be constructed in this way with the help of so-called open spectra.

If $\mathcal{A}$ is the collection of all *well-monotone* open covers (that is, covers well-ordered by set-theoretic inclusion), then $\mathcal{U}_\mathcal{A}$ is the *well-monotone quasi-uniformity* of X. Its conjugate is always hereditarily precompact.

2. Basic concepts

Recently various studies about the number of compatible (transitive) quasi-uniformities and quasi-proximities on topological spaces were conducted. Useful methods of constructing quasi-uniformities were found as the result of such investigations.

2.1. THEOREM (compare KÜNZI [1993]). *A topological space admits a unique quasi-uniformity if and only if each of its interior-preserving open collections is finite.*

A topological space (X,τ) admits a unique quasi-proximity if and only if its topology τ is the unique base of open sets for τ that is closed under finite unions and finite intersections. (Here the convention that $\bigcap\emptyset = X$ was used.)

The latter (strictly weaker) condition is implied by hereditary compactness and is equivalent to finiteness in Hausdorff spaces; in KÜNZI and WATSON [1996a] it was shown that a T_1-space satisfying it need not be hereditarily compact.

2.2. LEMMA (KÜNZI [1999]). *Each topological space (X, τ) such that for any entourage W belonging to its fine quasi-uniformity, the cover $\{(W \cap W^{-1})(x) : x \in X\}$ has a subcover of cardinality κ, admits at most $2^{|\tau|^\kappa}$ quasi-uniformities.*

2.3. COROLLARY. *A topological space X of network weight $nw(X)$ admits at most $2^{2^{nw(X)}}$ quasi-uniformities.*

2.4. THEOREM (KÜNZI and LOSONCZI [2000]). *A topological space possessing more than one compatible quasi-uniformity admits at least $2^{2^{\aleph_0}}$ nontransitive quasi-uniformities.*

The quasi-proximity class of a transitive quasi-uniformity contains at least $2^{2^{\aleph_0}}$ transitive quasi-uniformities if it contains more than one member.

If $(X, \mathcal{W})$ is a quasi-uniform space possessing an entourage W and a subspace A such that either $\{W(x) : x \in A\}$ or $\{W^{-1}(x) : x \in A\}$ does not have a subcollection of cardinality smaller than κ covering A (where κ denotes an infinite cardinal), then there are at least 2^{2^κ} quasi-uniformities belonging to the quasi-proximity class of $\mathcal{W}$.

2.5. COROLLARY. *If a quasi-proximity class of a quasi-uniformity possesses more than one member, then it contains at least $2^{2^{\aleph_0}}$ quasi-uniformities.*

2.6. REMARK. It is unknown whether each quasi-proximity class that contains more than one member contains at least $2^{2^{\aleph_0}}$ nontransitive quasi-uniformities.

Recently KÜNZI [200?a] showed that this is indeed the case for the Pervin quasi-proximity class of a topological space.

By a result of KÜNZI and LOSONCZI [2001] *for each nonzero cardinal κ there exists a T_0-space possessing exactly κ compatible totally bounded quasi-uniformities.*

In [2000] LOSONCZI mentioned that for any topological space there is a one-to-one correspondence between the compatible *transitive* totally bounded quasi-uniformities and the bases for the topology that are closed under finite unions and finite intersections.

2.7. THEOREM (KÜNZI and PÉREZ-PEÑALVER [2000]). *Each supersober (in particular, each T_2-)space having a discrete subspace of infinite cardinality κ admits at least 2^{2^κ} transitive totally bounded quasi-uniformities.*

A topological space which admits a nontransitive totally bounded quasi-uniformity admits at least $2^{2^{\aleph_0}}$ nontransitive totally bounded quasi-uniformities.

Here "supersober" means that the convergence set of each convergent ultrafilter is the closure of some unique singleton.

2.8. THEOREM (GERLITS, KÜNZI, LOSONCZI and SZENTMIKLÓSSY [2002]). *Each infinite Tychonoff space admits a nontransitive totally bounded quasi-uniformity.*

Several results about the semilattice $TB(X)$ of the compatible totally bounded quasi-uniformities of a topological space X were established in KÜNZI and LOSONCZI [2001]; e.g. $TB(X)$ can be a non-modular lattice.

KÜNZI (compare [1993]) characterized internally those topological spaces that admit a coarsest quasi-uniformity with the help of the concept of the *handy relation*. His condition is implied by local compactness and is equivalent to this property in supersober spaces; "local compactness" means that each point of the space has a neighborhood base consisting of compact sets.

2.9. THEOREM (LOSONCZI[1998]). *For a locally compact T_2-space X the quasi-proximity class of the coarsest compatible quasi-uniformity contains a unique element if and only if X is compact or nonLindelöf.*

2.10. DEFINITION. A topological space is called *transitive* if its fine quasi-uniformity is transitive.

The property is known to be neither finitely productive nor hereditary (compare KÜNZI [1993]). Kofner's plane provides an example of a nontransitive space, since it is quasi-metrizable, but not non-archimedeanly quasi-metrizable. Recently another such space based on combinatorics was constructed in KÜNZI and WATSON [1999].

Questions about transitivity motivated a lot of the early work on quasi-uniformities and various kinds of spaces were shown to be transitive (compare FLETCHER and LINDGREN [1982] and KÜNZI [1993]): for instance, the orthocompact semistratifiable spaces, the suborderable spaces, and the T_1-spaces having an orthobase (see KOFNER [1981]).

An interesting recent result is the following fact due to Junnila, Künzi and Watson.

2.11. THEOREM (JUNNILA, KÜNZI and WATSON [200?]). *For any neighbornet V of a hereditarily metacompact compact regular space there is a transitive neighbornet T such that $T \subseteq V^3$.*

2.12. COROLLARY. *Each locally compact hereditarily metacompact regular space is transitive.*

In fact, recent investigations revealed that the condition of local compactness can be weakened considerably in the latter result (see KÜNZI [200?b]).

A useful characterization of T_1-spaces having an orthobase by decreasing chains of transitive partial neighbornets was obtained in JUNNILA and KÜNZI [1993a].

Based on a similar idea, in KÜNZI [2000] a new proof of KOFNER's result [1981] was provided that for an arbitrary neighbornet V of a T_1-space X which possesses an orthopairbase, V^2 belongs to the fine quasi-uniformity of X.

2.13. REMARK. However, major problems concerning transitive spaces remain open (compare KÜNZI [1993]):

Is the fine quasi-uniformity of each (quasi-metrizable) Moore space or each non-archimedeanly quasi-metrizable space transitive?

Is there in ZFC a compact Hausdorff space that is not transitive?

We conclude this section with a discussion of some newly introduced symmetry properties and a concept of monotonicity for quasi-uniform spaces.

A quasi-uniformity whose induced quasi-proximity is a proximity is called *proximally symmetric*.

FLETCHER and HUNSAKER showed in [1992b] that each proximally symmetric quasi-uniform space is quiet (defined below).

Several related symmetry properties of quasi-uniform spaces are best understood by means of their induced quasi-proximity.

So DEÁK [1994a] called a quasi-uniform space $(X,\mathcal{U})$ *mixed-symmetric* if whenever A is a closed, B is an open subset of $(X,\mathcal{U})$ and there is $U \in \mathcal{U}$ such that $U(A) \cap B = \emptyset$, then there is $V \in \mathcal{U}$ such that $V(B) \cap A = \emptyset$.

Similarly a quasi-uniform space is called *closed symmetric* (resp. *open symmetric*) provided that the condition above holds for closed (resp. open) sets A and B.

He observed that mixed symmetry generalizes both open and closed symmetry. *Each mixed-symmetric uniformly regular* (defined below) *quasi-uniform space is quiet.*

2.14. DEFINITION. A quasi-uniformity $\mathcal{U}$ on a set X is called *monotonic* if there exists an operator $M : \mathcal{U} \to \mathcal{U}$ such that $M(U) \subseteq M(V)$ whenever $U, V \in \mathcal{U}$ and $U \subseteq V$, and such that $M(U)^2 \subseteq U$ whenever $U \in \mathcal{U}$.

2.15. THEOREM (GARTSIDE and MOODY [1993]). *A topological T_1-space admits a monotonic uniformity if and only if it is monotonically normal and has an orthobase.*

JUNNILA and KÜNZI [1993a] proved that the fine quasi-uniformity of the Sorgenfrey line is not monotonic. Furthermore they established that *each T_1-space with an orthobase admits a monotonic quasi-uniformity.* A countable space need not admit any monotonic quasi-uniformity.

3. Extensions and completions

Császár, Deák and Losonczi studied in detail the following problem about extensions of quasi-uniformities (see CSÁSZÁR [1999] and DEÁK [1993b]):

Let $\mathcal{U}$ be a quasi-uniformity on X, τ the topology induced by $\mathcal{U}$, and σ an extension of τ defined on the carrier set $Y \supseteq X$; describe those quasi-uniformities $\mathcal{W}$ extending $\mathcal{U}$ that are compatible with σ.

The special cases where $\mathcal{W}$ is transitive or totally bounded, or the remainder $Y \setminus X$ is finite, or X is dense, doubly dense (that is, dense in $(Y, \mathcal{W})$ as well as in its conjugate) or supdense (that is, dense in $(Y, \mathcal{W}^s)$) in Y have attracted special interest. DEÁK [1993a, 1993b] showed that several methods developed by Bing and Hausdorff for the case of extensions of (pseudo)metrics apply to the asymmetric setting as well.

Many investigations were devoted to the standard completion of quasi-uniform spaces, the so-called *bicompletion* (see e.g. FLETCHER and LINDGREN [1982]).

3.1. DEFINITION. A quasi-uniform space $(X,\mathcal{U})$ is called *bicomplete* provided that $\mathcal{U}^s$ is a complete uniformity.

Each quasi-uniform T_0-space $(X,\mathcal{U})$ has an (up to quasi-uniform isomorphism) unique T_0-bicompletion $(\widetilde{X},\widetilde{\mathcal{U}})$ in the sense that the space $(\widetilde{X},\widetilde{\mathcal{U}})$ is a bicomplete T_0-extension of $(X,\mathcal{U})$ in which $(X,\mathcal{U})$ is $\tau(\widetilde{\mathcal{U}}^s)$-dense.

The uniformities $(\widetilde{\mathcal{U}})^s$ and $\widetilde{\mathcal{U}^s}$ coincide. Furthermore if $f : (X,\mathcal{U}) \to (Y,\mathcal{V})$ is a quasi-uniformly continuous map into a bicomplete quasi-uniform T_0-space $(Y,\mathcal{V})$, then there exists a (unique) quasi-uniformly continuous extension $\widetilde{f} : \widetilde{X} \to Y$ of f.

Bicompletions of totally bounded T_0-quasi-uniformities $\mathcal{U}$ yield *joincompactifications*, that is, $\tau(\widetilde{\mathcal{U}^s})$ is compact.

KÜNZI (compare [1993]) constructed the Fell compactification of a locally compact T_0-space as the bicompletion of its coarsest compatible quasi-uniformity.

3.2. THEOREM (BRÜMMER and KÜNZI [2002]). *The quasi-proximity space induced by the bicompletion of a quasi-uniform T_0-space $(X,\mathcal{U})$ is a subspace of the quasi-proximity space induced by the Samuel bicompactification $(\widetilde{X},\widetilde{\mathcal{U}_\omega})$ of $(X,\mathcal{U})$.*

The class of *locally compact strongly sober spaces* (sometimes also called *skew compact*; see KOPPERMAN [1995, Comment 4.12]), is a suitable analogue in the category $\mathbf{Top_0}$ of T_0-spaces to the class of compact spaces in the category of Hausdorff spaces. Here "strongly sober" means "compact and supersober".

The strongly sober locally compact spaces (see also the articles of HÖTZEL ESCARDÓ [2001], LAWSON [1991], SALBANY and TODOROV [2000] and SMYTH [1992] for further details) can be characterized as the topological T_0-spaces that admit a totally bounded bicomplete quasi-uniformity. It is the coarsest compatible quasi-uniformity of such spaces.

3.3. DEFINITION. A *$*$-compactification* of a quasi-uniform T_1-space $(X,\mathcal{U})$ is a compact quasi-uniform T_1-space $(Y,\mathcal{V})$ that has a $\tau(\mathcal{V}^s)$-dense subspace quasi-uniformly isomorphic to $(X,\mathcal{U})$.

Continuing work of FLETCHER and LINDGREN [1986], Romaguera and Sánchez-Granero obtained the following results. (Recall that a quasi-uniformity $\mathcal{U}$ is called *point-symmetric* if $\tau(\mathcal{U}) \subseteq \tau(\mathcal{U}^{-1})$; see e.g. FLETCHER and LINDGREN [1982].)

3.4. THEOREM (ROMAGUERA and SÁNCHEZ-GRANERO [200?a, 200?b]). *A quasi-uniform T_1-space is $*$-compactifiable if and only if it is point-symmetric and its bicompletion is compact.*

If it exists, the $*$-compactification is unique and can be identified with the subspace of the bicompletion consisting of the closed singletons.

Each point-symmetric totally bounded quasi-uniform T_1-space is $*$-compactifiable.

A T_1-compactification $c(X)$ of a topological T_1-space X is of Wallman type if and only if X admits a point-symmetric totally bounded transitive quasi-uniformity $\mathcal{U}$ such that the $*$-compactification of $(X,\mathcal{U})$ is equivalent to $c(X)$.

Finally Romaguera and Sánchez-Granero showed that the $*$-compactification of the Pervin quasi-uniformity of any normal T_1-space X is the Čech-Stone compactification of X.

The search continues for concepts of completeness whose underlying idea is less symmetric than that of bicompleteness. The following notion was dealt with extensively in FLETCHER and LINDGREN [1982].

3.5. DEFINITION. A filter $\mathcal{F}$ on a quasi-uniform space $(X,\mathcal{U})$ is called a PS (that is, Pervin-Sieber)-Cauchy filter if for each $U \in \mathcal{U}$ there is $x \in X$ such that $U(x) \in \mathcal{F}$.

A quasi-uniform space $(X, \mathcal{U})$ is called *PS-complete* provided that each PS-Cauchy filter has a cluster point.

For locally symmetric (see e.g. FLETCHER and LINDGREN [1982]) quasi-uniform spaces this is equivalent to the condition that each PS-Cauchy filter converges. That property is called *convergence completeness* in FLETCHER and LINDGREN [1982].

In recent investigations on quasi-uniform spaces PS-completeness was often replaced by left K-completeness or S-completeness, which we shall discuss in the following. The latter concepts turned out to be more flexible than PS-completeness and allowed one to generalize many classical results (often under nontrivial modifications) to the asymmetric setting.

3.6. DEFINITION. A net $(x_d)_{d \in D}$ in a quasi-uniform space $(X, \mathcal{U})$ is called *left K-Cauchy* (resp. *right K-Cauchy*) if for any entourage $U \in \mathcal{U}$ there is $d \in D$ such that $d_2, d_1 \in D$ and $d_2 \geq d_1 \geq d$ imply that $(x_{d_1}, x_{d_2}) \in U$ (resp. $(x_{d_2}, x_{d_1}) \in U$).

The corresponding concept for filters was introduced by ROMAGUERA [1992, 1996].

3.7. DEFINITION. A filter $\mathcal{F}$ on a quasi-uniform space $(X, \mathcal{U})$ is said to be *left* (resp. *right*) *K-Cauchy* if for each $U \in \mathcal{U}$ there is an $F \in \mathcal{F}$ such that $U(x) \in \mathcal{F}$ (resp. $U^{-1}(x) \in \mathcal{F}$) whenever $x \in F$.

A quasi-uniformity is called *left (resp. right) K-complete* provided that each left (resp. right) K-Cauchy filter converges.

A filter $\mathcal{F}$ on a quasi-uniform space $(X, \mathcal{U})$ is called *stable* if $\bigcap_{F \in \mathcal{F}} U(F) \in \mathcal{F}$ whenever $U \in \mathcal{U}$.

We next mention some useful facts about the last concepts: *Each left (resp. right) K-Cauchy filter converges to its cluster points. Any right K-Cauchy filter is stable; the converse holds for ultrafilters. A filter is stable and left K-Cauchy if and only if it is Cauchy with respect to the supremum uniformity.*

In KÜNZI [1995a] resp. KÜNZI and ROMAGUERA [1997a] it was proved that left K-completeness (resp. right K-completeness) is equivalent to left K-completeness (resp. right K-completeness) defined with nets. PS-completeness implies left K-completeness.

3.8. THEOREM (ROMAGUERA [1992]). *A quasi-pseudometric space is left K-sequentially complete if and only if its induced quasi-uniformity is left K-complete (equivalently, each co-stable filter has a cluster point* (compare KÜNZI and RYSER [1995])).

Here a filter is called *co-stable* if it is stable in the conjugate space. Every regular left K-complete quasi-pseudometric space is a Baire space (see e.g. KÜNZI [2001]).

By a result of KÜNZI [1995a] a quasi-uniform space is compact if and only if it is precompact and left K-complete.

3.9. THEOREM (ROMAGUERA [1996]). *A uniformly regular quasi-uniform (T_0-)space is S-complete* (defined below) *if and only if it is left K-complete.*

Romaguera also proved that a topological space is compact if and only if each compatible quasi-uniformity is left (resp. right K-complete) and established that a quasi-uniform T_1-space $(X, \mathcal{U})$ has a T_1 left (resp. right) K-completion if and only if whenever $\mathcal{F}$ is a left (resp. right) K-Cauchy filter on $(X,\mathcal{U})$ that $\tau(\mathcal{U}^{-1})$-converges to some point $x \in X$, then $\mathcal{F}$ is $\tau(\mathcal{U})$-convergent to x.

Related results are also known to hold for quasi-metric spaces (compare e.g. ROMAGUERA [1992] and ROMAGUERA and SALBANY [1990]).

3.10. PROPOSITION (ROMAGUERA [1996]). *A quasi-uniform space is hereditarily precompact if and only if each ultrafilter is a left K-Cauchy filter.*

Similarly a quasi-pseudometric space is hereditarily precompact if and only if each sequence has a left K-Cauchy subsequence (see KÜNZI, MRŠEVIĆ, REILLY and VAMANAMURTHY [1993]).

A quasi-uniformity $\mathcal{U}$ is totally bounded if and only if both $\mathcal{U}$ and $\mathcal{U}^{-1}$ are hereditarily precompact; indeed hereditary precompactness is preserved under arbitrary products of quasi-uniform spaces (see KÜNZI, MRŠEVIĆ, REILLY and VAMANAMURTHY [1993]).

Similarly according to KÜNZI, MRŠEVIĆ, REILLY and VAMANAMURTHY [1994] the product of a hereditarily precompact and a hereditarily preLindelöf quasi-uniform space is hereditarily preLindelöf.

In KÜNZI and ROMAGUERA [1998] it was shown that a quasi-uniformity generated by a preorder $\leq$ is hereditarily precompact if and only if $\leq$ is a *well-quasi-order* (that is, every nonempty set has at least one but no more than a finite number of (nonequivalent) minimal elements).

3.11. THEOREM (KÜNZI [2000]). *The property that each co-stable filter clusters is preserved under quasi-uniformly open continuous surjections between quasi-uniform spaces.*

The latter property for uniform spaces is equivalent to supercompleteness, that is, completeness of the Hausdorff uniformity, and in quasi-uniform spaces implies left K-completeness, which need not be preserved under such maps.

It seems unlikely that a simple theory of left K-completions exists (but see LOWEN and VAUGHAN [1999] and the remarks on the S-completion below), since in general convergent filters are not left K-Cauchy: e.g. a regular quasi-metric space in which each convergent sequence has a left K-Cauchy subsequence is metrizable (see KÜNZI, MRŠEVIĆ, REILLY and VAMANAMURTHY [1993, Proposition 4]).

Although many classical results about completeness do not hold for right K-completeness (see e.g. KÜNZI [2001]), nevertheless, this property behaves much better than left K-completeness in function spaces and hyperspaces (see below).

SMYTH [1994] suggested endowing a quasi-uniform space $(X,\mathcal{U})$ with a topology τ that is not necessarily its standard topology $\tau(\mathcal{U})$, but is linked to $\mathcal{U}$ by some additional axioms. He also defined a concept of a Cauchy filter (now called *S-Cauchy filter*) on such *topological quasi-uniform spaces* and called them complete if every round S-Cauchy filter is the τ-neighborhood filter of a (unique) point.

In [1993] SÜNDERHAUF verified that Smyth's idea of the (now called) S-completion (Smyth completion) of topological quasi-uniform T_0-spaces yields a satisfactory theory from the categorical point of view. He called a quasi-uniform T_0-space equipped with its standard topology *S-completable* if its S-completion also carries the standard topology and proved that a quasi-uniformity is S-completable if and only if each (round) S-Cauchy filter is stable.

For S-completable spaces the construction of the S-completion coincides with the bicompletion.

3.12. PROPOSITION (KÜNZI [1995a]). *A quasi-uniform T_0-space $(X,\mathcal{U})$ is S-completable (S-complete) if and only if each left K-Cauchy filter is Cauchy in $(X,\mathcal{U}^s)$ (converges with respect to $\tau(\mathcal{U}^s)$).*

In particular, each S-complete quasi-uniform space is left K-complete and bicomplete. S-completeness of topological quasi-uniform spaces was also characterized by nets (see SÜNDERHAUF [1995a, 1997] and KÜNZI and SCHELLEKENS [2002]).

3.13. THEOREM (FERRARIO and KÜNZI [1991] and KÜNZI [1995a]). *If an arbitrary topological T_0-space is equipped with the well-monotone quasi-uniformity (and its standard topology), then we obtain the sobrification endowed with its well-monotone quasi-uniformity as its S-completion (= bicompletion).*

Recently FLAGG, KOPPERMAN and SÜNDERHAUF [1999] defined a certain kind of well-monotone quasi-uniformity $\mathcal{U}_\tau$ of a topological quasi-uniform T_0-space $(X,\mathcal{U},\tau)$ and showed that the S-completion of $(X,\mathcal{U},\tau)$ can be obtained via the bicompletion of $(X,\mathcal{U}_\tau)$.

In BONSANGUE, RUTTEN and VAN BREUGEL [1998], as well as in the papers FLAGG and SÜNDERHAUF [2002] and KÜNZI and SCHELLEKENS [2002] the (quasi-pseudometric) Yoneda completion of an extended quasi-pseudometric T_0-space was studied.

It is known that *the Yoneda completion (resp. sequential Yoneda completion) of an extended quasi-pseudometric T_0-space (X,d) yields the quasi-uniform space of the S-completion (resp. the subspace of the S-completion consisting of the round S-Cauchy filters associated with the left K-Cauchy sequences) of $(X,\mathcal{U}_d,\tau(d))$. Sequential Yoneda completion and Yoneda completion coincide for S-completable quasi-pseudometric T_0-spaces.*

An interesting extension of the bicompletion for the *subclass* of quiet quasi-uniform T_0-spaces was investigated by DOITCHINOV [1991].

3.14. DEFINITION. Call a filter $\mathcal{G}$ on a quasi-uniform space $(X,\mathcal{U})$ a *D-Cauchy filter* if there exists a co-filter $\mathcal{F}$ on X (that is, for each $U \in \mathcal{U}$ there are $F \in \mathcal{F}$ and $G \in \mathcal{G}$ such that $F \times G \subseteq U$). Then $(\mathcal{F},\mathcal{G})$ is said to be a *Cauchy filter pair*.

A quasi-uniformity is called *D-complete* provided that each D-Cauchy filter converges in $(X,\tau(\mathcal{U}))$.

Clearly, each convergence complete quasi-uniform space is D-complete.

3.15. DEFINITION. A quasi-uniform space $(X,\mathcal{U})$ is called *quiet* if for any entourage $U \in \mathcal{U}$ there exists an entourage $V \in \mathcal{U}$ such that for any Cauchy filter pair $(\mathcal{F},\mathcal{G})$ on $(X,\mathcal{U})$ and $V^{-1}(y) \in \mathcal{F}$ and $V(x) \in \mathcal{G}$ it follows that $(x,y) \in U$.

DOITCHINOV [1991] showed that each quiet quasi-uniform T_0-space $(X,\mathcal{U})$ has a standard D-completion, now called its *Doitchinov completion*, that is, there is an (up to quasi-uniform isomorphism) uniquely determined quiet D-complete quasi-uniform T_0-space $(X^+,\mathcal{U}^+)$ into which $(X,\mathcal{U})$ is embedded as a doubly dense subspace (compare DEÁK [1994b]); each quasi-uniformly continuous map $f : (X,\mathcal{U}) \to (Y,\mathcal{V})$ into an arbitrary quiet D-complete quasi-uniform T_0-space $(Y, \mathcal{V})$ has a unique quasi-uniformly continuous extension.

The Doitchinov completion of the conjugate of a quiet quasi-uniform T_0-space can be identified with the conjugate of its Doitchinov completion. Doitchinov's theory can readily be reformulated in terms of nets.

3.16. DEFINITION. A quasi-uniformity $\mathcal{U}$ on a set X is called *uniformly regular* if for any $U \in \mathcal{U}$ there is $V \in \mathcal{U}$ such that $\mathrm{cl}_{\tau(\mathcal{U})} V(x) \subseteq U(x)$ whenever $x \in X$.

Quiet quasi-uniform spaces are (doubly) uniformly regular. It follows from the results of FLETCHER, HEJCMAN and HUNSAKER [1990] and KÜNZI [1990b] that for a topological space the property of admitting a uniformly regular quasi-uniformity lies strictly between regularity and complete regularity.

Each D-complete uniformly regular quasi-uniformity is bicomplete (see FLETCHER and HUNSAKER [1992a]). In KÜNZI [2002] it was observed that *the fine quasi-uniformity of each regular preorthocompact semistratifiable space (or each regular T_1-space with an ortho-pairbase) is uniformly regular.*

3.17. THEOREM (FLETCHER and HUNSAKER [1990]). *Every point-symmetric uniformly regular D-complete quasi-uniform space is quiet.*

The fine transitive quasi-uniformity of a topological space is quiet and convergence complete if it is uniformly regular.

Quietness is a rather restrictive property; so each quiet totally bounded quasi-uniform space is uniform (compare FLETCHER and HUNSAKER [1998] and KÜNZI [1990c]). For this reason, other classes of quasi-uniform spaces that have canonical D-complete extensions were considered.

In [1999] DOITCHINOV said that a quasi-uniformity is *stable* if each D-Cauchy filter is stable. He constructed a canonical D-complete extension for an arbitrary stable quasi-uniform T_0-space. Quietness and stability are independent concepts for quasi-uniform spaces.

Stability is a strong property for quasi-pseudometrizable quasi-uniformities (see JUNNILA and KÜNZI [1993b]). Among other things they showed the following.

3.18. THEOREM (JUNNILA and KÜNZI [1993b]). *Each completely regular pseudocompact stable quasi-pseudometric space is compact, and for a stable quasi-pseudometric space the following conditions are equivalent: separable, ccc, weakly Lindelöf and pseudo-$\aleph_1$-compact.*

In the paper [1995] DEÁK and ROMAGUERA called a quasi-uniform space *co-stable* if each co-D-Cauchy filter is stable. They proved that *each mixed-symmetric quiet quasi-uniform space is co-stable*. Furthermore *D-completeness and bicompleteness coincide in co-stable quiet quasi-uniform spaces.*

In the literature several modifications of completeness properties were considered where "(co-)stable filters" were replaced by "weakly Cauchy filters" (that is, filters $\mathcal{F}$ satisfying the condition that for an arbitrary entourage U, $\bigcap_{F \in \mathcal{F}} U^{-1}(F) \neq \emptyset$).

Imitating the concept of S-completeness, authors also created further completeness properties by discussing supconvergence instead of ordinary convergence (see for instance

PÉREZ-PEÑALVER and ROMAGUERA [1999a] as well as the paper ROMAGUERA and SCHELLEKENS [2000]). Other authors tried to work under some weakened forms of quietness (see e.g. DEÁK [1994b] and ROMAGUERA [2000]).

4. Functorial quasi-uniformities

Methods from categorical topology have continued to play a prominent role in the theory of quasi-uniform spaces during the last years. Let T denote the (obvious) forgetful functor from the category **Quu** to the category **Top** of topological spaces and continuous maps.

4.1. DEFINITION. A *functorial admissible quasi-uniformity* on the category **Top** is a functor $F : \mathbf{Top} \to \mathbf{Quu}$ such that $TF = 1$.

Functorial admissible quasi-uniformities on subcategories of **Top** are defined similarly (see e.g. BRÜMMER [1999]). All quasi-uniformities constructed above by Fletcher's method are functorial.

In 1969 Brümmer showed that the coarsest functorial admissible quasi-uniformity on **Top** is the Pervin quasi-uniformity. He also popularized the spanning construction for building sections of the forgetful functor T : A T-section F is *spanned* by a class $\mathcal{A}$ of quasi-uniform spaces if for each X in **Top**, FX carries the initial quasi-uniformity determined by all continuous maps from X to spaces of $T\mathcal{A}$.

In the following paragraphs let K denote the bicompletion functor (on the category of quasi-uniform T_0-spaces).

Brümmer called a T-section F *lower K-true* (resp. *upper K-true*) if $KFX \leq FTKFX$ (resp. $KFX \geq FTKFX$) for any space X in $\mathbf{Top_0}$; T-sections that satisfy both conditions are called *K-true*. (Here $\leq$ corresponds to the coarser relation for quasi-uniformities.)

He showed that a T-section F is lower K-true if and only if F is spanned by a class of bicomplete quasi-uniform spaces.

4.2. THEOREM (KÜNZI [1993], BRÜMMER and KÜNZI [2002]). *A T-section F is upper K-true if and only if F is finer than the well-monotone quasi-uniformity W (equivalently, FX is bicomplete whenever X is sober).*

The well-monotone, the fine and the fine transitive quasi-uniformities are all K-true (see FERRARIO and KÜNZI [1991]); the Pervin and the point-finite quasi-uniformities are lower, but not upper K-true; the locally finite quasi-uniformity is neither lower nor upper K-true according to BRÜMMER and KÜNZI [2002].

If F is K-true, then TKF is a reflection. Kimmie (see BRÜMMER and KÜNZI [200?]) constructed an upper K-true T-section F with $TKF = TKW$ that is not lower K-true. Since TKW is the sobrification reflector in $\mathbf{Top_0}$ by a result of FERRARIO and KÜNZI [1991], TKF is idempotent.

In fact, for a T-section F, TKF is a reflection if and only if TKF is idempotent; each such F is upper K-true (see BRÜMMER and KÜNZI [200?]). The bicompletion of a lower K-true section gives rise to a monad on $\mathbf{Top_0}$ (compare BRÜMMER [1999]).

Many results dealing with completeness properties of functorial quasi-uniformities were obtained over the years. FLETCHER and KÜNZI [1985] studied PS-completeness of various functorial quasi-uniformities. They showed that *the fine quasi-uniformity of an arbitrary nontrivial Tychonoff Σ-product is not PS-complete.*

For many kinds of spaces (for instance Moore spaces) it remains unknown whether the fine quasi-uniformity is necessarily PS-complete. In KÜNZI and WATSON [1996b] a non-Hausdorff example of a quasi-metrizable space whose fine quasi-uniformity is not PS-complete was exhibited. The problem of constructing such a regular space remains open.

4.3. THEOREM (FERRARIO and KÜNZI [1991]). *A topological space admits a bicomplete quasi-uniformity if and only if its fine quasi-uniformity is bicomplete.*

The semicontinuous quasi-uniformity is bicomplete for any completely regular hereditarily realcompact space.

A topological space is hereditarily closed-complete if its semicontinuous quasi-uniformity is bicomplete.

The fine quasi-uniformity of each quasi-pseudometrizable and each quasi-sober space is bicomplete.

PÉREZ-PEÑALVER and ROMAGUERA [1999b] noted that the fine quasi-uniformity $\mathcal{U}$ of any quasi-pseudometrizable bispace X is (cofinally) bicomplete, since $\mathcal{U}^s$ coincides with the fine uniformity of the pseudometrizable supremum topology on X. KÜNZI and ROMAGUERA [1996a] established that the well-monotone quasi-uniformity of any (T_0-)space X is left K-complete; it is S-complete if and only if X is sober (see KÜNZI [2002]).

The fine quasi-uniformity of a topological space need neither be right K-complete nor D-complete, according to KÜNZI and RYSER [1995] resp. KÜNZI and ROMAGUERA [1996a]. However in KÜNZI [2002] it was shown that *the fine quasi-uniformity of each submetacompact space is right K-complete.*

In [2000] NAUWELAERTS described the Cartesian closed topological hull of the category of extended quasi-pseudometric spaces and nonexpansive maps as a subconstruct of the category of quasi-distance spaces and nonexpansive maps. NAUWELAERTS [2001] also provided a concrete description of the Cartesian closed topological hull of **Quu** inside the category of quasi-(semi)uniform limit spaces and quasi-uniformly continuous maps.

5. Quasi-pseudometric spaces

A T_1-space is a *γ-space* if it admits a local quasi-uniformity with a countable base (see e.g. FLETCHER and LINDGREN [1982]). A topological space is quasi-pseudometrizable if and only if it admits a local quasi-uniformity with a countable base whose conjugate filter is a local quasi-uniformity. A topological space is non-archimedeanly quasi-pseudometrizable if and only if it possesses a σ-interior-preserving base.

Examples (see FOX and KOFNER [1985] and e.g. FLETCHER and LINDGREN [1982]) show that these three concepts are distinct in the class of Tychonoff spaces. Junnila established that each developable γ-space is quasi-metrizable (compare e.g. FLETCHER and LINDGREN [1982]). Several kinds of γ-spaces are known to be non-archimedeanly quasi-metrizable, for instance, the suborderable spaces and the T_1-spaces having orthobases (compare FLETCHER and LINDGREN [1982] as well as KOFNER [1981]).

Only few new results about quasi-metrizability were discovered in the last two decades. In particular it remains unknown whether each developable quasi-metrizable space is non-archimedeanly quasi-metrizable.

Hung gave a characterization of quasi-pseudometrizability that illustrates the difference between quasi-metrizable spaces and γ-spaces.

5.1. THEOREM (HUNG [1998]). *Quasi-pseudometrizability of a topological space X is the availability on X of a decreasing neighborhood base $\langle U_n(x)\rangle$ at every $x \in X$, so constituted that, for every countable and relatively locally finite $A \subseteq X$ and $n \in \omega$, we have $U_m^2(A) \subseteq U_n(A)$ for some $m \in \omega$ (dependent on A and n).*

Related to an old, unpublished (see KÜNZI [2001]) bitopological criterion for quasi-metrizability due to Fox, KOPPERMAN [1993] characterized quasi-pseudometrizable topologies as topologies having a σ-self-cocushioned pairbase whose dual is σ-self-cocushioned. (Fox had characterized quasi-metrizable bispaces as the pairwise stratifiable doubly γ-bispaces.)

A connection between γ-spaces and annihilators was pointed out by SUZUKI, TAMANO and TANAKA [1989]. Recently BROWN [1999] obtained a bitopological version of a theorem by Arhangel'skiĭ that a Hausdorff space is metrizable if and only if it is fully normal and has a base of countable order.

A useful concept of paracompactness is still unknown for quasi-pseudometric bispaces. A natural stumbling block seems to be the observation made by Fox that there are pairwise stratifiable, pairwise developable bispaces which are *not* quasi-metrizable (compare KÜNZI [2001]).

Several interesting new results about completeness properties in quasi-(pseudo)metric spaces were discovered during the last years.

5.2. THEOREM (CIESIELSKI, FLAGG and KOPPERMAN [2002]). *A topological space X is Polish if and only if its topology can be induced by some quasi-metric the conjugate quasi-metric of which induces a compact (not necessarily Hausdorff) topology.*

More generally, KÜNZI [200?c] showed recently that a metrizable space is completely metrizable if and only if it admits a (totally bounded) quasi-uniformity $\mathcal{U}$ such that $\tau(\mathcal{U}^{-1})$ is compact.

5.3. THEOREM (KÜNZI [1992]). *Each Tychonoff sequentially PS-complete quasi-metric space is Čech complete.*

Each Tychonoff orthocompact Čech complete topological space with a G_δ-diagonal admits a convergence complete (non-archimedean) quasi-metric.

5.4. THEOREM (KÜNZI [2002]). *A quasi-metrizable space admits a left K-complete quasi-metric if and only if it possesses a λ-base in the sense of Wicke and Worrell.*

5.5. THEOREM (ROMAGUERA and SALBANY [1993]). *A quasi-pseudometrizable bispace admits a bicomplete quasi-pseudometric if and only if its supremum topology is completely pseudometrizable.*

5.6. THEOREM (JUNNILA and KÜNZI [1998]). *A metrizable space admits a bicomplete quasi-metric if and only if it is an $F_{\sigma\delta}$-subset of each metric space in which it is embedded.*

5.7. THEOREM (KÜNZI [2002]). *A quasi-metrizable Moore space admits a right K-complete quasi-metric if and only if it admits a bicomplete quasi-metric (equivalently, has a complete sequence — in the sense of Frolík — of σ-discrete closed covers).*

In KÜNZI and WAJCH [1997] the Junnila-Künzi result was generalized to higher Borel classes by weakening the bicompleteness condition accordingly. Related results on maps were presented in KÜNZI and WAJCH [1998]; e.g. it was shown that if $f : X \to Y$ is a map between metric spaces, then there exists a quasi-metric d on X inducing the topology of X such that f regarded as a map from $(X, \max\{d, d^{-1}\})$ to Y is continuous if and only if f in the original topology of X is a σ-discrete map of Borel class 1.

It is well known that each quasi-pseudometric T_0-space possesses a *quasi-pseudometric* bicompletion (see DI CONCILIO [1971]). DOITCHINOV [1988] developed a conjugate invariant quasi-metric completion theory for so-called *balanced quasi-metric* spaces that imitates the idea of quietness.

The induced quasi-uniformity $\mathcal{U}_d$ of any balanced quasi-metric space (X, d) is quiet and the topology $\tau(d)$ is completely regular. DEÁK [1990] constructed an example of a quiet quasi-metric space that is not completely regular. In DOITCHINOV [1991] it was shown that for quiet quasi-metric quasi-uniformities sequential D-completeness is equivalent to D-completeness.

In a study motivated by problems in Theoretical Computer Science, MATTHEWS [1994] introduced the concept of a *partial metric* and showed that an equivalent theory can be obtained by so-called weighted quasi-pseudometrics.

5.8. DEFINITION. A quasi-pseudometric (T_0-)space (X, q) is *weightable* if there exists a (so-called weight) function $|\cdot| : X \to [0, \infty)$ satisfying $q(x, y) + |x| = q(y, x) + |y|$ whenever $x, y \in X$.

KÜNZI and VAJNER [1994] observed that *the topology induced by a weightable quasi-pseudometric is quasi-developable*. Furthermore they developed a method of constructing weighted quasi-pseudometrics with the help of weighted paths.

By observations of KÜNZI [1995a] every totally bounded quasi-uniformity having a countable base can be induced by a weightable quasi-pseudometric and each weightable (T_0-)quasi-pseudometric induces an S-completable quasi-uniformity. Weightability of a quasi-pseudometric is preserved under the bicompletion (see the paper OLTRA, ROMAGUERA and SÁNCHEZ-PÉREZ [200?]). A canonical construction of weighted quasi-pseudometric spaces from metric spaces was discussed in VITOLO [1999].

Recent investigations studied the generation of partial metrics via certain (possibly negative) valuations and connections with dimension theories of modular lattices.In SCHELLEKENS [200?] the correspondence between partial metrics and semivaluations was explained.

HECKMANN [1999] used a variant of Matthews's partial metrics to investigate some properties of a computational model for Polish spaces (compare with EDALAT and HECKMANN [1998]), which is based on a concept of formal balls.

SCHELLEKENS [1995] provided a topological foundation for the complexity analysis of programs via his "complexity (distance) spaces". They are weightable and, thus, S-completable. Together with ROMAGUERA, SCHELLEKENS [1999] determined some of the quasi-pseudometric properties of his complexity spaces and their so-called duals.

6. Uniformizable ordered spaces

Nachbin (compare e.g. FLETCHER and LINDGREN [1982]) had observed that for any quasi-uniform T_0-space $(X,\mathcal{U})$, the triple $(X,\tau(\mathcal{U}^s),\cap\mathcal{U})$ is a topological T_2-ordered space (that is, the partial order $\cap\mathcal{U}$ is $\tau(\mathcal{U}^s)\times\tau(\mathcal{U}^s)$-closed in $X\times X$). We shall call the topological T_2-ordered spaces determined by this construction the *completely regularly ordered spaces.*

He characterized the topological ordered spaces that allow T_2-ordered compactifications as the completely regularly ordered spaces and constructed the largest such T_2-ordered compactification, which is now called the *Nachbin compactification.*

By one of his results any compact T_2-ordered space is determined by a unique quasi-uniformity $\mathcal{U}$. Furthermore, $\tau(\mathcal{U})$ is the upper and $\tau(\mathcal{U}^{-1})$ the lower topology of X. PRIESTLEY [1972] gave an application of Nachbin's theory in her duality theory which represents distributive lattices by those compact T_2-ordered spaces whose determining quasi-uniformity is transitive.

The T_2-ordered compactifications of a completely regularly ordered space $(X,\tau,\leq)$ can be constructed by bicompleting $\mathcal{U}_\omega$ where $\mathcal{U}$ is an arbitrary quasi-uniformity determining X. In [1993] MOONEY and RICHMOND studied cardinality and structure of semilattices of ordered compactifications.

We next mention a few recent results in this area.

6.1. THEOREM (KÜNZI [1990a]). *A completely regularly ordered space $(X,\tau,\leq)$ is strictly completely regularly ordered in the sense of* LAWSON [1991] *if and only if the bispace $(X,\tau^\sharp,\tau^\flat)$ is pairwise completely regular. Here $\tau^\sharp$ resp. $\tau^\flat$ denotes the upper resp. lower topology of X.*

A completely regularly ordered I-space is strictly completely regularly ordered if it satisfies at least one of the following conditions: it is locally compact, it is a C-space, or it is a topological lattice.

According to PRIESTLEY [1972] a topological ordered space is called an *I-space* (resp. *C-space*) if $A\uparrow$ and $A\downarrow$ are open (resp. closed) whenever A is open (resp. closed). Each topological lattice is known to be an I-space.

In KÜNZI [1990a] a completely regularly ordered C-space that is not strictly completely regularly ordered was constructed. An example of a completely regularly ordered I-space X which is not strictly completely regularly ordered such that the topology τ of X is metrizable and the bispace $(X,\tau^\sharp,\tau^\flat)$ is pairwise regular, but not pairwise completely regular, was defined in KÜNZI and WATSON [1994].

In [2001] NAILANA called a pairwise completely regular bispace $(X, \mathcal{P}, \mathcal{Q})$ *strictly completely regular* if its topologies are the upper and lower topologies of the completely regularly (T_2-)ordered space $(X, \mathcal{P} \vee \mathcal{Q}, \leq_{\mathcal{P}})$. (Here $\leq_{\mathcal{P}}$ denotes the specialization order of $\mathcal{P}$.)

He showed that the category of strictly completely regular bispaces is isomorphic to the category of strictly completely regularly ordered spaces. According to NAILANA and SALBANY [200?] for strictly completely regular bispaces the Čech-Stone bicompactification can be obtained from the Nachbin compactification of their associated topological ordered space.

For a recent investigation on the Čech-Stone bicompactification of a bispace we refer the reader to GARCÍA-FERREIRA, ROMAGUERA and SANCHIS [1999].

Some new applications of uniform topology to domain theory were discussed by KUMMETZ in [1999].

7. Hyperspaces and (multi)function spaces

Hypertopologies became increasingly important in the theory of quasi-uniform spaces during the last years.

7.1. DEFINITION. Let $(X, \mathcal{U})$ be a quasi-uniform space. On the set $\mathcal{P}_0(X)$ of nonempty subsets of X the *Hausdorff(-Bourbaki) quasi-uniformity* $\mathcal{U}_H$ is generated by the base $\{U_H : U \in \mathcal{U}\}$ where $U_H = \{(A, B) : B \subseteq U(A) \text{ and } A \subseteq U^{-1}(B)\}$.

KÜNZI and RYSER [1995] observed that $\mathcal{U}_H$ is precompact (resp. totally bounded, joincompact) if and only if $\mathcal{U}$ is precompact (resp. totally bounded, joincompact). Bicompleteness and left K-completeness on the other hand (even for quasi-metric spaces) often behave badly under the Hausdorff hyperspace construction.

Extending the Burdick-Isbell Theorem from the uniform to the quasi-uniform setting they established that *$\mathcal{U}_H$ is right K-complete if and only if each stable filter on $(X, \mathcal{U})$ has a cluster point.*

In KÜNZI and ROMAGUERA [1997b] it was observed that the condition that each co-stable filter clusters in $(X, \mathcal{U})$ is only necessary, but not sufficient for left K-completeness of $\mathcal{U}_H$. Furthermore it was proved that the Hausdorff quasi-uniformity of the well-monotone quasi-uniformity of any topological space is left K-complete.

7.2. THEOREM (KÜNZI and ROMAGUERA [1998]). *The topology $\tau(\mathcal{U}_H)$ induced by the Hausdorff quasi-uniformity $\mathcal{U}_H$ of a quasi-uniform space $(X, \mathcal{U})$ is compact if and only if $(X, \tau(\mathcal{U}))$ is compact and $(X_m, \mathcal{U}^{-1}|X_m)$ is hereditarily precompact, where X_m denotes the set of the minimal elements of X with respect to the (specialization) preorder $\bigcap\mathcal{U}$.*

Künzi and Romaguera also noted that the problem of characterizing hereditary precompactness of $\mathcal{U}_H$ is related to combinatorial results on better quasi-ordering.

CAO, KÜNZI, REILLY and ROMAGUERA [1998] extended a result of Morita from uniform to quasi-uniform spaces, by establishing that for a so-called compactly symmetric quasi-uniform space $(X, \mathcal{U})$, the Hausdorff quasi-uniformity on the set $\mathcal{K}_0(X)$ of nonempty compact sets of $(X, \mathcal{U})$ is PS-complete if and only if $\mathcal{U}$ is PS-complete (for related results see also SÁNCHEZ-GRANERO [2001]).

7.3. THEOREM (RODRÍGUEZ-LÓPEZ and ROMAGUERA [200?]). *For a quasi-uniform space $(X, \mathcal{U})$, the Hausdorff quasi-uniformity induces on $\mathcal{K}_0(X)$ [resp. on $\mathcal{P}_0(X)$] the Vietoris topology if and only if for each $K \in \mathcal{K}_0(X)$, $\mathcal{U}^{-1}|K$ is precompact [resp. $\mathcal{U}$ belongs to the Pervin quasi-proximity class of its induced topology and $\mathcal{U}^{-1}$ is hereditarily precompact].*

7.4. COROLLARY. *On $\mathcal{P}_0(X)$ the Hausdorff quasi-uniformity of a Pervin quasi-uniformity (resp. well-monotone quasi-uniformity) induces the Vietoris topology.*

7.5. COROLLARY. *For a quasi-metric space (X, d), the Hausdorff quasi-pseudometric on $\mathcal{K}_0(X)$ [resp. $\mathcal{P}_0(X)$] induces the Vietoris topology if and only if every convergent sequence in X has a right K-Cauchy subsequence [resp. $\mathcal{U}_d$ is the Pervin quasi-uniformity of X].*

RODRÍGUEZ-LÓPEZ [2001, 200?] studied several types of Fell topologies in the bitopological setting. For many classical results on hyperspaces he proposed bitopological variants. He also showed that for a quasi-pseudometrizable space (X, τ) the Vietoris topology on $\mathcal{P}_0(X)$ is the supremum of all Wijsman topologies associated with quasi-pseudometrics compatible with τ.

BURDICK [1999] defined hyperspace operators for biquasi-proximity spaces and for biquasi-uniform spaces.

Recently quasi-uniform function spaces also attracted a lot of attention (compare the paper SÜNDERHAUF [1995b]). In [1996b] KÜNZI and ROMAGUERA discussed various kinds of completeness properties of the quasi-uniformity of quasi-uniform convergence.

7.6. THEOREM (KÜNZI and ROMAGUERA [1996b]). *For a topological space X and a quasi-uniform space $(Y, \mathcal{U})$ the quasi-uniformity of quasi-uniform convergence on the set Y^X of all functions from X to Y is right K-complete (bicomplete) if and only if $\mathcal{U}$ is right K-complete (bicomplete). The corresponding result does not hold for left K-completeness.*

It followed from results of KÜNZI [1995b], PAPADOPOULOS [1994], RENDER [1998] and ROMAGUERA [1995] that the class of small-set symmetric quasi-uniform spaces is particularly useful in the study of function spaces. For instance *the quasi-uniformity of quasi-uniform convergence on compacta induces on the set $C(X, Y)$ of continuous functions from X to Y the compact-open topology if $\mathcal{U}$ is small-set symmetric, but not for arbitrary $\mathcal{U}$.*

Here a quasi-uniformity is called *small-set symmetric* if its conjugate is point-symmetric (compare KÜNZI, MRŠEVIĆ, REILLY and VAMANAMURTHY [1993]). A useful observation about continuous maps between quasi-uniform spaces is the following.

7.7. THEOREM (MARÍN and ROMAGUERA [1996a]). *Every continuous function from a Lebesgue quasi-uniform space* (defined e.g. in FLETCHER and LINDGREN [1982]) *to a small-set symmetric quasi-uniform space is quasi-uniformly continuous.*

In this context let us also mention that by a result of KÜNZI [2000] each open continuous map from a compact *uniform* space into an arbitrary quasi-uniform space is quasi-uniformly open.

RENDER [1998] gave an example of a locally symmetric compact quasi-uniform T_2-space X such that $\mathcal{C}(X,X)$ is not PS-complete for the quasi-uniformity of quasi-uniform convergence.

CAO [1996] (see also KÜNZI and ROMAGUERA [1997a]) proved that if X is a topological space and $(Y,\mathcal{U})$ is a quiet (or small-set symmetric) quasi-uniform space, then $\mathcal{C}(X,Y)$ need not be closed in Y^X equipped with the quasi-uniformity of quasi-uniform convergence; however, $\mathcal{C}(X,Y)$ is closed in Y^X if $\mathcal{U}$ is locally symmetric (see PAPADOPOULOS [1995]).

In [1995/7] ROMAGUERA and RUIZ-GÓMEZ studied bitopological versions of theorems of pointwise, compact and (quasi-)uniform compact convergence on spaces of continuous functions, extending classical theorems on function spaces to the bitopological case. Recently NAILANA [2000, 2001] developed a corresponding theory for topological ordered spaces by generalizing for instance the following result to these spaces:

Let (X,τ) and (Y,τ') be two topological spaces such that (Y,τ') contains a nontrivial path. The compact-open topology on $\mathcal{C}(X,Y)$ is completely metrizable if and only if (X,τ) is a hemicompact k-space and (Y,τ') is completely metrizable.

CAO, REILLY and ROMAGUERA [1998] investigated various properties of completeness of the *multifunction space* (equipped with its Hausdorff quasi-uniformity) in terms of properties of the range space.

KHANH [1989] presented open mapping theorems for families of multifunctions in right K-sequentially complete quasi-metric spaces. Related to his investigations, KÜNZI [2000] considered asymmetric versions of the fact that each almost uniformly open (multivalued) map with closed graph from a supercomplete uniform space into an arbitrary uniform space is uniformly open.

8. Topological algebra

In this final section we shall discuss several applications of methods from asymmetric topology to topological algebra.

8.1. DEFINITION. A *paratopological group* is a group equipped with a topology such that its group operation (multiplication) is continuous.

Clearly each paratopological group topology possesses an obvious conjugate topology that is defined by algebraic inversion, and an associated supremum topology which gives rise to a topological group. Many authors have determined conditions under which a paratopological group is a topological group (see e.g. KENDEROV, KORTEZOV and MOORS [2001]).

By definitions analogous to those used in the case of topological groups, on a paratopological group G canonical quasi-uniformities can be defined; for instance the two-sided quasi-uniformity is generated by $\{U_B : U \in \eta(e)\}$ where $\eta(e)$ is the neighborhood filter at the neutral element e and $U_B = \{(x,y) \in G \times G : y \in xU \text{ and } y \in Ux\}$ whenever $U \in \eta(e)$.

Generalizing many results known from the theory of topological groups, Marín and Romaguera conducted a systematic study of these quasi-uniformities.

8.2. THEOREM (MARÍN and ROMAGUERA [1996b]). *The ground set of the bicompletion of the two-sided quasi-uniformity of a paratopological (T_0-)group carries the structure of a paratopological group; moreover the quasi-uniformity of that bicompletion yields the two-sided quasi-uniformity of the constructed paratopological group.*

Each first-countable paratopological group admits a left-invariant quasi-pseudometric compatible with its left quasi-uniformity.

MARÍN and ROMAGUERA [1998] proved that *the ground set of the bicompletion of the left quasi-uniformity of a paratopological (T_0-)group G carries the structure of a topological semigroup*. They also characterized those paratopological groups for which the semigroup is a group H by the condition that *each filter which is Cauchy with respect to the left uniformity is Cauchy with respect to the right uniformity of the associated topological group on G*; under the latter condition the bicompleted left quasi-uniformity of G yields the left quasi-uniformity of the paratopological group H.

8.3. THEOREM (KÜNZI, ROMAGUERA and SIPACHEVA [1998]). *The two-sided quasi-uniformity of a regular paratopological group is quiet (in the sense of Doitchinov).*

The Doitchinov completion of the two-sided quasi-uniformity of an abelian regular paratopological group can be considered an abelian paratopological group.

However there is a regular paratopological group in which the product of two Cauchy filter pairs need not be a Cauchy filter pair (with respect to the two-sided quasi-uniformity) so that their result cannot be generalized to arbitrary paratopological groups.

In [200?] ROMAGUERA, SANCHIS and TKACHENKO showed that for any topological (or quasi-uniform) space X the free (resp. free abelian) paratopological group over X exists by constructing explicitly the extensions of quasi-pseudometrics.

Together with KÜNZI, MARÍN and ROMAGUERA [2001] generalized parts of their theory from paratopological groups to *topological monoids*. Among other things, they showed that *for a topological T_0-monoid X in which the left translations are open, the bicompletion of the left quasi-uniformity of X can be considered a topological monoid which contains the topological space X as a supdense submonoid.* For monoids which are (left-)cancellative or which are locally totally bounded, theorems similar to those known from the theory of (para)topological groups were established for the one-sided (resp. two-sided) quasi-uniformities.

In ROMAGUERA, SÁNCHEZ-PÉREZ and VALERO [200?] a method was presented to generate quasi-metrics from certain classes of subadditive functions defined on monoids.

PORTER [1993] continued the research of quasi-uniformities on homeomorphism groups which Fletcher and his students had begun three decades ago (see KÜNZI [1998]).

For instance these researchers had proved the following result: *If G is a group of self-homeomorphisms of a quasi-uniform space $(X, \mathcal{U})$ that is quasi-equicontinuous with respect to $\mathcal{U}$, then $(G, \circ, \tau(\mathcal{Q}_p))$ is a paratopological group, which is a topological group provided that $\mathcal{U}$ is point-symmetric.* (Here $\mathcal{Q}_p$ denotes the quasi-uniformity of pointwise convergence.)

Among other things Porter showed that the open-open topology on the group of all self-homeomorphisms of a topological space X coincides with the topology of quasi-uniform convergence transmitted by the Pervin quasi-uniformity of X.

In [1998] KÜNZI applied results about function spaces to quasi-uniform isomorphism groups.

8.4. THEOREM (KÜNZI [1998]). *If* $(X,\mathcal{U})$ *is a quasi-uniform space and* $\mathcal{Q}(X)$ *is the group of all quasi-uniform self-isomorphisms of* $(X,\mathcal{U})$ *equipped with the quasi-uniformity* $\mathcal{Q}_{\mathcal{U}}$ *of quasi-uniform convergence, then* $\mathcal{Q}_{\mathcal{U}}$ *is the right quasi-uniformity of the paratopological group* $(\mathcal{Q}(X),\circ,\tau(\mathcal{Q}_{\mathcal{U}}))$; *its two-sided quasi-uniformity is bicomplete if* $\mathcal{U}$ *is bicomplete.*

Künzi argued that the latter construction applied to the fine quasi-uniformity of a completely metrizable space yields a topological group, while this is not the case for the fine quasi-uniformity of the Sorgenfrey line.

An algebraic structure endowed with a topology is called *(quasi-)uniformizable* if there is a compatible (quasi-)uniformity such that the defining operations of the algebra are (quasi-)uniformly continuous; e.g. a paratopological group is quasi-uniformizable if and only if its left quasi-uniformity coincides with its right quasi-uniformity. So far only few publications have been devoted to this subject (see for instance the article of CORBACHO-ROSAS, DIKRANJAN and TARIELADZE [2001] dealing with semigroups).

In an application of quasi-uniformities to automata and language theory, PIN and WEIL [1999] proved an Eilenberg–like one-to-one correspondence between pseudovarieties of ordered semigroups and so-called varieties of quasi-uniformities.

WEHRUNG [1993] introduced an intrinsic subinvariant extended quasi-pseudometric on each positively ordered monoid which is defined in terms of the evaluation map from the monoid to its bidual and which turns out to be bicomplete for several important classes of monoids.

Recently quite a number of papers were written in the area of asymmetric functional analysis (see e.g. ALEGRE, FERRER and GREGORI [1999], and ALIMOV [2001]). Here the basic concept is that of an asymmetric (pre)norm, that is, a nonnegative and subadditive positively homogeneous function $\|\cdot\|$ defined on a (real) linear space.

Questions about extensions of such functions from cones of linear spaces to spanned subspaces were studied in GARCÍA-RAFFI, ROMAGUERA and SÁNCHEZ-PÉREZ [200?c]. In a related article of these authors [200?d] those asymmetric normed linear spaces were characterized whose induced topology is Hausdorff.

An asymmetric normed T_0-space whose induced quasi-pseudometric is bicomplete, is called a *biBanach space*. In GARCÍA-RAFFI, ROMAGUERA and SÁNCHEZ-PÉREZ [200?a] the bicompletion of an asymmetric normed linear T_0-space was constructed by verifying that each asymmetric normed linear T_0-space is isometrically isomorphic to a supdense subspace of a biBanach space that is unique (up to isometric isomorphism).

The dual cone of an asymmetric normed linear space was introduced in GARCÍA-RAFFI, ROMAGUERA and SÁNCHEZ-PÉREZ [200?b], where an asymmetric version of Alaoglu's theorem was also established.

ROMAGUERA and SCHELLEKENS [200?] showed that asymmetric normed spaces provide a suitable setting to carry out an analysis of the so-called dual-complexity space. The latter concept was extended from $p = 1$ to $p > 1$ in GARCÍA-RAFFI, ROMAGUERA and SÁNCHEZ-PÉREZ [200?e] in order to conduct complexity analysis of some kinds of exponential time algorithms.

Independently, an interesting concept of a *convex quasi-uniform structure* was introduced by KEIMEL and ROTH [1992] in their book "Ordered Cones and Approximation". For such cones Hahn-Banach type theorems were proved.

References

ALEGRE, C., J. FERRER and V. GREGORI
[1999] On the Hahn-Banach Theorem in certain linear quasi-uniform structures, *Acta. Math. Hungar.* **82**, 325–330.

ALIMOV, A. R.
[2001] On the structure of the complements of Chebyshev sets, *Functional Analysis Appl.* **35** (3), 176–182.

BENTLEY, H. L., H. HERRLICH and M. HUŠEK
[1998] The historical development of uniform, proximal, and nearness concepts in topology, in: *Handbook of the History of General Topology, Vol. 2*, Aull, C. E., and R. Lowen, eds., Dordrecht, Kluwer, pp. 577–629.

BONSANGUE, M.M., J.J.M.M. RUTTEN and F. VAN BREUGEL
[1998] Generalized metric spaces: completion, topology, and powerdomains via the Yoneda embedding, *Theoretical Computer Science* **193**, 1–51.

BRATTKA, V.
[200?a] Recursive quasi-metric spaces, *Theoretical Computer Science*, to appear.
[200?b] Generated quasi-metric hyper and function spaces, *Topology Appl.*, to appear.

BROWN, L.M.
[1999] $P - Q$-metrizability of bitopological spaces with a dual base of countable order, *Acta Math. Hungar.* **83**, 315–325.

BRÜMMER, G.C.L.
[1999] Categorical aspects of the theory of quasi-uniform spaces, *Rend. Ist. Mat. Univ. Trieste* **30** Suppl., 45–74.

BRÜMMER, G.C.L. and H.P.A. KÜNZI
[2002] Bicompletion and Samuel bicompactification, *Appl. Categ. Struct.* **10**, 317–330.
[200?] Idempotency of extensions via the bicompletion, in preparation.

BURDICK, B.S.
[1999] Uniform and proximal hyperspaces for bitopological spaces, *Topology Proc. Summer* **24**, 71–88.

CAO, J.
[1996] Answers to two questions of Papadopoulos, *Questions Answers Gen. Topology* **14**, 111–116.

CAO, J., H.P.A. KÜNZI, I.L. REILLY and S. ROMAGUERA
[1998] Quasiuniform hyperspaces of compact subsets, *Topology Appl.* **87**, 117–126.

CAO, J., I.L. REILLY and S. ROMAGUERA
[1998] Some properties of quasiuniform multifunction spaces, *J. Austral. Math. Soc. (Ser. A)* **64**, 169–177.

CIESIELSKI, K., R.C. FLAGG and R. KOPPERMAN

[2002] Polish spaces, computable approximations, and bitopological spaces, *Topology Appl.* **119**, 241–256.

CLEMENTINO, M.M. and D. HOFMANN

[200?] Topological features of lax algebras, preprint.

CORBACHO-ROSAS, E., D. DIKRANJAN and V. TARIELADZE

[2001] Absorption adjunctable groups, in: *Nuclear Groups and Lie Groups*, Martín-Peinador, E. and J. Núñez-García, eds., Research and Exposition in Mathematics **24**, Heldermann, Berlin, pp. 77–104.

CSÁSZÁR, Á.

[1999] Old and new results on quasi-uniform extension, *Rend. Ist. Mat. Univ. Trieste* **30** Suppl., 75–85.

DEÁK, J.

[1990] A non-completely regular quiet quasi-metric, *Math. Pannonica* **1** (2), 111–116.

[1993a] Extending a quasi-metric, *Studia Sci. Math. Hungar.* **28**, 105–113.

[1993b] A survey of compatible extensions (presenting 77 unsolved problems), *Colloq. Math. Soc. János Bolyai* **55**, (Top., Pécs Hungary, 1989), 127–175.

[1994a] A note on weak symmetry properties of quasi-uniformities, *Studia Sci. Math. Hungar.* **29**, 433–435.

[1994b] Extending and completing quiet quasi-uniformities, *Studia Sci. Math. Hungar.* **29**, 349–362.

DEÁK, J. and S. ROMAGUERA

[1995] Co-stable quasi-uniform spaces, *Ann. Univ. Sci. Budapest.* **38**, 55–70.

DEZA, M. and E. PANTELEEVA

[2000] Quasi-semi-metrics, oriented multi-cuts and related polyhedra, *Eur. J. Comb.* **21**, 777–795.

DI CONCILIO, A.

[1971] Spazi quasimetrici e topologie ad essi associate, *Rend. Accad. Sci. Fis. Mat. Napoli* **38** (IV), 113–130.

DOITCHINOV, D.

[1988] On completeness in quasi-metric spaces, *Topology Appl.* **30**, 127–148.

[1991] A concept of completeness of quasi-uniform spaces, *Topology Appl.* **38**, 205–217.

[1999] E-completions of quasi-uniform spaces, *Proc. Symposium on Categorical Topology*, University of Cape Town 1994, Banaschewski, B., C. R. A. Gilmour and H. Herrlich, eds., Dept. Math., University of Cape Town, 89–102.

EDALAT, A. and R. HECKMANN

[1998] A computational model for metric spaces, *Theoretical Computer Science* **193**, 53–73.

FERRARIO, N. and H.P.A. KÜNZI

[1991] Bicompleteness of the fine quasi-uniformity, *Math. Proc. Camb. Phil. Soc.* **109**, 167–186.

FERREIRA, M.J. and J. PICADO

[200?] The Galois approach to uniform structures, preprint.

FLAGG, R.C.

[1997] Quantales and continuity spaces, *Algebra Univ.* **37**, 257–276.

FLAGG, R.C. and R.D. KOPPERMAN

[1997] Continuity spaces: Reconciling domains and metric spaces, *Theoretical Computer Science* **177**, 111–138.

FLAGG, R.C., R. KOPPERMAN and P. SÜNDERHAUF
[1999] Smyth completion as bicompletion, *Topology Appl.* **91**, 169–180.

FLAGG, R.C. and P. SÜNDERHAUF
[2002] The essence of ideal completion in quantitive form, *Theoretical Computer Science* **278**, 141–158.

FLETCHER, P., J. HEJCMAN and W. HUNSAKER
[1990] A noncompletely regular quiet quasi-uniformity, *Proc. Amer. Math. Soc.* **108**, 1077–1079.

FLETCHER, P. and W. HUNSAKER
[1990] Uniformly regular quasi-uniformities, *Topology Appl.* **37**, 285–291.
[1992a] Completeness using pairs of filters, *Topology Appl.* **44**, 149–155.
[1992b] Symmetry conditions in terms of open sets, *Topology Appl.* **45**, 39–47.
[1998] A note on totally bounded quasi-uniformities, *Serdica Math. J.* **24**, 95–98.

FLETCHER, P., W. HUNSAKER and W. LINDGREN
[1995] A functional approach to uniform and quasi-uniform frames, in: *Topology with Applications*, Bolyai Soc. Math. Studies, Vol. 4, Szekszárd 1993, pp. 217–222.

FLETCHER, P. and H.P.A. KÜNZI
[1985] Extension properties induced by complete quasi-uniformities, *Pacific J. Math.* **120**, 357–384.

FLETCHER, P. and W.F. LINDGREN
[1982] *Quasi-Uniform Spaces*, Lecture Notes Pure Appl. Math. **77**, Dekker, New York.
[1986] Compactifications of totally bounded quasi-uniform spaces, *Glasgow Math. J.* **28**, 31–36.

FOX, R. and J. KOFNER
[1985] A regular counterexample to the γ-space conjecture, *Proc. Amer. Math. Soc.* **94**, 502–506.

GARCÍA-FERREIRA, S., S. ROMAGUERA and M. SANCHIS
[1999] On the Stone-Čech bicompactification of a bispace, *J. Austral. Math. Soc. (Ser. A)* **66**, 358–372.

GARCÍA-RAFFI, L.M., S. ROMAGUERA and E.A. SÁNCHEZ-PÉREZ
[200?a] The bicompletion of an asymmetric normed linear space, *Acta Math. Hungar.*, to appear.
[200?b] The dual space of an asymmetric normed linear space, preprint.
[200?c] Extensions of asymmetric norms to linear spaces, *Rend. Ist. Math. Univ. Trieste*, to appear.
[200?d] On Hausdorff asymmetric normed linear spaces, *Houston J. Math.*, to appear.
[200?e] Sequence spaces and asymmetric norms in the theory of computational complexity, *Math. Comput. Modelling*, to appear.

GARTSIDE, P.M. and P.J. MOODY
[1993] A note on proto-metrisable spaces, *Topology Appl.* **52**, 1–9.

GERLITS, J., H.P.A. KÜNZI, A. LOSONCZI and Z. SZENTMIKLÓSSY
[2002] The existence of compatible nontransitive totally bounded quasi-uniformities, *Topology Appl.* **117**, 139–147.

GOLAN, J.S.
[1998] Semiring-valued quasimetrics on the set of submodules of a module, *Math. J. Okayama Univ.* **40**, 33–38.

HECKMANN, R.

[1999] Approximation of metric spaces by partial metric spaces, *Appl. Categ. Struct.* **7**, 71–83.

HEITZIG, J.

[200?] Many familiar categories can be interpreted as categories of generalized metric spaces, *Appl. Categ. Struct.*, to appear.

HITZLER, P. and A.K. SEDA

[1999] Some issues concerning fixed points in computational logic: Quasi-metrics, multivalued mappings and the Knaster-Tarski theorem, *Topology Proc.* **24** (Summer), 223–250.

HÖTZEL ESCARDÓ, M.

[2001] The regular-locally compact coreflection of a stably locally compact locale, *J. Pure Appl. Algebra* **157**, 41–55.

HOWES, N.R.

[1995] *Modern Analysis and Topology*, Springer, Berlin.

HUNG, H.H.

[1998] Quasi-metrizability, *Topology Appl.* **83**, 39–43.

JUNNILA, H.J.K. and H.P.A. KÜNZI

[1993a] Ortho-bases and monotonic properties, *Proc. Amer. Math. Soc.* **119**, 1335–1345.

[1993b] Stability in quasi-uniform spaces and the inverse problem, *Topology Appl.* **49**, 175–189.

[1998] Characterizations of absolute $F_{\sigma\delta}$-sets, *Czechoslovak Math. J.* **48** (123), 55–64.

JUNNILA, H.J.K., H.P.A. KÜNZI and S. WATSON

[200?] On a class of hereditarily paracompact spaces, *Topology Proc.*, to appear. (Russian translation in *Fundam. Prikl. Mat.* **4** (1998), 141–154).

KEIMEL, K. and W. ROTH

[1992] *Ordered Cones and Approximation*, Lecture Notes Math. **1517**, Springer, Berlin.

KENDEROV, P.S., I. S. KORTEZOV and W.B. MOORS

[2001] Topological games and topological groups, *Topology Appl.* **109**, 157–165.

KHANH, P.Q.

[1989] On general open mapping theorems, *J. Math. Anal. Appl.* **144**, 305–312.

KOFNER, J.

[1980] On quasi-metrizability, *Topology Proc.* **5**, 111–138.

[1981] Transitivity and ortho-bases, *Canad. J. Math.* **33**, 1439–1447.

KOPPERMAN, R.D.

[1993] Which topologies are quasimetrizable? *Topology Appl.* **52**, 99–107.

[1995] Asymmetry and duality in topology, *Topology Appl.* **66**, 1–39.

KUMMETZ, R.

[1999] Continuous domains with approximating mappings and their uniformity, *Topology Proc.* **24** (Summer), 267–294.

KÜNZI, H.P.A.

[1990a] Completely regular ordered spaces, *Order* **7**, 283–293.

[1990b] A regular space without a uniformly regular quasi-uniformity, *Monatsh. Math.* **110**, 115–116.

[1990c] Totally bounded quiet quasi-uniformities, *Topology Proc.* **15**, 113–115.

[1992] Complete quasi-pseudo-metric spaces, *Acta Math. Hungar.* **59**, 121–146.

[1993] Quasi-uniform spaces — eleven years later, *Topology Proc.* **18**, 143–171.

[1995a] Nonsymmetric topology, in: *Topology with Applications*, Bolyai Soc. Math. Studies, Vol. 4, Szekszárd 1993, pp. 303–338.

[1995b] On quasi-uniform convergence and quiet spaces, *Questions Answers Gen. Topology* **13**, 87–92.

[1998] Quasi-uniform isomorphism groups, in: *Recent Progress in Function Spaces*, Di Maio, G. and L. Holá, eds., Quaderni di matematica, Dip. Mat. Sec. Univ. Napoli **3**, 187–220.

[1999] Nontransitive quasi-uniformities. *Publ. Math. Debrecen* **55**, 161–167.

[2000] Preservation of completeness under mappings in asymmetric topology, *Appl. Gen. Topology* **1**, 99–114.

[2001] Nonsymmetric distances and their associated topologies: about the origins of basic ideas in the area of asymmetric topology, in: *Handbook of General Topology, Vol. 3*, Aull, C. E. and R. Lowen, eds., Kluwer, Dordrecht, pp. 853–968.

[2002] Quasi-metrizable spaces satisfying certain completeness conditions, *Acta Math. Hungar.* **95**, 345–357.

[200?a] Nontransitive quasi-uniformities in the Pervin quasi-proximity class, *Proc. Amer. Math. Soc.*, to appear.

[200?b] Transitivity of hereditarily metacompact spaces, preprint.

[200?c] Cocompactness and quasi-uniformizability of completely metrizable spaces, preprint.

KÜNZI, H.P.A. and A. LOSONCZI

[2000] On some cardinal functions related to quasi-uniformities, *Houston J. Math.* **26**, 299–313.

[2001] Semilattices of totally bounded quasi-uniformities, *Topology Appl.* **114**, 273–284.

KÜNZI, H.P.A., J. MARÍN and S. ROMAGUERA

[2001] Quasi-uniformities on topological semigroups and bicompletion, *Semigroup Forum* **62**, 403–422.

KÜNZI, H.P.A., M. MRŠEVIĆ, I.L. REILLY and M.K. VAMANAMURTHY

[1993] Convergence, precompactness and symmetry in quasi-uniform spaces, *Math. Japonica* **38**, 239–253.

[1994] Prelindelöf quasi-pseudo-metric and quasi-uniform spaces, *Mat. Vesn.* **46**, 81–87.

KÜNZI, H.P.A. and M.J. PÉREZ-PEÑALVER

[2000] The number of compatible totally bounded quasi-uniformities, *Acta Math. Hungar.* **88**, 15–23.

KÜNZI, H.P.A. and S. ROMAGUERA

[1996a] Some remarks on Doitchinov completeness, *Topology Appl.* **74**, 61–72.

[1996b] Completeness of the quasi-uniformity of quasi-uniform convergence, in: *Papers on General Topology and Applications*, Eleventh Summer Conference, Southern Maine, 1995, Annals New York Acad. Sci. **806**, pp. 231–237.

[1997a] Spaces of continuous functions and quasi-uniform convergence, *Acta Math. Hungar.* **75**, 287–298.

[1997b] Left K-completeness of the Hausdorff quasi-uniformity, *Rostock. Math. Kolloq.* **51**, 167–176.

[1998] Well-quasi-ordering and the Hausdorff quasi-uniformity, *Topology Appl.* **85**, 207–218.

KÜNZI, H.P.A., S. ROMAGUERA and O.V. SIPACHEVA

[1998] The Doitchinov completion of a regular paratopological group, *Serdica Math. J.* **24**, 73–88.

KÜNZI, H.P.A. and C. RYSER

[1995] The Bourbaki quasi-uniformity, *Topology Proc.* **20**, 161–183.

KÜNZI, H.P.A. and M.P. SCHELLEKENS

[2002] On the Yoneda-completion of a quasimetric space, *Theoretical Computer Science* **278**, 159–194.

KÜNZI, H.P.A. and V. VAJNER

[1994] Weighted quasi-metrics, in: *Papers on General Topology and Applications*, Eighth Summer Conf. Queens College, 1992, Annals New York Acad. Sci., **728**, pp. 64–77.

KÜNZI, H.P.A. and E. WAJCH

[1997] Borel classification via quasi-metrics, *Topology Appl.* **77**, 183–192.

[1998] On σ-discrete Borel mappings via quasi-metrics, *Czechoslovak Math. J.* **48** (123), 439–455.

KÜNZI, H.P.A. and S. WATSON

[1994] A metrizable completely regular ordered space, *Comment. Math. Univ. Carol.* **35**, 773–778.

[1996a] A nontrivial T_1-space admitting a unique quasi-proximity, *Glasgow Math. J.* **38**, 207–213.

[1996b] A quasi-metric space without complete quasi-uniformity, *Topology Appl.* **70**, 175–178.

[1999] A nontransitive space based on combinatorics, *Boll. U.M.I.* **(8) 2-B**, 315–317.

LAWSON, J.D.

[1991] Order and strongly sober compactifications, in: *Topology and Category Theory in Computer Science*, G.M. Reed, A.W. Roscoe and R.F. Wachter, eds., Clarendon Press, Oxford, pp. 179–205.

LOSONCZI, A.

[1998] The cardinality of the coarsest quasi-proximity class of locally compact T_2-spaces, *Topology Proc.* **23** (Spring), 245–262.

[2000] Topological spaces with a coarsest compatible quasi-uniformity, *Quaestiones Math.* **23**, 67–75.

LOWEN, R. and D.J. VAUGHAN

[1999] A non quasi-metric completion for quasi-metric spaces, *Rend. Ist. Mat. Univ. Trieste* **30** Suppl., 145–163.

LOWEN, R. and B. WINDELS

[1998] **AUnif**: A common supercategory of **pMet** and **Unif**, *Internat. J. Math. Math. Sci.* **21**, 1–18.

MARÍN, J. and S. ROMAGUERA

[1996a] On quasi-uniformly continuous functions and Lebesgue spaces, *Publ. Math. Debrecen* **48**, 347–355.

[1996b] A bitopological view of quasi-topological groups, *Indian J. Pure Appl. Math.* **27**, 393–405.

[1998] Bicompleting the left quasi-uniformity of a paratopological group, *Arch. Math. (Basel)* **70**, 104–110.

MATTHEWS, S.G.

[1994] Partial metric topology, in: *Papers on General Topology and Applications*, Eighth Summer Conf. Queens College, 1992, Annals New York Acad. Sci. **728**, pp. 183–197.

MOONEY, D.D. and T.A. RICHMOND

[1993] Cardinality and structure of semilattices of ordered compactifications, in: *Papers on General Topology and Applications*, Ninth Summer Conf. Slippery Rock, PA, 1993, Annals New York Acad. Sci. **767**, pp. 188–193.

MURDESHWAR, M.G. and S.A. NAIMPALLY

[1966] *Quasi-Uniform Topological Spaces*, Noordhoff, Groningen.

NAILANA, K.R.

[2000] Ordered spaces and quasi-uniformities on spaces of continuous order-preserving functions, *Extracta Math.* **15**, 513–530.

[2001] Strict complete regularity in the categories of bitopological spaces and ordered topological spaces, *Publ. Math. Debrecen* **58**, 693–705.

[2001] Quasi-pseudometrizability of the point open ordered spaces and the compact open ordered spaces, *Internat. J. Math. Math.* **26**, 385–392.

NAILANA, K.R. and S. SALBANY

[200?] A note on the Stone-Čech bicompactification of strictly completely regular bitopological spaces, preprint.

NAUWELAERTS, M.

[2000] Cartesian closed hull for (quasi-)metric spaces (revisited), *Comment. Math. Univ. Carol.* **41**, 559–573.

[2001] The Cartesian closed topological hull of the category of (quasi-)uniform spaces (revisited), *Rend. Ist. Mat. Univ. Trieste* **32** Suppl. 2, 101–125.

OLTRA, S., S. ROMAGUERA and E.A. SÁNCHEZ-PÉREZ

[200?] Bicompleting weightable quasi-metric spaces and partial metric spaces, *Rend. Circ. Mat. Palermo II*, to appear.

PAPADOPOULOS, B.K.

[1994] Quasi-uniform convergence on function spaces, *Questions Answers Gen. Topology* **12**, 121–131.

[1995] A note on the paper "Quasi-uniform convergence on function spaces", *Questions Answers Gen. Topology* **13**, 55–56.

PÉREZ-PEÑALVER, M.J. and S. ROMAGUERA

[1999a] Weakly Cauchy filters and quasi-uniform completeness, *Acta Math. Hungar.* **82**, 217–228.

[1999b] Cofinal bicompleteness and quasi-metrizability, *Rend. Ist. Mat. Univ. Trieste* **30** Suppl., 165–172.

PIN, J.-E. and P. WEIL

[1999] Uniformities on free semigroups, *Internat. J. Algebra Comput.* **9**, 431–453.

PORTER, K.F.

[1993] The open-open topology for function spaces, *Internat. J. Math. Math. Sci.* **16**, 111–116.

PRIESTLEY, H.A.

[1972] Ordered topological spaces and the representation of distributive lattices, *Proc. London Math. Soc.* **24**, 507–530.

RENDER, H.

[1998] Generalized uniform spaces and applications to function spaces, *Studia Sci. Math. Hungar.* **34**, 429–443.

RODRÍGUEZ-LÓPEZ, J.

[2001] Fell type topologies of quasi-pseudometric spaces and the Kuratowski-Painlevé convergence, *Appl. Gen. Topology* **2**, 9–26.

[200?] Fell type topologies of quasi-uniform spaces, *New Zealand J. Math.*, to appear.

RODRÍGUEZ-LÓPEZ, J. and S. ROMAGUERA

[200?] The relationship between the Vietoris topology and the Hausdorff quasi-uniformity, *Topology Appl.*, to appear.

ROMAGUERA, S.
[1992] Left K-completeness in quasi-metric spaces, *Math. Nachr.* **157**, 15–23.
[1995] An example on quasi-uniform convergence and quiet spaces, *Questions Answers Gen. Topology* **13**, 169–171.
[1996] On hereditary precompactness and completeness in quasi-uniform spaces, *Acta Math. Hungar.* **73**, 159–178.
[2000] A new class of quasi-uniform spaces, *Math. Pannonica* **11** (1), 17–28.

ROMAGUERA, S. and M. RUIZ-GÓMEZ
[1995/7] Bitopologies and quasi-uniformities on spaces of continuous functions I, II, *Publ. Math. Debrecen* **47**, 81–93 and **50**, 1–15.

ROMAGUERA, S. and S. SALBANY
[1990] On countably compact quasi-pseudometrizable spaces, *J. Austral. Math. Soc. (Ser. A)* **49**, 231–240.
[1993] On bicomplete quasi-pseudometrizability, *Topology Appl.* **50**, 283–289.

ROMAGUERA, S. and M. SÁNCHEZ-GRANERO
[200?a] Completions and compactifications of quasi-uniform spaces, *Topology Appl.*, to appear.
[200?b] A quasi-uniform characterization of Wallman type compactifications, preprint.

ROMAGUERA, S., E.A. SÁNCHEZ-PÉREZ and O. VALERO
[200?] Quasi-normed monoids and quasi-metrics, preprint.

ROMAGUERA, S., M. SANCHIS and M. TKACHENKO
[200?] Free paratopological groups, preprint.

ROMAGUERA, S. and M.P. SCHELLEKENS
[1999] Quasi-metric properties of complexity spaces, *Topology Appl.* **98**, 311–322.
[2000] Cauchy filters and strong completeness of quasi-uniform spaces, *Rostock. Math. Kolloq.* **54**, 69–79.
[200?] Duality and quasi-normability for complexity spaces, *Appl. Gen. Topology*, to appear.

SALBANY, S. and T. TODOROV
[2000] Nonstandard analysis in topology: Nonstandard and standard compactifications, *Journal Symbol. Logic* **65**, 1836–1840.

SÁNCHEZ-GRANERO, M.A.
[2001] Weak completeness of the Bourbaki quasi-uniformity, *Appl. Gen. Topology* **2**, 101–112.

SCHELLEKENS, M.P.
[1995] The Smyth completion: a common foundation for denotational semantics and complexity analysis, in: *Proc. MFPS 11, Electronic Notes in Theoretical Computer Science* **1**. URL: http://www.elsevier.nl/locate/entcs/volume1.html
[200?] The correspondence between partial metrics and semivaluations, *Theoretical Computer Science.*, to appear.

SCHMITT, V.
[200?] Applying enriched categories to quasi-uniform spaces, preprint.

SEDA, A.K.
[1997] Quasi-metrics and the semantics of logic programs, *Fundamenta Informaticae* **29**, 97–117.

SMYTH, M.B.
[1992] Stable compactification I, *J. London Math. Soc.* **45**, 321–340.
[1994] Completeness of quasi-uniform and syntopological spaces, *J. London Math. Soc.* **49**, 385–400.

SÜNDERHAUF, P.
[1993] The Smyth-completion of a quasi-uniform space, in: *Languages and Model Theory, "Algebra, Logic and Applications"*, Droste, N. and Y. Gurevich, eds., Semantics of Programming Gordon and Breach Sci. Publ., New York, pp. 189–212.
[1995a] Quasi-uniform completeness in terms of Cauchy nets, *Acta Math. Hungar.* **69**, 47–54.
[1995b] Constructing a quasi-uniform function space, *Topology Appl.* **67**, 1–27.
[1997] Smyth completeness in terms of nets: the general case, *Quaestiones Math.* **20**, 715–720.

SUZUKI, J., K. TAMANO and Y. TANAKA
[1989] κ-metrizable spaces, stratifiable spaces and metrization, *Proc. Amer. Math. Soc.* **105**, 500–509.

VITOLO, P.
[1999] The representation of weighted quasi-metric spaces, *Rend. Ist. Mat. Univ. Trieste* **31**, 95–100.

WEHRUNG, F.
[1993] Metric properties of positively ordered monoids, *Forum Math.* **5**, 183–201.

CHAPTER 13

Function Spaces

Witold Marciszewski[1]

Institute of Mathematics, University of Warsaw, Banacha 2, 02–097 Warszawa, Poland
E-mail: wmarcisz@mimuw.edu.pl

Contents

[1] This article was written while the author held a temporary position at the Institute of Mathematics of the Polish Academy of Sciences in 2001/02.

RECENT PROGRESS IN GENERAL TOPOLOGY II
Edited by Miroslav Hušek and Jan van Mill

1. Introduction

The function spaces considered in this survey are the spaces $C_p(X)$ of continuous real-valued functions on topological spaces X equipped with the pointwise convergence topology. The systematic study of these function spaces was initiated by A.V. Arhangel'skii in the seventies; since then he was the motor vivendi for the development of the field. The successive stages of the progress in the theory of function space were the subject of his several excellent comprehensive survey articles [1978], [1987], [1988], [1990], [1992b], [1997], [1998a], and [1998b]. The books ARHANGEL'SKII [1992a] and, more recent, VAN MILL [2001] are also superb sources of information on the subject.

We shall present here some substantial developments of the main topics discussed in Arhangel'skii's articles and also some new areas of research, which emerged during the last ten years. The limited space prompted us to make a rather narrow selection from a vast material concerning $C_p(X)$; inevitable, our choice was heavily influenced by our own interest in the subject. The present article, to some extent, is an expansion and continuation of our survey article MARCISZEWSKI [1998b].

In general, our terminology concerning function spaces follows ARHANGEL'SKII [1992a]. For other notions that we are using, we refer the reader to ENGELKING [1989] (for the general topology notions), KURATOWSKI [1966] and KECHRIS [1995] (for the notions of descriptive set theory), and VAN MILL [2001] (for the notions of infinite-dimensional topology).

We consider only completely regular spaces. For such a space X, by $C_p^*(X)$ we denote the subspace of the function space $C_p(X)$ consisting of bounded functions. We say that the spaces X and Y are *t-equivalent* (resp., *t^*-equivalent*) if $C_p(X)$ is homeomorphic to $C_p(Y)$ (resp., $C_p^*(X)$ and $C_p^*(Y)$ are homeomorphic). In a similar way we define the relations of *l-equivalence*, *l^*-equivalence* (for linearly homeomorphic function spaces), *u-equivalence*, and *u^*-equivalence* (for uniformly homeomorphic function spaces). Recall that the map $\Phi : C_p(X) \to C_p(Y)$ is *uniformly continuous* if, for every neighborhood U of zero (i.e., the zero function) in $C_p(Y)$, there is a neighborhood V of zero in $C_p(X)$ such that $(\Phi(f) - \Phi(g)) \in U$ for every $f, g \in C_p(X)$ with $(f - g) \in V$. The map $\Phi : C_p(X) \to C_p(Y)$ is a *uniform homeomorphism* if both Φ and Φ^{-1} are uniformly continuous.

Let us recall some notions from the descriptive set theory that will be frequently used. Let α be a countable ordinal. We say that a metrizable space X is an *absolute Borel set* of additive class α (resp., multiplicative class α) if X embedded into an arbitrary metric space M is a Borel subset of M of this class, see KURATOWSKI [1966]. In particular, absolute Borel sets of multiplicative class 1, i.e., *absolute G_δ-sets*, are completely metrizable spaces. By $\mathcal{A}_\alpha$ (resp., $\mathcal{M}_\alpha$) we denote the class of metrizable spaces that are absolute Borel of additive (resp., multiplicative) class α. We use the notation Σ_n^1 and Π_n^1 for the classes of *projective spaces*, see KECHRIS [1995]. Σ_1^1 is the class of metrizable *analytic spaces*, i.e., metrizable spaces which are continuous images of the irrationals.

The ordinals, appearing as the examples of topological spaces, are always equipped with the standard order topology.

2. Function spaces on metrizable spaces

In this section we will discuss results concerning the function spaces $C_p(X)$ for metrizable spaces X. We shall concentrate on the problem of the classification of these function spaces with respect to (linear, uniform) homeomorphisms. The general problem which naturally emerges is the following:

2.1. PROBLEM. Which topological properties of the metrizable space X are preserved by (linear, uniform) homeomorphisms of spaces $C_p(X)$ (or $C_p^*(X)$)?

This problem has been extensively investigated also in the much wider class of completely regular spaces. For many standard topological properties of X, e.g., cardinal functions of X, it has been determined whether these properties are preserved by (linear, uniform) homeomorphisms of function spaces, i.e., by the relation of (l-equivalence, u-equivalence) t-equivalence, see ARHANGEL'SKII [1992a]. The simplicity of the behavior of cardinal functions on metrizable spaces makes this particular case different, in many respects, from the general one.

First, let us recall some basic facts on cardinal invariants of metrizable spaces X preserved by the t-equivalence. The weight of X is such an invariant since, for all spaces X, the *network weight* of X and $C_p(X)$ are equal, see ARHANGEL'SKII [1992a] (for completely regular spaces X, the weight of X is not preserved even by the l-equivalence). For all spaces X, the cardinality of X is preserved by the t-equivalence since it is equal to the weight of $C_p(X)$ (when X is infinite), or to the dimension of $C_p(X)$ (when X is finite). One should keep in mind that the metrizability is not preserved by the l-equivalence, see ARHANGEL'SKII [1992a]. However, any space u-equivalent to a metrizable compactum is metrizable.

It is well-known that many topological properties of metrizable spaces X of geometric character, like connectedness, local connectedness, or contractibility, are not preserved by the l-equivalence; e.g., the unit interval $[0,1]$, the circle S^1, and the product $[0,1]\times(\omega+1)$ are l-equivalent (see ARHANGEL'SKII [1991]). However, the fundamental notion of dimension behaves in a much better way in that respect:

2.2. THEOREM (PESTOV [1982]). *If $C_p(X)$ and $C_p(Y)$ are linearly homeomorphic then* $\dim X = \dim Y$.

While some topological invariants of the space X are preserved by surjections of function spaces (see 2.15, 2.16, 2.17, and 2.19), the dimension of X can be raised by a continuous linear surjection of $C_p(X)$.

2.3. THEOREM (LEIDERMAN, MORRIS and PESTOV [1997]). *For every finite dimensional metrizable compactum K there exists a continuous linear surjection*

$$\Phi : C_p([0,1]) \to C_p(K).$$

2.4. THEOREM (LEIDERMAN, LEVIN and PESTOV [1997]). *Let K be a finite dimensional metrizable compact space. Then there exist a 2-dimensional metrizable compactum L and a continuous open linear surjection $\Psi : C_p(L) \to C_p(K)$.*

Pestov's theorem 2.2 was generalized by Gul'ko:

2.5. THEOREM (GUL'KO [1992]). *If the spaces X and Y are u-equivalent then* $\dim X = \dim Y$.

Both theorems of Pestov and Gul'ko were proved for completely regular spaces X and Y, where $\dim X$ is the *covering dimension* of X (see ENGELKING [1989]). The key ingredient of the proof of Gulko's theorem was the following.

2.6. THEOREM (GUL'KO [1992]). *Let X and Y be u-equivalent (u^*-equivalent) separable metrizable spaces. Then Y is a countable union of closed subsets Y_n which can be embedded in X, and the same for X.*

Slightly modifying Gul'ko's argument (c.f. MARCISZEWSKI and PELANT [1997], MARCISZEWSKI [200?a]) it is possible to remove the separability assumption from the above result. In effect, one obtains the following:

2.7. COROLLARY. *Let $\mathcal{P}$ be the class of metrizable spaces with the following properties:*

(i) *if $X \in \mathcal{P}$ and Y is a subset of X then $Y \in \mathcal{P}$,*

(ii) *if X is a metrizable space which is a countable union of closed subsets $X_n \in \mathcal{P}$ then $X \in \mathcal{P}$.*

Then, for metrizable spaces X and Y which are u-equivalent, $X \in \mathcal{P}$ if and only if $Y \in \mathcal{P}$.

Observe that, for every $n \in \omega$, the class of metrizable spaces X with the *small inductive dimension* $\operatorname{ind} X \leq n$ (or the covering dimension $\dim X \leq n$) satisfies the conditions (i) and (ii) of the above corollary. Hence, from 2.7 we can easily derive the preservation of dimension ind and dim of metrizable spaces by the u-equivalence and the u^*-equivalence. Another consequence of Theorem 2.6 is the following example.

2.8. EXAMPLE. There exists a family $\{K_\alpha : \alpha < 2^\omega\}$ of 1-dimensional metrizable continua such that K_α and K_β are not u-equivalent for $\alpha \neq \beta$.

We can obtain the required family of continua by means of a 1-dimensional *hereditarily indecomposable continuum* M constructed by COOK in [1967]. The continuum M has the property that, for every subcontinuum $K \subset M$, every continuous map $f : K \to M$ is either the identity or is constant. Standard properties of hereditarily indecomposable continua (see KURATOWSKI [1968, §48.VI]) allow one to select a family $\{K_\alpha : \alpha < 2^\omega\}$ of nontrivial pairwise disjoint subcontinua of M. One can easily verify that this family satisfies the assertion of Example 2.8, using Theorem 2.6, the Baire category theorem, and Janiszewski's theorem (KURATOWSKI [1968, §47.III.1]).

We do not know if Gul'ko's theorem 2.5 can be generalized for the t-equivalence. This is one of the most interesting open problems in the theory of function spaces $C_p(X)$.

2.9. PROBLEM (ARHANGEL'SKII). Let X and Y be t-equivalent (metrizable, compact) spaces. Is $\dim X = \dim Y$?

A remarkable theorem of Cauty provides a partial affirmative answer to this question (see ENGELKING [1995] for the definition of *strongly infinite-dimensional spaces*):

2.10. THEOREM (CAUTY [1999]). *Let X and Y be metrizable compact spaces such that $C_p(Y)$ is an image of $C_p(X)$ under a continuous open mapping. Then if some finite power Y^n is strongly infinite-dimensional then X^k is also strongly infinite-dimensional, for some natural number k.*

This result should be compared with Theorem 2.4. From Cauty's theorem immediately follows:

2.11. COROLLARY (CAUTY). *The Hilbert cube $[0,1]^\omega$ is not t-equivalent to any finite dimensional metrizable compact space X.*

This was the first example of uncountable compact metrizable spaces with nonhomeomorphic function spaces. Modifying Cauty's technique and using an idea of GUL'KO from [1992] it is possible to prove the following:

2.12. THEOREM (MARCISZEWSKI [200?a]). *Let X and Y be t-equivalent (resp., t*-equivalent) metrizable spaces. Then X is countable dimensional if and only if Y is so.*

Recall that a space X is *countable dimensional* if X is a countable union of finite dimensional subspaces (we can consider here both the small inductive dimension ind and the covering dimension dim). Theorem 2.12 can be easily derived from the following general result (its proof can be also found in van Mill's book [2001, Ch. 6.11]).

2.13. THEOREM (MARCISZEWSKI [200?a]). *Let X and Y be t-equivalent (resp., t*-equivalent) metrizable spaces. Then Y is a countable union of G_δ-subsets Y_n which are homeomorphic to G_δ-subsets of X, and vice versa.*

Using this result we can formulate an appropriate modification of Theorem 2.7 for the t-equivalence. Unfortunately, Theorem 2.13 sheds no light on the t-equivalence between finite dimensional separable completely metrizable spaces. This follows from the fact that every pair X and Y of uncountable finite dimensional separable completely metrizable spaces satisfies the assertion of this theorem. In particular, we still cannot determine whether standard examples of uncountable metrizable compacta are t-equivalent (which is clearly related to Problem 2.9):

2.14. QUESTION. Is $C_p([0,1])$ homeomorphic to $C_p(2^\omega)$ ($C_p([0,1]^2)$)?

The descriptive set theory provides us with certain important topological invariants of metrizable spaces X preserved by maps of $C_p(X)$. First, let us recall the result of Uspenskiĭ concerning compact metrizable spaces, which may be viewed as spaces of the absolute class $\mathcal{M}_0$ (see the Introduction).

2.15. THEOREM (USPENSKIĬ [1982]). *Let X and Y be metrizable spaces and let $\Phi : C_p(X) \to C_p(Y)$ be a uniformly continuous surjection. If X is compact then Y is also compact.*

In general, the compactness is preserved by the u-equivalence (see USPENSKIĬ [1982]). This is not true for the t-equivalence, since GUL'KO and KHMYLEVA [1986] proved that $C_p([0,1])$ is homeomorphic to $C_p(\mathbb{R})$. OKUNEV [1989] proved that the σ-compactness of X is preserved by homeomorphisms of $C_p(X)$. For metrizable spaces X the situation is even better: Metrizable space X is σ-compact if and only if $C_p(X)$ (resp., $C_p^*(X)$) is analytic, see CHRISTENSEN [1974], DOBROWOLSKI and MARCISZEWSKI [1995], and ANDRETTA and MARCONE [2001]. From this characterization we immediately obtain:

2.16. THEOREM. *Let X and Y be metrizable spaces and let $C_p(Y)$ be a continuous image of $C_p(X)$ (resp., $C_p^*(Y)$ be a continuous image of $C_p^*(X)$). If X is σ-compact, so is Y.*

The space of the rationals $\mathbb{Q}$ is t-equivalent to the compact space $\omega + 1$, see Theorem 2.24. This shows that complete metrizability is also not preserved by the t-equivalence (recall that completely metrizable spaces are absolute G_δ-spaces). On the other hand, for linear maps we have the following theorem.

2.17. THEOREM (BAARS, DE GROOT and PELANT [1993]). *Let X and Y be metrizable spaces and let $\Phi : C_p(X) \to C_p(Y)$ (resp., $\Phi : C_p^*(X) \to C_p^*(Y)$) be a continuous linear surjection. If X is completely metrizable (i.e., X is an absolute G_δ), so is Y.*

We do not know if this result remains true for uniformly continuous maps:

2.18. PROBLEM (MARCISZEWSKI and PELANT). Let X and Y be (separable) metrizable spaces and let $\Phi : C_p(X) \to C_p(Y)$ (resp., $\Phi : C_p^*(X) \to C_p^*(Y)$) be a uniformly continuous surjection (uniform homeomorphism). Let X be completely metrizable. Is Y also completely metrizable?

A similar question for higher Borel classes has an affirmative answer, which can be viewed as an extension of Theorem 2.17:

2.19. THEOREM (MARCISZEWSKI and PELANT [1997]). *Let X and Y be metrizable spaces and let $\Phi : C_p(X) \to C_p(Y)$ (resp., $\Phi : C_p^*(X) \to C_p^*(Y)$) be a uniformly continuous surjection. If $X \in \mathcal{M}_\alpha$ (resp., $X \in \mathcal{A}_\alpha$), $\alpha > 1$, then also $Y \in \mathcal{M}_\alpha$ (resp., $Y \in \mathcal{A}_\alpha$).*

A counterpart of the above theorem holds true for all projective classes (MARCISZEWSKI and PELANT [1997, Thm. 3.5]).

Applying Theorem 2.13 we can prove the following result on the preservation of Borel and projective classes under the t-equivalence.

2.20. THEOREM (MARCISZEWSKI [200?a]). *Let X and Y be t-equivalent (resp., t^*-equivalent) metrizable spaces. Then we have:*

(i) *$X \in \mathcal{A}_\alpha$ if and only if $Y \in \mathcal{A}_\alpha$ for $\alpha \geq 2$,*

(ii) *$X \in \mathcal{M}_\alpha$ if and only if $Y \in \mathcal{M}_\alpha$ for $\alpha \geq 3$,*

(iii) *$X \in \Sigma_n^1$ (resp., $X \in \Pi_n^1$) if and only if $Y \in \Sigma_n^1$ (resp., $Y \in \Pi_n^1$) for $n \geq 1$.*

The preservation of analytic (Σ_1^1) spaces under the t-equivalence had been proved earlier by OKUNEV [1989] (the preservation of the classes Σ_n^1 can be also derived from his results). We do not know the answer to the following:

2.21. QUESTION. Let X and Y be t-equivalent (separable) metrizable spaces such that $X \in \mathcal{M}_2$ (i.e., X is an absolute $F_{\sigma\delta}$-space). Does Y belong to the class $\mathcal{M}_2$?

The following result of Baars exhibits yet another topological property of metrizable spaces preserved by the l-equivalence (in fact he proved this theorem for first countable paracompact spaces).

2.22. THEOREM (BAARS [1994]). *Let X and Y be l-equivalent metrizable spaces. Then X is scattered if and only if Y is scattered.*

As we mentioned before the spaces $\mathbb{Q}$ and $\omega + 1$ are t-equivalent. Hence Theorem 2.22 is not true for the t-equivalence. The question of whether 2.22 holds for u-equivalent metrizable spaces seems to be open.

Recall that for a scattered space X, the *scattered height* $ht(X)$ is the smallest ordinal α such that the α-th Cantor-Bendixson derivative $X^{(\alpha)}$ is empty.

2.23. THEOREM (BAARS, DE GROOT, VAN MILL and PELANT [1993]). *Let X and Y be l^*-equivalent metrizable spaces. Then $ht(X) < \omega$ if and only if $ht(Y) < \omega$.*

This theorem is applied to distinguish between the relations of l- and l^*-equivalence, see Example 3.14.

The following table summarizes main results on preservation of properties of metrizable spaces X by maps of function spaces $C_p(X)$.

Properties of metrizable spaces X preserved by l-, u-, and t-equivalence			
Property of X or cardinal invariant	l-equivalence	u-equivalence	t-equivalence
weight	+	+	+
covering dimension dim	+	+	?
small inductive dimension ind	+	+	?
countable dimensionality	+	+	+
compactness	+	+	-
σ-compactness	+	+	+
complete metrizability	+	?	-
$X \in \mathcal{A}_\alpha$ for $\alpha \geq 2$	+	+	+
$X \in \mathcal{M}_2$	+	+	?
$X \in \mathcal{M}_\alpha$ for $\alpha \geq 3$	+	+	+
$X \in \Sigma^1_n$	+	+	+
$X \in \Pi^1_n$	+	+	+
X is a Baire space	-	-	-
X is scattered	+	?	-

So far, we discussed the results to the effect that for some pairs of metrizable spaces X and Y the function spaces $C_p(X)$ and $C_p(Y)$ cannot be (linearly, uniformly) homeomorphic. Now, let us pass on to results establishing the existence of such homeomorphisms of function spaces. Typically, general results in this direction concern the case of function spaces on zero-dimensional spaces X, especially the function spaces on countable spaces X. The topological classification of $C_p(X)$ for countable metrizable X is particulary simple:

2.24. THEOREM (DOBROWOLSKI, MOGILSKI and GUL'KO [1990], CAUTY [1991]). *If X and Y are countable nondiscrete metrizable spaces, then X and Y are t-equivalent and t^*-equivalent.*

Some special cases of this theorem were proved earlier by VAN MILL [1987] and BAARS, DE GROOT, VAN MILL and PELANT [1989]. For uniform homeomorphisms we have the following result of Gul'ko.

2.25. THEOREM (GUL'KO [1988]). *All infinite countable compact spaces X are u-equivalent.*

In the same paper Gul'ko showed that, for countable compact spaces X, the linear topological classification of the spaces $C_p(X)$ coincides with the classification of Banach spaces $C(X)$ given by BESSAGA and PEŁCZYŃSKI in [1960] (the same result has been proved independently by BAARS and DE GROOT [1992]). For the case of uncountable zero-dimensional metrizable compacta we have:

2.26. THEOREM (BAARS and DE GROOT [1992]). *All uncountable metrizable compact zero-dimensional spaces X are l-equivalent.*

The results on linear classification of $C_p(X)$ (resp., $C_p^*(X)$), for locally compact zero-dimensional separable metrizable spaces X, can be found in BAARS and DE GROOT [1992] (resp., BAARS [1993]). Another result concerning zero-dimensional metrizable spaces was proved by Arhangel'skii.

2.27. THEOREM (ARHANGEL'SKII [1991]). *Let X and Y be separable zero-dimensional non-σ-compact completely metrizable spaces. Then X and Y are l-equivalent.*

The case of X of positive dimension is more complicated, and usually, substantial additional restrictions on the structure of X are imposed.

2.28. THEOREM (PAVLOVSKIĬ [1980]). *Let X and Y be finite polyhedra such that* $\dim X = \dim Y$. *Then X and Y are l-equivalent.*

Theorem of Pavlovskiĭ was generalized by Arhangel'skii for the class of so called *Euclidean-resolvable compacta* (see ARHANGEL'SKII [1991]). KAWAMURA and MORISHITA in [1996] proved that all compact manifolds of the same dimension are l-equivalent. In the same paper they gave a complete classification of CW-complexes with respect to the l-equivalence (some partial results concerning the l-equivalence of noncompact polyhedrons and CW-complexes were obtained earlier by DRANISHNIKOV [1986] and ARHANGEL'SKII [1991]).

The results concerning the l-equivalence of uncountable metrizable compacta we discussed above should be compared with the classical theorem of Milyutin (see SEMADENI [1971]) saying that all Banach spaces $C(K)$, for uncountable metrizable compact spaces K, are isomorphic. This result, together with the mentioned earlier theorem of Bessaga and Pełczyński, gives a complete isomorphic classification of Banach spaces $C(K)$ for metrizable compacta K. Examples 2.8 and 4.11 indicate considerable difficulties on the way to such complete linear (or uniform) classification of $C_p(K)$ for this class of spaces K. More feasible seems the problem of characterizing spaces K which are l-equivalent (resp.,

u-equivalent or t-equivalent) to a given standard metrizable compactum, like the Cantor set 2^ω, the n-dimensional cube $[0,1]^n$, the Hilbert cube $[0,1]^\omega$, etc.. A satisfactory solution to such a problem should be an internal characterization of spaces K, not referring to the structure of the function space $C_p(K)$. In this spirit we can reformulate the results of Gul'ko 2.25, and Baars and de Groot 2.26 as follows (in fact, we also need to use other results concerning the properties preserved by the u-equivalence).

2.29. THEOREM (GUL'KO). *A space X is u-equivalent to $\omega+1$ if and only if X is infinite countable and compact.*

From the mentioned earlier linear classification of $C_p(X)$ for countable compact X, it follows that X is l-equivalent to $\omega+1$ if and only if X is infinite countable, compact, and of finite scattered height.

2.30. THEOREM (BAARS and DE GROOT). *For a space X, the following are equivalent:*
(i) *X is l-equivalent to the Cantor set 2^ω,*
(ii) *X is u-equivalent to 2^ω,*
(iii) *X is uncountable, metrizable, compact, and zero-dimensional.*

For the Hilbert cube we have (a slight modification of) the result of VALOV [1991]:

2.31. THEOREM (VALOV). *For a space X, the following are equivalent:*
(i) *X is l-equivalent to the Hilbert cube $[0,1]^\omega$,*
(ii) *X is u-equivalent to $[0,1]^\omega$,*
(iii) *X is metrizable compact, and contains a copy of $[0,1]^\omega$.*

VALOV [1991] proved also a similar characterization for the n-dimensional *universal Menger compactum μ^n.*

2.32. THEOREM (VALOV). *For a space X, the following are equivalent:*
(i) *X is l-equivalent to μ^n,*
(ii) *X is u-equivalent to μ^n,*
(iii) *X is n-dimensional, metrizable, compact, and contains a copy of μ^n.*

The implication (iii)⇒(i) in Theorems 2.31 and 2.32 can be shown using factorization techniques (see Proposition 4.4). For the proof of the implication (ii)⇒(iii) one should use Theorems 2.5, 2.6, 2.15, the Baire Category Theorem, and the fact that every nonempty open subset of $[0,1]^\omega$ (resp., μ^n) contains a copy of $[0,1]^\omega$ (resp., μ^n).

But the following problem is still open.

2.33. PROBLEM (ARHANGEL'SKII). Find an internal characterization of spaces X which are l-equivalent to the cube $[0,1]^n$.

Some partial results in this direction can be found in ARHANGEL'SKII [1991], KAWAMURA and MORISHITA [1996], KOYAMA and OKADA [1987], MORISHITA [1999], PAVLOVSKIĬ [1980], and VALOV [1991].

Recently Górak has characterized spaces u-equivalent to the n-dimensional cube.

2.34. THEOREM (GÓRAK [200?]). *The space X is u-equivalent to $[0,1]^n$ if and only if X is metrizable compact, n-dimensional, and every non-empty closed subset A of X contains a non-empty relatively open subset U which can be embedded in $[0,1]^n$.*

3. Function spaces on countable spaces

Let us recall that the space $C_p(X)$ is metrizable if and only if X is countable. For infinite countable space X, the space $C_p(X)$ is a separable infinite-dimensional linear metric space - a dense linear subspace of the countable product of real lines $\mathbb{R}^X$. Such spaces appear as standard objects of research in infinite-dimensional topology and the methods from this field are very useful in analyzing the metrizable $C_p(X)$. One of most effective tools in this investigation was the technique of *absorbing sets* (see the article DIJKSTRA and VAN MILL [200?]). In particular, this technique was applied in the proofs of Theorems 2.24, 3.1, 3.2, 3.7, 3.8, and 3.9. In the previous section we described some results on function spaces on metrizable countable spaces X. In general, the class of function spaces on completely regular countable spaces X is much more rich. It is a well-known fact that there exist 2^{2^ω} many countable spaces X which are pairwise non-t-equivalent (see MARCISZEWSKI [1998b]). Hence, we cannot expect reasonable general results on classification of such function spaces without some restrictions on the spaces under consideration. For countable metrizable spaces X, both spaces $C_p(X)$ and $C_p^*(X)$ are absolute $F_{\sigma\delta}$-sets (see VAN MILL [1987]). Therefore, it seemed natural to investigate the metrizable space $C_p(X)$ (or $C_p^*(X)$) which are absolute $F_{\sigma\delta}$-sets, or more general - absolute Borel spaces and projective spaces. For function spaces of the class $F_{\sigma\delta}$, we have the following generalization of Theorem 2.24 (here the space σ is the subspace of $\mathbb{R}^\omega$ consisting of all eventually zero sequences).

3.1. THEOREM (DOBROWOLSKI, MARCISZEWSKI and MOGILSKI [1991]). *If X is a countable nondiscrete space and $C_p(X)$ is an absolute $F_{\sigma\delta}$-set, then $C_p(X)$ and $C_p^*(X)$ are homeomorphic to σ^ω.*

DIJKSTRA, GRILLIOT, LUTZER and VAN MILL in [1985] (see also VAN MILL [1999]) have proved that, for nondiscrete space X, the space $C_p(X)$ cannot be an absolute $G_{\delta\sigma}$-set. Therefore Theorem 3.1 completes the topological classification of spaces $C_p(X)$ which are absolute Borel sets of the class not greater than 2.

ARHANGEL'SKII in [1992b] considered the following stronger version of the t-equivalence. We say that the spaces X and Y are *absolutely t-equivalent* (shortly, *at-equivalent*) if the pairs of the spaces $(\mathbb{R}^X, C_p(X))$ and $(\mathbb{R}^Y, C_p(Y))$ are homeomorphic. If a map $\Phi : C_p(X) \to C_p(Y)$ is a uniform homeomorphism, then Φ can be extended to a homeomorphism between $\mathbb{R}^X$ and $\mathbb{R}^Y$, see ARHANGEL'SKII [1992b]. Therefore the u-equivalence implies the at-equivalence. In this context it is worth to mention a relative version of Theorem 3.1 (similar results on relative homeomorphisms can be also found in BAARS, GLADDINES and VAN MILL [1993]).

3.2. THEOREM (CAUTY, DOBROWOLSKI and MARCISZEWSKI [1993], DIJKSTRA and MOGILSKI [1996]). *Let X and Y be countable nondiscrete spaces such that $C_p(X)$ and $C_p(Y)$ are absolute $F_{\sigma\delta}$-sets. If both spaces X and Y are compact, or both are not compact, then the pairs $(\mathbb{R}^X, C_p(X))$ and $(\mathbb{R}^Y, C_p(Y))$ are homeomorphic.*

Let us mention here that compactness is preserved by the at-equivalence (see ARHANGEL'SKII [1992b]). In the case of compact spaces the above theorem can be derived from Gul'ko's theorem 2.25.

Theorems 3.1 and 3.2 allow us to give examples of countable spaces showing that certain topological properties are not preserved by the t-equivalence and the at-equivalence. Many important examples of metrizable $C_p(X)$ are provided by countable spaces X with exactly one nonisolated point. Such spaces are related in a natural way to filters on the set of natural numbers ω. Given a filter F on ω (we consider only free filters on ω, i.e., filters containing all cofinite subsets of ω), by ω_F we denote the space $\omega \cup \{\infty\}$, equipped with the following topology: All points of ω are isolated and the family $\{A \cup \{\infty\} : A \in F\}$ is a neighborhood base at ∞. We can also treat F as a topological space regarding F as a subset of the Cantor set 2^ω. The descriptive complexity of the function space $C_p(\omega_F)$ is strictly related to the descriptive class of the filter F, see LUTZER, VAN MILL and POL [1985], CALBRIX [1985], [1988], and DOBROWOLSKI, MARCISZEWSKI and MOGILSKI [1991]. For example the projective classes of $C_p(\omega_F)$ and F are the same and $C_p(\omega_F)$ is an absolute $F_{\sigma\delta}$-set if and only if F is $F_{\sigma\delta}$. There exist $F_{\sigma\delta}$ filters F such that the space ω_F does not contain any nontrivial convergent sequence (hence is not a *k-space*), or ω_F is not an $\aleph_0$-*space* (see DOBROWOLSKI, MARCISZEWSKI and MOGILSKI [1991]). By Theorems 3.1 and 3.2 such spaces ω_F are t-equivalent to the compact space $\omega + 1$ and at-equivalent to the metrizable locally compact space $\omega + \omega$.

Using the spaces ω_F, for suitable Borel and projective filters F on ω, we can show that the descriptive complexity of the spaces $C_p(X)$ can be arbitrarily high. In such a way, LUTZER, VAN MILL and POL [1985] and CALBRIX [1985], [1988] constructed examples of Borel spaces $C_p(X) \in \mathcal{M}_\alpha \setminus \mathcal{A}_\alpha$, for all $\alpha > 2$, and examples of spaces $C_p(X)$ of arbitrary projective classes. But Borel spaces $C_p(X)$ are always of the exact multiplicative class:

3.3. THEOREM (CAUTY, DOBROWOLSKI and MARCISZEWSKI [1993]). *Let X be a countable infinite space such that $C_p(X)$ is an absolute Borel set. Then there exists a countable ordinal $\alpha \geq 1$ such that $C_p(X) \in \mathcal{M}_\alpha \setminus \mathcal{A}_\alpha$.*

This result generalizes (for countable spaces X) the theorem of Dijkstra, Grilliot, Lutzer and van Mill mentioned before. The special case of Theorem 3.3, for the spaces of the form ω_F, had been proved earlier by CALBRIX [1988].

A rather surprising result of Cauty showed that Theorem 3.1 cannot be extended for higher Borel classes (see DIJKSTRA and VAN MILL [200?] or MARCISZEWSKI [1998b] for additional comments):

3.4. EXAMPLE (CAUTY [1998]). For every ordinal $\alpha \geq 3$ ($n \geq 1$) there exist countable spaces X and Y such that $C_p(X)$ and $C_p(Y)$ belong to the class $\mathcal{M}_\alpha \setminus \mathcal{A}_\alpha$ (resp., $\Sigma^1_n \setminus \Pi^1_n$, $\Pi^1_n \setminus \Sigma^1_n$) and are not homeomorphic.

Let us note that the examples of nonhomeomorphic spaces $C_p(X)$ and $C_p(Y)$ which are both analytic and not coanalytic (i.e., in $\Sigma^1_1 \setminus \Pi^1_1$) or coanalytic and not analytic had been previously constructed in MARCISZEWSKI [1993], under some additional set-theoretic assumptions. In a view of Cauty's example it seems natural to ask the following:

3.5. QUESTION. Do there exist infinitely many (continuum many) pairwise nonhomeomorphic spaces $C_p(X)$ of a given Borel class $\mathcal{M}_\alpha \setminus \mathcal{A}_\alpha$, $\alpha \geq 3$ (projective class)?

It is also natural to ask what is the relationship between the descriptive complexity of the space $C_p(X)$ and the space $C_p(A)$ of functions on a closed subset A of X. Clearly, if

$C_p(X)$ is analytic then $C_p(A)$, being a continuous image of $C_p(X)$, is also analytic. But the descriptive complexity of $C_p(A)$ can be higher than the complexity of $C_p(X)$:

3.6. THEOREM (MARCISZEWSKI [1995]). *Let A be a countable space such that $C_p(A)$ is analytic. Then A can be embedded as a closed subset in a countable space X with $C_p(X)$ which is an absolute $F_{\sigma\delta}$-set.*

An application of the absorbing set technique yields the following Cantor-Bernstein-type principle:

3.7. THEOREM (CAUTY, DOBROWOLSKI and MARCISZEWSKI [1993]). *Let X and Y be countable spaces such that $C_p(X)$ and $C_p(Y)$ are analytic. Then the function spaces $C_p(X)$ and $C_p(Y)$ are homeomorphic if $C_p(X)$ embeds as a closed subset in $C_p(Y)$ and vice versa.*

It is worth mentioning that, for a countable space X, the condition that $C_p(X)$ is analytic is equivalent to the existence of an embedding of X into the function space $C_p(\mathbb{P})$ on the irrationals $\mathbb{P}$ (see MARCISZEWSKI [1995]). The following counterpart of Theorem 3.7 was proved by Banakh and Cauty.

3.8. THEOREM (BANAKH and CAUTY [1997]). *For countable spaces X and Y, the function spaces $C_p^*(X)$ and $C_p^*(Y)$ are homeomorphic if each of these spaces contains a closed topological copy of the other.*

Comparing Theorems 3.7 and 3.8 it is natural to ask: What is the difference between the topological structure of the function spaces $C_p(X)$ and $C_p^*(X)$ for countable X? How different are the topological (resp., linear, uniform) classifications of these two classes of function spaces? It is well-known that, for infinite countable X, the space $C_p^*(X)$ is always of the first category. Clearly, the space ω demonstrates that this is not true for spaces $C_p(X)$. We also have examples of nondiscrete spaces X with $C_p(X)$ of the second category. Namely, the space $C_p(\omega_F)$ is of the first category if and only if the filter F is so (see VAN MILL [2001]). Hence, if F is an ultrafilter on ω then $C_p(\omega_F)$ is of the second category (actually, a Baire space), and is not homeomorphic to $C_p^*(\omega_F)$. On the other hand we have the following.

3.9. THEOREM (BANAKH and CAUTY [1997]). *For a countable infinite space X, the spaces $C_p(X)$ and $C_p^*(X)$ are homeomorphic if and only if $C_p(X)$ is a σZ-space.*

We refer the reader to VAN MILL [2001] or DIJKSTRA and VAN MILL [200?] for the definition of *σZ-spaces*. Let us note that, for nondiscrete countable X, analytic spaces $C_p(X)$ are σZ-spaces. Hence, an analytic space $C_p(X)$ is homeomorphic to $C_p^*(X)$ provided X is countable and nondiscrete. Therefore Theorem 3.8 may be viewed as a generalization of Theorem 3.7.

In general case we have the following relationship between the spaces $C_p(X)$ and $C_p^*(X)$.

3.10. THEOREM (BANAKH and CAUTY [1997]). *Let X be a countable nondiscrete space. Then the space $C_p^*(X)$ is homeomorphic to $C_p(X) \times \sigma$.*

From this theorem we derive the following:

3.11. COROLLARY (BANAKH and CAUTY [1997]). *If countable spaces X and Y are t-equivalent, then X and Y are t^*-equivalent.*

However, the reverse implication does not hold true:

3.12. EXAMPLE (MARCISZEWSKI and VAN MILL [1998]). There exists countable spaces X and Y which are t^*-equivalent and not t-equivalent.

Let us describe these spaces. First, recall that a filter F on ω is a *P-filter* if, for every sequence (U_n) of sets from F, there exists $A \in F$ which is *almost contained* in every U_n, i.e., $A \setminus U_n$ is finite. P-ultrafilters are also called *P-points*. We take an ultrafilter F on ω which is not a P-point (see VAN MILL [1984]). We put $X = \omega \times \omega_F$ and $Y = X \oplus (\omega + 1)$ — the discrete sum of spaces X and $\omega + 1$. The function spaces $C_p(X)$ and $C_p(Y)$ are topologically distinct, since the first is a Baire space and the second one is of the first category. The existence of the homeomorphism between the spaces $C_p^*(X)$ and $C_p^*(Y)$ can be established with a help of Theorems 3.8 and 3.1 and the following result:

3.13. THEOREM (MARCISZEWSKI [1998a]). *For a filter F on ω the following are equivalent:*

(i) *F is a second category P-filter,*

(ii) *F is a hereditary Baire space,*

(iii) *$C_p(\omega_F)$ is a hereditary Baire space,*

(iv) *$C_p(\omega_F)$ does not contain a closed copy of the space σ^ω (a copy of σ).*

Recall that the space X is *hereditary Baire* if every closed subset A of X is a Baire space. By a theorem of Hurewicz, for metrizable spaces X, this property is equivalent to the condition that X does not contain any closed copy of the rationals $\mathbb{Q}$. P-points (hence, second category P-filters) can be constructed under some additional set-theoretic assumptions, e.g., the continuum hypothesis (CH). The question of whether the existence of second category P-filters can be proved without any additional set-theoretic assumptions seems to be open. Note that the existence of such filters is equivalent to the existence of hereditary Baire spaces $C_p(X)$ for countable nondiscrete X (see MARCISZEWSKI [1998a]). Theorem 3.13 generalizes a similar result on ultrafilters proved earlier by GUL'KO and SOKOLOV [1998] (related results can be also found in MICHALEWSKI [1998] and BOUZIAD [2000a]). Let us also note that the Baire property of the function space $C_p(X)$ has been characterized in terms of the topological properties of the space X by PYTKEEV [1985], TKACHUK [1985] and van Douwen (unpublished). GUL'KO and SOKOLOV in [1998] stated a problem about similar characterization of hereditary Baire spaces $C_p(X)$.

Also the linear classifications of the spaces $C_p(X)$ and $C_p^*(X)$ are distinct:

3.14. EXAMPLE (BAARS, DE GROOT, VAN MILL and PELANT [1993])). The ordinal spaces ω^2 and ω^ω are l-equivalent but not l^*-equivalent.

The fact that $C_p^*(\omega^2)$ and $C_p^*(\omega^\omega)$ are not linearly homeomorphic follows from Theorem 2.23. We do not know the answer to the following:

3.15. PROBLEM. Let X and Y be (countable, metrizable) l^*-equivalent spaces. Are then X and Y l-equivalent?

4. Products of function spaces

Many typical constructions of (linear, uniform) homeomorphisms between spaces $C_p(X)$ or $C_p^*(X)$ rely on some factorization properties of these function spaces. Such techniques may be illustrated by the following version of a well-known Decomposition Scheme (see SEMADENI [1971]).

4.1. PROPOSITION. *Let $C_p(X)$ and $C_p(Y)$ be function spaces such that each of them is a (linear) factor of the other one, i.e., there exist (linear topological) spaces E and F such that $C_p(X)$ is (linearly) homeomorphic to $C_p(Y) \times E$ and $C_p(Y)$ is (linearly) homeomorphic to $C_p(X) \times F$. If $C_p(X)$ is (linearly) homeomorphic to $(C_p(X))^\omega$, then $C_p(X)$ is (linearly) homeomorphic to $C_p(Y)$.*

We refer the reader to ARHANGEL'SKII [1991] and [1992b] for another results employing factorization methods. The question of whether the space $C_p(Y)$ is a factor of $C_p(X)$ is connected to the problem of the existence of a continuous extender $e : C_p(A) \to C_p(X)$ for a closed subset $A \subset X$. Recall that a map $e : C_p(A) \to C_p(X)$ is an *extender* if $e(f)|A = f$ for all $f \in C_p(A)$. This connection is illustrated by the following standard fact (see ARHANGEL'SKII [1992a] or VAN MILL [2001]).

4.2. PROPOSITION. *Let A be a subset of a space X. If there exists a continuous (linear) extender $e : C_p(A) \to C_p(X)$ (resp., $e : C_p^*(A) \to C_p^*(X)$) then the spaces $C_p(X)$ and $C_p(A) \times \{f \in C_p(X) : f|A \equiv 0\}$ (resp., $C_p^*(X)$ and $C_p^*(A) \times \{f \in C_p^*(X) : f|A \equiv 0\}$) are (linearly) homeomorphic.*

Dugundji Extension Theorem provides continuous linear extenders $e : C_p(A) \to C_p(X)$ and $e' : C_p^*(A) \to C_p^*(X)$ for every closed subset A of a metrizable space X (see VAN MILL [2001]). Therefore, for metrizable spaces we can reformulate the Decomposition Scheme in the following way.

4.3. PROPOSITION. *Let X and Y be metrizable spaces such that each contains a closed topological copy of the other. If $C_p(X)$ is (linearly, uniformly) homeomorphic to $(C_p(X))^\omega$, then $C_p(X)$ is (linearly, uniformly) homeomorphic to $C_p(Y)$.*

Theorem 2.27 can be derived from this proposition. For the spaces of bounded function we can replace the countable product by the c_0-*product* of $C_p^*(X)$, i.e., the space $(C_p^*(X))_0^\omega = \{(f_n) \in (C_p^*(X))^\omega : \lim \|f_n\|_\infty = 0\}$ (where $\|\cdot\|_\infty$ is the supremum norm in $C_p^*(X)$):

4.4. PROPOSITION. *Let X and Y be metrizable spaces such that each contains a closed topological copy of the other. If $C_p^*(X)$ is (linearly, uniformly) homeomorphic to $(C_p^*(X))_0^\omega$, then $C_p^*(X)$ is (linearly, uniformly) homeomorphic to $C_p^*(Y)$.*

Proposition 4.4 is useful in obtaining results like Theorems 2.26, 2.31, and 2.32.

The Dugundji Extension Theorem holds true for the wider class of *stratifiable spaces*. Also, linear continuous extenders $e : C_p(A) \to C_p(X)$ (resp., $e' : C_p^*(A) \to C_p^*(X)$) exist, if either A is a compact metrizable subset of a completely regular space X, or A is a separable completely metrizable closed subset of a normal space X, or else A is a completely metrizable closed subset of a paracompact space X (see VAN MILL [2001]).

A standard example of a closed subset A of a space X without a continuous extender $e : C_p(A) \to C_p(X)$ is the remainder $\omega^* = \beta\omega \setminus \omega$ of the *Čech-Stone compactification* $\beta\omega$ of ω. Using cardinal invariants it is easy to verify that $C_p(\omega^*)$ cannot be embedded in $C_p(\beta\omega)$ (hence, $C_p(\omega^*)$ is not a factor of $C_p(\beta\omega)$ and there is no continuous extender for these function spaces). The *pseudocharacter* of $C_p(X)$ is equal to the density of X (see ARHANGEL'SKII [1992a]), therefore the pseudocharacter of $C_p(\omega^*)$ is greater than the pseudocharacter of $C_p(\beta\omega)$. But this phenomenon may also occur for countable spaces:

4.5. EXAMPLE (VAN MILL and POL [1993]). There exist a countable X and a closed $A \subset X$ with no continuous extender $e : C_p(A) \to C_p(X)$.

Actually, this example has a stronger property; there is no extender $e : C_p(A) \to C_p(X)$ measurable with respect to the σ-algebra generated by the Souslin sets, but $C_p(A)$ is a factor of $C_p(X)$. Yet, applying Theorem 3.6, for a countable space A with analytic non-Borel $C_p(A)$, we obtain an example of a countable X with a closed subset A such that $C_p(A)$ is not a factor of $C_p(X)$.

The following interesting question concerning continuous extenders remains open.

4.6. QUESTION (ARHANGEL'SKII). Does there exist a continuous extender
$e : C_p(\{0,1\}^{\omega_1}) \to C_p([0,1]^{\omega_1})$?

ARHANGEL'SKII and CHOBAN [1990] proved that no such extender can be linear.

For the application of the factorization techniques mentioned above it is important to determine when the function space $C_p(X)$ is (linearly, uniformly) homeomorphic to $(C_p(X))^\omega$, $C_p(X) \times C_p(X)$, or $C_p(X) \times \mathbb{R}$. These fundamental questions were asked by ARHANGEL'SKII in [1978], [1990], [1992b], [1997], and [1998a]. They are also related to another general problem by Arhangel'skii: which topological properties of $C_p(X)$ are shared by $C_p(X) \times C_p(X)$? These problems turned out to be difficult and there are only a few known positive general results concerning these factorization properties of $C_p(X)$.

4.7. THEOREM (ARHANGEL'SKII [1992b]). *If a space X contains a nontrivial convergent sequence, or X is not pseudocompact, then $C_p(X)$ is linearly homeomorphic to $C_p(X) \times \mathbb{R}$.*

In particular, the above theorem holds true for metrizable spaces, Lindelöf non-compact spaces, and spaces with countable network (hence, for countable spaces). However, this is not true for all compact spaces, see Example 4.15.

4.8. THEOREM (CAUTY, DOBROWOLSKI and MARCISZEWSKI [1993]). *Let F be a first category filter on ω. Then the space $C_p(\omega_F)$ is homeomorphic to $(C_p(\omega_F))^\omega$.*

We do not know if this result can be generalized for all infinite countable spaces.

4.9. PROBLEM. Is $C_p(X)$ homeomorphic to $(C_p(X))^\omega$ for every infinite countable space X? Is $C_p(\omega_F)$ homeomorphic to $(C_p(\omega_F))^\omega$ for every filter F on ω?

Observe that $C_p(X)$ cannot be uniformly homeomorphic to $(C_p(X))^\omega$ for any (nonempty) compact space X. This follows from Uspenskiĭ's result on preservation of compactness by the u-equivalence and the fact that the product $(C_p(X))^\omega$ can be easily identified with the space $C_p(X \times \omega)$.

In general, examples show that almost all factorization properties of $C_p(X)$ cannot be proved without essential restrictions on the class of spaces X.

4.10. EXAMPLE (GUL'KO [1990]). The space $C_p(\omega_1+1)$ is not homeomorphic to its own square $C_p(\omega_1+1) \times C_p(\omega_1+1)$.

Independently, MARCISZEWSKI [1988] constructed a separable compact space X of scattered height 3 such that $C_p(X)$ is not homeomorphic to $C_p(X) \times C_p(X)$. POL [1995] proved that for Cook's continuum M described in Section 2, the spaces $C_p(M)$ and $C_p(M) \times C_p(M)$ are not linearly homeomorphic. This paper contains also other examples of metrizable spaces X with the same property of the function space $C_p(X)$; one of them is a zero-dimensional subset X of the real line $\mathbb{R}$. Recently, it has been proved that, for Cook's continuum M, the space $C_p(M)$ is not uniformly homeomorphic to $C_p(M) \times C_p(M)$.

4.11. EXAMPLE (VAN MILL, PELANT and POL [200?]). There exist an infinite metrizable compact space X such that the spaces $C_p(X)$ and $C_p(X) \times C_p(X)$ are not uniformly homeomorphic.

But the following problems remain unsolved.

4.12. PROBLEM (ARHANGEL'SKII). Is $C_p(X)$ homeomorphic to $C_p(X) \times C_p(X)$ for every infinite (compact) metrizable space X?

4.13. PROBLEM (ARHANGEL'SKII). Does there exist a continuous map from $C_p(X)$ onto $C_p(X) \times C_p(X)$ for every (compact) space X?

The second question has an affirmative answer if X is compact and zero-dimensional or metrizable compact. This question is related to the following:

4.14. PROBLEM (ARHANGEL'SKII). Let X be a (compact) space such that $C_p(X)$ is Lindelöf. Is $C_p(X) \times C_p(X)$ also Lindelöf?

Recall that for a compact space X, the space $C_p(X)$ is Lindelöf if and only if it is normal. In general, $C_p(X)$ is Lindelöf if and only if it is paracompact (see, ARHANGEL'SKII [1992a]). The question of whether $C_p(X) \times C_p(X)$ is normal provided $C_p(X)$ is normal is also open.

4.15. EXAMPLE (MARCISZEWSKI [1997]). There exists an infinite compact space X with no continuous linear surjection from $C_p(X)$ onto $C_p(X) \times \mathbb{R}$. In particular, the space $C_p(X)$ is not linearly homeomorphic to the product $C_p(X) \times E$ for any nontrivial linear topological space E.

In MARCISZEWSKI [1997], two examples of spaces X with these properties are given. The first one is zero-dimensional compact and is constructed by transfinite induction, using the idea of "killing maps" invented by Kuratowski and Sierpiński. The second example X is non-compact, but much easier to describe: For every infinite subset A of ω we choose a weak P-point p_A in ω^* in such a way that $A \in p_A$ and p_A and $p_{A'}$ are not equivalent (via bijection of ω) for $A \neq A'$. We take $X = \omega \cup \{p_A : A \subset \omega,\ A \text{ is infinite}\}$ considered as a subspace of $\beta\omega$. Recall that a point $p \in \omega^*$ is a *weak P-point* if p is not in the closure of any countable set $D \subset \omega^* \setminus \{p\}$, see VAN MILL [1984].

We do not know if, for any of these examples, $C_p(X)$ is homeomorphic to $C_p(X) \times \mathbb{R}$. In general, the following problem is still open:

4.16. PROBLEM (ARHANGEL'SKII). Let X be an infinite (compact) space. Is $C_p(X)$ homeomorphic to $C_p(X) \times \mathbb{R}$?

5. Condensations of function spaces

A continuous bijection from a space X onto a space Y is called a *condensation* of X onto Y. We shall discuss some recent results concerning the following interesting

5.1. PROBLEM (ARHANGEL'SKII). When does there exist a condensation of $C_p(X)$ onto a compact (σ-compact) space?

In other words, we want to determine which function spaces $C_p(X)$ admit a weaker compact (σ-compact) topology ($C_p(X)$ is σ-compact only for finite X). The problem of the existence of such a weaker topology is related to an old question by S. Banach of whether every separable Banach space admits a weaker metrizable compact topology. Banach's question was solved in the affirmative by PYTKEEV [1976]. Pytkeev proved that every separable metrizable space which is an absolute Borel set and is not σ-compact, has a condensation onto the Hilbert cube (clearly, every σ-compact Banach space, i.e., topologically $\mathbb{R}^n$, can be condensed onto a metrizable compactum). Pytkeev's theorem was also used to prove the following result.

5.2. THEOREM (ARHANGEL'SKII [2000]). *For every σ-compact metrizable space X, $C_p(X)$ condenses onto a metrizable compactum.*

For metrizable compact spaces X this result was proved independently in CASARRUBIAS-SEGURA [2001].

For a dense subset D of the space X, by $C_D(X)$ we denote the space $\{f|D : f \in C_p(X)\}$ considered as a subspace of the product $\mathbb{R}^D$. Clearly $C_p(X)$ condenses onto $C_D(X)$. Hence if, for some countable dense subset $D \subset X$, the space $C_D(X)$ is absolute Borel and not σ-compact, then by Pytkeev's result $C_p(X)$ can be condensed onto the Hilbert cube. This happens for all infinite σ-compact metrizable spaces X. Also some non-metrizable compact spaces X possess this property, e.g., the *two arrows space*, the *Helly space*, or the Cantor cube $\{0,1\}^{2^\omega}$, see DOBROWOLSKI and MARCISZEWSKI [1995] and ARHANGEL'SKII and PAVLOV [200?]. Note that, by CHRISTENSEN's result [1974], the space $C_D(X)$ is not absolute Borel for metrizable non-σ-compact X. Yet, Arhangel'skii's theorem 5.2 has been generalized by Michalewski to the following effect.

5.3. THEOREM (MICHALEWSKI [200?]). *For every metrizable analytic space X, $C_p(X)$ condenses onto a metrizable compactum.*

However, it is not possible to generalize this result for all separable metrizable spaces.

5.4. EXAMPLE (MARCISZEWSKI [200?b]). Assuming that $\mathfrak{d} = 2^\omega$, there exists a subspace $X \subset \mathbb{R}$ such that there is no condensation of $C_p(X)$ onto a σ-compact space.

Let us recall that the above assumption $\mathfrak{d} = 2^\omega$ means that the minimal cardinality of a *dominating family* of functions $f : \omega \to \omega$ is equal to 2^ω (see VAN DOUWEN [1984]). We do not know if such an example can be constructed without any additional set-theoretic assumptions.

ARHANGEL'SKII and PAVLOV [200?] proved that for some classes of spaces X the function space $C_p(X)$ cannot be condensed onto a compact space. One of their most interesting results concerns *Corson compact spaces*, i.e., compact spaces which can be embedded into some Σ-product of real lines.

5.5. THEOREM (ARHANGEL'SKII and PAVLOV [200?]). *If K is a non-metrizable Corson compact space then $C_p(K)$ does not condense onto a compact space.*

Arhangel'skii also stated the following problem. Does $C_p(K)$ condense onto a σ-compact space for every compact space K? Quite recently Burke and Pol have shown that consistently this problem has a negative solution:

5.6. THEOREM (BURKE and POL [200?]). *Assuming the continuum hypothesis* (CH), *the space $C_p(\omega^*)$ cannot be condensed onto a Lindelöf space (in particular, onto a σ-compact space).*

Again, we do not know if such an example exists without any additional set-theoretic assumptions. Let us mention that CASARRUBIAS-SEGURA [2001] and ARHANGEL'SKII and PAVLOV [200?] proved that $C_p(\omega^*)$ has no condensation onto a compact space.

6. Miscellaneous results

Below we would like to present a brief overview of other interesting recent results on function spaces not fitting in topics discussed in previous sections.

We start with important results concerning the preservation of the *Lindelöf number* $l(X)$ by the l-equivalence (the first one was announced by Velichko in 1991).

6.1. THEOREM (VELICHKO [1998]). *The Lindelöf property is preserved by the l-equivalence.*

This result was generalized by Bouziad (for some classes of spaces X such generalizations were obtained earlier by BAARS [1994], BAARS and GLADDINES [1996], VALOV [1997], and VALOV and VUMA [2000]).

6.2. THEOREM (BOUZIAD [2000b]). *Let X and Y be l-equivalent spaces. Then the Lindelöf numbers of X and Y are equal.*

We do not know if the Lindelöf number is preserved by the t-equivalence (or by the u-equivalence).

Okunev proved several results concerning the preservation of the *hereditary Lindelöf number* $hl(X)$, the *hereditary density* $hd(X)$, and the *spread* $s(X)$ of the space X and its finite powers. For the l-equivalence, some partial results in this direction were obtained by TKACHUK [1997].

6.3. THEOREM (OKUNEV [1997]). *If X and Y are t-equivalent (or t*-equivalent) spaces, then for every $n \in \omega$, $hl(X^n) = hl(Y^n)$, $hd(X^n) = hd(Y^n)$, and $s(X^n) = s(Y^n)$.*

It is known that the *tightness* $t(X)$ of the space X is not preserved even by linear homeomorphisms of $C_p(X)$ (see OKUNEV [1990]). But for the class of compact spaces X the situation is different:

6.4. THEOREM (OKUNEV [2002]). *If X and Y are t-equivalent compact spaces, then $t(X) = t(Y)$.*

Let us also allude to some results characterizing certain covering properties of spaces X (or of all finite powers of X) in terms of sequential properties of the closure operator in function spaces. A classical example of such characterization is (a special case of) the theorem of Arhangel'skii and Pytkeev (see, ARHANGEL'SKII [1992a]) saying that $C_p(X)$ has countable tightness if and only if X^n is Lindelöf for every n. Another well-known characterization due to GERLITS and NAGY [1982] gives an equivalence between the *Fréchet property* of $C_p(X)$ and the γ*-property* of X. In such a way, ARHANGEL'SKII [1986] characterized the *Menger property* of all finite powers of X; SAKAI [1988] characterized the *Rothberger's property* C'' of X. More recent such results, connecting the *Hurewicz property* of X (or finite powers of X) with some sequential properties of $C_p(X)$, were proved in SCHEEPERS [1997], and KOČINAC and SCHEEPERS [200?]. Another characterization of this kind was given by SAKAI in [2000]. In this context it is also worth mentioning the results of FREMLIN [1994] concerning the properties of the *sequential closure* in spaces $C_p(X)$.

Some classes of compact spaces X, like *Eberlein* or Corson compact spaces can be characterized in terms of the properties of the function space $C_p(X)$ (see ARHANGEL'SKII [1992a]) Recently, KALENDA [2000] has proved that the class of *Valdivia compact spaces* can be also characterized in similar way.

Another interesting area of research focuses on the following general problem: Which spaces can be embedded in the spaces $C_p(X)$, where X is compact or Lindelöf? We refer the reader to ARHANGEL'SKII [1998a] for a survey on this topic.

References

ANDRETTA, A. and A. MARCONE

[2001] Pointwise convergence and the Wage hierarchy, *Comment. Math. Univ. Carolin.* **42**, 159–172.

ARHANGEL'SKII, A.V.

[1978] Structure and classification of topological spaces and their cardinal invariants, *Uspekhi Mat. Nauk* **33**, 29–84.

[1986] Hurewicz spaces, analytic sets and fan tightness of function spaces, *Soviet Math. Dokl.* **33**, 396–399.

[1987] A survey of C_p-Theory, *Q & A in General Topology* **5**, special issue.

[1988] Some results and problems in C_p-Theory, in *Proc. Sixth Prague Topological Sympos.*, Z.Frolík, ed., pp. 11–31.

[1990] Problems in C_p-Theory, in *Open Problems in Topology*, J.van Mill and G.M. Reed, eds., North-Holland, Amsterdam, pp. 601–615.

[1991] On linear topological classification of spaces of continuous function in the topology of pointwise convergence, *Math. USSR Sb.* **70**, 129–142.

[1992a] *Topological Function Spaces*, Kluwer Acad. Publ., Dordrecht.

[1992b] C_p-Theory, in *Recent Progress in General Topology*, M.Hušek and J.van Mill, eds., Elsevier, Amsterdam, pp. 1–56.

[1997] Some recent results and open problems in general topology (Russian), *Uspekhi Mat. Nauk* **52**, 45–70.

[1998a] Embeddings in C_p-spaces, *Topology Appl.* **85**, 9–33.

[1998b] Some observations on C_p-theory and bibliography, *Topology Appl.* **89**, 203–221.

[2000] On condensations of C_p-spaces onto compacta, *Proc. Amer. Math. Soc.* **128**, 1881–1883. Erratum: *Proc. Amer. Math. Soc.* **130**, (2002), 1875.

ARHANGEL'SKII, A.V. and M.M. CHOBAN

[1990] On the position of a subspace in the whole space, *Compt. Rend. Acad. Bulg. Sci.* **43**, 13–15.

ARHANGEL'SKII, A.V. and O.I. PAVLOV

[200?] A note on condensations of $C_p(X)$ onto compacta, *Comment. Math. Univ. Carolin.*, to appear.

BAARS, J.

[1993] On the l_p^*-equivalence of certain locally compact spaces, *Topology Appl.* **52**, 43–57.

[1994] Function spaces on first countable paracompact spaces, *Bull. Acad. Pol. Sci.* **42**, 29–35.

BAARS, J. and H. GLADDINES

[1996] On the linear invariance of Lindelöf numbers, *Canad. Math. Bull.* **39**, 129–137.

BAARS, J., H. GLADDINES and J. VAN MILL

[1993] Absorbing systems in infinite-dimensional manifolds, *Topology Appl.* **50**, 147–182.

BAARS, J. and J. DE GROOT

[1992] *On topological and linear equivalence of certain function spaces*, CWI Tracts 86, Stichting Mathematisch Centrum, Centrum voor Wiskunde en Informatica, Amsterdam.

BAARS, J., J. DE GROOT, J. VAN MILL and J. PELANT

[1989] On topological and linear homeomorphisms of certain function spaces, *Topology Appl.* **32**, 267–277.

[1993] An example of ℓ_p-equivalent spaces which are not ℓ_p^*-equivalent, *Proc. Amer. Math. Soc.* **119**, 963–969.

BAARS, J., J. DE GROOT and J. PELANT

[1993] Function spaces of completely metrizable spaces, *Trans. Amer. Math. Soc.* **340**, 871–879.

BANAKH, T. and R. CAUTY

[1997] Universalité forte pour les sous-ensambles totalement bornés. Applications aux espaces $C_p(X)$, *Colloquium Math.* **73**, 25–33.

BESSAGA, C. and A. PEŁCZYŃSKI

[1960] Spaces of continuous functions (IV) (On isomorphical classification of spaces of continuous functions), *Studia Math.* **19**, 53–62.

BOUZIAD, A.

[2000a] Filters, consonace and hereditary Baireness, *Topology Appl.* **104**, 27–38.

[2000b] Le degré de Lindelöf est l-invariant, *Proc. Amer. Math. Soc.* **129**, 913–919.

BURKE, D.K. and R. POL

[200?] On non-measurablility of ℓ_∞/c_0 in its second dual, preprint.

CALBRIX, J.

[1985] Classes de Baire et espaces d'applications continues, *C. R. Acad. Sc. Paris* **301**, 759–762.

[1988] Filtres boréliens sur l'ensemble des entiers et espaces des applications continues, *Rev. Roumaine Math. Pures Appl.* **33**, 655–661.

CASARRUBIAS-SEGURA, F.

[2001] On compact weaker topologies in function spaces, *Topology Appl.* **115**, 291–298.

CAUTY, R.

[1991] L'espace des fonctions continues d'un espace métrique dénombrable, *Proc. Amer. Math. Soc.* **113**, 493–501.

[1998] La classe borélienne ne détermine pas le type topologique de $C_p(X)$, *Serdica Math. J.* **24**, 307–318.

[1999] Sur l'invariance de la dimension infinie forte par t-équivalence, *Fund. Math.* **160**, 95–100.

CAUTY, R., T. DOBROWOLSKI and W. MARCISZEWSKI

[1993] A contribution to the topological classification of the spaces $C_p(X)$, *Fund. Math.* **142**, 269–301.

CHRISTENSEN, J.P.R.

[1974] *Topology and Borel Structure*, North-Holland, Amsterdam, London.

COOK, H.

[1967] Continua which admit only the identity mapping onto nondegenerate subcontinua, *Fund. Math.* **60**, 241–249.

DIJKSTRA, J., T. GRILLIOT, D. LUTZER and J. VAN MILL

[1985] Function spaces of low Borel complexity, *Proc. Amer. Math. Soc.* **94**, 703–710.

DIJKSTRA, J. and J. VAN MILL

[200?] Infinite-dimensional topology, this volume.

DIJKSTRA, J. and J. MOGILSKI

[1996] The ambient homeomorphy of certain function spaces and sequence spaces, *Comment. Math. Univ. Carolin.* **37**, 595–611.

DOBROWOLSKI, T., S.P. GUL'KO and J. MOGILSKI

[1990] Function spaces homeomorphic to the countable product of ℓ_f^2, *Topology Appl.* **34**, 153–160.

DOBROWOLSKI, T. and W. MARCISZEWSKI

[1995] Classification of function spaces with the pointwise topology determined by a countable dense set, *Fund. Math.* **148**, 35–62.

DOBROWOLSKI, T., W. MARCISZEWSKI and J. MOGILSKI

[1991] On topological classification of function spaces of low Borel complexity, *Trans. Amer. Math. Soc.* **328**, 307–324.

VAN DOUWEN, E.

[1984] The integers and topology, in *Handbook of Set-Theoretic Topology*, K.Kunen and J.Vaughan, eds., North-Holland, Amsterdam, pp. 111–167.

DRANISHNIKOV, A.N.

[1986] Absolute F-valued retracts and spaces of functions in the topology of pointwise convergence, *Siberian Math. J.* **27**, 366–376.

ENGELKING, R.
[1989] *General Topology*, Heldermann Verlag, Berlin.
[1995] *Theory of Dimension Finite and Infinite*, Helderman Verlag, Lemgo.

FREMLIN, D.H.
[1994] Sequential convergence in $C_p(X)$, *Comment. Math. Univ. Carolin.* **35**, 371–382.

GERLITS, J. and ZS. NAGY
[1982] Some properties of $C(X)$, I, *Topology Appl.* **14**, 151–161.

GÓRAK, R.
[200?] Spaces u-equivalent to the n-cube, preprint.

GUL'KO, S.P.
[1988] The space $C_p(X)$ for countable infinite compact X is uniformly homeomorphic to c_0, *Bull. Acad. Pol. Sci.* **36**, 391–396.
[1990] Spaces of continuous functions on ordinals and ultrafilters, *Mat. Zametki* **47**, 26–34.
[1992] On uniform homeomorphisms of spaces of continuous functions, *Proc. Steklov Inst. Math.* **3**, 87–93.

GUL'KO, S.P. and T.E. KHMYLEVA
[1986] Compactness is not preserved by the relation of t-eqivalence, *Math. Notes* **39**, 484–488.

GUL'KO, S.P. and G.A. SOKOLOV
[1998] P-points in N^* and the spaces of continuous functions, *Topology Appl.* **85**, 137–142.

KALENDA, O.
[2000] A characterization of Valdivia compact spaces, *Collectanea Math.* **51**, 59–81.

KAWAMURA, K. and K. MORISHITA
[1996] Linear topological classification of certain function spaces on manifolds and CW-complexes, *Topology Appl.* **85**, 137–142.

KECHRIS, A.S.
[1995] *Classical Descriptive Set Theory*, Springer-Verlag, New York.

KOČINAC, LJ. and M. SCHEEPERS
[200?] Combinatorics of open covers (VII): Groupability, preprint.

KOYAMA, A. and T. OKADA
[1987] On compacta which are l-equivalent to I^n, *Tsukuba J. Math.* **11**, 147–156.

KURATOWSKI, K.
[1966] *Topology* I, Academic Press and PWN, New York and London.
[1968] *Topology* II, Academic Press and PWN, New York and London.

LEIDERMAN, A.G., M. LEVIN and V.G. PESTOV
[1997] On linear continuous open surjections of the spaces $C_p(X)$, *Topology Appl.* **81**, 269–279.

LEIDERMAN, A.G., S.A. MORRIS and V.G. PESTOV
[1997] The free abelian topological group and the free locally convex space on the unit interval, *Jour. London Math. Soc* **56**, 529–538.

LUTZER, D., J. VAN MILL and R. POL
[1985] Descriptive complexity of function spaces, *Trans. Amer. Math. Soc.* **291**, 121–128.

MARCISZEWSKI, W.

[1988] A function space $C(K)$ not weakly homeomorphic to $C(K) \times C(K)$, *Studia Math.* **88**, 129–137.

[1993] On analytic and coanalytic function spaces $C_p(X)$, *Topology Appl.* **50**, 241–248.

[1995] A countable X having a closed subspace A with $C_p(A)$ not a factor of $C_p(X)$, *Topology Appl.* **64**, 141–147.

[1997] A function space $C_p(X)$ not linearly homeomorphic with $C_p(X) \times \mathbb{R}$, *Fund. Math.* **153**, 125–140.

[1998a] P-filters and hereditary Baire function spaces, *Topology Appl.* **89**, 241–247.

[1998b] Some recent results on function spaces $C_p(X)$, in *Recent Progress in Function Spaces* (Quaderni di Matematica, vol. 3), pp. 221–239.

[200?a] On properties of metrizable spaces X preserved by t-equivalence, *Mathematika*, to appear.

[200?b] A function space $C_p(X)$ without a condensation onto a σ-compact space, *Proc. Amer. Math. Soc.*, to appear.

MARCISZEWSKI, W. and J. VAN MILL

[1998] An example of t_p^*-equivalent spaces which are not t_p-equivalent, *Topology Appl.* **85**, 281–285.

MARCISZEWSKI, W. and J. PELANT

[1997] Absolute Borel sets and function spaces, *Trans. Amer. Math. Soc.* **349**, 3585–3596.

MICHALEWSKI, H.

[1998] Game-theoretic approach to the hereditary Baire property of $C_p(N_F)$, *Bull. Acad. Pol. Sci.* **46**, 135–140.

[200?] Condensations of projective sets onto compacta, *Proc. Amer. Math. Soc.*, to appear.

VAN MILL, J.

[1984] An introduction to $\beta\omega$, in *Handbook of Set-Theoretic Topology*, K.Kunen and J.Vaughan, eds., North-Holland, Amsterdam, pp. 503–567.

[1987] Topological equivalence of certain function spaces, *Compositio Math.* **63**, 159–188.

[1999] $C_p(X)$ is not $G_{\delta\sigma}$: a simple proof, *Bull. Acad. Pol. Sci.* **47**, 319–323.

[2001] *The Infinite-Dimensional Topology of Function Spaces*, North-Holland Mathematical Library 64, North-Holland, Amsterdam.

VAN MILL, J., J. PELANT and R. POL

[200?] Note on function spaces with the topology of pointwise convergence, *Archiv der Math.*, to appear.

VAN MILL, J. and R. POL

[1993] A countable space with a closed subspace without measurable extender, *Bull. Acad. Pol. Sci.* **41**, 279–283.

MORISHITA, K.

[1999] On spaces that are l-equivalent to a disk, *Topology Appl.* **99**, 111–116.

OKUNEV, O.G.

[1989] Weak topology of a dual space and a t-equivalence relation, *Math. Notes* **46**, 534–538.

[1990] A method for constructing examples of M-equivalent spaces, *Topology Appl.* **36**, 157–171, Erratum: *Topology Appl.* **49**, (1993), 191–192.

[1997] Homeomorphisms of function spaces and hereditary cardinal invariants, *Topology Appl.* **80**, 177–188.

[2002] Tightness of compact spaces is preserved by the t-equivalence relation, *Comment. Math. Univ. Carolin.* **43**, 335–342.

PAVLOVSKIĬ, D.S.

[1980] On spaces of continuous functions, *Dokl. Akad. Nauk SSSR* **253**, 38–41.

PESTOV, V.G.

[1982] The coincidence of the dimension dim of l-equivalent topological spaces, *Soviet Math. Dokl.* **26**, 380–383.

POL, R.

[1995] On metrizable E with $C_p(E) \not\cong C_p(E) \times C_p(E)$, *Mathematika* **42**, 49–55.

PYTKEEV, E.G.

[1976] Upper bounds of topologies, *Math. Notes* **20**, 831–837.

[1985] The Baire property of spaces of continuous functions (Russian), *Matem. Zametki* **38**, 726–740.

SAKAI, M.

[1988] Property C'' and function spaces, *Proc. Amer. Math. Soc.* **104**, 917–919.

[2000] Variations on tightness in function spaces, *Topology Appl.* **101**, 273–280.

SCHEEPERS, M.

[1997] A sequential property of $C_p(X)$ and a covering property of Hurewicz, *Proc. Amer. Math. Soc.* **125**, 2789–2795.

SEMADENI, Z.

[1971] *Banach Spaces of Continuous Functions*, PWN, Warsaw.

TKACHUK, V.V.

[1985] Characterization of Baire property in $C_p(X)$ by the properties of a space X (Russian), in *The Mappings and the Extensions of Topological Spaces*, Ustinov, pp. 21–27.

[1997] Some non-multiplicative properties are l-invariant, *Comment. Math. Univ. Carolin.* **38**, 169–175.

USPENSKIĬ, V.V.

[1982] A characterization of compactness in terms of uniform structure in a function space, *Uspekhi Matem. Nauk* **37**, 183–184.

VALOV, V.M.

[1991] Linear topological classification of certain function spaces, *Trans. Amer. Math. Soc.* **327**, 583–600.

[1997] Function spaces, *Topology Appl.* **81**, 1–22.

VALOV, V.M. and D. VUMA

[2000] Lindelöf degree and function spaces, in *Papers in honour of Bernhard Banaschewski*, (Cape Town, 1996), Kluwer Acad. Publ., Dordrecht, pp. 475–483.

VELICHKO, N.V.

[1998] The Lindelöf property is l-invariant, *Topology Appl.* **89**, 277–283.

CHAPTER 14

Topology and Domain Theory

Keye Martin
Oxford University Computing Laboratory, Oxford, UK
E-mail: Keye.Martin@comlab.ox.ac.uk

M. W. Mislove
Department of Mathematics, Tulane University, New Orleans, LA 70118, U.S.A.
E-mail: mislove@tulane.edu

G. M. Reed
Oxford University Computing Laboratory, Oxford, UK
E-mail: Mike.Reed@comlab.ox.ac.uk

Contents

RECENT PROGRESS IN GENERAL TOPOLOGY II
Edited by Miroslav Hušek and Jan van Mill

1. Introduction

Domain theory traces its history back to the need to define mathematical models of programming languages. The impetus was the introduction of a variety of high-level programming languages and the increasing complexity of their design and use in the 1960's. This led to an acknowledged need for models for programming languages that would support precise reasoning about program behavior. Such models were required both to give an unambiguous definition of a given programming language (i.e., one not dependent on a given compiler), and to allow the verification of assertions about programs written in that language (e.g., that two programs with different syntax have the same effect). The field of *Denotational Semantics* was introduced by Christopher Strachey at Oxford University in the mid-sixties to meet this need. STRACHEY [1973, 1974] and others were able to provide denotations for language constructs using higher order functions in some mathematical universe.

The techniques developed in denotational semantics were successful for procedural languages, functional languages, and later parallel languages. The initial problem was the lack of a theory for producing mathematical models that met all the requirements:

(a) Modelling recursion required functions to have fixed points.

(b) Modelling functional languages required a cartesian closed category so that the set of functions between objects was itself an object of the category.

(c) Modelling more complicated languages required solutions to recursive definitions of the universes themselves (e.g., $U \simeq (U \to U) + (U \times U) + B$).

In 1969, DANA SCOTT [1970] discovered an elegant theory that could provide a rigorous mathematical foundation for denotational semantics. This theory, called *Domain Theory*, has evolved to become not only an important tool for applications in computer science, but also an exciting field of ongoing research in pure mathematics. The essential ingredient was the precise definition of "universe" or "domain." This definition was given in terms of partial orders satisfying certain completeness conditions. Domains carry several intrinsic topologies: the most fundamental is the *Scott topology* which is crucial to the theory. The others – the Lawson topology and the μ-topology – also play important roles in the theory and in the applications of domain theory to computer science and to other areas. As but one example of the pervasiveness of topology in the theory, the original domain Scott devised to model the untyped lambda calculus (essentially, an abstract functional programming language without assignment) used an inverse limit construction based on the Scott topology. A reference on general relations between topology and domain theory is MISLOVE [1998].

Several works already have documented the evolution of domain theory as it is applied to programming language semantics – cf., e.g., AMADIO and CURIEN [1998] and GRIFFOR, *et al.* [1994]; ABRAMSKY [1994] and GIERZ, *et al* [1980] provide a more theoretical introduction to the area. Our goal in this article is to give an update to these presentations by outlining the major results that have occurred in domain theory in the 1990s; our presentation is admittedly somewhat biased toward the research interests of the authors.

At the beginning of the 1990s, the role of domain theory as a tool for modelling imperative and functional programming languages was reasonably well understood. In some areas, such as concurrency, it's application was still evolving, partly because concurrency itself was still maturing. During the 1990s, concurrency, including real-time concurrency, was successfully modelled. But perhaps the major advances in domain theory during that period really were in the growing applications of domain theory to other areas, both of computing and of mathematics, and in the change of focus these caused in the thrust of research in the area.

The 1990s began with the Abramsky's seminal work ABRAMSKY [1991a] intimately relating domain theory and logic as tools for modelling programming languages. His work on "domain theory in logical form" essentially provides a clear summary of the role of domain theory in semantics, as well as relating it to the obvious alternative method for reasoning about programs – logic. One could say that Abramsky's work showed that domain theory was one part of a three-part basis for reasoning about imperative and functional programming languages:

- *universal algebra* provides the abstract basis for presenting the syntax of such programming languages,
- *domain theory* provides the mathematical models for these languages, and
- *logic* provides the means for reasoning about these languages in terms of their models.

The link between domains and logic is via Stone Duality : one associates to a domain D the logic whose Lindenbaum algebra is the distributive lattice generated by the compact elements of the domain. Conversely, to an appropriate logic, one associates the domain whose compact elements are the sup-prime elements of the Lindenbaum algebra of the logic. This is essentially topological in nature: the compact elements of a domain D generate basic Scott open subsets of D which, on the logical side, are interpreted as observable properties that are expressible in the logic of the domain. Dually, each observable property expressible in the logic corresponds to a Scott open subset of the domain, and hence is generated by compact elements, since these form a basis for the domain.

Furthermore, this association only works for domains freely generated by the basic constructs of domain theory – lift, coalesced sum, product, function space, strict function space and the three standard power domains, so the language under study has to have a domain-theoretic model that uses only these constructs. Under this correspondence, each of these basic domain constructors corresponds to a specific logical combinator – for example, the function space corresponds to lattice maps between the associated Lindenbaum algebras, and the power domains correspond to modal operators.

What's "buried under the hood" here is the fact that domain theory brings with it a wide range of supporting category theory that is needed to reason about the higher order constructs that pervade modern programming languages.

Shortly after Abramsky's work appeared, new applications of domain theory were devised by Abbas Edalat. Edalat realized that domain theory provided tools for modelling a wide variety of applications, from fractals to neural networks to computational geometry. In a series of papers EDALAT [1995b, 1995a, 1997], Edalat showed how the constructs of domain theory could be used to provide models for these varying phenomena, often providing clearer and more succinct avenues to obtaining results than the original approaches. In the case of real analysis, domain theory also could be used to extend known results (cf. EDALAT [1995a]). Edalat's research program gave rise to one of the problems that grew to play a central role in the theory over the decade of the 1990s – modelling topological spaces within domains. We devote considerable time to a discussion of this problem in Section 3.

In addition to Edalat's work, there were other efforts to extend domain theory in other directions in the 1990s. Two of these efforts were presented in MISLOVE [1991] and MISLOVE, ROSCOE and SCHNEIDER [1995], where domain theory was extended to include more general structures. These extensions were driven by the need to provide models that didn't satisfy the usual completeness axioms of domains. It turned out that a satisfactory theory could be devised in both cases, results we outline in Subsection 2.6 We also discuss developments in the use of metric spaces, as opposed to partial orders, in providing models for denotational semantics.

One of the fundamental roles that domain theory plays is to provide a setting in which there is a wealth of functions having fixed points. Classically, every Scott continuous selfmap of a domain has a least fixed point. But this plethora of mappings with fixed points is not broad enough to capture all areas of application. For example, it is easy to encounter quite natural domain-theoretic models of algorithms which are not monotone, let alone Scott continuous. So, something more is needed to model these constructs. The work on measurement MARTIN [2000a] presents an approach that overcomes the shortcomings of traditional domain theory. Remarkably, this approach also provides a basis for reasoning about the computational complexity of algorithms in a domain-theoretic setting. This theory also gives rise to the third topology on domains listed above – the μ-topology. We touch on certain results this approach has spawned in Section 4.

AN HISTORICAL NOTE: On first thought, it might appear that there would be little interaction between the pure, continuous world of topology and the applied, discrete world of computer science. However, as the comments above make clear, this is certainly not the case. In fact, topology was there at the beginning. Perhaps the two names most often associated with the foundation of computer science are Turing and von Neumann. Both did early work in topology. Indeed it was a lecture by von Neumann in the topology seminar of Newman at Cambridge in 1935 that inspired Turing to work on Hilbert's problem about the decidability of arithmetic. This work led to his invention of the *Turing Machine*. Furthermore, the first actual computers were produced by teams including Turing and Newman at Manchester and von Neumann in Philadelphia. Of course, Turing and von Neumann were extremely clever people who worked in many areas, and WWII pushed their efforts into computer science. One might well ask if the connection with topology is only accidental. However, it is clear that any mathematical foundation of stored-program computation had to be based on the manipulation of functions, the finite approximation of infinite objects, and a notion of convergence. It is thus not surprising that topology and topologists have played a role in developing this foundation from the outset.

2. Domain theory

1. Continuous posets

A *poset* is a partially ordered set – i.e., a set together with a reflexive, antisymmetric and transitive relation.

2.1. DEFINITION. Let $(P, \sqsubseteq)$ be a partially ordered set. A nonempty subset $S \subseteq P$ is *directed* if $(\forall x, y \in S)(\exists z \in S)\, x, y \sqsubseteq z$. The *supremum* of a subset $S \subseteq P$ is the least of all its upper bounds provided it exists. This is written $\bigsqcup S$.

2.2. DEFINITION. For a subset X of a poset P, set

$$\uparrow\! X := \{y \in P : (\exists x \in X)\, x \sqsubseteq y\} \;\;\&\;\; \downarrow\! X := \{y \in P : (\exists x \in X)\, y \sqsubseteq x\}.$$

We write $\uparrow\! x = \uparrow\!\{x\}$ and $\downarrow\! x = \downarrow\!\{x\}$ for elements $x \in X$.

2.3. DEFINITION. For elements x, y of a poset, write $x \ll y$ iff for all directed sets S with a supremum,

$$y \sqsubseteq \bigsqcup S \Rightarrow (\exists s \in S)\, x \sqsubseteq s.$$

We set $\twoheaddownarrow x = \{a \in D : a \ll x\}$ and $\twoheaduparrow x = \{a \in D : x \ll a\}$.

For the symbol "$\ll$," read "approximates."

2.4. DEFINITION. A *basis* for a poset P is a subset B such that $B \cap \twoheaddownarrow x$ contains a directed set with supremum x for all $x \in P$. A poset is *continuous* if it has a basis.

2.5. EXAMPLE (EDALAT and HECKMANN [1998]). For a metric space (X, d),

$$\mathbf{B}X = X \times [0, \infty)$$

ordered by

$$(x, r) \sqsubseteq (y, s) \Leftrightarrow d(x, y) \leq r - s.$$

is a continuous poset with

$$(x, r) \ll (y, s) \Leftrightarrow d(x, y) < r - s.$$

If A is a dense subset of X, then $A \times \mathbb{Q}$ is basis for the poset $\mathbf{B}X$.

2. *Domains*

Domains are posets that carry intrinsic notions of approximation and completeness.

2.6. DEFINITION. A *dcpo* is a poset in which every directed subset has a supremum. A *continuous dcpo* is a continuous poset which is also a dcpo. A *domain* is a continuous dcpo.

2.7. EXAMPLE (GIERZ, *et al* [1981]). Let X be a locally compact Hausdorff space. Its *upper space*

$$\mathbf{U}X = \{\emptyset \neq K \subseteq X : K \text{ is compact}\}$$

ordered under reverse inclusion

$$A \sqsubseteq B \Leftrightarrow B \subseteq A$$

is a continuous dcpo:

- For directed $S \subseteq \mathbf{U}X$, $\bigsqcup S = \bigcap S$, and
- $A \ll B \Leftrightarrow B \subseteq \mathrm{int}(A)$.

2.8. EXAMPLE. For a metric space (X, d),

$$\mathbf{B}X \text{ is a domain} \Leftrightarrow d \text{ is complete.}$$

If (x_n, r_n) is an increasing sequence, then

$$\bigsqcup_{n \geq 1} (x_n, r_n) = (\lim x_n, \lim r_n).$$

Every directed subset S of $\mathbf{B}X$ contains an increasing sequence with the same set of upper bounds as S.

3. *Countably based domains*

2.9. DEFINITION. A poset is *ω-continuous* if it has a countable basis.

The next example is due to SCOTT [1970].

2.10. EXAMPLE. The collection of compact intervals of the real line

$$\mathbf{I}\mathbb{R} = \{[a, b] : a, b \in \mathbb{R} \ \& \ a \leq b\}$$

ordered under reverse inclusion

$$[a, b] \sqsubseteq [c, d] \Leftrightarrow [c, d] \subseteq [a, b]$$

is an ω-continuous dcpo:

- For directed $S \subseteq \mathbf{I}\mathbb{R}$, $\bigsqcup S = \bigcap S$,
- $I \ll J \Leftrightarrow J \subseteq \mathrm{int}(I)$, and
- $\{[p, q] : p, q \in \mathbb{Q} \ \& \ p \leq q\}$ is a countable basis for $\mathbf{I}\mathbb{R}$.

The domain $\mathbf{I}\mathbb{R}$ is called the *interval domain.*

2.11. EXAMPLE. $\mathbf{U}X$ is ω-continuous iff X is second countable.

2.12. EXAMPLE. $\mathbf{B}X$ is ω-continuous iff X is separable.

4. Algebraic domains

2.13. DEFINITION. An element x of a poset P is *compact* if $x \ll x$. The set of compact elements in P is written $K(P)$. A poset is *algebraic* if its compact elements form a basis; it is *ω-algebraic* if it has a countable basis of compact elements.

The next domain is of central importance in recursion theory (cf, ODIFREDDI [1989]).

2.14. EXAMPLE. The set of partial mappings on the naturals

$$[\mathbb{N} \rightharpoonup \mathbb{N}] = \{ f \mid f : \mathbb{N} \rightharpoonup \mathbb{N} \text{ is a partial map}\}$$

ordered by extension

$$f \sqsubseteq g \Leftrightarrow \mathrm{dom}(f) \subseteq \mathrm{dom}(g) \ \& \ f = g \text{ on } \mathrm{dom}(f)$$

is an ω-algebraic dcpo:

- For directed set $S \subseteq [\mathbb{N} \rightharpoonup \mathbb{N}]$, $\bigsqcup S = \bigcup S$,
- $f \ll g \Leftrightarrow f \sqsubseteq g \ \& \ \mathrm{dom}(f)$ is finite, and
- $K([\mathbb{N} \rightharpoonup \mathbb{N}]) = \{f \in [\mathbb{N} \rightharpoonup \mathbb{N}] : \mathrm{dom}(f) \text{ finite}\}$ is a countable basis for $[\mathbb{N} \rightharpoonup \mathbb{N}]$.

5. Scott domains

Historically speaking, the definition we adopt of Scott domain is somewhat nonstandard.

2.15. DEFINITION. A *Scott domain* is a domain with a least element $\bot$ in which any two elements bounded from above have a supremum.

2.16. EXAMPLE. The collection of functions

$$\Sigma^\infty = \{ s \mid s : \{1, \ldots, n\} \to \{0,1\}, 0 \le n \le \infty \}$$

ordered by extension

$$s \sqsubseteq t \Leftrightarrow |s| \le |t| \ \& \ (\forall\, 1 \le i \le |s|)\ s(i) = t(i),$$

where $|s|$ is written for the cardinality of $\mathrm{dom}(s)$, is an ω-algebraic Scott domain:

- For directed $S \subseteq \Sigma^\infty$, $\bigsqcup S = \bigcup S$,
- $s \ll t \Leftrightarrow s \sqsubseteq t \ \& \ |s| < \infty$,
- $\{s \in \Sigma^\infty : |s| < \infty\}$ is a countable basis for Σ^∞,
- The least element $\bot$ is the unique s with $|s| = 0$, and
- For any two elements x, y bounded from above, we have $x \sqsubseteq y$ or $y \sqsubseteq x$.

2.17. EXAMPLE. $[\mathbb{N} \rightharpoonup \mathbb{N}]$ is an ω-algebraic Scott domain.

6. The need for more general domains

We have listed a number of the most common classical domains, but there are others as well. For example, the SFP-objects, which are simultaneously limits and colimits of a countable sequence of finite posets, their generalization to bifinite domains in the algebraic case, and the retracts of these objects in the continuous case, as well as the so-called FS-domains also play major roles, because they include the maximal cartesian closed categories of domains (cf. ABRAMSKY and JUNG [1994] and JUNG [1989]). We close this section with a brief indication of why more general structures than domains also play a role in the theory and its applications. A domain is directed complete: every directed subset has a least upper bound. But some applications require this condition to be relaxed – either in the direction of continuous objects which are not directed complete, or in the direction of non-continuous objects which are only *locally bounded*. Here is a brief summary of how these objects arose.

ALGEBRAIC AND CONTINUOUS POSETS. Algebraic posets satisfy the property that each element is the directed supremum of compact elements, but they are not assumed to be directed complete. These objects were introduced in MISLOVE [1991], where they were used to give an alternate presentation of the failures model for the process algebra CSP of BROOKES, HOARE and ROSCOE [1984]. In the *failures model* for CSP, each process P has its meaning given as a set F_P of *failures* – pairs (s, X) consisting of a finite string $s \in \Sigma^*$ over the alphabet Σ of actions representing a possible trace of P, and a refusal set $X \subseteq \Sigma$ of actions that P might refuse after having executed the string s. There are several healthiness conditions that are imposed on process meanings, all aimed at presenting a compositional model for the process algebra CSP. The conditions which are assumed for a process meaning F_P are the following:

(a) $F_P \neq \emptyset$.

(b) $(s, X) \in F_P$ and $t \leq s$ imply $(t, \emptyset) \in F_P$.

(c) $(s, X) \in F_P$ and $Y \subseteq X$ imply $(s, Y) \in F_P$.

(d) $(s, X) \in F_P$ for all $X \subsetneq Y$ imply $(s, Y) \in F_P$.

(e) $(s, X) \in F_P$ and $(s\hat{\ }a, \emptyset) \notin F_P$ imply $(s, X \cup \{a\}) \in F_P$.

This family of sets is ordered by reverse inclusion – this is the so-called *order of nondeterminism*, since a process P is below a process Q iff $F_P \supseteq F_Q$. So, the higher a process is in this order, the more deterministic it is, since it has fewer failures.

2.18. THEOREM (MISLOVE [1991]). *Let* $S = (\Sigma^* \times \mathcal{P}(\Sigma), \sqsubseteq)$, *where the order* $\sqsubseteq$ *is given by*

$$(s, X) \sqsubseteq (t, Y) \quad \Leftrightarrow \quad (s < t \wedge X = \emptyset) \vee (s = t \wedge X \subseteq Y).$$

Then $X \in \Gamma(S) = \{X \subseteq S \mid \emptyset \neq X \text{ Scott closed}\}$ *iff* X *satisfies (1) – (4) above.*

All of the operators of CSP naturally define continuous operations on $\Gamma(S)$ with respect to usual set containment, and only the hiding operator fails to induce an operation on $\Gamma(S)$ which is continuous with respect to reverse set containment. We let FM denote the family of elements of $\Gamma(S)$ satisfying condition (5) above.

2.19. PROPOSITION (MISLOVE [1991]). *All of the standard operators of CSP define Scott-continuous operations on* $(FM, \supseteq)$.

So condition (5) – which asserts that "impossible events can be added to the refusal set" – is precisely what is needed for all the CSP operators to be continuous with respect to reverse set containment.

While the poset S above is algebraic – in fact, $K(S) = S$, the theory laid out in MISLOVE [1991] generalizes in a straightforward manner to the realm of continuous posets. This was first carried out in the thesis by HAN ZHANG [1993]. The following theorem summarizes the relevant results for this theory. Recall that a space X is sober iff the mapping $x \mapsto \overline{\{x\}}: X \to CSpec(\Gamma(X), \subseteq)$ is a homeomorphism onto the family of coprimes of $(\Gamma(X), \subseteq)$, the family of non-empty closed subsets of X. Here, the closed sets of $CSpec(\Gamma(X), \subseteq)$ are $\{A \cap CSpec(\Gamma(X)) \mid A \in \Gamma(X)\}$.

2.20. THEOREM (MISLOVE [1991], ZHANG [1993]). *The functor* $\Gamma\colon \mathsf{CPos} \to \mathsf{CD}$ *sending a continuous poset to its family of non-empty, Scott-closed subsets is left adjoint to the functor* $CSpec\colon \mathsf{CD} \to \mathsf{CPos}$ *sending a completely distributive lattice to its set of coprimes. Moreover,* $CSpec(\Gamma(P))$ *is the sobrification of the continuous poset* P *in the Scott topology, so a continuous poset is sober iff it is a dcpo.* □

Furthermore, the continuous poset P determines the way-below relation on $\Gamma(P)$, in that $\{\downarrow F \mid \emptyset \neq F \subseteq P \text{ finite}\}$ is a basis for $\Gamma(P)$.

LOCAL DCPOS. Another extension of domains was devised in the direction of non-continuous posets. The impetus again were models for CSP, but this time, the problem was how to devise models for an extension of the syntax of CSP that included infinitary operators. Here's some brief background on how this came about.

Process algebras have proved to be a fruitful way to study concurrency, and they have the added advantage that both a process and its specification can be encoded in the algebra. The brief description given above of the order of nondeterminism on the failures model for CSP serves to illustrate the point. Indeed, if P is a process and Q represents a specification that P is supposed to meet, then we can see whether P actually meets Q simply by testing whether $F_P \subseteq F_Q$: if this holds, then every behavior of P in the model is allowed by Q, which is precisely what is intended. So, the order of nondeterminism given above also serves as an order of specification, or, as it more commonly is called, it is an order of *refinement*. This is a very appealing situation – being able to write specifications in the same process algebra as the processes being specified, and then testing whether processes meet their specifications becomes a simple set containment question. In fact, this is exactly how the model-checker FDR (cf., ROSCOE [1998]) which has been developed for CSP works: one inputs a process and its specification, the tool FDR mechanically checks whether each behavior of P is allowed by Q. If not, it also gives a trace of the actions allowed by P which are not allowed by Q.

It also would be nice to extend this setting so that processes that involve unbounded choice could be included in the process algebra, and hence in its models. This takes us from the realm of finitary operators to infinitary ones. It is relatively straightforward to deal with the syntax issues so that the process algebra is a set – one puts a cardinal bound on the size of the set of processes one can combine using the infinitary operator intended to be included in the model. For example, it would be reasonable to include only countable nondeterminism, and then one could reason about specifications that would allow processes to execute an action any finite number of times, but would not allow them to execute infinitely many actions. This would amount to an assurance that the process in question would not fail to terminate.

The problems encountered in modelling infinitary operators have to do with the fixed point theory. In the setting of dcpos and Scott continuous maps, infinitary operators cause unwanted behaviors to appear in their meanings. For example, a process that could do an action a any finite number of times would, using continuity, also have the possibility of doing infinitely many a's, which is not what we want. In order to overcome this, BARRETT and ROSCOE [1990] devised a model for CSP including an unbounded choice operator by extending the model for finitary CSP to include infinite traces explicitly. A similar model was devised by Schneider to support unbounded nondeterminism in timed CSP, and both approaches were unified in MISLOVE, ROSCOE and SCHNEIDER [1995]. The problems and their common solution can be described as follows.

In the finitary setting, operators can be modelled as Scott-continuous mappings, and such operators naturally include infinite behaviors in their meanings. But, for infinitary operators one has to separate finite behaviors from infinite ones. At the same time, one can't simply resort to posets as models, since there are instances when one wants infinite behaviors to be included, and, in any case, fixed points are still needed if one is to model recursion. So, one needs a model in which some, but not all, selfmaps have fixed points.

The models that were devised in BARRETT and ROSCOE [1990] and, more generally, in MISLOVE, ROSCOE and SCHNEIDER [1995] rely on a simple idea to ensure that fixed points exist for those selfmaps that need them. First of all, recall that a *prefixed point* of a selfmap $f\colon P \to P$ is an element $x \in P$ satisfying $f(x) \le x$. A *local cpo* is a poset P in which $\downarrow x$ is a cpo for each $x \in P$. Then the result ensuring fixed points exist is called the *Dominated Convergence Theorem:*

2.21. THEOREM (MISLOVE, ROSCOE and SCHNEIDER [1995]). *Let E be a set and $\iota\colon E \to D$ be a mapping. Suppose that, for each $f\colon D \to D$ there is a mapping $\hat{f}\colon E \to E$ satisfying $\iota \circ f \le \hat{f} \circ \iota$. If $x \in E$ is a fixed point of $\hat{f}$, then $\iota(x)$ is a prefixed point for f. In particular, if D is a* local cpo, *then f has a least fixed point in $\downarrow x$.*

The gist of the modelling technique in BARRET and ROSCOE [1990] amounts to devising an extension of the failures model that supports infinite behaviors, and that satisfies the following property: each infinitary operator on the model supporting unbounded choice can be refined by a finitary operator on the failures model for CSP. Since the finitary operator leaves the failures model invariant, it has a fixed point on that model; it then follows from Theorem 2.21 that the infinitary operator has this fixed point as a prefixed point. The model supporting unbounded choice is a local cpo by construction, so the infinitary operator has a fixed point, which also can be shown to be a least fixed point.

Similar results supporting the same reasoning for models of timed CSP are presented in MISLOVE, ROSCOE and SCHNEIDER [1995]. Moreover, these results were extended in MISLOVE [1995] to a more abstract setting where it was shown that there is a cartesian closed category of local cpos and monotone mappings in which each infinitary operator from an abstract syntax process algebra has a least fixed point. This means the results are not confined to the realm of CSP and its models, but in fact can be generalized to a setting of an abstract process algebra for which one wants to add unbounded choice.

An open question here is the relation of local cpos to topology. These structures don't appear to have an obvious analogue in topological spaces – one could conjecture that there would be a category of "locally sober spaces" for which a subcategory would correspond to the local cpos that also are continuous posets. But it would be interesting to have a topological analogue of the Dominated Convergence Theorem in topological spaces.

METRIC SPACES AS MODELS FOR COMPUTATION The use of metric spaces as a mathematical foundation for denotational semantics was first introduced by Nivat in the late 1970s (e.g., NIVAT [1979], ARNOLD and NIVAT [1980]). The development of a comprehensive metric "domain" theory was accomplished by de Bakker and his colleagues and students in Amsterdam during the 1980s and 1990s (e.g., DE BAKKER and ZUCKER [1982], DE BAKKER and MEYER [1987], AMERICA and RUTTEN [1989], VAN BREGUEL [1990], and DE BAKKER and RUTTEN [1992]). Several researchers have sought to unify the metric and order approaches to domain theory (e.g., ROUNDS [1985], MATTHEWS [1986], and SMYTH [1988]).

There are several advantages in the use of metric models. For example, as mentioned in VAN BREGUEL [1990], the space of subsets in a metric space is much simpler and better understood than the powerdomain constructions in PLOTKIN [1976] and SMYTH [1978]. This space can be used to model many of the programming language constructs previously requiring powerdomains. Furthermore, the use of fixed point induction to prove that programs meet their specifications is much more general in a metric setting ROSCOE [1982].

A surprising development in the late 1980s and 1990s was the crucial need for metric space models to capture an extension of CSP to include real-time constructs. As outlined earlier in this section, untimed models of CSP have mostly been defined in terms of partial orders. However, several researchers have defined complete metric spaces over CSP and similar models of concurrency. These metrics were usually based on the number of initial steps over which a pair of processes behave indistinguishably. Recursion is defined via the Banach unique fixed point theorem for contraction mappings. Having a unique fixed point, as opposed to the least fixed point guaranteed by the Tarski fixed point theorem for continuous or monotone mappings on complete partial orders, often has advantages in developing a proof theory. However, not all desired recursions give rise to contraction mappings. Hence, the metric approach, while useful in conjunction with a partial order model, does not provide a general theory for untimed CSP. With the introduction of continuous timing constructs into CSP, it was found that the partial order approach led to immediate problems. Under the realistic assumption that only finitely many events can occur in finite time, it became clear that there could not exist a bottom element, representing the most nondeterministic process, for any appropriate partial order. Also, the requirement that increasing sequences in such orders have least upper bounds would cause additional problems.

Fortunately, the use of a complete metric based on the length of time in which a pair of processes behave indistinguishably gave rise to successful semantic models. Furthermore, under axioms to ensure the finitary assumption mentioned above, all desired mappings are contraction mappings, and we avoid the problems with the metric approach in the untimed models. The successful development of continuous timed models for CSP (which fully extend the standard untimed models) was accomplished in REED and ROSCOE [1986, 1988], REED [1990] and REED and ROSCOE [1999]. Although these models are highly more complex mathematically than the untimed models, a comprehensive theory of timed concurrency has now been built with several successful real-world applications in robotics, embedded systems, and network analysis (see REED and ROSCOE [1999]). Recently, using new metric models for untimed CSP, Ouaknine has been able to provide translation functions from the timed models into his new models which enable the model-checking of temporal behavior in CSP (OUAKNINE and REED [1999] and OUAKNINE [2001]).

3. Models of topological spaces

The order-theoretic structure of a domain allows for the derivation of several intrinsically defined topologies. The topology of interest in the study of models is the Scott topology.

3.1. DEFINITION. A subset U of a poset P is *Scott open* if

(i) U is an upper set: $x \in U \;\&\; x \sqsubseteq y \Rightarrow y \in U$, and

(ii) U is inaccessible by directed suprema: For every directed $S \subseteq P$ with a supremum,

$$\bigsqcup S \in U \Rightarrow S \cap U \neq \emptyset.$$

The collection of all Scott open sets on P is called the Scott topology.

Unless explicitly stated otherwise, all topological statements about posets are made with respect to the Scott topology.

3.2. PROPOSITION. *A function $f : P \to Q$ between posets is continuous iff*

(i) *f is monotone: $x \sqsubseteq y \Rightarrow f(x) \sqsubseteq f(y)$.*

(ii) *f preserves directed suprema: For every directed $S \subseteq P$ with a supremum, its image $f(S)$ has a supremum, and*

$$f(\bigsqcup S) = \bigsqcup f(S).$$

Now we need a way to understand the open sets themselves.

3.3. THEOREM (ZHANG [1993]). *The collection $\{\twoheaduparrow x : x \in P\}$ is a basis for the Scott topology on a continuous poset.*

In particular, this last result applies to domains, where it was known long before the result above.

3.4. EXAMPLE. A basic open set in $\mathbf{I}\mathbb{R}$ is

$$\twoheaduparrow[a, b] = \{x \in \mathbf{I}\mathbb{R} : x \subseteq (a, b)\}$$

while a basic open set in Σ^∞ is

$$\uparrow s = \{t \in \Sigma^\infty : (\exists u \in \Sigma^\infty)\, t = su\}$$

for s finite.

3.5. DEFINITION. An element x of a poset P is *maximal* if

$$(\forall y \in P)\, x \sqsubseteq y \Rightarrow x = y.$$

The set of maximal elements is $\max(P)$.

By the Hausdorff Maximality Principle, each element in a dcpo has a maximal element above it. A continuous poset, on the other hand, may have no maximal elements.

3.6. DEFINITION. A *model* of a space X is a continuous dcpo D and a homeomorphism

$$X \simeq \max(D)$$

from X onto the maximal elements of D in their relative Scott topology.

The *model problem* in domain theory calls for a characterization of those spaces which have a model.

1. Examples of models

The classical spaces of analysis all have natural models.

3.7. EXAMPLE. Models of classical spaces.

(i) $\max(\Sigma^\infty) \simeq \mathbb{C}$ (Cantor set).

(ii) $\max([\mathbb{N} \rightharpoonup \mathbb{N}]) \simeq \mathbb{R} \setminus \mathbb{Q}$ (the irrationals).

(iii) $\max(\mathbf{I}\mathbb{R}) \simeq \mathbb{R}$ (the reals).

(iv) $\max(\mathbf{U}X) \simeq X$ (X locally compact Hausdorff).

(v) $\max(\mathbf{B}X) \simeq X$ (X complete metric space).

Now for an example of a space without a model.

3.8. THEOREM (MARTIN[1999]). *Every space with a model is Baire.*

3.9. COROLLARY. *There is no model of the rationals.*

This also yields a new approach to unifying the Baire theorems of analysis.

3.10. COROLLARY. *Locally compact Hausdorff spaces are Baire.*

3.11. COROLLARY. *Complete metric spaces are Baire.*

We now turn to what are historically among the most important subclasses of domains: The compact domains and the countably based domains.

2. Compact models of spaces

3.12. DEFINITION. The *lower weak topology* on a domain P has for a basis the family $\{P \backslash \uparrow F \mid F \subseteq P \text{ finite}\}$. The *Lawson topology* is the common refinement of the Scott topology and the lower weak topology. It has $\{U \backslash \uparrow F \mid F \subseteq P\ finite\}$ as a basis, where U ranges over the Scott open sets.

3.13. DEFINITION. A domain is *compact* if it is Scott compact and the intersection of any two Scott compact upper sets is Scott compact.

The word compact is used because the condition above is equivalent to compactness in the Lawson topology (which also is the patch of the Scott), cf., ABRAMSKY and JUNG [1989].

3.14. THEOREM (MARTIN [1998]). *A space has a compact model iff it has a model that is a Scott domain.*

This equivalence remains valid for algebraic domains, as well as for countably based domains. Thus, to study compact models, we need only consider Scott domains.

3.15. THEOREM (KAMIMURA and TANG [1984], FLAGG and KOPPERMAN[1997]). *A space X is Polish and zero-dimensional iff there is an ω-algebraic Scott domain D with $X \simeq \max(D)$.*

3.16. THEOREM (LAWSON [1997], CIESIELSKI,FLAGG and KOPPERMAN [1999]). *A space X is Polish iff there is an ω-continuous Scott domain D with $X \simeq \max(D)$.*

The results above are indicative of a more general theme:

3.17. THEOREM (FLAGG and KOPPERMAN [1997], MARTIN [2000c]). *A metric space is completely metrizable and zero-dimensional iff there is an algebraic Scott domain D with $X \simeq \max(D)$.*

3.18. THEOREM (MARTIN [2000c]). *If the space of maximal elements in a Scott domain is developable, then it is Čech-complete.*

3.19. COROLLARY. *Any metric space with a model by a Scott domain is completely metrizable.*

Does every complete metric space have a model by a Scott domain? The answer must be yes, but there is currently no known proof. More generally, it is conjectured in MARTIN [2000c] that a developable space has a model by a Scott domain iff it is Čech-complete. Each Čech-complete developable space (1) has a model by a continuous dcpo, and (2) has a dense metrizable subset which has a model by a Scott domain; cf., MARTIN and REED [200?].

3. Countably based models of spaces

Scott domains D have an important property: For all x, the set $\uparrow x \cap \max(D)$ is a closed subset of the space $\max(D)$. Formally, this is expressed by saying that *the relative Scott and Lawson topologies on* $\max(D)$ *agree.*

3.20. PROPOSITION. *Let D be a compact domain. Then the relative Scott and Lawson topologies on* $\max(D)$ *agree.*

3.21. EXAMPLE. $\mathbf{B}X$ is a domain satisfying Lawson's condition that is *usually* not a Scott domain (even if we add a bottom element). Consider for instance $X = \mathbb{R}^2$.

3.22. THEOREM (LAWSON [1997]). *A space X is Polish iff there is an ω-continuous dcpo D such that*

(i) $X \simeq \max(D)$, *and*

(ii) *The relative Scott and Lawson topologies on* $\max(D)$ *agree.*

Thus far, all results on countably based models have either implicitly or explicitly involved assumptions that allow one to make use of classical completeness arguments from topology: In all cases, it happens that D with its Lawson topology λ_D is Čech-complete and that $\max(D)$ in its Scott topology is actually a G_δ subset of (D, λ_D). Thus, these results are more about the Lawson topology than they are about the Scott topology.

But in studying models, we seek an understanding of the true expressivity of the space of maximal elements in a domain. What makes this difficult is that D in its Scott topology is only a T_0-space in general, and it seems safe enough to say that topology is not replete with notions of completeness developed for spaces with such little separation.

Let T_3 denote the class of ω-continuous dcpo's D in which $\max(D)$ is regular. Then all of the domains studied by Lawson belong to T_3. But every Polish space which is not zero-dimensional gives rise to a natural model that violates Lawson's condition. Here is an example.

3.23. EXAMPLE (MARTIN [200?b]). Order the intervals

$$\mathbf{I}\mathbb{R}_{\frac{1}{2}} = \{[a,b] : a, b \in \mathbb{R} \;\&\; a \le b\}$$

by

$$[a,b] \sqsubseteq [c,d] \iff [c,d] \subseteq (a,b) \;\&\; \mu[c,d] \le \mu[a,b]/2$$

or $[a,b] = [c,d]$, where $\mu([a,b]) = |b-a|$. Then $\mathbf{I}\mathbb{R}_{\frac{1}{2}}$ is an ω-algebraic model of the real line with the property that every element is either compact or maximal – just like Σ^∞.

Why is this natural? It is natural in the same way that the upper space $\mathbf{U}X$ of a locally compact Hausdorff space X is natural – each is constructed using the idea that a space has a basis of closed neighborhoods in which certain filtered intersections are nonempty. In the case of a complete metric space, we have to know that the sets in such an intersection have diameters tending to zero, which is what the order on $\mathbf{I}\mathbb{R}_{\frac{1}{2}}$ ensures.

3.24. THEOREM (MARTIN [200?d]). *The space of maximal elements in an ω-continuous dcpo is regular iff Polish.*

This is a theorem about the Scott topology (the Baire Theorem 3.8 is another). Its proof is an exciting application of an idea due to CHOQUET [1969] – who had the vision to consider a notion of a completeness for spaces with no separation. These spaces are called *strong Choquet* by Kechris in classical descriptive set theory KECHRIS [1994], TELGÁRSKY [1987] identifies them by saying that player II has a winning (not necessarily stationary) strategy in the Choquet game, while the first author refers to them as *Choquet complete* in MARTIN [200?d].

Very recently, this result was extended to models of metric spaces in general MARTIN [200?e].

4. Measurement

The measurement formalism MARTIN [2000a] is a theory about uncertainty and its applications. The idea is that a domain D is a collection of informative objects, and that a measurement $\mu : D \to [0, \infty)^*$ assigns to each $x \in D$ its corresponding amount of information μx. This simple idea can be used to establish fixed point theorems not previously available (MARTIN [2000b, 2001c]), to measure the rate at which processes manipulate information MARTIN [200?c], and various other things in and out of computation (MARTIN [2001a, 2001b]). Surprisingly, it can also be used to define the next natural class in the hierarchy for countably based models.

Let $[0, \infty)^*$ denote the set of nonnegative reals ordered upside down.

4.1. DEFINITION. A *measurement* on a domain D is a Scott continuous mapping $\mu : D \to [0, \infty)^*$ such that for all $x \in D$ with $\mu x = 0$ and any sequence (x_n) with $x_n \ll x$,

$$\lim_{n \to \infty} \mu x_n = 0 \Rightarrow \bigsqcup_{n \geq 1} x_n = x,$$

and (x_n) is directed. The elements with measure zero

$$\ker \mu = \{x \in D : \mu x = 0\}$$

comprise the *kernel* of μ.

Intuition: μx is a measure of the *uncertainty* in x.

4.2. LEMMA (MARTIN [2000b]). *If μ is a measurement on D, then* $\ker \mu \subseteq \max(D)$.

In many important cases, but not all, we have the equality $\ker \mu = \max(D)$. Nevertheless, in the technical sense, studying the kernel of a measurement on a domain may be regarded an instance of the model problem.

4.3. THEOREM (MARTIN [200?b]). *If a nonempty space is the kernel of a measurement on a domain, then it has a model.*

4.4. EXAMPLE. Canonical measurements.

(i) $(\mathbf{I}\mathbb{R}, \mu)$, the interval domain with the length measurement $\mu[a, b] = b - a$.

(ii) $([\mathbb{N} \rightharpoonup \mathbb{N}], \mu)$, the partial functions on the naturals with

$$\mu f = |\mathrm{dom}(f)|$$

where $|\cdot| : \mathcal{P}\omega \to [0, \infty)^*$ is the measurement on the algebraic lattice $\mathcal{P}\omega$ given by

$$|x| = 1 - \sum_{n \in x} \frac{1}{2^{n+1}}.$$

(iii) $(\Sigma^\infty, 1/2^{|\cdot|})$, the Cantor set model where $|\cdot| : \Sigma^\infty \to [0, \infty]$ is the length of a string.

(iv) $(\mathbf{U}X, \mathrm{diam})$, the upper space of a locally compact metric space (X, d) with

$$\mathrm{diam}\, K = \sup\{d(x, y) : x, y \in K\}.$$

(v) $(\mathbf{B}X, \pi)$, the formal ball model of a complete metric space (X, d) with

$$\pi(x, r) = r.$$

In each case, we have $\ker \mu = \max(D)$.

Let T_μ be the class of all ω-continuous dcpo's D with a measurement μ such that $\ker \mu = \max(D)$.

4.5. THEOREM (MARTIN [200?d]). *Let D be an ω-continuous dcpo. If $\max(D)$ is regular, then there is a measurement $\mu : D \to [0, \infty)^*$ with $\ker \mu = \max(D)$.*

Thus, all countably based models of metric spaces belong to T_μ, i.e., $T_3 \subseteq T_\mu$.

4.6. DEFINITION. Let D be a continuous dcpo with a measurement μ. A monotone map $f : D \to D$ is a *contraction* if there is a constant $0 \le c < 1$ with

$$\mu f(x) \le c \cdot \mu x$$

for all $x \in D$.

In the following result, we assume that any two points $x, y \in \ker \mu$ are bounded below by some $z \sqsubseteq x, y$.

4.7. THEOREM (MARTIN [2000b]). *If $f : D \to D$ is a contraction on (D, μ) and there is a point $x \in D$ with $x \sqsubseteq f(x)$, then*

$$x^\star = \bigsqcup_{n \ge 0} f^n(x) \in \ker \mu$$

is the unique fixed point of f on D. Further, $x^\star$ is an attractor in two different senses:

(i) *For all $x \in \ker \mu$, $f^n(x) \to x^\star$ in the Scott topology on $\ker \mu$, and*

(ii) *For all $x \sqsubseteq x^\star$, $\bigsqcup_{n \ge 0} f^n(x) = x^\star$, and this supremum is a limit in the Scott topology on D.*

When a domain has a least element, the last result is easier to state.

4.8. COROLLARY. *Let D be a domain with least element $\bot$ and measurement μ. If a map $f : D \to D$ is a contraction, then*

$$x^\star = \bigsqcup_{n \geq 0} f^n(\bot) \in \ker \mu$$

is the unique fixed point of f on D. In addition, the other conclusions of Theorem 4.7 hold as well.

All domains considered in this paper thus far have the property that for all $x, y \in D$ there is $z \in D$ with $z \sqsubseteq x, y$. So one can freely apply Theorem 4.7 in each of these cases.

4.9. EXAMPLE. Let $f : X \to X$ be a contraction on a complete metric space X with Lipschitz constant $c < 1$. The mapping $f : X \to X$ extends to a monotone map on the formal ball model $\bar{f} : \mathbf{B}X \to \mathbf{B}X$ given by

$$\bar{f}(x, r) = (fx, c \cdot r),$$

which satisfies

$$\pi \bar{f}(x, r) = c \cdot \pi(x, r),$$

where $\pi : \mathbf{B}X \to [0, \infty)^*$ is the standard measurement on $\mathbf{B}X$, $\pi(x, r) = r$. Now choose r so that $(x, r) \sqsubseteq \bar{f}(x, r)$. By Theorem 4.7, $\bar{f}$ has a unique attractor which implies that f does also because $X \simeq \ker \pi$.

We can also use the upper space $(\mathbf{U}X, \text{diam})$ to prove the Banach contraction theorem for compact metric spaces by applying the technique of the last example. Here is an example from computation.

4.10. EXAMPLE. Consider a functional like

$$\phi : [\mathbb{N} \rightharpoonup \mathbb{N}] \to [\mathbb{N} \rightharpoonup \mathbb{N}]$$

$$\phi(f)(k) = \begin{cases} 1 & \text{if } k = 0, \\ kf(k-1) & \text{if } k \geq 1 \;\&\; k - 1 \in \text{dom}(f), \end{cases}$$

which is easily seen to be monotone. Applying $\mu : [\mathbb{N} \rightharpoonup \mathbb{N}] \to [0, \infty)^*$, we compute

$$\begin{aligned}
\mu\phi(f) &= |\text{dom}(\phi(f))| \\
&= 1 - \sum_{k \in \text{dom}(\phi(f))} \frac{1}{2^{k+1}} \\
&= 1 - \left(\frac{1}{2^{0+1}} + \sum_{k-1 \in \text{dom}(f)} \frac{1}{2^{k+1}} \right) \\
&= 1 - \left(\frac{1}{2} + \sum_{k \in \text{dom}(f)} \frac{1}{2^{k+2}} \right) \\
&= \frac{1}{2} \left(1 - \sum_{k \in \text{dom}(f)} \frac{1}{2^{k+1}} \right) \\
&= \frac{\mu f}{2}
\end{aligned}$$

which means ϕ is a contraction on the domain $[\mathbb{N} \rightharpoonup \mathbb{N}]$. By the contraction principle,

$$\bigsqcup_{n \in \mathbb{N}} \phi^n(\perp) = \text{fac}$$

is the unique fixed point of ϕ on $[\mathbb{N} \rightharpoonup \mathbb{N}]$, where $\perp$ is the function defined nowhere. This provides a formal justification for the usual recursive definition of the factorial function.

Because of applications like these, we would like to know when a space can be the kernel of a measurement on a domain. Major insights are provided by the following results.

4.11. THEOREM (MARTIN and REED [200?]). *A space is developable and T_1 iff it is the kernel of a measurement on a continuous poset.*

4.12. THEOREM (MARTIN and REED [200?]). *Each Čech-complete, developable space is the kernel of a measurement on a continuous dcpo.*

4.13. THEOREM (MARTIN and REED [200?]). *There exists a completeness condition C such that a T_1-space is developable with completeness condition C iff it is the kernel of a measurement on a continuous dcpo.*

4.14. THEOREM (MARTIN and REED [200?]). *There exists a continuous dcpo in which the maximal elements X form a G_δ-set in the Scott topology, but for which there exists no measurement having X as the kernel.*

In each of the above four theorems, the production of the desired poset from the given topological space is based on the generic Moore space construction in REED [1974]. An interesting open question is whether or not there exists a Scott domain in which the maximal elements X form a G_δ-set, but for which there exists no measurement with X as the kernel.

We close this discussion on topological aspects of measurement by mentioning one other series of results. In developing a domain theoretic foundation for the analysis of fractals MARTIN [200?a], one encounters the necessary class of *Lebesgue measurements.* It turns out that a space is metrizable iff it is the kernel of a Lebesgue measurement on a continuous poset, while it is completely metrizable iff it is the kernel of a Lebesgue measurement on a continuous *dcpo.* For the first result, a metrization theorem due to ARHANGEL'SKII [1995] figures prominently, while in the latter another appeal to Choquet completeness is made.

ACKNOWLEDGEMENT The authors express their thanks to the US Office of Naval Research for its support during the preparation of this paper. The second author also wishes to thank the NSF for support during his work on this article.

References

ABRAMSKY, S.

[1991a] Domain theory in logical form, *Annals of Pure and Applied Logic* **51**, 1–77.

[1991b] A domain equation for bisimulation, *Information and Computation* **92**, 161–218.

ABRAMSKY, S. and A. JUNG

[1994] Domain theory, in *Handbook of Logic in Computer Science* vol. III., S. Abramsky, D. M. Gabbay and T. S. E. Maibaum, editors, Oxford University Press, 1–168

AMADIO, R. and P.-L. CURIEN

[1998] *Domains and Lambda Calculi*, Cambridge Tracts in Theoretical Computer Science.

AMERICA, P. and J.J.M.M. RUTTEN

[1989] Solving reflexive domain equations in a category of complete metric spaces, *Journal of Computer and System Sciences* **39**, 345–375.

ARHANGEL'SKII, A.V.

[1995] *General topology iii: paracompactness, function spaces, descriptive theory*, A.V. Arhangelskii, ed., Encyclopedia of Mathematical Sciences, vol 51, Springer-Verlag.

ARNOLD, A. and M. NIVAT

[1980] Metric interpretations of infinite trees and semantics of nondeterministic recursive programs, *Theoretical Computer Science* **11**, 181–205.

DE BAKKER, J.W. and J.-J.CH. MEYER

[1987] Order and metric in the stream semantics of elemental concurrency, *Acta Information* **24**, 491–511.

DE BAKKER, J.W. and J.J.M.M. RUTTEN, (EDITORS)

[1992] *Ten years of Concurrency Semantics, selected papers of the Amsterdam Concurrency Group*, World Scientific (Singapore).

DE BAKKER, J. W. and J.I. ZUCKER

[1982] Processes and the denotational semantics of concurrency, *Information and Control* **54**, 70–120.

BARRETT, G. and A.W. ROSCOE

[1990] Unbounded nondeterminism in CSP, in *Lecture Notes in Comp. Sci.* **442**, 160–193.

VAN BREUGEL, F.C.

[1990] *Topological models in comparative semantics*, PhD thesis, Vrije Universiteit Amsterdam.

BROOKES, S., C.A.R. HOARE and A.W. ROSCOE

[1984] A theory of communicating sequential processes, *Journal of the ACM* **31**, 560–599.

CHOQUET, G.

[1969] *Lectures in Analysis*, W. A. Benjamin, vol. I, New York.

CIESIELSKI, K., R.C. FLAGG and R. KOPPERMAN

[1999] Characterizing topologies with bounded complete computational models, *Proceedings of MFPS XV, ENTCS* vol. 20.

EDALAT, A.

[1995a] Domain theory and integration, *Theoretical Computer Science*, **195**, 163–193.

[1995b] Dynamical Systems, measures and fractals via domain theory, *Information and Computation* **120**, 32–48.

[1997] When Scott is weak on the top, *Mathematical Structures in Computer Science* **7**, 401–417.

EDALAT, A. and R. HECKMANN

[1998] A computational model for metric spaces, *Theoretical Computer Science* **193**, 53–73.

ENGELKING, R.

[1977] *General Topology*, Polish Scientific Publishers, Warszawa.

FLAGG, B. and R. KOPPERMAN

[1997] Computational models for ultrametric spaces, *Proceedings of MFPS XIII, ENTCS* vol. 6, Elsevier Science.

GIERZ, G., K.H. HOFMANN, K. KEIMEL, J. LAWSON, M. MISLOVE and D. SCOTT

[1980] *A Compendium of Continuous Lattices*, Springer-Verlag, Berlin, Heidelberg, New York.

GRIFFOR, E.R., T. LINDSTROM and V. STOLTENBERG-HANSEN

[1994] *Mathematical Theory of Domains*, Cambridge University Press.

HOFMANN, K.H. and M.W. MISLOVE

[1981] Local compactness and continuous lattices, in: B. Banaschewski and R. E. Hofmann, editors, *Continuous Lattices*, Proceedings Bremen 1979, Lecture Notes in Mathematics (Springer-Verlag) vol. 871, pp. 209–248.

JUNG, A.

[1989] *Cartesian closed categories of domains*, CWI Tracts 66, Amsterdam.

HARRINGTON,L., A.S. KECHRIS and A. LOUVEAU

[1990] A Glimm-Effros Dichotomy For Borel Equivalence Relations, *Journal of the American Mathematical Society* **3** (4), 903–928.

KAMIMURA, T. and A. TANG

[1984] Total objects of domains, *Theoretical Computer Science* **34**, 275–288.

KECHRIS, A.

[1994] *Classical descriptive set theory*, Springer-Verlag.

LAWSON, J.

[1997] Spaces of Maximal Points, *Mathematical Structures in Computer Science* **7**, 543–555.

MARTIN, K.

[1998] Domain theoretic models of topological spaces, *Proceedings of Comprox III, Electronic Notes in Theoretical Computer Science* **13**.

[1999] Nonclassical techniques for models of computation, *Topology Proc., Summer Issue* **24**, 375–405.

[2000a] *A foundation for computation*, Ph.D. Thesis, Department of Mathematics, Tulane University.

[2000b] The measurement process in domain theory, in *Proceedings of the* 27th *International Colloquium on Automata, Languages and Programming (ICALP)*, Lecture Notes in Comp. Sci., Springer-Verlag, vol. 1853, 116–126.

[2000c] The space of maximal elements in a compact domain, *Electronic Notes in Theoretical Computer Science* **40**.

[2001a] A principle of induction, *Lecture Notes in Comp. Sci.*, Springer-Verlag, vol. 2142, 458–468.

[2001b] A renee equation for algorithmic complexity, *Lecture Notes in Comp.r Sci., Springer-Verlag* vol. 2215, 201-218.

[2001c] Unique fixed points in domain theory, *Electronic Notes in Theoretical Computer Science* **45**.

[200?a] Fractals and domain theory, submitted.

[200?b] Ideal models of spaces, *Theoretical Computer Science*, to appear.

[200?c] Powerdomains and zero finding, *Electronic Notes in Theoretical Computer Science*, to appear.

[200?d] The regular spaces with countably based models, *Theoretical Computer Science*, to appear.

[200?e] Topological games in domain theory, *Topology Appl.*, submitted.

MARTIN, K. and G.M. REED

[200?] Measurement and topology, to appear.

MATTHEWS, S.G.

[1986] *Metric domains for completeness*, PhD thesis, University of Warwick.

MISLOVE, M.

[1991] Algebraic posets, algebraic cpos and models of concurrency, in: *Proceedings of the Oxford Symposium on Topology*, G. M. Reed, A. W. Roscoe and R. Wachter, eds., Oxford University Press, 75–111.

[1995] Denotational models for unbounded nondeterminism, *Proceedings of MFPS XI, Electronic Notes in Theoretical Computer Science* **1**, Elsevier.

[1998] Topology, domain theory and theoretical computer science, *Topology Appl.* **89**, 3–59.

MISLOVE, M., A.W. ROSCOE and S. SCHNEIDER

[1995] Fixed points without completeness, *Theoretical Computer Science* **138**, 273–314.

NIVAT, M..

[1979] Infinite words, infinite trees, infinite computations, in: *Mathematical Centre Tracts* vol.109 Mathematical Center, Amsterdam, 3–52.

ODIFREDDI, P.

[1989] *Classical recursion theory*, Studies in Logic and the Foundations of Mathematics, vol. 125, Elsevier Science, North Holland.

OUAKNINE, J.

[2001] *Discrete analysis of continuous behaviour in real-time concurrent systems*, DPhil thesis, Oxford University.

OUAKNINE, J. and G.M. REED

[1999] Model-checking temporal behaviour in CSP, *Proceedings of PDFTA'99* (CSREA Press), 295–304.

PLOTKIN, G.D.

[1976] A powerdomain construction, *SIAM Journal of Computation* **5**, 452–487.

REED, G.M.

[1974] On chain conditions in Moore spaces, *General Topology Appl.* **4**, 255–267.

[1990] A hierarchy of models for real-time distributed computing, *Lecture Notes in Comp. Sci.* **442**, 80–128.

REED, G.M. and A.W. ROSCOE

[1986] A timed model for communicating sequential processes, *Theoretical Computer Science* **58**, 249–261.

[1988] Metric spaces as models for real-time concurrency, *Lecture Notes in Comp. Sci.* **298**, 331–343.

[1999] The timed failures-stability model for CSP, *Theoretical Computer Science* **211**, 85–127.

ROSCOE, A.W.

[1982] *A mathematical theory of communicating processes*, DPhil thesis, Oxford University.

[1998] *The Theory and Practice of Concurrency*, Prentice Hall.

ROUNDS, W.C.

[1985] On the relationship between Scott domains, synchronization trees, and metric spaces, *Information and Computation* **66**, 6–28.

SCOTT, D.
[1970] *Outline of a mathematical theory of computation*, Technical Monograph PRG-2, November 1970.

SMYTH, M.
[1978] Power domains, *Journal of Computer and System Sciences* **16**, 23–36.
[1988] Quasi-uniformities: reconciling domains with metric spaces, *Lecture Notes in Comp. Sci.* **298**, 236–253.

STRACHEY, C.
[1973] Varieties of Programming Language, in: *International Computing Symposium 1972*, 222–233; also as: Technical Monograph PRG-10, Programming Research Group, Oxford, 1973.

STRACHEY, C. and C.P. WADSWORTH
[1974] Continuations: A Mathematical Semantics for Handling Full Jumps, *Technical Report PRG-11, Programming Research Group*, Oxford University.

TELGÁRSKY, R.
[1987] Topological games: on the fiftieth anniversary of the Banach–Mazur game, *Rocky Mountain Journal of Mathematics* **17** (2), 227–276.

ZHANG, H.
[1993] *Dualities of domains*, Ph.D. thesis, Department of Mathematics, Tulane University.

CHAPTER 15

Topics in Dimension Theory

Roman Pol

Institute of Mathematics, University of Warsaw, Banacha 2, 02-097 Warszawa, Poland
E-mail: pol@mimuw.edu.pl

Henryk Toruńczyk

Institute of Mathematics, University of Warsaw, Banacha 2, 02-097 Warszawa, Poland
E-mail: torunczy@mimuw.edu.pl

Contents

RECENT PROGRESS IN GENERAL TOPOLOGY II
Edited by Miroslav Hušek and Jan van Mill

1. Introduction

We shall outline in this survey several areas in the dimension theory where some significant developments took place during the last several years. The dimension theory is a vast field, with methods ranging from a refined set-theory to the ones involving advanced homotopy theory and sophisticated geometric constructions. Trying to keep a balance in this diversified terrain, we had to neglect many valuable ideas and results, omitting more general subjects, and the topics where algebraic topology or specific geometric methods take over. We felt excused in some degree by an existence of excellent books and survey articles in the literature, covering various aspects of the subject. ENGELKING [1995] provides a comprehensive treatment of the general dimension theory up to 1995. This will be our basic source of terminology and references. The compactification problems in dimension theory are addressed in AARTS and NISHIURA [1993]. The monographs by CHIGOGIDZE and FEDORCHUK [1992] and CHIGOGIDZE [1998] give a good exposition of soft mappings and related topics concerning universal spaces. The important theory of Menger manifolds is outlined by CHIGOGIDZE, KAWAMURA and TYMCHATYN [1995]. Rich material, concerning infinite-dimensional topology and some subtle aspects of the dimension theory of separable metrizable spaces, is contained in the books by VAN MILL [1989], [2001]. A valuable source of information is a special issue of Uspekhi Mat. Nauk, devoted to the legacy of P.S.Urysohn. It contains in particular a survey article by BOGATYI [1998] on embeddings in Euclidean spaces, with many refinements of classical Hurewicz's theorems, and an article by SHCHEPIN [1998] presenting a broad spectrum of ideas linking the dimension theory with algebra and geometry. One should mention also comprehensive articles by DRANISHNIKOV and SHCHEPIN [1986] and DRANISHNIKOV [1988] on the homological dimension theory. The dimension theory of topological groups is discussed by SHAKHMATOV [1990].

2. Terminology

Our terminology follows ENGELKING [1995]. This is also the case with our notation, with one little exception - the transfinite extensions of the small and large inductive dimensions are denoted also by ind and Ind, respectively.

By a mapping we mean a continuous function. We say that a mapping $f : X \to Y$ covers essentially an n-cell C in Y, i.e., a topological copy of the cube $\mathbb{I}^n$, if each mapping $g : f^{-1}(C) \to C$ extending the restriction $f \mid f^{-1}(\partial C)$ is a surjection.

A *light mapping* is a perfect map with zero-dimensional fibers.

Most of this survey concerns metrizable spaces and speaking about a space without any additional explanation, we assume metrizability. By a *compactum* we understand a compact metrizable space.

A space X is *countable-dimensional* if it is a countable union of finite-dimensional subspaces, cf. ENGELKING [1995], 5.1.1.

For CW complexes K, L, their *join* $K * L$ is the union of mapping cylinders of the projections $p_K : K \times L \to K$ and $p_L : K \times L \to L$, intersecting along $K \times L$.

The *smash product* $K \wedge L$ of CW complexes K, L with base points is $K \times L / K \vee L$, where $K \vee L$ is the set of points (x, y) with at least one coordinate being a base point.

We shall denote by $\mathbb{I}$ the unit interval, $\mathbb{R}$ stands for the real line, and $\mathbb{S}^n$ is the unit n-sphere centered at 0 in $\mathbb{R}^{n+1}$.

3. Weakly infinite-dimensional spaces and Haver's property *C*

We shall consider in this section only metrizable spaces. Although the topic can be discussed in a more general setting, cf. ENGELKING [1995], 6.3.6, some fundamental problems in this area already concern compacta.

(A) Following HAVER [1974] we say that a metrizable space X has *property* C, or X is a C-space, if for any sequence $\mathcal{U}_1, \mathcal{U}_2, \ldots$ of open covers of X there exist disjoint open collections $\mathcal{V}_1, \mathcal{V}_2, \ldots$ such that $\mathcal{V}_i$ refines $\mathcal{U}_i$ and $\bigcup_i \mathcal{V}_i$ covers X.

Any C-space is *weakly infinite-dimensional*. Indeed, if one requires only that the collections $\mathcal{V}_i$ exist for any sequence of two-element covers $\mathcal{U}_i$, one gets a characterization of weakly infinite-dimensional spaces. No examples are known of metrizable weakly infinite-dimensional spaces without property C, and it is one of the most important problems concerning infinite-dimensional spaces, whether the two notions coincide for compacta. Some interesting connections of this problem with certain questions involving cohomological dimension and homotopy theory were established by DRANISHNIKOV [1992].

A natural transfinite extension of the covering dimension provides a stratification of the class of weakly infinite-dimensional compacta, cf. ENGELKING [1995], p.354 (a closely related Borst-Henderson index is considered in Section 10) and an analogous transfinite classification of compacta with property C is described by CHATYRKO [1991].

The celebrated Henderson theorem asserts that each strongly infinite-dimensional compactum contains a non-trivial continuum all of whose non-trivial subcontinua are strongly infinite-dimensional, cf. VAN MILL [2001], ENGELKING [1995], p.267. Analogously, any compactum without property C contains a non-trivial continuum all of whose non-trivial subcontinua fail property C, cf. POL [1996a] and LEVIN [1995a].

3.1. PROBLEM. Let $f : X \to Y$ be a continuous map between compacta with Y and all fibers $f^{-1}(y)$ weakly infinite-dimensional. Is then X weakly infinite-dimensional?

The answer is positive, if in addition Y is a C-space; if Y and the fibers are C-spaces, also X is a C-space, cf. ENGELKING [1995], 6.3.9. The Henderson theorem shows that any negative solution of Problem 3.1 gives raise to a light mapping of a strongly infinite-dimensional compactum onto a weakly infinite-dimensional one, and moreover, the mapping is not injective on any non-trivial continuum, cf. 7.3.

(B) Haver defined C-spaces to extract from countable-dimensionality a property useful in some selection problems. Interesting results along these lines were obtained recently by USPENSKIĬ [1998], and GUTEV and VALOV [2002].

Uspenskiĭ proved in particular the following

3.2. THEOREM. *A metrizable space X has property C if and only if for every Banach space E and any open set $V \subset X \times E$ with nonempty contractible vertical sections, V contains the graph of a continuous function $f : X \to E$.*

Another result of Uspenskiĭ emphasizes similarities between weak infinite-dimensionality and property C. For a compactum X, weak infinite-dimensionality means that X

has no universal map onto $\mathbb{I}^\infty$, i.e., for any mapping $f : X \to \mathbb{I}^\infty$ there is a mapping $g : X \to \mathbb{I}^\infty$ with $g(x) \neq f(x)$ for $x \in X$, cf. SEGAL and WATANABE [1991]. Let $\mathcal{Z}$ be the collection of all nonempty compact Z-*sets* in $\mathbb{I}^\infty$, i.e., for $Z \in \mathcal{Z}$, every mapping from $\mathbb{I}^\infty$ to itself can be approximated by mappings into $\mathbb{I}^\infty \setminus Z$. We shall consider $\mathcal{Z}$ with the Hausdorff metric. The Uspenskiĭ theorem reads as follows:

3.3. THEOREM. *A compactum X is a C-space if and only if for any mapping $F : X \to \mathcal{Z}$ there is a mapping $g : X \to \mathbb{I}^\infty$ with $g(x) \notin F(x)$, for $x \in X$.*

The next theorem, obtained by Gutev and Valov, is related to a problem formulated by MICHAEL [1990].

3.4. THEOREM. *Let X be a metrizable C-space, let G be a G_δ-set in a Banach space E, and let F be a lower-semicontinuous set-valued map with $F(x) \subset G$ being nonempty convex sets, relatively closed in G. Then there is a mapping $f : X \to E$ with $f(x) \in F(x)$, for $x \in X$.*

(C) We shall consider now mappings between compacta whose range has property C. USPENSKIĬ [2000] proved the following theorem, where "typical" refers to the Baire category in the space of continuous mappings into the unit interval.

3.5. THEOREM. *Let $f : X \to Y$ be a light mapping between compacta. If Y has property C, then for a typical mapping $u : X \to \mathbb{I}$, all sets $u(f^{-1}(y))$ are 0-dimensional.*

For countable-dimensional Y this was proved earlier by Toruńczyk, and it is an open problem if any restrictions on Y are necessary.

The next result is due to LEVIN and ROGERS [2000].

3.6. THEOREM. *Let $f : X \to Y$ be an open mapping between compacta with perfect fibers and let Y have property C. Then f maps some 0-dimensional compactum onto Y.*

Using this result, the authors proved also that, under the assumptions of Theorem 3.6, there exists a continuous surjection $g : X \to Y \times \mathbb{I}$ such that $f = p \circ g$, p being the projection onto Y. This yields instantly two disjoint compact sets F_0, F_1 in X, each mapped by f onto Y.

For finite-dimensional Y these results have been established by Bula, and extended to the case of countable-dimensional Y by Gutev. The sets F_0, F_1 may not exist for arbitrary Y, cf. DRANISHNIKOV [1990], KATO and LEVIN [2000]. Interestingly, Kato and Levin proved that if the fibers of the mapping f are non-trivial hereditarily indecomposable continua, one can always find the sets F_0, F_1, without any additional assumptions on Y.

(D) We shall close this section with recalling one more fundamental problem:

3.7. PROBLEM. Does there exist a compactum which is not countable-dimensional and whose subsets are all weakly infinite-dimensional?

4. Extension theory

We shall write $X\tau K$ (another standard notation is $K \in AE(X)$) if for any closed A in X and each mapping $f : A \to K$ there is a continuous extension $\overline{f} : X \to K$ of f.

By the Alexandroff theorem, $X\tau\mathbb{S}^n$ is equivalent to $\dim X \le n$. If $K = K(G,n)$ is the Eilenberg-Mac Lane complex for the group G, then $X\tau K$ means that the *cohomological dimension* of X with respect to G is not greater than n, $\dim_G X \le n$, cf. WALSH [1981].

During the several last years a general theory emerged, describing properties of metrizable spaces X in relation $X\tau K$ with certain CW complexes K. These developments had a great impact on the dimension theory. A prominent role in originating this process was played by WALSH [1981]. Some essential steps in shaping up the theory are outlined by DRANISHNIKOV and DYDAK [2001], Introduction. We shall concentrate on the results in this area which either extend or provide counterparts to certain basic facts in the classical dimension theory of metrizable spaces.

(A) We shall begin with a generalization of the *Menger-Urysohn formula*

$$\dim(A \cup B) \le \dim A + \dim B + 1\,.$$

Let us recall that $K * L$ is the join of CW complexes K, L, cf. Section 2.

The following theorem was obtained by DYDAK [1996].

4.1. THEOREM. *Let $X = A \cup B$ be a metrizable space. If for CW complexes K, L we have $A\tau K$ and $B\tau L$, then $X\tau(K * L)$.*

Setting $K = \mathbb{S}^k$, $L = \mathbb{S}^l$, and identifying $\mathbb{S}^k * \mathbb{S}^l$ with $\mathbb{S}^{k+l+1}$, one readily gets the Menger-Urysohn formula.

DYDAK [1996] derived also from Theorem 4.1 the Menger-Urysohn formula for cohomological dimension $\dim_R$ with respect to any ring R with unity (for $R = \mathbb{Z}$ this was earlier established by Rubin). However, Dranishnikov, Repovš and Shchepin proved that the formula fails for $\dim_G$ with respect to the group $G = \mathbb{Q}/\mathbb{Z}$. A discussion of this topic can be found in DYDAK [1994].

(B) The next theorem, proved by DRANISHNIKOV and DYDAK [2001] (for compact spaces, DRANISHNIKOV [1996]) is related to the *decomposition theorem*: any metrizable space X with $\dim X \le k + l + 1$ can be split into sets A, B such that $\dim A \le k$ and $\dim B \le l$.

4.2. THEOREM. *Let X be a metrizable separable space with $X\tau(K * L)$, where K and L are countable CW complexes. Then there is a decomposition $X = A \cup B$ with $A\tau K$ and $B\tau L$.*

Again, setting $K = \mathbb{S}^k$, $L = \mathbb{S}^l$ one gets (for separable spaces) the decomposition theorem.

(C) The *Eilenberg-Borsuk theorem* asserts that any mapping $f : A \to \mathbb{S}^k$ from a closed set in a metrizable space with $\dim X \le n$ extends continuously over a neighborhood U of A with $\dim(X \setminus U) \le n - k - 1$, cf. AARTS and NISHIURA [1993], Theorem 4.10.

This theorem was extended (for separable spaces) by DRANISHNIKOV and DYDAK [2001] (DRANISHNIKOV [1996] for compacta) to the following effect.

4.3. THEOREM. *Let $f : A \to K$ be a mapping from a closed subset of a metrizable separable space X into a CW complex K. Then for any countable CW complex L with $X\tau(K * L)$ there is a continuous extension $\overline{f} : U \to K$ of f over a neighborhood of A with $(X \setminus U)\tau L$.*

(D) Another basic fact about the covering dimension is the *enlargement theorem* that any metrizable X with $\dim X = n$ has a completion X^* with the same dimension. A related result is the following theorem due to OLSZEWSKI [1995b] (c.f. also section 13(A)):

4.4. THEOREM. *If $X\tau K$ where X is separable metrizable and K is a countable CW-complex, then there is a completion X^* of X with $X^*\tau K$.*

The clever idea of Olszewski's proof was subsequently exploited in several papers, cf. DRANISHNIKOV and DYDAK [2001], Section 3 and DRANISHNIKOV and KEESLING [2001].

The compactification problem in this general setting is, however, more complex than for the covering dimension. DYDAK and WALSH [1991] constructed separable metrizable X such that $X\tau K(\mathbb{Z}, n)$ but no compactification of X satisfies this relation. The general problem is discussed by CHIGOGIDZE [2000].

(E) The following theorem due to LEVIN, RUBIN and SCHAPIRO [2000] is related to the Mardešić *factorization theorem* for the covering dimension, cf. ENGELKING [1995], 3.4.1.

4.5. THEOREM. *Let $f : X \to Y$ be a continuous map between compact spaces and let $X_i \subset X$ be compact sets. There exists a compact space Z of weight not exceeding the weight of Y, and continuous maps $g : X \to Z$, $h : Z \to Y$ with $f = h \circ g$ such that for any CW complex K, $X_i\tau K$ implies $g(X_i)\tau K$, $i = 1, 2, \ldots$*

The factorization preserves in addition weak infinite-dimensionality, or property C.

RUBIN and SCHAPIRO [1999] obtained also the following extension of the Nagami theorem asserting that the *inverse limit* of an inverse sequence of metrizable n-dimensional spaces has dimension $\leq n$.

4.6. THEOREM. *Let X be the limit of an inverse sequence (X_i, p_i) of metrizable spaces. Then, for any CW complex K with $X_i\tau K$, $i = 1, 2, \ldots$, we have $X\tau K$.*

(F) A generalization of the *Hurewicz formula* $\dim X \leq \dim Y + \dim f$ is provided by the following result of LEVIN and LEWIS [200?].

4.7. THEOREM. *Let $f : X \to Y$ be a continuous map of a finite-dimensional compactum X onto Y. If K, L are CW complexes, with K countable, such that $f^{-1}(y)\tau K$ for any $y \in Y$, and $Y\tau L$, then $X\tau(K \wedge L)$.*

Since $\mathbb{S}^k \wedge \mathbb{S}^l = \mathbb{S}^{k+l}$, setting $K = \mathbb{S}^k$, $L = \mathbb{S}^l$, one gets the Hurewicz formula. The case $K = \mathbb{S}^0$ does not require the restriction on X. Indeed, DRANISHNIKOV and USPENSKIĬ [1988] proved that if $f : X \to Y$ is a light mapping between compacta and $Y\tau L$, then $X\tau L$.

Another interesting mapping theorem established by Dranishnikov and Uspenskiĭ asserts that if $f : X \to Y$ is a continuous map between compacta whose fibers are at most $(k+1)$-element, and $Y\tau L$ then $(X \times \mathbb{I}^k)\tau L$.

For the projection $f : Y \times Z \to Y$, Theorem 4.7 yields a result proved earlier by DRANISHNIKOV and DYDAK [2001].

(G) Let us close this section with a brief description of a new very interesting concept - the *extensional dimension* e-dim X, introduced by DRANISHNIKOV [1998].

Given CW complexes K, L, the inequality $K \preceq L$ means that $X\tau K$ implies $X\tau L$ for any compact space X. Let $[K]$ be the equivalence class with respect to the relation ($K \preceq L$ and $L \preceq K$). Then, for any compact space X, the collection of all CW complexes K with $X\tau K$ has the least element M with respect to the partial order $\preceq$, and the class $[M]$ is called the extensional dimension of X, e-dim $X = [M]$.

It is not known if e-dim X can be always represented as $[K]$ with a countable CW complex K. It is also unknown, if for any countable CW complex K there is a metrizable compact space X with e-dim $X = [K]$.

5. Products of compact spaces

(A) DRANISHNIKOV [1988] solved an old problem of Borsuk by constructing AR compacta X, Y with $\dim(X \times Y) < \dim X + \dim Y$. In this example $\dim X = \dim Y = 4$, $\dim(X \times Y) = 7$, and it is an open problem if there are compacta X, Y violating the *logarithmic law*, with X being a 3-dimensional AR.

(B) Some very interesting and deep research involving dimension of products has been stimulated by the following conjecture. We say that two mappings $f : X \to \mathbb{R}^n$ and $g : Y \to \mathbb{R}^n$ have *unstable intersection* if there are arbitrarily close continuous approximations $f' : X \to \mathbb{R}^n$, $g' : Y \to \mathbb{R}^n$ of f and g, respectively, with disjoint images.

5.1. PROBLEM. Let X, Y be compacta. Is it true that $\dim(X \times Y) < n$ if and only if every mappings $f : X \to \mathbb{R}^n$, $g : Y \to \mathbb{R}^n$ have unstable intersection?

A vital role in originating this subject has been played by a question of Yaki Sternfeld (related to his results cited in Section 9) about an existence of n-dimensional compacta whose every mapping into $\mathbb{R}^{2n}$ can be approximated by embeddings. A thorough analysis of disjoint membranes in Euclidean cubes and essential mappings, conducted by Mc Cullough and Rubin, and Krasinkiewicz and Lorentz, followed by a work by Spież, resulted in establishing in 1988 that the compacta X answering the Sternfeld question are exactly the ones with $\dim(X \times X) < n$. As demonstrated by KRASINKIEWICZ [1989], this property is equivalent to the condition that any two mappings $f, g : X \to \mathbb{R}^n$ have unstable intersection.

Further or independent significant contributions to Problem 5.1 were made by Dranishnikov, Repovš, Segal, Shchepin, Spież, Sternfeld, Toruńczyk and West. An outline of this work is given by SHCHEPIN [1998] and DRANISHNIKOV [2000], Introduction.

The implication $\Longleftarrow$ in Problem 5.1 was confirmed by Dranishnikov and West; an elegant proof was found by STERNFELD [1996].

Concerning the implication $\Longrightarrow$, DRANISHNIKOV [2000] proved the following

5.2. THEOREM. *Let* X, Y *be compacta with* $\max\{\dim X, \dim Y\} \le n-2$ *and let* $f : X \to \mathbb{R}^n$, $g : Y \to \mathbb{R}^n$ *be continuous maps. Then* f, g *can be approximated arbitrarily closely by mappings* $f' : X \to \mathbb{R}^n$, $g' : Y \to \mathbb{R}^n$ *with*

$$\dim(f'(X) \cap g'(Y)) \le \dim(X \times Y) - n.$$

These two theorems, combined with some earlier results, confirm the conjecture in Problem 5.1, except for the case $\dim X = \dim Y = n - 2 > 2$, which still remains open. An important aspect of DRANISHNIKOV [2000] is a development of some new methods involving the extensional dimension (cf. Section 4(G)), which are of independent interest.

Some other noteworthy outcome of the work stimulated by Problem 5.1 is a new construction of compacta X with $\dim(X \times X) < 2\dim X$, cf. KARNO and KRASINKIEWICZ [1989], and a characterization of compacta with this property in terms of extensions of mappings into Boltyanskiĭ - Kodama "bubbles", cf. SPIEŻ [1990]. Also, some useful conditions have been found such that for any pair of mappings $f : X \to \mathbb{R}^n$, $g : Y \to \mathbb{R}^n$ and $\varepsilon > 0$ satisfying these conditions, the maps f, g can be ε-approximated by mappings with disjoint images, cf. SPIEŻ and TORUŃCZYK [1991]. These conditions are pertinent to Problem 5.1, but require the assumption $2\dim X + \dim Y \le 2n - 2$; the role of this restriction is unclear.

(C) The following result of DRANISHNIKOV [2000] links in a remarkable way the dimension of products with the Menger-Urysohn formula: if a compactum X satisfies the condition $\dim(X \times X) = 2\dim X$, then for any decomposition $X = A \cup B$,

$$\dim X \le \dim(A \times B) + 1.$$

It is unknown, if the assumption about $X \times X$ is necessary. The weaker inequality, with 1 replaced by 2, is true without this restriction, cf. DRANISHNIKOV [2000].

6. Products of non-compact spaces

(A) KULESZA [1996] constructed striking examples illustrating the behavior of the dimension of finite powers of separable metrizable spaces. Among Kulesza's examples is a separable metrizable space X with

$$\dim X = \dim X^2 = 1 \text{ and } \dim X^3 = 2.$$

Kulesza conjectured that for any subadditive sequence d_n of natural numbers, $d_{n+m} \le d_n + d_m$, there is a separable metrizable X with $\dim X^n = d_n$ for $n = 1, 2, \ldots$

This conjecture was confirmed by Kulesza for some sequences d_n, including the sequence $m, m+1, m+2, \ldots$ or $2 < m, m+1, m+2, m+2, \ldots$

Concerning topological groups, HATTORI [1994] refined some earlier results of Keesling and Kulesza to the following effect: for each $n \le m$ there is a subgroup of $\mathbb{R}^{n+1}$ whose dimension is n and the dimension of its countable power is m.

It is a classical result that the multiplication of any metrizable space by the interval increases the dimension by 1. DRANISHNIKOV, REPOVŠ and SHCHEPIN [1993] showed that the interval can not be replaced here by an arbitrary curve: there exists a 2-dimensional set $X \subset \mathbb{R}^3$ and a 1-dimensional metrizable continuum K such that $\dim(X \times K) = 2$; independently, this result was obtained by Karno (unpublished manuscript).

(B) Recall that a metrizable separable space X is *weakly 1-dimensional* if $\dim X = 1$ but the kernel, i.e., the set of points in X without any local base consisting of clopen sets, is 0-dimensional. Although this is a classical notion, only recently it was noticed that the countable product of weakly 1-dimensional spaces is always 1-dimensional, cf. VAN MILL [2001], 3.11.12.

(C) Turning to infinite-dimensional spaces, it is a major open problem whether the product of two weakly infinite-dimensional compacta must be weakly infinite-dimensional (this is a particular case of Problem 3.1). However, the behavior of non-compact weakly infinite-dimensional spaces under the product operation is rather erratic: E. POL [1993] constructed for each n a metrizable separable Y such that Y^n is a C-space while Y^{n+1} is not weakly infinite-dimensional.

7. Hereditarily indecomposable continua in dimension theory

A continuum X is *hereditarily indecomposable* if for any subcontinua A, B in X with nonempty intersection, either $A \subset B$, or $B \subset A$. Hereditarily indecomposable continua of arbitrarily large dimension were constructed by Bing in 1951, but only recently such continua were recognized as essential tools in proving some interesting results in the dimension theory, whose statements do not involve the indecomposability; we shall present certain instances of this phenomenon.

The significance of hereditarily indecomposable continua is apparently related to their genericity. Let us call a compactum X a *Bing space* if all subcontinua of X are hereditarily indecomposable. KRASINKIEWICZ [1996] and LEVIN [1996] proved that for a typical (in the sense of Baire category in the function space) continuous map from a compactum to the unit interval, all fibers are Bing spaces.

(A) Passing to the applications in the dimension theory, we shall begin with a solution by LEVIN and STERNFELD [1997] of an outstanding problem concerning *hyperspaces*.

7.1. THEOREM. *For any 2-dimensional compactum X the space $\mathcal{C}(X)$ of all subcontinua of X with the Hausdorff metric is infinite-dimensional.*

A vital element in the proof by Levin and Sternfeld is that for any compactum X there is a hereditarily indecomposable continuum Z with $\dim Z = \dim X$ and a light mapping from Z into X.

Theorem 7.1 was improved by LEVIN and ROGERS [2000] to the effect that the hyperspace $\mathcal{C}(X)$ fails to have property C, whenever $\dim X \geq 2$. It is unknown, if $\mathcal{C}(X)$ must be strongly infinite-dimensional in this case.

(B) The next application is pertinent to the following closely related results, where (a) was obtained by PASYNKOV [1975] and (b) by TORUŃCZYK [1985].

7.2. THEOREM. *Let $f : X \to Y$ be a mapping between compacta with* $\dim f^{-1}(y) \le k$, *for $y \in Y$. If* $\dim Y < \infty$ *then*

(a) *for almost all mappings $g : X \to \mathbb{I}^k$ in the function space, the diagonal map $f\Delta g : X \to Y \times \mathbb{I}^k$ has 0-dimensional fibers;*

(b) *there exists a σ-compact set $A \subset X$ with* $\dim A \le k - 1$ *such that the restriction $f \mid (X \setminus A)$ has 0-dimensional fibers.*

It is unknown, whether the assumption $\dim Y < \infty$ is necessary. However, using Bing spaces, STERNFELD [1995] showed that, when raising 0 to 1 in the assertions (a) and (b), one can remove the restriction on Y. The first part of Sternfeld's theorem was improved by LEVIN [1996], who also exploited Bing spaces, to the effect that for arbitrary Y, one gets (a) with k replaced by $k+1$. Levin showed also that for any mapping $f : X \to Y$ between compacta, whose fibers are weakly infinite-dimensional, there is a weakly infinite-dimensional $A \subset X$ with $f \mid (X \setminus A)$ having 0-dimensional fibers.

Part (b) of Theorem 7.2 is connected to another result of TORUŃCZYK [1985] that for any light mapping $f : X \to Y$ between compacta, with finite-dimensional range, there is a 0-dimensional $A \subset X$ with $f \mid (X \setminus A)$ finite-to-one. Combining this result with a certain reasoning involving hereditarily indecomposable continua, one gets the following theorem, cf. POL [1996a].

7.3. THEOREM. *Let $f : X \to Y$ be a light mapping between compacta. If* $\dim X \ge 3$ *and* $\dim Y < \infty$, *then f is injective on a non-trivial continuum in X.*

Anderson's light open mapping from the Menger universal curve onto the Hilbert cube shows that the assumption $\dim Y < \infty$ is necessary in the result preceding Theorem 7.3. We do not know, however, of any example showing that this is also the case for the theorem.

(C) Hereditarily indecomposable continua of higher dimensions have a remarkable dimensional structure. Given an n-dimensional hereditarily indecomposable continuum X, let $B_r(X)$ be the set consisting of points in X that belong to some r-dimensional continuum, but avoid any non-trivial continuum of dimension less than r, $1 \le r \le n$. The elements of the top layer $B_n(X)$ are called *Bing points* in X.

The following results were obtained by R.POL and REŃSKA [2002].

7.4. THEOREM. *Let X be an n-dimensional metrizable hereditarily indecomposable continuum. Then, for any σ-compact set C with* $\dim C \le n - 2$,

$$\dim (B_n(X) \setminus C) = 1.$$

Also, $\dim B_r(X) = n - (r - 1)$, for $r \ge 2$ the layers $B_r(X)$ are never $G_{\delta\sigma}$-sets being always $G_{\delta\sigma\delta}$-sets, and no G_δ-set in X containing $B_r(X)$ with $1 < r < n$ is homogeneous. It is unclear, if the layers $B_r(X)$ are non-homogeneous.

For an infinite-dimensional hereditarily indecomposable continuum X, an analogue for the set of Bing points in X is the complement $B_\infty(X)$ of the union of all non-trivial finite-dimensional continua in X. E.POL and REŃSKA [200?] showed that for metrizable countable-dimensional hereditarily indecomposable continua X, the diversity among types of the sets $B_\infty(X)$ is much greater than in the finite-dimensional case. An interesting open problem is, however, whether in this case one can have $\dim B_\infty(X) = n$ for arbitrary n. E.Pol observed that a positive solution would provide an example of a countable-dimensional compactum which cannot be mapped onto a finite-dimensional compactum by any mapping with finite-dimensional fibers, cf. ENGELKING and E.POL [1983], Problem 7.18.

(D) Hereditarily indecomposable continua are also useful, as was demonstrated by LEVIN [1995b], in construction of *Henderson's continua*, i.e., continua whose each non-trivial subcontinuum is infinite-dimensional. Following Levin 's idea one can show that for any infinite cardinal κ there is a hereditarily indecomposable continuum X of weight κ, such that each non-trivial continuum in X admits an essential mapping onto the Tychonoff cube $\mathbb{I}^\kappa$, cf. HART, VAN MILL and POL [200?].

We refer the reader to BALL, HAGLER and STERNFELD [1998] and HART, VAN MILL and POL [200?] for some more information concerning general structure of hereditarily indecomposable continua.

8. Pushing compacta off affine manifolds in Euclidean spaces

Given a compactum X and a set M in $\mathbb{R}^m$, we shall say that X can be pushed off M if for any $\epsilon > 0$ there is a mapping $f : X \to \mathbb{R}^m \setminus M$ which is ϵ-close to the identity, i.e., $\| f(x) - x \| < \epsilon$ for $x \in X$.

A classical theorem of Alexandroff asserts that a compactum X in $\mathbb{R}^m$ which can be pushed off any $(m-n-1)$-dimensional polyhedron is at most n-dimensional. CHOGOSHVILI [1938] formulated some strengthenings of Alexandroff's Theorem, however recent striking examples of STERNFELD [1993a] and DRANISHNIKOV [1997] provided counterexamples to these claims.

Sternfeld proved the following.

8.1. THEOREM. *For any $n \geq 2$ and sufficiently large m, every n-dimensional hereditarily indecomposable continuum has a topological copy in $\mathbb{R}^m$ that can be pushed off any $(m-2)$-dimensional manifold $\{(x_1, \ldots, x_m) : x_i = a, x_j = b\}$, $i \neq j$.*

Some refinements of Sternfeld's ideas are presented by ANCEL and DOBROWOLSKI [1997]. The Dranishnikov theorem reads as follows:

8.2. THEOREM. *For each $k \geq 1$ there is a $2(k-1)$-dimensional compactum in $\mathbb{R}^{2k}$ which can be pushed off any k-dimensional affine manifold in $\mathbb{R}^{2k}$.*

On the other hand, DOBROWOLSKI, LEVIN and RUBIN [1997] imposed some conditions on compacta X which ensure that no topological copy of X in $\mathbb{R}^4$ can be pushed off any 2-dimensional affine manifold in $\mathbb{R}^4$. In particular, this is true for 2-cells. However, it is not known if there is an n-cell in $\mathbb{R}^m$, $n > 2$, which can be pushed off any $(m-n)$-dimensional affine manifold in $\mathbb{R}^m$.

9. Basic embeddings

Following Sternfeld, we shall say that a compactum $X \subset X_1 \times \ldots \times X_m$ is basically embedded in the product of compacta X_i, if for every continuous function $f : X \to \mathbb{R}$ there are continuous functions $f_i : X_i \to \mathbb{R}$ such that

$$(\star) \qquad f(x) = \sum_{i=1}^{m} f_i(x_i), \quad \text{whenever } x = (x_1, \ldots, x_m) \in X.$$

We shall also say that a continuous injection $h : X \to X_1 \times \ldots X_m$ of a compactum X is a *basic embedding* if $h(X)$ is basically embedded in the product.

A thorough investigation of this notion, introduced in connection with Hilbert's 13th problem, was carried out by STERNFELD [1989]. We shall concentrate only on some special aspects of this very interesting topic.

Sternfeld proved, refining the classical embedding theorem, that a typical mapping of an n-dimensional compactum into $\mathbb{R}^{2n+1}$ is a basic embedding. He proved also that no embedding of a compactum X with $\dim X = n$ into $\mathbb{R}^{2n}$ is basic.

Let us recall that locally connected continua without any simple closed curves are called *dendrites*, and the end-points of dendrites are the points of order one. Any dendrite embeds in the plane. Extending some results of Bowers, STERNFELD [1993b] proved the following

9.1. THEOREM. *Let D_i, $i = 1, \ldots, n$, be dendrites with dense sets of end-points. Then, for any compactum X with* $\dim X \le n$, *the set of basic embeddings of X into $\mathbb{I} \times \prod_{i=1}^{n} D_i$ is residual in the space of mappings of X into the product.*

Sternfeld showed also that the assertion of Theorem 9.1 characterizes compacta X with $\dim X \le n$ and that no n-dimensional compactum X, for $n \ge 2$, is basically embedded in any product of n dendrites. Some results concerning basic embeddings in the plane are discussed by REPOVŠ and SKOPENKOV [1999].

Interesting links between basic embeddings and the subject of Section 9 were explored by LEVIN and STERNFELD [1996], cf. also sec 14(K).

10. Transfinite dimensions

(A) The following question was formulated by Henderson, Kozlowski and Walsh in 1983.

10.1. PROBLEM. Is it possible to map a countable-dimensional compactum onto an uncountable-dimensional one by a mapping that is a hereditary shape equivalence?

A continuous map $f : X \to Y$ between compacta is a *hereditary shape equivalence* if for every closed A in Y, and each ANR-space Z, the restriction $f \mid f^{-1}(A) : f^{-1}(A) \to A$ induces a one-to-one correspondence between the homotopy classes of the mappings from A to Z and the homotopy classes of the mappings from $f^{-1}(A)$ to Z.

The problem is closely related to the question if hereditarily shape equivalences between countable-dimensional compacta can arbitrarily raise the transfinite dimensions, cf. DIJKSTRA and MOGILSKI [1997]. DIJKSTRA [1996] constructed the following relevant example.

10.2. EXAMPLE. There exists a hereditary shape equivalence $f : X \to Y$ between AR-compacta with $\operatorname{ind} X = \operatorname{Ind} X = \omega$ and $\operatorname{ind} Y = \operatorname{Ind} Y = \omega + 1$.

(B) We shall recall now Smirnov's *transfinite cubes* S_α and their modifications H_α by Henderson, playing a prominent role in the theory of infinite-dimensional spaces.

The compactum S_n is the Euclidean n-cube $\mathbb{I}^n$, $S_{\alpha+1} = S_\alpha \times \mathbb{I}$ and, for any limit $\alpha < \omega_1$, S_α is the one-point compactification of the free union of S_β with $\beta < \alpha$. Every S_α has countably many components, each being a closed cell. Henderson extended by transfinite induction every S_α to an AR-compactum H_α, and defined for each H_α a natural "boundary" ∂H_α. The "interior" $H_\alpha \backslash \partial H_\alpha$ is a disjoint union of open cells whose closures are the open components of Smirnov's compactum $S_\alpha \subset H_\alpha$. Henderson called a mapping $f : X \to H_\alpha$ essential if every mapping $g : X \to H_\alpha$ coinciding with f on $f^{-1}(\partial H_\alpha)$ is a surjection.

Let us associate to each compactum X the *Borst-Henderson* index $d_{BH}(X)$, letting $d_{BH}(X)$ be the supremum of ordinals α such that X admits an essential mapping onto H_α, if the supremum is countable, or $d_{BH}(X) = \infty$, otherwise. If $d_{BH}(X) \neq \infty$, then $d_{BH}(X) = \alpha$ where α is the maximal ordinal such that some compactum Z in X admits a mapping $f : Z \to S_\alpha$, which covers essentially each open component of S_α. In a slightly different setting, index $d_{BH}(X)$ was discussed by BORST [1988].

For any compactum X, $d_{BH}(X) \neq \infty$ means that X has no essential mappings onto $\mathbb{I}^\infty$, i.e., X is weakly infinite-dimensional. Henderson proved that for any compactum X, $d_{BH}(X) \leq \operatorname{Ind} X$. However, Dijkstra demonstrated, using his Example 10.2, that there is no characterization of transfinite dimension Ind in terms of essential mappings into fixed AR-compacta K_α with specified "boundaries" ∂K_α.

It is unknown, for what ordinals $\alpha < \omega_1$, whenever $f : X \to Y$ is a finite-to-one surjection and X is a compactum with $d_{BH}(X) \geq \alpha$, also $d_{BH}(Y) \geq \alpha$. This question is pertinent to Problem 3.1, cf. POL [1996b]. Related to this topic is the notion of the transfinite order of finite-to-one mappings between compacta, discussed by HATTORI and YAMADA [1998].

(C) For any Smirnov's compactum S_α, we have $\operatorname{Ind} S_\alpha = \alpha$, and also a natural transfinite extension of the covering dimension associates to S_α the ordinal α, cf. ENGELKING [1995], 7.3.N. However, the transfinite inductive dimensions may differ for Smirnov's compacta, and the evaluation of the transfinite dimension $\operatorname{ind} S_\alpha$ for all $\alpha < \omega_1$ is an interesting open problem. In this direction, CHATYRKO [1999] improved some earlier results of Luxemburg to the following effect.

10.3. THEOREM. *For any limit ordinal* $\lambda < \omega_1$, $\operatorname{ind}(S_{\lambda+2^m-1}) \leq \lambda + m$.

Chatyrko conjectures that for any limit $\lambda < \omega_1$, $\operatorname{ind}(S_{\lambda+2^m-1}) = \lambda + m$.

Some other noteworthy results of CHATYRKO and KOZLOV [2000] and CHATYRKO and HATTORI [2001] deal with transfinite dimensions of products. Among them is the evaluation

$$\operatorname{Ind}(S_\alpha \times S_\beta) = \alpha \oplus \beta, \quad \operatorname{ind}(S_\alpha \times S_\beta) = \operatorname{ind} S_{\alpha\oplus\beta},$$

where $\alpha \oplus \beta$ is the "natural sum" of ordinals, and the following theorem, where $\lambda(\alpha)$ is the maximal limit ordinal not greater than $\alpha < \omega_1$:

10.4. THEOREM. *Let X be a compactum with infinite* $\operatorname{ind} X = \alpha$ *such that the complement of the union of open sets U in X with* $\operatorname{ind} U < \lambda(\alpha)$ *is finite-dimensional. Then there exists n such that for any separable metrizable Y with* $n \leq \operatorname{ind} Y < \omega$, $\operatorname{ind}(X \times Y) < \operatorname{ind} X + \operatorname{ind} Y$.

It is also unknown if for every $\alpha < \omega_1$ there is a compactum X_α with $\operatorname{ind} X_\alpha = \operatorname{Ind} X_\alpha = \alpha$.

11. The gap between the dimensions

(A) We shall begin with metrizable spaces. Let us recall that ind and Ind coincide for metrizable separable spaces and dim and Ind are equal for any metrizable space.

The celebrated Roy space Δ is a completely metrizable space of weight $2^{\aleph_0}$ with $\operatorname{ind} \Delta = 0$ and $\operatorname{Ind} \Delta = 1$. Kulesza and Ostaszewski defined independently, about 1990, completely metrizable spaces of weight $\aleph_1$ with ind equal to 0, and Ind equal to 1. Kulesza's space K has a remarkably simple description. Let, for any ordinal $\alpha < \omega_1$, $\alpha = \lambda(\alpha) + n(\alpha)$, where $\lambda(\alpha)$ is the limit ordinal and $n(\alpha)$ is a nonnegative integer. Then the space K is the following subspace of the countable product ω_1^N of the space of countable ordinals with the order topology: a sequence $s : N \to \omega_1$ belongs to K if and only if $n(s(1)) > 0$ and, whenever $n(s(k)) = 0$, $\lambda(s(k+1)) = \lambda(s(k))$ and $n(s(k+i)) = k$ for $i = 1, 2, \ldots$ LEVIN [2000] gave a simplification of the original Kulesza's proof that $\operatorname{Ind} K > 0$.

Of great significance in this area are recent constructions of MRÓWKA [1997], [2000]. Mrówka defined a metrizable space M (in the original notation - $\mu\nu_0$) with $\operatorname{ind} M = 0$ and formulated a condition (S) which he used to prove that any completion of M^2 contains the Euclidean square. DOUGHERTY [1997] showed that (S) is consistent relative to a large cardinal, and conversely, consistency of (S) implies consistency of a large cardinal. In effect, under this hypothesis, Mrówka gave the first example of a metrizable space with the gap between ind and Ind greater than 1, which is also the first example of a metrizable space all whose completions increase the dimension ind of the space. This remarkable work was subsequently simplified by KULESZA [200?], who strengthened also Mrówka's results to the following effect: assuming (S), any completion of M^n contains $\mathbb{I}^n$.

Kulesza showed also that Mrówka's space M is N*-compact*, i.e., M embeds as a closed subspace into some product of natural numbers. Therefore, under (S), M is also the first metrizable N-compact space without N-compact completions.

MRÓWKA [2000] provides an assessment of some open problems in this topic. Let us recall in particular that no examples are known of metrizable groups G with $\operatorname{ind} G < \operatorname{Ind} G$.

(B) In another direction, DELISTATHIS and WATSON [2000] solved an outstanding open problem by constructing, using CH, a regular space X with $\dim X = 1$ and $\operatorname{ind} X = \operatorname{Ind} X = 2$, which is a continuous image of a separable metrizable space. No such examples are known in the realm of the usual set theory. It is also unclear if the space X is a quotient image of a separable metrizable space.

(C) We shall now turn to *n-manifolds*, i.e., connected Hausdorff spaces locally homeomorphic to $\mathbb{R}^n$. Evidently, for any n-manifold M, $\operatorname{ind} M = n$. However, answering a natural question, explicitly formulated by M.M.Postnikov in his textbook on differential geometry, FEDORCHUK and FILIPPOV [1992] constructed, using CH, normal manifolds with non-coinciding dimensions. In particular, they showed under CH, that there exists a normal, countably compact 3-manifold F with $\dim F = 3$ and $\operatorname{Ind} F = 4$. Subsequently, also using CH, FEDORČUK [1995] defined for any $m > 4$ a perfectly normal, separable 4-manifold F with $m - 1 \leq \dim F \leq m < \operatorname{Ind} F$. We refer the reader to FEDORCHUK [1998] for a comprehensive account on the dimension theory of nonmetrizable manifolds, including many interesting problems in this area.

12. Dimension-raising mappings with lifting properties

(A) EDWARDS - WALSH RESOLUTIONS AND DRANISHNIKOV'S SOLUTION OF ALEXANDROFF'S PROBLEM.

We have already indicated in Section 4 a great influence of WALSH [1981] on current research in dimension theory. Edwards and Walsh constructed, for each compactum X with $\dim_{\mathbb{Z}} X = n$, a cell-like surjection $f : \tilde{X} \to X$ defined on an n-dimensional compactum $\tilde{X}$. Let us recall that "cell-like" means that, upon embedding $\tilde{X}$ in $\mathbb{I}^\infty$, any fiber $f^{-1}(x)$ is contractible in each of its neighborhoods in $\mathbb{I}^\infty$, cf. also (D).

Dranishnikov combined the Edwards - Walsh approach with K-theory to get a spectacular solution of a fundamental Alexandroff's Problem, by constructing infinite-dimensional compacta X with $\dim_{\mathbb{Z}} X = 3$. Subsequently, DYDAK and WALSH [1993] constructed infinite-dimensional compacta X with $\dim_{\mathbb{Z}} X = 2$. In effect, one obtains also cell - like mappings of $\mathbb{I}^5$ onto infinite-dimensional compacta. As was demonstrated by Kozlowski and Walsh, there are no such mappings for $\mathbb{I}^3$, but the case of $\mathbb{I}^4$ remains open. Some recent cohomological refinements of the Edwards - Walsh construction are given by KOYAMA and YOKOI [2001].

(B) MAPPING SOFTLY MENGER COMPACTA ONTO THE HILBERT CUBE.

We shall say that $f : X \to Y$ is a *soft mapping* with respect to a class $\mathcal{C}$ of pairs of spaces (K, L), L closed in K, if for any $(K, L) \in \mathcal{C}$ and mappings $u : L \to X$, $v : K \to Y$ satisfying $f \circ u = v$, there is a continuous extension $\overline{u} : K \to X$ of u with $f \circ \overline{u} = v$.

We shall denote by μ_n the *Menger universal compactum* M_n^{2n+1}, cf. ENGELKING [1995], 1.11.

DRANISHNIKOV [1986] constructed continuous surjections $d_n : \mu_n \to \mathbb{I}^\infty$ with many softness properties, which became valuable tools in the dimension theory. Certain additional properties of the Dranishnikov resolutions were established by AGEEV, REPOVŠ and SHCHEPIN [1996].

We shall describe below some of the noteworthy properties of the maps d_n. We shall call a pair (K, L) of compacta n-admissible, if for any pair of continuous maps $u : L \to \mathbb{S}^n \times \mathbb{S}^n$, $v : K \to \mathbb{S}^n$ with $proj \circ u = v$, there is a continuous extension $\overline{u} : K \to \mathbb{S}^n \times S^n$ of u with $proj \circ \overline{u} = v$ and $\overline{u}^{-1}(s, s) \subset L$ for each $s \in \mathbb{S}^n$.

The pairs (K, L) with $\dim L \leq n - 2$ and $\dim K \leq n$, or $\dim K \leq n - 1$, or L being an $(n - 1)$-dimensional ANR-compactum, are n-admissible.

Ageev, Repovš and Shchepin showed that Dranishnikov's resolutions $d_n : \mu_n \to \mathbb{I}^\infty$ are soft with respect to the class of n-admissible pairs of compacta. They proved also that there exists a G_δ-set $C_n \subset \mu_n$, dense in each fiber of d_n, such that the restriction $d_n \mid C_n : C_n \to \mathbb{I}^\infty$ is soft with respect to the pairs (K, L) of completely metrizable separable spaces with $\dim K \leq n$ and L closed in K. Resolutions with properties similar to that of $d_n \mid C_n$ were first constructed by Chigogidze.

It is worth noticing that the idea of the construction of the resolution d_n was based on a very ingenious geometric proof by KOZLOVSKIĬ [1986] (announced in 1982) of the Keldysh theorem that the square is an image of a curve under an open light mapping.

(C) MAPPING SOFTLY σ-COMPACT COUNTABLE-DIMENSIONAL SPACES ONTO THE HILBERT CUBE.

ZARICHNYĬ [1995] obtained the following remarkable result: there exists a continuous surjection $f : X \to \mathbb{I}^\infty$, where $X \in AR$(metrizable) is a countable union of finite-dimensional compacta, which is soft with respect to the pairs (K, L) such that K is a countable union of finite-dimensional compacta and L is closed in K.

The ideas of Zarichnyĭ's construction were used by CAUTY [2001] in his proof of the famous Schauder conjecture asserting that any continuous map $f : C \to C$ of a convex set in a linear topological space with $\overline{f(C)}$ compact, has a fixed point.

Some other interesting applications of the Zarichnyĭ resolution can be found in BANAKH, RADUL and ZARICHNYĬ [1996]

(D) DIMENSION - RAISING UV^n - MAPPINGS.

A mappings $f : X \to Y$ between compacta is a *UV^n-mapping* if, upon embedding X in the Hilbert cube, for any neighborhood U of an arbitrary fiber $f^{-1}(y)$ there is a smaller neighborhood V of $f^{-1}(y)$ such that the inclusion of V into U induces trivial homomorphisms of the homotopy groups in dimensions $\leq n$. We refer the reader to DAVERMAN [1986], Section 16, for some important (approximate) lifting properties of UV^n-mappings.

ČERNAVSKIĬ [1985] gave a beautiful construction of UV^k-mappings of any cube $\mathbb{I}^n$ with $2k + 3 \leq n$ onto each cube of higher dimension. For $k = 0$ this yields the famous Keldysh theorem that $\mathbb{I}^3$ can be mapped onto $\mathbb{I}^4$ by a map with connected fibers.

Černavskiĭ's ideas were employed by Ferry in his elegant proof of the Bestvina–Walsh–Wilson theorem that any mapping of $\mathbb{S}^n$ into itself is homotopic to a UV^k-mapping, whenever $2k + 3 \leq n$.

DRANISHNIKOV [1987] proved (using methods different from the ones by Černavskiĭ) that any n-connected LC^n-compactum can be represented as an image of a UV^{n-1}-mapping defined on $\mathbb{I}^{2n+1}$, or on $\mathbb{I}^\infty$, or else on an $(n + 1)$-dimensional AR-compactum.

Earlier, Bestvina established similar results, where the domains of the UV^{n-1}-mappings were the Menger compacta M^{2n+3}_{n+1}. The Bestvina resolutions play an essential role in the theory of Menger manifolds, cf. Section 13(B).

13. Universal spaces

We say that X is a *universal space* for a class $\mathcal{C}$ of topological spaces if $X \in \mathcal{C}$ and any $Y \in \mathcal{C}$ embeds in X.

(A) UNIVERSAL SPACES IN THE EXTENSION THEORY.

The following theorem was proved by OLSZEWSKI [1995a] for separable spaces, and LEVIN [200?] in the general case.

13.1. THEOREM. *For any countable CW complex K and an infinite cardinal κ, the class of metrizable space X with $X\tau K$ and density $\leq \kappa$ contains a complete universal space.*

This result was refined by CHIGOGIDZE and VALOV [2001] to the following effect: given a countable CW complex and an infinite cardinal κ, there is a surjection $f : M \to l_2(\kappa)$ onto the Hilbert space of density κ which is soft with respect to the pairs (A, B), where A is metrizable, $A\tau K$ and B is closed in A, cf. Section 12(B).

The mapping f has also the property that for any complete metric space Z with density κ and $Z\tau K$, any mapping $u : Z \to M$ and any open cover $\mathcal{U}$ of M, there exists an embedding $v : Z \to M$ onto a closed subset, which is $\mathcal{U}$-close to g and satisfies $f \circ u = f \circ v$.

For the Eilenberg - Mac Lane complexes $K(\mathbb{Z}, n)$, a universal space in the class of separable metrizable spaces X with $X\tau K(\mathbb{Z}, n)$, i.e., such that $\dim_{\mathbb{Z}} X \leq n$, was constructed earlier by DYDAK and MOGILSKI [1994]. A key role in their construction was played by a theorem of RUBIN and SCHAPIRO [1987], extending the Edwards - Walsh resolution (cf. 12(A)) to non-compact spaces.

The question of existence of a universal space in the class of compacta with a given integral cohomological dimension is open.

A general problem of characterizing CW complexes K for which the class of separable metrizable spaces X with $X\tau K$ has a universal space is discussed by CHIGOGIDZE [2000].

(B) TOPOLOGICAL CHARACTERIZATION OF UNIVERSAL NÖBELING SPACES.
The Nöbeling spaces N_n^{2n+1}, cf. ENGELKING [1995], 1.11, have the following properties:

(i) N_n^{2n+1} is an n-dimensional completely metrizable separable absolute extensor for the class of n-dimensional separable metrizable spaces,

(ii) for any open cover $\mathcal{U}$ of N_n^{2n+1} and any mapping $f : X \to N_n^{2n+1}$ there is a $\mathcal{U}$-close to f embedding $h : X \to N_n^{2n+1}$ onto a closed subspace.

KAWAMURA, LEVIN and TYMCHATYN [1997] proved that for $n = 1$ the properties (i) and (ii) characterize topologically N_1^3. It was announced recently by Ageev that also for arbitrary n, properties (i) and (ii) determine N_n^{2n+1} up to a homeomorphism, cf. PEARL [2000], Problem 608.

Let us recall that topological characterizations of Menger universal compacta M_n^{2n+1} have been obtained in the early eighties by Bestvina, cf. CHIGOGIDZE, KAWAMURA and TYMCHATYN [1995]. The Nöbeling spaces can be considered in a natural way as pseudo-interiors of the corresponding Menger compacta, which provides some essential information about their structure, cf. CHIGOGIDZE, KAWAMURA and TYMCHATYN [1996].

14. Miscellaneous topics

(A) COLORING OF MAPPINGS.
Any fixed-point free autohomeomorphism $f : X \to X$ of an n-dimensional paracompact space can be colored with $n + 3$ colors, i.e., there are closed sets $A_1, \ldots, A_{n+3}$ covering X such that $f(A_i) \cap A_i = \emptyset$, $i = 1, \ldots, n + 3$. This interesting theorem was established by AARTS, FOKKING and VERMEER [1996] for metrizable spaces and proved in full

generality by VAN HARTSKAMP and VERMEER [1996]. The book by VAN MILL [2001] contains an exposition of this topic.

(B) ALMOST n-DIMENSIONAL SPACES.
LEVIN and TYMCHATYN [1999] call a separable metrizable space X *almost n-dimensional* if X has a countable base $\mathcal{B}$ such that each pair of elements of $\mathcal{B}$ with disjoint closures can be separated in X by at most $(n-1)$-dimensional closed set. Levin and Tymchatyn showed that for $n \geq 1$, weakly n-dimensional spaces are at most n-dimensional. Using this result for $n = 1$, they answered an old question of R.Duda, to the effect that any separable metrizable space whose all connected subsets are locally connected is 1-dimensional.

Earlier, OVERSTEEGEN and TYMCHATYN [1994] proved that almost 0-dimensional spaces are at most 1-dimensional and used this fact to demonstrate that the space of autohomeomorphisms of the Menger compactum M_n^k, $k > n$, is 1-dimensional, cf. also BRECHNER and KAWAMURA [2001] and VAN MILL [2001].

(C) COMPACTIFICATIONS AND THE COMPACTNESS DEGREE.
A space X is *rim-compact* if X has a base whose elements have compact boundaries. AARTS and COPLAKOVA [1993] solved a problem of Isbell, constructing a Tychonoff rim-compact space such that for any compactification $Y$of X, $\dim(Y \setminus X) \geq 1$.

Let us recall that the topological invariant *cmp* is a counterpart of the inductive dimension ind, where at the level 0, the class of rim-compact spaces replaces the class of spaces with a base consisting of clopen sets. It is an interesting problem, stated by de Groot and Nishiura, what is $\mathrm{cmp} J_n$, where J_n is the n-cube with removed geometric interior of the top. CHATYRKO and HATTORI [200?] proved that $\mathrm{cmp} J_n \leq m+1$, provided $n \leq 2^m - 1$. In particular, for $n \geq 5$, $\mathrm{cmp} J_n < \mathrm{def} J_n$, def being the minimal dimension of the remainder of a compactification, cf. AARTS and NISHIURA [1993].

Another noteworthy result concerning compactifications is a construction by CHARALAMBOUS [1997] of a normal space X with $\operatorname{ind} X = 0$, no compactification (in fact, no Lindelöf extension) of which has transfinite dimension ind.

(D) EMBEDDINGS INTO PRODUCTS OR TOPOLOGICAL GROUPS.
KOYAMA, KRASINKIEWICZ and SPIEŻ [200?] strengthened a result of Borsuk to the effect that the 2-sphere can not be embedded into the symmetric product of any 1-dimensional continuum, cf. ILLANES and NADLER [1999], Question 83.14.

KULESZA [1993] proved that the complete bipartite graph does not embed in any 1-dimensional group, and that any group which contains a copy of the hedgehog with $\aleph_1$ spines must be infinite-dimensional.

(E) DARBOUX PROPERTY AND DIMENSION.
A real-valued function $f : X \to \mathbb{R}$ has *Darboux property* if f maps connected sets onto connected sets. The following result was obtained by CIESIELSKI and WOJCIECHOWSKI [2001]: for any σ-compact metrizable X, the dimension of X is the minimal non-negative integer n such that any real-valued function on X is the algebraic sum of at most $(n+1)$ Darboux functions.

(F) THE n-DIMENSIONAL KERNEL.
Given an n-dimensional compactum X, let $K(X)$ be the union of all n-dimensional Cantor manifolds in X, cf. ALEXANDROV and PASYNKOV [1973]. JACKSON and MAULDIN

[1992] showed that for each n there is an n-dimensional compactum X in $\mathbb{R}^{n+2}$ for which $K(X)$ is a projection of a coanalytic set, i.e., Σ_2^1-set, but is not in a lower projective class. It is not known, however, if for any compactum X in $\mathbb{R}^3$, $K(X)$ must be analytic.

(G) EXTENSION PROPERTIES AND METRIZABILITY.
Let us write $X \in ANE(Y)$ if any mapping $u : A \to X$, with A closed in Y, extends to a continuous map $\overline{u} : U \to X$ over a neighborhood of A.

CHIGOGIDZE and ZARICHNYĬ [1998] proved that for any connected, non-contractible, countable CW complex K, if X is a compact space with $X \tau K$, and $X \in ANE(Y)$ for any compact $Y \tau K$, then X is metrizable.

For $K = \mathbb{S}^n$ this was established earlier by Dranishnikov, which in turn extended Shchepin's theorem that finite-dimensional compact ANE-spaces are metrizable.

(H) ON A THEOREM OF ČERNAVSKIĬ.
Let $f : M \to N$ be an open mapping with discrete fibers between n-manifolds, and let B_f be the set of branching points of f, i.e., the points in M at which f is not a local homeomorphism.

ČERNAVSKIĬ [1964] proved that $\dim B_f \leq n - 2$. He also conjectured that, whenever $B_f \neq \emptyset$, the upper bound is attained. This turned out not to be the case. However, recently MARTIO and RIAZANOV [1999] proved that, whenever $n \geq 2$, $n \neq 3$, and $B_f \neq \emptyset$, we have $\mathrm{dem} B_f = n - 2$, where dem stands for the *embedding dimension* introduced by Štanko.

(I) CLOSED IMAGES OF LOCALLY COMPACT METRIZABLE SPACES.
Let $L(n)$ be the class of n-dimensional closed images of locally compact metrizable separable spaces. KAWAMURA and TSUDA [1998] constructed an universal space in the class $L(n)$. Assuming GCH, the authors proved analogous results for the classes $L_\kappa(n)$, where the separability is replaced by density $\leq \kappa$ (the case $n = 0$ required no extra axioms).

(J) CANTOR MANIFOLDS FOR TRANSFINITE DIMENSION IND.
We say that a compactum X with $\mathrm{Ind}\, X = \alpha + 1$ is a *Cantor manifold* if for any closed set L separating X, $\mathrm{Ind} L \geq \alpha$. The first Cantor manifolds for each $\alpha + 1$ were constructed by OLSZEWSKI [1994]. REŃSKA [2001] gave simpler constructions to the following effect: for any infinite $\alpha < \omega_1$ there is a Cantor manifold X with $\mathrm{Ind}\, X = \alpha + 1$ such that X is a disjoint union of countably many closed cells and irrationals.

(K) BASIC EMBEDDINGS AND $C_p(X)$.
Let $C_p(X)$ be the space of continuous real-valued functions on the compactum X, endowed with the pointwise topology. If X is basically embedded in the product $X_1 \times \ldots \times X_m$ of compacta, formula $(\star)$ in Section 9 defines a linear continuous surjection $L : C_p(X_1 \oplus \ldots \oplus X_m) \to C_p(X)$. LEIDERMAN, LEVIN and PESTOV [1997] proved that, for $m = 2$, the map L is open (it is unclear if this is true for $m > 2$), and they derived from this result that for every n-dimensional compactum Y there exists a 2-dimensional compactum X and an open linear surjection from $C_p(X)$ onto $C_p(Y)$.

Let us recall that, by a result of Gul'ko, a uniform homeomorphism between $C_p(X)$ and $C_p(Y)$ yields $\dim X = \dim Y$.

References

AARTS, J.M. and E. COPLAKOVA
[1993] The dimension of remainders of rim-compact spaces, *Fund. Math.* **143**, 287–289.

AARTS, J.M., R.J. FOKKING and J. VERMEER
[1996] Variation on a theorem of Lusternik and Schnirelmann, *Topology* **35**, 1051–1056.

AARTS, J.M. and T. NISHIURA
[1993] *Dimension and Extensions*, North Holland, Amsterdam.

AGEEV, S.M., D. REPOVŠ and E.V. SHCHEPIN
[1996] On the softness of the Dranishnikov resolution (Russian), *Tr. Mat. Inst. Steklova* **212**, 7–32; *translation in Proceedings of the Steklov Inst. Math.* **212**, 3–27.

ALEXANDROFF, P.S. and B.A. PASYNKOV
[1973] *Introduction to dimension theory* (Russian), Nauka, Moskva.

ANCEL, F.D. and T. DOBROWOLSKI
[1997] On the Sternfeld-Levin counterexamples to a conjecture of Chogoshvili-Pontrjagin, *Topology Appl.* **80**, 7–19.

BALL, R.N., J.N. HAGLER and Y. STERNFELD
[1998] The structure of atoms (hereditarily indecomposable continua), *Fund. Math.* **156**, 261–278.

BANAKH, T., T. RADUL and M. ZARICHNYĬ
[1996] *Absorbing Sets in Infinite-Dimensional Manifolds*, VNTL Publishers, Lviv.

BOGATYI, S.A.
[1998] Geometry of mappings into Euclidean space (Russian), *Uspekhi Matem. Nauk* **53** (5), 27–72; *translation in Russian Math. Surveys* **53** (5), 893–920.

BORST, P.
[1988] Classification of weakly infinite-dimensional spaces, Part I: A transfinite extension of the covering dimension, *Fund. Math.* **130**, 1–25.

BRECHNER, B. and K. KAWAMURA
[2001] On the dimension of a homeomorphism group, *Proc. Amer. Math. Soc.* **129**, 617–620.

CAUTY, R.
[2001] Solution du problème de point fixe de Schauder (French), [Solution of Schauder's fixed point problem], *Fund. Math.* **170**, 231–246.

ČERNAVSKIĬ, A.V.
[1964] Finite-to-one open mappings on manifolds (Russian), *Mat. Sb.* **65**, 357–369.
[1985] Generalization of L. V. Keldysh's construction of monotone mappings of a cube onto a higher dimensional cube (Russian), *Uspiekhi Mat. Nauk* **40** (4), 209–211.

CHARALAMBOUS, M.G.
[1997] A normal space Z with $\operatorname{ind} Z = 1$ no compactification of which has transfinite dimension, *Topology Proceedings* **22**, 95-101.

CHATYRKO, V.A.
[1991] Weakly infinite-dimensional spaces (Russian), *Uspekhi Mat. Nauk* **46**, 161–177; *translation in Russian Math. Surveys* **46**, 191–210.
[1999] On finite sum theorems for transfinite inductive dimensions, *Fund. Math.* **162**, 91–98.

CHATYRKO, V.A. and Y. HATTORI

[2001] Estimations of small transfinite dimension in separable metrizable spaces, *Tsukuba J. Math.* **25**, 221–228.

[200?] On a question of de Groot and Nishiura, *Fund. Math.*, to appear.

CHATYRKO, V.A. and K.L. KOZLOV

[2000] On (transfinite) small inductive dimension of products, *Comment. Math. Univ. Carolin.* **41**, 597-603.

CHIGOGIDZE, A.

[1998] *Inverse spectra*, North-Holland, Amsterdam.

[2000] Compactifications and universal spaces in extension theory, *Proc. Amer. Math. Soc.* **128**, 2187–2190.

CHIGOGIDZE, A. and V.V. FEDORCHUK

[1992] *Absolute Retracts and Infinite-Dimensional Manifolds* (Russian), Nauka, Moscow.

CHIGOGIDZE, A., K. KAWAMURA and E.D. TYMCHATYN

[1995] Menger Manifolds, in: *Continua with the Houston problem book*, ed. H. Cook et al., Marcel Dekker Inc., New York, 37–88.

[1996] Nöbeling spaces and pseudo-interiors of Menger compacta, *Topology Appl.* **68**, 33–65.

CHIGOGIDZE, A. and V. VALOV

[2001] Universal metric spaces and extension dimension, *Topology Appl.* **113**, 23–27.

CHIGOGIDZE, A. and M.M. ZARICHNYĬ

[1998] On absolute extensors modulo a complex, *Topology Appl.* **86**, 169–178.

CHOGOSHVILI, G.

[1938] On a theorem in the theory of dimensionality, *Compositio Math.* **5**, 292–298.

CIESIELSKI, K. and J. WOJCIECHOWSKI

[2001] Topological dimension and sums of connectivity functions, *Topology Appl.* **112**, 181–191.

DAVERMAN, R.J.

[1986] *Decompositions of manifolds*, Academic Press, New York 1986.

DELISTATHIS, G. and S. WATSON

[2000] A regular space with a countable network and different dimensions, *Trans. Amer. Math. Soc.* **352**, 4095–4111.

DIJKSTRA, J.J.

[1996] A dimension raising hereditary shape equivalence, *Fund. Math.* **149**, 265–274.

DIJKSTRA, J.J. and J. MOGILSKI

[1997] Countable dimensionality and dimension raising cell-like maps, *Topology Appl.* **80**, 73–79.

DOBROWOLSKI, T., M. LEVIN and L. RUBIN

[1997] Certain 2-stable embeddings, *Topology Appl.* **80**, 81-90.

DOUGHERTY, R.

[1997] Narrow coverings of ω-ary product spaces, *Annals of Pure and Applied Logic* **88**, 47–91.

DRANISHNIKOV, A.N.

[1986] Universal Menger compacta and universal mappings (Russian), *Mat. Sbornik* **129** (1), 121 - 139.

[1987] On resolutions of LC^n compacta, *Lecture Notes in Math.* **1283**, 49-59.

[1988] Homological dimension theory, *Uspekhi Mat. Nauk* **43**, 11–55.

[1990] A fibration that does not accept two disjoint many-valued sections, *Topology Appl.* **35**, 71–73.

[1992] Stable cohomotopy dimension and weakly infinite-dimensional spaces, *Topology Appl.* **47**, 79–81.

[1996] On the mapping intersection problem, *Pacific J. Math.* **173**, 403–412.

[1997] On Chogoshvili's conjecture, *Proc. Amer. Math. Soc.* **125**, 2155–2160.

[1998] On the theory of extensions of mappings of compact spaces (Russian), *Uspekhi Mat. Nauk* **53**, 65–72; *translation in Russian Math. Surveys* **53**, 929–935.

[2000] On the dimension of the product of two compacta and the dimension of their intersection in general position in Euclidean space, *Trans. Amer. Math. Soc.* **352**, 5599–5618.

DRANISHNIKOV, A.N. and J. DYDAK

[2001] Extension theory of separable metrizable spaces with applications to dimension theory, *Trans. Amer. Math. Soc.* **353**, 133–156.

DRANISHNIKOV, A.N. and J. KEESLING

[2001] The countable extension basis theorem and its applications, Geometric topology: Dubrownik 1998, *Topology Appl.* **113**, 29–38.

DRANISHNIKOV, A.N., D. REPOVŠ and E.V. SHCHEPIN

[1993] Dimension of products with continua, *Topology Proceedings* **18**, 57–73.

DRANISHNIKOV, A.N. and E.V. SHCHEPIN

[1986] Cell-like mappings. The problem of raising the dimension (Russian), *Uspekhi Mat. Nauk* **41**, 49–90.

DRANISHNIKOV, A.N. and V.V. USPENSKIĬ

[1988] Light maps and extensional dimension, *Topology Appl.* **80**, 91–99.

DYDAK, J.

[1994] Union theorem for cohomological dimension: a simple counterexample, *Proc. Amer. Math. Soc.* **121**, 295–297.

[1996] Extension Theory: the interface between set-theoretic and algebraic topology. Proceedings of the International Conference on Set-theoretic Topology and its Applications (Matsuyama, 1994), *Topology Appl.* **74**, 225–258.

DYDAK, J. and J. MOGILSKI

[1994] Universal cell - like maps, *Proc. Amer. Math. Soc.* **122**, 943–948.

DYDAK, J. and J.J. WALSH

[1991] Spaces without cohomological dimension preserving compactifications, *Proc. Amer. Math. Soc.* **113**, 1155–1162.

[1993] Infinite-dimensional compacta having cohomological dimension two: an application of the Sullivan conjecture, *Topology* **32**, 93–104.

ENGELKING, R.

[1995] *Theory of Dimensions, Finite and Infinite*, Helderman Verlag, Lemgo.

ENGELKING, R. and E. POL

[1983] Countable-dimensional spaces: a survey, *Dissertationes Math.* **216**.

FEDORCHUK, V.V.

[1995] A differentiable manifold with noncoinciding dimensions for the continuum hypothesis (Russian), *Matem. Sb.* **186**, 149–160; *translation in Sb. Math.* **186**, 151–162.

[1998] The Urysohn identity and the dimension of manifolds (Russian), *Uspekhi Matem. Nauk* **53**, 73–113; *translation in Russian Math. Surveys* **53**, 937–974.

FEDORCHUK, V.V. and V.V. FILIPPOV

[1992] Manifolds with noncoinciding inductive dimensions (Russian), *Matem. Sb.* **183**, 29–44; *translation in Russian Acad. Sci. Sb. Math.* **77**, 25–36.

GUTEV, V. and V. VALOV

[2002] Continuous selections and C-spaces, *Proc. Amer. Math. Soc.* **130**, 233–242.

HART, K.P., J. VAN MILL and R. POL

[200?] Remarks on hereditarily indecomposable continua, *Topology Proceedings*, to appear.

VAN HARTSKAMP, M. and J. VERMEER

[1996] On coloring of maps, *Topology Appl.* **73**, 181–190.

HATTORI, Y.

[1994] Dimension and products of topological groups, *Yokohama Math. J.* **42**, 31–40.

HATTORI, Y. and K. YAMADA

[1998] Finite-to-one mappings and large transfinite dimension, *Topology Appl.* **82**, 181–194.

HAVER, W.E.

[1974] A covering property for metric spaces, *Lecture Notes in Math.* **375**, 108–113.

ILLANES, A. and S.B. NADLER

[1999] *Hyperspaces: Fundamentals and Recent Advances*, Marcel Dekker, New York.

JACKSON, S. and R.D. MAULDIN

[1992] Some complexity results in topology and analysis, *Fund. Math.* **141**, 75–83.

KARNO, Z. and J. KRASINKIEWICZ

[1989] On some famous examples in dimension theory, *Fund. Math.* **143**, 213-220.

KATO, H. and M. LEVIN

[2000] Open maps on manifolds which do not admit disjoint closed subsets intersecting each fiber, *Topology Appl.* **103**, 221–228.

KAWAMURA, K., M. LEVIN and E.D. TYMCHATYN

[1997] A characterization of 1-dimensional Nöbeling spaces, *Topology Proc.* **22**, Summer, 155–174.

KAWAMURA, K. and K. TSUDA

[1998] Universal spaces for finite-dimensional closed images of locally compact metric spaces, *Topology Appl.* **85**, 175–198.

KOYAMA, A., J. KRASINKIEWICZ and S. SPIEŻ

[200?] On embeddings of compacta into Cartesian products of compacta, preprint.

KOYAMA, A. and K. YOKOI

[2001] On Dranishnikov's cell-like resolutions, *Topology Appl.* **113**, 87–106.

KOZLOVSKIĬ, I.M.

[1986] Dimension-raising open mappings of compacta onto polyhedra as spectral limits of inessential mappings of polyhedra, (Russian), *Dokl. Akad. Nauk SSSR* **286** 535–538. English transl. in *Soviet Math. Dokl.* **33**, 118–121.

KRASINKIEWICZ, J.
[1989] Imbeddings into $\mathbb{R}^n$, *Fund. Math.* **133**, 247–253.
[1996] On mappings with hereditarily indecomposable fibers, *Bull. Pol. Acad. Sci.* **44**, 147–156.

KULESZA, J.
[1993] Spaces which do not embed in topological group of the same dimension, *Topology Appl.* **50**, 139–145.
[1996] The dimension of X^n where X is a separable metric space, *Fund. Math.* **150**, 43–54.
[200?] Some new properties of Mrówka's space $\nu\mu_0$, *Proc. Amer. Math. Soc.*, to appear.

LEIDERMAN, A., M. LEVIN and V. PESTOV
[1997] On linear continuous open surjections of the spaces $C_p(X)$, *Topology Appl.* **81**, 269–279.

LEVIN, M.
[1995a] Inessentiality with respect to subspaces, *Fund. Math.* **147**, 93–98.
[1995b] A short construction of hereditarily infinite-dimensional compacta, *Topology Appl.* **65**, 97–99.
[1996] Bing maps and finite-dimensional maps, *Fund. Math.* **151**, 47–52.
[2000] A remark on Kulesza's example, *Proc. Amer. Math. Acad.* **128**, 623–624.
[200?] On extension dimension of metrizable spaces, preprint.

LEVIN, M. and W. LEWIS
[200?] Some mapping theorems for extensional dimension, preprint.

LEVIN, M. and J.T. ROGERS, JR
[2000] A generalization of Kelley's theorem for C-spaces, *Proc. Amer. Math. Soc.* **128**, 1537–1541.

LEVIN, M., L. RUBIN and P.J. SCHAPIRO
[2000] The Mardešić factorization theorem for extension theory and C-separation, *Proc. Amer. Math. Soc.* **128**, 3099–3106.

LEVIN, M. and Y. STERNFELD
[1996] Monotone basic embeddings of hereditarily indecomposable continua, *Topology Appl.* **68**, 241–249.
[1997] The space of subcontinua of a 2-dimensional continuum is infinite dimensional, *Proc. Amer. Math. Soc.* **125**, 2771–2775.

LEVIN, M. and E.D. TYMCHATYN
[1999] On the dimension of almost n-dimensional spaces, *Proc. Amer. Math. Soc.* **127**, 2793–2795.

MARTIO, O. and V. RIAZANOV
[1999] On Černavskiĭ's theorem (Russian), *Dokl. Akad. Nauk* **368**, 595–598.

MICHAEL, E.
[1990] Some problems, *Open problems in Topology*, J. van Mill and J.M. Reed (Editors), North Holland, Amsterdam, 271–278.

VAN MILL, J.
[1989] *Infinite-Dimensional Topology*, North-Holland, Amsterdam.
[2001] *The Infinite-Dimensional Topology of Function Spaces*, North-Holland, Amsterdam.

MRÓWKA, S.
[1997] Small inductive dimension of completions of metric spaces, *Proc. Amer. Math. Soc.* **125**, 1545–1554.
[2000] Small inductive dimension of completions of metric spaces, II, *Proc. Amer. Math. Soc.* **128**, 1247–1256.

OLSZEWSKI, W.
[1994] Cantor manifolds in the theory of transfinite dimension, *Fund. Math.* **145**, 39–64.
[1995a] Universal separable metrizable spaces of given cohomological dimension, *Topology Appl.* **61**, 293–299.
[1995b] Completion theorem for cohomological dimensions, *Proc. Amer. Math. Soc.* **123**, 2261–2264.

OVERSTEEGEN, L.G. and E.D. TYMCHATYN
[1994] On the dimension of certain totally disconnected spaces, *Proc. Amer. Math. Soc.* **122**, 885–891.

PASYNKOV, B.A.
[1975] On dimension and geometry of mappings (Russian), *Dokl. AN SSSR* **221**, 543–546.

PEARL, E.
[2000] Open problems in topology, seventh status report, *Topology Appl.* **114**, 333–352.

POL, E.
[1993] Spaces whose nth power is weakly infinite-dimensional but whose $(n+1)$th power is not, *Proc. Amer. Math. Soc.* **117**, 871–876.

POL, E. and M. REŃSKA
[200?] On Bing points in infinite-dimensional hereditarily indecomposable continua, *Topology Appl.*, to appear.

POL, R.
[1996a] On light mappings without perfect fibers on compacta, *Tsukuba J. Math.* **20**, 11–19.
[1996b] Finite-to-one mappings on infinite-dimensional compacta, *Ann. New York Acad. Scien.* **788**, 175–183.

POL, R. and M. REŃSKA
[2002] On the dimensional structure of hereditarily indecomposable continua, *Trans. Amer. Math. Soc.* **354**, 2921–2932.

REŃSKA, M.
[2001] On Cantor manifolds for the large transfinite dimension, *Topology Appl.* **112**, 1–11.

REPOVŠ, D. and A.B. SKOPENKOV
[1999] New results on embeddings of polyhedra and manifolds in Euclidean spaces, *Uspekhi Mat. Nauk* **54**, 61–108; *translation in Russian Math. Surveys* **54**, 1149–1196.

RUBIN, L. and P.J. SCHAPIRO
[1987] Cell-like maps onto non-compact spaces of finite-dimensional cohomological dimension, *Topology Appl.* **27**, 221–244.
[1999] Limit theorem for inverse sequences of metric spaces in extension theory, *Pacific J. Math.* **187**, 177–186.

SHCHEPIN, E.V.
[1998] Arithmetic in dimension theory (Russian), *Uspekhi Mat. Nauk* **53**, 115–212; *translation in Russian Math. Surveys* **53**, 975–1069.

SEGAL, J. and T. WATANABE
[1991] Universal maps and infinite-dimensional spaces, *Bull. Acad. Pol. Sci.* **39**, 225–228.

SHAKHMATOV, D.

[1990] A survey of current researchs and open problems in the dimension theory of topological groups, *Q & A in General Topology* **8**, 101–126.

SPIEŻ, S.

[1990] The structure of compacta satisfying $\dim(X \times X) < 2 \dim X$, *Fund. Math.* **135**, 127–145.

SPIEŻ, S. and H. TORUŃCZYK

[1991] Moving compacta in $\mathbb{R}^n$ apart, *Topology Appl.* **41**, 193–204.

STERNFELD, Y.

[1989] Hilbert's 13th problem and dimension, *Geometric aspects of functional analysis (1987-88)*, 1–49, *Lecture Notes in Math.* **1376**, Springer, Berlin.

[1993a] Stability and dimension – a counterexample to a conjecture of Chogoshvili, *Trans. Amer. Math. Soc.* **340**, 243–251.

[1993b] Mappings in dendrites and dimension, *Houston J. Math.* **19**, 483–497.

[1995] On finite-dimensional maps and other maps with "small" fibers, *Fund. Math.* **147**, 127–133.

[1996] A short elementary proof of the Dranishnikov–West theorem on stable intersection of compacta in Euclidean spaces, Proceedings of the International Conference on Set-theoretic Topology and its Applications (Matsuyama, 1994), *Topology Appl.* **74**, 177–178.

TORUŃCZYK, H.

[1985] Finite-to-one restrictions of continuous functions, *Fund. Math.* **125**, 237–249.

USPENSKIĬ, V.V.

[1998] A selection theorem for C-spaces, 8th Prague Topological Symposium on General Topology and its Relations to modern Analysis and Algebra (1996), *Topology Appl.* **85**, 351–374.

[2000] A note on a question of R. Pol concerning light maps, *Topology Appl.* **103**, 291–294.

WALSH, J.J.

[1981] Dimension, cohomological dimension, and cell-like mappings, *Lecture Notes in Math.* **870**, 105–118.

ZARICHNYĬ, M.M.

[1995] Universal map of σ onto Σ and absorbing sets in the classes of absolute Borelian and projective finite-dimensional spaces, *Topology Appl.* **67**, 221–230.

CHAPTER 16

Continuous Selections of Multivalued Mappings

Dušan Repovš[1]

Institute of Mathematics, Physics and Mechanics, University of Ljubljana
Jadranska 19, P. O. Box 2964 Ljubljana, Slovenia 1001
E-mail: dusan.repovs@uni-lj.si

Pavel V. Semenov[2]

Department of Mathematics, Moscow City Pedagogical University
2nd Sel'skokhozyastvennyi pr. 4, Moscow, Russia 129226
E-mail: pavels@orc.ru

Contents

[1] Supported in part by the Ministry of Education, Science and Sport of the Republic of Slovenia research program No. 0101-509.
[2] Supported in part by the RFBR research grant No. 02-01-00014.

RECENT PROGRESS IN GENERAL TOPOLOGY II
Edited by Miroslav Hušek and Jan van Mill

In this paper we have collected a selection of recent results of theory of continuous selections of multivalued mappings. We have also considered some important applications of these results to other areas of mathematics. The first three parts of the paper are devoted to convex-valued mappings, to selectors on hyperspaces, and to links between selection theory for LSC mappings and approximation theory for USC mappings, respectively. The fourth part includes various other results.

Since our recent book REPOVŠ and SEMENOV [1998a] comprehensively covers most important work in this area approximately until the mid 1990's, we have therefore decided to focus in this survey on results which have appeared since then. As is often the case with surveys, due to the limitations of space, one has to make a selection. Therefore we apologize to all those authors whose results could not be included in this paper.

1. Solution of Michael's problem for *C*-domains

A singlevalued mapping $f : X \to Y$ between sets is said to be a *selection* of a given multivalued mapping $F : X \to Y$ if $f(x) \in F(x)$, for each $x \in X$. Note that by the Axiom of Choice selections always exist. We shall be working in the category of *topological* spaces and *continuous* singlevalued mappings. There exist many selection theorems in this category. However, the citation index of one of them is by an order of magnitude higher than for any other. This is the Michael selection theorem for convex-valued mappings:

1.1. THEOREM (MICHAEL [1956a]). *A multivalued mapping $F : X \to Y$ admits a continuous singlevalued selection, provided that the following conditions are satisfied:*

(1) *X is a paracompact space;*

(2) *Y is a Banach space;*

(3) *F is a lower semicontinuous (LSC) mapping;*

(4) *For every $x \in X$, $F(x)$ is a nonempty convex subset of Y; and*

(5) *For every $x \in X$, $F(x)$ is a closed subset of Y.*

A natural question arises concerning the necessity (essentiality) of any of the conditions (1)-(5). Here is a summary of known results:

Ad 1. With fixed conditions (2)-(5), condition (1) turned out to be necessary. This is a characterization of paracompactness in MICHAEL [1956a].

Ad 2. With fixed conditions (1), (3)-(5), condition (2) can easily be weakened to the following condition:

(2') Y is a Fréchet space.

However, the question about the necessity of condition (2') is in general still open. In many special cases (which cover the most important situations), the problem of complete metrizability of the space Y in which the images lie has already been solved in the affirmative.

MÄGERL [1978] has provided an affirmative answer in the case when Y is a compact subset of a topological linear space E, by proving that Y must be metrizable if every closed- and convex-valued LSC mapping from a paracompact domain X to Y admits a continuous singlevalued selection. Moreover, it suffices to take for the domain X only zero-dimensional compact spaces (in the sense of the Lebesgue covering dimension dim).

Nedev and Valov have shown that in Mägerl's theorem it suffices to require instead of a singlevalued continuous selection that there exists a multivalued USC selection. They also proved that Y must be completely metrizable if Y is a normal space (see NEDEV and VALOV [1984]).

VAN MILL, PELANT and POL [1996] have proved, without the convexity condition (4), that a metrizable range Y must be completely metrizable if for every 0-dimensional domain X, each closed-valued LSC mapping $F : X \to Y$ admits a singlevalued continuous selection.

Ad 3. Recall, that lower semicontinuity of a multivalued mapping $F : X \to Y$ between topological spaces X and Y means that for each $x \in X$ and $y \in F(x)$, and each open neighborhood $U(y)$, there exists an open neighborhood $V(x)$ such that $F(x') \cap U(y) \neq \emptyset$, whenever $x' \in V(x)$. Applying the Axiom of Choice to the family of nonempty intersections $F(x') \cap U(y)$, $x' \in V(x)$, we see that LSC mappings are exactly those, which admit local (noncontinuous) selections. In other words, the notion of lower semicontinuity is by definition very close to the notion of a selection.

Clearly, one can consider a mapping F which has an LSC selection G and then apply Theorem 1.1 to the mapping $\overline{conv}\ G \subset F$. For a metric space X, one of the the largest classes of such mappings was introduced by GUTEV [1993] under the name *quasi lower semicontinuous maps* (for more details see §3 of Part B in REPOVŠ and SEMENOV [1998a]).

Ad 4. This is essentially the only nontopological and nonmetric condition in (1)–(5). For $\dim X = n + 1 < \infty$ and Y completely metrizable it is possible (by MICHAEL [1956b]) to weaken the convexity restriction to the following purely topological condition:

(4') $F(x) \in C^n$ and $\{F(x)\}_{x \in X} \in ELC^n$.

In the infinite-dimensional case, it follows from the work PIXLEY [1974] and MICHAEL [1992] that there does not exist any purely topological analogue of condition (4) which would be sufficient for a selection theorem for an arbitrary paracompact domain.

In REPOVŠ and SEMENOV [1995, 1998b, 1998c, 1999] various possibilities were investigated to avoid convexity in metric terms. We exploited Michael's idea of paraconvexity in MICHAEL [1959a]. To every closed nonempty subset P of the Banach space B, a numerical function $\alpha_P : (0, \infty) \to [0, \infty)$ was associated. The identity $\alpha_P = 0$ is equivalent to convexity of P. Then all main selection theorems for convex-valued mappings remain valid if one replaces the condition $\alpha_{F(x)} = 0$ with the condition of the type $\alpha_{F(x)} < 1$, uniformly for all $x \in X$.

Ad 5. In general, one cannot entirely omit the condition of closedness of values of $F(x)$. However, if it is strongly needed then it can be done. For example, by MICHAEL [1989], in the finite-dimensional selection theorem, the closedness of $F(x)$ in Y can be replaced by the closedness of all $\{x\} \times F(x)$ in some G_δ-subset of the product $X \times Y$. Or, by MICHAEL [1956a], if X is perfectly normal and Y is separable, then it suffices to assume in Theorem 1.1 that the convex set $F(x)$ contains all interior (in the convex sense) points of its closure.

Around 1970 Michael and Choban independently showed that one can drop the closedness of $F(x)$ on any countable subset of the domain (for more details see Part B in RE-

POVŠ and SEMENOV [1998a]). Michael proposed the following way of uniform omission of closedness:

1.2. PROPOSITION (MICHAEL [1990]). *Let Y be any completely metrizable subset of a Banach space B, with the following property:*

(*) $K \subset C \subset Y \Longrightarrow \overline{conv}\ K \subset C$, *where K is a compactum and C is convex and closed (in Y).*

Then every LSC mapping $F : X \to Y$ defined on a paracompact space X with closed (in Y) and convex images has a continuous selection.

□ The compact-valued selection theorem guarantees, due to complete metrizability of Y, the existence of a compact-valued LSC selection $H : X \to Y$ of the mapping F. It remains to apply Theorem 1.1 to the multivalued selection $\overline{conv}\ H$ of the given mapping F. □

By the Aleksandrov theorem, such an Y must be a G_δ-subset of B. Property (*) is satisfied by any intersection of a countable number of open convex sets: it suffices to consider the corresponding Minkowski functionals.

However, there exist convex G_δ-sets which are not intersection of any countable number of open convex sets. For example, in the compactum $P[0,1]$ of all probability measures on the segment $[0,1]$ such is the convex complement of any absolutely continuous measure.

Hence at present, one of the central problems of selection theory is the following problem No. 396 from VAN MILL and REED [1990]:

1.3. PROBLEM (MICHAEL [1990]). Let Y be a G_δ-subset of a Banach space B. Does then every LSC mapping $F : X \to Y$ of a paracompact space X with convex closed values in Y have a continuous selection?

GUTEV [1994] proved that the answer is affirmative when X is a countably dimensional metric spaces or a strongly countably dimensional paracompact space. Problem 1.3 has recently been answered in the affirmative for domains having the so-called C-property:

1.4. THEOREM (GUTEV and VALOV [2002]). *The answer to Problem 1.3 above is affirmative for C-spaces X.*

□ We present an adaptation of the original Gutev-Valov argument. Of the C-property we shall need only the part of a theorem of USPENSKII [1998], to the effect that every mapping of such an X into a Banach space with open graph and aspherical values has a selection.

Hence let A_n be closed subsets of a Banach space B and

$$F : X \to Y = B \setminus (\bigcup_{n=1}^{\infty} A_n)$$

a convex-valued LSC mapping with values that are closed in Y. Let $\Phi(x) = Cl_B(F(x))$, $x \in X$. Apply Theorem 1.1 to the mapping $\Phi : X \to B$. Let $\mathbb{S}_\Phi$ be the set of all selections of Φ, endowed with the topology defined by the following local basis (fine topology):

$$\mathbb{O}(f, \varepsilon(\cdot)) = \{g : \|f(x) - g(x)\| < \varepsilon(x)\},$$

where $\varepsilon : X \to (0, \infty)$ runs through the set of all continuous mappings.

It is well-known that the space $C(X, B)$ of all singlevalued continuous mappings from X to B, endowed with such fine topology is a Baire space. Moreover, it contains the uniform topology. Clearly, $\mathbb{S}_\Phi$ is a uniformly closed subset of $C(X, B)$. Hence $\mathbb{S}_\Phi$ is also a Baire space.

With each closed $A_n \subset B$ one can naturally associate the set of selections, which avoid the set A_n. Namely, let $\mathbb{A}_n = \{f \in \mathbb{S}_\Phi : f(x) \notin A_n, \text{ for all } x \in X\}$. If $f \in \bigcap_n \mathbb{A}_n$ then $f : X \to B \setminus (\bigcup_n A_n) = Y$, i.e for every $x \in X$,

$$f(x) \in Y \cap \Phi(x) = Y \cap Cl_B(F(x)) = F(x),$$

because $F(x)$ is closed in Y.

It remains to verify that for every $n \in \mathbb{N}$, the families $\mathbb{A}_n$ of functions are open, nonempty and dense in $\mathbb{S}_\Phi$, and then apply the Baire property of $\mathbb{S}_\Phi$. Since we are dealing with a unique A_n, it is possible to simply delete the index n.

That $\mathbb{A} = \{f \in \mathbb{S}_\Phi : f(x) \notin A, \text{ for all } x \in X\}$ is open in $\mathbb{S}_\Phi$ for a closed $A \subset B$, is clear: if $f(X) \subset B \setminus A$, then for $\varepsilon(x) = \frac{1}{2} \operatorname{dist}(f(x), A)$, the inclusion $g \in \mathbb{O}(f, \varepsilon(\cdot)) \cap \mathbb{S}_\Phi$ implies $g(X) \subset B \setminus A$.

So far all proofs have been a repetition of the argument from MICHAEL [1988]. Formally speaking, that $\mathbb{A}$ is nonempty follows from density of $\mathbb{A}$ in $\mathbb{S}_\Phi$. However, we shall proceed in reverse order since it is more convenient to begin with the nonemptiness of $\mathbb{A}$.

Let us define a mapping $[\Phi < A] : X \to B$ as follows:

$$[\Phi < A]\ (x) = \{y \in B \ : \ y \text{ is closer to } \Phi(x) \text{ than to } A\}.$$

Clearly $F(x) \subset [\Phi < A](x)$. Hence our new mapping assumes nonempty values. Since the set A is closed and the mapping Φ is LSC, it follows that the graph $Gr[\Phi < A]$ is open.

Let us prove asphericity of each set $[\Phi < A]\ (x)$, $x \in X$. To this end we first deform $[\Phi < A]\ (x)$ into $\Phi(x) \setminus A$, and then we check the asphericity of the latter difference. For $y \in [\Phi < A]\ (x)$ we choose $r(y) > 0$ such that the closed ball $\overline{D}(y, r(y))$ intersects $\Phi(x)$ but does not intersect A. A simple selection (or separation in Dowker's spirit) arguments show that one can assume that $r(\cdot)$ is continuous. We apply Theorem 1.1 to the mapping $y \mapsto \Phi(x) \cap \overline{D}(y, r(y))$, i.e. we pick one of its continuous selections, say $z(\cdot)$.

It is geometrically evident that the entire segment $[z(y), y]$ lies in $[\Phi < A]\ (x)$ and it thus simply linear homotopy deforms $[\Phi < A]\ (x)$ into $\Phi(x) \setminus A$ (see Fig. 1).

Let us now verify that $\Phi(x) \cap A$ is a Z-subset of $\Phi(x)$ with respect to finite-dimensional domains. To this end, let us consider any mapping $\gamma : K \to \Phi(x)$ of a finite-dimensional K and for any $\delta > 0$ we associate to it the multivalued mapping $\Gamma : K \to Y = B \setminus (\bigcup_n A_n)$ given by:

$$\Gamma(k) = Cl_Y(F(x) \cap D(\gamma(k), \delta)).$$

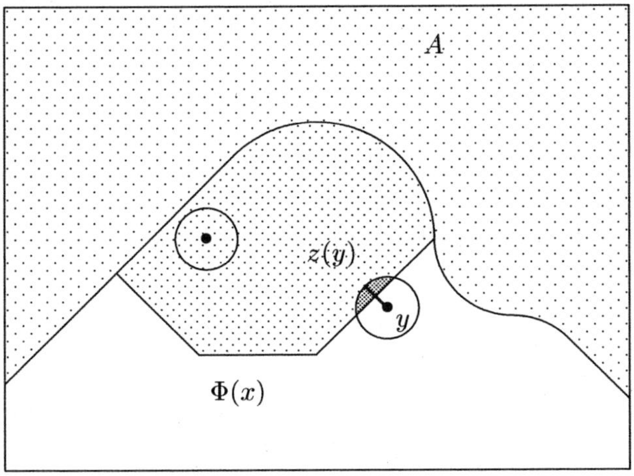

Figure 1

The finite-dimensional selection theorem applies to mapping Γ, due to complete metrizability of Y, convexity of $F(x)$ and continuity of γ. Hence the resulting selection γ' is δ-close to γ and avoids A. The asphericity of the difference $\Phi(x) \setminus A$ now follows by a standard argument (see USPENSKII [1998]).

Thus we can apply the Uspenskii selection theorem to the mapping $[\Phi < A] : X \to Y$ defined on the C-space X. Let g be a selection of $[\Phi < A]$. We repeat the previous proof, choosing closed balls $\overline{D}(g(x), r(x))$ intersecting $\Phi(x)$ but avoiding the set A, such that $r(\cdot) : X \to (0, \infty)$ is a continuous mapping. Then a selection (one more application of Theorem 1.1) of the mapping $x \mapsto \Phi \cap \overline{D}(g(x), r(x))$ is the desired selection of the mapping Φ, avoiding A. Therefore we have proved the nonemptiness of the set $\mathbb{A} \subset C(X, B)$.

In order to prove the density of $\mathbb{A} \subset \mathbb{S}_\Phi$ we pick $\phi \in \mathbb{S}_\Phi$ and a continuous mapping $\varepsilon : X \to (0, \infty)$. Then one can repeat the above argument on nonemptiness of the set of selections avoiding $\mathbb{A}$ for the mapping

$$\Psi(x) = \Phi(x) \cap \overline{D}(\varphi(x), \frac{\varepsilon(x)}{2}).$$

In other words, there is an element in $\mathbb{A}$ which is ε-close to φ. This proves the density of $\mathbb{A}$ in $\mathbb{S}_\Phi$. □

For the sake of completeness we reproduce here the complete statement of the result of Gutev and Valov.

1.5. THEOREM (GUTEV and VALOV [2002]). *For any paracompact space X the following conditions are equivalent:*

(a) *X is a C-space;*

(b) *Let Y be a Banach space and $F : X \to Y$ an LSC mapping with closed convex values. Then, for every sequence of closed-valued mappings $\Psi_n : X \to Y$ such that each Ψ_n has a closed graph and $\Psi_n(x) \cap F(x)$ is a Z_∞-set in $F(x)$ for every*

$x \in X$ and $n \in \mathbb{N}$, there exists a singlevalued continuous mapping $f : X \to Y$ with $f(x) \in F(x) \setminus \bigcup\{\Psi_n(x) : n \in \mathbb{N}\}$, for each $x \in X$; and

(c) *Let Y be a Banach space and $F : X \to Y$ be an LSC mapping with closed convex values. Then, for every closed $A \subset Y$ there exists a singlevalued continuous selection for F avoiding A, provided that $A \cap F(x)$ is a Z_∞-set in $F(x)$, for each $x \in X$.*

At present it is reasonable to expect that an affirmative solution of Problem 1.3 would yield a characterization of C-property of the domain X.

1.6. PROBLEM. Are the conditions $(a) - (c)$ from Theorem 1.5 equivalent to the following condition:

(d) Let Y be any G_δ-subset of a Banach space and $F : X \to Y$ an LSC mapping with convex values which are closed in Y. Then F admits a singlevalued continuous selection.

As a continuation of the above technique, Valov has given a selectional characterization of paracompact spaces, having the so-called *finite* C-property. To introduce it we do not use the original definition given by BORST [200?], but its characterization via the C-property. Namely, a paracompact space X has finite C-property if there exists a C-subcompact $K \subset X$ such that $\dim A < \infty$, for each closed $A \subset X \setminus K$.

1.7. THEOREM (VALOV [2002]). *For any paracompact space X the following conditions are equivalent:*

(a) *X has finite C-property;*

(b) *For any space Y and any infinite aspherical filtration $\{F_n : X \to Y\}_{n=1}^\infty$ of strongly LSC mappings there exists $m \in \mathbb{N}$ such that F_m admits a singlevalued continuous selection; and*

(c) *For any space Y and any infinite aspherical filtration $\{F_n : X \to Y\}_{n=1}^\infty$ of open-graph mappings there exists $m \in \mathbb{N}$ such that F_m admits a singlevalued continuous selection.*

Here, *strong lower semicontinuity* of a mapping $F : X \to Y$ means that the set $\{x \in X : K \subset F(x)\}$ is open for each subcompactum $K \subset X$. As for a filtration of mappings, we have that for each $x \in X$ and for each natural $n \in \mathbb{N}$,

$$F_1(x) \subset F_2(x) \subset F_3(x) \subset \ldots$$

and the inclusion $F_n(x) \subset F_{n+1}(x)$ is homotopically trivial up to dimension n (compare with SHCHEPIN and BRODSKY [1996]).

The coincidence of the class of spaces having finite C-property with the class of weakly infinitely dimensional spaces (in the sense of Smirnov) is a necessary condition for the affirmative solution of one of the main problems of infinite dimensional theory: Does every weakly infinite-dimensional compact metric space have the C-property?

2. Selectors for hyperspaces

2.A. Given a Hausdorff space X and the family $\mathcal{F}(X)$ of all nonempty closed subsets of X, we say that a singlevalued mapping $s : \mathcal{F}(X) \to X$ is a *selector* on $\mathcal{F}(X)$, provided that $s(A) \in A$, for every $A \in \mathcal{F}(X)$. From the formal point of view, a selector is simply a selection of the multivalued evaluation mapping, which associates to each $A \in \mathcal{F}(X)$ the same A, but as a subset of X. However, historically the situation was converse. Fifty years ago, in his fundamental paper MICHAEL [1951] proposed a splitting of the problem about existence of a selection $g : Y \to 2^X$ into two separate problems: first, to check that g is continuous and second, to prove that there exists a selector on 2^X. Hence, the selection problem was originally reduced to a certain selector problem.

Subsequently, the situation has stabilized to the present state. Namely, selectors are a special case of selections, but with an important exception: as a rule, no general selection theorem can be directly applied for resolving a specific problem on selectors. Specific tasks require specific techniques. Well-known papers ENGELKING, HEATH and MICHAEL [1968], CHOBAN [1970], and NADLER and WARD [1970] illustrate the point.

From early 1970's to mid 1990's the best result on continuous selectors was due to VAN MILL and WATTEL [1981], who characterized the orderable Hausdorff compacta as the compacta having a continuous selector for the family of at most two-points subsets (hence it was the extension of the similar result for the class of continua MICHAEL [1951]).

In the last five years the interest in theory of selectors has sharply increased – perhaps the monograph of BEER [1993] was one of the reasons. Over thirty papers have been published or are currently in print. We have chosen the results of HATTORI and NOGURA [1995] and VAN MILL, PELANT and POL [1996] as the starting point of this part of the survey.

2.B. For a subset $\mathbb{S} \subset \mathcal{F}(X)$ a *selector* is a mapping $s : \mathbb{S} \to X$ which selects a point $s(A) \in A$ for each $A \in \mathbb{S}$. Here, hyperspaces $\mathcal{F}(X)$ and their subsets are endowed with the *Vietoris* topology τ_V which is generated by all families of the type

$$\{A \in \mathcal{F}(X) : A \subset \bigcup_{i=1}^{n} V_i, \ \ A \cap V_i \neq \emptyset \},$$

over all finite collections of open subsets V_i of X. It is well-known that for metric spaces the Vietoris topology and the Hausdorff distance topology coincide if and only if the space X is compact.

By HUREWICZ [1928], for each metrizable space X the absence of a closed subspace of X homeomorphic to the rationals $\mathbb{Q}$ is equivalent to X being a *hereditarily Baire* space, i.e. every nonempty closed subspace of X is a Baire space. Due to the absence of continuous selectors for $\mathcal{F}(\mathbb{Q})$ (see ENGELKING, HEATH and MICHAEL [1968]), every metrizable space admitting a continuous selector is hereditarily Baire. This implication holds in the class of all regular spaces.

2.1. THEOREM (HATTORI and NOGURA [1995]). *Let X be a regular space having a continuous selector for $\mathcal{F}(X)$. Then X is a hereditarily Baire space.*

□ For a representation of a first Baire category space $X = \bigcup_{n=1}^{\infty} X_n$ as a union of closed nowhere dense subsets $X_n \subset X$ and for any continuous selector $s : \mathcal{F}(X) \to X$ it is possible to inductively construct a sequence of pairs $\{(A_n, F_n)\}_{n=1}^{\infty}$ such that for each n:

(0) A_n is a regular closed subset of X and F_n is a finite subset of the interior of A_n;

(1) $F_{n-1} \subset F_n \subset IntA_n \subset A_n \subset A_{n-1}$;

(2) $s(A) \notin X_n$, whenever $F_n \subset A \subset A_n$; and

(3) $s(A_n) \notin F_n$.

Having done such a construction, we see that $s(\bigcap_{n=1}^{\infty} A_n) \notin X_n$. This contradicts the fact that $X = \bigcup_{n=1}^{\infty} X_n$. □

Note that GUTEV, NEDEV, PELANT and VALOV [1992] proved (in a somewhat similar manner) that a metric space X is hereditarily Baire whenever every LSC mapping from the Cantor set to $\mathcal{F}(X)$ admits a USC compact-valued selection. Moreover, they showed that under such hypotheses either X is *scattered* (i.e. every closed subset has an isolated point) or X contains a homeomorphic copy of the Cantor set.

Thus we have the following facts for hyperspaces of the rationals:

2.2. THEOREM (ENGELKING, HEATH and MICHAEL [1968] HATTORI and NOGURA [1995]). *There exist no continuous selectors for $\mathcal{F}(\mathbb{Q})$, for the family of all closed nowhere dense subsets of $\mathbb{Q}$ or for the family of all clopen subsets of $\mathbb{Q}$. There exists a continuous selector on the family of subsets of $\mathbb{Q}$ of the form $C \cap \mathbb{Q}$, where C is connected subset of the real line.*

A natural question concerning existence of selectors for the family of all discrete closed subsets of $\mathbb{Q}$ arises immediately. A negative answer is a direct corollary of the following theorem in which $\mathcal{C}(M)$ denotes the family of all discrete closed subsets of a metric space M, which admits a representation as the value of some Cauchy sequence having no limit.

2.3. THEOREM (VAN MILL, PELANT and POL [1996]). *Let $\mathcal{C}(M)$ have a singlevalued continuous selector s. Then M is a completely metrizable space. Moreover, one can assume that every selector s is USC and finite-valued.*

Nogura and Shakhmatov investigated spaces with a "small" number of different continuous selectors. Recall that *orderability* of a topological space X means the existence of a linear order, say $<$, on X such that the family of intervals and rays with respect to $<$ constitutes a base for topology of X.

2.4. THEOREM (NOGURA and SHAKHMATOV [1997a]). *Let X be an infinite connected Hausdorff space. Then there are exactly two continuous selectors for $\mathcal{F}(X)$ if and only if X is compact and orderable.*

As a corollary, the topological (up to a homeomorphism) description of the interval is as follows: this is an infinite, separable, connected, Hausdorff space, admitting exactly two selectors for $\mathcal{F}(X)$. They also proved the following result:

2.5. THEOREM (NOGURA and SHAKHMATOV [1997b]). *A Hausdorff space X has finitely many selectors if and only if X has finitely many components of connectedness and there exists a compatible linear order $<$ on X such that every closed subset A of X has a minimal element with respect to $<$.*

Moreover, the total numbers of different selectors over all X in Theorem 2.5 constitutes a sufficiently scattered subsequence of the natural numbers $\mathbb{N}$. The first seven members are: $1, 2, 4, 24, 576, 720$ and 4096. This total number is a function of two natural parameters: the number n of all components of connectedness and the number m of all compact, nonsingleton components of connectedness (see NOGURA and SHAKHMATOV [1997b] for the precise formula).

2.C. There exist different topologies on the set $\mathcal{F}(X)$ whose restrictions on $X \subset \mathcal{F}(X)$ are compatible with the original topology of X. The Vietoris topology τ_V is only one of them. Hence different topologies on hyperspaces give different problems on selectors, continuous with respect to these topologies. Gutev was the first to systematically study this subject. For a metric space (X, d) the *d-proximal topology* $\tau_{\delta(d)}$ on $\mathcal{F}(X)$ is defined as the Vietoris topology τ_V but with the following additional "boundary" restriction

$$\{A \in \mathcal{F}(X) : A \subset \bigcup_{i=1}^{n} V_i, \quad A \cap V_i \neq \emptyset, \quad dist(A, X \setminus \bigcup_{i=1}^{n} V_i) > 0\ \},$$

over the all finite collections of open subsets V_i of X.

Thus $\tau_{\delta(d)} \subset \tau_V$ and it can easily be seen that $\tau_{\delta(d)} \subset \tau_{H(d)}$, where $\tau_{H(d)}$ is the topology on $\mathcal{F}(X)$ generated by the Hausdorff metric.

2.6. THEOREM (GUTEV [1996]). *For every complete nonarchimedean metric d on the space X there exists a $\tau_{\delta(d)}$-continuous selector for $\mathcal{F}(X)$.*

This theorem improves earlier results in ENGELKING, HEATH and MICHAEL [1968] and CHOBAN [1970] because each completely metrizable space X with $\dim X = 0$ admits a complete nonarchimedean metric d. The assumption in Theorem 2.6 that the metric d is nonarchimedean cannot be simply replaced by $\dim X = 0$. Namely, on the space $\mathbb{I}$ of irrationals there exists a complete compatible metric d such that $(\mathcal{F}(\mathbb{I}), \tau_{\delta(d)})$ has no continuous selectors – see COSTANTINI and GUTEV [200?].

Theorem 2.6 holds for separable spaces X for the so-called *d-ball proximal* topology $\tau_{\delta B(d)} \subset \tau_{\delta(d)}$, which is defined as topology $\tau_{\delta(d)}$ with the additional restriction that the union $\bigcup_{i=1}^{n} V_i$ can be represented as a union of a finite number of closed balls of (X, d) (for details see GUTEV [1996], GUTEV and NOGURA [2000]).

Bertacchi and Costantini unified separability of the domain with the nonarchimedean restriction on metric d.

2.7. THEOREM (BERTACCHI and COSTANTINI [1998]). *Let (X, d) be a separable complete metric space with a nonarchimedean metric d. Then the hyperspace $(\mathcal{F}(X), \tau_W(d))$ admits a selector if and only if it is totally disconnected.*

Here, $\tau_{W(d)}$ in Theorem 2.7 stands for the *Wijsman* topology which is the weakest topology on $\mathcal{F}(X)$ with continuous distance functions $dist(x, \cdot) : \mathcal{F}(X) \to \mathbb{R}$, $x \in X$. Note, that $\tau_{W(d)} \subset \tau_{\delta B(d)}$.

To end this list of the most important recent hyperspace topology results, recall that the *Fell* topology τ_F has the base:

$$\{A \in \mathcal{F}(X) : A \subset \bigcup_{i=1}^{n} V_i, \;\; A \cap V_i \neq \emptyset\}$$

over all finite collections of open subsets V_i of X with compact complement of the union $\cup V_i$.

Gutev and Nogura presented the most comprehensive up-to-date view of selectors for Vietoris-like topologies.

2.8. THEOREM (GUTEV and NOGURA [2000]). *Let X be a completely metrizable space which has a clopen $\mathbb{D}$-orderable base for some $\mathbb{D} \subset \mathcal{F}(X)$. Then $\mathcal{F}(X)$ admits a $\tau_{V(\mathbb{D})}$-continuous selector.*

Here, the base of the topology $\tau_{V(\mathbb{D})}$, which is called a $\mathbb{D}$-*modification* of the Vietoris topology τ_V, constitutes the base of the Vietoris topology neighborhoods:

$$\{A \in \mathcal{F}(X) : A \subset \bigcup_{i=1}^{n} V_i, \;\; A \cap V_i \neq \emptyset\},$$

with the additional property that the complement $X \setminus \cup V_i$ can be represented as a union of a finite number of elements of the family $\mathbb{D}$.

The following examples show, that $\tau_{V(\mathbb{D})}$-continuous selectors with respect to various $\mathbb{D} \subset \mathcal{F}(X)$ are continuous selectors with respect to Vietoris-like topologies $\tau_{\delta(d)}$, $\tau_{\delta B(d)}$, $\tau_{B(d)}$, $\tau_{W(d)}$, and τ_F. Below, the *d-clopeness* of $A \subset X$ means that $dist(A, X \setminus A) > 0$ and *d-strongly clopeness* of $A \subset X$ means the existence of a finite $F \subset A$ and a positive number $d(x) < dist(x, X \setminus A)$, for every $x \in F$, such that A is the union of closed balls of radii $d(x)$, centered at $x \in F$.

2.9. THEOREM (GUTEV and NOGURA [2000]). *Let (X, d) be a metric space.*

(a) *If $\mathbb{D}$ is a family of d-clopen subsets of (X, d) then $\tau_{V(\mathbb{D})} \subset \tau_{\delta(d)}$;*

(b) *If $\mathbb{D}$ is a family of d-clopen subsets of (X, d) which are finite unions of closed balls then $\tau_{V(\mathbb{D})} \subset \tau_{B(d)}$;*

(c) *If $\mathbb{D}$ is a family consisting of finite unions of closed balls then $\tau_{V(\mathbb{D})} \subset \tau_{B(d)}$;*

(d) *If $\mathbb{D}$ is a family of strongly d-clopen subsets of (X, d) then $\tau_{V(d)} \subset \tau_{W(d)}$;*

(e) *If $\mathbb{D}$ is a family of compact subsets of X then $\tau_{V(\mathbb{D})} \subset \tau_F$.*

The proof of Theorem 2.9 looks like a sophisticated modification (or "$\mathbb{D}$-modification") of the well-known method of coverings based on the existence of a suitable sieve (p, γ) on X.

Here, a pair (p, γ) is called a *sieve* on X if $p = \{(p_n, A_n\}$ is a countable spectrum of a discrete pairwise disjoint index sets A_n and surjections $p_n : A_{n+1} \to A_n$ and $\gamma = \{\gamma_n\}$ is a sequence of open coverings $\gamma_n = \{V_{\alpha,n} : \alpha \in A_n\}$ which are linked together with the property that $V_\alpha = \bigcup\{V_\beta : \beta \in p_n^{-1}(\alpha)\}$, $\alpha \in A_n$.

So the values of a $\tau_{V(\mathbb{D})}$-continuous selector on $\mathcal{F}(X)$ are constructed as the kernels of p-chains of some precise and $\mathbb{D}$-orderable sieve on X. A series of results on selectors follows from Theorems 2.8 and 2.9. For instance, as an improvement of Theorem 2.6, one can see that every completely metric space (X, d) admits a $\tau_{\delta(d)}$-continuous selector, whenever d-clopen subsets of X constitute a base of topology of X. An interesting approach was proposed for the Fell topology and ball proximal topology. Thus Theorems 2.10 and 2.11 below are a part of a result of Gutev and Nogura.

For the moment, let us say that a surjection $l : (X, d) \to (Y, p)$ between metric spaces is a *fiber-isometry* if $d(x, y) = p(l(x), l(y))$, for all x, y with different $l(x)$ and $l(y)$.

2.10. THEOREM (GUTEV and NOGURA [2000]). *Let Y be the hedgehog of the countable weight over a convergent sequence of reals and X a strongly zero-dimensional metrizable nonlocally compact space. Then there exists a surjection $l : X \to Y$ such that one can associate to each compatible metric p on Y a compatible metric d on X such that:*

(a) *l is a fiber-isometry with respect to d and p; and*

(b) *X has a clopen base of closed d-balls.*

In particular, the associated mapping $l_{\mathcal{F}} : \mathcal{F}(Y) \to \mathcal{F}(X)$ is continuous with respect to ball proximal topologies. By virtue of Theorem 2.10 and the absence of ball proximal-continuous selectors for hyperspaces of hedgehog Y, the following characterization theorem can be proved:

2.11. THEOREM (GUTEV and NOGURA [2000]). *For a strongly zero-dimensional metrizable space X the following conditions are equivalent:*

(a) *X is locally compact and separable; and*

(b) *There exists a τ_F-continuous selector on $\mathcal{F}(X)$.*

As a continuation of this result and in the spirit of the van Mill-Wattel theorem, GUTEV and NOGURA [2001, 200?a] have recently proved that a Hausdorff space X is topologically well-orderable if and only if $\mathcal{F}(X)$ admits a τ_F-continuous selector. ARTICO and MARCONI [2001] and GUTEV [2001] have generalized this characterization to $\mathcal{F}_2(X) = \{A \in \mathcal{F}(X) : |A| \le 2\}$.

2.D. Let us return to the Vietoris topology on $\mathcal{F}(X)$. Another characterization of the van Mill-Wattel type was obtained by Fujii and Nogura. See also MIYAZAKI [2001a] for extending van Mill-Wattel result to the class of almost compact spaces.

2.12. THEOREM (FUJII and NOGURA [1999]). *Let X be a compact Hausdorff space. The following two conditions are equivalent:*

(a) *X is homeomorphic to an ordinal space; and*

(b) *There exists a continuous selector $s : \mathcal{F}(X) \to X$, whose values are isolated points of a closed subset of X.*

Artico, Marconi, Moresco and Pelant proposed one more selector description of certain topological properties. A topological space is said to be *nonarchimedean* if for some base of open sets for an arbitrary pair of nondisjoint members of the base one of the members is a subset of the other one. A nonarchimedean space is a P-space if and only if all of its countable subsets are closed (and hence discrete).

2.13. THEOREM (ARTICO, MARCONI, MORESCO and PELANT [2001]). *Let X be a nonarchimedean P-space. Then the following conditions are equivalent:*

(a) *X is scattered;*

(b) *X is topologically well-orderable space; and*

(c) *There exists a continuous selector on $\mathcal{F}(X)$.*

FUJII, MIYAZAKI and NOGURA [2002] have recently showed that any countable regular space X admits a continuous selector if and only if it is scattered.

Now, let $s : \mathcal{F}(X) \to X$ be a continuous selector and let $G \subset X$ be nonempty and clopen. One can define another selector, say s_G which associates to each A outside of G exactly $s(A)$ and to each A meeting G, the value $s(A \cap G)$. It easy to see that s_G is also continuous and that $s_G(X) \in G$ (see Fig. 2).

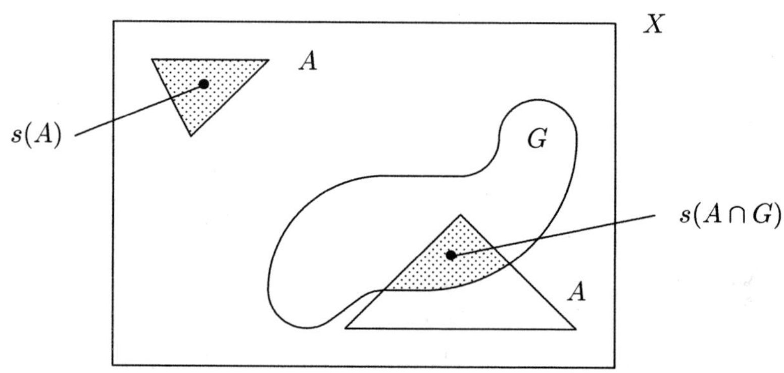

Figure 2

This observation allowed GUTEV and NOGURA [2001] to prove that if a zero-dimensional (in the ind sense) Hausdorff space X admits a continuous selector then the set $\{f(X) : f$ is a continuous selector$\}$ is dense in X. Conversely, if the set $\{f(X) : f$ is a continuous selector$\}$ is dense in X, then X is totally disconnected. A local "countable" version of such observation yields the following characterization:

2.14. THEOREM (GUTEV and NOGURA [2001]). *For any first countable Hausdorff space X admitting a continuous selector the following conditions are equivalent:*

(a) ind $X = 0$; *and*

(b) *For each point $x \in X$ there exists a selector $f_x : \mathcal{F}(X) \to X$ which is "maximal" with respect to x, i.e. which selects the point x for every closed $A \subset X$, containing x.*

As a corollary, a locally compact Hausdorff space X admitting a continuous selector is zero-dimensional in the ind sense if and only if the set $\{f(X) : f$ is a continuous selector$\}$ is dense in X.

Subsequently, Garcia-Ferreira, Gutev, Nogura, Sanchis and Tomita examined more closely the notion of x-maximal and minimal selectors. Hence in the spirit of Michael theory they proved that for a Hausdorff space X, admitting a continuous selector every such selector is maximal with respect to some point if and only if X has at most two different continuous selectors on $\mathcal{F}(X)$:

2.15. THEOREM (GARCIA-FERREIRA ET AL. [200?]). *For any countable space X the following two conditions are equivalent:*

(a) *X is a scattered metrizable space; and*

(b) *For every point $x \in X$ there exists a continuous x-maximal selector on $\mathcal{F}(X)$.*

3. Relations between *U*- and *L*-theories

3.A. Everyone understands the notion of continuity of a singlevalued mapping between topological spaces in a unique sense. For a multivalued mapping the term "continuity" has a "multivalued" interpretation, because of many different topologies on the hyperspace $\mathcal{F}(X)$ which coincide on $X \subset \mathcal{F}(X)$ with the original topology. In this part we consider links between two most useful types of continuity - the upper semicontinuity and the lower semicontinuity of multivalued mappings. In both cases one can try to find some suitable relations between multivalued and singlevalued mappings.

In the first case (upper semicontinuity) the notion of *approximation* by singlevalued mappings arises naturally. For lower semicontinuity the notion of a *selection* is a starting point. Hence we shall for the moment talk about U*-theory* and about L*-theory*, respectively.

On the one hand, the main techniques and facts of U-theory and L-theory look very similar. For example, UV^n-properties of values $F(x)$ and $ELC^n \& C^n$-properties in the case $\dim X < \infty$ (or the nontopological restriction of convexity of values for an arbitrary domain, which occurs as a paracompact space in both theories).

Moreover, in both cases there are principal obstructions to purely topological passage from finite dimensional to the infinite dimensional cases - see examples of PIXLEY [1974], TAYLOR [1975] and DRANISHNIKOV [1993].

On the other hand, no theorems of U-theory follow directly from theorems of L-theory and vice versa. For example, clearly an ε-approximation f of F can be defined as a selection of a double ε-enlargement $F_\varepsilon : x \mapsto D_Y(F(D_X(x, \varepsilon)), \varepsilon)$. Many authors have observed this fact. But in general, there is no information about topological or convexity-like properties of values of such an enlargement. Hence the selection theory cannot be applied directly. In brief, we had two closed but "parallel" theories for multivalued mappings.

3.B. Shchepin and Brodsky have proposed a unified approach of simultaneously using U- and L-theories in order to find a new proof of the finite-dimensional Michael selection theorem together with its recent generalization in MICHAEL [1989]. The key ingredient of the their considerations is the notion of a *filtration* of a multivalued mapping. Two different kinds of filtrations were used.

For a topological spaces X and Y a finite sequence $\{F_i\}_{i=0}^n$ of a multivalued mappings $F_i : X \to Y$ is said to be an *L-filtration* if: (1) F_i is a selection of F_{i+1}, $0 \le i < n$; (2) the identity inclusions $F_i(x) \subset F_{i+1}(x)$ are *i-apolyhedral* for all $x \in X$, i.e. for each polyhedron P with $\dim P \le i$ every continuous mapping $g : P \to F_i(x)$ is null-homotopic in $F_{i+1}(x)$; (3) the families $\{\{x\} \times F_i(x)\}_{x \in X}$ are ELC^{i-1} families of subsets of the Cartesian product $X \times Y$, $0 \le i \le n$; (4) for every $0 \le i \le n$ there exists a G_δ-subset of $X \times Y$ such that $\{x\} \times F_i(x)$, $x \in X$, are closed in this G_δ-subset.

3.1. THEOREM (SHCHEPIN and BRODSKY [1996]). *Let X be a paracompact space with $\dim X \le n$, Y a complete metric space and $\{F_i\}_{i=0}^n$ an L-filtration of maps $F_i : X \to Y$. Then the mapping $F_n : X \to Y$ admits a singlevalued continuous selection.*

It is easy to see that Theorem 3.1 applies to the constant filtration $F_i = F$ and hence we obtain the finite-dimensional selection theorem as a corollary of the filtered Theorem 3.1. Moreover, for a constant filtration $F_i = F$ with F having the properties (3) and (4) from the definition of L-filtration above, we obtain the generalization of the finite-dimensional selection theorem which was proposed in MICHAEL [1989]. Note, that almost the same filtered approach for open-graph mappings was earlier proposed by BIELAWSKI [1989], but with no detailed argumentation.

Next, we define the notion of a *U-filtration*. For topological spaces X and Y a finite sequence $\{H_i\}_{i=0}^n$ of compact-valued upper semicontinuous mappings is said to be a *U-filtration* if: (1') H_i is a selection of H_{i+1}, $0 < i < n$; (2') the identity inclusions $H_i(x) \subset H_{i+1}(x)$ are *UV^i-aspherical* for all $x \in X$, i.e. for every open $U \subset H_{i+1}(x)$ there exists a smaller open $V \subset H_i(x)$ such that every continuous mapping $g : S^i \to V$ is null-homotopic in U; where S^i is the standard i-dimensional sphere. Clearly, condition (2') in the definition of a U-filtration looks like an approximate version of condition (2) in the definition of an L-filtration.

We shall now formulate the notion of (graphic) approximations of multivalued mappings $F : X \to Y$. Let $\mathcal{V} = \{\mathcal{V}_\alpha\}_{\alpha \in A}$ be a covering of X and $\mathcal{W} = \{W_\gamma\}_{\gamma \in \Gamma}$ a covering of Y. A singlevalued mapping $f : X \to Y$ is said to be a $(\mathcal{V} \times \mathcal{W})$-approximation of F if for every $x \in X$ there exist $\alpha \in A$ and $\gamma \in \Gamma$ and points $x' \in X$, $y' \in F(x')$ such that x and x' belong to V_α, $f(x)$ and y' belongs to W_γ. In other words, the graph of f lies in the neighborhood of the graph F with respect to the covering $\{V_\alpha \times W_\gamma\}_{\alpha \in A, \gamma \in \Gamma}$ of the Cartesian product $X \times Y$.

For metric spaces X and Y and for coverings $\mathcal{V}$ and $\mathcal{W}$ of X and Y by $\varepsilon/2$-open balls we obtain the more usual notion of (graphic) ε-approximations AUBIN and CELLINA [1984]. Namely, a singlevalued mapping $f_\varepsilon : X \to Y$ between metric spaces (X, ρ) and (Y, d) is said to be *ε-approximation* of a given multivalued mapping $F : X \to Y$ if for every $x \in X$ there exist points $x' \in X$ and $y' \in F(x')$ such that $\rho(x, x') < \varepsilon$ and $d(y', f_\varepsilon(x)) < \varepsilon$.

We recall the well-known Cellina approximation theorem (see AUBIN and CELLINA [1984]) which states that each convex-valued USC $F : X \to Y$ from a metric space (X, ρ) into a normed space $(Y, \|\cdot\|)$ with convex values $F(x)$, $x \in X$, is *approximable*, i.e. that for every $\varepsilon > 0$ there exists a singlevalued continuous ε-approximation of F. The following theorem is a natural finite-dimensional version of Cellina's theorem:

3.2. THEOREM (SHCHEPIN and BRODSKY [1996]). *Let X be a paracompact space with $\dim X \le n$, Y an ANE for the class of all paracompacta and $\{H_i\}_{i=0}^n$ a U-filtration of* *mappings $H_i : X \to Y$. Then for every open in $X \times Y$ neighborhood G of the graph $\Gamma(H_n)$ of the mapping H_n there exists a continuous singlevalued mapping $h : X \to Y$ such that the graph $\Gamma(h)$ lies in G.*

The following theorem gives an intimate relation between L-filtrations and U-filtrations. This theorem can also be regarded as a *filtered* analogue of the compact-valued selection theorem:

3.3. THEOREM (SHCHEPIN and BRODSKY [1996]). *Let X be a paracompact space, Y be a complete metric space and $\{F_i\}_{i=0}^n$ an L-filtration of maps $F_i : X \to Y$. Then there exists a U-filtration $\{H_i\}_{i=0}^n$ such that H_n is a selection of F_n. Moreover each H_i is a selection of F_i, $0 \leq i \leq n$ and the inclusions $H_i(x) \subset H_{i+1}(x)$ are UV^i-apolyhedral.*

We point out some discordance: in the definition of an L-filtration we talked about *apolyhedrality* and in the definition of a U-filtration about *asphericity*. In view of Theorem 3.3, the UV^i-asphericity of $H_i(x) \subset H_{i+1}(x)$ cannot be directly derived from the UV^i-asphericity of inclusions $F_i(x) \subset F_{i+1}(x)$, $x \in X$, of given L-filtration $\{F_i\}_{i=0}^n$. Moreover, there is a gap in the original proof of Theorem 3.1 — the authors in fact need an L-filtration $\{F_i\}_{i=0}^{n^2}$ of length n^2, not n. BRODSKY [2000] later partially filled this gap by considering *singular* filtrations (see below). Recently, BRODSKY, CHIGOGIDZE and KARASEV [2002] have completely solved the problem (see Theorem 4.22 below).

We now formulate the crucial technical ingredient of the whole procedure. The following theorem asserts the existence of another U-filtration $\{H_i'\}_{i=0}^n$ accompanying a given L-filtration $\{F_i\}_{i=0}^n$. Here, we drop the conclusions that $H_0(x) \subset F_0(x), \ldots, H_{n-1}(x) \subset F_{n-1}(x)$ and add the property that the sizes of values $H_n(x)$ can be chosen to be less than arbitrary given $\varepsilon > 0$.

3.4. THEOREM (SHCHEPIN and BRODSKY [1996]). *Let X be a paracompact space with $\dim X \leq n$, Y a Banach space and $\varepsilon > 0$. Then for every L-filtration $\{F_i\}_{i=0}^n$, every U-filtration $\{H_i\}_{i=0}^n$ with H_n being a selection of F_n and every open in $X \times Y$ neighborhood G of the graph $\Gamma(H_n)$ of the mapping H_n there exists another U-filtration $\{H_i'\}_{i=0}^n$ such that:*

(1) *H_n' is a selection of F_n;*

(2) *The graph $\Gamma(H_n')$ lies in G; and*

(3) diam $H_n'(x) < \varepsilon$, *for each $x \in X$.*

The proof of Theorem 3.4 is divided, roughly speaking into two steps. One can begin by the application of Theorem 3.2 to the given a U-filtration $\{H_i\}_{i=0}^n$ with $H_n \subset F_n$. Hence we obtain some singlevalued continuous mapping $h\colon X \to Y$ which is an approximation of H_n. Then we perform a "thickening" procedure with h, in order to obtain a new L-filtration $\{F_i'\}_{i=0}^n$ with small sizes of values $F_n'(x)$, $x \in X$. Such an L-filtration $\{F_i'\}_{i=0}^n$ naturally arises from the ELC^{n-1} properties of the values of the final mapping F_n of a given L-filtration $\{F_i\}_{i=0}^n$. Finally, we use the "filtered" compact-valued selection Theorem 3.3 exactly for the new L-filtration $\{F_i'\}_{i=0}^n$. The result of such an application gives the desired U-filtration $\{H_i'\}_{i=0}^n$ with small sizes of values $H_n'(x)$, $x \in X$. This is the strategy of the proof of Theorem 3.4. The inductive repetition for some series $\sum_k \varepsilon_k < \varepsilon$ involves in each fiber $F_n(x)$ a sequence of subcompacta $H_n^k(x)$ which turns out to be fundamental with respect to the Hausdorff metric. Its limit point gives the value $f(x)$ of the desired selection f of F.

3.C. For resolving some difficulties with the proof of Theorem 3.1 and in order to find new applications of the filtered approach, BRODSKY [2002, 200?] introduced the notion of a *singular* filtration of a multivalued mapping.

For two multivalued mappings $\Phi : X \to Y$ and $\Psi : X \to Z$ with the same base, their *fiberwise transformation* is defined as a singlevalued continuous mapping $\hat{f} : \Gamma_\Phi \to \Gamma_\Psi$ between their graphs such that $\hat{f}(\{x\} \times \Phi(x)) \subset \{x\} \times \Psi(x), x \in X$.

For a multivalued mapping $F : X \to Y$ its *singular* n-length filtration is defined as a triple $\mathbb{F} = \{\hat{f}_i, F_i, f_i\}$ where $F_i : X \to Y_i$ are multivalued mappings, $f_i : F_i \to F$ and $\hat{f}_i : F_i \to F_{i+1}$ are fiberwise transformations such that $f_i = f_{i+1} \circ \hat{f}_i$ (see Fig. 3).

A singular filtration $\mathbb{F}$ is said to be:

(1) *simple* if all fiberwise transformations are fiberwise inclusions;

(2) *complete* if all spaces Y_i are completely metrizable and all fibers $\{x\} \times F_i(x)$ are closed in some G_δ-subset of $X \times Y_i$;

(3) *contractible* if inclusions $\hat{f}_i(\{x\} \times F_i(x)) \subset \{x\} \times F_{i+1}(x), x \in X$ are homotopically trivial;

(4) *connected* if inclusions $\hat{f}_i(\{x\} \times F_i(x)) \subset \{x\} \times F_{i+1}(x), x \in X$ are i-aspherical for all i;

(5) *lower continuous* if all mappings F_i are LSC and family $\{\{x\} \times F_i(x)\}_{x \in X}$ is ELC^{i-1}; and

(6) *compact* if all mappings F_i all compact-valued and USC.

3.5. THEOREM (BRODSKY [2000]). *For each complete, connected and lower continuous n-length filtration $\mathbb{F} = \{\hat{f}_i, F_i, f_i\}$ of mappings from a metrizable space X with* $\dim X \leq n$ *into a completely metrizable space Y there exists a singlevalued continuous selections of F_n.*

Theorem 3.5 is based on the following theorem which, briefly speaking, reduces a singular filtration to some simple filtration with nice topological properties. Note, that in the following theorem n can be equal to infinity.

3.6. THEOREM (BRODSKY [2000]). *For each complete, connected and lower continuous n-length filtration $\mathbb{F} = \{\hat{f}_i, F_i, f_i\}$ of mappings from a metrizable space X with* $\dim X \leq n$ *into a completely metrizable space Y there exist compact, contractible, simple n-filtration $\mathbb{G} = \{\hat{g}_i, G_i, g_i\}$ and fiberwise transformation $\hat{h} : \mathbb{F} \to \mathbb{G}$.*

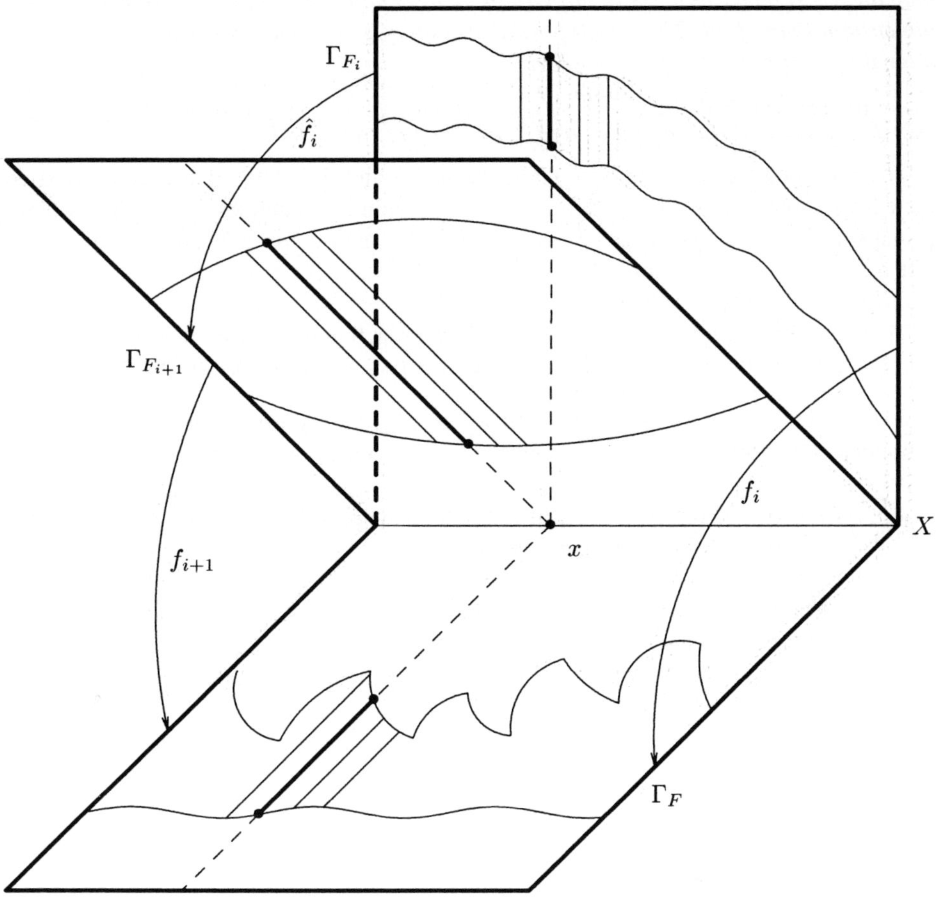

Figure 3

Clearly, Theorem 3.6 is a singular version of Theorem 3.3. Some words about its proof are in order. Metrizable domain X is the image of some zero-dimensional metric space Z under some perfect mapping $p : Z \to X$. The inductive procedure of extensions of $\hat{h}$ reduces to a selection problem for a suitable multivalued mapping from Z to a space of continuous singlevalued mappings from the graph of fiberwise join $\Gamma_{F * p^{-1}}$. The latter functional space is endowed by some asymmetric (not metric) and the analogue of standard zero-dimensional selection theorem done here "by hands", following known covering technique. In Theorem 3.6 metrizability of X is needed because of necessity of certain asymmetry in a suitable functional space.

We formulate two possibilities for applications which give the first known positive step towards a solution of the two-dimensional Serre fibration problem (see Problem 5.12 below).

3.7. THEOREM (BRODSKY [2002]). *Let $f : X \to Y$ be a mapping from a metrizable space onto an ANR metrizable space with all preimages homeomorphic to a fixed compact two-dimensional manifold. Then each partial section of f over closed subsets $A \subset X$ admits a local continuous extension, whenever:*

(1) *f is homotopically 0-regular; or*

(2) *f is a Serre fibration and X and Y are locally connected.*

The intermediate result between Theorems 3.5 and 3.7 states that a connected complete lower continuous 2-length singular filtration admits a continuous selection whenever it maps an ANR metric space X into a complete metric space Y and all values of the last member of the filtration are hereditary aspherical. Note, that any two-dimensional manifold is a locally hereditary aspherical space.

There are many applications of the filtered approach in the approximation theory which we shall omit here for the lack of space (see BRODSKY [1999, 2002]).

3.D. In Sections 1 and 2 above problems from L-theory were reduced to problems in U-theory. Here we consider a converse reduction which was proposed in REPOVŠ and SEMENOV [200?].

A family $\mathcal{L}$ of nonempty subsets of a topological space Y is said to be *selectable* with respect to a pair (X, A) if for each LSC mapping $F : X \to Y$ with values from $\mathcal{L}$ (i.e. $F(x) \in \mathcal{L}$ for every $x \in X$) and each selection $s : A \to Y$ of the restriction $F|_A$ there exists a selection $f : X \to Y$ of F which extends s (shortly, $\mathcal{L} \in \mathbb{S}(X, A)$).

For a positive r and for a family $\mathcal{L}$ of nonempty subsets of a metric space Y we denote by $\mathcal{L}_r$ the family of all subsets of Y which are r-close (with respect to the Hausdorff distance) to the elements of the family.

A family $\mathcal{L}$ of a nonempty subsets of a metric space Y is said to be *nearly selectable* with respect to a pair (X, A) if for every $\varepsilon > 0$ there exists $\delta > 0$ such that for each LSC mapping $F : X \to Y$ with values from $\mathcal{L}_\delta$ and for each selection $s : A \to Y$ of the restriction $F|_A$ there exists an ε-selection $f : X \to Y$ of F which extends s. Shortly, $\mathcal{L} \in \mathbb{NS}(X, A)$. Below, we shall consider *only* hereditary families of sets.

3.8. THEOREM (REPOVŠ and SEMENOV [200?]). *Let $H : X \to Y$ be a USC mapping between metric spaces and A a closed subset of X, and let all values of H be in some family $\mathcal{L}$ which is nearly selectable with respect to the pair (X, A). Then for every covering ω of X, every $\varepsilon > 0$ and every selection $s : A \to Y$ of the restriction $H|_A$ there exists an $(\omega \times \varepsilon)$-approximation of H which extends s.*

□ For a given $\varepsilon > 0$ we choose $\varepsilon > \delta > 0$ with respect to the definition of near selectability of the family $\mathcal{L}$. Next we construct a new multivalued mapping $F : X \to Y$ such that:

(a) F is an LSC mapping and $H(x) \subset F(x)$, for every $x \in X$; and

(b) For every $x \in X$, there exists $z_x \in X$ such that $H(z_x) \subset F(x) \subset D(H(z_x), \delta)$.

In particular, $F(x) \in \mathcal{L}_\delta$. By paracompactness of the domain X, one can find a locally finite open star-refinement ν of the given covering of X. For each $x \in X$, let V_x be an arbitrary element of the covering ν such that $x \in V_x$. Now the covering of the domain arises naturally. Namely,

$$U_x = H_{-1}(D(H(x), \delta)) \bigcap V_x, \; x \in X.$$

We shall need the following lemma:

3.9. LEMMA (SEMENOV [2000b]). *For each positive τ and each open covering $\{U_x\}_{x\in X}$, $x \in U_x$, of a metric space X there exists a lower semicontinuous numerical function $l : X \to (0, \tau/2]$ with the following property: for every $x \in X$ there exists $z \in X$ such that $z \in D(x, l_x) \subset U_z$.*

We now apply this lemma to the covering $\{U_x\}_{x\in X}$ chosen above, for $\tau = \varepsilon$, and we define $F : X \to Y$, by setting $F(x) = H(D(x, l_x))$. □

A typical example of a nearly selectable family with respect to arbitrary paracompact domains and their closed subsets is the family of all nonempty convex subsets of a normed space. As a special case we have for the empty set A:

3.10. COROLLARY (REPOVŠ and SEMENOV [200?]). *Let $H : X \to Y$ be an USC mapping between metric spaces and let all values of H belong to some family $\mathcal{L}$ which is nearly selectable with respect to X. Then H is approximable.*

As a concrete application we have the following relative approximation fact:

3.11. COROLLARY (REPOVŠ and SEMENOV [200?]). *Let $H : X \to Y$ be a convex-valued USC mapping from a metric space X into a normed space Y and $A \subset X$ a closed subset. Let $\varepsilon > 0$ be given. Then:*

(1) *Each $(\varepsilon/4)$-selection $s : A \to Y$ of the restriction $H|_A$ can be extended to some ε-approximation $h : X \to Y$ of H; and*

(2) *There exists a continuous function $\delta : X \to (0, \infty)$ such that each $\delta(\cdot)$-approximation $\alpha : A \to Y$ of H can be extended to an ε-approximation $h : X \to Y$ of H.*

In the nonconvex situation the same technique was applied in SEMENOV [2000b] for resolving approximation problem for paraconvex-valued mappings. For a nonempty closed subset $P \subset Y$ of a Banach space $(Y, \|\cdot\|)$ and for an open ball $D \subset Y$ of radius r, one defines:

$$\delta(P, D) = \sup\{dist(q, P)/r \mid q \in conv(P \cap D)\},$$

and the value of its *function of nonconvexity* α_P at a point $r > 0$ is defined as $\alpha_P(r) = \sup\{\delta(P, D)\}$, where sup is taken over the set of all open balls of radius r. Next, a subset of a Banach space is said to be *α-paraconvex* if its function of nonconvexity majorates by the preassigned constant $\alpha \in [0, 1)$.

A direct calculation shows that for every $\alpha < 1$ and $R > 0$ the family $\mathcal{P}_{\alpha,R}$ of all α-paraconvex subsets P of a Banach space Y with $diam\, P < R$ is nearly selectable with respect to paracompact spaces. Moreover, one can check that in this case $\delta = \frac{\varepsilon}{12(1+\alpha)R}$ is a suitable answer for $\delta = \delta(\varepsilon)$ in the definitions above. Hence for any $\alpha \in [0, 1)$ and any USC mapping $F : X \to Y$ from a metric space to a normed space we see that F is approximable if all values $F(x)$, $x \in X$, are α-paraconvex in Y.

As for other unified U- and L- facts we conclude by the following result.

3.12. THEOREM (BEN-EL-MECHAIEKH and KRYZSZEWSKI [1997]). *Let $F : X \to Y$ be a LSC mapping and $H : X \to Y$ a USC mapping from a paracompact space X into a Banach space Y. Suppose that both mappings are convex-valued, F is closed-valued and $F(x) \cap H(x) \neq \emptyset, x \in X$. Then for every $\varepsilon > 0$ there exists a continuous singlevalued mapping $f : X \to Y$ which is a selection of F and ε-approximation of H.*

4. Miscellaneous results

4.A. Problem 1.3 deals with generalized ranges of multivalued mappings: substitution of a G_δ-subset instead of a Banach space. The following problem is related to generalized domains (as a variation, one can consider reflexive Banach spaces instead of Hilbert space).

4.1. PROBLEM (CHOBAN, GUTEV and NEDEV). Does every LSC closed- and convex-valued mapping from a collectionwise normal and countably paracompact domain into a Hilbert space admit a singlevalued continuous selection?

As a continuation of the result in NEDEV [1987], Choban and Nedev considered more complicated, in general *nonparacompact* domains of an LSC mappings. They extended a given LSC mapping to some paracompactification (Dieudonné completion) of an original domain and then applied Theorem 1.1. Recall that GO-spaces are precisely the subspaces of linearly ordered spaces. Their result is a step towards resolving (still open) Problem 4.1.

4.2. THEOREM (CHOBAN and NEDEV [1997]). *Every LSC closed- and convex-valued mapping* $F : X \to Y$ *from a generalized ordered space* X *to a reflexive Banach space* Y *has a singlevalued continuous selection.*

Shishkov obtained similar results for domains which are σ-products of metric spaces. Such a product of uncountably many copies of reals is collectionwise normal and countably paracompact but not pseudoparacompact.

4.3. THEOREM (SHISHKOV [2001]). *Every closed- and convex-valued LSC mapping of a* σ*-product of a metric spaces into a Hilbert space has a singlevalued continuous selection.*

Initially, Shishkov result dealt with separable metric spaces. He had earlier proved that the same selection result holds for any reflexive range and any collectionwise normal, countably paracompact and pseudoparacompact domain. Recently Shishkov has strengthened the Choban-Nedev theorem above because of paracompactness of the Dieudonné completition of GO-spaces.

4.4. THEOREM (SHISHKOV [2002]). *Each LSC closed- and convex-valued mapping of a normal and countably paracompact domain into a reflexive Banach space admits a LSC closed- and convex-valued extension over the Dieudonne completition of the domain.*

It is interesting that the property of the domain to be collectionwise normal and countably paracompact admits a characterization via multivalued selections. It turns out that for such purpose it suffices to consider in the assumption of the classical compact-valued Michael's selection theorem (see MICHAEL [1959c]) not only an LSC mapping, but such a mapping together with its a USC selection.

4.5. THEOREM (MIYAZAKI [2001b]). *For a* T_1*-space* X *the following conditions are equivalent:*

(a) X *is a collectionwise normal and countably paracompact space; and*

(b) *For every completely metrizable space Y, every LSC mapping $F : X \to Y$ with $F(x)$ either compact or $F(x) = Y$ for all $x \in X$ and every compact-valued USC selection $H : X \to Y$ of F there exist a compact-valued USC mapping $\Phi : X \to Y$ and a compact-valued LSC mapping $G : X \to Y$ such that $H(x) \subset G(x) \subset \Phi(x) \subset F(x)$, $x \in X$.*

See MIYAZAKI [2001b] for other conditions which are equivalent to $(a), (b)$. Inside the class of all normal spaces the collectionwise normality property has the following multivalued extension-type description.

4.6. THEOREM (MIYAZAKI [2001b]). *For a normal space X the following conditions are equivalent:*

(a) *X is collectionwise normal; and*

(b) *For every finite-dimensional completely metrizable space Y and every USC mapping $H : X \to Y$ with values consisting of at most n points there exist a compact-valued USC mapping $\Phi : X \to Y$ and a compact-valued LSC mapping $G : X \to Y$ such that $H(x) \subset G(x) \subset \Phi(x)$, $x \in X$.*

Recall also the following generalization of the compact-valued Michael's selection theorem to the class of Čech-complete spaces.

4.7. THEOREM (CALBRIX and ALLECHE [1996]). *For each paracompact space X, each regular AF-complete space Y admitting a weak k-development and each closed-valued LSC mapping $F : X \to Y$ there exist a compact-valued USC mapping $\Phi : X \to Y$ and a compact-valued LSC mapping $G : X \to Y$ such that $G(x) \subset \Phi(x) \subset F(x)$, $x \in X$.*

Note that every AF-complete submetrizable space X has a weak k-development. A space is called *AF-complete* if it is Hausdorff and has a sequence of open coverings which is complete. The class of all Čech-complete spaces coincides with the class of all completely regular AF-complete spaces and completely metrizable spaces are exactly metrizable AF-complete spaces.

4.B. Künzi and Shapiro used Theorem 1.1 to prove the uniform version of the Dugundji extension theorem for partially defined mappings:

4.8. THEOREM (KÜNZI and SHAPIRO [1997]). *For each metrizable space X there exists a continuous mapping $E : C_{vc}(X) \to C_b(X)$ such that $E(f)|_{dom\ f} = f$ for all maps $f \in C_{vc}(X)$ and for every $K \in exp_c(X)$ the restriction $E|_{p^{-1}(K)}$ is a linear positive operator with $E(id_K) = id_X$.*

Here $C_{vc}(X)$ and $C_b(X)$ are sets of all continuous numerical mappings f with compact domain $dom\ f \subset X$ and all continuous bounded numerical mappings on the whole space X. Elements of $C_{vc}(X)$ are identified with their graphs and topology is induced by the Vietoris topology on $\mathcal{F}(X \times \mathbb{R})$, where $C_b(X)$ is endowed with the usual sup-norm topology. One can associate to each $f \in C_{vc}(X)$ its domain and obtain the projection p onto $exp_c(X)$ - the compact exponent of X.

A sketch of the proof goes as follows. For a Banach space B and every $K \in exp_c(B)$ one must consider the subset $R(K) \subset C_b(B, B)$ consisting of all $r : B \to B$ with

$r(B) \subset \overline{conv}(K)$ and $r|_K = id_K$. Clearly, $R(K)$ is a closed convex subset of $C_b(B,B)$ which is nonempty, due to the Dugundji theorem. It turns out that Theorem 1.1 is applicable to the mapping $R : exp_c(B) \to C_b(B,B)$. Hence the desired operator of simultaneous extension can be given by the formula

$$E(f)(x) = \int f d\mu(dom\ f)(x),$$

where X is embedded into the conjugate space of the Banach space $B = BL(X,d)$ of all bounded Lipshitz numerical mappings on the metric space (X,d).

Moreover, the above formula works for mappings not only to reals, but to Banach spaces and Cartesian products of Banach spaces. Metrizability of the domain X can be weakened to the restriction that X is one-to-one continuous preimage of a metric space. Note that the one-point-Lindelöfication of an uncountable discrete space admits such an operator E, although it is not a submetrizable space.

STEPANOVA [1993] had earlier characterized preimages of metric spaces under perfect mappings as spaces X for which a continuous mapping $E : C_{vc}(X) \to C_b(X)$ exists with $E(f)|_{dom\ f} = f$ and $\sup_{x \in dom\ f} |f(x)| = \sup_{x \in X} |E(f)(x)|$.

4.C. Filippov and Drozdovsky introduced new types of semicontinuity which unify lower and upper semicontinuity.

4.9. DEFINITION. A multivalued mapping $F : X \to Y$ is said to be *mixed semicontinuous* at a point $x \in X$ if for each open sets U and V with $F(x) \subset U$ and $F(x) \cap V \neq \emptyset$, respectively, there exists an open neighborhood W of x such that for every $x' \in W$ one of the following holds:

$$F(x') \subset U \qquad \text{or} \qquad F(x') \cap V \neq \emptyset.$$

They proved the following theorem concerning USC selections for mappings which are mixed semicontinuous at each point of domain:

4.10. THEOREM (FILIPPOV and DROZDOVSKY [1998, 2000]). *Let X be a hereditary normal paracompact space and Y a completely metrizable space. Then every compact-valued mixed semicontinuous mapping $F : X \to Y$ has a USC compact-valued selection.*

Considering the case $Y = \{0; 1\}$, it easy to see that the domain X must be hereditarily normal whenever Theorem 4.10 holds for each mixed semicontinuous mapping. Theorem 4.10 is useful in theory of differential equations with multivalued right-hand sides because of the well-known conditions in DAVY [1972] for the original mapping F imply the same conditions for USC selection of F. Hence, the inclusion $y' \in F(t,y)$ admits a solutions for a mixed semicontinuous right-hand side.

In their proof the authors used the idea of universality of the zero-dimensional selection theorem (see Part A of REPOVŠ and SEMENOV [1998a]): they considered the projection $\pi_X : A(X) \to X$ of the absolute of the domain over the domain. This is a perfect mapping and $A(X)$ is a paracompact space, because X is such. The hereditary normality of domain, extremal disconnectedness of the absolute and mixed continuity of F show that the mapping $G : A(X) \to Y$ defined by $G(x) = \liminf_{y \to x} F(\pi_X(y))$, is an LSC selection of the composition $F \circ \pi_X$ with nonempty closed values. By the compact-valued selection theorem we find an USC compact-valued selection H of G and finally $H \circ \pi_X^{-1}$ gives the desired selection of F.

A simple example of a mixed continuous mapping $F : X \to Y$ is given by the mapping $F(x) = \Psi(x), x \in A; F(x) = \Phi(x), x \in X \setminus A$ where $\Phi : X \to Y$ is compact-valued LSC mapping and $\Psi : A \to Y$ is its USC compact-valued selection over closed subset $A \subset X$. Hence, as a corollary of Theorem 4.7 we see that the selection Ψ admits an USC extension over whole X.

FRYSZKOWSKI and GORNIEWICZ [2000] introduced somewhat different type of mixed continuity. They considered mappings which are lower semicontinuous at some points of the domain and upper semicontinuous at all remaining points of the domain. General theorems on multivalued selections are proved together with various applications in theory of differential inclusions.

Finally, we mention here one more "unified" selection result, which has recently been proved by Arutyunov. The following theorem looks like a mixture of theorems of Kuratowski–Ryll-Nardzewski and Michael-Pixley:

4.11. THEOREM (ARUTYUNOV [2001]). *Let $F : X \to Y$ be a measurable mapping from a metric space X endowed by a σ-additive, regular measure, $dim_X Z \leq 0$ and $A \subset X$ such that all values $F(x), x \in A$, are convex and F is LSC over $Z \cup Cl(A)$. Then F admits a singlevalued measurable selection which is continuous over $Z \cup A$.*

Moreover as usual, the set of all such selections is pointwise dense in values of multivalued mapping F. For applications in the optimal control theory see ARUTYUNOV [2000].

4.D. Cauchy problem for differential inclusions $x' \in F(t, x)$, $x(0) = 0$ was first reduced in ANTOSIEWICZ and CELLINA [1975] to a selection problem for some multivalued mapping $\hat{F} : K \to L_1(I, \mathbb{R}^n)$. Here I is a segment of reals, K is some suitable convex compactum of continuous functions $u : I \to \mathbb{R}^n$ and

$$\hat{F}(u) = \{v \in L_1(I, \mathbb{R}^n) | v(t) \in F(t, u(t)) \text{ a.e. in } I\}.$$

The mapping $\hat{F}$ is LSC whenever F is also such. But the values of $\hat{F}$ are in general nonconvex. They are decomposable subsets of $L_1(I, \mathbb{R}^n)$.

4.12. DEFINITION. A set Z of a measurable mappings from a measurable space $(T, \mathbb{A}, \mu)$ into a topological space E is said to be *decomposable* if for every $f, g \in Z$ and for every $A \in \mathbb{A}$, the mapping defined by $h(t) = f(t)$, when $t \in A$ and $h(t) = g(t)$, when $t \notin A$, belongs to Z.

The intersection of all decomposable sets, containing a given set S, is called the *decomposable hull $Dec(S)$* of the set. For spaces of numerical functions on nonatomic domains, the decomposable hull of the two-point set is homeomorphic to the Hilbert space. Hence, it is a very unusual convexity-like property. Thus it is impossible to adapt the proof of the convex-valued selection theorem directly to decomposable-valued mappings. One of the reasons is a big difference between the mapping which associates to each set its convex hull (it is continuous in the Hausdorff metric on subsets), and the one which associates to each set its decomposable hull (it fails to be continuous). For example, the decomposable hull $Dec(D(f,a))$ of any ball $B(f,a)$ coincides with the entire space $L_1(T,E)$.

FRYSZKOWSKI [1983] (resp. BRESSAN and COLOMBO [1988]) proved selection theorems for a decomposable-valued LSC mappings with compact metric domains (resp. with separable metric domains). In both cases the so-called Lyapunov convexity theorem or its generalizations were used. The principal obstruction for a similar proof of the selection theorem for any paracompact domains is that the Lyapunov theorem fails for infinite number nonatomic real-valued measures.

In an attempt to return to the original idea of the Theorem 1.1, AGEEV and REPOVŠ [2000] introduced the notions of *dispersibly decomposable* sets and *dispersibly decomposable* hulls $Disp(A) \subset Dec(A)$. All decomposable sets are dispersibly decomposable sets and also (which is more important) all open and closed balls are dispersibly decomposable. Precisely the latter fact enables one to apply the usual techniques developed for the Michael convex-valued selection Theorem 1.1. Thus they proved the following selection theorem for the multivalued mappings with uniformly dispersed values (the so called *dispersible* multivalued mappings):

4.13. THEOREM (AGEEV and REPOVŠ [2000]). *Let $(T, \mathbb{A}, \mu)$ be a separable measurable space, E a Banach space, X a paracompact space and $L_1(T,E)$ the space of all Bochner integrable functions. Then each dispersible closed-valued mapping $F : X \to L_1(T,E)$ admits a continuous selection.*

The main technical step was the following lemma on a dividing of segment onto disjoint measurable subsets.

4.14. LEMMA. *For every $\sigma > 0$ and every point $s = (s_0, s_1, ..., s_n) \in \Delta^n$ of the standard n-dimensional simplex Δ^n there exists a partition $P = \{P_i\}_{i=0}^n$ of the interval I such that*

$$|m(P_i \cap J) - s_i \cdot m(J)| < \sigma,$$

for each $0 \leq i \leq n$ and each subinterval $J \subset I$.

The partitions from Lemma 4.14 are called σ-approximatively s-dispersible.

4.15. DEFINITION. A multivalued mapping $F : X \to L_1(I,E)$ is said to be *dispersible* if for each $x_0 \in X$, $\varepsilon > 0$, $s \in \Delta^n$ and each functions $u_0, u_1, \ldots, u_n \in F(x_0)$ there exist a neighborhood $V(x_0)$ of the point x_0 and a number $\sigma > 0$ such that for any σ-approximatively s-dispersible partition $P = \{P_i\}_{i=0}^n$, the function $\sum_{i=0}^{n} u_i \cdot \chi_{P_i}$ is contained in $D(F(x),\varepsilon)$, for every point $x \in V(x_0)$.

After checking that a LSC mapping F is dispersible whenever for each point $x \in X$ the value $F(x)$ is a decomposable set, one can obtain the generalization of the Fryszkowski, Bressan and Colombo theorems to arbitrary paracompact domains.

The following theorem substantially generalizes GONCHAROV and TOLSTONOGOV [1994] to paracompact domains:

4.16. THEOREM (AGEEV and REPOVŠ [2000]). *Let $(T, \mathbb{A}, \mu)$ be a separable measurable space and X be a paracompact space. Let $F : X \to L_1(T, E)$ be a dispersible closed-valued mapping and $\{ G_i : X \to L_1(T, E)\}_{i \in \mathbb{N}}$ be a sequence of dispersible multivalued mappings with open graphs such that $D(G_i(x); \varepsilon_i) \subset G_{i+1}(x)$, where the sequence $\{\varepsilon_i\}$ does not depend on $x \in X$. If for every point $x \in X$ the intersection $\Phi(x) = F(x) \cap G(x)$ is nonempty, where $G(x) = \bigcup_{i=1}^{\infty} G_i(x)$, then the multivalued mapping $\Phi : X \to L_1(T, E)$, $x \mapsto \Phi(x)$, admits a continuous selection.*

In a series of papers TOLSTONOGOV [1999a, 1999b, 1999c] studied selections passing through fixed points of multivalued contractions, depending on a parameter, with decomposable values. In particular, such parametric fixed points sets are absolute retracts and the sets of such selections are dense in the set of all continuous selections of the convexified mappings. Earlier, GORNIEWICZ and MARANO [1996] proposed unified approach for proving such nonparametric facts as for convex-valued contraction and also for decomposable-valued contractions (see also GORNIEWICZ, MARANO and SLOSARSKI [1996]).

4.E. Continuing the subject of unusual convexities, let us say something about some papers which are related to different kinds of such structures. Saveliev proposed a relaxation of Michael's axiomatic structure $\{(M_n, k_n)\}$ on a metric space M (see MICHAEL [1959b]) in the following three directions. First, he assumed M to be uniform. This reminds one of the approaches of GEILER [1970] and VAN DE VEL [1993b]. Second, the convex combination functions k_n were assumed to be multivalued. Recall that we have met such a situation earlier for decomposable-valued mappings. Moreover the sequence of mappings k_n was replaced by a multivalued (and partially defined) mapping C from the set $\Delta(M)$ of all formal convex combinations of elements of M into M. This repeats the approach of HORVATH [1991]. Briefly, a convexity on M is defined as a triple (M, C, Z) where Z is a topology on M which may be different from the uniform topology of M.

4.17. THEOREM (SAVELIEV [2000]). *Let X be a normal space, M a complete uniform space, and (M, C, Z) a continuous convexity with a countable convex uniform base and with uniform topology of M which is finer than the topology Z. Then every LSC closed-valued and convex-valued mapping from X to Z admits a selection whenever $p(X) \geq l_u(M)$.*

Here $l_u(M)$ denotes the Lindelöf number of the uniformity and $p(X)$ is the largest cardinal number $\mu \leq \kappa$ (where κ is a cardinal much bigger than cardinalities of all sets considered) such that each open cover of X whose cardinality is less than μ has a locally finite open refinement. Note that $p(X) = \kappa$ for a paracompact space X. Hence by putting $Z = M$ in Theorem 4.17 one can obtain the selection theorem for paracompact domains. Such a theorem includes as a special cases the convex-valued selection theorems of MICHAEL [1959b], CURTIS [1985], HORVATH [1991], and VAN DE VEL [1993a, 1993b].

JI-CHENG HOU [2001] proved a selection theorem for mappings into spaces having H-structure (in the sense of Horvath) which are ball-locally-uniformly LSC, but in general not LSC (see Part B of REPOVŠ and SEMENOV [1998a]).

Colombo and Goncharov considered a specific type of convexity in Hilbert spaces.

4.18. DEFINITION. A closed subset K of a Hilbert space is called *ϕ-convex* if there exists a continuous function $\phi : K \to [0, \infty)$ such that

$$\langle v, y - x \rangle \ \leq \ \phi(x)\|v\| \cdot \|y - x\|^2$$

for all $y, x \in K$ and all v proximally normal to K at x.

All convex sets as well as sets with sufficiently smooth boundary are ϕ-convex. In such sets one can obtain a kind of geodesic between two points which allows a convexity type structure.

4.19. THEOREM (COLOMBO and GONCHAROV [2001]). *Each continuous mapping from a metric space into a finite-dimensional Euclidean space admitting as values closed simply connected C^2-manifolds with negative sectional curvature, uniformly bounded from below, has a dense family of continuous selections.*

Continuous singlevalued selections f of a given multivalued mapping F are usually constructed as uniform limits of sequences of certain approximations $\{f_n\}$ of F. Practically all known selection results have been obtained by using one of the following two approaches for a construction of $\{f_n\}$. In the first (and the most popular) one, the method of outside approximations, mappings f_n are continuous ε_n-selections of F, i.e. $f_n(x)$ all lie *near* the set $F(x)$ and all mappings f_n are *continuous*. In the second one, the method of inside approximations, f_n are δ_n-continuous selections of F, i.e. $f_n(x)$ all lie *in* the set $F(x)$, however f_n are discontinuous.

In REPOVŠ and SEMENOV [1999] continuous selections were constructed as uniform limits of a sequence of δ-continuous ε-selections. Such a method was needed in order to unify different kinds of selection theorems. Namely, one forgets about closedness of values $F(x)$ over a countable subset C of domain and restricts nonconvexity of values $F(x)$ outside a zero-dimensional subset of domain. The density theorem holds as well.

4.20. THEOREM (REPOVŠ and SEMENOV [1999]). *Let $\beta : (0, \infty) \to (0, \infty)$ be a weakly g-summable function and $F : X \to Y$ a lower semicontinuous mapping from a paracompact space X into a Banach space Y. Suppose that $C \subset X$ is a countable subset of the domain such that values $F(x)$ are closed for all $x \in X \setminus C$ and that $Z \subset X$ with $dim_X Z \leq 0$. Then F has a singlevalued continuous selection, whenever $\beta(\cdot)$ is a pointwise strong majorant of the function* $(\sup\{\alpha_{Cl(F(x))}(\cdot) \mid x \in X \setminus Z\})^+$.

In REPOVŠ and SEMENOV [2001] we compared the nonconvexity of the set and nonconvexity of its ε-neighborhoods. The answers depend on smoothness properties of a unit sphere of a Banach range space. Hence on the one hand there exist a 4-dimensional Banach space B, its 1-dimensional subset P and a sequences ε_n, t_n of positive reals tending to zero such that the function of nonconvexity $\alpha_P(\cdot)$ always is less than some $q \in [0, 1)$ while $\alpha_{D(P,\varepsilon_n)}$ are identically equal to 1 on the intervals $(0, t_n)$ (i.e., nonconvexity of neighborhoods in principle differs from nonconvexity of the set). On the other hand, for uniformly convex Banach spaces, the inequality $\alpha_P(\cdot) < q < 1$ always implies the inequality $\alpha_{D(P,\varepsilon)}(\cdot) < p < 1$. Note, that even on the Euclidean plane there are examples with $q < p$.

4.F. Recently a new series of papers on continuous selections has appeared. All of them, are related to the substitution in selection theory of Lebesgue dimension dim by *extension dimension*. The extension dimension of a topological space equals to a class of CW-complexes, not a natural number. This notion was introduced by DRANISHNIKOV [1995] (see also DRANISHNIKOV and DYDAK [1996]). A comprehensive survey of the subject can be found in CHIGOGIDZE [2002].

KARASEV [200?] proved an $extdim$-analogue of Michael's finite-dimensional theorem. BRODSKY, CHIGOGIDZE and KARASEV [2002] found a unified "filtered" approach to both selection and approximation results with respect to $extdim$-theory.

For a CW-complexes L and K we say that $L \leq K$ if

$$(L \in AE(X)) \quad \Longrightarrow \quad (K \in AE(X))$$

for each X from a suitable class of spaces. First, authors were interested in the separable and metrizable situation. It now seems that proofs remain valid in general position, i.e. for paracompact domains and completely metrizable ranges. Thus $L \sim K$ if $L \leq K$ and $K \leq L$ and $[L]$ denotes the equivalence class.

4.21. DEFINITION. A space X is said to have *extension dimension* $\leq [L]$ (notation: $ed(X) \leq [L]$) if $L \in AE(X)$.

Clearly, $\dim X \leq n$ is equivalent to $ed(X) \leq [S^n]$ and $\dim_G X \leq n$ is equivalent to $ed(X) \leq [K(G,n)]$, where $K(G,n)$ is the Eilenberg-MacLane complex. One can develop homotopy and shape theories specifically designed to work for at most $[L]$-dimensional spaces (see CHIGOGIDZE [2002]). The theories were developed mostly for *finitely* dominated complexes L. Absolute extensors for at most $[L]$-dimensional spaces in a category of continuous maps are precisely $[L]$-soft mappings. And compacta of trivial $[L]$-shape are precisely $UV^{[L]}$-compacta.

As for selection theorems, they are presently known to be true for *finite* complexes L only. All notions used in the filtered approach to selection theorem (see Subsection 3.3 above) admit natural extensional analogues. For example, a pair of spaces $V \subset U$ is said to be $[L]$-*connected* if for every paracompact space X of extension dimension $ed(X) \leq [L]$ and for every closed subspace $A \subset X$ any mapping of A into V can be extended to a mapping of X into U. Or, a multivalued mapping $F : X \to Y$ is called $[L]$-*continuous at a point* $(x, y) \in \Gamma_F$ of its graph if for every neighborhood Oy of the point $y \in Y$, there is a neighborhood $O'y$ of the point y and a neighborhood Ox of the point $x \in X$ such that for all $x' \in Ox$, the pair $F(x') \cap O'y \subset F(x') \cap Oy$ is $[L]$-connected. Hence the above filtered Theorems 3.1 and 3.5 admit the following generalization.

4.22. THEOREM (BRODSKY, CHIGOGIDZE and KARASEV [2002]). *Let L be a finite CW-complex such that $[L] \leq [S^n]$ for some n. Let X be a paracompact space of extension dimension $ed(X) \leq [L]$. Let a complete lower $[L]$-continuous multivalued mapping Φ of X into a complete metric space Y contain an n-$UV^{[L]}$-filtered compact submapping Ψ which is singlevalued on some closed subset $A \subset X$. Then any neighborhood U of the graph Γ_Ψ in the product $X \times Y$ contains the graph of a singlevalued continuous selection s of the mapping Φ which coincides with $\Psi|_A$ on the set A.*

In particular, under the assumptions of Theorem 4.22 each complete lower $[L]$-continuous multivalued mapping $F : X \to Y$ into a complete metric space has a continuous selection.

There exists an extension dimensional version of Uspenskii's selection theorem for C-domains which we used in Section 1 above.

4.23. THEOREM (BRODSKY and CHIGOGIDZE [200?]). *Let L be a finite CW-complex and $F : X \to Y$ a multivalued mapping of a paracompact C-space X of extension dimension $ed(X) \le [L]$ to a topological space Y. If F admits infinite fiberwise $[L]_c$-connected filtration of strongly LSC multivalued mappings, then F has a singlevalued continuous selection.*

Recall, that a multivalued mapping F is said to be *strongly lower semicontinuous* if for any point $x \in X$ and any compact set $K \subset F(x)$ there exists a neighborhood V of x such that $K \subset F(z)$ for every $z \in V$.

As for an applications, we mention two facts concerning the so-called *Bundle problem* (see Problem 5.12. below):

4.24. THEOREM (BRODSKY, CHIGOGIDZE and SHCHEPIN [200?]). *Let $p : E \to B$ be a Serre fibration of LC^0-compacta with a constant fiber which is a compact two-dimensional manifold. If $B \in ANR$, then any section of p over closed subset $A \subset B$ can be extended to a section of p over some neighborhood of A.*

4.25. THEOREM (BRODSKY, CHIGOGIDZE and SHCHEPIN [200?]). *Let $p : E \to B$ be a topologically regular mapping of compacta with fibers homeomorphic to a 3-dimensional manifold. If $B \in ANR$, then any section of p over closed subset $A \subset B$ can be extended to a section of p over some neighborhood of A.*

5. Open problems

5.1. PROBLEM (MICHAEL). Let Y be a G_δ-subset of a Banach space B. Does then every LSC mapping $F : X \to Y$ of a paracompact space X with convex closed (in Y) values have a continuous selection?

We wish to emphasize that this problem is infinite-dimensional by its nature. Indeed, for $\dim X < \infty$ one can simply apply the finite-dimensional selection theorem. For C-domains X see proof in Section 1. Gutev has observed that using so-called selection-factorization technique of Choban and Nedev the problem for $\dim B < \infty$ reduces to metrizable domains and then one can apply the Michael selection theorem for perfectly normal domains and nonclosed-valued mappings into a separable range space. Maybe, one should first attempt it for Hilbert spaces B or for reflexive spaces B.

5.2. PROBLEM. Is it true that the affirmative answer to Problem 5.1 for an arbitrary Banach space B characterizes C-property of the domain?

5.3. PROBLEM (CHOBAN, GUTEV and NEDEV). Does each LSC closed- and convex-valued mapping from a collectionwise normal and countably paracompact domain into a Hilbert space admit a singlevalued continuous selection?

The following problems due to van de Vel relate to an axiomatic definition of the convexity notion. For detailed discussion see VAN DE VEL [1993a, 1993b].

5.4. PROBLEM (VAN DE VEL). Does every LSC compact- and convex-valued mapping from a normal domain into a metric space Y endowed by an uniform convex system admit a singlevalued continuous selection?

5.5. PROBLEM (VAN DE VEL). Problem 5.4 for paracompact domains and closed-valued mappings.

Recall, that a *convex system* on a set Y means a collection of subsets of Y which is closed for intersection and for chain union. The difference from convex structures is that the set Y itself needs not be convex. Subsets of convex sets are called admissible - they are the sets which have a convex hull. This definition includes the structures defined earlier by MICHAEL [1959b] and CURTIS[1985]. A *polytope* is the convex hull of an admissible finite set. Of course, singletons are assumed to be convex. The uniformity of the convex system means that all polytopes are compact, all convex sets are connected and the (partial) convex hull operator is uniformly continuous i.e. for all $\varepsilon > 0$ there is $\delta > 0$ such that if two finite sets A and B are δ-close in the Hausdorff metric, then $hull(A)$ and $hull(B)$ are ε-close. The answers are affirmative for uniform convex systems with real parameters, as defined by Michael and Curtis. It also holds for uniform metric convex systems which can be extended to uniform metric convex structures. Such an extension problem is in general also unsolved.

RICCERI [1987] proved that a multivalued mapping G from an interval $I \subset \mathbb{R}$ into a topological space Y admits an LSC multivalued selection H whenever the graph of G is connected and locally connected and for every open set $\Omega \subset Y$, the set $G^{-}(\Omega) \cap \operatorname{int}(I)$ has no isolated points. He stated the following factorization problem.

5.6. PROBLEM (RICCERI). Let X and Y be any topological spaces and let $F : X \to 2^Y$. Find suitable conditions under which there exist an interval $I \subset \mathbb{R}$, a continuous function $h : X \to I$ and a mapping $G : I \to 2^Y$, satisfying the following properties:

(1) $G(h(x)) \subset F(x)$ for all $x \in X$;

(2) The graph of G is connected and locally connected; and

(3) For every open set $\Omega \subset Y$, the set $G^{-}(\Omega) \cap \operatorname{int}(I)$ has no isolated points.

The motivation for this problem comes from the fact that each time it has a positive answer, the multifunction F admits a lower semicontinuous multiselection with nonempty values.

Next, we reproduce some problems proposed by GUTEV and NOGURA [200?b]. Below, $Sel(X)$ and $Sel_n(X)$ means the set of all continuous (with respect to Vietoris topology) selectors for closed subsets of X and for subsets, consisting of $\leq n$ elements. And a space X is *zero-dimensional* if it has a base of clopen sets, i.e. if $\operatorname{ind} X = 0$.

5.7. PROBLEM. Does there exist a space X such that $Sel_2(X) \neq \emptyset$ but $Sel_n(X) = \emptyset$ for some $n > 2$?

Recall that $Sel_2(X) \neq \emptyset \implies X$ is linearly ordered topological space $\implies Sel(X) \neq \emptyset$ for compact space X, but $Sel_2(\mathbb{R}) \neq \emptyset$ while $Sel(\mathbb{R}) = \emptyset$. Having also $Sel_2(X) \neq \emptyset \implies \operatorname{ind} X \leq 1$ for compact X it is naturally ask the following:

5.8. PROBLEM. Does there exist a space X such that $Sel_2(X) \neq \emptyset$ and $\operatorname{ind} X > 1$?

As mentioned above (see Section 2), $(Sel(X) \neq \emptyset \;\&\; \operatorname{ind} X = 0) \Longrightarrow \{f(X) : f$ is a continuous selector$\}$ is dense in X. In the other direction X is totally disconnected whenever $\{f(X) : f$ - continuous selector$\}$ is dense in X.

5.9. PROBLEM. Does there exist a space X which is not zero-dimensional but $\{f(X) :$ f is a continuous selector$\}$ is dense in X?

In comparison with results in ENGELKING, HEATH and MICHAEL [1968] and CHOBAN [1970], the following question seems to be interesting:

5.10. PROBLEM (GUTEV and NOGURA). Does there exist a zero-dimensional metrizable space X such that $\mathcal{F}(X)$ has a continuous selector but $\dim X \neq 0$?

The following *Bundle problem* has a negative answer for $n > 4$, a positive answer for $n = 1$ and partially positive solution for $n = 2$ (see Theorem 3.7 above). That for mappings between finite-dimensional compacta and for $n > 4$ the answer is affirmative.

5.11. PROBLEM (SHCHEPIN). Let $p : E \to B$ be a Serre fibration with a constant fiber which is an n-dimensional manifold. Is p a locally trivial fibration?

Shchepin has proposed Problems 5.12-5.15 related to the Bundle problem below:

5.12. PROBLEM. Does every open mapping of a locally connected continuum onto arc have a continuous section?

5.13. PROBLEM. Is any piecewise linear n-soft mapping of compact polyhedra a Serre n-fibration?

5.14. PROBLEM. Does every Serre fibration with a compact locally connected base have a global section if all of its fibers are contractible compact 4-manifolds with boundary?

5.15. PROBLEM. Is the complex-valued mapping $z_1^3 + z_2^3$ of $\mathbb{C}^2$ (2-dimensional complex space) onto $\mathbb{C}^1$ a Serre 1-fibration?

One of the approaches to a possible solution of the Bundle problem for $n = 2$ relates to a convexity-like structures in the space $H_0(D^2)$ of all autohomeomorphisms of two-dimensional disk which act identically over the boundary of the disk. Hence we can ask the following problems concerning the function of nonconvexity of H_0 considered as a subset of the Banach space $C(D^2; \mathbb{R}^2)$. Recall, that H_0 is a contractible ANR and moreover it is homeomorphic to the Hilbert space (Mason's theorem). The following three problems concerning paraconvexity of the set $H_0(D^2)$ are due to Shchepin and Semenov.

5.16. PROBLEM. Let $f \in H_0(D^2)$ and $dist(f, id|_{D^2}) = 2r$. Estimate the distance $dist(\frac{f+id}{2}; H_0)$. Is $0,5r$ the correct answer?

Passing to higher dimensional simplices we obtain:

5.17. PROBLEM. Let $f_1, f_2, ..., f_n \in H_0(D^2)$ and $f \in conv\{f_1, f_2, ..., f_n\}$. Is it true that $dist(f; H_0) \leq 0,5r$ where r is minimal radius of a ball which covers all $f_1, f_2, ..., f_n$?

As a preliminary step in attacking Problem 5.17 one can change autohomeomorphisms of disk to embeddings of a segment into the plane.

5.18. PROBLEM. Let f and g be two embeddings of the segment $[0, 1]$ into the Euclidean plane and $dist(f, g) = 2r$. Estimate the distance between the mapping $\frac{f+g}{2}$ and the set of all embeddings of this segment to the plane.

The two following problems are due to Semenov. The Nash's embedding theorem asserts that for each Riemannian metric ρ on a smooth compact manifold M^n there exists an isometric embedding of $(M^n; \rho)$ to $\mathbb{R}^N$ where $N \sim n^2$. Let us consider embeddings into the infinite-dimensional Hilbert space H.

5.19. PROBLEM. For each Riemannian metric ρ on M define $F(\rho)$ as the set of all isometric embeddings of $(M, \rho) \hookrightarrow H$. Does then the multivalued mapping F admit a continuous selection?

Note, that each value $F(\rho)$ is invariant under the action of the orthogonal group $O(H)$ which is homotopically trivial (due to the Kuiper theorem).

The following problem arises as a possible unification of proofs of Michael selection Theorem 1.1 and Fryszkowski selection theorem. For a Banach space B consider a subset S of all continuous linear operators on B with the property that $A \in S \implies (id - A) \in S$ and define S-convex subsets $C \subset B$ as the sets with the property that $x \in C, y \in C, A \in S \implies (id - A)x + Ay \in C$.

5.20. PROBLEM. Find a suitable axiomatic restrictions for S under which Theorem 1.1 holds for mappings with S-convex values.

Observe, that for $S = \{t \cdot id | t \in [0, 1]\}$ we obtain the standard convexity in Banach space B and for $B = L_1(T, \mu)$ and $S =$ the set of all operators of multiplications on characteristic functions of measurable sets S-convexity coincides with decomposability.

Recall that a subset M of a Banach space B is said to be *proximinal* if for each $x \in B$ the set $P_M(x)$ of points of M which are nearest to x is non-empty.

5.21. PROBLEM (DEUTSCH). Is there a semicontinuity condition on the metric projection P_M onto a proximinal subspace M in a Banach space that is both necessary and sufficient for the metric projection to admit a continuous selection?

FISCHER [1988] proved that for a Banach space of continuous functions on compacta and for any finite-dimensional subspace M an affirmative answer is given by the so-called *almost lower semicontinuity*. This last notion was introduced for general multivalued mappings by DEUTSCH and KENDEROV [1983] and characterizes the property of a multivalued mapping to have continuous ε-selections for any positive ε. It should be mentioned that LI [1991] has given an *intrinsic* characterization of those finite-dimensional subspaces M of $C_0(T)$ such that P_M admits a continuous selection. (These are the so-called *weakly regularly interpolating* subspaces.)

Recently BROWN, DEUTSCH, INDUMATHI AND KENDEROV [2002] have established a geometric characterization of those Banach spaces X in which P_M admits a continuous selection for any one-dimensional subspace M.

References

AGEEV, S.M. and D. REPOVŠ

[2000] On selection theorems with decomposable values, *Topol. Meth. Nonlin. Anal.* **15**, 385–399.

ANTOSIEWICZ, H.A. and A. CELLINA

[1975] Continuous selections and differential relations, *J. Diff. Eq.* **19**, 386–398.

ARTICO, G. and U. MARCONI

[2001] Selections and topologically well-ordered spaces, *Topol. Appl.* **115**, 299–303.

ARTICO, G., U. MARCONI, R. MORESCO and J. PELANT

[2001] Selectors and scattered spaces, *Topol. Appl.* **111**, 35–48.

ARUTYUNOV, A.V.

[2000] *Optimality conditions: Abnormal and Degenerate Problems*, Math. and Its Appl., vol. 526, Kluwer, Dordrecht.

[2001] Special selectors of multivalued mappings (Russian), *Dokl. Akad. Nauk* **377** (3), 298–300.

AUBIN, J.-P. and A. CELLINA

[1984] *Differential Inclusions. Set-Valued Maps and Viability Theory*, Grundl. der Math. Wiss. vol. 264, Springer-Verlag, Berlin, xiii+342 pp.

BEER, G.

[1993] *Topologies on Closed and Closed Convex Sets*, Math. and Its Appl., vol. 268, Kluwer, Dordrecht.

BEN-EL-MECHAIEKH, H. and W. KRYSZEWSKI

[1997] Equilibria of set-valued maps on nonconvex domains, *Trans. Amer. Math. Soc.* **349**, 4159–4179.

BERTACCHI, D. and C. COSTANTINI

[1998] Existence of selections and disconnectedness properties for the hyperspace of an ultrametric space, *Topol. Appl.* **88**, 179–197.

BIELAWSKI, R.

[1989] A selection theorem for open-graph multifunctions, *Fund. Math.* **133**, 97–100.

BORST, P.

[200?] Some remarks concerning C-spaces, preprint.

BRESSAN, A. and G. COLOMBO

[1988] Extensions ans selections of maps with decomposable values, *Studia Math.* **90**, 69–86.

BRODSKY, N.B.

[1999] On extension of compact-valued mappings, *Uspekhi Mat. Nauk* **54**, 251-252. English transl. in *Russian Math. Surveys* **54**, 256–257.

[2000] *Selections of filtered multivalued mappings* (Russian), Ph. D. Thesis., Moscow State University, Moscow.

[2002] Sections of maps with fibers homeomorphic to a two-dimensional manifold, *Topol. Appl.* **120**, 77–83.

[200?] Selections and approximations, preprint.

BRODSKY, N.B. and A. CHIGOGIDZE

[200?] Extension dimensional approximation theorem, preprint.

BRODSKY, N.B., A. CHIGOGIDZE and A. KARASEV

[2002] Approximations and selections of multivalued mappings of finite-dimensional space, *JP Journal of Geometry and Topology* **2**, 29–73.

BRODSKY, N.B., A. CHIGOGIDZE and E. SHCHEPIN

[200?] Sections of Serre fibrations with low-dimensional fibers, in preparation.

BROWN, A.L., F. DEUTSCH, V. INDUMATHI and P. KENDEROV

[2002] Lower semicontinuity concepts, continuous selections and set-valued metric projections, *J. of Approx. Theory* **115**, 120–143.

CALBRIX, J. and B. ALLECHE

[1996] Multifunctions and Čech-complete spaces, *Proc. of the 8-th Prague Top. Symp.*, 30–36.

CAUTY, R.

[1999] Sur les sélections continus des collections d'arcs, *Rend. Istit. Mat. Univ. Trieste* **31**, 135–142.

CHIGOGIDZE, A.

[2002] Infinite dimensional topology and shape theory, in: *Handbook of Geometric Topology*, eds. R. Daverman and R. Sher, North-Holland, Amsterdam, 307–371.

CHOBAN, M.M.

[1970] Manyvalued mappings and Borel sets, I (Russian), *Trudy Mosk. Mat. Obsch.* **22**, 229–250. Engl. transl. in: *Trans. Moscow Math. Soc.* **22**, 258–280.

CHOBAN, M.M. and S.Y. NEDEV

[1997] Continuous selections for mappings with generalized ordered domain, *Math. Balkanica (N. S.)* **11**, 87–95.

COLOMBO, G. and V.V. GONCHAROV

[2001] Continuous selections via geodesics, *Topol. Methods Nonlin. Anal.* **18**, 171–182.

COSTANTINI, C. and V. GUTEV

[200?] Recognizing special metrics by topological properties of the "metric"-proximal hyperspace, *Tsukuba J. Math*, to appear.

CURTIS, D.W.

[1985] Application of selection theorem to hyperspace contractibility, *Can. J. Math.* **37**, 747–759.

DAVY, J.L.

[1972] Properties of the solution set of generalized differential equation, *Bull. Austral. Math. Soc.* **6**, 379–398.

DEUTSCH, F. and P. KENDEROV

[1983] Continuous selections and approximate selection for set-valued mappings and applications to metric projections, *SIAM J. Math. Anal.* **14**, 185–194.

DRANISHNIKOV, A.N.

[1993] K-theory of Eilenberg-MacLane spaces and cell-like mapping problem, *Trans. Amer. Math. Soc.* **335**, 91–103.

[1995] The Eilenberg-Borsuk theorem for mappings in an arbitrary complex (Russian), *Mat. Sb.* **185** (4), 81–90. English transl. in: *Russian Acad. Sci. Sb. Math.* **81**, 467–475.

DRANISHNIKOV, A.N. and J. DYDAK

[1996] Extension dimension and extension types (Russian), *Trudy Mat. Inst. Steklova* **212**, 61–94. English transl. in: *Proc. Steklov Inst. Math.* **212**, 55–88.

DROZDOVSKY, S.A. and V.V. FILIPPOV

[1998] A selection theorem for a new class of multivalued mappings (Russian), *Uspekhi Mat. Nauk* **53** (5), 235–236. Engl. transl. in: *Russian Math. Surveys* **53**, 1089–1090.

[2000] A selection theorem for a new class of multivalued mappings (Russian), *Mat. Zam.* **66**, 503–507. Engl. transl. in: *Math. Notes* **66**, 411–414.

ENGELKING, R., R.W. HEATH and E. MICHAEL

[1968] Topological well-ordering and continuous selections, *Invent. Math.* **6**, 150–158.

FISCHER, T.

[1988] A continuity condition for the existence of a continuous selection for a set-valued mapping, *J. Approx. Theory* **49**, 340–345.

FRYSZKOWSKI, A.

[1983] Continuous selections for a class of nonconvex multivalued maps, *Studia Math.* **76**, 163–174.

FRYSZKOWSKI, A. and L. GORNIEWICZ

[2000] Mixed semicontinuous mappings and their applications to differential inclusions, *Set-Valued Anal.* **8**, 203–217.

FUJII, S., K. MIYAZAKI and T. NOGURA

[2002] Vietoris continuous selections on scattered spaces, *J. Math. Soc. Japan* **54**, 273–281.

FUJII, S. and T. NOGURA

[1999] Characterizations of compact ordinal spaces via continuous selections, *Topol. Appl.* **91**, 65–69.

GARCIA-FERREIRA, S., V. GUTEV, T. NOGURA, M. SANCHIS and A. TOMITA

[200?] Extreme selectors for hyperspaces of topological spaces, *Topology Appl.*, to appear.

GEILER, V.A.

[1970] Continuous selectors in uniform spaces (Russian), *Dokl. Akad. Nauk* **195** (1), 17–19.

GONČAROV, V.V. and A.A. TOLSTONOGOV

[1994] Continuous selections of a family of nonconvex-valued mappings with a noncompact domain (Russian), *Sibir. Mat. Zh.* **35**, 537–553. English transl. in: *Siberian Math. J.* **35**, 479–494.

GORNIEWICZ, L. and S. MARANO

[1996] On the fixed point set of multivalued contractions, *Rend. Circ. Mat. Palermo* **40**, 139–145.

GORNIEWICZ, L., S. MARANO and M. SLOSARSKI

[1996] Fixed points of contractive multivalued maps, *Proc. Amer. Math. Soc.* **124**, 2675–2683.

GUTEV, V.G.

[1993] Selection theorems under an assumption weaker than lower semicontinuity, *Topol. Appl.* **50**, 129–138.

[1994] Continuous selections, G_δ-subsets of Banach spaces and upper semicontinuous mappings, *Comm. Math. Univ. Carol.* **35**, 533–538.

[1996] Selections and hyperspace topologies via special metrics, *Topol. Appl.* **70**, 147–153.

[2000a] Weak factorizations of continuous set-valued mappings, *Topol. Appl.* **102**, 33–51.

[2000b] Generic extension of finite-valued u.s.c. selections, *Topol. Appl.* **104**, 101–118.

[2000c] An exponential mapping over set-valued mappings, *Houston J. Math.* **26**, 721–739.

[2001] Fell-continuous selections and topologically well-orderable spaces II, *Proceedings of the Ninth Prague Topological Symposium*, 147–153

GUTEV, V.G. and S.Y. NEDEV
[2001] Continuous selections and reflexive Banach spaces, *Proc. Amer. Math. Soc.* **129**, 1853–1860.

GUTEV, V.G., S.Y. NEDEV, J. PELANT and V.M. VALOV
[1992] Cantor set selectors, *Topol. Appl.* **44**, 163–166.

GUTEV, V.G. and T. NOGURA
[2000] Selections for Vietoris-like hyperspace topologies, *Proc. London. Math. Soc.* **80**, 235–256.
[2001] Vietoris continuous selections and disconnectedness-like properties, *Proc. Amer. Math. Soc.* **129**, 2809–2815.
[200?a] Fell continuous selections and topologically well-orderable spaces, preprint.
[200?b] Some problems on selections for hyperspace topologies, preprint.

GUTEV, V.G. and V. VALOV
[2002] Continuous selections and C-spaces, *Proc. Amer. Math. Soc.* **130**, 233–242.

HATTORI, Y. and T. NOGURA
[1995] Continuous selections on certain spaces, *Houston J. Math.* **21**, 585–594.

HORVATH, C.D.
[1991] Contractibility and generalized convexity, *J. Math. Anal. Appl.* **156**, 341–357.
[2000] On the existence of constant selections, *Topol. Appl.* **104**, 119–139.

HOU, JI-CHENG
[2001] Michael's selection theorem under an assumption weaker than lower semicontinuity in H-spaces, *J. Math. Anal. Appl.* **259**, 501–508.

HUREWICZ, W.
[1928] Relativ perfekte Teile von Punktmengen and Mengen (A), *Fund. Math.* **12**, 78–109.

KARASEV, A.
[200?] Continuous selections with respect to extension dimension, preprint.

KÜNZI, H.P.A. and L.B. SHAPIRO
[1997] On simultaneous extension of continuous partial functions, *Proc. Amer. Math. Soc.* **125**, 1853–1859.

LI, WU.
[1991] *Continuous Selections for Metric Projections and Interpolating Subspaces*, Peter Lang, Frankfurt.

MÄGERL, G.
[1978] Metrizability of compact sets and continuous selections, *Proc. Amer. Math. Soc.* **72**, 607–612.

MICHAEL, E.A.
[1951] Topologies on spaces of subsets, *Trans. Amer. Math. Soc.* **71**, 152–182.
[1956a] Continuous selections, I, *Ann. of Math.* **63**, 361–382.
[1956b] Continuous selections, II, *Ann. of Math.* **64**, 562–580.
[1959a] Paraconvex sets, *Math. Scand.* **7**, 372–376.
[1959b] Convex structures and continuous selections, *Can. J. Math.* **11**, 556–575.
[1959c] A theorem on semi-continuous set-valued functions, *Duke. Math. J.* **26**, 647–651.
[1988] Continuous selections avoiding a set, *Topol. Appl.* **28**, 195–213.
[1989] A generalization of a theorem on continuous selections, *Proc. Amer. Math. Soc.* **105**, 236–243.
[1990] Some problems, in *Open Problems in Topology*, J. van Mill and G. M. Reed, eds., North Holland, Amsterdam, 271–278.

[1992] Selection theorems with and without dimensional restrictions, (Int. Conf. in Memory of F. Hausdorff), *Math. Research* **67**, 218–222.

MICHAEL, E.A. and C.P. PIXLEY

[1980] A unified theorem on continuous selections, *Pacific J. Math.* **87**, 187–188.

VAN MILL, J., J. PELANT and R. POL

[1996] Selections that characterize topological completeness, *Fund. Math.* **149**, 127–141.

VAN MILL, J. and G.M. REED, editors

[1990] *Open Problems in Topology*, North-Holland, Amsterdam.

VAN MILL, J. and E. WATTEL

[1981] Selections and orderability, *Proc. Amer. Math. Soc.* **83**, 601–605.

MIYAZAKI, K.

[2001a] Continuous selections on almost compact spaces, *Sci. Math. Jap.* **53**, 489–494.

[2001b] Characterizations of paracompact-like properties by means of set-valued semi-continuous selections, *Proc. Amer. Math. Soc.* **129**, 2777–2782.

NADLER, JR., S.B. and L.E. WARD, JR.

[1970] Concerning continuous selections, *Proc. Amer. Math. Soc.* **25**, 369–374.

NEDEV, S.Y.

[1987] A selection example, *C. R. Acad. Bulg. Sci.* **40** (11), 13–14.

NEDEV, S.Y. and V.M. VALOV

[1983] On metrizability of selectors, *C. R. Acad. Bulg. Sci.* **36**, 1363–1366.

[1984] Normal selectors for the normal spaces, *C. R. Acad. Bulg. Sci.* **37**, 843–846.

NOGURA, T. and D. SHAKHMATOV

[1997a] Characterizations of intervals via continuous selections, *Rend. Circ. Mat. Palermo* **46**, 317–328.

[1997b] Spaces which have finitely many continuous selections, *Boll. Un. Mat. Ital.* **A 11**, 723–729.

PIXLEY, C. P.

[1974] An example concerning continuous selections on infinite dimensional spaces, *Proc. Amer. Math. Soc.* **43**, 237–244.

REPOVŠ, D. and P.V. SEMENOV

[1995] On functions of nonconvexity for graphs of continuous functions, *J. Math. Anal. Appl.* **196**, 1021–1029.

[1998a] *Continuous Selections of Multivalued Mappings.* Math. and Its Appl., vol. 455, Kluwer, Dordrecht.

[1998b] On nonconvexity of graphs of polynomials of several real variables, *Set-Valued. Anal.* **6**, 39–60.

[1998c] Continuous selections of nonlower semicontinuous nonconvex-valued mappings, *Diff. Incl. and Opt. Contr.* **2**, 253–262.

[1999] Continuous selections as uniform limits of δ-continuous ε-selections, *Set-Valued Anal.* **7**, 239–254.

[2001] On a relation between nonconvexity of the set and nonconvexity of its ε-neiborhood (Russian), *Matem. Zam.* **70**(2), 246–259. Engl. transl. in: Math. Notes **70**, 221–232.

[200?] On a relative approximation theorem, *Houston J. of Math.*, to appear.

RICCERI, B.

[1987] On multiselections of one real variable, *J. Math. Anal. Appl.* **124**, 225–236.

SAVELIEV, P.

[2000] Fixed points and selections of set-valued maps on spaces with convexity, *Int. J. Math. Math. Sci.* **24**, 595–612.

SEMENOV, P.V.

[2000a] Nonconvexity in problems of multivalued calculus, *J. Math. Sci. (N. Y.)* **100**, 2682–2699.

[2000b] On the Lebesgue function of open coverings, *Topol. Appl.* **107**, 147–152.

SHCHEPIN E.V. and N.B. BRODSKY

[1996] Selections of filtered multivalued mappings (Russian), *Trudy Mat. Inst. Steklova* **212**, 220–240. Engl. Transl. in: *Proc. Steklov Inst. Math.* **212**, 218–239.

SHISHKOV, I.

[2001] σ-products and selections of set-valued mappings, *Comment. Math. Univ. Carol.* **42**, 203–207.

[2002] Extensions of l.s.c. mappings into reflexive Banach spaces, *Set-Valued Anal.* **10**, 79–87.

STEPANOVA, E.N.

[1993] Extension of continuous functions and metrizability of paracompact p-spaces (Russian), *Mat. Zametki* **53**, 92–101. English transl. in: *Math. Notes* **53**, 308–314.

TAYLOR, J.L.

[1975] A counterexample in shape theory, *Bull. Amer. Math. Soc.* **81**, 629–632.

TOLSTONOGOV, A.A.

[1999a] L_p-continuous selectors of fixed points of multivalued mappings with decomposable values. I: Existence theorems (Russian), *Sibir. Mat. Zhur.* **40**, 695–709. Engl. transl. in: *Sib. Math. J.* **40**, 595–607.

[1999b] L_p-continuous selectors of fixed points of multivalued mappings with decomposable values. II: Relaxation theorems (Russian), *Sibir. Mat. Zhur.* **40**, 1167–1181. Engl. transl. in: *Sib. Math. J.* **40**, 991–1003.

[1999c] L_p-continuous selectors of fixed points of multivalued mappings with decomposable values. III: Relaxation theorems (Russian), *Sibir. Mat. Zhur.* **40**, 1380–1396. Engl. transl. in: *Sib. Math. J.* **40**, 1173–1187.

USPENSKII, V.V.

[1998] A selection theorem for C-spaces, *Topol. Appl.* **85**, 351–374.

VALOV, V.M.

[2002] Continuous selections and finite C-spaces, *Set-Valued Anal.* **10**, 37–51.

VAN DE VEL, M.L.J.

[1993a] A selection theorem for topological convex structures, *Trans. Amer. Math. Soc.* **336**, 463–496.

[1993b] *Theory of Convex Structures*, North-Holland Math Library, vol. 50, North-Holland, Amsterdam.

CHAPTER 17

Convergence in the Presence of Algebraic Structure

Dmitri Shakhmatov

Department of Mathematical Sciences, Faculty of Science, Ehime University, Matsuyama 790-8577, Japan
E-mail: dmitri@dpc.ehime-u.ac.jp

Contents

RECENT PROGRESS IN GENERAL TOPOLOGY II
Edited by Miroslav Hušek and Jan van Mill

We present a survey of selected results intended to demonstrate that the behavior of convergence properties (such as Fréchet-Urysohn property, sequentiality, tightness and α_i-properties) tends to improve as one passes from general topological spaces to topological groups and function spaces. In our "Miscellaneous" section we also include some recent results about topological groups in which convergent sequences play a prominent role.

Unless stated explicitly to the contrary, a term "space" or "topological space" means "Tychonoff space", i.e. most spaces in this survey are assumed to be Tychonoff by default.

1. Definitions of main convergence properties

Let X be a topological space. For $A \subseteq X$ we use $\overline{A}$ to denote the closure of A in X. A *sequence converging to* $x \in X$ is a countable infinite set $S \subseteq X$ such that $S \setminus U$ is finite for every open neighborhood U of x. A space X is *Fréchet-Urysohn* provided that for each set $A \subseteq X$ and every point $x \in \overline{A}$ there exists a sequence $S \subseteq A$ converging to x.

Recall that a subset A of a topological space X is called *sequentially closed* in X provided that A contains the limit of any sequence of points of A that converges in X. A space X is *sequential* provided that all sequentially closed subsets of X are closed in X.

As usual, for a set A and a cardinal τ, $[A]^{\leq\tau}$ denotes the set of all subsets of A of size less or equal than τ. According to ARHANGEL'SKIĬ [1971] the *tightness* $t(X)$ of a topological space X is defined as the smallest cardinal τ such that $\overline{A} = \bigcup\{\overline{B} : B \in [A]^{\leq\tau}\}$ for every $A \subseteq X$.

It is well-known (and easy to see) that

$$\text{first countable} \to \text{Fréchet-Urysohn} \to \text{sequential} \to t(X) = \omega. \tag{17.1}$$

1.1. DEFINITION (ARHANGEL'SKIĬ [1972, 1979]). Let X be a topological space. For $i = 1, 2, 3$ and 4 we say that X is an *α_i-space*[1] provided that for every countable family $\{S_n : n \in \omega\}$ of sequences converging to some point $x \in X$ there exists a (kind of diagonal) sequence S converging to x such that:

(α_1) $S_n \setminus S$ is finite for all $n \in \omega$,

(α_2) $S_n \cap S$ is infinite for all $n \in \omega$,

(α_3) $S_n \cap S$ is infinite for infinitely many $n \in \omega$,

(α_4) $S_n \cap S \neq \emptyset$ for infinitely many $n \in \omega$.

1.2. DEFINITION (NYIKOS [1992]). We say that a space X is an *$\alpha_{3/2}$-space* provided that for every countable family $\{S_n : n \in \omega\}$ of sequences converging to some point $x \in X$ such that $S_n \cap S_m = \emptyset$ for $n \neq m$, there exists a sequence S converging to x such that $S_n \setminus S$ is finite for infinitely many $n \in \omega$.

One has

$$\text{metric} \to \text{first countable} \to \alpha_1 \to \alpha_{3/2} \to \alpha_2 \to \alpha_3 \to \alpha_4 \tag{17.2}$$

where the only non-trivial implication $\alpha_{3/2} \to \alpha_2$ is due to NYIKOS [1992].

---

[1] In [1972, 1979] ARHANGEL'SKIĬ used a different terminology. We adopt here the term α_i-space which is currently being used.

1.3. DEFINITION. A space X has the *Ramsey property* if, for every matrix $\mathcal{M} = \{x_{ij} : i,j \in \omega\}$ of points in X such that $\lim_{i\to\infty} \lim_{j\to\infty} x_{ij} = x$ for some point $x \in X$, there exists an infinite set $M \subseteq \omega$ with the following property: For every U, an open neighborhood of x, one can find $k \in \omega$ such that $x_{mn} \in U$ whenever $m, n \in M$ and $k \leq m < n$.

A somewhat weaker property than in the above definition first appeared in the classical paper of RAMSEY [1929].

1.4. THEOREM (NOGURA and SHAKHMATOV [1995]).

(i) *A space with the Ramsey property is* α_3.

(ii) *There exists an* α_1*-space without the Ramsey property.*

It is unclear if a space with the Ramsey property must be an α_2-space.

2. Convergence properties in topological spaces

Let us first mention that Fréchet-Urysohn spaces need not be α_4:

2.1. EXAMPLE. Let X be the *countable Fréchet-Urysohn fan*. That is, $X = (\omega\times\omega)\cup\{*\}$, where $* \notin \omega \times \omega$, all points of the set $\omega \times \omega$ are isolated and the family

$$\{\{*\} \cup \{(n,m) : m \leq f(n) \text{ for all } n \in \omega\} : f \in \omega^\omega\}$$

serves as the family of neighborhoods of the only non-isolated point $*$. Then X is Fréchet-Urysohn but is not an α_4.

It is well-known that none of the arrows in (17.1) can be reversed even for compact spaces. We now turn to a question of whether some arrows in (17.2) can be reversed.

2.2. THEOREM (SIMON [1980]). *There exists a compact Fréchet-Urysohn* α_4*-space that is not* α_3.

2.3. THEOREM (REZNICHENKO [1986], GERLITS and NAGY [1988], NYIKOS [1989]). *There exists a compact Fréchet-Urysohn* α_3*-space that is not* α_2.

For $f, g \in \omega^\omega$ we write $f <^* g$ if $f(n) < g(n)$ for all but finitely many $n \in \omega$. A family $\mathcal{F} \subseteq \omega^\omega$ is *unbounded* if for every function $g \in \omega^\omega$ there exists $f \in \mathcal{F}$ such that $g <^* f$. We define $\mathfrak{b}$ to be the smallest cardinality of an unbounded family in $(\omega^\omega, <^*)$.

2.4. THEOREM (NYIKOS [1992]). *If* $\mathfrak{b} = \omega_1$ *holds, then there exists a countable Fréchet-Urysohn* α_2*-space that is not* α_1.

2.5. THEOREM (NYIKOS [1992]). *The existence of the following spaces is consistent with ZFC:*

(i) *a compact Fréchet-Urysohn* α_2*-space that is not* $\alpha_{3/2}$,

(ii) *a compact Fréchet-Urysohn* $\alpha_{3/2}$*-space that is not* α_1.

2.6. THEOREM (DOW [1990]). *α_2 implies α_1 in the Laver model for the Borel conjecture.*

It follows that α_2, $\alpha_{3/2}$ and α_1 properties coincide in the Laver model for the Borel conjecture.

The following result is part of the folklore.

2.7. THEOREM. *Let $G = \{f \in \{0,1\}^{\omega_1} : |\{\beta \in \omega_1 : f(\beta) = 1\}| \leq \omega\}$. Then G is a countably compact Fréchet-Urysohn topological group that is α_1 but is not first countable.*

The group G in the previous example is not countable. In fact, all countable subsets of G are metrizable (and thus first countable). The question of the existence of a *countable* Fréchet-Urysohn α_1-space that is not first countable turns out to be delicate. We start with two results of independent interest.

2.8. THEOREM (NYIKOS [1989]). *Every space of character $< \mathfrak{b}$ is α_1.*

2.9. THEOREM (MALYHIN and SHAPIROVSKIĬ [1974]). *If* MA *holds, then every countable space of character $< 2^\omega$ is Fréchet-Urysohn.*

Let G be a subgroup of $\{0,1\}^{\omega_1}$ algebraically generated by any countable dense subset of $\{0,1\}^{\omega_1}$. Clearly G is countable. Since $\omega_1 < \mathfrak{b} = 2^\omega$ under MA $+\neg$ CH, G is a Fréchet-Urysohn α_1-space by Theorems 2.9 and 2.8. Thus one gets

2.10. COROLLARY (MALYHIN, 1978). MA $+\neg$ CH *implies the existence of a countable Fréchet-Urysohn group that is an α_1-space but is not first countable.*

It turns out that the existence of a countable Fréchet-Urysohn α_1-space that is not first countable is not only consistent with ZFC but also independent of ZFC:

2.11. THEOREM (DOW and STEPRĀNS [1992]). *There is a model of ZFC in which all countable Fréchet-Urysohn α_1-spaces are first countable.*

The next result demonstrates that one cannot get the natural strengthening of both Theorems 2.6 and 2.11 simultaneously:

2.12. THEOREM (GERLITS and NAGY [1988], NYIKOS [1989]). *There exists a countable Fréchet-Urysohn α_2-space that is not first countable.*

3. Convergence properties in topological groups

Example 2.1 shows that our next theorem is specific for topological groups.

3.1. THEOREM (NYIKOS [1981]). *Every Fréchet-Urysohn topological group is α_4.*

It turns out that α_4 is precisely the property which in conjunction with sequentiality gives the Fréchet-Urysohn property in topological groups:

3.2. THEOREM (NYIKOS [1981]). *A sequential topological group is Fréchet-Urysohn if and only if it is* α_4.

As a corollary of Theorem 3.1 NYIKOS [1981] notices that the Fréchet-Urysohn space from Example 2.1 cannot be embedded as a subspace of a Fréchet-Urysohn topological group. It is unclear if every sequential space (or even Fréchet-Urysohn space) can be realized as a (preferably closed) subspace of some sequential group. (We note that the space from Example 2.1 is a closed subspace of a sequential group, namely, its free topological group.)

3.3. QUESTION (NYIKOS [1981]). Let G be a group equipped with a Fréchet-Urysohn topology with respect to which multiplication is continuous. Is then G an α_4-space?

We next examine whether some arrows in (17.2) can be reversed in the class of topological groups.

3.4. THEOREM (SHAKHMATOV [1990]). *Let* M *be a model of ZFC obtained by adding* ω_1 *many Cohen reals to an arbitrary model of ZFC. Then* M *contains a countable Fréchet-Urysohn topological group* G *that is not* α_3. (Note that G is α_4 by Theorem 3.1.)

3.5. THEOREM (SHIBAKOV [1999a]). CH *implies the existence of a countable Fréchet-Urysohn topological group that is* α_3 *but is not* α_2.

3.6. THEOREM (SHAKHMATOV [1990]). *Let* M *be a model of ZFC obtained by adding* ω_1 *many Cohen reals to an arbitrary model of ZFC. Then* M *contains a countable Fréchet-Urysohn topological group* G *that is* α_2 *but is not* $\alpha_{3/2}$.

3.7. THEOREM (SHIBAKOV [1999a]). *A Fréchet-Urysohn topological group that is an* $\alpha_{3/2}$*-space is* α_1. *Thus* $\alpha_{3/2}$ *and* α_1 *are equivalent for Fréchet-Urysohn topological groups.*

It seems unclear if $\alpha_{3/2}$ and α_1 are equivalent for all (i.e. not necessarily Fréchet-Urysohn) topological groups.

3.8. QUESTION (SHAKHMATOV [1990]). Is it consistent with ZFC that every Fréchet-Urysohn topological group is α_3? What about countable Fréchet-Urysohn topological groups?

3.9. QUESTION. Is it consistent with ZFC that every Fréchet-Urysohn topological group that is an α_3-space is automatically α_2? What about countable Fréchet-Urysohn topological groups?

3.10. QUESTION (SHAKHMATOV [1990]). Is it consistent with ZFC that every *countable* Fréchet-Urysohn topological group that is an α_2-space is first countable?

The reader may have noticed that all of the examples of Fréchet-Urysohn topological groups distinguishing α_i-properties presented so far are of consistency nature. This is because even the following fundamental problem is still open:

3.11. PROBLEM (MALYHIN, 1978). Without any additional set-theoretic assumptions beyond ZFC, does there exist a *countable* Fréchet-Urysohn topological group that is not first countable?

A consistent example was given in Corollary 2.10. The word "countable" is essential in view of Theorem 2.7.

We will now present a possible approach to constructing a ZFC example. First let us remind ourselves the following folklore construction of topological groups. Let $[\omega]^{<\omega}$ denote the set of all finite subsets of ω. For $A, B \in [\omega]^{<\omega}$ define $A{\cdot}B = (A \setminus B) \cup (B \setminus A) \in [\omega]^{<\omega}$. This operation makes $[\omega]^{<\omega}$ into an Abelian group with $\emptyset$ as the identity element such that $A \cdot A = \emptyset$ (thus A coincides with its own inverse, and all elements of $[\omega]^{<\omega}$ have order 2). Let $\mathcal{F}$ be a filter on ω. Let $G(\mathcal{F})$ be the group $([\omega]^{<\omega}, \cdot, \emptyset)$ equipped with the topology whose base of open neighborhoods of $\emptyset$ is given by the family $\{[F]^{<\omega} : F \in \mathcal{F}\}$. It is easy to see that $G(\mathcal{F})$ is Hausdorff if and only if $\mathcal{F}$ is free (i.e. $\bigcap \mathcal{F} = \emptyset$), and $G(\mathcal{F})$ is first countable if and only if the filter $\mathcal{F}$ is countably generated. Finally, let $\omega_{\mathcal{F}}$ be the space obtained by adding to the discrete copy of ω a single point $*$ whose filter of open neighborhoods is $\{F \cup \{*\} : F \in \mathcal{F}\}$. Observe that $\omega_{\mathcal{F}}$ is naturally homeomorphic to a subspace of $G(\mathcal{F})$.

3.12. THEOREM (NYIKOS [1992]).

Let $\mathcal{F}$ be a free filter on ω and let $G(\mathcal{F})$ and $\omega_{\mathcal{F}}$ be as described above. Then:

(i) α_4 is equivalent to α_2 for $G(\mathcal{F})$.

(ii) If $G(\mathcal{F})$ is Fréchet-Urysohn, then $G(\mathcal{F})$ (and hence $\omega_{\mathcal{F}}$) is α_2.

(iii) If $G(\mathcal{F})$ is $\alpha_{3/2}$, then it is α_1.

(iv) $G(\mathcal{F})$ is α_1 if and only if $\omega_{\mathcal{F}}$ is α_1.

REZNICHENKO and SIPACHEVA [1999] say that a filter $\mathcal{F}$ on ω is a *FUF-filter* provided that the following property holds: If $\mathcal{K} \subseteq [\omega]^{<\omega}$ is a family of finite subsets of ω such that for every $F \in \mathcal{F}$ there exists $K \in \mathcal{K}$ with $K \subseteq F$, then there exists a sequence $\{K_n : n \in \omega\} \subseteq \mathcal{K}$ so that for every $F \in \mathcal{F}$ one can find $n \in \omega$ with $K_m \subseteq F$ for all $m \geq n$. They observe that $G(\mathcal{F})$ is Fréchet-Urysohn if and only if $\mathcal{F}$ is an FUF-filter. Therefore, the existence of a free FUF-filter on ω that is not countably generated implies the existence of a countable Fréchet-Urysohn topological group that is not first countable, and so a positive answer to the following question would imply a positive answer to Problem 3.11:

3.13. QUESTION (REZNICHENKO and SIPACHEVA [1999]). Is there, in ZFC only, a free FUF-filter on ω that is not countably generated?

Item (ii) of Theorem 3.12 implies that a positive answer to Question 3.13 would provide a strengthening of the example from Theorem 2.12. Note also that item (iii) of the same theorem shows that a group of the form $G(\mathcal{F})$ is not suitable for getting a counterexample to the question mentioned right after Theorem 3.7. Some additional information can be found in SIPACHEVA [2002].

Comparison with item (ii) of Theorem 1.4 shows that the next theorem is specific for topological groups:

3.14. THEOREM (NOGURA and SHAKHMATOV [1995]). *A topological group that is an $\alpha_{3/2}$-space has the Ramsey property.*

3.15. QUESTION. Let G be a topological group.
(i) If G is an α_2-space, must G have the Ramsey property?
(ii) If G has the Ramsey property, is G an α_2-space?
What if G is additionally assumed to be Fréchet-Urysohn?

Finally, we turn to metrizability of topological groups. The classical Birkhoff-Kakutani theorem says that a topological group is metrizable if (and only if) it is first countable. Arkhangel'skiĭ generalized this theorem by showing that bisequential topological groups are metrizable (see ARHANGEL'SKIĬ and MALYKHIN [1996] for a more general result, and recall that first countable spaces are bisequential). NOGURA, SHAKHMATOV and TANAKA [1993] generalize these results further by showing that a topological group is metrizable if it has the weak topology with respect to a point-finite cover consisting of bisequential spaces. (They say that a space X has the *weak topology with respect to a cover* $\mathcal{C}$ of X provided that a subset F of X is closed in X if and only if its intersection $F \cap C$ is closed in C for every member of the cover $C \in \mathcal{C}$.) In addition, NOGURA, SHAKHMATOV and TANAKA [1993] establish the following theorem, each item of which also generalizes both Birkhoff-Kakutani's and Arkhangel'skiĭ's results:

3.16. THEOREM. *Let G be a topological group which has the weak topology with respect to a point-countable cover $\mathcal{C}$ consisting of bisequential spaces. Then G is metrizable in each of the following cases:*
(i) *G is an α_4-space,*
(ii) *$\mathcal{C}$ consists of closed subspaces and G does not contain a closed subspace homeomorphic to the countable Fréchet-Urysohn fan* (see Example 2.1) *or, equivalently, a closed subspace homeomorphic to S_2,*
(iii) *$\mathcal{C}$ is countable and increasing, and G contains no closed subspace homeomorphic to the countable Fréchet-Urysohn fan (equivalently, no closed subspace homeomorphic to S_2).*

Recall that S_2 is the Arens space, the standard sequential space of sequential order 2.

4. Convergence properties in topological groups with additional compactness conditions

Recall that a topological group is *totally bounded* (or *precompact*) if it is a subgroup of some compact group. Pseudocompact groups are totally bounded (COMFORT and ROSS [1966]). Therefore

compact → countably compact → pseudocompact → totally bounded

represents a chain of compactness-like conditions on a topological group.

Since (locally) compact groups have a nice structure, it seems natural to expect that many convergence properties should coincide for (locally) compact groups. The next theorem supports this thesis:

4.1. THEOREM (NOGURA and SHAKHMATOV [1995]). *All α_i properties for $i = 1, 3/2, 2,$ $3, 4$, as well as the Ramsey property, coincide for locally compact topological groups.* [2]

This result justifies the following

4.2. QUESTION. Are there any "new" implications between α_i-properties and the Ramsey property in a topological group G satisfying one of the following compactness conditions:
(i) (locally) countably compact,
(ii) (locally) pseudocompact,
(iii) totally bounded?

"New" above means an implication that does not hold for topological groups in general but holds for topological groups satisfying the additional compactness condition.

We note that in Theorem 4.1 we do not assume our locally compact group to be Fréchet-Urysohn. Moreover, the main value of this result lies outside of the class of Fréchet-Urysohn spaces. Indeed, a *locally compact group G with $t(G) = \omega$ is metrizable*, and so the result of Theorem 4.1 becomes trivial if the group in question is countably tight. However, countably compact Fréchet-Urysohn groups need not be first countable (see Theorem 2.7), and thus the next variation of Question 4.2 makes sense:

4.3. QUESTION. What is the answer to Question 4.2 if one additionally assumes in it that G is Fréchet-Urysohn?

A particular version of Questions 3.15 and 4.2 deserves special attention:

4.4. QUESTION. What is the answer to Question 3.15 if one additionally assumes that the group G has one of the following compactness properties:
(i) (locally) countably compact,
(ii) (locally) pseudocompact,
(iii) totally bounded?

NOGURA and SHAKHMATOV [1995] also obtained a couple of unexpected "metrization criteria" for locally compact groups:

4.5. THEOREM. *The following conditions are equivalent:*
(i) *every compact group that is an α_1-space is metrizable,*
(ii) *every locally compact group that is an α_4-space is metrizable,*
(iii) $\mathfrak{b} = \omega_1$.

4.6. COROLLARY. *Under* CH, *a locally compact group is metrizable if and only if it is α_4.*

Theorem 2.7 demonstrates that "local compactness" cannot be replaced by "countable compactness" in the above corollary.

---

[2] In fact, the same proof given in NOGURA and SHAKHMATOV [1995] works to prove a slight generalization of Theorem 4.1 with "locally compact group" replaced by "regular locally dyadic space". A space X is *locally dyadic* if every point x of X has an open neighborhood U such that the closure $\overline{U}$ of U is a dyadic compact space; that is $\overline{U}$ is a continuous image of $\{0,1\}^\tau$ for some cardinal τ. Locally compact groups are locally dyadic (CHOBAN [1977]; see also USPENSKIĬ [1988] for an "easy" proof).

5. Convergence properties in function spaces $C_p(X)$

For a topological space X let $C(X)$ be the set of all real-valued continuous functions on X. Let $\mathcal{F}$ be a collection of subsets of X. For $f \in C(X)$, $\varepsilon > 0$ and $F \in \mathcal{F}$ define $O(f, \varepsilon, F) = \{g \in C(X) : |g(x) - f(x)| < \varepsilon \text{ for all } x \in F\}$. The topology $\tau(\mathcal{F})$ on $C(X)$ generated by the family $\{O(f, \varepsilon, F) : f \in C(X), \varepsilon > 0, F \in \mathcal{F}\}$ as a subbase is called the *topology of uniform convergence on* $\mathcal{F}$. The topology of uniform convergence on finite subsets of X is called the *topology of pointwise convergence* on $C(X)$, or simply the *pointwise topology* on $C(X)$. The topology of uniform convergence on compact subsets of X is called the *compact-open topology* on $C(X)$.

We use $C_p(X)$ to denote the space $C(X)$ equipped with the topology of pointwise convergence. One can easily see that the pointwise topology on $C(X)$ is precisely the topology $C(X)$ inherits from $\mathbb{R}^X$, the latter space having the Tychonoff product topology.

For every space X and any family $\mathcal{F}$ of subsets of X that contains all finite subsets of X, the space $(C(X), \tau(\mathcal{F}))$ is both a (locally convex) topological vector space and a topological ring. In particular, $C_p(X)$ have these properties. It is perhaps due to this fact that convergence properties in function spaces exhibit even more peculiarities that in the case of topological groups. The following fundamental result of PYTKEEV [1992] demonstrates this thesis very well:

5.1. THEOREM. *Let X be a topological space and let $\mathcal{F}$ be a family of compact subsets of X containing all finite subsets of X. Then the following three properties are equivalent:*
(i) *$(C(X), \tau(\mathcal{F}))$ is Fréchet-Urysohn,*
(ii) *$(C(X), \tau(\mathcal{F}))$ is sequential,*
(iii) *$(C(X), \tau(\mathcal{F}))$ is a k-space.*

Two extreme cases of the above theorem, for the pointwise topology and the compact-open topology, were obtained earlier by GERLITS [1983] and PYTKEEV [1982].

An *ω-cover* of a space X is a cover $\mathcal{U}$ of X such that very finite subset of X is contained in some element $U \in \mathcal{U}$. A space X has *property* (γ) if every open ω-cover $\mathcal{U}$ of X contains a sequence $\{U_n : n \in \omega\} \subseteq \mathcal{U}$ such that $X = \bigcup_{n \in \omega} \bigcap_{k \geq n} U_k$. The equivalence of items (i), (ii), (iii) of our next theorem is due to GERLITS [1983], while the equivalence of items (i) and (iv) was proved by GERLITS and NAGY [1982].

5.2. THEOREM. *For every space X the following properties are equivalent:*
(i) *$C_p(X)$ is Fréchet-Urysohn,*
(ii) *$C_p(X)$ is sequential,*
(iii) *$C_p(X)$ is a k-space,*
(iv) *X has property (γ).*

We now turn to α_i-properties of $C_p(X)$.

5.3. THEOREM (SCHEEPERS [1998]). *Let X be a topological space. Then $C_p(X)$ is α_2 if and only if $C_p(X)$ is α_4. Therefore, all three properties α_4, α_3 and α_2 coincide for spaces of the form $C_p(X)$.*

Since $C_p(X)$ is a topological group, combining Theorems 3.1 and 5.3 yields:

5.4. COROLLARY (GERLITS and NAGY [1982]). *If $C_p(X)$ is Fréchet-Urysohn, then $C_p(X)$ is α_2.*

5.5. THEOREM (SCHEEPERS [1998]). *It is consistent with ZFC that there exists a subset of real numbers $X \subseteq \mathbb{R}$ such that $C_p(X)$ is Fréchet-Urysohn (and thus α_2) but is not α_1.*

Note that the existence of the above space X is not only consistent with ZFC but also independent of ZFC by Theorem 2.6.

SCHEEPERS [1998] also found a consistent example of a subspace X of the real line $\mathbb{R}$ such that $C_p(X)$ is α_1 but is not Fréchet-Urysohn.

CSÁSZÁR and LACZKOVICH [1975] and BUKOVSKÁ [1991] say that a sequence of real-valued functions $\{f_n : n \in \omega\}$ defined on a set X *quasi-normally converges* to a real-valued function f provided that there exists a sequence $\{\varepsilon_n : n \in \omega\}$ of positive real numbers such that: (i) $\lim_{n\to\infty} \varepsilon_n = 0$, and (ii) for each $x \in X$, $|f - f_n(x)| < \varepsilon_n$ for all but finitely many n. BUKOVSKÝ, RECŁAW and REPICKÝ [1991] say that a space X is a *QN-space* provided that, whenever a sequence $\{f_n : n \in \omega\}$ of continuous real-valued functions defined on X converges pointwise to the continuous function f, this convergence is automatically quasi-normal.

5.6. THEOREM (SCHEEPERS [1998]). *If $C_p(X)$ is an α_1-space, then X is a QN-space.*

5.7. QUESTION (SCHEEPERS [1998]). Does the converse hold? I. e., is it true that $C_p(X)$ is an α_1-space if and only if X is a QN-space?

5.8. QUESTION (SCHEEPERS [1998]). Find necessary and sufficient conditions on X for its function space $C_p(X)$ to be α_1.

Since $C_p(X)$ is a topological group, from Theorems 3.14, 1.4(i) and 5.3 it follows that $\alpha_{3/2} \to$ Ramsey property $\to \alpha_3 \to \alpha_2$ for the function space $C_p(X)$. This justifies the following:

5.9. QUESTION. Let X be a space.

(i) Are the Ramsey property and α_2-property equivalent for $C_p(X)$?

(ii) Is the Ramsey property equivalent to $\alpha_{3/2}$-property for $C_p(X)$?

Of course, item (i) of the above question is a particular version of item (i) of Question 3.15.

6. Convergence properties in products

1. Products of general spaces

The countable Fréchet-Urysohn fan from Example 2.1 demonstrates that the square of Fréchet-Urysohn space need not be Fréchet-Urysohn. Moreover, SIMON [1980] gave even a stronger counter-example:

6.1. THEOREM. *There exists a compact Fréchet-Urysohn space X such that $X \times X$ is not Fréchet-Urysohn.*

It is this failure of preservation of the Fréchet-Urysohn property that was the primary motivation for Arhangel'skiĭ when he introduced α_i-spaces. He also proved the following:

6.2. THEOREM (ARHANGEL'SKIĬ [1979]). *If X is a Fréchet-Urysohn α_3-space and Y is a (countably) compact Fréchet-Urysohn space, then $X \times Y$ is Fréchet-Urysohn.*

Note that Theorems 6.1 and 6.2 imply Theorem 2.2.

6.3. THEOREM (NOGURA [1985]).

(i) *For $i = 1, 2, 3$, if X and Y are α_i-spaces, then $X \times Y$ is also an α_i-space.*

(ii) *There exist compact Fréchet-Urysohn α_4-spaces X and Y such that $X \times Y$ is neither Fréchet-Urysohn nor α_4.*

6.4. THEOREM (COSTANTINI and SIMON [2000]). *There exist two countable Fréchet-Urysohn α_4-spaces X and Y such that $X \times Y$ is α_4 but fails to be Fréchet-Urysohn.*

Earlier a consistent example of such spaces X and Y was constructed by COSTANTINI [1999].

6.5. THEOREM. (i) *Under* CH, *there exist two countable Fréchet-Urysohn α_4-spaces X and Y such that $X \times Y$ is Fréchet-Urysohn but is not α_4* (SIMON [1998]).

(ii) *Under Open Colouring Axiom OCA, if X and Y are Fréchet-Urysohn α_4-spaces and $X \times Y$ is Fréchet-Urysohn, then $X \times Y$ is α_4* (TODORČEVIĆ [2002]).

Some strengthenings of Theorem 6.2 were obtained by DOLECKI and NOGURA [1999]. We refer the reader to DOLECKI and NOGURA [2002] and MYNARD [2000] for further recent results related to products.

2. *Products of topological groups*

Recall that a topological group G is *compactly generated* provided that there exists a compact set $K \subseteq G$ such that the smallest subgroup of G that contains K coincides with G.

6.6. THEOREM (TODORČEVIĆ [1993]). *There exist, in ZFC, two (compactly generated) Fréchet-Urysohn groups G and H such that $t(G \times H) > \omega$ (in particular, $G \times H$ is not Fréchet-Urysohn). Moreover, every countable subset of G and H is metrizable, and so both G and H are α_1.*

It is far from clear if two different groups in the above theorem can be replaced by a *single* group:

6.7. QUESTION. In ZFC, is there a Fréchet-Urysohn group G such that:

(i) $G \times G$ is not Fréchet-Urysohn, or even

(ii) $t(G \times G) > \omega$?

Theorems 6.8 and 6.10(i) provide consistent examples of such a group G.

6.8. THEOREM (MALYHIN and SHAKHMATOV [1992]). *Add a single Cohen real to a model of* MA $+\neg$ CH. *Then, in the generic extension, the exists a (hereditarily separable) Fréchet-Urysohn topological group G such that $t(G \times G) > \omega$ (in particular, $G \times G$ is not Fréchet-Urysohn). Moreover, G is an α_1-space.*

The above results do not help in getting *countable* groups for which the Fréchet-Urysohn property is not preserved by products.

6.9. THEOREM (SHIBAKOV [1996]). *Under* CH, *there exist two countable Fréchet-Urysohn Abelian groups G and H such that their product $G \times H$ is not (even) sequential.*

6.10. THEOREM. (i) *Under* CH, *there exists a countable Fréchet-Urysohn Abelian group G such that $G \times G$ is sequential but is not Fréchet-Urysohn* (SHIBAKOV [1999a]).

(ii) *Under Open Colouring Axiom OCA, if G and H are Fréchet-Urysohn groups, then the product $G \times H$ is Fréchet-Urysohn if and only if it is sequential* (TODORČEVIĆ [2002]).

In a private email communication to the author (dated May 24, 2002) SHIBAKOV notes the following reduction principle:

6.11. THEOREM. *Let G be a countable Fréchet-Urysohn group such that $G \times G$ is sequential but is not Fréchet-Urysohn. Let $H = G \times \mathbb{Q}$ be the product of G with the rationals $\mathbb{Q}$. Then H is a countable Fréchet-Urysohn group such that $H \times H$ is not sequential.*

□ Clearly H is countable. Since G is Fréchet-Urysohn, it is α_4 by Theorem 3.1. Thus G is strongly Fréchet-Urysohn, and so H is Fréchet-Urysohn. Note that $X = G \times G$ is a sequential space in which all points are G_δ. Since $\mathbb{Q}$ is not locally countably compact and $X \times \mathbb{Q}$ is not Fréchet-Urysohn (because it contains a closed copy of non-Fréchet-Urysohn space $G \times G$), by result of TANAKA [1976] the product $X \times \mathbb{Q}$ is not sequential. Since $H \times H$ contains a closed copy of $X \times \mathbb{Q}$, it follows that $H \times H$ is also not sequential. □

Applying this reduction principle to the group G from Theorem 6.10(i) one obtains

6.12. THEOREM (SHIBAKOV, 2002). *CH implies the existence of a countable Fréchet-Urysohn Abelian group H such that $H \times H$ is not (even) sequential.*

Whether the assumption of CH can be dropped in Theorems 6.9 and 6.12 remains an open question:

6.13. QUESTION. (i) In ZFC only, is there a countable Fréchet-Urysohn topological group G such that $G \times G$ is not Fréchet-Urysohn (not sequential)?

(ii) In ZFC only, does there exist two countable Fréchet-Urysohn topological groups G and H such that $G \times H$ is not Fréchet-Urysohn (not sequential)?

SHIBAKOV [1999a] notices that the group G from Theorem 6.10(i) cannot be α_3, thereby proving the existence of a countable Fréchet-Urysohn group that is not α_3 under CH. (Compare this with Theorem 3.4.) Indeed, suppose that G is α_3. Then $G \times G$ must be α_3 by Theorem 6.3(i). In particular, $G \times G$ is α_4. Being a sequential topological group, $G \times G$ must be Fréchet-Urysohn by Theorem 3.2, a contradiction.

Theorem 6.6 justifies the following

6.14. QUESTION. In ZFC only, is there a Fréchet-Urysohn topological group G such that G is α_1 but $G \times G$ is not Fréchet-Urysohn?

In view of Theorem 2.11, G must be uncountable. Since $G \times G$ is α_1 by Theorem 6.3, Theorem 3.2 implies that $G \times G$ cannot be sequential.

3. Products of function spaces $C_p(X)$

6.15. THEOREM (TKAČUK [1984]). *If $C_p(X)$ is Fréchet-Urysohn, then even its countable power $C_p(X)^\omega$ is Fréchet-Urysohn (and thus so are all finite powers $C_p(X)^n$).*

6.16. THEOREM (TODORČEVIĆ [1993]). *There exist two spaces X and Y such that both $C_p(X)$ and $C_p(Y)$ are Fréchet-Urysohn but $t(C_p(X) \times C_p(Y)) > \omega$ (in particular, $C_p(X) \times C_p(Y)$ is not Fréchet-Urysohn). Moreover, every countable subset of $C_p(X)$ and $C_p(Y)$ is metrizable, and so both $C_p(X)$ and $C_p(Y)$ are α_1.*

7. Sequential order in topological groups and function spaces

Let X be a topological space, and let A be a subset of X. Define a transfinite sequence $\{[A]_\alpha : \alpha \leq \omega_1\}$ of subsets of X as follows. Let $[A]_0 = A$. If α is a limit ordinal, then $[A]_\alpha = \bigcup\{[A]_\beta : \beta < \alpha\}$. Finally, define $[A]_{\alpha+1}$ be the set consisting of limits of all convergent sequences of points in $[A]_\alpha$. One can easily see that the set $[A]_{\omega_1}$ is sequentially closed in X. Therefore, if X is sequential, then $\overline{A} = [A]_{\omega_1}$ for every subset A of X. The minimal ordinal $\alpha \leq \omega_1$ such that $[A]_{\omega_1} = [A]_\alpha$ for all subsets A of X is called the *sequential order* of a space X. Using this terminology, a sequential space is Fréchet-Urysohn if and only if it is of sequential order one.

NYIKOS [1981] asked whether the sequential order of a sequential topological group is ω_1 if the group is not Fréchet-Urysohn. Answering this question, SHIBAKOV [1996a] used CH to give an example of a sequential group topology on the group of rational numbers whose sequential order is between 2 and ω. Refining his methods even further SHIBAKOV [1998a] obtained the following

7.1. THEOREM. *Assume* CH. *Then for any ordinal $\alpha < \omega_1$ there exists a countable sequential Abelian group of sequential order α.*

The existence of such a group in ZFC remains an open problem. However, the following two related examples were constructed in ZFC.

7.2. THEOREM (DOLECKI and PEIRONE [1992]). *For every countable ordinal α there exists a Hausdorff (but not regular) sequential topology of sequential order α on a countable Abelian group G such that the multiplication is jointly continuous.*

The inverse operation in G above must be discontinuous, because otherwise G would be a topological group and thus, being Hausdorff, would be Tychonoff.

7.3. THEOREM (PEIRONE [1994]). *For every countable ordinal α there exists a regular sequential topology of sequential order α on a countable Abelian group G such that the group multiplication is separately continuous and the inverse operation is continuous.*

FOGED [1981] constructed (in ZFC) a countable homogeneous sequential space of sequential order α, for every $\alpha < \omega_1$.

7.4. QUESTION. Does there exist, for every ordinal $\alpha \leq \omega_1$, a sequential topological group G_α of sequential order α which in addition has one of the following properties:
(i) totally bounded,
(ii) pseudocompact,
(iii) countably compact?

Question 7.4(iii) obviously goes in the opposite direction to the next question the author asked in 1990 (see Problems Section, Topology Proc. 15 (1990)):

7.5. QUESTION. Is a countably compact sequential group Fréchet-Urysohn?

SHIBAKOV [1998] proved that a sequential topological group with a point-countable k-network is metrizable if and only if its sequential order is less than ω_1.

7.6. THEOREM (FREMLIN [1994]). *For any space X, the sequential order of $C_p(X)$ is either 1 or ω_1.*

Note that if $C_p(X)$ is sequential, then $C_p(X)$ is Fréchet-Urysohn by Theorem 5.2, and so the last result is only "non-trivial" when $C_p(X)$ is *not* sequential.

8. Miscellaneous

1. Sequential continuity versus continuity

A map $f : X \to Y$ between spaces X and Y is called *sequentially continuous* provided that for every sequence $\{x_n : n \in \omega\} \subseteq X$ converging to $x \in X$ the sequence $\{f(x_n) : n \in \omega\} \subseteq Y$ converges to $f(x) \in Y$. Recall that a cardinal number κ is called *Ulam-measurable* if there exist an ultrafilter $\mathcal{F}$ on κ such that $\bigcap \mathcal{F} = \emptyset$ and $\bigcap \mathcal{E} \in \mathcal{F}$ whenever $\mathcal{E}$ is a countable subfamily of $\mathcal{F}$. The following classical theorem about automatic continuity of homomorphisms is of special importance:

8.1. THEOREM (VAROPOULOS [1964]). *Let G and H be two locally compact groups and $f : G \to H$ be a sequentially continuous group homomorphism. If the cardinality $|G|$ of G is not Ulam-measurable, then f is continuous.*

COMFORT and REMUS [1994] found a partial converse to this theorem, and a full converse was announced by Uspenskiĭ (however no proof have appeared in print yet).

HUŠEK [1996] proves that the following three cardinal numbers coincide: the smallest cardinal s such that there exists a sequentially continuous mapping $f\colon \{0,1\}^s \to \mathbb{R}$ that is not continuous; the smallest cardinal u such that there exists a uniformly sequentially continuous mapping $h\colon \{0,1\}^u \to \mathbb{R}$ which is not (uniformly) continuous; and the smallest cardinal g such that there exists a sequentially continuous group homomorphism from $\mathbb{Z}_2^g$ to some topological group that is not continuous. (Here $\mathbb{Z}_2 = \mathbb{Z}/2\mathbb{Z}$.) It is known that s is not greater than the first real-measurable cardinal m, and $s = m$ under MA. Some further related results can be found in BALCAR and HUŠEK [2001].

HUŠEK [1996] says that a topological group G is an s-group if every sequentially continuous homomorphism from G to any topological group is continuous, and he proves that if κ is a non-measurable cardinal, then every sequentially continuous homomorphism form a product of κ-many s-groups into a compact group is continuous.

ARHANGEL' SKIĬ, JUST and PLEBANEK [1996] construct, for a real-measurable cardinal κ, a sequentially continuous homomorphism of $\mathbb{Z}_2^\kappa$ onto a non-pseudocompact metrizable topological group (thus the homomorphism is not continuous). Additional results about (continuity of) sequentially continuous group homomorphisms can be found in ARHANGEL'SKIĬ [1994] and ARHANGEL'SKIĬ and JUST [1995].

2. *Sequentially complete groups*

A topological group G is called *sequentially complete* if G is sequentially closed in any other topological group that contains G as a subgroup, or equivalently, if G is sequentially closed in its Raikov completion. The class of sequentially complete groups is closed with respect to Cartesian products and passage to closed subgroups. Clearly, countably compact groups are sequentially complete.

8.2. THEOREM (DIKRANJAN and TKACHENKO [2001a]). *For every space X the following conditions are equivalent:*

(i) *the free topological group $F(X)$ of X is sequentially complete,*
(ii) *the free Abelian topological group $A(X)$ of X is sequentially complete,*
(iii) *the free precompact Abelian group $A^*(X)$ of X is sequentially complete,*
(iv) *X is sequentially closed in βX.*

In particular, if a space X is countably compact, then $F(X)$, $A(X)$ and $A^*(X)$ are all sequentially complete (DIKRANJAN and TKACHENKO [2000]).

DIKRANJAN and TKACHENKO [2000] introduced two strengthenings of sequential completeness. They call a topological group G *sequentially h-complete* (*sequentially q-complete*) if every continuous homomorphic image of G (every quotient group of G, respectively) is sequentially complete. Clearly, countably compact $\to$ sequentially h-complete $\to$ sequentially q-complete $\to$ sequentially complete. Totally bounded sequentially h-complete groups are pseudocompact, while totally bounded sequentially q-complete groups need not be pseudocompact (DIKRANJAN and TKACHENKO [2000]). Every pseudocompact Abelian group is a quotient of a pseudocompact Abelian sequentially complete group by a closed pseudocompact subgroup (DIKRANJAN and TKACHENKO [2001a]).

DIKRANJAN and TKACHENKO [2001] found non-trivial connections between sequential completeness, minimality and properties related to connectedness and disconnectedness. For example, the properties of being sequentially complete, pseudocompact and minimal are independent, i.e., the conjunction of any two of them does not imply the third one even in the class of Abelian groups. DIKRANJAN and TKACHENKO [2001] prove that a connected sequentially complete minimal Abelian group of non-measurable size is compact. They also show that each minimal sequentially complete hereditarily disconnected Abelian group has covering dimension zero.

3. *Topologies on groups determined by sequences*

A sequence S of points in a group G is called a *T-sequence* if there exists a group topology on G in which S converges to the identity element of G. A straightforward application of Zorn's lemma yields that for every T-sequence S there exists a maximal topology $\tau(S)$

on G which satisfies this property, in which case $\tau(S)$ is said to be *determined by the sequence* S. A comprehensive monograph by PROTASOV and ZELENYUK [1999] is devoted to a systematic study of the interconnections between the arithmetical properties of the T-sequence S and the topologico-algebraic properties of the topology $\tau(S)$. We mention only three results here.

8.3. THEOREM. *A sequence $S \subseteq G$ is a T-sequence if and only if there exists a topological group H containing G as an (algebraic) subgroup such that S converges to some element $h \in H$ such that G has trivial intersection with the cyclic group generated by h.*

8.4. THEOREM. *The topology $\tau(S)$ determined by a T-sequence S in a group G is always complete.*

The next theorem provides a nice "description" of T-sequences in the group $\mathbb{Z}$ of integers in number-theoretic terms:

8.5. THEOREM.

(i) *If $\{z_n : n \in \omega\} \subseteq \mathbb{Z}$ and $\lim_{n\to\infty} z_{n+1}/z_n$ is either infinite or finite and transcendental, then $\{z_n : n \in \omega\}$ is a T-sequence in $\mathbb{Z}$.*

(ii) *For every algebraic number r, there exists a sequence $\{z_n : n \in \omega\} \subseteq \mathbb{Z}$ which is not a T-sequence in $\mathbb{Z}$ such that $\lim_{n\to\infty} z_{n+1}/z_n = r$.*

Among numerous applications of T-sequences found in this monograph are the following:

(i) Every maximal topological group contains a countable open Boolean subgroup.

(ii) Under CH, every nondiscrete metrizable group topology on an arbitrary group has a refinement in which every nowhere dense subset is closed.

(iii) Every countable topologizable group admits a complete sequential group topology of sequential order ω_1.

(iv) Every infinite Abelian group admits a complete group topology for which characters do not separate points.

(v) Under CH, every infinite Abelian group admits a nondiscrete group topology in which all nowhere dense subsets are closed.

4. Suitable sets as super-sequences

Let us call an infinite compact space with a single non-isolated point a *super-sequence*. Clearly, convergent sequences are precisely countable super-sequences. Let us say that a subset X of a topological group G *topologically generates* G if G is the smallest closed subgroup of G containing X. HOFMANN and MORRIS [1990] call a subset X of a topological group G *suitable* if X topologically generates G, $e \notin X$ and X is discrete and closed in $G \setminus \{e\}$, where e is the identity element of G. One can easily see that an infinite suitable subset of a compact group is a super-sequence, which explains the relevance of this notion to our survey. HOFMANN and MORRIS [1990] proved that every locally compact group has a suitable set. It follows that every infinite compact group can be topologically generated by a super-sequence, thereby justifying the following cardinal invariant $seq(G) = \min\{|S| : S \subseteq G$ is a super-sequence topologically generating $G\}$ for an infinite compact group G.

8.6. THEOREM (DIKRANJAN and SHAKHMATOV [200?]). *Let G be an infinite compact group and $c(G)$ be the connected component of G. Then $seq(G) = w(G/c(G)) \cdot \sqrt{w(c(G))}$, where $\sqrt{\tau} = \min\{\sigma \geq \omega : \sigma^\omega \geq \tau\}$.*

We refer the reader to TKAČENKO [1997] for a comprehensive survey on suitable sets in topological groups.

5. *None of the above but in this section*

NOGURA, SHAKHMATOV and TANAKA [1997] give a series of examples demonstrating that the A-property (due to E. Michael) and α_4-property behave independently from each other in general spaces and groups. These two properties are known to coincide for Fréchet-Urysohn spaces but may differ in sequential spaces. They prove that the A-property and α_4-property coincide for hereditarily normal, sequential topological groups. as well as general sequential spaces in which every point is a G_δ-set. KABANOVA [1992] proves that every countable field admits a non-discrete field topology without non-trivial (i.e. infinite) convergent sequences (see also SHIBAKOV [1999] for a different proof). Additional information about convergence properties in groups can be found in SHAKHMATOV's [1999] survey.

Acknowledgement: The author would like to thank Dikran Dikranjan, Szymon Dolecki, Peter Nyikos, Alexander Shibakov and Stevo Todorčević for their valuable comments on the preliminary version of this manuscript.

References

ARHANGEL'SKIĬ, A.V.

[1971] Bicompacta that satisfy the Suslin condition hereditarily. Tightness and free sequences (Russian), *Dokl. Akad. Nauk SSSR* **199**, 1227–1230.

[1972] The frequency spectrum of a topological space and the classification of spaces (Russian), *Doklady Akad. Nauk SSSR* **206**, 265–268.

[1979] The spectrum of frequencies of a topological space and the product operation (Russian), *Trudy Moskov. Mat. Obshch.* **40**, 171–206.

[1994] On countably compact topologies on compact groups and on dyadic compacta, *Topology Appl.* **57** (2-3), 163–181.

ARHANGEL'SKIĬ, A.V. and W. JUST

[1995] Dense, sequentially continuous maps on dyadic compacta, *Topology Appl.* **64** (1), 95–99.

ARHANGEL' SKIĬ, A.V., W. JUST and G. PLEBANEK

[1996] Sequential continuity on dyadic compacta and topological groups, *Comment. Math. Univ. Carolin.* **37** (4), 775–790.

ARKHANGEL'SKIĬ, A.V. and V.I. MALYKHIN

[1996] Metrizability of topological groups (Russian), *Vestnik Moskov. Univ. Ser. I Mat. Mekh.* no. 3, 13–16.

BALCAR, B. and M. HUŠEK
[2001] Sequential continuity and submeasurable cardinals, *Topology Appl.* **111** (1-2), 49–58.

BUKOVSKÁ, Z.
[1991] Quasinormal convergence, *Math. Slovaca* **4**, 137–146.

BUKOVSKÝ, L., I. RECŁAW and M. REPICKÝ
[1991] Spaces not distinguishing pointwise and quasinormal convergence of real functions, *Topol. Appl.* **41**, 25–40.

CHOBAN, M.M.
[1977] Topological structure of subsets of topological groups and their quotients (Russian), in *Topological Structures and Algebraic Systems* (Stiinca, Kishinev), 117–163.

COMFORT, W.W. and D. REMUS
[1994] Compact groups of Ulam-measurable cardinality: partial converses to theorems of Arhangel'skiĭand Varopoulos, *Math. Japon.* **39** (2), 203–210.

COMFORT, W.W. and K.A. ROSS
[1966] Pseudocompactness and uniform continuity in topological groups, *Pacific J. Math.* **16**, 483–496.

COSTANTINI, C.
[1999] On a problem of Nogura about the product of Fréchet-Urysohn $\langle\alpha_4\rangle$-spaces, *Comment. Math. Univ. Carolin.* **40** (3), 537–549.

COSTANTINI, C. and P. SIMON
[2000] An α_4, not Fréchet product of α_4 Fréchet spaces, *Topology Appl.* **108** (1), 43–52.

CSÁSZÁR, Á. and M. LACZKOVICH
[1975] Discrete and equal convergence, *Studia Sci. Math. Hungarica,* **10**, 463–472.

DIKRANJAN, D. and D. SHAKHMATOV
[200?] The weight of closed subspaces algebraically generating a dense subgroup of a compact group, to appear.

DIKRANJAN, D. and M. TKAČENKO
[2000] Sequential completeness of quotient groups, *Bull. Austral. Math. Soc.* **61** (1), 129–150.
[2001] Sequentially complete groups: dimension and minimality, *J. Pure Appl. Algebra* **157** (2-3), 215–239.
[2001a] Weakly complete free topological groups, *Topology Appl.* **112** (3), 259–287.

DOLECKI, S. and T. NOGURA
[1999] Two-fold theorem on Fréchetness of products, *Czechoslovak Math. J.* **49**(124) no. (2), 421–429.
[2002] Countably infinite products of sequential topologies, *Sci. Math. Jpn.* **55** (1), 121–127.

DOLECKI, S. and R. PEIRONE
[1992] Topological semigroups of every countable sequential order, in *Recent developments of general topology and its applications* (Berlin, 1992), Math. Res. 67, Akademie-Verlag, Berlin, 80–84.

DOW, A.
[1990] Two classes of Fréchet-Urysohn spaces, *Proc. Amer. Math. Soc.* **108**, 241–247.

DOW, A. and J. STEPRĀNS
[1992] Countable Fréchet α_1-spaces may be first-countable, *Arch. Math. Logic* **32**, 3–50.

FOGED, L.
[1981] Sequential order of homogeneous and product spaces, *Topology Proc.* **6**, 287–298.

FREMLIN, D.H.

[1994] Sequential convergence in $C_p(X)$, *Comment. Math. Univ. Carolin.* **35** (2), 371–382.

GERLITS, J.

[1983] Some properties of $C(X)$, II, *Topology Appl.* **15** (3), 255–262.

GERLITS, J. and ZS. NAGY

[1982] Some properties of $C(X)$. I, *Topology Appl.* **14** (2), 151–161.

[1988] On Fréchet spaces, *Rend. Circolo Matem. Palermo, Ser. II,* **18**, 51–71.

HOFMANN, K.H. and S.A. MORRIS

[1990] Weight and c, *J. Pure Appl. Algebra* **68** (1-2), 181–194.

HUŠEK, M.

[1996] Sequentially continuous homomorphisms on products of topological groups, *Topology Appl.* **70** (2-3), 155–165.

KABANOVA, E.I.

[1992] Countable nondiscrete fields without nontrivial converging sequences (Russian), *Mat. Zametki* **52** (2), 62–65.

MALYHIN, V.I. and D.B. SHAKHMATOV

[1992] Cartesian products of Fréchet topological groups and functions spaces, *Acta Math. Hung.* **60** (3-4), 207–215.

MALYHIN, V.I. and B.E. SHAPIROVSKIĬ

[1974] Martin's Axiom and properties of topological spaces (Russian), *Dokl. Akad. Nauk SSSR* **213**, 532–535.

MYNARD, F.

[2000] Strongly sequential spaces, *Comment. Math. Univ. Carolinae* **41** (1), 143-153.

NOGURA, T.

[1985] The product of $\langle\alpha_i\rangle$-spaces, *Topol. Appl.* **21**, 251–259.

NOGURA, T. and D. SHAKHMATOV

[1995] Amalgamation of convergent sequences in locally compact groups, *C. R. Acad. Sci. Paris Sér. I Math.* **320**, 1349–1354.

NOGURA, T., D. SHAKHMATOV and Y. TANAKA

[1993] Metrizability of topological groups having weak topologies with respect to good covers, *Topology Appl.* **54** (1-3), 203–212.

[1997] α_4-property versus A-property in topological spaces and groups, *Studia Sci. Math. Hungar.* **33** (4), 351–362.

NYIKOS, P.J.

[1981] Metrizability and the Fréchet-Urysohn property in topological groups, *Proc. Amer. Math. Soc.* **83**, 793–801.

[1989] The Cantor tree and the Fréchet-Urysohn property, *Ann. New York Acad. Sci.* **552**, 109–123.

[1992] Subsets of ${}^{\omega}\omega$ and the Fréchet-Urysohn and α_i-properties, *Topology Appl.* **48**, 91–116.

PEIRONE, R.

[1994] Regular semitopological groups of every countable sequential order, *Topology Appl.* **58** (2), 145–149.

PROTASOV, I. and E. ZELENYUK

[1999] *Topologies on Groups Determined by Sequences,* Mathematical Studies Monograph Series, 4. VNTL Publishers, L'viv, 1999, 111 pp, ISBN 966-7148-66-1.

PYTKEEV, E.G.

[1982] Sequentiality of spaces of continuous functions (Russian), *Uspekhi Mat. Nauk* **37** 5(227), 197–198.

[1992] On a property of Fréchet-Urysohn spaces of continuous functions (Russian), *Trudy Mat. Inst. Steklov.* **193**, 156–161.

RAMSEY, F.P.

[1929] On a problem of formal logic, *Proc. London Math. Soc.* **30**, 264–286.

REZNICHENKO, E.A.

[1986] On the number of countable Fréchet-Urysohn spaces (Russian), in: *Continuous functions on topological spaces*, Latv. Gos. Univ., Riga, pp. 147–154.

REZNICHENKO, E.A. and O.V. SIPACHEVA

[1999] Properties of Fréchet-Urysohn type in topological spaces, groups and locally convex spaces (Russian), *Vestnik Moskov. Univ. Ser. I Mat. Mekh.* no. 3, 32–38.

SCHEEPERS, M.

[1998] $C_p(X)$ and Arhangel'skiĭ's α_i-spaces, *Topol. Appl.* **89**, 265–275.

SHAKHMATOV, D.

[1990] α_i-properties in Fréchet-Urysohn topological groups, *Topology Proc.* **15**, 143–183.

[1999] A comparative survey of selected results and open problems concerning topological groups, fields and vector spaces, *Topology Appl.* **91** (1), 51–63.

SHIBAKOV, A.

[1996] Examples of sequential topological groups under the continuum hypothesis, *Fund. Math.* **151** (2), 107–120.

[1996a] Sequential group topology on rationals with intermediate sequential order, *Proc. Amer. Math. Soc.* **124** (8), 2599–2607.

[1998] Metrizability of sequential topological groups with point-countable k-networks, *Proc. Amer. Math. Soc.* **126** (3), 943–947.

[1998a] Sequential topological groups of any sequential order under CH, *Fund. Math.* **155** (1), 79–89.

[1999] A countable nondiscrete topological field without nontrivial convergent sequences, *Proc. Amer. Math. Soc.* **127** (10), 3091–3094.

[1999a] Countable Fréchet topological groups under CH, *Topology Appl.* **91**, 119–139.

SIMON, P.

[1980] A compact Fréchet space whose square is not Fréchet, *Comment. Math. Univ. Carolinae,* **21**, 749–753.

[1998] A hedgehog in a product, *Acta Univ. Carolin. Math. Phys.* **39** (1-2), 147–153.

SIPACHEVA, O.V.

[2002] Spaces Fréchet-Urysohn with respect to families of subsets, *Topology Appl.* **121** (1-2), 305–317.

TANAKA, Y.

[1976] Products of sequential spaces, *Proc. Amer. Math. Soc.* **54**, 371–375.

TKAČENKO, M.

[1997] Generating dense subgroups of topological groups, *Topology Proc.* **22** (Summer), 533–582.

TKACHUK, V.V.

[1984] The multiplicativity of some properties of mapping spaces in the topology of pointwise convergence (Russian), *Vestnik Moskov. Univ. Ser. I Mat. Mekh.* no. 6, 36–39.

TODORČEVIĆ, S.

[1993] Some applications of S and L combinatorics, *Ann. New York Acad. Sci.*, **705**, 130–167.

[2002] A proof of Nogura's conjecture, *Proc. Amer. Math. Soc.*, to appear.

USPENSKIĬ, V.V.

[1988] Why compact groups are dyadic, in *General topology and its relations to modern analysis and algebra*, VI (Prague, 1986), Res. Exp. Math., 16, Heldermann, Berlin 1988, 601–610.

VAROPOULOS, N.TH.

[1964] A theorem on the continuity of homomorphisms of locally compact groups, *Proc. Cambridge Philos. Soc.* **60**, 449–463.

CHAPTER 18

Descriptive Set Theory in Topology

Sławomir Solecki

Department of Mathematics, 1409 W. Green St., University of Illinois, Urbana, IL 61801, USA
E-mail: ssolecki@math.uiuc.edu

Contents

RECENT PROGRESS IN GENERAL TOPOLOGY II
Edited by Miroslav Hušek and Jan van Mill

1. Introduction

This paper is a survey of some applications, discovered during the last decade, of descriptive set theory in various topological contexts.

Section 1 contains several results on actions of Polish topological groups in which structural assumptions on the acting group or definability assumptions on the action imply results on the complexity of the action.

In Section 2, we turn the table and show how, assuming some kind of upper bounds on the complexity of all the actions of a given group, one can deduce consequences concerning the group topology on the group. We present results on the structure of ideals of subsets of $\mathbb{N}$, the set of all non-negative integers, which turn out to be related to the questions about group actions.

In Section 3, we study indecomposable continua, in particular, the global structure of the family of all composants of an indecomposable continuum, and we obtain a classification of such families.

In Section 4, we present results on classifying certain types of topological objects, metric spaces and complex manifolds, up to a natural notion of isomorphism.

This short summary makes it clear that the topics covered in this survey are very different from each other. However, they do share an important feature aside from the fact that descriptive set theoretic methods are used in their proofs. Even though results obtained are of obvious topological relevance, the objects that are studied directly are not themselves reasonable topological spaces but rather arise from such spaces by dividing them by equivalence relations. The topological spaces being divided are in most situations Polish spaces, and equivalence relations by which the division is being done are definable (analytic or Borel).

A topological space is called *Polish* if it is completely metrizable and separable. The compact metric, and therefore Polish, space $2^{\mathbb{N}}$, which is simply the Cantor set $\{0,1\}^{\mathbb{N}}$ with the product topology, will be frequently used. A subset of a Polish space is said to be *analytic* if it is the image of a Borel subset B of a Polish space via a Borel function defined on B or, equivalently, on the whole ambient Polish space. Clearly all Borel subsets of Polish spaces are analytic. The family of all Borel subsets of a Polish space can be naturally represented as an increasing union of ω_1 (= the first uncountable ordinal number) many subfamilies. These are defined as follows. Let Σ^0_1 be the family of all open sets and, for $1 < \alpha < \omega_1$, let Σ^0_α consist of countable unions of sets each of which has the complement in some Σ^0_ξ for some $\xi < \alpha$. So, for example, Σ^0_2 sets are identical with F_σ sets, and G_δ and $F_{\sigma\delta}$ sets are the complements of sets in Σ^0_2 and Σ^0_3, respectively. Also

$$\Sigma^0_\alpha \subseteq \Sigma^0_\beta, \text{ for } 1 \leq \alpha < \beta < \omega_1, \quad \text{and} \quad \bigcup_{1 \leq \alpha < \omega_1} \Sigma^0_\alpha = \text{Borel sets}.$$

Therefore, the families Σ^0_α form a convenient scale for measuring complexity of Borel sets with simpler Borel sets residing in Σ^0_α with smaller α's. For these and other notions of classical descriptive set theory the reader is referred to KECHRIS [1995].

We introduce now some fundamental notions relevant to the study of equivalence relations. Given two equivalence relations E and F defined on Polish spaces X and Y, respectively, we say that E is *Borel reducible to* F, in symbols $E \leq F$, if there exists a Borel function $f : X \to Y$ such that

$$xEy \Leftrightarrow f(x)Ff(y). \tag{18.1}$$

If this relation holds, we think of E as simpler than F. The relation $E \leq F$ means that there exists an injection from X/E to Y/E, the families of all equivalence classes of E and F, respectively, which has a lifting to a Borel mapping from X to Y, namely the Borel function f. We say that E and F are *Borel bireducible*, in symbols $E \sim F$, if $E \leq F$ and $F \leq E$, and we write $E < F$ if $E \leq F$ and $E \not\sim F$. The relation $\leq$ is sometimes modified to suit better a situation at hand. For example, f from (18.1) is assumed to be injective, or continuous, or measurable, etc. We will not be concerned with these modifications except for one. The equivalence relations E and F are said to be *Borel isomorphic* if there exists a Borel bijection $f : X \to Y$ for which (18.1) holds. Borel bireducibility does not imply Borel isomorphism. All these notions can be easily generalized to apply to equivalence relations defined only on some Borel subsets of Polish spaces.

An equivalence relation E on a Polish space is called *universal*, for a class of equivalence relations $\mathcal{F}$, if it belongs to $\mathcal{F}$ and $F \leq E$ for each $F \in \mathcal{F}$.

If E is an equivalence relation defined on X and $A \subseteq X$, let $E|A$ stand for the equivalence relation on A making two points $x, y \in A$ equivalent exactly when xEy.

An important class of equivalence relations is given by group actions. Let G be a group acting on a set X. The *orbit equivalence relation* E_G^X on X is defined by

$$xE_G^X y \Leftrightarrow y = gx \text{ for some } g \in G.$$

Thus, the equivalence classes of E_G^X are precisely the orbits of the action. A subset A of X is called *G-invariant* if $x \in A$ implies $gx \in A$ for any $g \in G$. If G is a topological group and X is a topological space, an action of G on X is called continuous (Borel) if it is continuous (Borel) as a function from $G \times X$ to X.

The first two sections of the present survey deal with actions of groups on Polish spaces. These two sections roughly parallel each other but they approach the subject from two different directions. In Section 2, we will introduce a few natural notions of simplicity of orbit equivalence relations. These notions are getting more relaxed as the section progresses, but all of them can be expressed by a requirement of the form $E_G^X \leq E$ for some fixed equivalence relation E. In this section, we will survey results which say that, under some definability assumptions on E_G^X or algebraic or topological assumptions on G, E_G^X is not simple only if there exists a canonical obstacle for it. In Section 3, we will change our perspective and, fixing a group G, we will look at all continuous actions of G on Polish spaces and, assuming that the orbit equivalence relations of these actions are simple, in the appropriate senses, we will deduce consequences for the topology on G.

2. Polish topological group actions

By a *Polish group* we mean a topological group whose topology is Polish. (For information on general topological groups see KELLY [1955].) Examples of Polish groups include all locally compact, second countable groups; separable Banach spaces with addition; the group S_∞ of all permutations of $\mathbb{N}$ with composition as the group operation and with the topology inherited from the inclusion $S_\infty \subseteq \mathbb{N}^\mathbb{N}$; the group of all unitary operators on a separable Hilbert space with composition as the group operation and with the strong topology; the group of all homeomorphisms of a compact metric space with the uniform topology and with composition as the group operation. For more on these examples of Polish groups see KECHRIS [1995].

There are two important classes of Polish groups whose definitions have to do with the type of metrics these groups carry. All metrics considered below are assumed to be compatible with the topology on the group they are defined on. Each Polish group has a complete metric, since it is Polish, and has a left-invariant metric, since all metrizable groups have such metrics by the Birkhoff-Kakutani theorem, see KELLY [1955]. (A metric d is *left-invariant* if $d(gx, gy) = d(x, y)$ for all $g, x, y \in G$.) However, it is not the case that each Polish group carries a complete left-invariant metric. The first class of groups which will be important later on is the class of Polish groups which do have *left-invariant complete metrics.* (In fact, if a group has a left-invariant complete metric, then all left-invariant metrics on it are complete, see BECKER [1998].) This class includes, for example, all locally compact Polish groups and all solvable Polish groups, see HJORTH and SOLECKI [1999]. For basic facts about groups with left-invariant complete metrics see BECKER [1998].

A metric d on a Polish group G is called *two-sided invariant* if $d(gxh, gyh) = d(x, y)$ for all $g, h, x, y \in G$. The second important class of groups defined in terms of metrics is the class of all Polish groups with two-sided invariant metrics. By Klee's theorem these are precisely those Polish groups which admit an open basis at 1 each element of which is invariant under inner automorphisms, see KELLY [1955]. Two-sided invariant metrics on Polish groups are always complete, therefore, all Polish groups with two-sided invariant metrics have left-invariant complete metrics. All Abelian Polish and all compact Polish groups have two-sided invariant metrics and there are non-Abelian, non-locally compact Polish groups with two-sided invariant metrics. The class of groups with two-sided invariant metrics was investigated in the context of locally compact groups. For such groups the existence of a two-sided invariant metric places interesting structural restrictions on the group, see for example GROSSER and MOSKOWITZ [1971].

2.1. EXAMPLES. 1. S_∞ and the group of all homeomorphisms of the interval $[0, 1]$ are Polish groups without left-invariant complete metrics.

2. Let $\mathbb{Z}$ be the group of the integers, let $\mathbb{Z}_2$ be the two element cyclic group, and let $(\mathbb{Z}_2)^\mathbb{Z}$ be the group of infinite in two directions sequences of elements of $\mathbb{Z}_2$. The group $\mathbb{Z} \ltimes (\mathbb{Z}_2)^\mathbb{Z}$, which is the semidirect product of $\mathbb{Z}$ with $(\mathbb{Z}_2)^\mathbb{Z}$ with $\mathbb{Z}$ acting on $(\mathbb{Z}_2)^\mathbb{Z}$ by the shift of coordinates, is an example of a locally compact group which, like all locally compact groups, has a left-invariant complete metric but does not have a two-sided invariant metric. See MORAN and WILLIAMSON [1978].

3. Countable products of countable groups, where the countable groups are taken with the discrete topology, are groups with two-sided invariant metrics. If all of the factor groups are infinite and one of them is not Abelian, we get an example of a non-Abelian, non-locally compact group with an invariant metric.

In this and the next section, we will study continuous actions of Polish groups on Polish spaces. This is a rather general setting. In particular, a theorem from BECKER and

KECHRIS [1996] shows that we do not gain any generality if we allow the action to be Borel, rather than continuous, and the space on which the action is done to be a Borel subset of a Polish space rather than a Polish space.

First we discuss descriptive set theoretic complexity of orbit equivalence relations. Let G be a Polish group acting continuously on a Polish space X. We think of E_G^X as a subset of $X \times X$ consisting of all pairs (x, y) with $xE_G^X y$. A direct calculation shows that E_G^X is analytic. The equivalence relation E_G^X need not be Borel. We will see in the sequel that in many situations the assumption that E_G^X is Borel has consequences for the structure of the equivalence relation as well as for the structure of the group G. BECKER and KECHRIS [1996] proved the theorem below showing that each orbit equivalence relation can be pieced together from ω_1 Borel equivalence relations. Recall that a mapping is called *C-measurable* if it is measurable with respect to the smallest σ-algebra containing the analytic sets. For the definition of "in the codes" in the theorem below we refer the reader to BECKER and KECHRIS [1996].

2.2. THEOREM. *Let G be a Polish group acting continuously on a Polish space X. There exists a family of pairwise disjoint, G-invariant Borel sets A_ξ, $\xi < \omega_1$, which cover X and have the following properties:*

(i) *$E_G^X | A_\xi$ is Borel for each $\xi < \omega_1$;*

(ii) *If A is a Borel G-invariant subset of X with $E_G^X | A$ Borel, then $A \subseteq \bigcup_{\xi < \xi_0} A_\xi$ for some $\xi_0 < \omega_1$.*

Moreover, the function mapping $x \in X$ to the unique $\xi < \omega_1$ with $x \in A_\xi$ is "simply definable" or, more precisely, C-measurable in the codes.

We will state now some useful results on universal orbit equivalence relations. Given a Polish group G there exists a universal equivalence relation among orbit equivalence relations induced by continuous actions of G on Polish spaces. In fact, more is true as seen from the following theorem proved by BECKER and KECHRIS [1996].

2.3. THEOREM. *Let G be a Polish group. There exists a Polish space X_0 and a continuous action of G on X_0 such that for any continuous action of G on a Polish space X there exists a Borel injection $f : X \to X_0$ with $f(gx) = gf(x)$ for all $g \in G$ and $x \in X$.*

In particular, there exists a Borel G-invariant subset Z of X_0 such that E_G^X is Borel isomorphic to $E_G^{X_0} | Z$. Thus, $E_G^X \leq E_G^{X_0}$.

An equivalence relation $E_G^{X_0}$ as in the theorem is unique only up to Borel bireducibility. Frequently, we fix one such equivalence relation, denote it by E_G^∞, and call it the *universal orbit equivalence relation of G*.

USPENSKII [1986] proved that there exists a Polish group G_0 with the property that each Polish group is topologically isomorphic with a closed subgroup of G_0. In fact, the group of all homeomorphisms of the Hilbert cube $[0, 1]^{\mathbb{N}}$ with the compact-open topology can be taken as G_0. Let us fix a universal orbit equivalence relation $E_{G_0}^\infty$ of G_0. Now it follows from a theorem of Mackey on extension of actions, see BECKER and KECHRIS [1996], that for any Polish group G and any action of G on a Polish space X, $E_G^X \leq E_{G_0}^\infty$. Therefore, $E_{G_0}^\infty$ is universal among all orbit equivalence relations induced by continuous actions of Polish groups on Polish spaces. We will denote it by E_P^∞ and call it the *universal orbit equivalence relation of Polish group actions*.

1. Topological Vaught conjecture

The most fundamental question about an action of a group is the question about the number of orbits, that is, about the number of equivalence classes in E_G^X. Let $=_{2^{\mathbb{N}}}$ stand for the equivalence relation of equality on $2^{\mathbb{N}}$, that is, for $x, y \in 2^{\mathbb{N}}$,

$$x =_{2^{\mathbb{N}}} y \Leftrightarrow x = y. \tag{18.2}$$

(Since any uncountable Polish space is Borel isomorphic with $2^{\mathbb{N}}$, equality on any such space is Borel isomorphic with $=_{2^{\mathbb{N}}}$.) The simplest problem making the above question precise is whether the following dichotomy is always true:

$$E_G^X \text{ has countably many equivalence classes or } \quad =_{2^{\mathbb{N}}} \leq E_G^X. \tag{18.3}$$

Note that $=_{2^{\mathbb{N}}} \leq E_G^X$ implies that there are $2^{\aleph_0}$ orbits. Dichotomy (18.3) is called the *topological Vaught conjecture* and it is a generalization of the Vaught conjecture from model theory. In fact, the model theoretic Vaught conjecture (for $L_{\omega_1\omega}$ sentences) is equivalent, by BECKER and KECHRIS [1996], to the topological Vaught conjecture for orbit equivalence relations of continuous actions of S_∞ on Polish spaces. For a thorough discussion of connections with model theory and various versions of the topological Vaught conjecture see BECKER and KECHRIS [1996].

A theorem of BURGESS [1979] says that if we augment dichotomy (18.3) by adding the option that there can be $\aleph_1$ E_G^X equivalence classes, then the resulting trichotomy is true. In fact, the only property of E_G^X relevant to this trichotomy is the fact that it is an analytic equivalence relation. Thus, the topological Vaught conjecture amounts to eliminating the possibility of E_G^X having $\aleph_1$ equivalence classes. An exhaustive study of hypothetical realizations of this possibility was carried out by BECKER [1994].

One should mention that if E_G^X happens to be Borel, then the topological Vaught conjecture holds for this E_G^X for completely general reasons which do not involve the fact that E_G^X is an orbit equivalence relation. The following theorem is due to SILVER [1980].

2.4. THEOREM. *Let X be a Polish space, and let E be a Borel equivalence relation on X. Then either E has countably many equivalence classes or $=_{2^{\mathbb{N}}} \leq E$.*

Since all the continuous actions of locally compact Polish groups on Polish spaces induce Borel orbit equivalence relations, it follows from Silver's theorem that the topological Vaught conjecture holds for actions of such groups. We will see a far reaching strengthening of this result in Theorem 2.9(iii). SAMI [1994] proved that the topological Vaught conjecture holds for continuous actions of Polish Abelian groups on Polish spaces. Sami's result was the first one which goes beyond Silver's theorem since, as proved in SOLECKI [1995], orbit equivalence relations of actions of Polish Abelian groups need not be Borel. The strongest result towards proving the topological Vaught conjecture at this point is the following theorem due to HJORTH [2001].

2.5. THEOREM. *Let G be a Polish group acting continuously on a Polish space X. Assume G does not have a closed subgroup which can be mapped by a continuous homomorphism onto S_∞. Then either E_G^X has countably equivalence classes or $=_{2^{\mathbb{N}}} \leq E_G^X$.*

The class of groups in this theorem contains all Polish groups with left-invariant complete metrics and many more. The topological Vaught conjecture for arbitrary Polish groups remains open. However, if any substantial further progress is made, that is, if one is able to prove the topological Vaught conjecture for the orbit equivalence relations of all continuous actions on Polish spaces of a Polish group G, which is not already subject to Theorem 2.5, then, by using Mackey's theorem on extending actions of Polish groups, it will follow that the topological Vaught conjecture holds for all continuous actions of S_∞ on Polish spaces, see BECKER and KECHRIS [1996]. As mentioned above, this would solve the model theoretic Vaught conjecture which has been open (in full generality) for over 40 years.

2. *The Glimm-Effros dichotomy*

There are good reasons for considering smooth equivalence relations as being simple. These are the equivalence relations whose equivalence classes can be distinguished by assigning to them in a Borel fashion real numbers or elements of the Cantor set $2^{\mathbb{N}}$. Here is a precise definition. An equivalence relation E on a Polish space X is called *smooth* if there exists a Borel function $f : X \to 2^{\mathbb{N}}$ such that xEy exactly when $f(x) = f(y)$ for $x, y \in X$. This is equivalent to saying that E is smooth if, and only if, $E \leq =_{2^{\mathbb{N}}}$. Thus, $=_{2^{\mathbb{N}}}$ is a universal smooth equivalence relation.

2.6. EXAMPLE. To give a nontrivial and familiar example of a smooth equivalence relation consider the Polish space of complex $n \times n$ matrices $\mathcal{M}_n$. (Topologically this is simply $\mathbb{R}^{2n^2}$.) Make two matrices M_1 and M_2 in $\mathcal{M}_n$ equivalent if they are similar, that is, if for some invertible matrix P, $PM_1P^{-1} = M_2$. Call this Borel equivalence relation F_n. Now, F_n turns out to be smooth. To see this, recall that each matrix M in $\mathcal{M}_n$ has finitely many Jordan canonical forms which differ from each other by permutations of their elementary Jordan matrices. Now given such an M, code in a Borel way the finite set of Jordan canonical forms of M by an element of $2^{\mathbb{N}}$. This is a Borel function from $\mathcal{M}_n$ to $2^{\mathbb{N}}$ with the property that two matrices are F_n equivalent if, and only if, the elements of $2^{\mathbb{N}}$ assigned to them are equal. Thus, this Borel assignment witnesses that $F_n \leq =_{2^{\mathbb{N}}}$.

There is a simple non-smooth equivalence relation. Define E_0 on $2^{\mathbb{N}}$ by the formula

$$xE_0y \Leftrightarrow \exists N \forall n > N \; x(n) = y(n). \tag{18.4}$$

where, of course, we treat $x, y \in 2^{\mathbb{N}}$ as binary sequences $(x(n))$ and $(y(n))$. We have $E_0 \not\leq =_{2^{\mathbb{N}}}$ since each Borel function $f : 2^{\mathbb{N}} \to 2^{\mathbb{N}}$ constant on E_0 equivalence classes is constant on a comeager subset of $2^{\mathbb{N}}$. In fact, E_0 is Borel isomorphic with Vitali's equivalence relation on the interval $[0, 1]$ identifying two reals if they lie in the same coset of $\mathbb{Q}$ and used to find a Lebesgue non-measurable subset of $[0, 1]$. In this form, $E_0 \not\leq =_{2^{\mathbb{N}}}$ is a consequence of the well-known fact that any Borel function from $[0, 1]$ to $2^{\mathbb{N}}$, or to any other Polish space, invariant under translations by elements of $\mathbb{Q}$ is constant on a comeager subset of $[0, 1]$. It is not difficult to see that we actually have $=_{2^{\mathbb{N}}} < E_0$.

The Glimm-Effros dichotomy states that Borel embedding E_0 is the canonical obstacle to being smooth. To be more precise, we say that the *Glimm-Effros dichotomy* holds for a continuous action of a Polish group G on a Polish space X if

$$E_G^X \text{ is smooth or } E_0 \leq E_G^X.$$

This type of dichotomy was first established by Glimm for continuous actions of Polish locally compact groups and then extended by EFFROS [1965] to continuous actions of Polish groups on Polish spaces with F_σ orbit equivalence relations. The Glimm-Effros dichotomy is stronger than the topological Vaught conjecture; it implies the topological Vaught conjecture for any action for which it holds. Again, like in the case of the topological Vaught conjecture, if E_G^X happens to be Borel, the Glimm-Effros dichotomy holds for it for very general reasons because of the following theorem due to HARRINGTON, KECHRIS and LOUVEAU [1990].

2.7. THEOREM. *Let X be a Polish space and let E be a Borel equivalence relation on X. Then either E is smooth or $E_0 \leq E$.*

Unlike the topological Vaught conjecture, the Glimm-Effros dichotomy is known to fail for some continuous actions of Polish groups, for example S_∞, on Polish spaces, see HJORTH and KECHRIS [1995].

The following fact can be deduced from results of BECKER and KECHRIS [1996]. Let E_G^X be the orbit equivalence relation of a continuous action of a Polish group G on a Polish space X. Then the following two conditions are equivalent.

(I) E_G^X is smooth.

(II) There exists a Polish topology τ on X which includes the original one, keeps the action of G on X continuous, and is such that E_G^X is G_δ in the product topology $\tau \times \tau$ on $X \times X$.

Point (II) gives a quantization of smoothness of E_G^X. Since, as can be easily proved, the Polish topology τ as in (II) consist of Borel subsets of the old topology, there exists $\alpha < \omega_1$ with the property that τ is included in Σ_α^0. Now, the existence of a small such α indicates that smoothness is achieved easily; if the smallest such α is large, smoothness is difficult to achieve. Thus, the simplest way in which E_G^X can be smooth is when $\alpha = 1$, that is, when τ is the original topology. We can, therefore, formulate the *strong Glimm-Effros dichotomy*:

$$E_G^X \text{ is } G_\delta \text{ or } E_0 \leq E_G^X.$$

The strong Glimm-Effros dichotomy turns out to be true if we make some assumptions on the action itself or on the structure of the acting group. The following theorem is a result of the first type. Point (i) was established by EFFROS [1965] with a more stringent assumption that E_G^X is F_σ. (The present version follows from results in BECKER and KECHRIS [1996]). In fact, Effros' result formed a root out of which all the other results of this section grew. Point (ii) was proved by HJORTH and SOLECKI [1999]. Recall that the *stabilizer* of a point x is defined to be the group $\{g \in G : gx = x\}$.

2.8. THEOREM. *Suppose G is a Polish group acting continuously on a Polish space X. If one of the following assumptions holds*

(i) *each orbit of the action is F_σ,*

(ii) *G has a left-invariant complete metric and the stabilizer of each $x \in X$ is compact, then we have that E_G^X is G_δ or $E_0 \leq E_G^X$.*

It should be pointed out that both conditions in the above theorem imply that E_G^X is Borel ((i) by a result of SAMI [1994] and (ii) by the theorem that Borel sets with compact fibers have Borel projections, see KECHRIS [1995]), so the Glimm-Effros dichotomy, but not the strong form of it, for such E_G^X is immediate from the Harrington-Kechris-Louveau theorem stated above.

The next result shows that one can prove the strong Glimm-Effros dichotomy by making only assumptions about the structure of the acting Polish group. If G is Polish locally compact, then its continuous actions on Polish spaces fulfill condition (i) in Theorem 2.8. Therefore, all continuous actions of such groups satisfy the strong Glimm-Effros dichotomy. The first result going beyond locally compact groups was established by Solecki in HJORTH and SOLECKI [1999] and constitutes point (i) of the theorem below. Points (ii) and (iii) are due to Hjorth and come from HJORTH and SOLECKI [1999] and HJORTH [2000b], respectively.

2.9. THEOREM. *Let G be a Polish group acting on a Polish space X. In the following situations either E_G^X is G_δ or $E_0 \leq E_G^X$.*

(i) *G has a two-sided invariant metric.*

(ii) *G is nilpotent.*

(iii) *G is a countable product of Polish locally compact groups.*

In the proofs of the above theorem, one actually shows that the groups G from (i), (ii), and (iii) have a certain property which is interesting in its own right. After HJORTH [2000b], call a Polish group a *Glimm-Effros group* if whenever G act continuously on a Polish space X and the action has a non-meager, dense orbit, then the action is transitive, that is, there is only one orbit. Now, as essentially proved in HJORTH and SOLECKI [1999] and explicitly stated in HJORTH [2000b], if a Polish group G is a Glimm-Effros group, then all its continuous actions on Polish spaces satisfy the strong Glimm-Effros dichotomy. HJORTH asked in [2000b] if this is the only reason for the strong Glimm-Effros dichotomy to hold for all continuous actions of a given Polish group.

2.10. QUESTION. Is it true that if all continuous actions of a Polish group G on Polish spaces satisfy the strong Glimm-Effros dichotomy, then G is a Glimm-Effros group?

We will come back to the Glimm-Effros groups later.

It turns out that the strong Glimm-Effros dichotomy is a delicate property and it can fail for actions just a bit more complicated than the ones in Theorem 2.8 and for Polish groups just a bit more complicated than the ones in Theorem 2.9. HJORTH and SOLECKI [1999] exhibit an example of a Polish solvable group G with a left-invariant complete metric and a continuous action of G on a Polish space X with all stabilizers locally compact and with each orbit F_σ or G_δ. This action has two orbits (so $E_0 \not\leq E_G^X$) both of which are dense (so, by the Baire Category Theorem, at least one of them is not a G_δ). Another example in HJORTH and SOLECKI [1999] shows that one cannot remove the assumption of the existence of a left-invariant complete metric on G in Theorem 2.8(ii).

It was proved in HJORTH and SOLECKI [1999] that all solvable and, therefore, nilpotent Polish groups admit complete left-invariant metrics. (For a more direct proof see GAO [1998].) It is easy to see that any left-invariant metric on locally compact groups is complete and that possession of a left-invariant complete metric is preserved under countable products. Therefore, Polish groups with left-invariant complete metrics constitute a natural class of groups including the groups in Theorem 2.9(i)–(iii). As the above example shows, the strong Glimm-Effros dichotomy may fail for such groups, however, BECKER [1998] proved that the Glimm-Effros dichotomy does hold for them. In fact, in the case when E_G^X is smooth, a topology making it G_δ as in (II) can be found which is not too complicated. Here is Becker's theorem.

2.11. THEOREM. *Let G be a Polish group with a complete left-invariant metric acting continuously on a Polish space X. Then either E_G^X is smooth or $E_0 \leq E_G^X$.*

Moreover, if E_G^X is smooth, one can find a Polish topology τ on X as in (II) *such that each set in τ is in Σ_ω^0 with respect to the original topology.*

The following question asked by BECKER [1998] remains open.

2.12. QUESTION. Does there exist a Polish group without a left-invariant complete metric such that all its continuous actions on Polish spaces induce orbit equivalence relations fulfilling the Glimm-Effros dichotomy?

The next question also has to do with Theorem 2.11. Its weaker version can be found in BECKER [1998].

2.13. QUESTION. Let G have a left-invariant complete metric, and let it act continuously on a Polish space X. Is it true that either $E_0 \leq E_G^X$ or there is a Polish topology τ on X containing the original one, keeping the action continuous, making E_G^X G_δ and such that all sets in τ are F_σ in the old topology?

There exists a weaker form of the Glimm-Effros dichotomy in which the role of $2^{\mathbb{N}}$ is played by countable subsets of ω_1 and which holds for all continuous Polish group actions on Polish spaces. Its relationship to the Glimm-Effros dichotomy is, in a sense, the same as the relationship of Burgess' theorem mentioned in Subsection 2.1 to the topological Vaught conjecture. This weaker form of the dichotomy is included in the following theorem due to Becker, and Hjorth and Kechris, see HJORTH and KECHRIS [1995]. C-measurability was defined before the statement of Theorem 2.2; for the meaning of "in the codes," we refer the reader to HJORTH and KECHRIS [1995]. $\mathcal{P}_{\aleph_0}(\omega_1)$ stands for the set of all countable subsets of ω_1.

2.14. THEOREM. *If G acts continuously on X, G and X are Polish, then either there exists a C-measurable in the codes function $f : X \to \mathcal{P}_{\aleph_0}(\omega_1)$ such that*

$$xE_G^X y \Leftrightarrow f(x) = f(y)$$

or $E_0 \leq E_G^X$.

3. *Actions of* S_∞ *and turbulence*

We will consider here the Polish group S_∞ of all permutations of $\mathbb{N}$ and its continuous actions. This group plays a special role for two reasons. First, as seen in Subsections 2.1 and 2.2, it is in some situations the simplest group for which a general approach does not work. The topological Vaught conjecture is not known to hold precisely for actions of those Polish groups which contain closed subgroups which can be mapped by continuous homomorphisms onto S_∞. Also S_∞ is an example of a Polish group without a left-invariant complete metric; therefore, the results in Subsection 2.2 do not apply to it. In fact, the Glimm-Effros dichotomy is known to fail for certain continuous actions of S_∞. The second reason for the particular interest in continuous actions of S_∞ is the fact that the orbit equivalence relations induced by such actions are, up to Borel bireducibility, precisely the equivalence relations of isomorphism on countable models of $L_{\omega_1\omega}$ sentences, BECKER and KECHRIS [1996]. The way these are defined is explained in an example in the next paragraph. These equivalence relations are of obvious importance in model theory.

2.15. EXAMPLE. It follows from the remark in the introduction to Section 2, that there exists a universal equivalence relation among orbit equivalence relations induced by continuous actions of S_∞ on Polish spaces. As proved by BECKER and KECHRIS [1996] this universal relation can be chosen to have a particularly nice form. It is the equivalence relation on undirected graphs whose domain is $\mathbb{N}$. We make two such graphs equivalent if they are isomorphic. Here is a formal definition. Consider $\mathcal{P}(\mathbb{N}\times\mathbb{N})$, the family of all subsets of $\mathbb{N}\times\mathbb{N}$, identified via indicator functions with the space $\{0,1\}^{\mathbb{N}\times\mathbb{N}}$, which is compact metric. Let $\mathcal{G}$ be the subspace of it consisting of all x which code an undirected graph relation on $\mathbb{N}$, that is, for all $n,m\in\mathbb{N}$,

$$(n,n)\notin x \text{ and } (n,m)\in x \Rightarrow (m,n)\in x.$$

Then $\mathcal{G}$ is a G_δ subset of $2^{\mathbb{N}\times\mathbb{N}}$ so it is a Polish space with the inherited topology. We define a continuous action of S_∞ on $\mathcal{G}$ by letting

$$gx = \{(n,m) : (g^{-1}(n), g^{-1}(m)) \in x\}$$

for $g\in S_\infty$ and $x\in\mathcal{G}$. Then $E^{\mathcal{G}}_{S_\infty}$ is a universal relation for orbit equivalence relations induced by continuous actions of S_∞ on Polish spaces. We will denote it by $E^{\infty}_{S_\infty}$ and call it the *graph isomorphism equivalence relation.* Note that two graphs represented by $x,y\in\mathcal{G}$ are isomorphic precisely when $y=gx$ for some $g\in S_\infty$. In fact, this g is an isomorphism between the graphs coded by x and y.

The above example indicates that being Borel reducible to the orbit equivalence relation of a continuous action of S_∞ on a Polish space has an interesting interpretation. If this happens for an equivalence relation E on a Polish space X, then $E\leq E^{\mathcal{G}}_{S_\infty}=E^{\infty}_{S_\infty}$, that is, we can assign in a Borel way graphs on $\mathbb{N}$ to points of X so that two points lie in the same E equivalence class precisely when the graphs assigned to them are isomorphic. This amounts to saying that we can classify E equivalence classes by graphs or, in other words, we can assign graphs as complete invariants to equivalence classes of E. This is analogous to, though more complicated than, smoothness which is defined as the ability to assign in a Borel way elements of $2^{\mathbb{N}}$ to equivalence classes as complete invariants. So in this section by simple we will regard those orbit equivalence relations which are Borel reducible to $E^{\infty}_{S_\infty}$.

It is important, however, to keep in mind that actions of S_∞ can be complicated in ways different than anything we encountered so far. In fact, there exist simple to define actions of S_∞ on Polish spaces whose orbit equivalence relations are not Borel reducible to orbit equivalence relations induced by continuous actions of Polish groups with complete left-invariant metrics. HJORTH [1999] proved that the action of S_∞ on $\mathbb{R}^{\mathbb{N}}$ by permuting the coordinates is of this sort.

On the other hand, there are actions of relatively simple (Abelian) Polish groups whose orbit equivalence relations cannot be Borel reduced to orbit equivalence relations induced by actions of S_∞. This was first discovered by Friedman whose proof appeared only years later in FRIEDMAN [2000]. One would like to have a canonical obstacle for an orbit equivalence relation to admit classification of its equivalence classes by graphs, that is, to admit Borel reducibility to an orbit equivalence relation of S_∞. In order to find such an obstacle, Hjorth defined a dynamical property of an action called turbulence. Let G be a Polish group acting continuously on a Polish space X. The action is called *turbulent* if

(i) all orbits are meager and dense;

(ii) for any $x \in X$ and any open sets $x \in U \subseteq X$ and $1 \in V \subseteq G$ the set

$$\{g_1 \cdots g_k x : g_1, \ldots, g_k \in V \text{ and, for each } i \leq k, g_1 \cdots g_i x \in U\}$$

is dense in some open non-empty subset of U.

2.16. EXAMPLES. 1. It will follow from the next theorem that no continuous action of S_∞ on a Polish space is turbulent. In particular, the action from Example 2.15 is not turbulent.

2. Let G be c_0, the group of all sequences of reals tending to 0, with the metric $d((a_n),(b_n)) = \sup_n |a_n - b_n|$. The group G with the topology induced by this metric is Polish. Define the action of G on $\mathbb{R}^{\mathbb{N}}$ by letting $(a_n) \in G$ applied to $(x_n) \in \mathbb{R}^{\mathbb{N}}$ be the sequence $(a_n + x_n)$. This action is turbulent. A verification of this and a general result in which similar actions are proved to be turbulent can be found in HJORTH [2000b].

Let G be a Polish group acting continuously on a Polish space X. Let Y be a G-invariant Borel subset of X. We say that the action of G on Y is *potentially turbulent*, if there exists a Polish topology τ on Y which is stronger than the original subspace topology on Y, keeps the action continuous and makes it turbulent on Y. For thorough expositions of the theory of turbulence the reader is referred to HJORTH [2000b] and KECHRIS [200?].

HJORTH [2000b] proved that potential turbulence is incompatible with Borel reducibility to orbit equivalence relations induced by continuous actions of S_∞ on Polish spaces.

2.17. THEOREM. *Let G act continuously on a Polish space X. If there exists a Borel G-invariant subspace of X on which the action is potentially turbulent, then E^X_G is not reducible to the orbit equivalence relation induced by a continuous action of S_∞ on a Polish space.*

Furthermore, Hjorth proved that, under certain assumptions, potential turbulence is the only possible obstacle to Borel reduction to an action of S_∞. As in some situations seen before, these assumptions are of two kinds: definability assumptions on the orbit equivalence relation or structural assumptions on the acting Polish group. Hjorth's results are included in the following theorem: point (i) comes from HJORTH [200?a] and point (ii) from HJORTH [2000b].

2.18. THEOREM. *Let G be a Polish group acting on a Polish space X. Either on some G-invariant Borel subset of X the action is potentially turbulent or E_G^X is Borel reducible to the orbit equivalence relation induced by a continuous action of S_∞ on a Polish space, if one of the following conditions holds:*

(i) *E_G^X is Borel.*

(ii) *G is a Glimm-Effros group.*

The definition of Glimm-Effros groups was given after Theorem 2.9. As mentioned there, this class of groups includes Polish groups which have two-sided invariant metrics, are nilpotent, or are countable products of locally compact Polish groups. In fact, as proved in HJORTH [2000b], for products of locally compact groups the second possibility from Theorem 2.18 ($E_G^X \leq E_{S_\infty}^\infty$) always holds.

The following question is due to Hjorth.

2.19. QUESTION. Does the dichotomy from Theorem 2.18 hold for arbitrary continuous actions of Polish groups on Polish spaces?

Potential turbulence of E_G^X is a dynamical condition on the action. In the context of equivalence relations, it would be desirable to prove that it is equivalent to saying that $E \leq E_G^X$ where E comes from some fixed (not depending on E_G^X) finite family of turbulent orbit equivalence relations. These finitely many turbulent equivalence relations would form a simple "basis for turbulence." The possibility of existence of such a finite basis was, however, disproved by Farah. The theorem below, whose proof uses some ideas from the theory of Banach spaces concerning Tsirelson spaces, comes from FARAH [2001a].

2.20. THEOREM. *Let $\mathcal{F}$ be a family of turbulent orbit equivalence relations with the property that for each turbulent orbit equivalence relation E induced by a continuous action of some Polish group on a Polish space there is an element of $\mathcal{F}$ which Borel reduces to E. Then $\mathcal{F}$ has cardinality continuum.*

Another way in which the failure of a finite basis to exist can be established is described by FARAH [2001b].

4. *Beyond Polish group actions*

The final problem we face in this section is when a given analytic equivalence relation is, or can be Borel reduced to, the orbit equivalence relation induced by a continuous action of a Polish group on a Polish space. One structural obstacle for the existence of such a Borel reduction was discovered by Kechris and Louveau. Define first an equivalence relation on $(2^\mathbb{N})^\mathbb{N}$ denoted by E_1 and given by the formula

$$(x_n)E_1(y_n) \Leftrightarrow \exists N \forall n \geq N \; x_n = y_n. \tag{18.5}$$

The following theorem is from KECHRIS and LOUVEAU [1997].

2.21. THEOREM. *Let X be a Polish space with a Polish group G acting continuously on it. Then $E_1 \not\leq E_G^X$.*

Since $\leq$ is transitive, it follows from the above result that if E is Borel reducible to the orbit equivalence relation of a continuous action of a Polish group on a Polish space, then $E_1 \not\leq E$. This suggests the following question/conjecture which was formulated by Kechris and Louveau.

2.22. QUESTION. Let E be a Borel, or even analytic, equivalence relation on a Polish space X. Is it true that either $E_1 \leq E$ or $E \leq E_G^Y$ for some continuous action of a Polish group G on a Polish space Y?

This problem is open but some progress has been made and will be described in Subsection 3.2.

3. Topologies on groups and ideals and complexity of their actions

In this section, we will study group actions from a different perspective. Given a group G, we will consider the problem of whether placing assumptions on the complexity of orbit equivalence relations of all continuous actions of G on Polish spaces implies that G is in some sense topologically simple.

1. Actions and compactness and local compactness of Polish groups

Since trivially all Polish groups induce orbit equivalence relations Borel bireducible with $=_{2^{\mathbb{N}}}$, the first problem here is to inquire for what Polish groups all continuous actions on Polish spaces give rise only to smooth orbit equivalence relations. It is an easy folklore fact that all orbit equivalence relations induced by continuous actions of Polish compact groups are smooth. SOLECKI [2000] proved that the converse holds, therefore, we have the following theorem. Recall that an action of G on X is *free* if $gx = x$ for some $x \in X$ implies $g = 1$.

3.1. THEOREM. *Let G be a Polish group. The following two conditions are equivalent.*

(i) *G is compact.*

(ii) *For each Polish space X on which G acts continuously, E_G^X is smooth.*

Moreover in (ii) *one can replace "acts continuously" by "acts continuously and freely" and obtain an equivalent condition.*

An analogous question arises for local compactness. To formulate it precisely, we will need first to define another subclass of Borel equivalence relations. A Borel equivalence relation E on a Polish space X is called *countable* if each equivalence class of E is countable. Examples of such equivalence relations include $=_{2^{\mathbb{N}}}$ and E_0. By a theorem of FELDMAN and MOORE [1977] each Borel countable equivalence relation is the orbit equivalence relation of a Borel action of a countable (discrete) group. Since, by a change of topology on the space acted on, we can make sure that the action is continuous, BECKER and KECHRIS [1996], we see that countable Borel equivalence relations are among orbit equivalence relations induced by continuous actions of Polish locally compact groups on

Polish spaces. As was the case with previously studied classes of equivalence relations, it turns out that there exists a universal countable equivalence relation. Such an equivalence relation can be defined as follows.

3.2. EXAMPLE. Let F_∞ be the free group on countably infinite number of generators. Consider the space of all subsets of F_∞ identified via indicator functions with $\{0,1\}^{F_\infty}$ with the product topology which is metric compact. We make two such subsets x and y equivalent if there exists a $g \in F_\infty$ with $y = gx = \{gf : f \in x\}$. We denote this equivalence relation by E_∞.

KECHRIS in [1992] and [1994] established the following precise description of orbit equivalence relations induced by continuous actions of Polish locally compact groups. This theorem gives a purely Borel version of results of FELDMAN, HAHN and MOORE [1979] who studied actions of locally compact Polish groups on Polish spaces with quasi-invariant Borel probability measures. To state Kechris' result, we will need the notion of the product of equivalence relations. If E and F are equivalence relations on X and Y, respectively, let $E \times F$ be the equivalence relation on $X \times Y$ identifying pairs (x_1, y_1) and (x_2, y_2) if, and only if, $x_1 E x_2$ and $y_1 F y_2$. Define also $I_{2^{\mathbb{N}}}$ to be the equivalence relation on $2^{\mathbb{N}}$ which makes any two points equivalent.

3.3. THEOREM. *Let G be a locally compact Polish group acting continuously on a Polish space X. There exist G-invariant, disjoint Borel sets C, U with $C \cup U = X$ and such that*

(i) *$E_G^X | C$ is countable,*

(ii) *$E_G^X | U$ is Borel isomorphic to $F \times I_{2^{\mathbb{N}}}$ where $F = E_G^X | Z$ for a Borel set $Z \subseteq U$ which has non-empty countable intersection with each orbit included in U.*

It is an immediate consequence of this theorem that orbit equivalence relations induced by continuous actions of locally compact Polish groups on Polish spaces are Borel reducible to countable Borel equivalence relations and, therefore, to E_∞. This suggests that to obtain a result analogous to Theorem 3.1 for local compactness one should replace smoothness in point (ii) of this theorem by Borel reducibility to a Borel countable equivalence relation. This gives the following question due to Kechris.

3.4. QUESTION. Let G be a Polish group. Assume that all continuous actions of G on Polish spaces induce orbit equivalence relations which are $\leq E_\infty$. Is it true that G is locally compact?

Some partial results towards the affirmative answer to this question were found in SOLECKI [2000]. In particular, the answer is "yes" if G is a separable Banach space or if it is a product of countably many locally compact Abelian Polish groups. Moreover, HJORTH [200?b] has isolated a dynamical condition with the property that if a Polish group G acts on a Polish space X and E_G^X is Borel, then either E_G^X is Borel reducible to a countable Borel equivalence relation or the action of G on a Borel G-invariant subset of X fulfills this dynamical condition. In the latter case, E_G^X cannot be Borel reduced to E_∞.

Apart from comparing an orbit equivalence relation via Borel reducibility with other equivalence relations, there is another way of measuring how complicated such an equivalence relation is. This is done by evaluating its descriptive set theoretic complexity. The following question is most natural in this context. Does the fact that all orbit equivalence relations induced by continuous actions of a Polish group G are Borel imply that the topology on G is simple? If G is Polish locally compact, then all its continuous actions on Polish spaces have Borel orbit equivalence relations. What about the converse? This question has been studied for certain subclasses of the class of Polish groups.

It turns out that the converse is false in general. SOLECKI [1995] investigated which groups of the form $\prod_n H_n$, where each H_n, $n \in \mathbb{N}$, is countable Abelian with discrete topology, are capable of inducing non-Borel orbit equivalence relations. Note that under the condition that infinitely many of the H_n's are infinite, that is, when the question is worth studying at all, these groups are topologically identical. Yet, among such groups there are groups of both types: those inducing non-Borel orbit equivalence relations and those incapable of doing so. The dividing line is defined by algebraic, rather than topological, conditions on the H_n's.

Another class of groups for which the question was studied is the class of separable Banach spaces with vector addition as group operation. It turns out that among such groups the converse does hold. HJORTH [2000a] proved that a number of classical separable, infinite dimensional Banach spaces have continuous actions on Polish spaces with non-Borel orbit equivalence relations. Building on his methods, SOLECKI [2000] proved that, in fact, all separable infinite dimensional Banach spaces have such actions which gives the following theorem.

3.5. THEOREM. *Let G be a separable Banach space. Then G is locally compact* (*that is, finitely dimensional*) *if, and only if, for each continuous action of G on a Polish space X, E_G^X is Borel.*

The next result is a digression from the theme of this subsection but, after reading Section 2, the reader may wonder what are the Polish groups whose continuous actions on Polish spaces induce orbit equivalence relations which are Borel reducible to the graph isomorphism equivalence relation $E_{S_\infty}^\infty$. The following theorem of GAO and KECHRIS [200?], which generalizes a theorem of HJORTH [2000b], is a relevant result.

3.6. THEOREM. *Let G be the group of all isometries of a Polish locally compact metric space with the compact-open topology.* (*Such a group is Polish.*) *Then for any continuous action of G on a Polish space X, $E_G^X \leq E_{S_\infty}^\infty$.*

The isometry groups of Polish locally compact metric spaces include, for example, S_∞ and countable products of locally compact Polish groups. A description of such groups can be found in GAO and KECHRIS [200?].

2. *Polishability of Borel groups and Borel ideals*

This subsection is related to Subsection 2.4. We ask here the most general question: can the existence of *any* natural Polish group topology on a group be recognized by studying complexity of actions of the group? It is natural to consider this question for metric Borel groups. A group is called *metric Borel* if it is a metrizable separable topological group which is absolutely Borel, that is, it is a Borel subset of some, or equivalently all, of its metric completions. After KECHRIS and LOUVEAU [1997], a metric Borel group G is called *Polishable* if there exists a Polish group topology on it which has the same Borel sets as the original topology. This is equivalent to saying that there exists a Polish group topology on G which is stronger than the original one. Such a Polish group topology on a metric Borel group is canonical; if it exists, it is unique. In fact, it is frequently the natural topology with which a given group is considered.

3.7. EXAMPLES. 1. Let G be the group of all sequences (x_n) in $\mathbb{R}^{\mathbb{N}}$ with $\sum_n |x_n| < \infty$. It is a metric Borel group with the topology inherited from $\mathbb{R}^{\mathbb{N}}$. (It is an F_σ subgroup of its completion $\mathbb{R}^{\mathbb{N}}$.) The metric $d((x_n),(y_n)) = \sum_n |x_n - y_n|$ induces a stronger topology on G and this topology happens to be a Polish group topology. Therefore, G is Polishable.

2. Let G be the subgroup of the Polish group H of all homeomorphisms of the interval $[0,1]$ consisting of all f for which both f and f^{-1} are absolutely continuous. This is an $F_{\sigma\delta}$ subgroup of H and, therefore, metric Borel with the subspace topology inherited from H.

Elements of G have derivatives almost everywhere which are summable with respect to the Lebesgue measure. The metric

$$d(f,g) = \int_0^1 |f'(x) - g'(x)|\, dx$$

induces a Polish group topology on G which is stronger than the one coming from the inclusion in H. This makes G a Polishable group. For proofs see SOLECKI [1999b].

3. Let G be the group of all sequences (x_n) in the Polish group $\mathbb{R}^{\mathbb{N}}$ with the property that $x_n = 0$ for large enough n. Then G is a metric Borel group which is not Polishable. If it were Polishable, by the Baire Category Theorem, for some $N \in \mathbb{N}$, the subgroup $G_N = \{(x_n) \in G : x_n = 0 \text{ for all } n \geq N\}$ would be non-meager in the Polish group topology on G, which would make it open by Pettis' theorem (see KECHRIS [1995]), implying that the index of G_N in G is countable. This can be easily verified to be false.

Thus, the main problem here is whether Polishability of a Borel metric group G can be recognized by the orbit equivalence relations continuous actions of G on Polish spaces give rise to. This brings us back to Theorem 2.21 of Kechris and Louveau which can be reformulated as follows. Let G be a Polishable metric Borel group. If G acts continuously on a Polish space X, then $E_1 \not\leq E_G^X$. Therefore, Polishable groups are not capable of inducing orbit equivalence relations above E_1. The natural conjecture here is that it is indeed the distinguishing feature of Polishable groups among metric Borel groups. Thus, we are lead to the question of whether a metric Borel group G which is not Polishable can have a continuous action on a Polish space X with $E_1 \leq E_G^X$.

If G is a metric Borel group, there exists a canonical free continuous action of G on a Polish space which may be particularly relevant in this context. Namely, there exists a Polish group H such that G is a dense subgroup of H and the topology on G inherited from the inclusion in H is the original topology on G. In fact, the group H with these properties is unique up to isomorphism over G, that is, if H' is another Polish group for which the above conditions hold, then there exists a homeomorphic isomorphism $f : H \to H'$ which is the identity mapping on G. In Examples 3.7 it is easy to guess what the Polish groups H are. (For more details on such a "group completion" of G see SOLECKI [1999b].)

Now consider the action of G on H given by

$$G \times H \ni (g, h) \to hg^{-1} \in H. \tag{18.6}$$

This is a continuous action. Its orbits are the left cosets of G in H. The orbit equivalence relation E_G^H of this action will be denoted simply by $E_{H/G}$ and will be called the *coset equivalence relation*. Coset equivalence relations were considered by MILLER [1977] who, generalizing an earlier result of MACKEY [1957], showed that if G is not closed in H, then $E_{H/G}$ is not smooth. (The converse is obvious.) One may hope that non-Polishability of a Borel metric group will be revealed by the coset equivalence relation as well. This suggests the following question.

3.8. QUESTION. Let G be a Borel subgroup of a Polish group H. Is it true that G is Polishable if, and only if, $E_1 \not\leq E_{H/G}$?

Obviously, the direction $\Rightarrow$ follows from Kechris and Louveau's theorem. In full generality this question is still open. There are, however, some interesting cases in which it is known that the answer to it is "yes". We will consider these cases below.

Before we do that, let us record a theorem on the structure of Polishable subgroups of Polish groups. Examples of Polishable subgroups described in Examples 3.7 are quite simple Borel subsets (F_σ and $F_{\sigma\delta}$) of the ambient Polish group but, as proved by SAINT-RAYMOND [1976], Polishable subgroups can occur arbitrarily high in the hierarchy Σ^0_α of Borel subsets of the Polish group in which they are contained. However, by the following theorem, a Polishable subgroup G of a Polish group H is recoverable by a procedure in which simple ($F_{\sigma\delta}$) Polishable groups play a major role. The special case of the result below (for H a separable Fréchet topological vector space and G a linear subspace of H which is separable Fréchet with a topology stronger than the one inherited from the inclusion in H) was proved by SAINT-RAYMOND [1976]. The theorem was established in full generality by SOLECKI [1999b].

3.9. THEOREM. *Let H be a Polish group, and let G be a Borel subgroup of H. If G is Polishable, then there exists a countable ordinal $\alpha_0 < \omega_1$ and Polishable subgroups G_α of H for $\alpha \leq \alpha_0$ with Polish group topologies τ_α with the following properties:*

(i) $G_0 = \overline{G}$, $G_{\alpha_0} = G$.

(ii) *$G_{\alpha+1} < G_\alpha$ for each $\alpha < \alpha_0$ and $G_\lambda = \bigcap_{\alpha<\lambda} G_\alpha$ for $\lambda \leq \alpha_0$ limit.*

(iii) *$G_{\alpha+1}$ is $F_{\sigma\delta}$ in the topology τ_α on G_α.*

(iv) *If $A \subseteq G_\alpha$ is $F_{\sigma\delta}$ in τ_α and $G \subseteq A$, then $A \cap G_{\alpha+1}$ is comeager in $\tau_{\alpha+1}$.*

Moreover, α_0 and the sequence $(G_\alpha)_{\alpha\leq\alpha_0}$ with properties (i)–(iv) are unique.

Now we turn our attention back to the dichotomy between Polishability and Borel reduction of E_1 or, more precisely, to Question 3.8. A particular case of this question was first considered in KECHRIS and LOUVEAU [1997] where the authors showed that the answer is "yes" if G is the increasing union of countably many closed subgroups of H. We will present a generalization of this result later.

The first class of metric Borel groups we will discuss here are Borel ideals of subsets of $\mathbb{N}$. Establishing the dichotomy for such ideals involves proving a structure theorem for them which, in turn, has various other applications.

We identify here the family $\mathcal{P}(\mathbb{N})$ of all subsets of $\mathbb{N}$ with the compact metric space $2^{\mathbb{N}}$ by identifying a subset of $\mathbb{N}$ with its indicator function. The space $2^{\mathbb{N}}$ is also a compact metric group when considered with pointwise addition modulo 2. By an *ideal* we mean a family of subsets of $\mathbb{N}$ which is closed under taking finite unions and under taking subsets. All ideals are assumed to contain the singletons of natural numbers. We say that an ideal is Borel if it is a Borel subset of $2^{\mathbb{N}}$. Note that ideals are subgroups, in fact, dense subgroups of $2^{\mathbb{N}}$. Therefore, if an ideal is Borel, it is a Borel metric group. It acts on $2^{\mathbb{N}}$ by translations as described by (18.6) and produces the coset equivalence relation. It turns out that for the coset equivalence relations induced by Borel ideals the main question of this subsection can be answered affirmatively. In fact, a much more precise result can be proved. We will need a few new definitions.

First we define a natural way of comparing ideals which is more refined than comparing the coset equivalence relations via $\leq$. Let I and J be two ideals. We say that I is *Rudin-Blass below* J, in symbols $I \leq_{RB} J$, if there exists a finite-to-one function $f : \mathbb{N} \to \mathbb{N}$ such that, for each $x \subseteq \mathbb{N}$,

$$x \in I \Leftrightarrow f^{-1}(x) \in J.$$

If this happens, then the function $2^{\mathbb{N}} \ni x \to f^{-1}(x) \in 2^{\mathbb{N}}$ shows that

$$I \leq_{RB} J \Rightarrow E_{2^{\mathbb{N}}/I} \leq E_{2^{\mathbb{N}}/J}.$$

It turns out that the equivalence relation E_1 defined by (18.5) is the coset equivalence relation of a Borel ideal. Define the ideal I_1 of subsets of $\mathbb{N} \times \mathbb{N}$ as follows

$$I_1 = \{x \subseteq \mathbb{N} \times \mathbb{N} : \exists n \in \mathbb{N}\ x \subseteq \{0, \ldots, n\} \times \mathbb{N}\}. \tag{18.7}$$

By enumerating $\mathbb{N} \times \mathbb{N}$ by $\mathbb{N}$, we can consider I_1 to be an ideal of subsets of $\mathbb{N}$. It is easy to see that

$$E_{2^{\mathbb{N} \times \mathbb{N}}/I_1} \text{ is Borel isomorphic to } E_1.$$

There are submeasures behind ideals which are Polishable groups. A function $\varphi : \mathcal{P}(\mathbb{N}) \to \mathbb{R}$ is a *submeasure* if φ attains only nonnegative values, $\varphi(\emptyset) = 0$, and, for any two subsets x, y of $\mathbb{N}$,

$$\varphi(x) \leq \varphi(x \cup y) \leq \varphi(x) + \varphi(y).$$

Define

$$\mathrm{Exh}(\varphi) = \{x \in \mathcal{P}(\mathbb{N}) : \lim_n \varphi(x \setminus \{0, \ldots, n\}) = 0\}.$$

It is not difficult to check that $\mathrm{Exh}(\varphi)$ is an ideal. We call a submeasure lower semicontinuous if it is lower semicontinuous as a function from $2^{\mathbb{N}}$ (identified with $\mathcal{P}(\mathbb{N})$) to $\mathbb{R}$. If φ is a lower semicontinuous submeasure, then $d(x, y) = \varphi((x \setminus y) \cup (y \setminus x))$ is a metric and it induces a topology on $\mathrm{Exh}(\varphi)$ which shows that this ideal is a Polishable group.

3.10. EXAMPLES. 1. The ideal I_1 defined in (18.7) is not Polishable and, therefore, not of the form $\mathrm{Exh}(\varphi)$ for a lower semicontinuous submeasure ϕ.

2. Let $\varphi : \mathcal{P}(\mathbb{N}) \to \mathbb{R}$ be defined by

$$\varphi(x) = \sup_n \frac{1}{n} |x \cap \{0, \ldots, n-1\}|$$

where $|\cdot|$ stands for the number of elements in a set. Then φ is a lower semicontinuous submeasure and, as is easy to verify, $\text{Exh}(\varphi)$ is the of all subsets of $\mathbb{N}$ which have density 0, that is, all those $x \subseteq \mathbb{N}$ with $\lim_n (1/n)|x \cap \{0, \ldots, n-1\}| = 0$.

The following theorem was proved by SOLECKI [1999a]. Note that the dichotomy "$E_1 \leq E_{2^{\mathbb{N}}/G}$ or G is Polishable" follows immediately from the main part of the theorem by the discussion above.

3.11. THEOREM. *Let G be a Borel ideal of subsets of $\mathbb{N}$. Then precisely one of the following two conditions holds.*

(i) *$G = Exh(\varphi)$ for a lower semicontinuous submeasure φ.*

(ii) *$I_1 \leq_{RB} G$.*

Therefore, either G is Polishable or $E_1 \leq E_{2^{\mathbb{N}}/G}$.

As a byproduct of the above theorem, we obtain a characterization of the important class of P-ideals among Borel ideals. An ideal G is called a *P-ideal* if for any $x_n \in G$, $n \in \mathbb{N}$, there exists $x \in G$ with $x_n \setminus x$ finite for all n. Point (i) in Theorem 3.11 is easily seen to be incompatible with G being a P-ideal, therefore, all Borel P-ideals are of the form $\text{Exh}(\varphi)$ for a lower semicontinuous submeasure ϕ. It is worth pointing out that ideals of this form are $F_{\sigma\delta}$.

Groups somewhat similar to ideals were considered by Casevitz. He studied linear subspaces G of $\mathbb{R}^{\mathbb{N}}$ with the following additional property: if $(x_n) \in G$, and $(y_n) \in \mathbb{R}^{\mathbb{N}}$ is such that $|y_n| \leq |x_n|$ for all n, then $(y_n) \in G$. Such subspaces of $\mathbb{R}^{\mathbb{N}}$ are called *hereditary*. CASEVITZ [2000] established the following theorem.

3.12. THEOREM. *Let G be a Borel hereditary linear subspace of $\mathbb{R}^{\mathbb{N}}$. Then either G is Polishable or $E_1 \leq E_{\mathbb{R}^{\mathbb{N}}/G}$.*

In fact, as for ideals, both cases in the dichotomy have strong consequences for the structure of G. For example, a Borel hereditary linear subspace of $\mathbb{R}^{\mathbb{N}}$ which is Polishable is $F_{\sigma\delta}$.

The following theorem goes beyond subgroups of Polish groups with additional algebraic and combinatorial structure (like $2^{\mathbb{N}}$ and $\mathbb{R}^{\mathbb{N}}$). It was proved by SOLECKI [200?]. Point (ii) generalizes a result from KECHRIS and LOUVEAU [1997] where the G_n's are assumed to be closed in H.

3.13. THEOREM. *Let H be a Polish group and let G be its Borel subgroup. Then under either one of the following conditions*

(i) *G is Abelian and $F_{\sigma\delta}$,*

(ii) *$G = \bigcup_n G_n$ where each G_n is a Polishable subgroup of H and $G_n \subseteq G_{n+1}$ for each $n \in \mathbb{N}$,*

we have that G is Polishable or $E_1 \leq E_{H/G}$.

4. Composants in indecomposable continua

The results in this section concern indecomposable continua and have little to do with Polish group actions. However, Borel equivalence relations again play a prominent role. By a *continuum* we understand a compact connected metric space. A continuum is *indecomposable* if it is not the union of two proper subcontinua. A *hereditarily indecomposable continuum* is a continuum whose all subcontinua are indecomposable. Such continua are rather complicated but they form a dense G_δ in the family of all subcontinua of $[0,1]^{\mathbb{N}}$. (The family of all subcontinua of $[0,1]^{\mathbb{N}}$ is a compact subset of the space of all compact subsets of $[0,1]^{\mathbb{N}}$ with the Vietoris topology.)

4.1. EXAMPLES. 1. The interval $[0,1]$ is *not* an indecomposable continuum.

2. Let $f_n : \mathbb{T} \to \mathbb{T}$ be given by $f_n(z) = z^2$ where $\mathbb{T} = \{z \in \mathbb{C} : |z| = 1\}$ is the unit circle in the complex plane. Then the inverse limit of the system $(\mathbb{T}, f_n)_{n\in\mathbb{N}}$ is an indecomposable continuum called the dyadic solenoid.

3. The pseudo-arc, whose description can be found in NADLER [1992], is a hereditarily indecomposable continuum.

For more examples of indecomposable continua and their natural occurrence in dynamical systems see, for instance, NADLER [1992] and KENNEDY and YORKE [1995].

Let C be an indecomposable continuum. For $x \in C$, by the *composant* of x we understand the union of all proper subcontinua containing x. Composants are an important feature of the structure of indecomposable continua. They have been studied, both individually and as a whole family, since the 1920's. The first result about the global structure of the family of all composants was proved by MAZURKIEWICZ [1927] who showed that there exists a Cantor subset of C which picks *at most* one element from each composant. A natural question arose: does there exist a Borel subset of C which picks *precisely* one point from each composant? Such a set is called a *Borel transversal*. There were several notable partial results related to this question. COOK [1964] showed that a Borel transversal cannot be F_σ. KRASINKIEWICZ [1974] showed that a Borel transversal does not exist in the Knaster buckethandle continuum and other similar continua, EMERYK [1980] showed the same for the pseudo-arc, and ROGERS [1986] for solenoids and solenoids of pseudo-arcs.

As we will see the question about the existence of a Borel transversal can be formulated in the language of Borel equivalence relations and fully resolved in this context. Our main goal, however, will be a classification of the family of all composants of indecomposable continua. The equivalence relations $=_{2^{\mathbb{N}}}$, E_0, and E_1, defined by (18.2), (18.4), (18.5), will play a role in achieving this goal. Recall also the notion of smoothness from Subsection 2.2.

From indecomposability of C it follows that, for $x, y \in C$, the composants of x and y are either equal or disjoint. Thus, we obtain the *composant equivalence relation* E_C on C defined by

$$xE_Cy \Leftrightarrow \text{the composant of } x \text{ is equal to the composant of } y.$$

Clearly, xE_Cy precisely when x and y belong to a proper subcontinuum of C.

The equivalence relation E_C was considered in ROGERS [1986] where it was proved to be F_σ. This was strengthened by SOLECKI [2002] to the following theorem.

4.2. THEOREM. *Let C be an indecomposable continuum. Then there exist compact equivalence relations F_n, $n \in \mathbb{N}$, such that $F_n \subseteq F_{n+1}$, for each n, and $E_C = \bigcup_n F_n$.*

Since it is not difficult to see that compact equivalence relations are smooth, it follows from Theorem 4.2 that E_C is the increasing union of smooth equivalence relations. Such equivalence relations are called *hypersmooth.* They admit a classification up to Borel bireducibility. This classification is contained in the following theorem which is the conjunction of Theorem 2.4, Theorem 2.7, and a theorem proved by KECHRIS and LOUVEAU [1997].

4.3. THEOREM. *Let E be a hypersmooth equivalence relation. Then precisely one of the following is true.*

(i) *E has countably many equivalence classes;*
(ii) *$E \sim =_{2^{\mathbb{N}}}$;*
(iii) *$E \sim E_0$;*
(iv) *$E \sim E_1$.*

The equivalence relations are listed in points (i)–(iv) in the order of increasing complexity: if E has countably many classes, then $E <=_{2^{\mathbb{N}}}$ (this is obvious) and $=_{2^{\mathbb{N}}}< E_0 < E_1$ (the first of these inequalities was justified in Subsection 2.2, the second one follows, for example, from Theorem 2.21).

Now Theorem 4.2 and Theorem 4.3 open a possibility of completely classifying the composant equivalence relations. The equivalence relation E_C is as in (i)–(iv) in Theorem 4.3. But can all these possibilities be attained? This question is related to the question about the existence of a Borel set which picks precisely one point from each composant, that is, from each equivalence class of E_C. The existence of such a Borel transversal is equivalent for a σ-compact equivalence relation (and E_C is σ-compact) to smoothness. On the list in Theorem 4.3, E in points (i) and (ii) is smooth and E in points (iii) and (iv) is not. Therefore, to see that Borel transversals do not exist, it is sufficient to show that E_C is not as in (i) and (ii), for example, by proving that $E_0 \leq E_C$.

SOLECKI [2002] established the following two theorems, the first of which gives a sufficient condition for a σ-compact equivalence relation to Borel reduce E_0.

4.4. THEOREM. *Let E be a σ-compact equivalence relation on a Polish space X. Assume that the set of all $x \in X$ such that the equivalence class of x is not locally closed at x is non-meager. Then $E_0 \leq E$.*

Since it is easy to check that E_C satisfies the assumptions of this theorem, combining it with Theorem 4.3, we immediately get the following result.

4.5. THEOREM. *Let C be an indecomposable continuum. Then either $E_C \sim E_0$ or $E_C \sim E_1$. In particular, there exists no Borel subset of C which has precisely one point in common with each composant.*

The next theorem shows that Theorem 4.5 cannot be improved, that is, that both cases, $E_C \sim E_0$ and $E_C \sim E_1$, are possible. Actually, $E_C \sim E_1$ holds for the whole class of hereditarily indecomposable continua C. Theorem 4.6(ii) was proved by SOLECKI [2002].

To see point (i) of Theorem 4.6, it is enough to show that $E_1 \not\leq E_C$. But, as noticed by ROGERS [1986], if C is a solenoid, E_C is the orbit equivalence relation induced by a continuous action of $\mathbb{R}$ on C. Therefore, by Theorem 2.21, $E_1 \not\leq E_C$. (In fact, for solenoids as well as for the Knaster buckethandle continuum, it is not difficult to prove directly that $E_C \sim E_0$.)

4.6. THEOREM.

(i) $E_C \sim E_0$ *for* C *a solenoid (in particular, the dyadic solenoid from Examples 4.1).*

(ii) $E_C \sim E_1$ *for any hereditarily indecomposable continuum* C.

Note that as an immediate consequence of Theorem 4.6(ii) and Theorem 2.21, we see that, for a hereditarily indecomposable continuum C, E_C is not the orbit equivalence relation as those studied in Sections 2 and 3, that is, induced by a continuous action of a Polish group on C. As mentioned above, E_C for a solenoid C is such an orbit equivalence relation.

Theorems 4.5 and 4.6 give a full classification of composant equivalence relations. They demonstrate that the class of all indecomposable continua splits naturally into two subclasses: simple indecomposable continua ($E_C \sim E_0$) and complicated ones ($E_C \sim E_1$).

There are many questions open about these subclasses. The most important problem, however, seems to be that of determining the "generic" behavior of composants in an indecomposable continuum. Let me explain what I mean by that. The next theorem, proved by SOLECKI [2002], shows that whether $E_C \sim E_0$ or $E_C \sim E_1$ holds may be determined by the behavior of E_C on a small subset of C and the "generic" behavior of E_C may be different from that on the small set. We say that a set $A \subseteq C$ is E_C*-invariant* if for any $x \in U$ and $y \in C$ with xE_Cy, we have $y \in U$.

4.7. THEOREM. *There exists an indecomposable continuum* C *and an* E_C*-invariant dense* G_δ*-set* $U \subseteq C$ *such that*

(i) $E_C \sim E_1$,

(ii) $E_C|U \sim E_0$.

Therefore, we have the following problem. The affirmative answer to this question would give a classification of indecomposable continua into two classes according to the generic behavior of composants. Recall that $I_{2^{\mathbb{N}}}$ is the equivalence relation on $2^{\mathbb{N}}$ which makes any two points equivalent. The product of two equivalence relations is defined in Subsection 3.1.

4.8. QUESTION. Let C be an indecomposable continuum. Does there exist a comeager E_C-invariant Borel subset U of C with the property that $E_C|U$ is Borel isomorphic with E_1 via an isomorphism for which images and preimages of meager sets are meager or $E_C|U$ is Borel isomorphic with $E_0 \times I_{2^{\mathbb{N}}}$ via a Borel isomorphism as above?

Note that since composants are uncountable and equivalence classes of E_0 are countable, in the above question, we need to consider $E_0 \times I_{2^{\mathbb{N}}}$ rather than E_0. These two equivalence relations are, however, Borel bireducible. A positive answer to the above question would imply a positive answer to the question of KURATOWSKI [1932] which can be reformulated as follows.

4.9. QUESTION. Is it true that each Borel E_C-invariant subset of an indecomposable continuum C is meager or comeager?

As noted by MAULDIN [1990], the nonexistence of a Borel transversal for E_C on an indecomposable continuum C, which is a consequence of Theorem 4.5, would also follow from the positive answer to Kuratowski's question. Therefore, this question can be thought of as a strengthening of the question on the existence of Borel transversals.

5. Classifications of topological objects

The notions of definable equivalence relations and Borel reducibility are very helpful in formalizing various classification problems. In many situations in mathematics, we want to classify objects from a certain class up to a certain notion of isomorphism. Frequently in such situations, we are able to find a Polish space which consists of objects in question and, moreover, contains an isomorphic copy of each object in the class being considered. Now, the isomorphism relation becomes an equivalence relation on the Polish space. Frequently it happens to be analytic or even Borel. We can now ask precise mathematical questions about the complexity of this equivalence relation in terms of Borel bireducibility with some known and better understood equivalence relations. This approach has been successfully applied recently to many classification problems in topology, algebra, and analysis. Below we will survey several pieces of work that have to do with classifying topological objects.

1. Polish metric spaces

We consider here the class of all Polish metric spaces, that is, sets equipped with a complete separable metric. The main problem is to classify such spaces, or spaces from some subclasses of Polish metric spaces, up to isometry. These type of questions were first considered by GROMOV [1999] for compact metric spaces. VERSHIK [1998] proposed to study the general problem of classifying the isometry equivalence relation on all Polish metric spaces and remarked that the isometry on this large class of spaces is more complicated than on compact metric spaces. This general problem was studied in depth by GAO and KECHRIS [200?] and by CLEMENS [2001].

An important role here is played by the Urysohn space $\mathbb{U}$. This is a Polish metric space which contains an isometric copy of any Polish metric space. (If we add that the Urysohn space is also ultrahomogeneous, that is, that any isometry between finite subsets of $\mathbb{U}$ can be extended to an isometry of the whole space, then it becomes a unique, up to isometry, Polish metric space with these two properties. For more information on the Urysohn space see GROMOV [1999].) Note that each isometric copy of a Polish metric space in $\mathbb{U}$ is closed. Therefore, the space of closed subsets of $\mathbb{U}$, denoted by $\mathcal{F}(\mathbb{U})$, consists of Polish metric spaces and each Polish metric space is represented in it by an isometric copy. Now, $\mathcal{F}(\mathbb{U})$ carries a canonical σ-algebra, the so called Effros Borel structure, which is generated by sets of the form $\{F \in \mathcal{F}(\mathbb{U}) : F \cap W \neq \emptyset\}$ for $W \subseteq \mathbb{U}$ open. Moreover, there exists a (non-unique) Polish topology on $\mathcal{F}(\mathbb{U})$ whose Borel sets coincide with the Effros Borel structure. We consider this Polish space with the equivalence relation of isometry between its elements. This equivalence relation turns out to be analytic. We call it the *isometry equivalence relation on Polish metric spaces.*

A summary of results from CLEMENS [2001] and from GAO and KECHRIS [200?] along with background information can be found in CLEMENS, GAO and KECHRIS [2001]. The authors obtain a number of theorems on the complexity of the isometry equivalence relation on subclasses of the class of Polish metric spaces defined by topological, metric, metric homogeneity, and metric rigidity conditions or a mixture of such conditions. I will present below only a sample of these results.

The following theorem was proved by GAO and KECHRIS [200?]. Independently, CLEMENS [2001] established the part of this result which gives the lower estimate on the complexity of the isometry equivalence relation on Polish metric spaces. The equivalence relation E_P^∞ was defined in the introduction to Section 2.

5.1. THEOREM. *The equivalence relation of isometry on Polish metric spaces is Borel bireducible with the universal orbit equivalence relation of Polish group actions* E_P^∞.

Among locally compact Polish metric spaces the restrictions of the isometry equivalence relation to various topologically defined subclasses give very different complexities. This is illustrated by the following theorem. Point (i) is due to GROMOV [1999], point (ii) to Hjorth, see GAO and KECHRIS [200?], and point (iii) to GAO and KECHRIS [200?]. Recall that the equivalence relations $E_{S_\infty}^\infty$ and E_∞ were defined in Subsections 2.3 and 3.1, respectively. For the record let us state that $E_\infty < E_{S_\infty}^\infty < E_P^\infty$.

5.2. THEOREM.

(i) *The isometry equivalence relation on compact metric spaces is smooth, in fact, is Borel bireducible with* $=_{2^{\mathbb{N}}}$.

(ii) *The isometry equivalence relation on connected locally compact Polish metric spaces is Borel bireducible with the universal countable equivalence relation* E_∞.

(iii) *The isometry equivalence relation on zero dimensional locally compact Polish metric spaces is Borel bireducible with the graph isomorphism equivalence relation* $E_{S_\infty}^\infty$.

The following question asked in CLEMENS, GAO and KECHRIS [2001] remains unanswered.

5.3. QUESTION. What is the exact complexity of the isometry equivalence relation on Polish locally compact metric spaces? Is it $E_{S_\infty}^\infty$?

The problem of classifying compact metrizable spaces up to homeomorphism also remains open. This problem is formalized as follows. Consider the space $\mathcal{K}$ of all compact subsets of the Hilbert cube $[0,1]^{\mathbb{N}}$. With the Vietoris topology this is a Polish, in fact compact metric, space which contains a homeomorphic copy of each compact metrizable space as an element. Make two elements of $\mathcal{K}$ equivalent if they are homeomorphic. This is an equivalence relation called the *homeomorphism equivalence relation*. CLEMENS, GAO and KECHRIS [2001] pose the following problem.

5.4. QUESTION. What is the exact complexity of the homeomorphism equivalence relation among compact metrizable spaces?

Only very partial results are known. Kechris and Solecki proved, using Z-sets, that this equivalence relation is Borel reducible to E_P^∞. Hjorth showed that it is strictly above E_∞.

2. Complex manifolds

We will consider here complex manifolds of dimension $n \in \mathbb{N}$, $n \geq 1$, with the equivalence relation of being biholomorphic. If $n = 1$, such manifolds are called Riemann surfaces, and the biholomorphic mappings among Riemann surfaces are called conformal mappings. The study of complexity of these equivalence relations was initiated by BECKER, HENSON and RUBEL [1980] who proved, in a language different from the language of Borel reducibility, that the conformal equivalence relation on complex domains, and therefore on Riemann surfaces, is not smooth. HJORTH and KECHRIS [2000] found the precise complexity of the conformal equivalence relation on Riemann surfaces. (The definition of the Polish space which consists of n dimensional complex manifolds and contains a biholomorphic copy of each such manifold is somewhat complicated. We refer the reader to HJORTH and KECHRIS [2000] for a precise description.)

The following classification result for one dimensional complex manifolds was obtained by HJORTH and KECHRIS [2000]. One direction of their proof uses Theorem 3.3.

5.5. THEOREM. *The relation of conformal equivalence on Riemann surfaces is Borel bireducible with the universal countable Borel equivalence relation* E_∞.

Furthermore, HJORTH and KECHRIS [2000] proved that for higher dimensional complex manifolds the situation is more complicated. The proof of this result uses the notion of turbulence from Subsection 2.3.

5.6. THEOREM. *The relation of biholomorphic equivalence among n dimensional complex manifolds, for $n \geq 2$, is not Borel reducible to the graph isomorphism equivalence relation* $E^\infty_{S_\infty}$.

ACKNOWLEDGMENTS. I thank Alekos Kechris for his helpful comments. This work was supported by NSF grants DMS-9803676 and DMS-0102254. A part of this paper was written while I was visiting the Mathematics Department at Caltech. I thank it for the support during this time.

References

BECKER, H.

[1994] The topological Vaught conjecture and minimal counterexamples, *J. Symb. Logic* **59**, 757–784.

[1998] Polish group actions: dichotomies and generalized elementary embeddings, *J. Amer. Math. Soc.* **11**, 397–449.

BECKER, J., C.W. HENSON and L.A. RUBEL

[1980] First-order conformal invariants, *Ann. of Math.* (2) **112**, 123–178.

BECKER, H. and A.S. KECHRIS

[1996] *The Descriptive Set Theory of Polish Group Actions*, London Mathematical Society Lecture Note Series 232, Cambridge University Press.

BURGESS, J.P.

[1979] A reflection phenomenon in descriptive set theory, *Fund. Math.* **104**, 127–139.

CASEVITZ, P.

[2000] Dichotomies pour les espaces de suites reelles, *Fund. Math.* **165**, 249–284.

CLEMENS, J.D.

[2001] Ph.D. Thesis, University of California, Berkeley.

CLEMENS, J.D., S. GAO and A.S. KECHRIS

[2001] Polish metric spaces: their classification and isometry groups, *Bull. Symb. Logic* **7**, 361–375.

COOK, H.

[1964] On subsets of indecomposable continua, *Colloq. Math.* **13**, 37–43.

EFFROS, E.G.

[1965] Transformation groups and C^*-algebras, *Ann. of Math.* (2) **81**, 38–55.

EMERYK, A.

[1980] Partitions of indecomposable continua into composants, in *Proceedings of the International Conference in Geometric Topology*, Polish Scientific Publisher, pp. 137–140.

FARAH, I.

[2001a] Basis problem for turbulent actions, I. Tsirelson submeasures, *Ann. Pure Appl. Logic* **108**, 189–203.

[2001b] Basis problem for turbulent actions, II. c_0-equalities, *Proc. London Math. Soc.* **82**, 1–30.

FELDMAN, J., P. HAHN and C.C. MOORE

[1979] Orbit structure and countable sections for actions of continuous groups, *Adv. Math.* **26**, 186–230.

FELDMAN, J. and C.C. MOORE

[1977] Ergodic equivalence relations, cohomology and von Neumann algebras, I, *Trans. Amer. Math. Soc.* **234**, 289–324.

FRIEDMAN, H.

[2000] Borel and Baire reducibility, *Fund. Math.* **164**, 61–69.

GAO, S.

[1998] On automorphism groups of countable structures, *J. Symb. Logic* **63**, 891–896.

GAO, S. and A.S. KECHRIS

[200?] *On the Classification of Polish Metric Spaces Up To Isometry*, Mem. Amer. Math. Soc., to appear.

GROMOV, M.

[1999] *Metric Structures for Riemannian and Non-Riemannian Spaces*, Progress in Mathematics 152, Birkhauser.

GROSSER, S. and M. MOSKOWITZ

[1971] Compactness condition in topological groups, *J. Reine Angew. Math.* **246**, 1–40.

HARRINGTON, L., A.S. KECHRIS and A. LOUVEAU

[1990] A Glimm-Effros dichotomy for Borel equivalence relations, *J. Amer. Math. Soc.* **3**, 903–928.

HJORTH, G.

[1999] Orbit cardinals: on the definable cardinalities arising as quotient spaces of the form X/G where G acts on a Polish space X, *Israel J. Math.* **111**, 221–261.

[2000a] Actions by the classical Banach spaces, *J. Symb. Logic* **65**, 392–420.

[2000b] *Classification and Orbit Equivalence Relations*, Mathematical Surveys and Monographs 75, American Mathematical Society.

[2001] Vaught conjecture on analytic sets, *J. Amer. Math. Soc.* **14**, 125–143.

[200?a] A dichotomy theorem for turbulence, *J. Symb. Logic*, to appear.

[200?b] A dichotomy theorem for being essentially countable, to appear.

HJORTH, G. and A.S. KECHRIS

[1995] Analytic equivalence relations and Ulm-type classifications, *J. Symb. Logic* **60**, 1273–1300.

[2000] The complexity of the classification of Riemann surfaces and complex manifolds, *Illinois J. Math.* **44**, 104–137.

HJORTH, G. and S. SOLECKI

[1999] Vaught conjecture and the Glimm-Effros property for Polish transformation groups, *Trans. Amer. Math. Soc.* **351**, 2623–2641.

KECHRIS, A.S.

[1992] Countable sections for locally compact group actions, *Ergodic Theory Dynam. Systems* **12**, 283–295.

[1994] Countable sections for locally compact group actions. II, *Proc. Amer. Math. Soc.* **120**, 241–247.

[1995] *Classical Descriptive Set Theory*, Springer.

[200?] Actions of Polish groups and classification problems, in *Analysis and Logic*, London Mathematical Society Lecture Note Series, Cambridge University Press, to appear.

KECHRIS, A.S. and A. LOUVEAU

[1997] The classification of hypersmooth Borel equivalence relations, *J. Amer. Math. Soc.* **10**, 215–242.

KELLY, J.L.

[1955] *General Topology*, Van Nostrad.

KENNEDY, J.A. and J.A. YORKE

[1995] Bizarre topology is natural in dynamical systems, *Bull. Amer. Math. Soc. (N.S.)* **32**, 309–316.

KRASINKIEWICZ, J.

[1974] On a class of indecomposable continua, *Colloq. Math.* **32**, 72–75.

KURATOWSKI, K.

[1932] Sur un probleme topologique concernant les systems strictement transitifs, *Fund. Math.* **19**, 252–256.

MACKEY, G.W.

[1957] Borel structure on groups and their duals, *Trans. Amer. Math. Soc.* **85**, 134–165.

MAULDIN, R.D.

[1990] Problems in topology arising from analysis, in *Open Problems in Topology*, J. Van Mill and G.M. Reed, editors, North-Holland, 617–629.

MAZURKIEWICZ, S.

[1927] Sur les continus indecomposables, *Fund. Math.* **10**, 305–310.

MILLER, D.E.

[1977] On the measurability of orbits in Borel actions, *Proc. Amer. Math. Soc.* **63**, 165–170.

MORAN, W. and J.H. WILLIAMSON
[1978] Isotone measures on groups, *Math. Proc. Cambridge Philos. Soc.* **84**, 89–107.

NADLER, JR., S.B.
[1992] *Continuum Theory. An Introduction*, Marcel Dekker.

ROGERS, JR., J.T.
[1986] Borel transversals and ergodic measures on indecomposable continua, *Top. Appl.* **24**, 217–227.

SAINT-RAYMOND, J.
[1976] Espaces a modele separable, *Ann. Inst. Fourier (Grenoble)* **26**, 211–256.

SAMI, R.
[1994] Polish group actions and the Vaught conjecture, *Trans. Amer. Math. Soc.* **341**, 335–353.

SILVER, J.
[1980] Counting the number of equivalence classes of Borel and coanalytic equivalence relations, *Ann. Math. Logic* **18**, 1–28.

SOLECKI, S.
[1995] Equivalence relations induced by actions of Polish groups, *Trans. Amer. Math. Soc.* **347**, 4765–4777.
[1999a] Analytic ideals and their applications, *Ann. Pure. Appl. Logic* **99**, 51–72.
[1999b] Polish group topologies, in *Sets and Proofs*, London Mathematical Society Lecture Note Series 258, Cambridge University Press, pp. 339–364.
[2000] Actions of non-compact and non-locally compact Polish groups, *J. Symb. Logic* **65**, 1881–1894.
[2002] The space of composants of an indecomposable continuum, *Adv. Math.* **166**, 149–192.
[200?] Subgroups of Polish groups, in preparation.

USPENSKII, V.V.
[1986] A universal topological group with a countable basis, *Funct. Anal. Appl.* **20**, 86–87.

VERSHIK, A.M.
[1998] The universal Urysohn space, Gromov triples and random metrics on the natural numbers, *Russ. Math. Surveys* **53**, 921–928.

CHAPTER 19

Topological Groups: Between Compactness and $\aleph_0$-boundedness

Mikhail Tkachenko

Departamento de Matemáticas, Universidad Autónoma Metropolitana, Av. San Rafael Atlixco 186
Col. Vicentina, Del. Iztapalapa, C.P. 09340, México D.F.
E-mail: mich@xanum.uam.mx

Contents

RECENT PROGRESS IN GENERAL TOPOLOGY II
Edited by Miroslav Hušek and Jan van Mill

1. Introduction

In the last ten years, several survey articles on topological groups appeared (see COMFORT, HOFMANN and REMUS [1992], TKACHENKO [1999, 2000]). Our aim here is to complement these articles, as well as ARHANGEL'SKII's paper [2002] in this volume, and present three special topics: countably compact groups, (strictly) o-bounded groups and $\mathbb{R}$-factorizable groups. This choice is motivated by recent interesting contributions to these areas and, of course, by the author's personal preferences.

Each of the three classes contains the class of compact topological groups, but the permanence properties of these classes (as well as the methods applied for their study) differ substantially. We shall see below that countably compact groups are both strictly o-bounded and $\mathbb{R}$-factorizable. However, the study of countably compact groups presents major difficulties, especially when constructing examples with preassigned properties. The problem of (finite) productivity of the class of countably compact topological groups is especially intriguing and difficult. All known counterexamples have been constructed with the use of extra set-theoretic tools such as Martin's Axiom (MA, for short) or its weaker versions. In addition, MA is very useful when characterizing the algebraic structure of small countably compact groups (see Section 2).

The classes of o-bounded and strictly o-bounded were introduced in 1997 by Okunev and the author, respectively, with the idea of extending the class of subgroups of σ-compact groups. It turned out that these two classes of groups are sufficiently wide, but, under certain hypotheses, (strictly) o-bounded groups are topologically isomorphic to subgroups of σ-compact groups. In Section 3, we discuss the relations between o-bounded, strictly o-bounded and σ-compact groups as well as the stability of these classes with respect to taking direct products, continuous homomorphic images and subgroups.

One defines $\mathbb{R}$-factorizable groups as the groups G with the property that every continuous real-valued function on G is refined by a continuous homomorphism of G to a second countable group. The introduction of this class of groups is motivated by the fact established in the thirties in PONTRYAGIN [1939] (without using the term): every compact topological group is $\mathbb{R}$-factorizable. Then it was proved in COMFORT and ROSS [1966] that every continuous real-valued function on a pseudocompact group G extends to a continuous function over the Raĭkov completion ϱG of G. Since the group ϱG is compact, all pseudocompact (hence, countably compact) topological groups are $\mathbb{R}$-factorizable. The reader will see in Section 4 that the class of $\mathbb{R}$-factorizable groups is considerably wider: it contains, for example, all precompact groups and all Lindelöf groups. We also formulate a number of open problems about $\mathbb{R}$-factorizable groups.

It is worth mentioning that countably compact groups, o-bounded groups and $\mathbb{R}$-factorizable groups are $\aleph_0$-*bounded* in the sense that each of these groups can be covered by countably many translates of any neighborhood of the identity. The notion of an $\aleph_0$-bounded group was introduced in GURAN [1981] and plays an important role here. This explains, in part, the title of the article.

2. Countably compact groups

Countable compactness occupies a place between compactness and pseudocompactness. Each of the three properties has a strong impact on the structure of topological groups, especially in the Abelian case. It suffices to mention a description of the algebraic structure of compact Abelian groups given in HARRISON [1959] and HULANICKI [1958]. An almost complete characterization of the algebraic structure of pseudocompact topological groups was given in DIKRANJAN and SHAKHMATOV [1998]. In this section, we discuss the topological and algebraic aspects of countably compact topological groups, including the algebraic structure of small countably compact groups.

1. Weight and size of countably compact groups

It is well known that a compact topological group G of weight $\kappa \geq \omega$ admits a continuous map onto the Tychonoff cube I^κ and, hence, $|G| = 2^\kappa$ (see Theorem 1.45 in COMFORT [1984] or SHAKHMATOV [1994]). In particular, $w(G) < |G|$ for every infinite compact group G. Another important fact was established in IVANOVSKII [1958] and KUZ'MINOV [1959]: all compact groups are dyadic. Hence every infinite compact group contains lots of non-trivial convergent sequences.

None of the above results is valid for countably compact topological groups (at least, consistently). Indeed, let $\kappa = 2^{\mathfrak{c}}$. Then the density of $\mathbb{T}^\kappa$ is not greater than $\mathfrak{c}$ by the Hewitt–Marczewski–Pondiczery theorem, where $\mathbb{T}$ is the circle group. Therefore, $\mathbb{T}^\kappa$ contains a dense subgroup H of the size $\mathfrak{c}$, and the standard closing off argument implies that there exists a countably compact subgroup G of $\mathbb{T}^\kappa$ of the size $\mathfrak{c}$ containing H. Then G is dense in $\mathbb{T}^\kappa$, so that $w(G) = w(\mathbb{T}^\kappa) = \kappa$. We thus have $|G| = \mathfrak{c} < 2^{\mathfrak{c}} = w(G)$.

The existence of non-trivial convergent sequences in infinite countably compact groups is a more subtle problem. The first counterexample was constructed in HAJNAL AND JUHÁSZ [1976] under the CH. Later on, numerous examples of countably compact groups without non-trivial convergent sequences (having certain additional properties) were presented in VAN DOUWEN [1980a], HART AND VAN MILL [1991], TKACHENKO [1990] and TOMITA [1997a, 1997b, 1999]. However, all known constructions of such groups depend on extra set-theoretic assumptions (see Subsections 2 and 3 of this section).

The study of cardinal characteristics of pseudocompact and countably compact groups was initiated in VAN DOUWEN [1980b]. It was noted there that every infinite pseudocompact group G has no isolated points and, hence, satisfies $|G| \geq \mathfrak{c}$. This simple fact naturally led van Douwen to considering all possible sizes of countably compact groups. He also observed that, for every infinite cardinal κ with $\kappa^\omega = \kappa$, there exists a countably compact subgroup G of $\mathbb{Z}(2)^\kappa$ of the size κ; one can take G to be the Σ-product lying in $\mathbb{Z}(2)^\kappa$. In the same article, van Douwen conjectured that $|G|^\omega = |G|$ or, at least, $cf(|G|) > \omega$ for every infinite countably compact group G. The next result proved in VAN DOUWEN [1980b] shows that this conjecture is consistent with ZFC. We abbreviate the Generalized Continuum Hypothesis to GCH.

2.1. THEOREM. *If GCH holds, then every infinite countably compact group G satisfies $|G|^\omega = |G|$ and, hence, $cf(|G|) > \omega$.*

In fact, Theorem 2.1 remains valid for all pseudocompact homogeneous spaces.

It turns out that van Douwen's conjecture is independent of ZFC. This follows from a result obtained in TOMITA [200?]:

2.2. THEOREM. *It is consistent with ZFC that* $\aleph_1 = \mathfrak{c} < \aleph_\omega$ *and there exists a countably compact topological group of size* $\aleph_\omega$.

An interesting complement to Theorem 2.1 was proved in MALYKHIN AND SHAPIRO [1985]. It was shown that, under GCH, the weight of a pseudocompact topological group without non-trivial convergent sequences cannot have countable cofinality. However, it is proved in TOMITA [200?] that the existence of a countably compact group of size $\aleph_\omega$ without non-trivial convergent sequences is consistent with ZFC. So, both the weight and size of a countably compact topological group can (consistently) have countable cofinality, even if the group contains no convergent sequences other than trivial ones.

2. *Products of countably compact groups*

It is interesting to compare the permanence properties of compact, countably compact and pseudocompact topological groups. Clearly, the class of compact groups is closed under taking arbitrary direct products, closed subgroups and continuous homomorphic images. The class of pseudocompact groups is also productive by the celebrated theorem in COMFORT and ROSS [1966], it is closed under taking continuous homomorphic images, but a closed subgroup of a pseudocompact group can fail to be pseudocompact. In fact, every precompact group is topologically isomorphic to a closed subgroup of a pseudocompact group (see COMFORT and ROBERTSON [1988]). Quite differently, closed subgroups and continuous homomorphic images of countably compact groups are countably compact, but under additional set-theoretic assumptions, the product operation can destroy the countable compactness of topological groups. The first example to this effect was produced under Martin's Axiom in VAN DOUWEN [1980a]. The argument of van Douwen was as follows. First, he constructed, under MA, a countably compact subgroup G of $\mathbb{Z}(2)^{\mathfrak{c}}$ of size $\mathfrak{c}$ without non-trivial convergent sequences. Then, in ZFC, he defined two countably compact subgroups H_1 and H_2 of G with a countably infinite intersection $K = H_1 \cap H_2$. This immediately implies that $\{(x, x) : x \in K\}$ is a countably infinite closed subgroup of the product $H_1 \times H_2$, so that the group $H_1 \times H_2$ is not countably compact. Later on, van Douwen's result was improved in HART and VAN MILL [1991], where a single countably compact group H of size $\mathfrak{c}$ with a non countably compact square was presented under $\mathrm{MA}_{countable}$ (a weaker version of MA restricted to countable partially ordered sets). Their example is a subgroup $H = C \oplus K$ of $\mathbb{Z}(2)^{\mathfrak{c}}$, where C is a countable group and K is a group with the property that the closure of every countable subset of K is compact. In particular, H contains infinite compact subgroups and, therefore, non-trivial convergent sequences.

Much earlier, in HAJNAL and JUHÁSZ [1976], the Continuum Hypothesis (CH, for short) was used to construct a countably compact hereditarily separable subgroup G of $\mathbb{Z}(2)^{\mathfrak{c}}$ of size $\mathfrak{c}$ without non-trivial convergent sequences. That construction, combined with the ZFC part of van Douwen's argument, also implies the existence of two countably compact topological groups whose product is not countably compact.

The extension operation in the class of topological groups looks very much like the product operation. A topological group G is said to be an *extension of a group H via N*

provided that N is a closed normal subgroup of G and the quotient group G/N is topologically isomorphic to H. The properties of the group G are determined, to some extent, by those of the groups N and G/N. Indeed, it is well known that if both groups N and G/N are compact, pseudocompact, complete, connected or metrizable, then so is G. Very recently, BRUGUERA and TKACHENKO [200?] constructed in ZFC a dense pseudocompact subgroup G of $\mathbb{Z}(2)^{\mathfrak{c}}$ and a closed subgroup N of G with the following properties:

- the closure of every countable set in N is compact;
- the quotient group G/N is compact and metrizable;
- G contains a sequence converging to a point of $\mathbb{Z}(2)^{\mathfrak{c}} \setminus G$.

In particular, both groups N and G/N are countably compact but G is not. So, extensions of topological groups do not preserve countable compactness.

All groups mentioned so far are subgroups of $\mathbb{Z}(2)^{\mathfrak{c}}$ and, hence, they are Boolean and zero-dimensional. This makes the problem of characterizing the algebraic structure of countably compact topological groups especially interesting. In TKACHENKO [1990], CH was applied to construct a countably compact Hausdorff group topology on a free Abelian group with $\mathfrak{c}$ generators. The idea of the construction was to find a dense countably compact subgroup G of $\mathbb{T}^{\mathfrak{c}}$ algebraically isomorphic to the free Abelian group of size $\mathfrak{c}$, where $\mathbb{T}$ is the circle group with its usual compact topology. Additionally, G was connected, locally connected, hereditarily separable and contained no convergent sequences other than the trivial ones. The choice of $\mathbb{T}$ instead of $\mathbb{Z}(2)$ for the construction of the group G is forced by the fact that every countably compact Hausdorff group topology on a free Abelian group has to be infinite dimensional (see TKACHENKO [1990, Note 1]).

Further progress in the study of countably compact groups is due to A. Tomita who clarified both the topological properties and the algebraic structure of these groups. It was shown in TOMITA [1998] that a non-trivial free Abelian group does not admit a sequentially compact Hausdorff group topology. Since sequential compactness is countably productive and implies countable compactness, this fact follows from a more general result also proved in TOMITA [1998]:

2.3. THEOREM. *Let G be an infinite free Abelian group endowed with a Hausdorff group topology. Then the group G^{ω} is not countably compact.*

Let p be a free ultrafilter on ω. A Hausdorff space X is called *p-compact* if every sequence $\{x_n : n \in \omega\} \subseteq X$ has a p-limit point in X. It is easy to verify that p-compactness is productive and implies countable compactness. Therefore, Theorem 2.3 implies that an infinite free Abelian group admits a p-compact Hausdorff group topology for no free ultrafilter p on ω. It is unknown, however, whether the existence of a Hausdorff group topology on an infinite free Abelian group with countably compact square is consistent with ZFC.

In the same article, TOMITA [1998], the author presented a construction of a countably compact Hausdorff group topology on a free Abelian group with $\mathfrak{c}$ generators which makes use of MA$_{(\sigma\text{-}centered)}$. Another interesting result proved by Tomita concerns *initially ω_1-compact* topological groups (i.e., groups with the property that every open cover of size at most ω_1 has a finite subcover):

2.4. THEOREM. *The existence of an initially ω_1-compact Hausdorff group topology on some infinite free Abelian group is independent of* $\mathfrak{c} = \aleph_2$.

The proof of Theorem 2.4 is naturally split in two parts. First, it is shown in TOMITA [1998], under MA(*σ-centered*) + $\mathfrak{c} = \aleph_2$, that a free Abelian group with $\mathfrak{c}$ generators admits an initially ω_1-compact Hausdorff group topology. The second part requires the use of Kunen's axiom KA which says that there exists a free ultrafilter p on ω with a base of cardinality ω_1. It is known that $KA + \mathfrak{c} = \aleph_2$ is consistent with ZFC. It is easy to verify that, for Kunen's ultrafilter p, an arbitrary product of initially ω_1-compact spaces is p-compact and, hence, countably compact. Therefore, Theorem 2.3 implies that, under $KA + \mathfrak{c} = \aleph_2$, no infinite free Abelian group admits an initially ω_1-compact Hausdorff group topology.

By Theorem 2.3, the ω-power of a topological group is never countably compact if the underlying group is free Abelian. It turns out that a free Abelian group of size $\mathfrak{c}$ admits $\mathfrak{c}$ many distinct countably compact Hausdorff group topologies whose countable products remain countably compact (see TOMITA 19-To97b):

2.5. THEOREM. *Under MA(σ-centered), there exists a family* $\{G_\alpha : \alpha < \mathfrak{c}\}$ *of countably compact topological groups, each of which algebraically coincides with a free Abelian group of size* $\mathfrak{c}$*, such that the product* $\prod_{\alpha<\beta} G_\alpha$ *is countably compact for each* $\beta < \mathfrak{c}$.

Combining Theorems 2.5 and 2.3, we conclude that the family $\{G_\alpha : \alpha < \mathfrak{c}\}$ contains $\mathfrak{c}$ many pairwise topologically non-isomorphic groups. This conclusion is improved in TOMITA [1999] where it is shown, under the same assumption MA(*σ-centered*), that a free Abelian group of size $\mathfrak{c}$ admits at least $\mathfrak{c}^+$ pairwise non-homeomorphic countably compact Hausdorff group topologies.

Summing up, there are many examples which witness, under CH or MA, that countable compactness is not productive in the class of topological groups. Therefore, the next question could be of whether something is left of countable compactness when taking direct products of countably compact topological groups. The answer is definitely "yes": any direct product of countably compact groups is *sequentially complete.* We recall that a topological group G is sequentially complete if no sequence from G converges to a point of $\varrho G \setminus G$, where ϱG is the Raĭkov completion of G. It is clear that every countably compact topological group is sequentially complete, any direct product of sequentially complete groups is sequentially complete and a closed subgroup of a sequentially complete group is also sequentially complete (see DIKRANJAN and TKACHENKO [2000, 2001]). In particular, closed subgroups of direct products of countably compact topological groups are sequentially complete. However, extensions of topological groups do not preserve sequential completeness as it follows from the example in BRUGUERA and TKACHENKO [200?] mentioned above. Since a first countable topological group is sequentially complete if and only if it is Raĭkov complete, we conclude that first countable closed subgroups of direct products of countably compact groups are compact. More information about sequentially complete groups can be found in SHAKHMATOV's article [2002] in this volume.

Surprisingly, almost all precompact Abelian groups can be obtained, starting from countably compact Abelian groups, if we are allowed to take direct products, closed subgroups and continuous isomorphic images (see DIKRANJAN and TKACHENKO [2002b]). First, we recall an important notion.

An infinite cardinal κ is said to be *Ulam-measurable* if there exists a free countably complete ultrafilter on a set of cardinality κ. Ulam-measurable cardinals are known to be very large; it is consistent with ZFC that no cardinal is Ulam-measurable.

2.6. THEOREM. *Assume that MA holds. Then every precompact Abelian group of non Ulam-measurable cardinality is a continuous isomorphic image of a closed subgroup of a direct product of countably compact Abelian topological groups.*

Since continuous homomorphisms and taking closed subgroups preserve countable compactness, Theorem 2.6 means that, in a sense, direct products almost completely destroy countable compactness. The proof of this result is based on the following fact: If an Abelian group G of non Ulam-measurable cardinality is endowed with the maximal precompact group topology, then it is topologically isomorphic to a closed subgroup of a direct product of countably compact Abelian groups. It is unknown, however, whether Theorem 2.6 remains valid in the non-Abelian case.

3. Algebraic structure of countably compact Abelian groups

It is well known that precompactness does not imply any constraints on the algebraic structure of an Abelian topological group. For example, every abstract Abelian group admits a Hausdorff precompact group topology generated by the family of all homomorphisms to the circle group $\mathbb{T}$. On the other hand, a complete description of the algebraic structure of compact Abelian groups (obtained as a solution to Halmos' problem) is given in Section 25 of HEWITT and ROSS [1979].

The counterpart of Halmos' problem for pseudocompact groups has been almost completely solved in DIKRANJAN and SHAKHMATOV [1998] after an important series of results by van Douwen, Comfort, Remus, Robertson, and van Mill, among others. The reader can find the pertinent information in the survey article COMFORT, HOFMANN and REMUS [1992].

The first restriction on the cardinality of an infinite pseudocompact group G was given in VAN DOUWEN [1980b]: $|G| \geq \mathfrak{c}$. In particular, infinite countably compact groups have cardinality at least $\mathfrak{c}$. The results of the previous section show how little we know about countably compact groups: even the problem of the existence (in ZFC) of a countably compact Hausdorff group topology on a free Abelian group of size $\mathfrak{c}$ remains open. It is unknown (in ZFC only) whether free Abelian groups of cardinality greater than $\mathfrak{c}$ admit a countably compact Hausdorff group topology. So, very simple Abelian groups present serious difficulties when searching for countably compact group topologizations. On the other hand, it was shown in DIKRANJAN and SHAKHMATOV [1998] that pseudocompact topologizations are much easier to achieve due to the well-known description of pseudocompact groups as G_δ-dense subgroups of compact groups given in COMFORT and ROSS [1966].

Surprisingly, Martin's Axiom makes it possible to completely describe the algebraic structure of countably compact Abelian groups of size $\mathfrak{c}$. We consider the case of torsion groups first. From now on, all group topologies are assumed to be Hausdorff. The next result was proved in DIKRANJAN and TKACHENKO [2002a].

2.7. THEOREM. *Assume that MA holds. Then, for a torsion Abelian group G of cardinality $\mathfrak{c}$, the following conditions are equivalent:*

(a) *G admits a pseudocompact group topology;*

(b) *G admits a countably compact group topology;*

(c) *G admits a countably compact group topology without non-trivial convergent sequences;*

(d) *G has finite exponent n and dG is either finite or has cardinality $\mathfrak{c}$, for every proper divisor d of n.*

Therefore, if a torsion Abelian group of size $\mathfrak{c}$ admits a pseudocompact group topology, it also admits a countably compact group topology (this conclusion requires MA).

For non-torsion groups, the situation is different. Given an Abelian group G, let $tor(G)$ be the torsion subgroup of G. If n is an integer, we denote by $G[n]$ the subgroup of G which consists of all elements $x \in G$ satisfying $nx = 0$. If n, d are non-zero integers, $d|n$ means that d divides n.

2.8. THEOREM. *Martin's Axiom implies that, for an Abelian non-torsion group G of cardinality $\mathfrak{c}$, the following conditions are equivalent:*

1) *G admits a countably compact group topology;*

2) *G admits a countably compact group topology without non-trivial convergent sequences;*

3) *$|G/tor(G)| = \mathfrak{c}$ and, for all $d, n \in \mathbb{N}$ with $d|n$, the group $dG[n] \cong G[n]/G[d]$ is either finite or has cardinality $\mathfrak{c}$.*

The above theorem was also established in DIKRANJAN and TKACHENKO [2002a]. It implies that the existence of a countably compact group topology on an Abelian non-torsion group G of size $\mathfrak{c}$ is a considerably stronger condition than the existence of a pseudocompact group topology. Indeed, from results of DIKRANJAN and SHAKHMATOV [1998] it follows that such a group G admits a pseudocompact group topology if and only if $|G/tor(G)| = \mathfrak{c}$. Denote by $Z(p)^{(\omega)}$ the direct sum of ω copies of the cyclic group $\mathbb{Z}(p)$ of a prime order p. Then the group $G = \mathbb{R} \oplus \mathbb{Z}(p)^{(\omega)}$ admits a pseudocompact group topology while no group topology on G is countably compact by Theorem 2.8. In fact, this conclusion does not require MA because the subgroup $H = \{0\} \times \mathbb{Z}(p)^{(\omega)}$ of G is unconditionally closed in G (being the kernel of the homomorphism $\varphi_p \colon G \to G$, $\varphi_p(x) = px$ for each $x \in G$). Therefore, if G had a countably compact group topology τ, H would be closed in (G, τ), hence countably compact. However, every infinite countably compact topological group has to have size $\geq \mathfrak{c}$, a contradiction.

Two strong results have recently been obtained in KOSZMIDER, TOMITA AND WATSON [200?]. It is shown there, improving earlier Tomita's results, that $\mathrm{MA}_{countable}$ implies the existence of a countably compact Hausdorff group topology on a free Abelian group of size $\mathfrak{c}$ which does not contain non-trivial convergent sequences . The second fact established by Koszmider, Tomita and Watson is even more striking. They construct by forcing a model of ZFC where CH holds and there exists a countably compact Hausdorff group topology on a free Abelian group of size $2^{\mathfrak{c}}$. This is the first step towards the classification of countably compact groups having size greater than $\mathfrak{c}$.

4. The Wallace problem

It is well known that every compact topological semigroup with two-sided cancellation is a topological group (see HEWITT and ROSS [1979]). In 1953, Wallace asked whether this result could be extended to countably compact topological semigroups with two-sided cancellation. Since then, the question has been known as *Wallace's problem.*

In several special cases, the answer to the Wallace problem is "yes". For instance, MUKHERJEA and TSERPES showed in [1972] that a first countable sequentially compact two-sided cancellative semigroup is a topological group. Then PFISTER proved in [1985] that if an abstract group is endowed with a countably compact semigroup topology, then it is a topological group. Therefore, a counterexample to Wallace's question cannot be algebraically a group. GRANT gave in [1993] the positive answer to Wallace's question in the case of a completely regular sequentially compact cancellative topological semigroup. YUR'EVA [1993] complemented both Grant's and Mukherjea–Tserpes's results as follows: every countably compact sequential topological semigroup is a topological group. Finally, ROBBIE and SVETLICHNY [1996] used a countably compact torsion-free Abelian group G from TKACHENKO [1990] to construct a countably compact subsemigroup of G which fails to be a group, thus answering the Wallace question in the negative. However, their construction makes use of CH. Recently, TOMITA [1997a] presented a construction of a subsemigroup of $\mathbb{T}^{\mathfrak{c}}$ with similar properties that depends on $\mathrm{MA}_{countable}$. In addition, the square of that semigroup is not countably compact. It is unknown whether such a subsemigroup can be constructed in ZFC or there exists a model in which the answer to Wallace's question is "yes".

5. Finer countably compact group topologies

It is shown in COMFORT, REMUS and ROBERTSON [1988, 1993] that, under suitable hypotheses, a non-metrizable compact topological group admits a strictly finer pseudocompact group topology. A construction of such a group topology can almost always be carried out within ZFC. The study of strictly finer countably compact group topologies on compact groups was started in ARHANGEL'SKII [1994].

The following result proved in ARHANGEL'SKII [1994] explains the importance of Ulam-measurable cardinals when finer countably compact group topologies are involved:

2.9. THEOREM. *If a compact topological group G admits a strictly finer countably compact group topology, then the cardinality of G is Ulam-measurable.*

Theorem 2.9 means, in particular, that the following assertion is consistent with ZFC: every continuous isomorphism $f: G \to H$ of a countably compact group G onto a compact group H is a homeomorphism. A substantial part of the proof of Theorem 2.9 given by Arhangel'skii was based on a result in VAROPOULOS [1964]: every sequentially continuous homomorphism $f: K \to L$ of locally compact topological groups K and L is continuous provided that the cardinality of K is not Ulam-measurable.

Arhangel'skii's theorem was complemented in COMFORT and REMUS [1994], where strictly finer countably compact group topologies were constructed on many compact groups of Ulam-measurable cardinality:

2.10. THEOREM. *Let G be a compact topological group of Ulam-measurable cardinality. If G is either Abelian or connected, then it admits a strictly finer countably compact group topology.*

The argument given by Comfort and Remus goes as follows. Let m be the first Ulam-measurable cardinal. First, they show that F^m admits a strictly finer countably compact group topology for every compact metrizable group F with $|F| \geq 2$. It is well known that a compact group G as in Theorem 2.10 admits a continuous homomorphism onto F^m, where F is a non-trivial compact metrizable group. Finally, it is proved that if $p\colon K \to L$ is a continuous homomorphism of an arbitrary compact group K onto a group L which admits a strictly finer countably compact group topology, then K also admits a strictly finer countably compact group topology. This implies the required conclusion.

Recently, Theorem 2.10 was improved in USPENSKIJ [200?]: every compact topological group of Ulam-measurable cardinality admits a strictly finer countably compact group topology.

3. O-bounded and strictly o-bounded groups

The notion of an $\aleph_0$-bounded topological group arose when trying to find an internal characterization of subgroups of Lindelöf topological groups (but the original problem is still open). A similar problem of characterizing subgroups of σ-compact topological groups has quite a satisfactory solution: a topological group G is topologically isomorphic to a subgroup of a σ-compact group iff G is *σ-precompact*, i.e., G is the union of countably many sets each of which is precompact in G. There are, however, several wider classes of topological groups which look very close to being subgroups of σ-compact groups. Here we consider two of them: o-bounded and strictly o-bounded groups.

A topological group G is called *o-bounded* (see TKACHENKO [1998], HERNÁNDEZ [2000], and HERNÁNDEZ, ROBBIE and TKACHENKO [2000]) if, for every sequence $\{U_n : n \in \omega\}$ of neighborhoods of the identity in G, there exists a sequence $\{F_n : n \in \omega\}$ of finite sets in G such that $G = \bigcup_{n\in\omega} F_n \cdot U_n$. It is clear that σ-precompact groups are o-bounded, and all o-bounded groups are $\aleph_0$-bounded. The class of o-bounded groups is closed under taking arbitrary subgroups and continuous homomorphic images HERNÁNDEZ [2000]. Not every second countable topological group is o-bounded: a usual diagonal argument implies that the groups $\mathbb{R}^\omega$ and $\mathbb{Z}^\omega$ are not o-bounded (see HERNÁNDEZ [2000]). Therefore, o-bounded groups form a proper subclass of $\aleph_0$-bounded groups.

Strictly o-bounded groups are defined in terms of the *OF-game*. Suppose that G is a topological group and that two players, say I and II, play as follows. Player I chooses an open neighborhood U_1 of the identity in G, and player II responds choosing a finite subset F_1 of G. In the second turn, player I chooses another neighborhood U_2 of the identity in G and player II chooses a finite subset F_2 of G. The game continues this way until we have the sequences $\{U_n : n \in \mathbb{N}\}$ and $\{F_n : n \in \mathbb{N}\}$. Player II wins if $G = \bigcup_{n\in\mathbb{N}} F_n \cdot U_n$. Otherwise, player I wins. The group G is called *strictly o-bounded* if player II has a winning strategy in the OF-game.

Clearly, σ-precompact groups are strictly o-bounded and strictly o-bounded groups are o-bounded. It is shown in HERNÁNDEZ [2000] that subgroups of strictly o-bounded groups are strictly o-bounded and the same is valid for continuous homomorphic images of strictly o-bounded groups.

We say that a topological group G is a *P-group* if countable intersections of open sets in G are open. By a result in HERNÁNDEZ [2000], every Lindelöf P-group is o-bounded. In fact, an argument similar to that of HERNÁNDEZ [2000] shows that every $\aleph_0$-bounded P-group is o-bounded. Recently, in KRAWCZYK and MICHALEWSKI [200?], the authors constructed a Lindelöf P-group which is not strictly o-bounded. This implies, in particular, that there is an o-bounded group which fails to be strictly o-bounded.

Lindelöf P-groups play an important role in the sequel. One of the peculiar properties of these groups is given in the next theorem.

3.1. THEOREM. *Every Lindelöf P-group is Raĭkov complete.*

Let $G = \prod_{i\in I} G_i$ be the direct product of countable discrete groups endowed with the ω-box topology. Denote by H the subgroup of G (known as the *σ-product* of the groups G_i's) consisting of all points of G almost all coordinates of which coincide with those of the neutral element of G. Then G and its subgroup H are P-groups. By a theorem in COMFORT [1975], the group H is Lindelöf. In what follows, we call any subgroup of such a group H a *Comfort-like group*. It is proved in HERNÁNDEZ [2000] that every Comfort-like group is strictly o-bounded. Note that the Lindelöf P-group H is uncountable if $|I| \geq \omega_1$ and $|G_i| \geq 2$ for each $i \in I$. Since every Lindelöf P-group is Raĭkov complete by Theorem 3.1, the group H is strictly o-bounded but not σ-precompact (otherwise it would be countable). We conclude that the σ-precompact groups form a proper subclass of strictly o-bounded groups.

As we mentioned above, not every second countable group is o-bounded — the group $\mathbb{Z}^\omega$ is a counterexample. It turns out that it suffices to know all second countable homomorphic images of a given $\aleph_0$-bounded group in order to decide whether the group is o-bounded or not (see HERNÁNDEZ [2000]):

3.2. THEOREM. *An $\aleph_0$-bounded group G is o-bounded if and only if all second countable continuous homomorphic images of G are o-bounded.*

An analogous assertion for strictly o-bounded groups is false. Indeed, let G be any Lindelöf P-group. Then every second countable continuous homomorphic image of G is countable and, hence, strictly o-bounded. Since Lindelöf P-groups need not be strictly o-bounded by a result of KRAWCZYK and MICHALEWSKI [200?], our claim follows. Earlier, the same conclusion was obtained in HERNÁNDEZ, ROBBIE and TKACHENKO [2000] under $\diamondsuit$.

One of the most important problems concerning (strict) o-boundedness is to find additional conditions on a (strictly) o-bounded group which imply that the group is σ-precompact. The following interesting result in this direction was recently proved in BANAKH [2000]:

3.3. THEOREM. *Suppose that a topological group G is a continuous homomorphic image of a second countable Weil complete group. Then G is o-bounded if and only if it is σ-precompact.*

Similarly, Banakh shows in the same article that if a strictly o-bounded group G is *analytic* (i.e., a continuous image of a complete separable metric space), then G is σ-precompact. In fact, the relation between (strict) o-boundedness and σ-precompactness in topological groups is considerably more complicated and profound. Let us denote by $itb(G)$ the minimal cardinality of a cover of G by precompact subsets. First, it is shown in BANAKH [2002] that every analytic topological group G satisfies $itb(G) \in \{1, \omega, \mathfrak{d}\}$, where $\mathfrak{d}$ is the minimal number of compact sets which cover $\mathbb{N}^\omega$. Then he shows in ZFC that every metrizable strictly o-bounded group G satisfies $itb(G) \leq \omega_1$. Surprisingly, the latter result can be complemented as follows (see BANAKH [2002, Theorem 4]):

3.4. THEOREM. *If there exists an Ulam-measurable cardinal, then every metrizable strictly o-bounded group is σ-precompact.*

The space $C_p(X)$ of all continuous real-valued functions on a Tychonoff space X with the pointwise convergence topology is a locally convex linear space and, hence, a topological group. The property of $C_p(X)$ being a (strictly) o-bounded group can be completely characterized in terms of the space X (see HERNÁNDEZ [2000]):

3.5. THEOREM. *For a Tychonoff space X, the following are equivalent:*

(a) *the group $C_p(X)$ is o-bounded;*

(b) *the group $C_p(X)$ is strictly o-bounded;*

(c) *the group $C_p(X)$ is σ-precompact;*

(d) *X is pseudocompact.*

In general, the class of o-bounded groups is not productive: KRAWCZYK and MICHALEWSKI [2001] constructed, under CH, two linear metric spaces (thus topological groups) which are o-bounded but their product is not. In fact, they mentioned in the same article that a construction of such a pair of linear spaces can be carried out under MA. It is also known that the product of an o-bounded group with a σ-precompact group is o-bounded (see HERNÁNDEZ [2000]). This makes it natural to ask the following. Suppose that the product of a topological group G with every (strictly) o-bounded group is (strictly) o-bounded. Is then G σ-precompact? As is shown in HERNÁNDEZ, ROBBIE and TKACHENKO [2000], the product of an $\aleph_0$-bounded P-group with every o-bounded group is o-bounded. Since there are uncountable Lindelöf P-groups, the answer to the question is "no" for o-bounded groups. Similarly, the product of every Comfort-like group with a strictly o-bounded group is strictly o-bounded (see HERNÁNDEZ, ROBBIE and TKACHENKO [2000]). Again, this implies that the answer to the above question is negative in the case of strictly o-bounded groups. It is unknown, however, whether the product of two strictly o-bounded groups is strictly o-bounded.

Let us turn back to the OF-game and say that a topological group G is *OF-determined* if one of the two players, I or II, has a winning strategy in this game on G. Otherwise, we say that the group G is *OF-undetermined.* It is easy to verify that every OF-undetermined group is o-bounded. The first example of an OF-undetermined group was constructed in HERNÁNDEZ, ROBBIE and TKACHENKO [2000] under the assumption of $\diamondsuit$; in fact, it was a P-group. Then, in BANAKH [2002], MA was used to show that every complete metric Abelian group which is not locally compact contains an OF-undetermined subgroup. Banakh also proved in the same article that every analytic group is OF-determined.

Very recently, an important step towards understanding the properties of o-bounded groups was made in BANAKH, NICKOLAS and SANCHIS [200?]. It is shown there without extra set-theoretic assumptions that the group $\mathbb{R}^\omega$ contains many OF-undetermined subgroups. In the same article, the authors construct, under MA, two o-bounded linear subspaces of $\mathbb{R}^\omega$ whose product fails to be o-bounded. This construction looks very close to that in KRAWCZYK and MICHALEWSKI [2001].

4. $\mathbb{R}$-factorizable groups

There are two different ways to express the idea that the topology of a topological group can be generated by continuous homomorphisms to second countable groups. The first one goes back to Guran's notion of $\aleph_0$-bounded groups. By a theorem in GURAN [1981], a topological group is $\aleph_0$-bounded if and only if it embeds as a subgroup into a direct product of second countable topological groups. Two different proofs of this important result are available in USPENSKIJ [1988] and TKACHENKO [1998]. The class of $\aleph_0$-bounded groups is stable with respect to taking direct products, arbitrary subgroups and continuous homomorphic images, etc. This makes $\aleph_0$-bounded groups look, within topological groups, very much like Tychonoff spaces do in the realm of topological spaces.

A more restrictive approach is to require that continuous homomorphisms of a topological group G to second countable topological groups generate *all continuous real-valued functions* on G. The first result in this direction was obtained in PONTRYAGIN [1939]: if $f: G \to \mathbb{R}$ is a continuous function on a compact topological group G, then there exists a closed normal subgroup N of type G_δ in G such that f is constant on every coset of N in G. Let $\pi: G \to G/N$ be the quotient homomorphism of G onto the group G/N. Since f is constant on xN for each $x \in G$, there exists a function $h: G/N \to \mathbb{R}$ such that $f = h \circ \pi$.

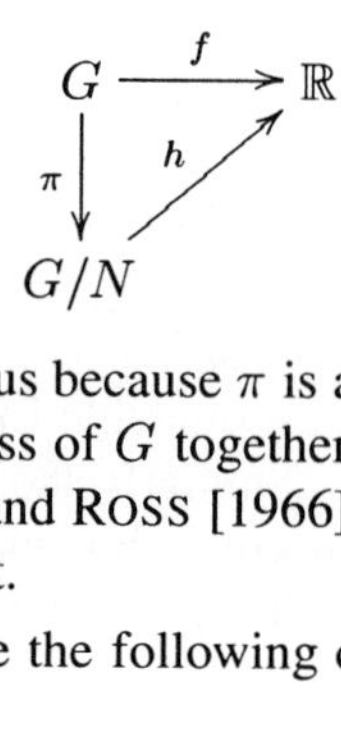

Note that the function h is continuous because π is an open homomorphism. In addition, the choice of N and the compactness of G together imply that the quotient group G/N is second countable. In COMFORT and ROSS [1966], this result was extended to the case when the group G is pseudocompact.

Motivated by these facts, we give the following definition (see TKACHENKO [1991a, 1991b]):

4.1. DEFINITION. A topological group G is $\mathbb{R}$-*factorizable* if, for every continuous function $f: G \to \mathbb{R}$, one can find a continuous homomorphism $p: G \to H$ onto a second countable topological group H and a continuous function $h: H \to \mathbb{R}$ such that $f = h \circ p$.

If f, p and h are as in Definition 4.1, we say that p *factorizes* f or, in symbols, $p \prec f$.

The class of $\mathbb{R}$-factorizable groups has proved to be useful in the study of Dieudonné and Hewitt–Nachbin completions of topological groups, dimensional properties of topological groups, etc. The reader can find many results about $\mathbb{R}$-factorizable groups in the survey article TKACHENKO [1998]. Our aim here is to mention briefly the most important properties of these groups and present new results obtained in the years 1998–2001.

1. General properties of $\mathbb{R}$-factorizable groups

Since continuous real-valued functions generate the topology of a topological group, every $\mathbb{R}$-factorizable group is $\aleph_0$-bounded. It turns out that the class of $\mathbb{R}$-factorizable groups is considerably wider than the class of compact or pseudocompact groups. In the next theorem, we collect several results from TKACHENKO [1991a, 1991b].

4.2. THEOREM. *The class of $\mathbb{R}$-factorizable groups contains:*

(a) *all precompact groups;*

(b) *all Lindelöf groups;*

(c) *arbitrary subgroups of σ-compact groups;*

(d) *direct products of second countable groups and their dense subgroups.*

Since every precompact group is a subgroup of a compact group, (a) is a special case of (c) in the list of the above theorem.

It is very natural to compare permanence properties of $\aleph_0$-bounded and $\mathbb{R}$-factorizable groups. The results of TKACHENKO [1991a, 1991b, 1998] revealed the following main problems in the field:

4.3. PROBLEM. Let G be an arbitrary $\mathbb{R}$-factorizable group.

(A) Is every locally finite family of open sets in G countable?

(B) Must all continuous homomorphic images of G be $\mathbb{R}$-factorizable?

(C) Is the class of $\mathbb{R}$-factorizable groups closed under taking direct products?

(D) What are the relations between $\aleph_0$-bounded and $\mathbb{R}$-factorizable groups?

Almost all results we discuss here concern one of the items (A)–(D) listed in Problem 4.3. In many special cases, we answer (A), (B) and (C) in the affirmative, while our answer to (D) is almost complete. The next theorem presents a good piece of information about (D) taken from TKACHENKO [1991a, 1991b].

4.4. THEOREM. *The following assertions hold:*

(a) *Every $\mathbb{R}$-factorizable group is $\aleph_0$-bounded;*

(b) *Every $\aleph_0$-bounded group is topologically isomorphic to a closed normal subgroup of an $\mathbb{R}$-factorizable group;*

(c) *There exist $\aleph_0$-bounded groups which fail to be $\mathbb{R}$-factorizable.*

Summarizing, $\mathbb{R}$-factorizable groups form a proper subclass of the class of $\aleph_0$-bounded groups. To see the difference between the two classes of groups and clarify (c) of Theorem 4.4, we present a simple example of an $\aleph_0$-bounded group which is not $\mathbb{R}$-factorizable.

4.5. EXAMPLE. Let $\mathbb{Z}(2) = \{0, 1\}$ be the discrete two-point group. Denote by G the subgroup of $\mathbb{Z}(2)^{\omega_1}$ consisting of all points which have only finitely many non-zero coordinates. Suppose that the group $\mathbb{Z}(2)^{\omega_1}$ carries the ω-box topology. Then both $\mathbb{Z}(2)^{\omega_1}$ and its subgroup G are P-groups. In addition, the group G is Lindelöf by a theorem in COMFORT [1975]. Let H be the subgroup of G which consists of all elements which have an even number of non-zero coordinates. It is easy to see that H is a proper dense subgroup of G and the weight of G and H is equal to ω_1. Applying these simple facts, one can verify that the group H is not $\mathbb{R}$-factorizable (it suffices to represent H as the union of two disjoint open sets which have a common cluster point in G).

Hence, we obtain the following:

4.6. COROLLARY. *Dense subgroups of Lindelöf groups need not be $\mathbb{R}$-factorizable.*

The group H in Example 4.5 is not Raĭkov complete since it is a proper dense subgroup of the group G. Hence, one may conjecture that Raĭkov complete $\aleph_0$-bounded groups are $\mathbb{R}$-factorizable. A counterexample to this conjecture has recently been presented in TKACHENKO [2001], but the construction of such a group is considerably more complicated than the one just described. Curiously enough, the group constructed in TKACHENKO [2001] is also a P-group.

Corollary 4.6 implies that dense subgroups of $\mathbb{R}$-factorizable groups can fail to be $\mathbb{R}$-factorizable. This gives rise to the following open problem.

4.7. PROBLEM. Suppose that a topological group G contains a dense $\mathbb{R}$-factorizable subgroup. Is then G $\mathbb{R}$-factorizable?

Clearly, the group G in Problem 4.7 is $\aleph_0$-bounded because it contains a dense $\aleph_0$-bounded subgroup (see (a) of Theorem 4.4). In Subsection 5 of this section, we present some cases when this problem has a positive solution.

$\mathbb{R}$-factorizable groups can be characterized by means of embeddings into larger topological groups. Let us say that a subset Y of a space X is *z-embedded* in X if, for every zero-set F in Y, there is a zero-set P in X such that $F = Y \cap P$. Obviously, every C^*-embedded subset is z-embedded, but not vice versa. The following result was proved in HERNÁNDEZ, SANCHIS and TKACHENKO [2000].

4.8. THEOREM. *Let H be a subgroup of a topological group G. If the group H is $\mathbb{R}$-factorizable, then it is z-embedded in G.*

Theorem 4.8 has a counterpart established in HERNÁNDEZ and TKACHENKO [1998]:

4.9. THEOREM. *If a subgroup H of an $\mathbb{R}$-factorizable group G is z-embedded in G, then H is $\mathbb{R}$-factorizable.*

Combining Theorems 4.8 and 4.9, we conclude that an $\aleph_0$-bounded group H is $\mathbb{R}$-factorizable iff it is z-embedded in *every* topological group which contains H as a subgroup. Note that discrete subgroups of an arbitrary topological group G are C-embedded in G, so one cannot omit "$\aleph_0$-bounded" in this characterization of $\mathbb{R}$-factorizable groups. In fact, every $\aleph_0$-bounded group H admits a "test" embedding into a single topological group G which "decides" whether H is $\mathbb{R}$-factorizable or not. Indeed, by Guran's theorem, H is topologically isomorphic to a subgroup of a direct product $G = \prod_{i\in I} K_i$ of second countable topological groups. Then G is $\mathbb{R}$-factorizable by (d) of Theorem 4.2, so H is $\mathbb{R}$-factorizable iff it is z-embedded in G.

Note that (b) of Theorem 4.2 and Theorem 4.9 imply that every z-embedded subgroup of a Lindelöf topological group is $\mathbb{R}$-factorizable. We do not know, however, whether such a subgroup is pseudo-ω_1-compact (i.e., whether every locally finite family of open sets in a subgroup is countable, see (A) of Problem 4.3):

4.10. PROBLEM. Is every z-embedded subgroup of a Lindelöf topological group pseudo-ω_1-compact? What about C^*-embedded subgroups?

Surprisingly, (A) of Problem 4.3 has a positive solution in the class of locally connected groups (see TKACHENKO [1991b]):

4.11. THEOREM. *Every locally connected $\mathbb{R}$-factorizable group is pseudo-ω_1-compact.*

Several cases when $\mathbb{R}$-factorizable groups turn out to be pseudo-ω_1-compact were found in TKACHENKO [1991b, 1998]:

4.12. THEOREM. *Let G be an $\mathbb{R}$-factorizable group of weight τ. Then G is pseudo-ω_1-compact in each of the following cases:*

(a) $\tau^\omega < 2^{\omega_1}$;

(b) $\tau = 2^\omega < 2^{\omega_1}$;

(c) *all continuous homomorphic images of G of weight $\leq 2^\omega$ are pseudo-ω_1-compact;*

(d) *every quotient of G of countable pseudocharacter is pseudo-ω_1-compact.*

Clearly, (b) of Theorem 4.12 follows from (a). In addition, (d) implies that if there exists an $\mathbb{R}$-factorizable group which is not pseudo-ω_1-compact, then there exists such a group of countable pseudocharacter (we also apply Theorem 4.18 here). Combining (b) and (c), we also see that if all continuous homomorphic images of an $\mathbb{R}$-factorizable group G are $\mathbb{R}$-factorizable, then G is pseudo-ω_1-compact (this requires the assumption $2^\omega < 2^{\omega_1}$).

It was recently shown in TKACHENKO [200?] that (A), (B) and (C) of Problem 4.3 have positive solutions in the class of P-groups. It turns out that the key property of $\mathbb{R}$-factorizable P-groups is pseudo-ω_1-compactness. We will discuss this in Subsections 2 and 3 of this section.

2. $\mathbb{R}$-factorizability of P-groups

It is easy to see that a Tychonoff space X is pseudo-ω_1-compact if and only if every continuous metrizable image of X is separable. Therefore, all countably cellular spaces, all Lindelöf spaces and all pseudocompact spaces are pseudo-ω_1-compact. The following result of TKACHENKO [200?] solves (A) of Problem 4.3 in the affirmative for P-groups.

4.13. THEOREM. *For a P-group G, the following conditions are equivalent:*

(a) *G is $\mathbb{R}$-factorizable;*

(b) *G is pseudo-ω_1-compact;*

(c) *G is $\aleph_0$-bounded and every its continuous homomorphic image H with $w(H) \leq \omega_1$ is Lindelöf.*

In fact, the list of equivalent conditions in Theorem 4.13 can be extended. Let τ be an infinite cardinal. Following Arhangel'skii, we say that a space X is *τ-stable* if every continuous image Y of X which admits a coarser Tychonoff topology of weight $\leq \tau$ satisfies $nw(Y) \leq \tau$. If X is τ-stable for each $\tau \geq \omega$, then X is said to be *stable*. It is known that arbitrary products and σ-products of second countable spaces are ω-stable, and every Lindelöf P-space is ω-stable by a result in ARHANGEL'SKII [1984]. It is not difficult to show that, for a P-space, ω-stability and pseudo-ω_1-compactness are equivalent. Therefore, one can add ω-stability to the list in Theorem 4.13.

It is worth mentioning that Theorem 4.13 cannot be extended to $\aleph_0$-bounded P-groups. Indeed, there exists an $\aleph_0$-bounded P-group H that is neither $\mathbb{R}$-factorizable, nor ω-stable, nor pseudo-ω_1-compact. One can take H to be a proper dense subgroup of a Lindelöf P-group of weight $\aleph_1$ (see Example 4.5).

It is shown in TKACHENKO [200?] that a Lindelöf P-group is τ-stable for each $\tau < \aleph_\omega$. We do not know, however, whether Lindelöf P-groups are stable. More generally, it would be interesting to solve the problem below.

4.14. PROBLEM. Find out which of the following assertions are valid:

(a) Every Lindelöf ω-stable topological group is stable.

(b) Every Lindelöf P-group is stable.

(c) Every $\mathbb{R}$-factorizable P-group is τ-stable for each $\tau < \aleph_\omega$.

It is easy to verify that every C-embedded subspace of a Lindelöf space is pseudo-ω_1-compact. In addition, every G_δ-dense z-embedded subset of a space X is C-embedded in X. Therefore, we can apply (b) of Theorem 4.2 and the equivalence of (a) and (b) in Theorem 4.13 to deduce the next result which solves Problem 4.10 in the special case of P-groups:

4.15. PROPOSITION. *If a P-group G is topologically isomorphic to a z-embedded subgroup of a Lindelöf group, then G is pseudo-ω_1-compact and $\mathbb{R}$-factorizable.*

One cannot drop "z-embedded" in the above result: every proper dense subgroup of a Lindelöf P-group of weight ω_1 is a counterexample. In fact, since Lindelöf groups are $\mathbb{R}$-factorizable, Proposition 4.15 can be given a stronger symmetric form that follows from Theorems 4.8, 4.9 and 4.13.

4.16. THEOREM. *Let a P-group H be a subgroup of an $\mathbb{R}$-factorizable group G. Then the following are equivalent:*

(a) *H is $\mathbb{R}$-factorizable;*

(b) *H is pseudo-ω_1-compact;*

(c) *H is z-embedded in G.*

Proposition 4.15 suggests the following tempting conjecture:

4.17. CONJECTURE. *Every $\mathbb{R}$-factorizable P-group is topologically isomorphic to a subgroup of a Lindelöf P-group.*

It follows from Theorem 4.13 that $\mathbb{R}$-factorizable P-groups of weight ω_1 are Lindelöf, so a counterexample to Conjecture 4.17 should be of weight $\geq \omega_2$. On the other hand, one can try to prove Conjecture 4.17 by taking the Raĭkov completion ϱH of an $\mathbb{R}$-factorizable P-group H. Then ϱH is also a P-group, so it suffices to prove that the group ϱH is Lindelöf. Since H admits a unique dense embedding (up to topological isomorphisms fixing points of H) into a Raĭkov complete topological group, Theorem 3.1 implies, in particular, that Conjecture 4.17 is equivalent to asking whether the group ϱH is Lindelöf.

Apart from completeness, Lindelöf P-groups have other peculiar properties. For example, every continuous homomorphism $f\colon G \to H$ of a Lindelöf P-group onto a P-group H is open (see TKACHENKO [200?]). We do not know under which conditions Lindelöf P-groups are monolithic. Very recently, ITZKOWITZ AND TKACHUK [200?] showed that the existence of non-monolithic Lindelöf groups is consistent with ZFC.

It is an open problem whether an analog of Theorem 4.4 (b) remains valid for P-groups. In other words, the problem is to embed an arbitrary $\aleph_0$-bounded P-group into an $\mathbb{R}$-factorizable P-group. HERNÁNDEZ and TKACHENKO [200?] established that if H is an arbitrary subgroup of an Abelian $\mathbb{R}$-factorizable P-group, then H is topologically isomorphic to a closed subgroup of another Abelian $\mathbb{R}$-factorizable P-group. Therefore, the classes of *all* subgroups and *closed* subgroups of Abelian $\mathbb{R}$-factorizable P-groups coincide. This result and Example 4.5 together imply that closed subgroups of $\mathbb{R}$-factorizable P-groups need not be $\mathbb{R}$-factorizable.

Let us call a group G *hereditarily $\mathbb{R}$-factorizable* if all subgroups of G are $\mathbb{R}$-factorizable. Many natural questions about hereditarily $\mathbb{R}$-factorizable groups are open. It is not known, for example, whether the group $\mathbb{Z}^{\omega_1}$ (or $\mathbb{Z}^{\mathfrak{c}}$) is hereditarily $\mathbb{R}$-factorizable or whether every hereditarily $\mathbb{R}$-factorizable group is pseudo-ω_1-compact. However, all hereditarily $\mathbb{R}$-factorizable P-groups are countable by a theorem in HERNÁNDEZ and TKACHENKO [200?].

3. *Continuous homomorphic images of $\mathbb{R}$-factorizable groups*

One of the most interesting (and, we believe, difficult) open questions is whether continuous homomorphisms preserve $\mathbb{R}$-factorizability (see (B) of Problem 4.3). The following result of TKACHENKO [1991b] answers the question in the special case of open homomorphisms.

4.18. THEOREM. *A quotient group of an $\mathbb{R}$-factorizable group is $\mathbb{R}$-factorizable.*

Since every continuous homomorphism $p\colon G \to H$ can be represented in the form $p = i \circ \pi$, where $\pi\colon G \to K$ is an open continuous homomorphism and $i\colon K \to H$ is a continuous isomorphism, Theorem 4.18 implies that (B) of Problem 4.3 is equivalent to asking whether a continuous isomorphic image of an $\mathbb{R}$-factorizable group is $\mathbb{R}$-factorizable.

In several special cases, the answer to (B) of Problem 4.3 is affirmative simply because a certain property (implying $\mathbb{R}$-factorizability of a group) is invariant under continuous homomorphisms. For example, we can go along the list (a)–(d) of Theorem 4.2 and note that precompactness, the Lindelöf property, and σ-precompactness are invariant under taking continuous homomorphic.

The situation with (d) of Theorem 4.2 is a bit different. We cannot claim that a continuous homomorphic image of a dense subgroup of a direct product of second countable groups is of the same type. Nevertheless, we have the following result (see TKACHENKO [200?]):

4.19. THEOREM. *Let G be a dense subgroup of a direct product of second countable topological groups. Then every continuous homomorphic image of G is $\mathbb{R}$-factorizable.*

The proof of Theorem 4.19 requires several facts. One of them is the next result established in SHCHEPIN [1976] and which is important in itself.

4.20. THEOREM. *For every continuous function $f\colon G \to \mathbb{R}$ on a topological group G satisfying $c(G) \le \tau$, there exists an open continuous homomorphism $\pi\colon G \to K$ onto a topological group K with $\psi(K) \le \tau$ and a continuous function $h\colon K \to \mathbb{R}$ such that $f = h \circ \pi$.*

Shchepin's theorem implies that a countably cellular topological group G is, in a sense, close to being $\mathbb{R}$-factorizable. Indeed, the group K in the above theorem has countable cellularity (as a continuous image of G), and every topological group of countable cellularity is $\aleph_0$-bounded (see TKACHENKO [1998]). Therefore, though K need not be second countable, it admits a coarser second countable Hausdorff group topology by a result in ARHANGEL'SKII [1980] or TKACHENKO [1998, Prop. 4.5]. Curiously, we do not know whether countably cellular topological groups are $\mathbb{R}$-factorizable (but the converse is false because Lindelöf topological groups can have uncountable cellularity).

To sketch the proof of Theorem 4.19, we recall that an *Oz space* (in another terminology, a *perfectly κ-normal space*) is a space with the property that the closure of every open set is a zero set. It is easy to verify that a dense subspace of an Oz space is also Oz. The first step is to show that every continuous homomorphic image of a direct product $P = \prod_{i\in I} G_i$ of second countable topological groups is an Oz-space (and $\mathbb{R}$-factorizable). This requires Theorems 4.2, 4.20 and a theorem on factorization of continuous functions defined on a product of separable spaces with values in a space of countable pseudocharacter (see JUHÁSZ [1971]). Once this is proved, the rest is easy. Indeed, suppose that G is an arbitrary dense subgroup of the direct product $P = \prod_{i\in I} G_i$. Extend a given continuous epimorphism $\varphi\colon G \to H$ to a continuous homomorphism $\tilde{\varphi}\colon P \to \varrho H$, where ϱH is the Raĭkov completion of H. By the above claim, the subgroup $H_0 = \tilde{\varphi}(P)$ of ϱH is $\mathbb{R}$-factorizable and an Oz space. Clearly, $H \subseteq H_0 \subseteq \varrho H$ and H is dense in H_0. By a result in BLAIR [1976], every dense subset of an Oz space is z-embedded, so it remains to apply Theorem 4.9 to conclude that the group H is $\mathbb{R}$-factorizable.

The conclusion of Theorem 4.19 remains valid for continuous homomorphic images of dense subgroups of direct products of topological groups with a countable network (the argument is the same). A similar assertion is also valid in the case when the factors are σ-compact groups or, even more generally, *Lindelöf Σ-groups*. In the latter case, one has to apply additionally a couple of results from TKACHENKO [1991a].

Another instance of preservation of $\mathbb{R}$-factorizability by continuous homomorphisms found in TKACHENKO [200?], is the case of P-groups.

4.21. THEOREM. *Every continuous homomorphic image of an ℝ-factorizable P-group is ℝ-factorizable.*

Since continuous homomorphic images of P-groups need not be P-groups, one cannot directly apply the equivalence (a) $\Leftrightarrow$ (b) of Theorem 4.13 to deduce the above preservation theorem. Nevertheless, the cellularity argument helps. To put it clearly, we present the next proposition (which is also an essential part of the proof of Theorem 4.13 given in TKACHENKO [200?]).

4.22. PROPOSITION. *Let G be an ℝ-factorizable P-group. Then:*

(a) *G is pseudo-ω_1-compact;*

(b) *$c(G) \le \omega_1$;*

(c) *if additionally $\psi(G) \le \omega_1$, then G is Lindelöf and satisfies $w(H) \le \omega_1$.*

The proof of (b) of Proposition 4.22 is based on the fact that every ℝ-factorizable P-group has many open continuous homomorphisms onto groups of weight $\le \omega_1$, and the family of such homomorphisms is ω_1-complete. The proof of (c) leans on (a) (so that $\psi(G) \le \omega_1$ implies $\chi(G) \le \omega_1$ and, hence, $w(G) \le \omega_1$) and the fact that every regular P-space of weight $\le \omega_1$ is zero-dimensional and paracompact.

Here is a brief sketch of the proof of Theorem 4.21. Let $p\colon G \to H$ be a continuous homomorphism of an ℝ-factorizable P-group G onto a group H and f be a continuous real-valued function on H. Since $c(G) \le \omega_1$ by (b) of Proposition 4.22 and $c(H) \le c(G)$, we can apply Theorem 4.20 to find a continuous homomorphism $\pi\colon H \to K$ onto a topological group K with $\psi(K) \le \omega_1$ and a continuous real-valued function g on K such that $f = g \circ \pi$. Denote by K^* the underlying group K endowed with the group topology whose base consists of G_δ-sets in K. The homomorphism $\lambda = \pi \circ p\colon G \to K$ is continuous and, since G is a P-group, the homomorphism $\lambda^*\colon G \to K^*$ pointwise coinciding with λ remains continuous. Denote by i the identity isomorphism of K^* onto K.

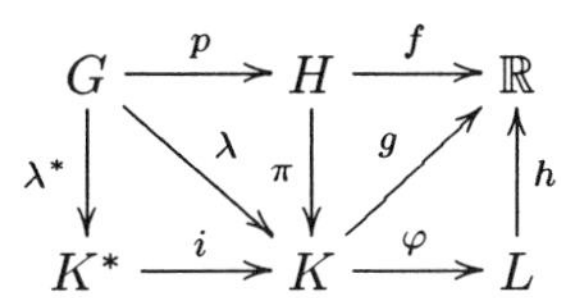

The group G is pseudo-ω_1-compact by (a) of Proposition 4.22, and so is K^* as a continuous image of G. Therefore, Theorem 4.13 implies that K^* is ℝ-factorizable. In addition, from the continuity of i it follows that $\psi(K^*) \le \psi(K) \le \omega_1$. By (c) of Proposition 4.22, the group K^* is Lindelöf. Clearly, the group $K = i(K^*)$ is also Lindelöf. Since every Lindelöf group is ℝ-factorizable, we can find a continuous homomorphism $\varphi\colon K \to L$ onto a second countable topological group L and continuous function $h\colon L \to \mathbb{R}$ such that $g = h \circ \varphi$. So, the homomorphism $\psi = \varphi \circ \pi$ of H onto L factorizes f. This proves the ℝ-factorizability of H.

All known ℝ-factorizable groups have the property that their continuous homomorphic images remain ℝ-factorizable. However, this does not exclude the possibility that *every* $\aleph_0$-bounded group could be a continuous homomorphic image of an ℝ-factorizable group. This is the best we can say up to the moment.

4. Products of $\mathbb{R}$-factorizable groups

Another mysterious problem is whether the class of $\mathbb{R}$-factorizable groups is productive (see (C) of Problem 4.3). The situation here is even less clear than that in the case of homomorphic images. Almost every attempt to verify productivity of certain $\mathbb{R}$-factorizable groups (even from a special class) presents serious difficulties or simply fails. Here is an open problem which shows the limitations of our knowledge.

4.23. PROBLEM.

(a) Is any product of $\mathbb{R}$-factorizable groups $\mathbb{R}$-factorizable?

(b) Is the product of two $\mathbb{R}$-factorizable groups $\mathbb{R}$-factorizable?

(c) Is the product of a compact group with any $\mathbb{R}$-factorizable group $\mathbb{R}$-factorizable?

(d) Is the product of two Lindelöf groups $\mathbb{R}$-factorizable?

(e) Is the product of a Lindelöf P-group with any $\mathbb{R}$-factorizable group $\mathbb{R}$-factorizable?

Clearly, the first item of the above problem is exactly (C) of Problem 4.3, while (b)–(e) tend to specify the general problem for different special cases.

It is easy to see that any direct product of precompact topological groups is $\mathbb{R}$-factorizable — this immediately follows from the productivity of the class of precompact groups and (a) of Theorem 4.2. Arbitrary subgroups of σ-compact groups form another class with this property:

4.24. PROPOSITION. *Let $\{G_i : i \in I\}$ be a family of σ-precompact topological groups. Then the product group $G = \prod_{i \in I} G_i$ is $\mathbb{R}$-factorizable.*

A similar result remains valid for direct products of arbitrary subgroups of Lindelöf Σ-groups; this follows from TKACHENKO [1998, Theorem 5.10]. The latter fact is one of the most general results about productivity in the class of $\mathbb{R}$-factorizable groups. Another special case of such a productivity has been recently obtained in TKACHENKO [200?]:

4.25. THEOREM. *Any direct product of $\mathbb{R}$-factorizable P-groups is $\mathbb{R}$-factorizable.*

Clearly, an infinite product of non-trivial topological groups is never a P-group, so one cannot directly apply Theorem 4.13 here. Nevertheless, it plays an important role in the corresponding argument in TKACHENKO [200?] which goes as follows. Let $G = \prod_{i \in I} G_i$ be a product of $\mathbb{R}$-factorizable P-groups. For every $J \subseteq I$, let $G_J = \prod_{i \in J} G_i$. We divide the proof into several steps.

Fact 1. *If $J \subseteq I$ is countable and $\varphi\colon G_J \to K$ is a continuous homomorphism onto a topological group K with $\psi(K) \leq \omega_1$, then K is Lindelöf.*

This requires Theorem 4.18, Proposition 4.22 and Noble's theorem which states that a countable product of Lindelöf P-spaces is Lindelöf. Further, we have:

Fact 2. *Every continuous real-valued function on G depends on at most countably many coordinates.*

Indeed, by a theorem in GLICKSBERG [1959], it suffices to verify that the product group G is pseudo-ω_1-compact. Note that, for every finite $J \subseteq I$, G_J is an $\aleph_0$-bounded P-group. Fact 1 and the equivalence of (b) and (c) in Theorem 4.13 together imply that G_J is pseudo-ω_1-compact. Then the Delta lemma argument applies to conclude that so is G.

Finally, let f be a continuous real-valued function on G. Fact 2 enables us to assume that the index set I is countable. The P-group G_J is $\mathbb{R}$-factorizable for each finite $J \subseteq I$, so (b) of Proposition 4.22 implies that $c(G_J) \leq \omega_1$. Hence, $c(G) \leq \omega_1$ by a theorem in NOBLE and ULMER [1972]. Apply Theorem 4.20 to find a continuous homomorphism $\varphi\colon G \to K$ onto a group K with $\psi(K) \leq \omega_1$ and a continuous function $g\colon K \to \mathbb{R}$ such that $f = g \circ \varphi$. The group K is Lindelöf by Fact 1 and, hence, $\mathbb{R}$-factorizable by (b) of Theorem 4.2.

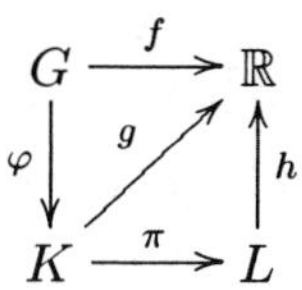

It remains to find a continuous homomorphism $\pi\colon K \to L$ onto a second countable group L and a continuous function $h\colon L \to \mathbb{R}$ satisfying $g = h \circ \pi$. Clearly, the homomorphism $p = \pi \circ \varphi$ of G to L factorizes f. So, G is $\mathbb{R}$-factorizable.

In fact, Theorems 4.21 and 4.25 admit a more general form: every continuous homomorphic image of a product of $\mathbb{R}$-factorizable P-groups is $\mathbb{R}$-factorizable (see TKACHENKO [200?]).

Now we present some comments about (b), (c) and (e) of Problem 4.23. Let us start with (c).

Suppose that a topological group G contains an uncountable locally finite family $\{U_\alpha : \alpha < \omega_1\}$ of non-empty open sets. Let $\mathbb{Z}(2) = \{0, 1\}$ be the discrete group with two elements. In the compact group $K = \mathbb{Z}(2)^{\omega_1}$, choose a family $\{V_\alpha : \alpha < \omega_1\}$ of proper canonical clopen sets, where each V_α depends only on the coordinate α. Then the family $\{U_\alpha \times V_\alpha : \alpha < \omega_1\}$ is locally finite in the product group $G \times K$, and one easily defines a continuous real-valued function f on $G \times K$ which depends on uncountably many coordinates related to the second factor. Therefore, f does not admit a factorization via a continuous homomorphism of $G \times K$ to a second countable topological group (all such homomorphisms depend on at most countably many coordinates). Therefore, if there exists an $\mathbb{R}$-factorizable group G which fails to be pseudo-ω_1-compact, then the product of G with the compact group K is not $\mathbb{R}$-factorizable. It turns out, however, that pseudo-ω_1-compactness of G is exactly the property which makes the product of G with every compact group $\mathbb{R}$-factorizable (see TKACHENKO [1998]):

4.26. THEOREM. *The following conditions are equivalent for an $\mathbb{R}$-factorizable group G:*

(a) *G is pseudo-ω_1-compact;*

(b) *$G \times \mathbb{Z}(2)^{\omega_1}$ is $\mathbb{R}$-factorizable;*

(c) *$G \times L$ is $\mathbb{R}$-factorizable for every compact group L.*

Thus, the answer to (c) of Problem 4.23 requires a solution to (A) of Problem 4.3 and vice versa. This explains, in part, the difficulties arising in the study of products of $\mathbb{R}$-factorizable groups. On the other hand, a product of an arbitrary $\mathbb{R}$-factorizable group with a second countable compact group is $\mathbb{R}$-factorizable TKACHENKO [1991b].

In the case when both factors in the product $G \times H$ are non-compact, the situation is even less clear. To give a first idea of what type of results one may expect, we present two facts in this direction. The first of them was proved in TKACHENKO [1991b].

4.27. PROPOSITION. *If G is a Lindelöf group and a group H is pseudocompact, then the product $G \times H$ is $\mathbb{R}$-factorizable.*

A space X is said to be *weakly Lindelöf* if every open cover of X contains a countable subfamily whose union is dense in X. Clearly, all Lindelöf spaces and all countably cellular spaces are weakly Lindelöf. The notion of a weakly Lindelöf group appears in the next result.

4.28. PROPOSITION. *Let G and H be $\mathbb{R}$-factorizable groups. If G is a P-group and H is weakly Lindelöf, then the product $G \times H$ is $\mathbb{R}$-factorizable.*

In particular, the product of a Lindelöf P-group with a Lindelöf group is $\mathbb{R}$-factorizable, thus answering (d) of Problem 4.23 in the special case when one of the factors is a P-group. We do not know, however, whether the product of a Lindelöf group with a precompact group is $\mathbb{R}$-factorizable.

Finally, an argument similar to that presented before Theorem 4.26 shows that if there exists an $\mathbb{R}$-factorizable K which fails to be pseudo-ω_1-compact, then $G \times K$ is not $\mathbb{R}$-factorizable for a certain Lindelöf P-group G (one can take G as in Example 4.5). This might be the answer to (e) of Problem 4.23. It is also unknown whether Proposition 4.28 remains valid in the case when H is pseudo-ω_1-compact (even if the P-group G is Lindelöf).

5. *Topological completions of $\mathbb{R}$-factorizable groups*

In general, the Hewitt–Nachbin and Dieudonné completions of a topological group need not be topological groups (see ARHANGEL'SKII's article [2002] in this volume). It turns out that $\mathbb{R}$-factorizable groups are much better in this respect. For a Tychonoff space X, let υX and μX be the Hewitt–Nachbin and Dieudonné completions of X, respectively. Given a topological group G, we define the *index of boundedness* of G, denoted by $ib(G)$, as the least infinite cardinal number τ such that G can be covered by at most τ translates of any neighborhood of the identity.

The next result shows that almost every topological group G satisfies $\mu G = \upsilon G$ (see TKACHENKO [200?]).

4.29. PROPOSITION. *Let G be a topological group such that its index of boundedness $ib(G)$ is not Ulam-measurable. Then $\mu G = \upsilon G$. In particular, every $\aleph_0$-bounded group G satisfies $\mu G = \upsilon G$.*

Let us say that a subgroup K of a topological group L is *G_δ-dense* in L if K meets every non-empty G_δ-set in L. It is well known that a precompact group K is pseudocompact iff it is a G_δ-dense subgroup of the Raĭkov completion ϱK. Similarly, the interaction between the properties of a group K and a larger topological group $L \supseteq K$ improves if K is G_δ-dense in L (see TKACHENKO [200?]):

4.30. PROPOSITION. *Let K be a G_δ-dense $\mathbb{R}$-factorizable subgroup of a topological group L. Then L is $\mathbb{R}$-factorizable and K is C-embedded in L.*

The above result gives a partial answer to Problem 4.7. It is easy to verify that if a P-group K is dense in a topological group L, then L is a P-group and K is G_δ-dense in L. Therefore, the next corollary to Proposition 4.30 is immediate.

4.31. COROLLARY. *Suppose that a P-group K is a dense subgroup of a topological group L. If K is $\mathbb{R}$-factorizable, then so is L.*

A topological group K is said to be a *PT-group* if the group multiplication and inversion in K can be continuously extended over the Dieudonné completion μK of K in such a way that μK becomes a topological group which contains K as a dense subgroup (see ARHANGEL'SKII [2002]). Clearly, all Raĭkov complete groups are PT-groups. Denote by $\varrho_\omega K$ the G_δ-closure of a topological group K in its Raĭkov completion ϱK. Clearly, $\varrho_\omega K$ is a dense subgroup of ϱK and K is G_δ-dense in $\varrho_\omega K$. The following result implies, in particular, that all $\mathbb{R}$-factorizable groups are PT-groups.

4.32. THEOREM. *Every $\mathbb{R}$-factorizable group K satisfies $\upsilon K = \mu K = \varrho_\omega K$ and, hence, is a PT-group. In addition, the group $\varrho_\omega K$ is $\mathbb{R}$-factorizable.*

In a slightly weaker form, Theorem 4.32 was proved in HERNÁNDEZ, SANCHIS and TKACHENKO [2000] and TKACHENKO [1998]. We do not know, however, whether the Raĭkov completion ϱK of an $\mathbb{R}$-factorizable group K is $\mathbb{R}$-factorizable. The question remains open even if the group K is countable. This means, in particular, that the $\mathbb{R}$-factorizability of separable groups is an open problem.

Corollary 4.6 implies that dense subgroups of $\mathbb{R}$-factorizable groups can fail to be $\mathbb{R}$-factorizable. In fact, Example 4.5 shows that even G_δ-dense subgroups of $\mathbb{R}$-factorizable groups need not be $\mathbb{R}$-factorizable. The explanation of this phenomenon comes if we combine Proposition 4.30 and Theorem 4.9:

4.33. PROPOSITION. *The following conditions are equivalent for a G_δ-dense subgroup K of an $\mathbb{R}$-factorizable group L:*

(1) *K is $\mathbb{R}$-factorizable;*

(2) *K is C-embedded in L.*

Finally, we note that $\mathbb{R}$-factorizable P-groups behave in a more predictable way (see TKACHENKO [200?]).

4.34. THEOREM. *The equalities $\upsilon H = \mu H = \varrho H$ are valid for every $\mathbb{R}$-factorizable P-group H. In particular, a Dieudonné complete $\mathbb{R}$-factorizable P-group is Raĭkov complete.*

Indeed, let ϱH be the Raĭkov completion of an $\mathbb{R}$-factorizable P-group H. Then ϱH is a P-group, so H is G_δ-dense in ϱH. By Proposition 4.33, H is C-embedded in ϱH and, hence, $\varrho H \subseteq \upsilon H$. Since H is $\mathbb{R}$-factorizable, from Theorem 4.32 it follows that $\upsilon H = \mu H$ is a topological group which contains H as a dense subgroup. Therefore, $\mu H \subseteq \varrho H$. This proves that $\upsilon H = \mu H = \varrho H$. If, in addition, the group H is Dieudonné-complete, then $H = \mu H$ and the above equalities imply that $H = \mu H = \varrho H$. So, H is Raĭkov complete.

Example 4.5 shows that there is an $\aleph_0$-bounded P-group H which does not satisfy the equality $\upsilon H = \varrho H$. Indeed, since the completion ϱH of the P-group H in Example 4.5 is Lindelöf (hence, $\mathbb{R}$-factorizable) and H is not $\mathbb{R}$-factorizable, from Proposition 4.33 it follows that H is not C-embedded in ϱH. Therefore, $\upsilon H \neq \varrho H$. In fact, $\upsilon H = H$ since H is a P-space of weight ω_1.

References

ARHANGEL'SKII, A.V.

[1980] Relations between invariants of topological groups and their subspaces, (in Russian), *Trudy Mosk. Mat. Obsch.* **47** (1984), 3–21.

[1984] Continuous mappings, factorization theorems and function spaces, (in Russian), *Trudy Mosk. Mat. Obsch.* **47**, 3–21.

[1994] On countably compact topologies on compact groups and on dyadic compacta, *Topology Appl.* **57**, 163–181.

[2002] Topological invariants in algebraic environment, in *Recent Progress in General Topology II*, this volume.

BANAKH, T.

[2000] Locally minimal topological groups and their embeddings into products of o-bounded groups, *Comment. Math. Univ. Carolin.* **41**, 811–815.

[2002] On index of total boundedness of (strictly) o-bounded groups, *Topology Appl.* **120**, 427–439.

BANAKH, T., P. NICKOLAS and M. SANCHIS

[200?] Application of filter games for constructing pathological examples of o-bounded topological groups, submitted.

BLAIR, R.L.

[1976] Spaces in which special sets are z-embedded, *Canad. J. Math.* **28**, 673–690.

BRUGUERA, M. and M. TKACHENKO

[200?] Extensions of topological groups do not respect countable compactness, preprint.

COMFORT, W.W.

[1975] Compactness-like properties for generalized weak topological sums, *Pacific J. Math.* **60**, 31–37.

[1984] Topological groups, in *Handbook of Set-theoretic Topology*, K. Kunen and J.E. Vaughan, eds.,North-Holland, Amsterdam, 1143–1263.

COMFORT, W.W., K.H. HOFMANN and D. REMUS

[1992] Topological Groups and Semigroups, in: *Recent Progress in General Topology*, M. Hušek and J. van Mill, eds., North-Holland, Amsterdam; pp. 239–274.

COMFORT, W.W. and D. REMUS

[1993] Pseudocompact refinements of compact group topologies, *Mathematische Zeitschrift* **215** (3), 337–346.

[1994] Compact groups of Ulam-measurable cardinality: partial converse to theorems of Arhangel'skiĭ and Varopoulos, *Math. Japonicae* **39**, 203–210.

COMFORT, W.W. and L. ROBERTSON

[1988] Extremal phenomena in certain classes of totally bounded groups, *Dissert. Math.* **272**, 1–48.

COMFORT, W.W. and K.A. ROSS

[1966] Pseudocompactness and uniform continuity in topological groups, *Pacific J. Math.* **16**, 483–496.

DIKRANJAN, D. and D. SHAKHMATOV

[1998] *Algebraic structure of the pseudocompact groups*, Memoirs Amer. Math. Soc., 133/633, pp. viii+83.

DIKRANJAN, D. and M. TKACHENKO

[2000] Sequential completeness of quotient groups, *Bull. Austral. Math. Soc.* **61**, 129–151.

[2001] Sequentially complete groups: dimension and minimality, *J. Pure Appl. Algebra* **157**, 215–239.

[2002a] Algebraic structure of small countably compact groups, *Forum Math.* **14**.

[2002b] Varieties generated by countably compact Abelian groups, *Proc. Amer. Math. Soc.* **130**, 2487–2496.

DIKRANJAN, D., M.G. TKAČENKO and V.V. TKACHUK

[2000] Topological groups with thin generating sets, *J. Pure Appl. Algebra* **145**, 123–148.

DOUWEN, E.K. VAN

[1980a] The product of two countably compact topological groups, *Trans. Amer. Math. Soc.* **262**, 417–427.

[1980b] The weight of a pseudocompact (homogeneous) space whose cardinality has countable cofinality, *Proc. Amer. Math. Soc.* **80**, 678–682.

GLICKSBERG, I.

[1959] Stone–Čech compactifications of products, *Trans. Amer. Math. Soc.* **90**, 369–382.

GRANT, D.L.

[1993] Sequentially compact cancellative topological semigroups: some progress on the Wallace problem, in: *Papers on General Topology and Applications, VIIth Summer Conference at the University of Wisconsin* (Madison, 1991), Annals of the New York Acad. Sci. **704**, 150–154.

GURAN, I.

[1981] On topological groups close to being Lindelöf, *Soviet Math. Dokl.* **23**, 173–175.

HAJNAL, A. and I. JUHÁSZ

[1976] A separable normal topological group need not be Lindelöf, *General Top. Appl.* **6**, 199–205.

HARRISON, D.

[1959] Infinite abelian groups and homological methods, *Ann. Math.* (2) **69**, 366–391.

HART, K.P. and J. VAN MILL

[1991] A countably compact group H such that $H \times H$ is not countably compact, *Trans. Amer. Math. Soc.* **323**, 811–821.

HERNÁNDEZ, C.

[2000] Topological groups close to being σ-compact, *Topology Appl.* **102**, 101–111.

HERNÁNDEZ, C., D. ROBBIE and M.G. TKACHENKO

[2000] Some properties of o-bounded and strictly o-bounded groups, Applied General Topology **1** (1), 29–43.

HERNÁNDEZ, C. and M.G. TKACHENKO

[1998] Subgroups of $\mathbb{R}$-factorizable groups, *Comment. Math. Univ. Carolin.* **39**, 371–378.

[200?] Subgroups and products of $\mathbb{R}$-factorizable P-groups, preprint.

HERNÁNDEZ, S., M. SANCHIS and M. TKACHENKO

[2000] Bounded sets in spaces and topological groups, *Topology Appl.* **101**, 21–43.

HEWITT, E. and K. ROSS

[1979] *Abstract Harmonic Analysis, Volume I,* Springer-Verlag, Berlin-Göttingen-Heidelberg.

HULANICKI, A.

[1958] Algebraic structure of compact abelian groups, *Bull. Acad. Polon. Sci. Sér. Sci. Math. Astr. Phys.* **6**, 71–73.

ITZKOWITZ, G.L. and V.V. TKACHUK

[200?] Any pseudo-pG-compact group G is fine, preprint.

IVANOVSKIĬ, L.N.

[1958] On a hypothesis of P.S. Alexandrov, (in Russian), *Dokl. AN SSSR* **123**, 785–786.

JUHÁSZ, I.

[1971] *Cardinal Functions in Topology*, Math. Centre Tracts 34, Amsterdam.

KOSZMIDER, P., A. TOMITA and S. WATSON

[200?] Forcing countably compact group topologies on a larger free Abelian group, *Topology Proc.*, to appear.

KRAWCZYK, A. and H. MICHALEWSKI

[2001] Linear metric spaces close to being σ-compact, preprint 46, Warsaw Univ.

[200?] An example of a topological group, *Topology Appl.*, to appear.

KUZ'MINOV, V.

[1959] On a hypothesis of P.S. Alexandrov in the theory of topological groups, (in Russian), *Doklady Akad. Nauk SSSR* **125**, 727–729.

MALYKHIN, V. AND L. SHAPIRO

[1985] Pseudocompact groups without converging sequences, *Math. Notes* **37**, 59–62.

MUKHERJEA, A. and N.A. TSERPES

[1972] A note on countably compact semigroups, *J. Austral. Math. Soc.* **13**, 180–184.

NOBLE, N. and M. ULMER

[1972] Factoring functions on Cartesian products, *Trans. Amer. Math. Soc.* **163**, 329–339.

PFISTER, H.

[1985] Continuity of the inverse, *Proc. Amer. Math. Soc.* **95**, 312–314.

PONTRYAGIN, L.S.

[1939] *Continuous Groups*, Moscow, 1938. English translation: *Topological groups*, Princeton University Press, Princeton 1939.

ROBBIE, D. and S. SVETLICHNY

[1996] An answer to A. D. Wallace's question about countably compact cancellative semigroups, *Proc. Amer. Math. Soc.* **124**, 325–330.

SHCHEPIN, E.V.

[1976] Real-valued functions and canonical sets in Tychonoff products and topological groups, *Russian Math. Surveys* **31**, 19–30.

SHAKHMATOV, D.B.

[1994] A direct proof that every infinite compact group G contains $\{0,1\}^{w(G)}$, in Papers on General Topology and Applications (8th Summer Conference at Queens College), *Annals of the New York Academy of Sciences* **728**, 276–283.

[2002] Convergence in the presence of algebraic structure, in *Recent Progress in General Topology II*, this volume.

TKACHENKO, M.G.

[1990] Countably compact and pseudocompact topologies on free Abelian groups, *Soviet Math. (Izv. VUZ)* **34**, 79–86.

[1991a] Factorization theorems for topological groups and their applications, *Topology Appl.* **38**, 21–37.

[1991b] Subgroups, quotient groups and products of $\mathbb{R}$-factorizable groups, *Topology Proc.* **16**, 201–231.

[1998] Introduction to topological groups, *Topology Appl.* **86**, 179–231.

[1999] Topological groups for topologists: Part I, *Bol. Soc. Mat. Mexicana* (3) **5**, 735–753.

[2000] Topological groups for topologists: Part II, *Bol. Soc. Mat. Mexicana* (3) **6**, 1–45.

[2001] Complete $\aleph_0$-bounded groups need not be $\mathbb{R}$-factorizable, *Comment. Math. Univ. Carolin.* **42** (3), 551-559.

[200?] $\mathbb{R}$-factorizable groups and subgroups of Lindelöf P-groups, submitted.

TOMITA, A.H.

[1997a] On the square of Wallace semigroups and topological free Abelian groups, *Topology Proc.* **22**, 331–349.

[1997b] On infinite products of countably compact groups, in *Proc. VIIIth Prague Topol. Symposium, P. Simon, ed.*, Prague, Czech Republic, August 18–24, 1996, Topology Atlas, 362–370.

[1998] The existence of initially ω_1-compact group topologies on free Abelian groups is independent of ZFC, *Comment. Math. Univ. Carolin.* **39**, 401–413.

[1999] On the number of countably compact group topologies on a free Abelian group, Topology Appl. **98**, 345–353.

[200?] Two countably compact topological groups: one of size $\aleph_\omega$ and the other of weight $\aleph_\omega$ without non-trivial convergent sequences, preprint.

USPENSKIJ, V.V.

[1988] Why compact groups are dyadic, in *Proc. of the 6th Prague Topological Symposium* 1986, Frolík Z., ed., Heldermann Verlag, Berlin, pp. 601–610.

[200?] On sequentially continuous homomorphisms of topological groups, preprint.

VAROPOULOS, N.TH.

[1964] A theorem on the continuity of homomorphisms of locally compact groups, *Proc. Camb. Phil. Soc.* **60**, 449–463.

YUR'EVA, A.A.

[1993] Countably compact sequential topological semigroup is a topological group, (in Russian), *Math. Stud.* **2**, 23–24.

CHAPTER 20

Essays

Contents

RECENT PROGRESS IN GENERAL TOPOLOGY II
Edited by Miroslav Hušek and Jan van Mill

The Early Development of Infinite Dimensional Topology

R. D. Anderson

By chance or good fortune, infinite dimensional topology got its start at almost the same time as the Prague Symposium. Its' origins came from a question posed by Vic Klee to the author in about 1960 or 1961: Is the product of a triod, T, and the Hilbert cube, Q, homeomorphic to the Hilbert cube? The Hilbert cube, Q, is the countable infinite product of closed intervals, a generalization of finite dimensional cubes. There are many similarities with finite dimensional cubes and many differences. Since we want the Hilbert cube to be metric, we cannot just think of it as the product of unit intervals since the distance between $(0, 0, \dots)$ and $(1, 1, \dots)$ must be finite which means that higher indexed coordinate intervals must be regarded as small, the ith interval could be i^{-1} in length, for example. Unlike a finite dimensional cube $\mathbb{I}^N$ which has a closed $(N - 1)$-dimensional topological sphere as its boundary, the Hilbert cube has a pseudo-boundary, the set of all points with at least one coordinate an end point of the coordinate interval. The pseudo-boundary of Q is dense in Q as is the pseudo-interior, the complement of the pseudo-boundary. Indeed, unlike a finite dimensional cube, Q is *homogeneous*, any point is like any other point. The fact that higher indexed coordinate factors must be regarded as small, makes it possible to pass many problems down the coordinate scale to infinity, which cannot be done in finite dimensional cubes.

The question about Q and $Q \times T$ supposedly dates back to Borsuk in the early half of the 20th century. The author had never thought deeply about such a question and had to develop special techniques for dealing with infinite dimensional topological objects. In the summer of 1962, he and his family had rented a house near Bath, England and while his wife and children went to Ireland, he paced the floor and the yard thinking about the problem and finally proved that Q and $T \times Q$ are homeomorphic (for a generalization, see [1]). This got him and others started. Later during Christmas vacation in Baton Rouge in 1964, he showed in [2] that ℓ_2 and s (the countable infinite product of lines) were homeomorphic, settling a question believed to have been posed originally by Fréchet. Thus infinite-dimensional topology was off and running. The intuition and techniques leading to these results were rather different than finite-dimensional topological intuition and techniques and led a number of geometric topologists to many new and sometimes surprising results.

A useful infinite-dimensional topological concept introduced early was that of a Z-set. A Z-set in s, thought of as the product of open intervals and thus as the pseudo interior of the Hilbert cube, was a closed subset Z of s (as a space) which was negligible, i.e. s was homeomorphic to $s \setminus Z$ by an arbitrarily small homeomorphism. In s as a subset of Q, an arbitrarily small motion could push Z off s to the pseudo boundary of Q, the set of all points of Q not in s. Any countable union of Z-sets in s was also negligible.

Among the early contributors to or stimulators of infinite-dimensional topology were the Polish mathematicians, Bessaga and Pełczyński, Henryk Toruńczyk, and Andrew Lelek,

2954 Fritchie Drive, Baton Rouge, LA 70809, USA; E-mail: randerso@regents.state.la.us

RECENT PROGRESS IN GENERAL TOPOLOGY II
Edited by Miroslav Hušek and Jan van Mill

and David Henderson (a student of Bing) and several of my doctoral students at LSU, including Raymond Wong, Jim West, and Tom Chapman. In 1970, I gave an invited address at the ICM (in Nice), as did Chapman, West, and Toruńczyk in the next three ICM's.

A major factor in the rather rapid growth of infinite dimensional topology was the occurrence of ICM's every four years and the presence of frequent national and international conferences on topology including the Prague Symposium starting in 1961 and repeated every five years. In the U.S. there were almost yearly NSF sponsored general and geometric topology conferences with lots of bright young people participating. The stimulating effect of the opportunity to share thoughts and results at frequent conferences is illustrated by a topology conference in Hercig Novi in southern Yugoslavia in 1968, which David Henderson and I both attended. As we crossed the Bay of Kotor on an excursion, I recall posing the issue to him as to whether any ℓ_2-manifold could be openly embedded in ℓ_2. Dave proved the open-embedding theorem a short time later, [3]. In the summer of 1970, several of us organized and held a special infinite dimensional topology conference at Oberwolfach.

In the latter 1970's, Henryk Toruńczyk gave an elegant characterization and classification of infinite dimensional manifolds in [4], [5], which in some sense settled a number of open questions and was probably the culmination of almost 20 years of very active work by many people. It was the basis for his invited ICM talk in the early 1980's.

References

[1] Anderson, R. D., *The Hilbert cube as a product of dendrons*, Notices Amer. Math. Soc. **11** (1964), 572.

[2] Anderson, R. D., *Hilbert space is homeomorphic to the countable infinite product of lines*, Bull. Amer. Math. Soc. **72** (1966), 515–519.

[3] Henderson, D. W., *Infinite-dimensional manifolds are open subsets of Hilbert space*, Topology **9** (1970), 312–318.

[4] Toruńczyk, H., *On CE-images of the Hilbert cube and characterizations of Q–manifolds*, Fund. Math. **106** (1980), 31–40.

[5] Toruńczyk, H., *Characterizing Hilbert space topology*, Fund. Math. **111** (1981), 247–262.

Topological Combinatorics: A Peaceful Pursuit

W. W. Comfort

The diversion of science to destructive purposes is as old as science itself. The observation that a club, knife, spear or arrow useful to bring down an animal could be used to similar effect against "the enemy" in the next village was readily perverted to the design of weapons strictly for anti-personnel purposes. Even while enhancing the human condition, science has degraded and brutalized it—early on with the misappropriation of potentially neutral or useful discoveries (gunpowder, the airplane) and more recently and egregiously by the development of malevolent technologies dedicated exclusively to killing: rockets, the H-bomb, biological warfare. The history of science is replete with examples of bright intellects in three categories: those who, driven either by a savage spirit or by ambition, lead and participate willingly in the project to kill; those who participate minimally and without enthusiasm; and those who refuse, often at personal risk. (For a carefully researched, illuminating account of the actions of German mathematicians in the first half of the twentieth century, see [11]. I know of no parallel comparably detailed study of a significant sector of the scientific community in the United States, where pressures to participate were much less strong and penalties for defiance commensurably less severe.)

The Irish number theorist Henry J. S. Smith is said [10] to have proposed this toast, perhaps on the occasion (1874) of his inauguration as President of the London Mathematical Society: "Here's to Mathematics—may she never be of use to anybody." In contrast to Smith, who evidently recognized at least the possibility of the (mis)use of mathematics, G. H. Hardy [5] was able, remarkably, to write in 1940 that "real mathematics has no effects on war . . . Mathematics is . . . a harmless and innocent occupation." Perhaps one may excuse this inexcusable pronouncement on the grounds that Hardy, surely a "real mathematician" by any workable standard, had somehow satisfied himself that the pertinent moral and ethical issues could be safely declared irrelevant, in effect defined out of existence. To him, applicable mathematics is disjoint from the "real"; indeed, "real mathematics must be justified as art if it can be justified at all" [5].

If the knowable universe it not quite Euclidean, surely it is metrizable and σ-compact, hence separable (and Hausdorff). This puts an upper bound (perhaps $\mathfrak{c}$, or $2^{\mathfrak{c}}$ to be safe) on the set of cardinal numbers which may reasonably be associated with any real-world problem. Thus a set-theoretic topologist working with cardinal invariants is almost surely living a professional life consistent with the familiar dictum of Hippocrates (which does not, incidentally, appear in his famous Oath) "First, do no harm." As a goal in itself, to "do no harm" seems pitifully unproductive and defensive, but as a point of departure it is an admirable counsel. To an emerging scholar approaching mathematical pubescence and seeking a rich and challenging outlet for emerging creative energies, I can think of no path less susceptible to deflection into destructive purposes than infinitary combinatorics. (For centuries, prime numbers were viewed deservedly as the purest of the pure. But Number Theory lost its innocence some decades ago with the advent of computer-driven encryption.)

Dept. of Math., Wesleyan University, Middletown, CT 06459, USA; E-mail: wcomfort@wesleyan.edu

RECENT PROGRESS IN GENERAL TOPOLOGY II
Edited by Miroslav Hušek and Jan van Mill

The price of clean hands is admittedly nontrivial. According to the logic here being advanced, the cost of leaving the world uncorrupted is to leave it unimproved.

It would stretch the truth beyond credibility to assert that set-theoretic topology *is* combinatorics. But examples abound which establish the suggestion that combinatorics is the bedrock upon which much of topology lies. I select here six combinatorial principles (abbreviated **CP**), all in ZFC, with in each case one or more topological consequences (**TC**). I use standard topological notation as given in the references cited, as well as these conventions. Always η is an ordinal, α and κ are infinite cardinals, and $U_\kappa(\alpha)$ is the set of κ-uniform ultrafilters on α, that is, those $p \in \beta(\alpha)$ for which $A \in p \Rightarrow |A| \geq \kappa$; we write $U(\alpha) := U_\alpha(\alpha)$. For a space $X = (X, \mathcal{T})$, $S(X)$ is the *Souslin number* of X, that is, the least cardinal not the cardinality of a cellular family in X; and X_α is X with the topology generated by $\{\cap\mathcal{U} : \mathcal{U} \in [\mathcal{T}]^{<\alpha}\}$. For a set $\{X_i : i \in I\}$ of spaces, X_I denotes $\Pi_i X_i$ with the usual product topology.

Must an ω-indexed family $\{A_n : n < \omega\}$ of finite sets have an infinite subfamily whose pairwise intersections coincide? Surely not. It is enough to take $A_n = n = \{0, 1, \cdots, n-1\}$; here, no three sets have pairwise identical intersection. That trivial example introduces and lends interest to the following combinatorial principle.

CP1 [Erdős and Rado] (1960, 1969). If $[\beta < \alpha \text{ and } \lambda < \kappa] \Rightarrow \beta^\lambda < \alpha$, then for every α-indexed family $\{A_\eta : \eta < \alpha\}$ of sets of cardinality $< \kappa$ there is a subfamily $\{A_\eta : \eta \in A\}$ with $A \in [\alpha]^\alpha$ whose pairwise intersections coincide; and conversely.

TC1 $S(X_I) = \sup\{S(X_F) : F \in [I]^{<\omega}\}$.

CP2 [Erdős and Rado] (1956), [Kurepa] (1959). $(\beth_n(\alpha))^+ \to (\alpha^+)^{n+1}_\alpha$; in particular, $(2^\alpha)^+ \to (\alpha^+)^2_\alpha$.

TC2.1 If each $S(X_i) \leq \alpha^+$, then $S(X_F) \leq (2^\alpha)^+$ ($|F| < \omega$), and hence $S(X_I) \leq (2^\alpha)^+$ by **TC1**.

So in particular, the anomalous behavior with respect to cellularity of a Souslin Line L is "bounded". One has $\omega^+ = S(L) < S(L \times L)$, but $\mathfrak{c}^+ = (2^\omega)^+$ is a universal upper bound for all numbers of the form $S(X_0 \times X_1)$ where the spaces X_i satisfy the countable chain condition. The world of mathematics is sometimes hectic, but it is not totally out of control.

TC2.2 A compact space X with $S(X) \leq \alpha^+$ satisfies $S(X_{\alpha^+}) \leq (2^\alpha)^+$.

Similarly here, upon adjoining all "$G_{\leq\alpha}$" sets to the topology of X, the "Souslin jump" is limited to a single exponentiation.

The remarkable and unexpected improvements of **TC2.1** and **TC2.2** achieved by Negrepontis and his school in the period 1980 $\pm$ 2, using combinatorial generalizations of **CP1** and **CP2**, seem inadequately known and appreciated: When $Y = X_I$ (or even $Y = (X_I)_{\alpha^+}$) in **TC2.1** or $Y = X_{\alpha^+}$ in **TC2.2**, then not only is $S(Y) \leq (2^\alpha)^+$ but $(2^\alpha)^+$ is a pre-calibre for Y. That means: Not only is no $(2^\alpha)^+$-indexed family $\{U_\eta : \eta < (2^\alpha)^+\}$ of nonempty open subsets of Y a pairwise disjoint family, but each such indexed family has a subfamily $\{U_\eta : \eta \in A\}$ with $|A| = (2^\alpha)^+$ with the finite intersection property.

We remark in passing that although a compact topological group G necessarily satisfies $S(G) \leq \omega^+$, this question of van Douwen apparently remains unsolved: Is there an upper bound on numbers of the form $S(X)$ with X a compact, homogeneous space? Is $\mathfrak{c}^+$ such a number?

CP3 Given $|S| = 2^\alpha$, there is $D \subseteq \alpha^S$ with $|D| = \alpha$ such that: for each $f \in \alpha^S$ and $F \in [S]^{<\omega}$ there is $g \in D$ such that $g|F = f|F$.

Here **CP3** is, of course, a statement in combinatorial language of the topological fact that the product of 2^α-many copies of the discrete space α has a dense subset of cardinality α. This is turn yields the Hewitt-Marczewski-Pondiczery theorem:

TC3.1 If $|I| \leq 2^\alpha$ and each $d(X_i) \leq \alpha$ ($i \in I$), then $d(X_I) \leq \alpha$.

The relation between **CP3** and topology is symbiotic, in that **CP3** itself is provable from elementary topological properties of the space $S = \{0,1\}^\alpha$. There follows in turn a familiar topological corollary to **TC3.1**.

TC3.2 [Marczewski] (1941) for $\alpha = \omega$, X_i metrizable; [Shanin] (1946). If each $d(X_i) \leq \alpha$ ($|I|$ arbitrary), then $S(X_I) \leq \alpha^+$. Hence each $f \in C(X_I, \mathbb{R})$ factors as $f = g \circ \pi_J$ for suitable $J \in [I]^{\leq\alpha}$, $g \in C(X_J, \mathbb{R})$.

The special case of **TC3.1** asserting that $[0,1]^\mathfrak{c}$ (alternatively, $\mathbb{R}^\mathfrak{c}$) is separable, always a surprise to undergraduates, is rendered transparent by Mrówka [8] *via* the observation that there are (only) countably many polynomials with rational coefficients. **TC3.3** with $\alpha = \omega$ then follows. Similarly, Oxtoby [9] contributed a nifty *ad hoc* proof that $S(X_I) \leq \omega^+$ when each X_i is separable: One easily associates with each $X_i = (X_i, \mathcal{T}_i)$ a probability measure μ_i such that $\mu_i(U) > 0$ whenever $\emptyset \neq U \in \mathcal{T}_i$, and since the product measure assigns positive measure to each basic open set in X_I, each cellular family there is countable. My efforts to find a generalization of either of these special arguments to the full statements given in **TC3.1** and **TC3.2** respectively have been unsuccessful.

When Čech (1937) described the fundamental properties of $\beta(X)$, whose existence he generously attributed to Tychonoff (1929/30), he left the cardinal number $|\beta(\alpha)|$ undetermined. Since **TC3.1** shows that $\beta(\alpha)$ maps continuously onto $\{0,1\}^{2^\alpha}$, **CP3** gives the answer.

TC3.3 [Hausdorff] (1936), [Pospíšil] (1937). $|\beta(\alpha)| = 2^{2^\alpha}$.

CP4 [Erdős] (1934). Given $|S| = \alpha^{<\kappa} := \Sigma_{\lambda<\kappa}\alpha^\lambda$, there is $\mathcal{A} \in [\mathcal{P}(S)]^{\alpha^\kappa}$ such that $A_0 \neq A_1$ (in $\mathcal{A}$) implies $|A_0 \cap A_1| < \kappa$. In particular [Tarski] (1928), there is $\mathcal{A} \in [\mathcal{P}(\alpha)]^{\alpha^\omega}$ such that $A_0 \neq A_1$ (in $\mathcal{A}$) implies $|A_0 \cap A_1| < \omega$.

TC4 $S(\beta(\alpha)\backslash\alpha) = (\alpha^\omega)^+$; $S(U_\kappa(\alpha^{<\kappa})) = (\alpha^\kappa)^+$; $S(U(\alpha)) = (2^\alpha)^+$ when $\alpha = 2^{<\alpha}$; and $S(U_{\alpha^+}(2^\alpha)) = (2^{(\alpha^+)})^+$.

In the following principle, a set $\mathcal{A}$ chosen maximal with respect to the relevant property is shown, by an argument reminiscent of Cantor's famous diagonalization proof that $|\mathbb{R}| > \omega$, not to have cardinality α. Choosing $\mathcal{A}$ so that $|\mathcal{A}| \geq \alpha$, one concludes that $|\mathcal{A}| > \alpha$, a strict inequality reflected in the strict inequalities of **TC5.1** and **TC5.2**. In the absence of additional hypotheses, for example GCH, the more precise (and desirable) result $|\mathcal{A}| = 2^\alpha$ in **CP5** is not available. Indeed Baumgartner [1] gives a model in which every $\mathcal{A}$ as in **CP5** (with $\alpha = \omega^+$) has $|\mathcal{A}| < 2^{(\omega^+)}$ (and hence $S(U(\omega^+)) \leq 2^{(\omega^+)}$).

CP5 There is $\mathcal{A} \in [\alpha^\alpha]^{\alpha^+}$ such that $f \neq g$ (in $\mathcal{A}$) implies $|\{\eta \in \alpha : f(\eta) = g(\eta)\}| < \alpha$. Equivalently: there is $\mathcal{A} \in [\mathcal{P}(\alpha)]^{\alpha^+}$ such that $A_0 \neq A_1$ (in $\mathcal{A}$) implies $|A_0 \cap A_1| < \alpha$.

TC5.1 $S(U(\alpha)) > \alpha^+$.

TC5.2 [Frayne, Morel and Scott] (1962) For $p \in U(\alpha)$, the ultrapower α^α/p satisfies $|\alpha^\alpha/p| > \alpha$.

A principle related to **CP5**, due to Sierpiński (1949) and developed in [1] and with many topological consequences, is that the partially ordered set $(\mathcal{P}(2^{<\alpha}), \subseteq)$ contains a chain of length 2^α; hence $\mathcal{P}(\alpha)$ contains a chain of length α^+.

CP6 (The Disjoint Refinement Lemma). For each α-indexed family $\{A_\eta : \eta < \alpha\}$ of sets, with each $|A_\eta| = \alpha$, there are pairwise disjoint sets $B_\eta \in [A_\eta]^\alpha$.

Proving **CP6** is a nice training exercise in well-ordering and the use of ordinals at the introductory graduate level. As with **CP1**, an appealing feature of the principle is its triviality at the extremes: When the given sets A_η coincide or are pairwise disjoint. The applications of **CP6** touch many topological subdisciplines. For a space $X = (X, \mathcal{T})$ write $\Delta(X) = \min\{|U| : \emptyset \neq U \in \mathcal{T}\}$, and recall that X is *maximally resolvable* if it has $\Delta(X)$-many pairwise disjoint dense subsets.

TC6.1 (Ceder [2]) If $\pi w(X) \leq \Delta(X)$, then X is maximally resolvable.

TC6.2 (a) $|U(\alpha)| = 2^{2^\alpha}$, and (b) $p \in U(\alpha) \Rightarrow \chi(p, U(\alpha)) > \alpha$.

Another appeal to **CP3** strengthens **TC6.2**(a):

TC6.3 [Pospíšil] (1939) $|\{p \in U(\alpha) : \chi(p, U(\alpha)) = 2^\alpha\}| = 2^{2^\alpha}$; indeed [Juhász] (1969) $|\{p \in U(\alpha) : \chi(p, U(\alpha)) < 2^\alpha\}| < 2^{2^\alpha}$.

Summary. Many aspects of general (set-theoretic) topology relating to cardinal invariants are rooted in infinitary combinatorics. In a world where benign science is too easily turned to malign purposes, this direction of inquiry is recommended as a safe haven, invitingly free of such applications.

Bibliographic Remarks. **CP1** and **CP2** are discussed and proved in full, together with the consequences mentioned here and many others, in [6] and [7]. Somewhat the same material is addressed in [3] and [4], as are **CP3–6** and a wealth of other combinatorial principles and topological consequences. The bibliographic citations given in truncated form above are available in [3] and [4] in the familiar extended format.

References

[1] Baumgartner, J.E., *Almost-disjoint sets, the dense set problem and the partition calculus*, Ann. Math. Logic **10** (1976), 401–439.

[2] Ceder, J. G., *On maximally resolvable spaces*, Fund. Math. **55** (1964), 87–93.

[3] Comfort, W. W. and S. Negrepontis, *The Theory of Ultrafilters*. Springer-Verlag, Berlin, 1974.

[4] Comfort, W. W. and S. Negrepontis, *Chain Conditions in Topology*. Cambridge University Press, Cambridge, 1982.

[5] Hardy, G. H., *A Mathematician's Apology*. Cambridge University Press, Cambridge, 1940.

[6] Juhász, I., *Cardinal Functions in Topology*. Mathematisch Centrum, Amsterdam, 1971.

[7] Juhász I., *Cardinal Functions in Topology—Ten Years Later*. Mathematisch Centrum, Amsterdam, 1980.

[8] Mrówka, S., *On the potency of subsets of βN*, Colloq. Math. **7** (1959), 23–25.

[9] Oxtoby, J.C., *Cartesian products of Baire spaces*, Fund. Math. **49** (1961), 157—166.

[10] Rosenbaum, R.A., *Another vicious versus*, an address to the Freshman class, pp. 1–8. Wesleyan University Press, Connecticut, 1982.

[11] Segal, S. L., *Topologists in Hitler's Germany*, In: History of Topology (I. M. James, ed.), pp. 849–861. Elsevier Science B. V., Amsterdam, 1999.

Topology Related to Rings of Real-Valued Continuous Functions. Where it has been and where it might be going

M. Henriksen

There is no book of Genesis for the subject of the title, and even the earliest papers devoted to it depended on the work of predecessors less than completely aware that they were helping to start a new area. The first clearly recognizable contribution to this subject is M.H. Stone's monumental *Applications of Boolean rings to general topology* [St] published in 1937, and a second prize close enough to be considered a tie must be awarded to E. Čech for [Ce]. Both of these authors depended on the work of Tychonoff [T], who showed that a space is a completely regular Hausdorff space if and only if it is a closed subspace of a product of copies of the closed interval [0,1], and that this latter product is compact. What is now called the product topology in the case of infinite products appears for the first time in [T], but the fact that an arbitrary product of compact spaces is compact, which is usually called the *Tychonoff theorem,* appeared first in [Ce]. As was pointed out by A. Shields in [Sh], what we call the Stone-Čech (in the United States and the Čech-Stone in Europe) compactification βX of a completely regular Hausdorff space (now called a *Tychonoff space)* X was not considered important enough by M.H. Stone to mention in the introduction of [St], and its algebraic significance is not discussed in [Ce]. The first definite development of the maximal compactification βX as the space of maximal ideals of the algebra $C^*(X)$ of bounded real-valued continuous functions appears in [GK] in 1939. The first paper devoted completely to our subject was E. Hewitt's monumental paper [Hew1] This brilliant paper sent the subject on its true course despite being marred by a number of serious errors. At this point, Hewitt withdrew from the field, and concerned himself mostly with functional analysis. He did not re-enter the area again until 1976 in the form of a high quality paper written with three co-authors was concerned with residue class fields of $C(X)$ mod maximal ideals. See [ACCH]. This was his last contribution to the field. He made many others to both topology and analysis. See the memorials dedicated to Hewitt written by W.W. Comfort and K. Ross in Topological Commentary, a part of Topology Atlas that can be found on the internet at the url: http://at.yorku.ca/topology/.

A lot of papers were written on this subject in the 1950s, and the state of the art up until 1960 is summarized in the Gillman-Jerison [GJ] published first in 1960. Reading this excellently written book is a must for anyone wishing to do research in this area. Published now by Springer-Verlag, it remains in print after 42 years. Missing definitions in what follows may be found in [GJ]. It helped to create another wave of publication that continued more or less through the mid 1980's summarized incompletely in [Hen1] and in [V1]. In the interim, three books appeared [Wa] on the Stone-Čech compactification, [We] on realcompactness (alias Hewitt-Nachbin completeness), and [PW], while concerned primarily with H-closed extensions, contains a large amount of material pertinent to the theme of this article. A conference dedicated to the forthcoming 25th anniversary

Harvey Mudd College, Claremont CA 91711, USA; E-mail: Henriksen@hmc.edu

RECENT PROGRESS IN GENERAL TOPOLOGY II
Edited by Miroslav Hušek and Jan van Mill

of the publication of [GJ] was held in 1982 in Cincinnatti Ohio. Its proceedings appeared in 1985; see [A].

Since that time, the volume of papers on rings of continuous functions in journals devoted mainly to general topology has diminished in the United States while increasing in some other countries. The scope of this article is too small to permit a more up to date survey, but will permit the mention only of a few sample papers. For example, work in this area has been done recently in Iran (see, for example, [AKR]), in Russia (see, for example, [V2] and [Z]), and in Spain (see, for example, [BM] and [GM]).

The hope that theorems in the main stream of general topology coupled be proved more easily by studying properties of rings of real-valued continuous functions is dashed by translation difficulties. Even though the ring of continuous functions on a realcompact space (one homeomorphic to a closed subspace of a product of real lines), "nice" topological properties translate into complicated ones algebraically, and vice versa. Also, the ring $C(X)$ cannot distinguish between a Tychonoff space X and its Hewitt real compactification υX, a distinction more naturally algebraic than topological. The latter is mitigated in part because metrizable and Lindelöf spaces are realcompact as are arbitrary products of realcompact spaces, but still creates difficulties. While this disappoints those mainly concerned with "pure" topology, it is intriguing to those of us fascinated by the translation process, and enables one to apply topological techniques to the study of certain kinds of ordered algebraic systems. For example, characterizing rings of all continuous functions on various classes of topological spaces within certain classes of lattice-ordered rings and algebras that are subdirect products of totally ordered rings or algebras. These are almost always called *f-rings.* An (incomplete) survey of activity in this area is given in [Hen2] where connections with other areas such as real semi-algebraic geometry, homological algebra, spectral theory, point-free topology, and non-standard analysis are given at least brief mention. In addition, it is of interest to set-theorists that answers to many questions which arise in the study of $C(X)$ seem to be independent in Zermelo-Fraenkel set theory together with the axiom of choice [ZFC]. For example, in [vDvM], it is shown that the Parovičenko characterization of $\beta\omega \setminus \omega$ depends essentially on the continuum hypothesis, and in [Wi], E.Wimmers give an exposition of how S. Shelah and W. Rudin show that whether there are P-points in $\beta\omega \setminus \omega$ cannot be determined in [ZFC].

Those of us who pursue the study of rings of real-valued continuous functions does not seem to have a happy home anywhere. The hypotheses of its theorems often seem unnatural both to topologists for the reasons given above and to commutative algebraists because they usually fail to obey "natural" chain conditions, and because any such integral domain is a field. Reducing mod a prime or maximal ideal yields either the real field or objects full of set-theoretic difficulties. Set-theorists examine questions in this area primarily as a possible source for undecidable problems. Functional analysts will not admit us to their fraternity because we ask and answer different questions about $C(X)$ from the ones they do, and are often not concerned about topologies on $C(X)$ itself.

Mathematicians usually admire depth over breadth, and few of us have the ability to prove deep theorems that require skills in several fields to understand. Despite this, aficionados of this kind of endeavor continue to persist. Going to Full Search on Math. Sci. Net and entering "Rings of continuous functions" in the slot labelled Anywhere yielded 439 entries at the end of January 2002, many of which appeared in the last decade. There are, without doubt, many more papers related to this subject to be found using different key words. So, while examining the interplay between a Tychonoff space X and the ring $C(X)$ may not yield instant fame, it will continue to fascinate enough mathematicians to keep it alive and well for some time to come.

References

[ACCH] Antonovskij, A., D. Chudnovsky, G. Chudnovsky, and E. Hewitt, *Rings of real-valued continuous functions II,* Math. Zeit. 176 (1981), 151-186

[A] Aull, C.E., *Rings of Continuous Functions,* Lecture notes in Pure and Applied Mathematics 95, Marcel Dekker Inc., New York 1985

[AKR] Azarpanah, F., O. Karamzadeh, A. Rezai-Aliabad, *On z^0-ideals in C(X),* Fund. Math. 160 (1999), 15–25.

[C] Čech, E., *On bicompact spaces,* Ann. of Math. 38 (1937), 823-844.

[BM] Bustamante, J. and F. Montalvo, *Stone-Weierstrass theorems in C*(X),* J. Approx. Theory 107 (2000), 143–159.

[GM] Garrido, I. and F. Montalvo,. *Algebraic properties of the uniform closure of spaces of continuous functions,* Papers on general topology and applications (Amsterdam, 1994), 101–107, Ann. New York Acad. Sci. 788, New York Acad. Sci., New York, 1996

[GK] Gelfand, I. and A. Kolmogoroff, *On rings of continuous functions on topological spaces,* Dokl. Akad. Nauk SSSR 22 (1939), 11-15

[GJ] Gillman, L. and M. Jerison, *Rings of Continuous Functions,* D. Van Nostrand Publ. Co., Princeton N.J. 1960.

[Hen1] Henriksen, M., *Rings of continuous functions from an algebraic point of view,* Ordered Algebraic Structures, Kluwer Academic Publishers, Dordrecht 1989, 144-174.

[Hen2] Henriksen, M., *A survey of f-rings and their generalizations,* ibid. 1997, 1-26

[Hew] Hewitt, E., *Rings of real-valued continuous functions I,* Trans. Amer.Math. Soc. 64 (1948), 54-99.

[M] Mulero, M., *Algebraic properties of rings of continuous functions*, Fund. Math. 149 (1996), 55–66

[PW] Porter, J. and R.G..Woods, *Extensions and Absoluters of Hausdorff Spaces,* Springer-Verlag, New York 1987.

[Sh] Shields, A., *Years ago,* Math. Intelligencer 9 (1987), 61-63.

[St] Stone, M.H., *Applications of the theory of Boolean rings to general topology,* Trans. Amer. Math. Soc. 41 (1937), 375-481.

[T] Tychonoff, A., *Über die topologische Erweiterung von Räume,* Math. Ann. 102 (1929), 544-561.

[V1] Vechtomov, E., *Rings of continuous functions, Algebraic aspects,* (Russian), Itogi Nauki i Tekhniki, Algebra. Topology. Geometry, Vol. 29 , 119–191, Akad. Nauk SSSR, Vsesoyuz. Inst. Nauchn. i Tekhn. Inform., Moscow, 1991. (Translated in J. Math. Sci. 71 (1994), no. 2, 2364–2408),

[V2] Vechtomov, E., *Rings of continuous functions with values in a topological division ring*, Topology, 2. J. Math. Sci. 78 (1996), 702–753.

[Wa] Walker, R., , *The Stone-Čech Compactification*, Springer-Verlag, New York, 1975.

[We] Weir, M., *Hewitt-Nachbin Spaces.* North Holland Publishing Company, Amsterdam 1975.

[Wi] Wimmers, E., *The Shelah P-point independence theorem.* Israel J. Math. 43 (1982), 28–48

[Z] Zakharov, V., *Classical extensions of the ring of continuous functions and the corresponding preimages of a completely regular space*, Algebra. 1. J. Math. Sci. 73 (1995), 114–139

Shape Theory

Sibe Mardešić

Topology, as we know it today, was founded around the beginning of the past century. By the end of the century it constituted an impressive and essential body of mathematics. Looking back at its development, we see general and algebraic topology as two of its main branches. Of its many subfields this author found particular interest in shape theory, an area where general and algebraic topology meet. The scope of shape theory is to extend homotopy theory, a vital part of algebraic topology which studies global properties of CW-complexes, to the study of analogous properties of more general spaces, especially metric compacta, which are objects of study of general topology. An extension of the theory to such spaces should not be neglected, because they appear naturally in many areas of mathematics. Examples are provided by fibers of mappings, sets of fixed points, boundaries of groups, attractors of dynamical systems, spectra of linear operators, fractal sets, etc.

When H. Poincaré founded algebraic topology in his basic paper on Analysis situs (1895), he still used intuitive arguments concerning manifolds and homology, just as did his predecessors B. Riemann and E. Betti. Efforts to give a rigorous treatment of this topic led to simplicial complexes and raised the problem of proving the topological invariance of their homology. J.W. Alexander solved the problem by introducing singular homology (1915). It applied to arbitrary topological spaces X and was based on considering mappings of polyhedra P (in particular, simplices) into X. The same idea is at the basis of homotopy theory, because the homotopy groups $\pi_n(X,*)$ are defined by considering mappings of the n-sphere S^n to the pointed space $(X,*)$. However, this approach is not adequate if the local behavior of X is not sufficiently regular.

This explains why at the basis of shape theory is the dual approach, which consists in considering mappings of spaces into polyhedra. This approach has its origins in the early work of P.S. Alexandroff on inverse systems and nerves $N(\mathcal{U})$ of coverings $\mathcal{U}$ of spaces X (1926, 1927) and on Kuratowski's notion of canonical mappings $\phi\colon X \to N(\mathcal{U})$ (1933). Therefore, Čech homology, introduced in various forms by P.S. Alexandroff (1927), L. Vietoris (1927), E. Čech (1932) and others, can be considered as the beginning of shape theory.

The essential step in founding modern shape theory was done by K. Borsuk in his basic paper on the homotopy properties of compacta (1968), where he defined the shape category Sh(CM). Its objects are all metric compacta X, embedded in the Hilbert cube I^ω, and the morphisms $F\colon X \to Y$ are defined using particular sequences of mappings $f_n\colon I^\omega \to I^\omega$, called fundamental sequences. Two compacta are said to be of the same shape provided they are isomorphic objects in Sh(CM). Borsuk also defined the shape functor $S\colon H(CM) \to Sh(CM)$, whose domain is the homotopy category H(CM) of metric compacta X. In the case when Y is a polyhedron (or an ANR), every shape morphism $F\colon X \to Y$ is of the form $F = S[f]$, where the homotopy class $[f]\colon X \to Y$ is unique.

Dept. Math., University of Zagreb, 41001 Zagreb, Croatia; E-mail: smardes@bde.math.hr

RECENT PROGRESS IN GENERAL TOPOLOGY II
Edited by Miroslav Hušek and Jan van Mill

Borsuk's theory was readily extended to metric spaces (R. Fox 1972) and to arbitrary topological spaces (S. Mardešić 1973). The objects of the shape category Sh(Top) are topological spaces and the morphisms $F\colon X \to Y$ are functions which with every homotopy class $[\psi]\colon Y \to P$ to a polyhedron P associate a homotopy class $[\phi]\colon X \to P$ in such a way that, whenever $[\psi']\colon Y \to P'$ and $[q]\colon P' \to P$ are homotopy classes such that $[\psi] = [q][\psi']$, then $[\phi] = [q][\phi']$. In 1970 S. Mardešić and J. Segal noticed that polyhedral and ANR-inverse systems represent the right tool for founding shape theory. Their approach, originally applicable to compact Hausdorff spaces, was extended to arbitrary spaces by K. Morita (1975). The general philosophy consisted in associating with every space X adequate inverse systems $\mathbf{X} = (X_\lambda, p_{\lambda\lambda'}, \Lambda)$ in the topological or the homotopical category of polyhedra and of generalizing homotopy of polyhedra to a homotopy of polyhedral systems. The freedom in the choice of the systems $\mathbf{X}$ proved to be of great practical value.

In the next ten years, under Borsuk's leadership, shape theory attracted a large number of researchers all over the world. They successfully extended a number of basic results of homotopy theory of CW-complexes to analogous results in shape theory. These results applied to spaces not subject to any local regularity conditions and included the classification of overlays (the shape-theoretic version of covering mappings) (R.H. Fox 1972), the Hurewicz and the Whitehead theorems (M. Moszyńska 1973, K. Morita 1974, J.E. Keesling 1976), Freudenthal's suspension theorem (Š. Ungar 1976), the Vietoris-Smale theorem (J. Dydak 1979), etc. Of course the standard tools of algebraic topology, like the homology and the homotopy groups, had to be replaced by adequate shape-theoretic notions, i.e., by homology progroups $\mathbf{H}_n(\mathbf{X}) = (H_n(X_\lambda), p_{\lambda\lambda'*}, \Lambda)$ and homotopy progroups $\underline{\pi}_n(\mathbf{X}) = (\pi_n(X_\lambda), p_{\lambda\lambda'\#}, \Lambda)$, respectively.

Closer to ideas of general topology were the newly introduced shape invariant classes of metric compacta, in particular, the class of fundamental absolute neighborhood retracts (FANR's) and the broader class of movable and n-movable compacta (Borsuk 1969). FANR's generalize compact ANR's and are their shape analogues. Connected FANR's coincide with metric continua which have the shape of a (possibly noncompact) ANR (D.A. Edwards and R. Geoghegan 1976),

A different direction in shape theory was inaugurated by T.A. Chapman in 1972. Using the theory of I^ω-manifolds, he proved that two Z-sets in I^ω have the same shape if and only if their complements are homeomorphic. Analogous complement theorems were also obtained for compacta suitably embedded in euclidean spaces (I. Ivanšić, R.B. Sher and G. Venema 1981). Embedding compacta in euclidean spaces up to shape is another topic of shape theory with strong geometric flavor (L.S. Husch and I. Ivanšić 1981). Other useful topics studied at that time were approximate fibrations (D.S. Coram and P.F. Duvall, 1977), shape fibrations (S. Mardešić and T.B. Rushing 1978; T. Yagasaki 1986) and shape dimension (S. Nowak 1981, S. Spież 1983).

Beside ordinary shape theory, a finer and more geometric theory was founded. It assumes an intermediate position between homotopy and ordinary shape. It was first defined for metric compacta (J.B. Quigley 1973, D.A. Edwards and H.M. Hastings 1976). It turned out that the founding of a strong shape theory for topological spaces requires rather sophisticated techniques (localization, coherent homotopy). Its development has been of interest to a number of shape-theorists during the last twenty years (F.W. Bauer, Yu.T. Lisitsa and

S. Mardešić, B. Günther, J. Dydak and S. Nowak).

There are several areas of general topology where shape theory has already proved to be very useful. In continua theory and in the theory of hyperspaces, 1-movability plays an important role (J. Krasinkiewicz 1978, H. Kato 1986). Shape theory was used in an essential way in the study of various compactifications (J. Keesling 1987). It found applications also in the theory of fixed points (K. Borsuk 1975; J. Segal and T. Watanabe 1992; Z. Čerin 1993).

Shape theory suggested new techniques of approximating spaces by polyhedra. It is well known that compact spaces can be represented as inverse limits of compact polyhedra. This most elegant way of defining complicated examples has been amply used by general topologists. It suffices to recall the standard definitions of the Cantor set or the dyadic solenoid. We now also have at our disposal approximate inverse systems of compacta (S. Mardešić and L.R. Rubin 1989). In dealing with noncompact spaces, a new tool are resolutions and approximate resolutions (S. Mardešić and T. Watanabe 1989). The latter were recently used in the theory of Lipschitz mappings and fractal dimensions (T. Miyata and T. Watanabe).

An area of mathematics closely related to general topology, where shape theory is already playing an essential role, is dynamical systems. E.g., the Conley index $h(S)$ of an isolated invariant set S of a flow is, by definition, the homotopy type of a certain space associated with S. In 1988 J.W. Robbin and D. Salamon generalized the Conley index so as to apply to differentiable flows and diffeomorphisms on smooth manifolds. Their index $s(S)$ is the shape of a certain space associated with S. In the study of the Conley index J.R.M. Sanjurjo (2000) has successfully used the shape-theoretic Lusternik-Schnirelmann category. B. Günther and J. Segal (1993) proved that a finite-dimensional compactum A is the attractor of a flow on a euclidean space if and only if it has the shape of a compact polyhedron. These are examples of important results, which can be stated only in terms of shape theory.

Shape-theoretic techniques have also proved useful in proper homotopy and the theory of ends (Z. Čerin 1978, M.L. Mihalik 1980). Suitably adapted they gave rise to a shape theory of C^*-algebras (B. Blackadar 1985; E.G. Effros and J. Kaminker 1986, M. Dadarlat and A. Némethi 1990). Strong shape theory stimulated new research concerning Steenrod ordinary and extraordinary homology, now called strong homology (D.A. Edwards and H.M. Hastings, T. Porter, A. Miminoshvili, Yu.T. Lisitsa and S. Mardešić, A.V. Prasolov).

This author believes that shape theory does have a future and will become part of the standard equipment of researchers in areas where global properties of irregular spaces matter.

References

[1] Mardešić, S. and J. Segal, History of shape theory and its application to general topology, Handbook of the History of General Topology, Aull, C.E. and R. Lowen, eds., Volume 3, Kluwer Academic Publishers, 2001, Dordrecht, The Netherlands.

Looking Back at Modern General Topology in the Last Century

Jun-iti Nagata

In the present essay we are going to review some of the topics that occurred in modern general topology.

At the beginning of history of general topology there was the era of great pioneers like F. Hausdorff, M. Fréchet, K. Kuratowski, P. Urysohn, A. Tychonoff, P. Alexandroff, E. Čech, R. Moore and so on. After the end of World War II (1945) general topology entered into the era of new developments, which we call modern general topology. What is the difference between classical general topology (old g.t. for short) and modern general topology (new g.t. for short)?

It seems to the author that, besides separation axioms, compactness and metrizability are the most basic conditions assumed for topological spaces in g.t., and among other significant topological properties are connectedness, dimension, cardinal invariants etc. In old g.t. separable metric spaces and compact spaces were often in the center of arguments while they had some difficulty in going beyond the wall of separable metrizability and compactness. The situation is well represented in the following statement quoted from the introduction of W.Hurewicz and H.Wallman's famous book Dimension Theory (1941): Throughout this book all spaces are separable metric .... This limitation is made because there arise grave difficulties in extending dimension theory to more general spaces.

In new g.t. separable metrizability is no more a condition of primary importance. Compactness is still an important condition, but it is not necessarily in the center of arguments. We should say that metrizability has taken over the position of separable metrizability, and paracompactness has almost taken over compactness' position. This remarkable change was initiated by A.H. Stone's epoch-making paper of 1948, that asserted the equivalence between paracompactness and full-normality for T_2-spaces and also paracompactness of every metric space. (Original references will be given only for less known results. For other results see some of the textbooks listed in References.) The transition of era became clear when J.Nagata & Yu.M.Smirnov's metrization theorem and R.H.Bing's metrization theorem were published (1950-'51), C.H.Dowker characterized countable paracompactness in terms of normality of product (1951), M.Katětov and K.Morita established a satisfactory dimension theory for general metric spaces (1952-'54), E.Micheal studied paracompactness and selection theory (1953-'56), and H.Tamano and K.Morita characterized paracompactness in terms of normality of product (1960-'61). Those results brought new developments of old aspects like metrization theory and dimension theory as well as the rise of new aspects like studies of generalized metric spaces, generalized paracompact spaces, normality of product spaces and so on.

A number of long-standing questions were answered in the late 20-th century, among which the most remarkable ones are: R.D.Anderson's result that Hilbert space is homeomorphic to the countable product of real lines (1966), D.Henderson's construction of an infinite-dimensional compact metric space with no positive-dimensional compact sub-

---

Uzumasa Higashiga-oka 13-2, Neyagawa, Osaka, 572-0841 Japan

RECENT PROGRESS IN GENERAL TOPOLOGY II
Edited by Miroslav Hušek and Jan van Mill

sets (1967), A.V.Arhangel'skii's positive answer to P.S.Alexandroff's problem that the cardinality of every first countable Lindelöf T_2-space is at most $\mathfrak{c}$ (1969) and M.E.Rudin's construction of a normal space whose product with $I = [0,1]$ is not normal (1971). Following Henderson and Rudin's examples many interesting pathological spaces were constructed. Anderson's result raised the new aspect of topology called infinite-dimensional topology. Arhangel'skii's work established the cardinal invariants as a subject of systematic studies. In fact results obtained on cardinal invariants before Arhangel'skii were somewhat sporadic.

In the meanwhile we should note another striking difference between old g.t. and new g.t. that exists in the foundation of arguments. Although F.B.Jones assumed CH to prove his metrization theorem (1937), people in olden times believed (or rather wanted) that most problems in g.t. could be eventually answered on ZFC basis. But this belief crumbled when many problems began to be solved on set theory assumptions like Martin's axiom, negation of CH and so on. Especially memorable was J.H.Silver and F.Tall's result that there is a separable normal non-metrizable Moore space on the assumption of negation of CH and Martin's axiom (1969), which combined with F.B.Jones's theorem implies that the existence of such a space is independent from ZFC. Nowadays problems are often discussed on consistency and independence basis. This makes a problem more complicated on one hand and easier on the other hand since one can assume various alternative answers other than yes and no in ZFC. Anyway this is a feature of new g.t.

Another distinctive feature of new g.t. is categorical method. Various topological properties like metrizability, compactness, paracompactness etc. have been generalized, and in this context categorical topology may be regarded as a generalization of theories like extensions of topological structures. This method certainly helps us to look at g.t. from a wider point of view, but the author hopes categorical terms will not be used in unnecessary circumstances.

Yet another feature of new g.t. is C_p-theory that was established by A.V.Arhangel'skii and his school around 1980 to study properties of a topological space X in relation with the topological ring (lattice and linear space) $C_p(X)$ of all real-valued continuous functions on X with the topology of point-convergence. (See Arhangel'skii's article in M.Hušek & J.van Mill [1992].) As is well-known, the ring structure of $C^*(X)$ (C(X)) of all bounded continuous functions (continuous functions) on a compact T_2-space X (realcompact space X) determines the topology of X. But if the space is not compact (not realcompact), then the same does not hold. Even if we regard $C^*(X)$ (C(X)) as a topological ring vested with the topology of uniform convergence, the situation does not change. Thus the topological ring $C^*(X)$ gives little information on topological properties of a non-compact space X. In fact we cannot tell by $C^*(X)$ if X is compact (metrizable, first countable etc.) or not. On the other hand if we regard $C(X)$ as a topological ring with the topology of point-convergence which we denote by $C_p(X)$, then it completely determines the topology of a Tychonoff space X (J.Nagata [1949]). Thus it is no wonder that C_p-theory has proved to be a very ample field, where many interesting and some astonishing results are being obtained. As an example, we can mention V.G.Pestov's result [1982] that if $C_p(X)$ and $C_p(Y)$ are linearly homeomorphic, then $\dim X = \dim Y$. This aspect of g.t. will develop further in the 21-st century. Some remarkable results were obtained also in the study of the hyperspace 2^X of all non-empty closed sets in a regular space X with the finite topology.

E.g., J.Keesling proved under CH that 2^X is normal iff X is compact, which implies that 2^X is normal iff it is compact (1970).

Now, let us turn our attention to media of communication like conferences, journals and books that enhanced the developments of g.t. in the last century. There were many conferences and symposia on g.t. in different places, Europe, Russia, U.S., Japan and so no. Perhaps the trend was initiated by the Prague Topological Symposium, whose 1-st meeting was organized by the organizing committee consisting of J.Novák (Chairman), K.Kuratowski, M.Katětov, P.S.Alexandroff, Z.Frolík and others, and the first session took place on September 1, 1961. Since then the Prague Symposium has been held every five years, and the 9-th met just in 2001. The first Prague Symposium was especially impressive, because both leading figures in the olden time like P.S.Alexandroff, K.Borsuk, K.Kuratowski and M.H.Stone and new leaders to come like R.D.Anderson, A.V.Arhangel'skii, R.H.Bing, C.H.Dowker, R.Engelking, J. de Groot, E.Hewitt, E.Michael and Yu.M.Smirnov met together in a room, which seemed to symbolize the opening of a new era.

As for journals, "General Topology and its Applications" was established by J. de Groot and others in 1971, and its title was changed in 1980 to "Topology and its Applications", where a number of significant papers on g.t. were published. In 1983 a journal specialized in g.t. was established by M.Atsuji, J.Nagata and others under the title "Questions and Answers in General Topology". This is the age of electronic communication, and no wonder "Topology Atlas" was established around 1995 by D.Shakhmatov, S.Watson and others to communicate various news on topology and publish electronic versions of proceedings of conferences and symposia. The author hopes that a global consensus will be established on the delicate issue about if an electronically published paper in Topology Atlas can be accepted by another printed journal.

A number of books were written on g.t. during the late 20-th century. Some were designed for general readers and others for specialists. Some were textbooks to cover a wider range of topics, and others concentrated on one or a few topics. Among popular textbooks for general readers are J.L.Kelly [1955], K.Kunen & J.Vaughan [1984], J.Nagata [1985] and R.Engelking [1989]. K.Morita & J.Nagata [1989] and M.Hušek & J. van Mill [1991] are not designed for general readers but cover well newer developments in various areas. Lastly "Encyclopedia of General Topology" edited by K.P.Hart, J.Nagata, J.E.Vaughan and others will be published probably in 2002 to summarize those results obtained in the last century.

We should like to conclude this essay with looking forward to g.t. in the new century. However, because the pages allowed for us are limited, just a couple of problems will be briefly mentioned while leaving detailed discussions to other occasions.

A fundamental task of general topology is to characterize topological properties of spaces, identify their topological structures and eventually classify all spaces. In other words the problems are: Find a nice property (or a collection of properties) $\mathcal{P}(X)$ of a space X such that (1) $\mathcal{P}(X)$ is equivalent to an important property of X, (2) $\mathcal{P}(X)$ is equivalent to that X is homeomorphic to a particular space (or $\mathcal{P}(X)$ characterizes the topological structure of X), or (3) all spaces belonging to certain class are topologically classified in accordance with $\mathcal{P}(X)$. It is desirable that $\mathcal{P}(X)$ be simple, practical and universal. We mean by 'practical' that the property is easy to handle. We mean by 'universal' that $\mathcal{P}(X)$'s are of similar kinds even for different kinds of spaces X's. Most of generalized metric spaces are defined or characterized in terms of base, network etc. It would be nice if they all could be characterized in terms of, say, base. (See J.Nagata [1999].)

For the author, what the term "space" means is an extensive vacancy, whose fundamental attribute consists of distance and dimension. A general topological space has no distance (metric), but still the concept of neighborhood implies a sense of nearness and farness, whose origin is the concept of distance. It seems an interesting theme to characterize topological properties of a metrizable space in terms of special metrics compatible with its topology. (See Y.Hattori & J.Nagata's article in Hušek & van Mill [1992].) Aside from classical results, a metrizable space of dim $\leq n$ was characterized by J.Nagata and P.A.Ostrand by special metrics (1958-'65). A strongly metrizable space was characterized by Y.Hattori [1986] by a special metric. Z.Balogh & G.Gruenhage [200?] also got interesting results in this aspect. Perhaps possibilities are not exhausted yet, and exploration should be pushed further (to wider classes of spaces like submetrizable spaces, too.)

As for the problem (2), we recall J. de Groot's result (1969), who characterized specific spaces like I^n and I^ω by use of a simple subbase property. It would be worthwhile to try to further that direction of study. Perhaps the problem (3) is the hardest while there are an infinite variety of topological spaces. In this aspect there are only very few examples of successful results for very small classes of spaces. But it should be tried further since the problem could be an eventual goal of general topology.

References

BALOGH, Z. AND G. GRUENHAGE

[200?] When the collection of ε-balls is locally finite, *Topology Appl.*, to appear.

ENGELKING, R.

[1989] *General Topology*, 2nd edition, Heldermann Verlag.

HATTORI, Y.

[1986] On special metrics characterizing topological properties, Fund. Math. **126**, 135–145.

HUŠEK, M. AND J. VAN MILL, ED.

[1992] *Recent Progress in General Topology*, North-Holland.

KELLY, J.L.

[1955] *General Topology*, D. van Nostrand

KUNEN, K. AND J. VAUGHAN, ED.

[1984] *Handbook of Set-Theoretic Topology*, North-Holland.

MORITA, K. AND J. NAGATA, ED.

[1989] *Topics in General Topology*, North-Holland.

NAGATA, J.

[1949] On lattices of functions on topological spaces and of functions on uniform spaces, *Osaka Math. J.* **1**, 166–181.

[1985] *Modern General Topology, 2nd revised ed.*, North-Holland.

[1999] Remarks on metrizability and generalized metric spaces, *Topology Appl.* **91**, 71–77.

PESTOV, V.G.

[1982] Coincidence of the dimension dim of l-equivalent spaces, *Soviet Math. Dokl.* **26**, 380–383.

Topology in the 20th Century

Mary Ellen Rudin

It is entertaining and perhaps humbling for a topologist to consider a subject like "Topology in the 20th Century". I received my PhD in 1949. My major professor, R. L. Moore, received his PhD in 1905. We were both topologists, actively involved in topological research as well as the wide community of mathematicians for our entire adult lives. Together we almost spanned the century. Our special topological interests were quite different but never trivial and were the specialties of many other topologists of our time from all over the world. These experiences, however, do *not* prepare me to deal with this topic.

The difficulty is that topology is not, and really never has been, one subject. Over the course of the 20th century topology has become everywhere dense in mathematics. The basic assumptions and definitions, the theorems which are considered classic and necessary for every student and educated mathematician to understand, the theorems which a particular topologist thinks are important or hopes to prove, the tools he expects to be used in proofs, the very meaning of the word topology, all vary so widely that large active groups of topologists can hardly speak to each other because their languages are so different. Topology has a remarkable talent for melding with other areas of mathematics. Two themes have really dominated 20th century mathematics: the building of complex technical structures and the erasing of boundaries between fields. Topology has played a significant roll in both of these trends.

The sheer increase in the size of the mathematical community and the improved ease of travel and communication, as well as more specialized journals and conferences, have led to substantial worldwide groups working on similar problems. These groups build more and more sophisticated and technical structures using specialized languages and tools from many other areas of mathematics, seeking applications outside of their own area. Generalized topological concepts lend themselves both as tools and applications for many areas of mathematics. The topologists working in such a group naturally use the language of the group and work on the problems basic to the group. Thus topology becomes ever more diverse and more technical.

A trivial illustration from my personal experience is found in topology seminars at the University of Wisconsin. There were 3 topologists here when I arrived 42 years ago, one algebraic, one geometric, one set theoretic. We had a weekly seminar which we all attended with our various students all of whom occasionally spoke in the seminar which was organized by R. H. Bing, the geometric topologist. Our concerns were different but our common geometric interests gave us a common language. There are now 10 topologists at Wisconsin. Topology seminars tend to be intense mini courses taught by outside experts in some specialized area and attended by 3 or 4 of the topologists and some of their students. Almost all of the seminars recently have been on topological applications to problems in physics using techniques from algebra, especially Lie algebra, differential equations, differential geometry, and, of course, some highly technical topology. All of

Dept. Math., University of Wisconsin, Madison, WI 53706, USA); E-mail: mrudin@math.wisc.edu

RECENT PROGRESS IN GENERAL TOPOLOGY II
Edited by Miroslav Hušek and Jan van Mill

the topologists here are active in research and attend seminars but often ones not called topology although topological concepts are fundamental tools or applications.

For instance, I attend the Logic seminar where I am very much at home. Two set theorists in this seminar regularly have graduate students writing theses in set theoretic topology and both of them also do research in this area and are frequent invited lecturers at topological conferences. I find set theoretic ideas beautiful and am drawn to those topological problems which are set theoretic in nature. However, most mathematicians have little knowledge of mathematical logic and one obvious consequence is that most topologists do not work in set theoretic topology.

Topology was a natural outgrowth of analysis toward the end of the 19th century. The motivation for much of analysis had been found in physical problems and proofs tended to be intuitive. Coming out of Fourier analysis, seeking a characterization of the real numbers, Georg Cantor's discovery of the use of one-to-one correspondences as an equivalence relation between sets pointed out among other things the need for a more clearly defined proof theory. Cantor and others then gave what we now see as topological characterizations of some specific Euclidean spaces. The possibility for an interesting and less rigid geometry was becoming clear. Two dimensional Riemann surfaces had been defined by Riemann already in the mid 19th century and he had suggested that this idea could be generalized to higher dimensions and used for defining function spaces. It was an idea that attracted a number of leading analysts and a clear exposition of how this could be used to define manifolds was given by Hermann Weyl in 1913, opening up possibilities for geometric, algebraic, and differential topology. Thus, at the beginning of the 20th century there were two principal groups of topologists heading in different directions. One is called "general topology".

Since it is there that I have personal experience I will begin with a brief review of general topology as I see it. General topologists were trying to build topology up from the bottom, moving hand in hand with their understanding of set theory. The simplest, most abstract and set theoretic definitions were made. Topological properties were thought of an axioms. Each axiom was tested for interesting pathologies and pathological interactions with other axioms. Each theorem was bounded by often more interesting counterexamples and conjectures. Hausdorff gave the simple, now customary, definition of a topological space in 1914. The very words general topology bring to mind Hausdorff, Sierpiński, Alexandroff and Urysohn, the whole Polish school, the theorems one found in Fundamenta during the 1920's and 1930's. The right definitions were found; the basic properties of separable metric spaces were clarified. For the most part the theorems were simple and elegant and they came with a variety of examples and hard questions. Much of the motivation came from functional analysis. Connectivity and low dimensional geometry were basic interests.

At the University of Texas following World War II I was in a position to observe a number of breakthroughs in general topology causing separate areas of more sophisticated and specialized work to emerge. The construction of a pseudoarc in 1948 (a continuum in the plane homeomorphic to each of its proper subcontinua), was one factor in "continua theory" becoming an even more active distinct field which over the years has partially evolved into "dynamical systems". "Dimension theory" was hardly a new field but Anderson's universal one dimensional curve did lead a new attack which today is primarily concerned with infinite dimensional manifolds [2]. R. H. Bing's more purely geometric proof of the Hauptvermutung for which Ed Moise had given a very complex algebraic proof in 1952, started a drive by Bing, his students, and others to settle the poincaré conjecture in dimension 3 (as yet unsolved) and, more generally, to understand the geometry of 3 manifolds. "Geometric topology" as this area was called became a vital separate area that included many algebraic topologists as well as more geometric types from general topology. The area of "rings of continuous functions", Stone-Čech compactifications, ultrafilters, Boolean algebras, a richly set theoretic area of interest to mathematicians in many fields, blossomed with the publication of a book in the field by Gillman and Jerison in 1954.

The definition of paracompactness by Dieudonné in 1944 together with the proof that metric spaces are paracompact led naturally to the "right" characterization of metrizable spaces discovered almost simultaneously by Bing in the United States, Nagata in Japan, and Smirnov in Russia by 1952. This accentuated the fact that little was understood about normal nonparacompact spaces or other classes of perhaps nonmetrizable spaces. The years from 1950 to 1970 were among the least beautiful in this type of general topology. The theorems tended to be unattractive, set theoretic type translations, of problems involving all sorts of complex conditions. The questions were often quite elegant, but the tools for solving them were unavailable. However, in 1963, Paul Cohen gave a model of the usual axioms for set theory in which the continuum hypothesis failed whose methods could be used to construct all sorts of models of set theory having different properties. Then using these models and other techniques from mathematical logic as well as just a more sophisticated knowledge of set theory, quite a few of the nice old topological problems were solved and new conjectures were made and solved. This area where the tools are mostly set theoretic is now called "set theoretic topology" and it is the area which has the closest intersection with computer science. It has been very active for the last quarter of the century. An excellent review of this kind of topology can be found in [2].

General topology has become a term used to cover a large number of specialized topological topics especially those which are quite abstract or set theoretic as well as some which are almost purely geometric. However the dominant topological theme throughout the century has been toward algebraic and differential topology, fields more immediately applicable outside of topology and mathematics.

The attack here is different. Assuming the spaces to be considered are manifolds, which removes many obstructions and almost all of the set theoretic complications, additional algebraic or differential structures are added. Not only may this make the problems more amenable to attack, the real interest is often in these more restricted classes. Tools then come from algebra, differential equations, differential geometry, ..., and the topologist becomes involved in improving these tools.

As a nonexpert but interested observer here, I enjoyed reading an article by Sir Michael Atiyah [1] on mathematics in the 20th century. Atiyah has been one of the leading algebraic and differential topologists for many years and he admits ignoring the significant advances in mathematical logic and computer science due to lack of knowledge. One is conscious as one reads that he views 20th century mathematics as vital because it is (and to whatever extent it is) geometric, topological, and applicable to physics. I recommend reading the article which I find well written and pertinent to our subject, so I will review it here.

Atiyah lists 8 "themes" which he sees in 20th century mathematics and he gives illustrations of each especially from complex variables, differential equations, differential geometry, number theory, and physics. Geometry, algebra, and topology are *assumed* to be part of all of them. The themes are:

1. local to global
2. increase in dimension
3. commutative to noncommutative
4. linear to nonlinear
5. geometry versus algebra
6. techniques in common
    (a) homology theory
    (b) K-theory
    (c) Lie-groups
7. finite groups
8. impact on physics

I certainly agree that these are themes one can see. I might comment that when discussing:

(5) Atiyah suggests "historic lines": Newton, Poincaré, Arnol'd as a geometric line and Leibnitz, Hilbert, Bourbaki as an algebraic line. He views geometric lines as the vital ones, but views algebra as both helpful and even necessary in the understanding of geometry.

(6) That there have almost ceased to be borders between fields with techniques coming from everywhere seems clear to me. But I find Atiyah's list of techniques really just a reflection of his field.

(7) Only the monster group seems interesting to Atiyah.

(8) For Atiyah, the really important thing for mathematics and in particular for topology is for it to lead to (8).

In his summary he points out that the first half of the 20th century was an "era of specialization" and the second half was an "era of unification".

There are no stone walls separating mathematical fields and no virtues in staying inside some imaginary wall. The geometry of manifolds is a basic concern of *all* topological fields and 3-dimensional manifolds are still far from being understood. Effective tools may come from anywhere and often involve the building of elaborate structures. Topology has become all too intricate to be understood even in its broader outlines by any one person.

References

[1] , M. Atiyah, *Mathematics in the 20th century*, Amer. Math. Monthly **108** (2001) 654–666.

[2] T. Koetsier, J. van Mill, *By their fruits ye shall know them: some remarks on the intersection of general topology with other areas of mathematics*, History of topology, edited by I. M. James, North-Holland (1999) 199–239.

Compact Extensions

Yuriĭ Mikhailovich Smirnov

The first compactifications were, apparently, the plane of complex numbers extended by one point and also projective plane and space, as the usual plane (space) extended by improper points. In that time, there was practically neither topology nor (needed and serious) concept of compactness.

As a start point of compactifications one should consider a method of a construction of conform extensions of plane regions, by means of the so called *ends*, described by Carathéodory in [1913], which lead later to Stoilow-Kerékjártó compactifications. His procedure was convenient in extensions having properties defined by purposes given earlier. Especially in that situation many topologists were interested in the last ten years. But many years ago already Hurewicz in [1927] and Tumarkin in [1927] proved that every normal space X having countable weight has a compactification with the same weight and dimension as X has. The same result for any weight and any dimension dim was proved later by Skljarenko in [1957]. For the dimension ind the result is not true (Smirnov). Freudenthal [1951], [1942], Morita [1952] and Skljarenko [1958], [1963] investigated spaces having punctiform or zero-dimensional remainders. In connection with that investigation, Skljarenko [1961], [1962a], [1962b] introduced a concept of perfect compactification.

Alexandrov had in mind abstract aims when he defined one-point compactifications for locally compact spaces in [1923]. A functional approach brought Čech in [1937] to a construction of maximal compactifications for completely regular spaces, and problems of lattices brought Wallman in [1938] to the same result for normal spaces.

Uniform spaces of A. Weil [1937] became in [1948] a basis for Samuel's isotone bijection between all precompact structures and all compactifications of a given Tychonov space. Later on (and independently), Smirnov [1952] proved an analogous result for proximity spaces and compactifications.

A connection to algebra was shown to be strong and interesting: Gel'fand, Raĭkov and Shilov in [1960] found an isotone bijection between all compactifications and all closed subrings (or subalgebras) of the ring $C^*(X)$ of bounded continuous functions on X distinguishing points and closed sets of X.

A question that appeared to be very important and interesting is that of compactifications for spaces having additional structures, or equivalently, extensions of such structures to compactifications. For proximity spaces (precompact uniform spaces) the question was solved by the above mentioned results of Samuel and Smirnov. For totally bounded (precompact) metrizable spaces an answer is in the affirmative although neither the author nor E.G. Skljarenko could find it in literature:

Take a space X with a totally bounded metric $d : X \times X \to \mathbb{R}$ and its compactification cX defined by its metric proximity (precompact uniform structure). Define a metric d_c on cX by $d_c(x,y) = \lim d(x_i, y_i)$, where $\{x_i\}$ and $\{y_i\}$ are sequences in X converging to

Mech-mat.fak., Moscow State Univ., 119899 Moscow, Russia; E-mail: smyrnoff@mech.math.msu.su

RECENT PROGRESS IN GENERAL TOPOLOGY II
Edited by Miroslav Hušek and Jan van Mill

x or y, resp. The defined function d_c does not depend on the choice of sequences and is really a metric.

For G-structures compatible with a given topology of X (i.e., if the corresponding action is continuous), compactifications were constructed by Palais in [1960] (for compact groups) and de Vries [1978] (for any group). A different principle was used by Smirnov in [1981] (Theorem C2 – the assumption of local compactness is not necessary there): a G-space has an equivariant compactification iff it is G-Tychonov (i.e., it is a T_0-space and all its continuous equiuniform functions separate points and closed sets in X). Antonjan [1979] extended to G-spaces the Gel'fand-Shilov theorem about bijection between compactifications and rings of equivariant functions. The assumption to be G-Tychonov is really necessary as shown by Megrelishvili in [1988] who constructed a Tychonov space without G-compactifications. It is easy to prove, using the standard procedure, that among G-compactifications on a space X (if they exist) there exists a maximal one (one can use Antonjan's result, too). In topological case the maximal compactifications are not simple at all, but in equivariant case even for such simple spaces like sphere or ball it is possible to find such actions on $\mathbb{R}^n$ that the sphere and ball are maximal G-compactifications of $\mathbb{R}^n$ (Smirnov [1994]). The same paper contains a proof that for projective spaces P^n a corresponding result does not hold, and some necessary and sufficient conditions for a G-compactification to be maximal.

References

ALEXANDROV, P.S. and P.S. URYSOHN

[1923] Sur les espaces topologiques compacts, *Bull. Acad. Polon. Sci., Sér. A*, 5–8.

ANTONJAN, S.A.

[1979] Classification of bicompact G-extensions by means of rings of equivariant mappings, (Russian), *Doklad AN Arm.SSR* **59**, 260–264.

ANTONJAN, S.A. and YU.M. SMIRNOV

[1981] universal objects and bicompact extensions for topological transformation groups, (Russian), *Doklady AN SSSR* **257**, 521–525.

CARATHÉODORY, C.

[1913] Über die Begrenzung einfach zusammenhängender Gebiete, *Math. Ann* **73**, 323–370.

ČECH, E.

[1937] On bicompact spaces, *Ann. Math.* **38**, 823–845.

FREUDENTHAL, H.

[1942] Neuaufbau der Endentheorie, *Ann. Math.* **43**, 261–279.

[1951] Kompaktisierungen und Bikompaktisierungen, *Indag. Math.* **13**, 184–192.

GEL'FAND, I.M., D.A. RAĬKOV and G.E. SHILOV

[1960] *Commutative Normed Rings*, (Russian), Izd. Fiz.-Mat.Liter., Moscow.

HUREWICZ, W.

[1927] Über das Verhältnis separabler Räume zu kompakten Räumen, *Proc. Acad. Wetensch. Amsterdam* **30**, 425–430.

MEGRELISHVILI, M.G.

[1988] Tychonov G-space without bicompact G-extension and G-linearization, (Russian), *Uspechi Mat. Nauk* **43**, 145–146.

MORITA, K.

[1952] On bicompactifications of semibicompact spaces, *Sei. Rep. Tokyo Bur. Dai.* **4A** , 222–229.

PALAIS, R.

[1960] *The Classification of G-Spaces*, Mem. Amer. Math. Soc. vol.36.

SAMUEL, P.

[1948] Ultrafilters and compactifications of uniform spaces, *Trans. Amer. Math. Soc.* **64**, 110–132.

SKLJARENKO, E.G.

[1957] On embedding of normal spaces into bicompacta of the same weight and dimension, (Russian), *Doklady AN SSSR* **117**, 36.

[1958] Bicompact extensions of semibicompact spaces, (Russian), *Doklady AN SSSR* **120**, 1200.

[1961] On perfect bicompact extensions, (Russian), *Doklady AN SSSR* **137**, 39–41.

[1962a] On perfect bicompact extensions, II, (Russian), *Doklady AN SSSR* **146**, 1031–1034.

[1962b] Some questions of theory of bicompact extensions, (Russian), *Izv. AN SSSR* **26**, 427–452.

[1963] Bicompact extensions with punctiform remainders and cohomology groups, (Russian), *Izv. AN SSSR* **27**, 1165–1180.

SMIRNOV, JU.M.

[1952] On proximity spaces, (Russian), *Matem. Sb.* **31**, 543–574.

[1994] Can simple geometric objects be maximal compact extensions of $\mathbb{R}^n$), Russian, *Uspechi Mat. Nauk* **49**, 213–214.

TUMARKIN, L.A.

[1927] On some new results and questions in general dimension theory, (Russian), in *Trudy Russ. Math. Conf.*, p. 239.

DE VRIES, J.

[1978] On the existence of G-compactifications, *Bull. Acad. Polon. Sci.* **26**, 275–280.

WALLMAN,

[1938] Lattices and topological spaces, *Ann. Math.* **40**, 112–127.

WEIL, A.

[1937] *Sur les espaces a structures uniforme et la topologie générale*, Actualités Sci. Industr., Paris, vol. 551.

Reminiscences

of L. Vietoris

(1) Aus einem Video

Gespräch mit Leopold Vietoris
Innsbruck, 9. Juni 1994
G. Helmberg ÖMG 1994:

Zur Entstehungsgeschichte des Enzyklopädieartikels von H. Tietze und L. Vietoris über die verschiedenen Zweige der Topologie [T-V]

Um 1925 hat H. Tietze auf Anregung von F. Klein einen Artikel über die verschiedenen Zweige der Topologie verfasst. Die Fahnenkorrekturen wurden von Klein auch an L. Brouwer nach Amsterdam gesandt – damals ein Zentrum der Topologie: Dort anwesend waren unter anderen P. Alexandrov, K. Menger und L. Vietoris, der sich im SS 1925 und im WS 1925/26 als Rockefeller-Stipendiat und anschließend als Assistent bei Brouwer aufhielt, wohl auf Empfehlung von Weitzenböck, seinem früheren Chef in Graz.

Brouwer war mit Tietzes Beitrag nicht recht zufrieden, insbesondere da die Dimensionstheorie darin zu kurz kam. In einer Strategiebesprechung mit Alexandrov, Weitzenböck, Menger und Vietoris wurde letzterer dazu "verurteilt", bei Tietze die Dimensionstheorie ordentlich zu vertreten. Vietoris war dazu bestens geeignet, hatte er doch als erste Tätigkeit im Rahmen seiner Assistentenstelle in Wien mit Menger dessen Dimensionstheorie zu diskutieren.

F. Klein war mit diesem "Amsterdamer Vorschlag" einverstanden und Vietoris, dem dies überaus peinlich war, wurde Tietze sozusagen aufgezwungen. Tietze, der lange vor Vietoris in Wien studiert hatte, war zu der Zeit bereits Professor in München, verhielt sich dem jungen Kollegen gegenüber aber sehr vornehm. Vietoris hat schon bei der ersten Zusammenkunft Tietzes Vertrauen gewonnen, es entwickelte sich ein gutes Arbeitsverhältnis – Vietoris hat sogar eine Woche bei der Familie Tietze in München gewohnt, wo er von Frau Tietze bestens versorgt wurde.

(2) Es entwickelte sich eine lebenslange Freundschaft (vgl. Nachruf [V], Reitberger [R]):

"... the main credit belongs to Tietze. The structure, the main parts of the text, in particular the bibliographical data was given by Tietze. The hours and days I spent together

Since, in the time of preparation of this book, L.Vietoris (1891–2002) was unable to write his own essay, we decided to place here several of his recent reminiscences. Prof. H.Reitberger from Innsbruck university kindly offered (with agreement of G.Helmberg) the parts 1 and 2, and Prof. E.Briescorn and Prof. W.Purkert from Bonn university kindly offered some letters by L.Vietoris (see the parts 3 and 4).

RECENT PROGRESS IN GENERAL TOPOLOGY II
Edited by Miroslav Hušek and Jan van Mill

with H. Tietze creating this work are unforgettable not only because I learned a lot during these consultations but also because I could stay as a guest in his house and I was taken care of by him and his charming wife. Beside his perspicacity, his enormous insight and his great knowledge of the literature which I admired, I appreciated his cheerfulness and thoughtful humor at the table. It was a special pleasure to play chess with him. In those days I got to know his love for nature ... Seefeld (near Innsbruck) used to be his summer resort for many years, a welcome opportunity for me to enjoy his and his wife's company."

(3) From a letter to Prof. Dr. Egbert Brieskorn (University of Bonn), dated Innsbruck, December 27, 1994.

"... Bei Ausbruch des ersten Weltkrieg im August 1914 hatte ich gerade 8 Semester meines Universitätsstudiums (Universität Wien) hinter mir und rückte zum österreichischen Heer ein, wurde im September 1915 auf dem russischen Kriegsschauplatz verwundet und kam nach meiner Ausheilung zu Anfang des Februar 1916 an die Südtiroler Front. Dort wurde ich zum Bergführer ausgebildet und diente als solcher bis zum Kriegsende an der Südtiroler Front. Dort geriet ich am Kriegsende (Nov. 1918) in italienische Kriegsgefangenschaft. Aus ihr kehrte ich in August 1919 nach Wien zurück.

Während diese fünf Jahre hatte ich viel Zeit, über Mengenlehre nachzudenken. Ich hatte sogar zur Vollendung meines Studiums für das Sommersemester 1916 und das Sommersemester 1918 Studienurlaube.

Alle diese Umstände ermöglichten mir, Úber meine wissenschaftlichen Probleme, d.h. über mengentheoretische Topologie nachzudenken und die einschlägige Literatur, besonders das 1914 erschienene Buch von Hausdorff, zu studieren, sodaß ich von der italienischen Gefangenschaft, in der wir anständig behandelt wurden, mit meiner fast fertigen Dissertation "Stetige Mengen" heimkehrte. Ich reichte sie im Dezember 1919 bei der Universität Wien ein und promovierte im Juni (Juli?) 1920.

Gegen Ende meines zweiten Studienurlaubs schrieb ich am 27.6.1918 an Hausdorff ein langen Brief, in den ich mitteile, was ich damals von meiner Dissertation hatte. ..."

(4) From a letter to Prof. Dr. Egbert Brieskorn (University of Bonn), dated Innsbruck, February 20, 1995.

"... *Es ist mir wichtig, festzustellen*, daß der in diesen Briefen vorkommende Begriff der *orientierten* Menge in meiner Dissertation "Stetige Mengen" (Monatshefte 31, S.184) eine *Präzisierung* erfahren hat.

Ihre hohe Meinung von Hausdorffs menschlicher Güte ist sicher berechtig. Auch *P.Alexandroff* und *P.Urisohn* haben sie, wie ich von Alexandroff erfahren habe, erlebt. Ich kann mich nicht erinnern, Hausdorff gesehen zu haben, obwohl ich, von 1922 an, etliche Jahresversammlungen der Deutschen Mathematikervereinigung besucht habe. Daß ich *Fraenkel* auf der Versammlung in Marburg (1923 oder 1924) gesehen habe, kann ich nich erinnern. *Zermelo* habe ich in Oberwolfach zweimal erlebt, einmal als eben Verschieden. ..."

The letters between Vietoris and Hausdorff will be published, with notes, in a forthcoming volume [H]. Here we remark that Vietoris wrote about his concept of connectedness (defined independently of Hausdorff), ordered topological spaces, compactness, and other topological concepts.

References

[H] *Felix Hausdorff, Gesammelte Werke*, E.Brieskorn, W.Purkert et al, eds., Springer Verlag, in preparation

[R] Reitberger, H., *The contributions of L.Vietoris and H.Tietze to the foundations of general topology*, Handbook of the History of General Topology I (ed. by C.E.Aull and R.Lowen), Kluwer 1997, 31–40.

[T-V] Tietze, H., Vietoris, L., *Beziehungen zwischen den verschiedenen Zweigen der Topologie*, Encyklopädie der Math. Wiss. III.1.2.13., Teubner, Leipzig, 1914-1931, 144–237.

[V] Vietoris,L.,*Heinrich Tietze (Nachruf)*, Almanach Österr. Akad. Wiss. 114 (1964), 360–369.

CHAPTER 21

List of Open Problems and Questions

from the contributions of the following authors:

Contents

Problems, Questions and Conjectures (with one or two exceptions) are copied from the same environments in the contributions. *Unknowns* are extracted from the text of the contributions.

RECENT PROGRESS IN GENERAL TOPOLOGY II
Edited by Miroslav Hušek and Jan van Mill

A.V. Arhangel'skii

PROBLEM **3.2.** *Is every regular paratopological group G Tychonoff? What if, in addition, G is first countable?*

PROBLEM **3.10.** *Is every first countable semitopological (paratopological) group subparacompact?*

PROBLEM **3.11.** *Can every first countable paratopological ((semitopological) group be condensed onto a metrizable space?*

PROBLEM **3.17.** *Suppose that G is a bisequential paratopological group such that $G \times G$ is Lindelöf. Must G have a countable base?*

PROBLEM **3.18.** *Is every regular bisequential paratopological group with a countable network first countable?*

PROBLEM **3.21.** *Suppose that F is a compact G_δ subspace of a (regular, Tychonoff) paratopological group G. Is then F dyadic?*

PROBLEM **4.4.** *Is every compact Mal'tsev space a retract of a compact topological group?*

PROBLEM **4.5.** *Is every metrizable Mal'tsev space retral? Is every countable Mal'tsev space retral?*

PROBLEM **4.6.** *Is every Mal'tsev Lindelöf Σ-space retral?*

PROBLEM **4.18.** *Is every T_0 rectifiable space Tychonoff?*

PROBLEM **4.21.** *Is every rectifiable paratopological group a topological group?*

PROBLEM **4.22.** *Suppose that G is a paratopological group and a Mal'tsev space. Is then G a topological group? Is G homeomorphic to a topological group?*

PROBLEM **4.24.** *Is every Tychonoff rectifiable space retral?*

PROBLEM **5.13.** *Is there in ZFC an example of a non-discrete extremally disconnected topological group?*

PROBLEM **5.14.** *Let G be an extremally disconnected quasitopological group. Is then true that there exists an open and closed Abelian subgroup of G?*

PROBLEM **5.16.** *Is there in ZFC a non-discrete submaximal topological group?*

PROBLEM **5.38.** *Is the product of two arbitrary non-discrete topological groups resolvable?*

PROBLEM **5.39.** *Is every extremally disconnected (regular) paratopological group a topological group?*

PROBLEM **5.40.** *Is there an example in ZFC of a nondiscrete extremally disconnected regular paratopological group?*

PROBLEM **6.5.** *Is for every Moscow group G true that $\mu G = \rho_\omega G$?*

PROBLEM **6.7.** *Is every C-embedded subgroup of a Moscow group Moscow?*

PROBLEM **6.8.** *Can every topological group be embedded in a Moscow group?*

PROBLEM **6.12.** *Is every Rajkov complete group projectively Moscow?*

PROBLEM **6.13.** *Is every $\aleph_0$-bounded group a PT-group?*

PROBLEM **6.14.** *Is every topological group with the countable Souslin number R-factorizable?*

PROBLEM **6.18.** *Let G_i be a topological group such that $\mu G_i = \rho_\omega G_i$, for each $i \in \omega$, and G the product of these groups. Assume also that G is a PT-group. Is then true that $\mu G = \Pi\{\mu G_i : i \in \omega\}$?*

PROBLEM **6.21.** *Let G be a topological group of countable tightness. Is then $G \times G$ a Moscow group? A PT-group? Is then the g-tightness of $G \times G$ countable?*

PROBLEM **6.22.** *Suppose G is an extremally disconnected topological group. Is then $G \times G$ Moscow? Is $G \times G$ a PT-group? Is the g-tightness of $G \times G$ countable?*

PROBLEM **6.23.** *Suppose G is an extremally disconnected group and B a compact group. Is then $G \times B$ a Moscow group?*

PROBLEM **6.24.** *Is the g-tightness of every Moscow group countable?*

PROBLEM **6.25.** *Suppose that G is a topological group of the countable g-tightness, and H is a dense subgroup of G. Is the g-tightness of H is countable?*

PROBLEM **7.8.** *Is it true that, for every metrizable space X, $A(X)$ is a $k_{\mathbb{R}}$-space? In particular, is the free Abelian topological group of the space Q of rational numbers a $k_{\mathbb{R}}$-space?*

PROBLEM **7.9.** *Is the o-tightness of the free (Abelian) topological group of a metrizable space countable?*

PROBLEM **7.10.** *Characterize metrizable spaces X such that $A(X)$ is a $k_{\mathbb{R}}$-space.*

PROBLEM **7.21.** *When the free topological group of a space X is Moscow?*

PROBLEM **7.22.** *Is the free (Abelian) topological group of a first countable space Moscow?*

PROBLEM **7.27.** *When $F(X)$ is C-embedded in $F(\mu X)$?*

PROBLEM **7.28.** *Suppose that X is a first countable space. Is $F(X)$ C-embedded in $F(\mu X)$?*

PROBLEM **7.31.** *Is $F(X)$ $(A(X))$ paracompact, for every paracompact p-space? What if $X = M \times B$, where M is metrizable and B is compact?*

PROBLEM **10.1.** *Given a class $\mathcal{P}$ of topological spaces, when every $X \in \mathcal{P}$ can be embedded as a closed subspace into a topological group $G \in \mathcal{P}$?*

PROBLEM **10.2.** *Given a class $\mathcal{P}$ of topological spaces, when, for every topological group G in $\mathcal{P}$, the square $G \times G$ belongs to $\mathcal{P}$?*

PROBLEM **10.3.** *Can Sorgenfrey line be embedded as a closed subspace into some Lindelöf topological group? Can every Lindelöf space be so embedded?*

PROBLEM **10.4.** *Can arbitrary Tychonoff space of countable tightness be embedded as a closed subspace into a topological group of countable tightness? What if we drop the word "closed" in this question?*

PROBLEM **10.5.** *Is there a countably compact topological group G such that the square $G \times G$ is not countably compact?*

PROBLEM **10.6.** *Is the product of any two topological groups of countable tightness a Moscow group? A topological group of countable o-tightness?*

PROBLEM **10.7.** *Is the square of a Moscow group a PT-group?*

PROBLEM **10.8.** *Is the square of a PT-group a PT-group?*

PROBLEM **10.10.** *Is μG homogeneous for every topological group G?*

PROBLEM **10.11.** *Is the product of arbitrary family of pseudocompact quasitopological groups pseudocompact?*

PROBLEM **10.12.** *Is there a Dowker topological field?*

PROBLEM **10.13.** *Is there in ZFC a countable non-metrizable Fréchet-Urysohn topological group?*

PROBLEM **10.14.** *Is every regular Fréchet-Urysohn paratopological group first countable?*

PROBLEM **10.15.** *Is every topological field with the countable Souslin number separable?*

PROBLEM **10.17.** *Is every countably compact sequential topological group a Fréchet-Urysohn space?*

PROBLEM **10.18.** *Is every countably compact sequential topological group G ω-monolithic, that is, is the closure of arbitrary countable subset of G metrizable? Compact?*

PROBLEM **10.19 .** *Is there a ZFC-example of a non-metrizable Rajkov complete Fréchet-Urysohn topological group?*

PROBLEM **10.20.** *Is every ω-monolithic Rajkov complete Fréchet-Urysohn topological group metrizable?*

PROBLEM **10.21.** *Is it true that every compact topological group contains a dense (or a G_δ-dense) subspace of countable tightness? A dense Fréchet-Urysohn subspace?*

PROBLEM **10.22.** *Is every topological group topologically isomorphic to a closed subgroup of a minimal topological group?*

PROBLEM **10.23.** *Is every topological group a quotient of a minimal topological group?*

PROBLEM **10.24.** *Is every topological group a retract of a minimal topological group?*

PROBLEM **10.26.** *Is every minimal topological group a PT-group? Is every minimal topological group Moscow?*

PROBLEM **10.27.** *When is a Lindelöf topological group G metrizable at infinity? Is it true in this case that the Souslin number of G is countable and G is a p-space?*

PROBLEM **10.28.** *Is it true that every topological group that is paracompact at infinity is a p-space?*

H.R. Bennett and D.J. Lutzer

QUESTION **3.4.** *Suppose X is an arbitrary weakly perfect GO-space. Must X be hereditarily weakly perfect?*

QUESTION **3.5.** *Is there a ZFC example of a perfect GO-space that does not have a σ-closed-discrete dense subset?*

QUESTION **3.6.** *Is there a ZFC example of a perfect GO-space that has a point-countable base and is not metrizable?*

QUESTION **3.8.** *In ZFC, is there an example of a perfect non-Archimedean space that is not metrizable?*

QUESTION **3.14.** *Is it true that any perfect GO-space can be topologically embedded in some perfect LOTS?*

QUESTION **4.14.** *For a GO-space X, find a topological property that solves the equation*
X is quasi-developable + (?) if and only if X has a $< \omega$-WUB.

QUESTION **5.6.** *Suppose X is a Lindelöf GO-space with a small diagonal that can be p-embedded in some LOTS. Must X be metrizable?*

QUESTION **5.21.** *In ZFC, does S^* have a continuous separating family?*

QUESTION **5.22.** *In ZFC, is there a non-metrizable perfect LOTS with a continuous separating family?*

QUESTION **5.23.** *In ZFC, is there an example of a GO-space X that has a continuous separating family, but whose LOTS extension X^* does not?*

QUESTION **6.4.** *Suppose A is a closed subset of a perfect LOTS. Is there a linear cch-extender from $C(A)$ to $C(X)$?*

UNKNOWN (following **9.2.**) It is an open question in ZFC whether every countably paracompact subspace of $[0, \omega_1)^2$ is normal.

J. Dijkstra and J. van Mill

UNKNOWN (following **3.9.**) Is every *compact* convex subset of a metrizable vector space an absolute retract?

G. Godefroy

UNKNOWN (following **2.3.**) Is the property "K Corson" determined by the space $\mathcal{C}(K)$ in ZFC?

UNKNOWN (following **2.3.**) It is not known whether a continuous image of a Radon-Nikodym compact set is Radon-Nikodym, and not even whether the class is stable under (non disjoint) union of two sets.

UNKNOWN (following **3.3.**) Is there in ZFC a scattered compact set K with the Namioka property such that the space $\mathcal{C}(K)$ has no equivalent norm and the weak and norm topologies coincide on the unit sphere?

PROBLEM **3.4.** *Does there exist a Baire space E, a compact set K, and a separately continuous function $f : E \times K \to \mathbf{R}$ with no point of joint continuity?*

UNKNOWN (following **3.4.**) In ZFC, has every Asplund space an equivalent norm with the Mazur intersection property?

PROBLEM **3.5.** *Let X be a Banach space which has an equivalent F-smooth norm. Does there exist an equivalent LUR norm on X?*

UNKNOWN (following **3.5.**) It is not known whether every Banach space which has an equivalent F-smooth norm has $\mathcal{C}^1$-smooth partitions of unity.

UNKNOWN (following **3.7.**) It is not known whether there exists a Banach space X which is a Borel subset of (X^{**}, w^*) without actually being a $K_{\sigma\delta}$ in that space.

PROBLEM **4.1.** *Let X and Y be two separable Banach spaces, such that there exists a bi-Lipschitz homeomorphism between X and Y. Does it follow that X and Y are linearly isomorphic?*

UNKNOWN (following **4.4.**) Let K be a countable compact set, and X be a Banach space which is Lipschitz-isomorphic to $\mathcal{C}(K)$. Is then X linearly isomorphic to $\mathcal{C}(K)$?

UNKNOWN (following **4.4.**) If a Banach space Y contains a subset which is Lipschitz-isomorphic to $c_0(\mathbf{N})$, does it contain a linear copy of $c_0(\mathbf{N})$?

UNKNOWN (following **4.7.**) Is the space $c_0(\mathbf{N})$ determined by its uniform structure?

PROBLEM **4.8.** *Let X be a Banach space which is uniformly homeomorphic to $c_0(\mathbf{N})$. Is the space X linearly isomorphic to $c_0(\mathbf{N})$?*

UNKNOWN (following **4.8.**) It is not known whether an isomorphic predual of $l_1(\mathbf{N})$ with summable Szlenk index is isomorphic to $c_0(\mathbf{N})$.

G. Gruenhage

UNKNOWN (in **2.**) Has every metrizable space with no compact open sets a Tychonoff connectification?

UNKNOWN (in **2.**) Is there a universal space (in ZFC) for ultrametric spaces of cardinality τ and of weight τ?

UNKNOWN (in **2.**) Does every metrizable space admit a metric d such that X has a σ-discrete base consisting of open d-balls?

UNKNOWN (in **2.**) Is every separable metrizable space having the UMP homeomorphic to a subset of the real line?

UNKNOWN (in **3.**) It is not known if $C_p(X)$ Lindelöf Σ and ω_1 a caliber for $C_p(X)$ implies X cosmic.

UNKNOWN (in **3.**) Does $C_p(X)$ a σ-space imply that X and $C_p(X)$ are cosmic?

UNKNOWN (in **5.**) Are all stratifiable spaces μ-spaces?

UNKNOWN (following **5.2.**) When $C_k(X)$ are M_1 spaces (in particular, $C_k(\mathbb{P})$, where $\mathbb{P}$ is the space of irrationals)?

UNKNOWN (following **5.2.**) When $C_k(X)$ are stratifiable (in particular, $C_k(\mathbb{Q})$?

UNKNOWN (following **5.2.**) Is the free group of a stratifiable space stratifiable?

UNKNOWN (in **6.**) Are ω_μ-stratifiable spaces ultraparacompact for $\mu > 0$?

UNKNOWN (in **9.**) Is every stratifiable space, or stratifiable μ-space, LF-netted?

UNKNOWN (in **9.**) It is open if there are ZFC examples of non-normal Σ-products of Lašnev spaces. In particular, it is not known if $S(2^{\mathfrak{c}})^2 \times \omega_1$ is non-normal in ZFC.

UNKNOWN (in **10.**) Can infinite metrizability numbers of locally compact (or compact) spaces be raised by perfect mappings? (In particular, the case $m(X) = \omega$ is unsettled).

UNKNOWN (in **10.**) Is $m(X) \leq \mathfrak{c}$ whenever X has a point-countable base?

PROBLEM **11.1.** *Are all stratifiable spaces M_1 spaces?*

PROBLEM **11.2.** *Is it consistent that there are no symmetrizable L-spaces?*

PROBLEM **11.3.** (a) *Is there a symmetrizable Dowker space?*

(b) *Suppose X is normal, and the union of countably many open metrizable subspaces. Must X be metrizable?*

(c) *Is every normal space with a σ-disjoint base paracompact?*

PROBLEM **11.4.** *Does a space X have a point-countable base iff X has a countable open point-network?*

PROBLEM **11.5.** *If every $\aleph_1$-sized subspace of a first-countable space X is metrizable, must X be metrizable?*

PROBLEM **11.6.** *Is Arhangel'skii's class MOBI preserved by perfect mappings?*

PROBLEM **11.7.** *Is there a class of spaces (and if so, describe it) which:*

(i) *contains all metrizable spaces;*

(ii) *is closed under the taking of closed subspaces, closed images, and countable products; and*

(iii) *is contained in the class of paracompact spaces?*

UNKNOWN (following **11.7.**) It is not known if X, Y paracompact $\Sigma^{\#}$ implies $X \times Y$ is paracompact (it is $\Sigma^{\#}$).

PROBLEM **11.8.** *Is there in ZFC a non-metrizable perfectly normal non-archimedean space?*

PROBLEM **11.9.** *Is there in ZFC a regular perfect first-countable space with no σ-discrete dense subset?*

PROBLEM **11.10.** *Is there a non-metrizable compact space with a small diagonal?*

N. Hindman and D. Strauss

UNKNOWN (following **3.2.**) It is not known whether extremally disconnected non-discrete topological groups can be defined in ZFC.

UNKNOWN (following **3.8.**) It is an open problem if there exists a topological group G, which is not totally bounded, for which uG has precisely one minimal left ideal.

UNKNOWN (following **4.6.**) It is an open question whether the assumption that $p = p + p$ in Theorem 4.6 can be replaced by the weaker assumption that $p \in \mathbb{N}^*$.

K. Kawamura

CONJECTURE **2.4.** *If a finite dimensional locally compact separable ANR X is topologically homogenous, then X is a topological manifold.*

CONJECTURE **2.5.** *If a generalized manifold X has the DDP, then X is topologically homogenous.*

CONJECTURE **2.6.** *Every finite dimensional G-space is a topological manifold.*

PROBLEM **3.3.** *Let $f : M^4 \to X$ be a cell-like map of a topological 4-manifold M onto a compact metric space X. Does X have a finite covering dimension?*

PROBLEM **3.6.** *Does the above theorem hold for arbitrary abelian group? Is it possible to choose above Z so that $\mathrm{c} - \dim_G Z \le n$?*

CONJECTURE **3.9.** *For each $n \ge 2$, there does not exist a universal space for the class of all compacta of integral cohomological dimension at most n.*

QUESTION **4.1.** *Let X be a locally compact polyhedron and suppose that $X \times Q$ admits a $\mathcal{Z}$-compactification. Then does X itself admits a $\mathcal{Z}$-compactification?*

CONJECTURE **4.7.** *If P and Q are closed aspherical manifolds with isomorphic fundamental groups, then their universal covers are homeomorphic.*

CONJECTURE **4.8.** *If P are Q are closed aspherical manifolds with isomorphic fundamental groups, then P and Q are homeomorphic.*

CONJECTURE **4.21.** *Let Γ_1 and Γ_2 be two isomorphic Coxeter groups (with possibly different presentations). Then $\partial\Gamma_1$ and $\partial\Gamma_2$ are homeomorphic.*

PROBLEM **5.4.** *If a closed manifold N has the hopfian fundamental group, is N hopfian?*

UNKNOWN (following **5.5.**) It is an open problem as to whether every hopfian manifold with the hyperhopfian fundamental group is a codimension 2 fibrator.

QUESTION **6.1.** *Let X be a Polish space with the following properties:*

(1) $\dim X = n$, *X is locally $(n-1)$-connected and $(n-1)$-connected, and*

(2) *for each Polish space Z with* $\dim Z \le n$, *every map $f : Z \to X$ is approximated arbitrarily closely by closed embeddings.*

Is then X homeomorphic to the n-dimensional Nöbeling space

$$N_n^{2n+1} = \{(x_i) \in \mathbf{R}^{2n+1} \mid \text{at most } n \text{ coordinates } x_i\text{'s are rational}\}?$$

CONJECTURE **6.3.** *Let X be a compactum in $\mathbf{R}^n$. If* $\dim X \ge k$, *then there exist an $(n-k)$-dimensional affine subspace L of $\mathbf{R}^n$ and an $\epsilon > 0$ such that, for each map $f : X \to \mathbf{R}^n$ with $d(f, id) < \epsilon$, $f(X)$ intersects with L.*

H.-P. Künzi

PROBLEM **2.6.** *It is unknown whether each quasi-proximity class that contains more than one member contains at least $2^{2^{\aleph_0}}$ nontransitive quasi-uniformities.*

PROBLEM **2.13.** (a) *Is the fine quasi-uniformity of each (quasi-metrizable) Moore space or each non-archimedeanly quasi-metrizable space transitive?*

(b) *Is there in ZFC a compact Hausdorff space that is not transitive?*

PROBLEM **4.2.** *Is there an example of a regular quasi-metrizable space whose fine quasi-uniformity is not PS-complete?*

PROBLEM **5.** *It remains unknown whether each developable quasi-metrizable space is non-archimedeanly quasi-metrizable.*

W. Marciszewski

PROBLEM **2.1.** *Which topological properties of the metrizable space X are preserved by (linear, uniform) homeomorphisms of spaces $C_p(X)$ (or $C_p^*(X)$)?*

PROBLEM **2.9.** *Let X and Y be t-equivalent (metrizable, compact) spaces.*
Is $\dim X = \dim Y$*?*

QUESTION **2.14.** *Is $C_p([0,1])$ homeomorphic to $C_p(2^\omega)$ ($C_p([0,1]^2)$)?*

PROBLEM **2.18.** *Let X and Y be (separable) metrizable spaces and let $\Phi : C_p(X) \to C_p(Y)$ (resp., $\Phi : C_p^*(X) \to C_p^*(Y)$) be a uniformly continuous surjection (uniform homeomorphism). Let X be completely metrizable. Is Y also completely metrizable?*

QUESTION **2.21.** *Let X and Y be t-equivalent (separable(metrizable spaces such that $X \in \mathcal{M}_2$ ((i.e., X is an absolute $F_{\sigma\delta}$-space). Does Y belong to the class $\mathcal{M}_2$?*

PROBLEM **2.33.** *Find an internal characterization of spaces X which are l-equivalent to the cube $[0,1]^n$.*

QUESTION **3.5.** *Do there exist infinitely many ((continuum many) pairwise nonhomeomorphic spaces $C_p(X)$ of a given Borel class $\mathcal{M}_\alpha \setminus \mathcal{A}_\alpha$, $\alpha \geq 3$ (projective class)?*

PROBLEM **3.15.** *Let X and Y be (countable, metrizable) l^*-equivalent spaces. Are then X and Y l-equivalent?*

QUESTION **4.6.** *Does there exist a continuous extender*
$e : C_p(\{0,1\}^{\omega_1}) \to C_p([0,1]^{\omega_1})$?

PROBLEM **4.9.** *Is $C_p(X)$ homeomorphic to $(C_p(X))^\omega$ for every infinite countable space X? Is $C_p(\omega_F)$ homeomorphic to $(C_p(\omega_F))^\omega$ for every filter F on ω?*

PROBLEM **4.12.** *Is $C_p(X)$ homeomorphic to $C_p(X) \times C_p(X)$ for every infinite (compact) metrizable space X?*

PROBLEM **4.13.** *Does there exist a continuous map from $C_p(X)$ onto $C_p(X) \times C_p(X)$ for every (compact) space X?*

PROBLEM **4.14.** *Let X be a (compact) space such that $C_p(X)$ is Lindelöf. Is $C_p(X) \times C_p(X)$ also Lindelöf?*

PROBLEM **4.16.** *Let X be an infinite (compact) space. Is $C_p(X)$ homeomorphic to $C_p(X) \times \mathbb{R}$?*

PROBLEM **5.1.** *When does there exist a condensation of $C_p(X)$ onto a compact space (σ-compact space)?*

K. Martin, M.W. Mislove and G.M. Reed

UNKNOWN (following **4.14.**) Open question is whether or not there exists a Scott domain in which the maximal elements X form a G_δ-set, but for which there exists no measurement with X as the kernel.

R. Pol and H. Toruńczyk

UNKNOWN (in **3.A.**) No examples are known of metrizable weakly infinite-dimensional spaces without property C, and it is one of the most important problems concerning infinite-dimensional spaces, whether the two notions coincide for compacta.

PROBLEM **3.1.** *Let $f : X \to Y$ be a continuous map between compacta with Y and all fibers $f^{-1}(y)$ weakly infinite-dimensional. Is then X weakly infinite-dimensional?*

UNKNOWN (in **3.C.**) Let $f : X \to Y$ be a light mapping between compacta and u a typical mapping $X \to \mathbb{I}$. Are all the sets $u(f^{-1}(y))$ 0-dimensional?

PROBLEM **3.7.** *Does there exist a compactum which is not countable-dimensional and whose subsets are all weakly infinite-dimensional?*

UNKNOWN (in **4.G**). It is not known if e-dim X can be always represented as $[K]$ with a countable CW complex K. It is also unknown, if for any countable CW complex K there is a metrizable compact space X with e-dim $X = [K]$.

UNKNOWN (in **5.A.**) It is an open problem if there are compacta X, Y violating the *logarithmic law*, with X being a 3-dimensional AR.

PROBLEM **5.1.** *Let X, Y be compacta. Is it true that* $\dim (X \times Y) < n$ *if and only if every mappings $f : X \to \mathbb{R}^n$, $g : Y \to \mathbb{R}^n$ have unstable intersection?*

UNKNOWN (in **5.C**). Let a compactum X be decomposed into A and B. Is $\dim X \leq \dim (A \times B) + 1$?

CONJECTURE *(in **6.A.**) For any subadditive sequence d_n of natural numbers, $d_{n+m} \leq d_n + d_m$, there is a separable metrizable X with $\dim X^n = d_n$ for $n = 1, 2, \ldots$*

UNKNOWN (in **7.B.**) Is the assumption $\dim Y < \infty$ necessary in Theorem 7.2?

UNKNOWN (in **7.C.**) For metrizable countable-dimensional hereditarily indecomposable continua X, can one have $\dim B_\infty(X) = n$ for arbitrary n?

UNKNOWN (end of Section **8**). It is not known if there is an n-cell in $\mathbb{R}^m$, $n > 2$, which can be pushed off any $(m - n)$-dimensional affine manifold in $\mathbb{R}^m$.

UNKNOWN (in **10.B.**) For what ordinals $\alpha < \omega_1$, whenever $f : X \to Y$ is a finite-to-one surjection and X is a compactum with $d(X) \geq \alpha$, also $d(Y) \geq \alpha$?

UNKNOWN (in **10.C.**) The evaluation of the transfinite dimension $\operatorname{ind} S_\alpha$ for all $\alpha < \omega_1$ is an interesting open problem.

UNKNOWN (in **10.C.**) It is unknown if for every $\alpha < \omega_1$ there is a compactum X_α with $\operatorname{ind} X_\alpha = \operatorname{Ind} X_\alpha = \alpha$.

UNKNOWN (in **11.A.**) No examples are known of metrizable groups G with $\operatorname{ind} G < \operatorname{Ind} G$.

UNKNOWN (in **11.B.**) Under CH, there is a regular space X with $\dim X = 1$ and $\operatorname{ind} X = \operatorname{Ind} X = 2$, which is a continuous image of a separable metrizable space. No such examples are known in the realm of the usual set theory. It is also unclear if the space X is a quotient image of a separable metrizable space.

UNKNOWN (in **12.A.**) Are there cell - like mappings of $\mathbb{I}^4$ onto infinite-dimensional compacta?

UNKNOWN (in **13.A.**) The question of existence of a universal space in the class of compacta with a given integral cohomological dimension is open.

UNKNOWN (in **14.F.**) It is not known if for any compactum X in $\mathbb{R}^3$, $K(X)$ must be analytic.

D. Repovš and P.V. Semenov

UNKNOWN (in **1.1, Ad 2.**) The question about the necessity (for the Michael's Theorem 1.1) of condition (2') is in general still open.

PROBLEM **1.3, 5.1.** *Let Y be a G_δ-subset of a Banach space B. Does then every LSC mapping $F : X \to Y$ of a paracompact space X with convex closed values in Y have a continuous selection?*

PROBLEM **1.6.** *Are the conditions* $(a) - (c)$ *from Theorem* 1.5 *equivalent to the following condition:*

(d) *Let Y be any G_δ-subset of a Banach space and $F : X \to Y$ an LSC mapping with convex values which are closed in Y. Then F admits a singlevalued continuous selection.*

PROBLEM (end of Section **1.**) *Does every weakly infinite-dimensional compact metric space have the C-property?*

PROBLEM **4.1, 5.3.** *Does every LSC closed- and convex-valued mapping from a collectionwise normal and countably paracompact domain into a Hilbert space admit a singlevalued continuous selection?*

PROBLEM **5.2.** *Is it true that the affirmative answer to Problem 5.1 for an arbitrary Banach space B characterizes C-property of the domain?*

PROBLEM **5.4.** *Does every LSC compact- and convex-valued mapping from a normal domain into a metric space Y endowed by an uniform convex system admit a singlevalued continuous selection?*

PROBLEM **5.5.** *Problem 5.4 for paracompact domains and closed-valued mappings.*

PROBLEM **5.6.** *Let X and Y be any topological spaces and let $F : X \to 2^Y$. Find suitable conditions under which there exist an interval $I \subset \mathbb{R}$, a continuous function $h : X \to I$ and a mapping $G : I \to 2^Y$, satisfying the following properties:*

(1) *$G(h(x)) \subset F(x)$ for all $x \in X$;*

(2) *The graph of G is connected and locally connected; and*

(3) *For every open set $\Omega \subset Y$, the set $G^{-}(\Omega) \cap$ int(I) has no isolated points.*

PROBLEM **5.7.** *Does there exist a space X such that $Sel_2(X) \neq \emptyset$ but $Sel_n(X) = \emptyset$ for some $n > 2$?*

PROBLEM **5.8.** *Does there exist a space X such that $Sel_2(X) \neq \emptyset$ and* $\operatorname{ind} X > 1$*?*

PROBLEM **5.9.** *Does there exist a space X which is not zero-dimensional but $\{f(X) :$ f is a continuous selector$\}$ is dense in X?*

PROBLEM **5.10.** *Does there exist a zero-dimensional metrizable space X such that $\mathcal{F}(X)$ has a continuous selector but* $\dim X \neq 0$*?*

PROBLEM **5.11.** *Let $p : E \to B$ be a Serre fibration with a constant fiber which is an n-dimensional manifold. Is p a locally trivial fibration?*

PROBLEM **5.12.** *Does every open mapping of a locally connected continuum onto arc have a continuous section?*

PROBLEM **5.13.** *Is any piecewise linear n-soft mapping of compact polyhedra a Serre n-fibration?*

PROBLEM **5.14.** *Does every Serre fibration with a compact locally connected base have a global section if all of its fibers are contractible compact 4-manifolds with boundary?*

PROBLEM **5.15.** *Is the complex-valued mapping $z_1^3 + z_2^3$ of $\mathbb{C}^2$ (2-dimensional complex space) onto $\mathbb{C}^1$ a Serre 1-fibration?*

PROBLEM **5.16.** *Let $f \in H_0(D^2)$ and* $\operatorname{dist}(f, id|_{D^2}) = 2r$. *Estimate the distance* $\operatorname{dist}(\frac{f+id}{2}; H_0)$. *Is $0,5r$ the correct answer?*

PROBLEM **5.17.** *Let $f_1, f_2, ..., f_n \in H_0(D^2)$ and $f \in conv\{f_1, f_2, ..., f_n\}$. Is it true that* $\operatorname{dist}(f; H_0) \leq 0,5r$ *where r is minimal radius of a ball which covers all $f_1, f_2, ..., f_n$?*

PROBLEM **5.18.** *Let f and g be two embeddings of the segment $[0, 1]$ into the Euclidean plane and* $\operatorname{dist}(f, g) = 2r$. *Estimate the distance between the mapping $\frac{f+g}{2}$ and the set of all embeddings of this segment to the plane.*

PROBLEM **5.19.** *For each Riemannian metric ρ on M define $F(\rho)$ as the set of all isometric embeddings of $(M, \rho) \hookrightarrow H$. Does then the multivalued mapping F admit a continuous selection?*

PROBLEM **5.20.** *Find a suitable axiomatic restrictions for S under which Theorem* 1.1 *holds for mappings with S-convex values.*

PROBLEM **5.21.** *Is there a semicontinuity condition on the metric projection P_M onto a proximinal subspace M in a Banach space that is both necessary and sufficient for the metric projection to admit a continuous selection?*

D. Shakhmatov

UNKNOWN (following **3.2.**) It is unclear if every sequential space (or even Fréchet-Urysohn space) can be realized as a (preferably closed) subspace of some sequential group.

QUESTION **3.3.** *Let G be a group equipped with a Fréchet-Urysohn topology with respect to which multiplication is continuous. Is then G an α_4-space?*

UNKNOWN (following **3.7.**) It seems unclear if $\alpha_{3/2}$ and α_1 are equivalent for all (i.e. not necessarily Fréchet-Urysohn) topological groups.

QUESTION **3.8.** *Is it consistent with ZFC that every Fréchet-Urysohn topological group is an α_3-space? What about countable Fréchet-Urysohn topological groups?*

QUESTION **3.9.** *Is it consistent with ZFC that every Fréchet-Urysohn topological group that is an α_3-space is automatically α_2? What about countable Fréchet-Urysohn topological groups?*

QUESTION **3.10.** *Is it consistent with ZFC that every countable Fréchet-Urysohn topological group that is an α_2-space is first countable?*

PROBLEM **3.11.** *Without any additional set-theoretic assumptions beyond ZFC, does there exist a countable Fréchet-Urysohn topological group that is not first countable?*

QUESTION **3.13.** *Is there, in ZFC only, a free FUF-filter on ω that is not countably generated?*

QUESTION **3.15.** *Let G be a topological group.*
(i) *If G is an α_2-space, must G have the Ramsey property?*
(ii) *If G has the Ramsey property, is G an α_2-space?*
What if G is additionally assumed to be Fréchet-Urysohn?

QUESTION **4.2.** *Are there any "new" implications between α_i-properties and the Ramsey property in a topological group G satisfying one of the following compactness conditions:*
(i) *((locally) countably compact,*
(ii) *(locally) pseudocompact,*
(iii) *totally bounded?*

QUESTION **4.3.** *What is the answer to Question 4.2 if one additionally assumes in it that G is Fréchet-Urysohn?*

QUESTION **4.4.** *What is the answer to Question 3.15 if one additionally assumes that the group G has one of the following compactness properties:*
(i) *(locally) countably compact,*
(ii) *(locally) pseudocompact,*
(iii) *totally bounded?*

QUESTION **5.7.** *Is it true that $C_p(X)$ is an α_1-space if and only if X is a QN-space?*

QUESTION **5.8.** *Find necessary and sufficient conditions on X for its function space $C_p(X)$ to be α_1.*

QUESTION **5.9.** *Let X be a space.*
(i) *Are the Ramsey property and α_2-property equivalent for $C_p(X)$?*

(ii) *Is the Ramsey property equivalent to $\alpha_{3/2}$-property for $C_p(X)$?*

QUESTION **6.7.** *In ZFC, is there a Fréchet-Urysohn group G such that:*
(i) *$G \times G$ is not Fréchet-Urysohn, or even*
(ii) *$t(G \times G) > \omega$?*

QUESTION **6.13.** (i) *In ZFC only, is there a countable Fréchet-Urysohn topological group G such that $G \times G$ is not Fréchet-Urysohn (not sequential)?*
(ii) *In ZFC only, does there exist two countable Fréchet-Urysohn topological groups G and H such that $G \times H$ is not Fréchet-Urysohn (not sequential)?*

QUESTION **6.14.** *In ZFC only, is there a Fréchet-Urysohn topological group G such that G is α_1 but $G \times G$ is not Fréchet-Urysohn?*

UNKNOWN (following **7.1.**) In ZFC, does there, for any ordinal $\alpha < \omega_1$, exist a countable sequential Abelian group of sequential order α?

QUESTION **7.4.** *Does there exist, for every ordinal $\alpha \leq \omega_1$, a sequential topological group G_α of sequential order α which in addition has one of the following properties:*
(i) *totally bounded,*
(ii) *pseudocompact,*
(iii) *countably compact?*

QUESTION **7.5.** *Is a countably compact sequential group Fréchet-Urysohn?*

S. Solecki

QUESTION **2.10.** *Is it true that if all continuous actions of a Polish group G on Polish spaces satisfy the strong Glimm-Effros dichotomy, then G is a Glimm-Effros group?*

QUESTION **2.12.** *Does there exist a Polish group without a left-invariant complete metric such that all its continuous actions on Polish spaces induce orbit equivalence relations fulfilling the Glimm-Effros dichotomy?*

QUESTION **2.13.** *Let G have a left-invariant complete metric, and let it act continuously on a Polish space X. Is it true that either $E_0 \leq E_G^X$ or there is a Polish topology τ on X containing the original one, keeping the action continuous, making E_G^X G_δ and such that all sets in τ are F_σ in the old topology?*

QUESTION **2.19.** *Does the dichotomy from Theorem* 2.18 *hold for arbitrary continuous actions of Polish groups on Polish spaces?*

QUESTION **2.22.** *Let E be a Borel, or even analytic, equivalence relation on a Polish space X. Is it true that either $E_1 \leq E$ or $E \leq E_G^Y$ for some continuous action of a Polish group G on a Polish space Y?*

QUESTION **3.4.** *Let G be a Polish group. Assume that all continuous actions of G on Polish spaces induce orbit equivalence relations which are $\leq E_\infty$. Is it true that G is locally compact?*

QUESTION **3.8.** *Let G be a Borel subgroup of a Polish group H. Is it true that G is Polishable if, and only if, $E_1 \not\leq E_{H/G}$?*

QUESTION **4.8.** *Let C be an indecomposable continuum. Does there exist a comeager E_C-invariant Borel subset U of C with the property that $E_C|U$ is Borel isomorphic with E_1 via an isomorphism for which images and preimages of meager sets are meager or $E_C|U$ is Borel isomorphic with $E_0 \times I_{2^{\mathbb{N}}}$ via a Borel isomorphism as above?*

QUESTION **4.9.** *Is it true that each Borel E_C-invariant subset of an indecomposable continuum C is meager or comeager?*

QUESTION **5.3.** *What is the exact complexity of the isometry equivalence relation on Polish locally compact metric spaces? Is it $E_{S_\infty}^\infty$?*

QUESTION **5.4.** *What is the exact complexity of the homeomorphism equivalence relation among compact metrizable spaces?*

M. Tkachenko

UNKNOWN (following **2.3.**) It is unknown whether the existence of a Hausdorff group topology on an infinite free Abelian group with countably compact square is consistent with ZFC.

UNKNOWN (following **2.6.**) Is, under MA, every precompact group of non Ulam-measurable cardinality a continuous isomorphic image of a closed subgroup of a direct product of countably compact topological groups?

UNKNOWN (beginning of Subsection 3 in Section **2.**) It is unknown (in ZFC only) whether free Abelian groups of cardinality greater than $\mathfrak{c}$ admit a countably compact Hausdorff group topology.

UNKNOWN (end of Subsection 4 in Section **2.**) Does there exist a model in which the answer to Wallace's question is "yes"?

UNKNOWN (following **3.5.**) It is unknown whether the product of two strictly o-bounded groups is strictly o-bounded.

PROBLEM **4.3.** *Let G be an arbitrary $\mathbb{R}$-factorizable group.*
(A) *Is every locally finite family of open sets in G countable?*
(B) *Must all continuous homomorphic images of G be $\mathbb{R}$-factorizable?*
(C) *Is the class of $\mathbb{R}$-factorizable groups closed under taking direct products?*

PROBLEM **4.7.** *Suppose that a topological group G contains a dense $\mathbb{R}$-factorizable subgroup. Is then G $\mathbb{R}$-factorizable?*

PROBLEM **4.10.** *Is every z-embedded subgroup of a Lindelöf topological group pseudo-ω_1-compact? What about C^*-embedded subgroups?*

PROBLEM **4.14.** *Find out which of the following assertions are valid:*
(a) *Every Lindelöf ω-stable topological group is stable.*
(b) *Every Lindelöf P-group is stable.*
(c) *Every $\mathbb{R}$-factorizable P-group is τ-stable for each $\tau < \aleph_\omega$.*

CONJECTURE **4.17.** *Every $\mathbb{R}$-factorizable P-group is topologically isomorphic to a subgroup of a Lindelöf P-group.*

UNKNOWN (following **4.17.**) Can arbitrary $\aleph_0$-bounded P-group be embedded into an $\mathbb{R}$-factorizable P-group?

UNKNOWN (following **4.17.**) It is not known whether the group $\mathbb{Z}^{\omega_1}$ (or $\mathbb{Z}^{\mathfrak{c}}$) is hereditarily $\mathbb{R}$-factorizable or whether every hereditarily $\mathbb{R}$-factorizable group is pseudo-ω_1-compact.

UNKNOWN (beginning of Subsection 3 in Section **4.**) Do continuous homomorphisms preserve $\mathbb{R}$-factorizability?

PROBLEM **4.23.** (a) *Is any product of $\mathbb{R}$-factorizable groups $\mathbb{R}$-factorizable?*
(b) *Is the product of two $\mathbb{R}$-factorizable groups $\mathbb{R}$-factorizable?*
(c) *Is the product of a compact group with any $\mathbb{R}$-factorizable group $\mathbb{R}$-factorizable?*
(d) *Is the product of two Lindelöf groups $\mathbb{R}$-factorizable?*
(e) *Is the product of a Lindelöf P-group with any $\mathbb{R}$-factorizable group $\mathbb{R}$-factorizable?*

UNKNOWN (following **4.28.**) Let G and H be $\mathbb{R}$-factorizable groups. If G is a P-group and H is pseudo-ω_1-compact, is then the product $G \times H$ is $\mathbb{R}$-factorizable?

UNKNOWN (following **4.32.**) We do not know whether the Raĭkov completion ϱK of an $\mathbb{R}$-factorizable group K is $\mathbb{R}$-factorizable.

UNKNOWN (following **4.32.**) The $\mathbb{R}$-factorizability of separable groups is an open problem.

Author Index

The numbers point to pages containing author's mathematical results or references in Chapters 1–19; the bold face page numbers correspond to lists of publications.

Special Symbols

Subject Index

This index concerns Chapters 1–19. Items in their usual order (like *Baire space*) provide often more information than those in reversed order (like *space – Baire*); if a notion occurs more often in the text, the latter case (i.e., *space – Baire*) may contain the arrow ⇝ only, referring to the item in its usual order (i.e., to *Baire space*). Sometimes there is no item for a property of group or semigroup (like *Baire group, Baire semigroup*) — those items are included under the corresponding property of "space" (i.e., under *Baire space* in our example). The page numbers in bold face point to places where the notion is defined.